Molecular Biology Biochemistry and Biophysics

31

Membrane Spectroscopy

Edited by Ernst Grell

With Contributions by
D. F. Bocian   S. I. Chan   M. C. Foster   N. P. Franks
U. P. Fringeli   E. Grell   Hs. H. Günthard   J. Heesemann
Y. K. Levine   M. M. Long   R. C. Lord   D. Marsh
R. Mendelsohn   N. O. Petersen   H. Ruf   D. W. Urry
J. Yguerabide   H. P. Zingsheim

With 146 Figures

Springer-Verlag
Berlin Heidelberg New York 1981

Dr. Ernst Grell
Max-Planck-Institut für Biophysik
Heinrich-Hoffmann-Straße 7
6000 Frankfurt 71

ISBN 3-540-10332-5 Springer-Verlag Berlin Heidelberg New York
ISBN 0-387-10332-5 Springer-Verlag New York Heidelberg Berlin

Library of Congress Cataloging in Publication Data

Main entry under title: Membrane spectroscopy. (Molecular biology, biochemistry, and biophysics ; 31) 1. Membranes (Biology) 2. Spectroscopy. I. Grell, Ernst, 1941- II. Series. [DNLM: 1. Membranes. 2. Spectrum analysis. W1 MO195T no. 31 / QH 601 M533] QH601.M4688 574.87'5'028 80-24524

Typesetting and Offsetprinting: Julius Beltz, Hemsbach/Bergstr.
Bookbinding: Brühlsche Universitätsdruckerei, Gießen.
2131/3130-543210

Foreword

The last 10 years have seen an enormous growth in our understanding of the molecular organisation of biological membranes. Experimental methods have been devised to measure the translational and rotational mobility of lipids and proteins, thereby furnishing a quantitative basis for the concept of membrane fluidity. Likewise, the asymmetry of bilayer membranes as evidenced by the asymmetric insertion of proteins and lipids has been put on firm experimental ground. At higher molecular resolution it has been possible to provide a detailed picture of the molecular conformation and dynamics of lipids and, to some extent, even of small peptides embedded in a bilayer matrix. Many of these achievements would not have been possible without the application of modern spectroscopic methods. Since these techniques are scattered in a variety of specialized textbooks the present monograph attempts to describe the key spectroscopic methods employed in present-day membrane research at an intermediate level. There is no question that the elusive detailed structure of the biological membrane demands a multiplicity of experimental approaches and that no single spectroscopic method can cover the full range of physical phenomena encountered in a membrane. Much confusion in the literature has arisen by undue generalizations without considering the frequency range or other limitations of the methods employed. It is to be hoped that the present monograph with its comprehensive description of most modern spectroscopic techniques, will contribute towards a further convergence of views among the spectroscopic specialists and will enhance the understanding of membrane structure.

Basel, September 1980 — Joachim Seelig

Preface

The aim of this book is to introduce the reader to the application of spectroscopic techniques to the study of membranes. The following principal methods are covered: magnetic resonance, optical and low-angle X-ray spectroscopy and chemical relaxation spectrometry.

Each chapter summarizes the experimental and theoretical principles of a particular technique and the special applications of that technique to the investigation of membranes. In addition, the contributions will critically review the current exploitation of the technique, from the point of the view of the authors, by considering the results obtained on membrane constituents, simple model membranes and on biological membrane systems of a highly complex nature. A common aspect in all the contributions is the intensive search for a detailed understanding of the structures and functions of biological membranes at a molecular level.

This book will help to facilitate and to stimulate future studies in this interesting field.

Frankfurt, Summer 1980

Ernst Grell

Table of Contents

Nuclear Magnetic Resonance Studies of the Phospholipid Bilayer Membrane

S.I. Chan, D.F. Bocian and N.O. Petersen

I. Introduction

The cell membrane has been the focus of much recent biochemical and biophysical research, primarily because of its role in cellular phenomena. Numerous efforts have been directed toward determining the motional state of the lipid components of the membrane, the motivation being the contention that the lipid bilayer is the basic matrix in which membrane proteins are embedded to form the biological membrane. As such, it is likely that such diverse phenomena as maintenance of ionic gradients and transmembrane potentials, activities of membrane-bound enzymes, interactions between membrane proteins, transmembrane signal transmission, intercellular communication, and manifestations of cellular development and cell transformation all depend on the structure and fluidity of the lipid bilayer.

The working hypothesis for the structure of the cell membrane is the fluid mosaic model (Singer and Nicolson, 1972), where the steady state arrangements of lipids and proteins are intrinsically dynamic in nature. Thus, the structural arrangement of membrane components contains essential information concerning membrane function. Similarly, understanding the dynamics of membrane components both in terms of internal freedom of motion and in terms of lateral movements of the components in the lipid phase is important for revealing how the various time-dependent membrane functions are controlled (Edidin, 1974). It is particularly in this latter area of membrane research that various spectroscopic techniques such as magnetic resonance are useful.

Biological membranes are multicomponent systems composed of many different membrane proteins and lipids. One of the primary lipid constituents of most biological membranes is phospholipids, which occur naturally with a variety of headgroups and acyl chains. Aqueous dispersions of phospholipids exhibit many of the physical properties of natural membranes, and accordingly these dispersions have been used extensively as membrane models for magnetic resonance studies (Lee et al., 1974; James, 1975; Tiddy, 1975, 1977; Seelig, 1977; Wennerström and Lindblom, 1977; Bocian and Chan, 1978). Phospholipids spontaneously aggregate to form bilayer structures, and the interactions between different bilayers result in a thermodynamically stable multilamellar superstructure (Fig. 1).

Pure phospholipid dispersions exist in a number of phases depending on temperature and water content. The two phases of primary importance as biological membrane models are the gel and the liquid crystalline phases. In multicomponent phospholipid systems,

A.A. Noyes Laboratory of Chemical Physics (Contribution No. 5853), California Institute of Technology, Pasadena, California 91125, USA

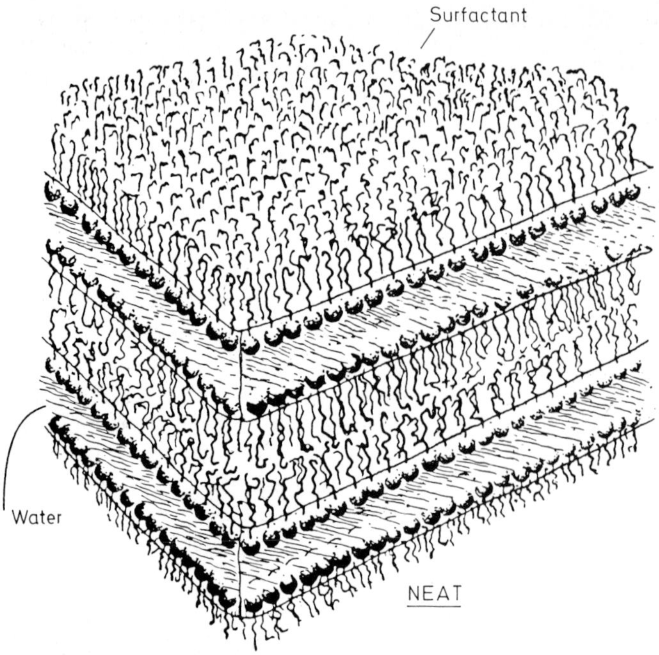

Fig. 1. A representation of the multilamellar superstructure formed by phospholipid/water dispersions. The individual lipid bilayers are of infinite extent and are separated by the interstitial water (Khetrapal et al., 1975)

species immiscibility and phase separations can occur and the phase diagrams become exceedingly complicated. This phenomenon has been investigated extensively by electron spin resonance (ESR) spin-label studies (Wu and McConnell, 1975). However, most nuclear magnetic resonance (NMR) studies have been performed on single-component systems and, in general, at temperatures above the gel → liquid crystalline phase transition temperature. In principle, NMR could be used to study multicomponent systems if the components could be suitably labeled. Also, sensitivity and abundance have made lipids the principal membrane components investigated to date by NMR. However, the direct observation of nonlipid membrane components is now becoming possible through the use of specific labeling of these components, and by the use of lipid components which do not contain the nucleus being observed (Feigenson et al., 1977).

For many years, much of our knowledge about the motional state of the bilayer has come from ESR measurements on spin-labeled lipids incorporated into the bilayer (Hubbell and McConnell, 1971; Jost et al., 1971; Schindler and Seelig, 1974). Recent efforts have been directed toward developing spectroscopic probes which would circumvent the possible perturbations in the bilayer produced by the bulky nitroxide radical. Nuclear magnetic resonance spectroscopy is a natural tool for this purpose and there has been considerable effort using ^{1}H, ^{2}H, ^{13}C, and ^{31}P probes which are either endogenous or incorporated into the lipid molecule (Lee et al., 1974; James, 1975; Tiddy, 1975, 1977; Seelig, 1977; Wennerström and Lindblom, 1977; Bocian and Chan, 1978). Both ESR and NMR are sensitive to molecular motions, but on different time-scales. With ESR, molecular processes occurring on the order of 10^{-7} s or faster can be sampled, while NMR in principle can be used to monitor motions occurring from 10^{-12} s to several seconds, depending on the parameter being monitored. Thus, the two methods complement each other and it is expected that together the two approaches give a more representative picture of the dynamics of the system.

In this chapter, we describe how NMR has been utilized to characterize the structural and dynamical properties of the phospholipid bilayer. We first delineate the special features of the lipid bilayer membrane which render the application of NMR to this system nonstandard. We then outline those elements of NMR theory that pertain to partially oriented molecules. Finally, we demonstrate how the measurement of NMR parameters such as chemical shift anisotropies, quadrupolar splittings, dipolar interactions, and spin-lattice and spin-spin relaxation times can be analyzed and interpreted to yield motional information about lipid molecules. Examples will be taken from the current literature to illustrate these techniques, both their utility as well as their limitations.

II. Partially Oriented Molecules

A phospholipid bilayer is a lyotropic liquid crystal, and as such its NMR properties are intermediate between those of an immobile solid and a nonviscous liquid. In effect we are dealing with partially oriented molecules. We therefore outline those unusual features of molecules undergoing restricted anisotropic motion.

In the usual application of NMR, in particular to organic chemistry, the molecules observed are either in the liquid phase or in solution. As a consequence, there is much rotational freedom, and a complete averaging of dipolar and quadrupolar interactions as well as anisotropy in the magnetic shielding occurs (Abragam, 1961). Narrow, well-resolved resonances result and the NMR theory of the molecular motions simplifies considerably, resulting in straightforward interpretation of the data.

In the other extreme, the solids, no motion is present, and all the aforementioned manifestations of the interactions in the NMR spectrum depend solely on the spatial orientation of the molecules. In this case, directional dependence of the parameters can be extracted, although it is not necessarily easy to relate them to the molecular frame without independent structural information. To complicate matters, there is usually strong interaction among the magnetic nuclei in the sample, creating a many-body problem. Nonetheless, the NMR theory is well understood and directly applicable (Abragam, 1961).

Partially oriented molecules, such as slowly reorienting rigid macromolecules or liquid crystalline phases, encompass both solid-like and liquid-like features in the manifestation of their NMR behavior. In order to facilitate the description of motion in such systems, it is helpful to introduce the following classes of motion:

1. *Isotropic motion* takes the molecule over *all* directions in space with *equal* probability *and* is uniquely defined by a single characteristic rotational time (Fig. 2a).
2. *Anisotropic* motion takes the molecule over *all* directions in space with equal probability but rotation about some molecular axes occurs at a different rate than about

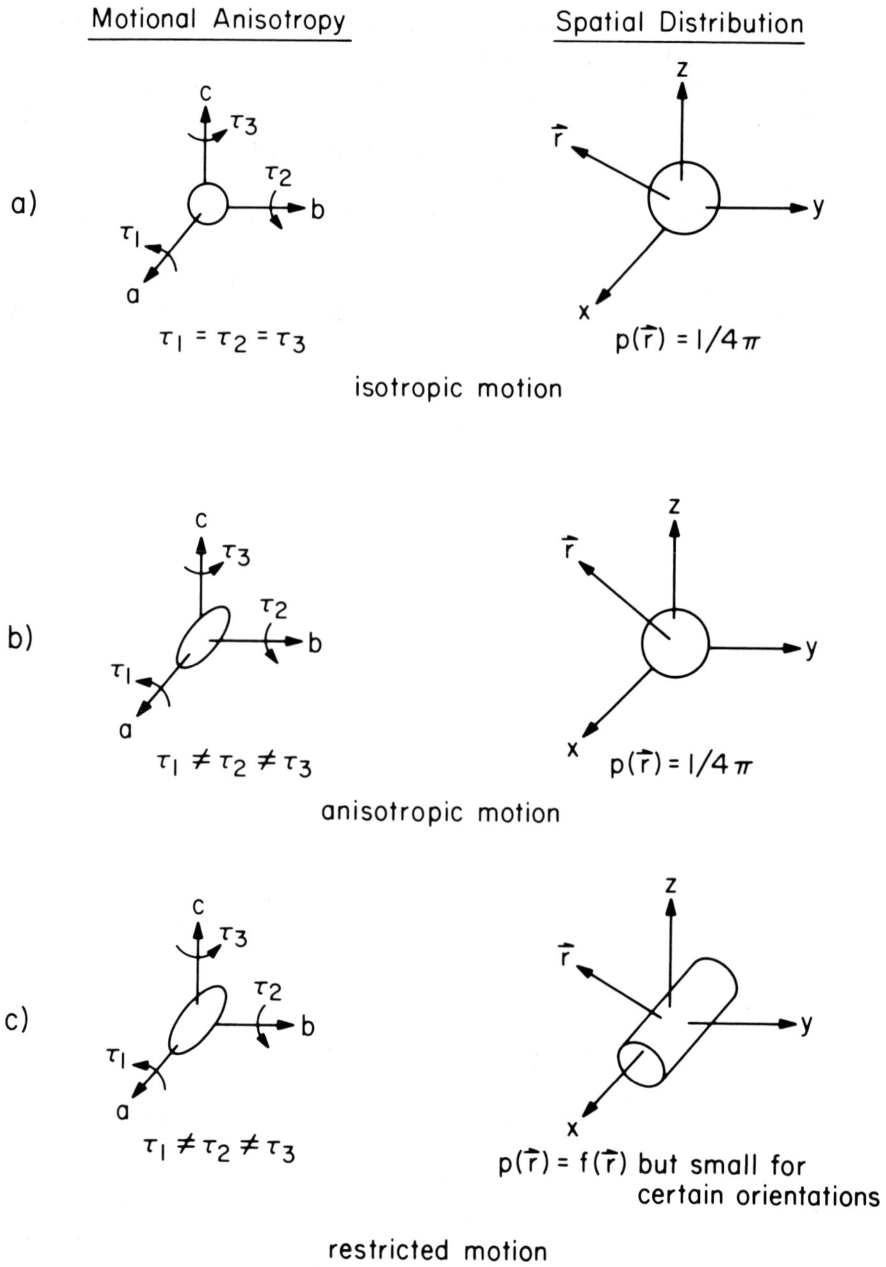

Fig. 2. A representation of (a) isotropic motion, (b) anisotropic motion, and (c) restricted motion. The motional anisotropy is defined in terms of a coordinate system ***a, b, c*** which is fixed in the individual molecules. Here the ellipsoids represent the temporal distribution of the three axes, whose rotational correlation times are given by τ_1, τ_2, and τ_3. The spatial distribution of the molecules is defined in terms of a laboratory-fixed coordinate system x, y, z. Here the ellipsoids represent the spatial distribution for the vector $\vec{r}$, whose orientational distribution function is given by $p(\vec{r})$.

others, so that *at least* two characteristic rotational times must be used to characterize the motion completely (Fig. 2b).

3. *Restricted motion* takes the molecule over *some* directions in space with high (and possibly equal) probability but over *other* directions in space with *no* or *low* probability. It is thus *not* the same as anisotropic motion in that it requires, in addition to the rotational times (usually two or three), the determination of the spatial angles through which the motion is allowed (Fig. 2c), that is the orientational distribution function.

In Figure 3, we define the laboratory and molecular axes which we will use to describe the orientation and motion of molecules in the phospholipid bilayer. The axis system a, b, c is defined with respect to the individual molecules, while axes x, y and z refer to a laboratory-fixed axis system where the external magnetic field $\vec{H}_0$ is directed along z. The axis system 1, 2, 3 is defined with respect to the entire bilayer unit, with the 3 axis directed along the bilayer normal $\vec{n}$.

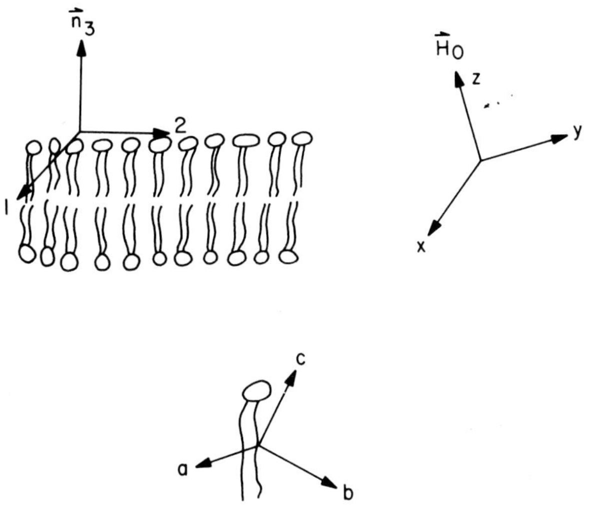

Fig. 3. The definition of three coordinate systems, appropriate for describing the orientation and motion of molecules in the phospholipid bilayer. For details see text

The importance of these definitions becomes evident when we consider partially oriented molecules. In the liquid crystal phases of many molecules the degree of order is high. In some nematic phases it has become clear that rapid orientation about the long axis of the molecules takes place, but that the motion of this axis is restricted to certain angular ranges (Pirs et al., 1976). The consequence of this is that, in spite of some fast motions, chemical shift anisotropies and dipolar and quadrupolar interactions which may exist are not completely averaged. We must keep in mind, however, that restricted motion may degenerate into unrestricted motion if sampling times are made sufficiently long. Thus, the timescale of observation determines whether the motion appears restricted or not. Of course, if one samples for longer times, more characteristic times must be introduced to define the motional state of the system.

At this point it is appropriate to consider the motions expected for lipid molecules in membrane systems. We can crudely divide these motions into three categories: (1) motions within the molecule, (2) motions of the molecule, and (3) motions of the membrane.

1. Motions within the lipid molecules are all associated with bond movements. Bond vibrations may conveniently be ignored because they are both low amplitude and high frequency motions, and therefore not detectable by NMR methods. Fast, low amplitude bond rotation oscillations likewise will have little detectable effect on the NMR spectra. On the other hand, high amplitude, low frequency bond rotations, such as trans-gauche rotations, can contribute significantly to the parameters monitored by NMR methods.
2. Motions of the molecule can be visualized to occur in several manners: (a) rotational diffusion about the axis $\vec{n}$ normal to the bilayer plane, (b) translational diffusion in the plane of the bilayer, corresponding to a displacement perpendicular to $\vec{n}$, and (c) translation perpendicular to the bilayer, i.e., a bobbing motion along $\vec{n}$. Other motions include phospholipid exchange between the two monolayers within the same bilayer, the so-called flip-flop motion, and phospholipid exchange between two different bilayers.
3. This third category represents the motions of the entire bilayer unit, collective motions whose observable effects on the NMR spectra can be similar to motion of individual molecules. Thus tumbling of sheets or bilayer fragments will have the same observable effect in many respects as lateral diffusion on a curved surface. Cooperative movement perpendicular to the bilayer may be visualized as lateral wave motions of the bilayer, a phenomenon not uncommon in certain liquid crystals.

We now proceed to consider the effects of the spatial ordering of the lipid molecules on the various NMR measurements, discussing the effects of restricted motion on the first order averaging of the secular terms in the Hamiltonian appropriate to the parameter being monitored. Here we focus particularly on the chemical shifts and the dipolar interactions, and briefly discuss quadrupolar interactions. Subsequently we will consider the effects of restricted motions on relaxation phenomena, observed through spin-lattice and spin-spin relaxation times.

III. Effects of Motional Restriction on NMR Spectra

The Hamiltonian for a motionally restricted system can be written formally as

$$\mathcal{H} = \overline{\mathcal{H}} + (\mathcal{H}(t) - \overline{\mathcal{H}}). \tag{1}$$

Terms in the Hamiltonian which are not motionally averaged to zero contribute to $\overline{\mathcal{H}}$, and lead to first order dispersion in the energy of the NMR transitions. In principle, it is possible to obtain information about the orientational order of the system from a knowledge of $\overline{\mathrm{H}}$. The second term enters in second order only and contributes to the various relaxation processes. Interpretation of the relaxation rates in terms of motional processes allows one to determine the timescales of those motional processes modulating $\mathcal{H}(t)$.

We now consider the effects of motional restriction on the averaging of the various terms in $\mathcal{H}$. We examine in turn the magnetic shielding, quadrupolar interactions, and dipolar interactions.

A. Magnetic Shielding

Magnetic shielding arises from the screening of a given nucleus by its electronic environment. Due to electronic currents induced in the molecule by the applied magnetic field $\vec{H}_0$, the effective magnetic field at the nucleus differs from the applied field by a small but measurable amount. The resonance position of the nucleus is a sensitive measure of the nature of the bonding in which it is involved. In general, the electronic environment is not symmetric about the nucleus, and the magnetic shielding will depend on the orientation of the molecule with respect to $\vec{H}_0$. This orientational dependence gives rise to the well-known "chemical shift anisotropy" observed in crystals and powders.

The magnetic shielding is generally described by

$$\mathcal{H}_z = -\vec{\mu} \cdot (1 - \underset{\sim}{\sigma}) \cdot \vec{H}_0 \tag{2}$$

where $(1 - \underset{\frown}{\sigma})$ is a symmetric tensor whose components depend on the orientation of the molecule with respect to the magnetic field. It is always possible to diagonalize the tensor by appropriate choice of a molecular coordinate system; thus, in what is usually called the principal axis system, we may write

$$\underset{\frown}{\sigma} = \begin{pmatrix} \sigma_{aa} & 0 & 0 \\ 0 & \sigma_{bb} & 0 \\ 0 & 0 & \sigma_{cc} \end{pmatrix} . \tag{3}$$

Since the magnetic moment, $\vec{\mu}$, of the nucleus under consideration is quantized along $\vec{H}_0$, only the component of the shielding tensor along the magnetic field, σ_z, is measured experimentally. In terms of the principal values of $\underset{\sim}{\sigma}$,

$$\sigma_z = \sigma_{aa}\sin^2\theta\cos^2\phi + \sigma_{bb}\sin^2\theta\sin^2\phi + \sigma_{cc}\cos^2\theta \tag{4}$$

where θ and ϕ are defined in Figure 4. The expression for the Zeeman energy (Eq. (2)) then simplifies to

$$\mathcal{H}_z = -\mu_z (1 - \sigma_z) H_0 . \tag{5}$$

Whenever possible, the principal values of $\underset{\frown}{\sigma}$ are determined by single crystal studies. When the crystal structure is known, these values of $\underset{\frown}{\sigma}$ can be referred to the molecular axes system. The ^{31}P NMR spectrum of a single crystal of phosphorylethanolamine (PE) in an arbitrary orientation is shown in Figure 5 (Kohler and Klein, 1976). The two observed signals correspond to the two molecules in the unit cell. The orientational dependence of the resonance frequency is shown in Figure 6. From analysis of the orientational dependence, the following principal values of $\underset{\frown}{\sigma}$ have been obtained: $\sigma'_{aa} = -67$ ppm,

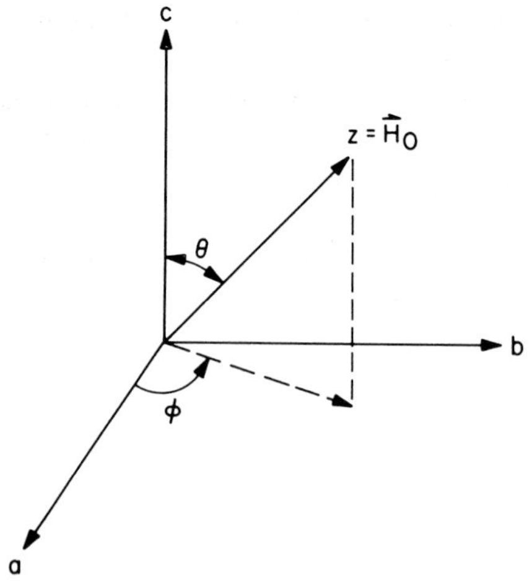

Fig. 4. The relationship between the laboratory- and molecule-fixed coordinate systems (cf. Fig. 3)

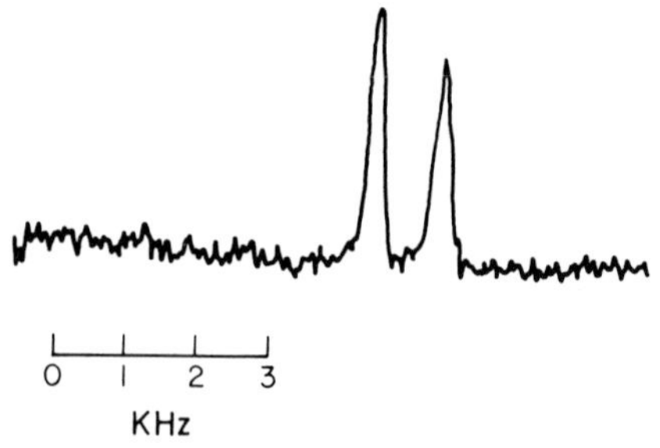

Fig. 5. The ^{31}P NMR spectrum (24.3 MHz) of a single crystal of phosphorylethanolamine at 25°C in an arbitrary orientation, in the presence of an 8.6 G ^1H decoupling field. The two resonances correspond to the two molecules in the unit cell (Kohler and Klein, 1976)

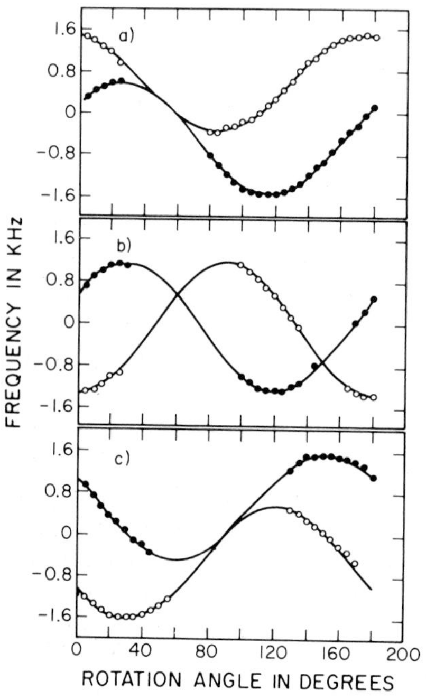

Fig. 6 a-c. The angular dependence of the ^{31}P resonance frequency for three orientations of a crystal of phosphorylethanolamine. The frequency is expressed relative to the resonance frequency of H_3PO_4 (Kohler and Klein, 1976)

σ'_{bb} = 13 ppm, σ'_{cc} = 69 ppm, where the prime here is used to indicate that $\underset{\frown}{\sigma}$ is referred to the average value of $\underset{\frown}{\sigma}$ for H_3PO_4 measured in solution.

Frequently the principal components of $\underset{\frown}{\sigma}$ can only be obtained from measurements on powder samples. In this case, the observed composite spectrum arises from all possible orientations of the molecule relative to $\vec{H}_0$. While the principal values of $\underset{\frown}{\sigma}$ can usually be ascertained from analysis of the powder spectrum, the location of the principal axes of the shielding tensor is not known relative to the molecular axes. The powder spectrum of anhydrous dipalmitoyllecithin (DPL) is shown in Figure 7a (Griffin, 1976). A similar

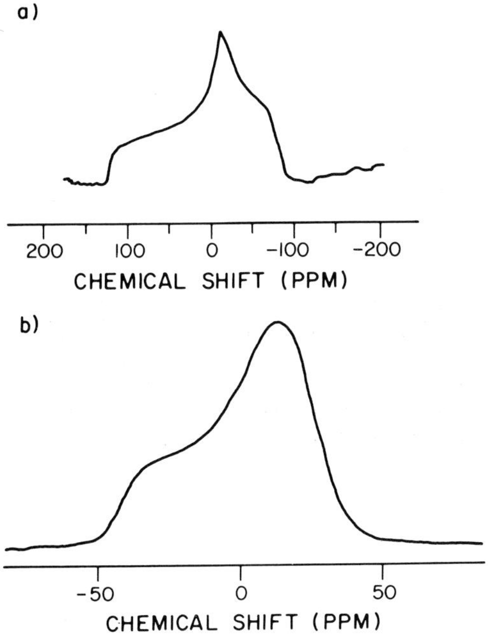

Fig. 7. The ^1H decoupled ^{31}P NMR spectra of powder samples of (a) anhydrous dipalmitoyllecithin (DPL) at 15°C (Griffin, 1976) and (b) a 1:1 mixture (by weight) of DPL and water at 21°C (Habercorn et al., 1978)

spectrum is obtained for a powder sample of PE, and analysis of this spectrum gives the same principal values of $\underset{\frown}{\sigma}$ as those obtained from single-crystal studies (Kohler and Klein, 1976). It is evident that for PE $\underset{\frown}{\sigma}$ is not axially symmetric. This is also the case for DPL, as can be seen by comparison of the spectrum of anhydrous powder with that of a 50% hydrated sample (Griffin, 1976; Habercorn et al., 1978), for which $\underset{\frown}{\sigma}$ is axially symmetric (Figure 7b).

If there is molecular motion, the chemical shift tensor components will be modulated by this motion. To determine how molecular motion can contribute to the averaging of $\underset{\frown}{\sigma}$, consider the case where $\underset{\frown}{\sigma}$ is axially symmetric. The principal values of $\underset{\frown}{\sigma}$ are then $\sigma_{aa} = \sigma_{bb} = \sigma_\perp$ and $\sigma_{cc} = \sigma_\parallel$. If the motion is much slower than the inverse of the chemical shift dispersion $(\nu_\parallel - \nu_\perp)$, there will be no observable effect on the spectrum. If, however, the motion is faster than $(\nu_\parallel - \nu_\perp)^{-1}$, typically 10^{-4}-10^{-6} s, averaging of the tensor will occur. For this reason $(\nu_\parallel - \nu_\perp)^{-1}$ is referred to as the timescale of observation of this experiment.

We now consider the effects of motional averaging on σ_z. The result, of course, depends on the details of the motion involved. We can examine the details if Eq. (4) is rewritten in terms of the so-called order parameters (Saupe, 1965) for the principal axes relative to the applied magnetic field. These order parameters are defined by

$$S_{pq} = \tfrac{1}{2} < 3 \cos \theta_p \cos \theta_q - \delta_{pq} > \qquad (6)$$

where θ is the angle between the p principal axis and the applied magnetic field and δ_{pq} is the Kronecker delta. Because only the principal values of $\underset{\frown}{\sigma}$ are included in Eq. (4), only the diagonal elements of the order parameter matrix need be considered, and σ_z can be rewritten as

$$\sigma_z = \tfrac{2}{3} \left[(\sigma_{aa} - \sigma_{cc}) S_{aa} + (\sigma_{bb} - \sigma_{cc}) S_{bb} \right] + \tfrac{1}{3} \mathrm{Tr} (\underset{\frown}{\sigma}) \ . \qquad (7)$$

To obtain this result we have made use of the well-known relationship

$$S_{aa} + S_{bb} + S_{cc} = 0. \qquad (8)$$

The last term in Eq. (7) gives the value of σ_z in liquids and solutions, since S_{aa} and S_{bb} are zero when the molecules undergo unrestricted motion. This quantity is frequently given the symbol $\bar{\sigma}$. For restricted motion, however, S_{aa} and S_{bb} do not vanish. Note that two independent order parameters are then needed to define the effects of motion on σ_z in general. Furthermore, when the static shielding tensor is axially symmetric only one order parameter can be measured in this experiment.

In the special case where the motion takes a molecular axis symmetrically about a director, Eq. (7) simplifies to

$$\sigma_z = \bar{\sigma} + \tfrac{1}{3} \left[(\sigma_{aa} - \sigma_{cc}) S_{\alpha\alpha} - (\sigma_{bb} - \sigma_{cc}) S_{\beta\beta} \right] (3 \cos^2 \theta' - 1) \qquad (9)$$

where θ' is the angle between the applied field and the director and the order parameters $S_{\alpha\alpha}$ and $S_{\beta\beta}$ are now referred to the director, that is

$$\begin{aligned} S_{\alpha\alpha} &= \tfrac{1}{2} < 3 \cos^2 \alpha - 1 > \\ S_{\beta\beta} &= \tfrac{1}{2} < 3 \cos^2 \beta - 1 > \end{aligned} \qquad (10)$$

where α and β are the angles between the a and b principal axes of the shielding tensor and the director. As we shall see, the case where the molecular motion results in effective axial symmetry is appropriate for lipid molecules in the liquid crystalline phase of phospholipid bilayers.

B. Quadrupolar Interactions

Nuclei with $I > \frac{1}{2}$ do not usually possess a spherical charge distribution within the nucleus. This nuclear charge distribution takes on either a prolate or oblate ellipsoidal shape. Accordingly, the energy of the nucleus depends on the orientation of this ellipsoid relative to the electronic charge distribution surrounding the atom. This interaction, frequently referred to as the electric quadrupolar interaction, lifts the $|m_I|$ degeneracy of the nuclear spin levels. Inasmuch as these energy spacings measure the orientational dependence of the nuclear spin energy in the electric field, it is clear that only the electric field gradient at the nucleus is important. Thus, the degeneracy of the nuclear spin levels of a quadrupolar nucleus is not removed in a spherical field.

The Hamiltonian for the quadrupolar interaction is given by

$$\mathcal{H}_Q = \frac{eQ}{4I(2I-1)} [V_{cc}(3I_c^2 - I^2) + (V_{aa} - V_{bb})(I_a^2 - I_b^2)] \tag{11}$$

where Q is the quadrupole moment of the nucleus, and V_{cc}, V_{aa}, V_{bb} denote the principal components of the electric field gradient tensor. Since the quadrupolar interaction is frequently monitored in the presence of an applied magnetic field, there is competition between the alignment of the spin $\vec{I}$, along the applied magnetic field and the electric field gradient. When the applied magnetic field is sufficiently large for the Zeeman energy to be dominant in the combined Zeeman-Quadrupolar Hamiltonian, it is convenient to write the Hamiltonian in the coordinate system defined by $\vec{H}_0$. The combined first order Hamiltonian is then

$$\mathcal{H} = -\gamma\hbar H_0 I_z + \frac{eQ}{4I(2I-1)} [3 I_z^2 - I^2] V_{zz} \tag{12}$$

where V_{zz} is the zz component of electric field tensor in the laboratory frame and is related to the principal values in the molecular frame by

$$V_{zz} = V_{aa} \sin^2\theta \cos^2\phi + V_{bb} \sin^2\theta \sin^2\phi + V_{cc} \cos^2\theta \,. \tag{13}$$

If we define the field gradient $V_{cc} = eq$ and the asymmetry parameter $\eta = (V_{aa} - V_{bb})/V_{cc}$ then Eq. (12) can be rewritten as

$$\mathcal{H} = -\gamma\hbar H_0 I_z + \frac{e^2 qQ}{8I(2I-1)} [3I_z^2 - I^2] [(3\cos^2\theta - 1) + \eta\sin^2\theta\cos 2\phi] \tag{14}$$

which predicts the following energy levels

$$E(m_I) = -\gamma\hbar H_0 m_I + \frac{e^2 qQ}{8I(2I-1)} [3m_I^2 - I(I+1)] \cdot [(3\cos^2\theta - 1) + \eta\sin^2\theta \cos 2\phi] \,. \tag{15}$$

The results for the case of I = 1 are shown in Figure 8. The resultant NMR spectrum can be seen to be a doublet displaced symmetrically about the Zeeman frequency by

$$\nu - \nu_0 = \pm \tfrac{3}{8} \frac{e^2 qQ}{h} [(3 \cos^2\theta - 1) + \eta\sin^2\theta \cos 2\phi] \,. \tag{16}$$

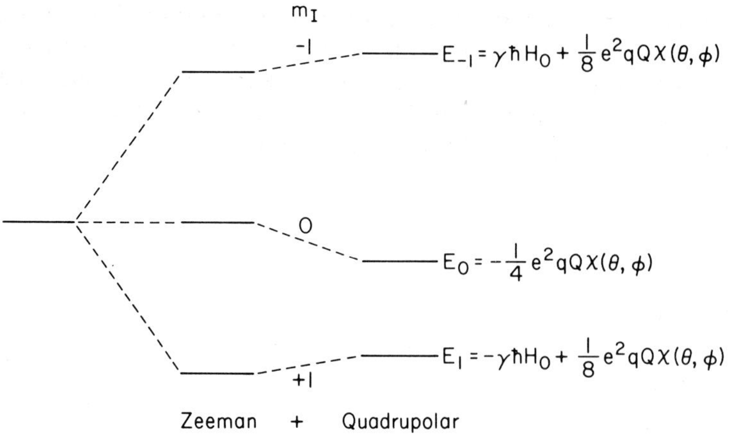

Fig. 8a. Nuclear energy levels of an I = 1 nucleus in a strong magnetic field. The angular function $\chi(\theta, \phi)$ is defined in Eq. (15). Relative contributions to the splitting from the nuclear Zeeman effect and the nuclear quadrupolar interaction are not drawn to scale

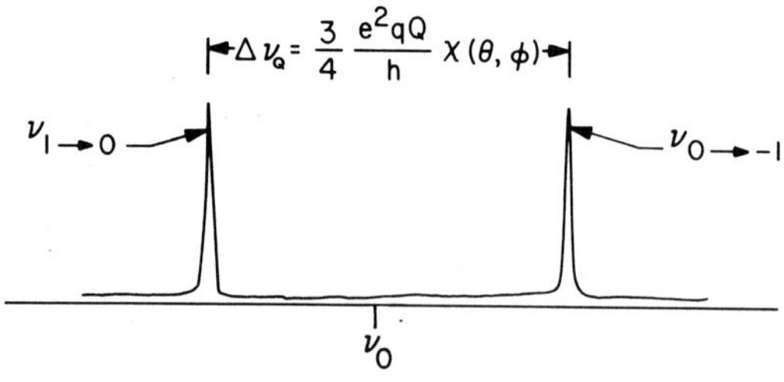

Fig. 8b. The NMR spectrum expected for a single crystal of a molecule with an isolated I = 1 nucleus at an arbitrary orientation in the magnetic field

As in the case of the magnetic shielding anisotropy, a detailed orientational study of the splitting in single crystals yields the molecular parameters of the Hamiltonian, which are e^2qQ and η for quadrupolar interactions.

Typical quadrupolar nuclei in membrane research include ^2H and ^{14}N. For a C-D bond in an alkyl chain e^2qQ/h is typically 170 kHz and η is near zero (Burnett and Muller, 1971); that is, the electric field gradient is essentially axially symmetric about the C-D bond. Since the ^2H NMR frequency is 13.8 MHz at 2.1 tesla, the typical field used in these experiments, the Zeeman interaction clearly dominates over the quadrupolar interaction and the theory outlined above is applicable. This may not be the case for ^{14}N. Here the NMR frequency is 6.5 MHz at the same field but e^2qQ/h can be several MHz (Schempp and Bray, 1970). Consequently, the first order Hamiltonian should not be used without some caution. Also, since p-orbitals are involved in nitrogen bonding, η can be quite large.

A typical ^2H NMR spectrum of specifically deuterated fatty acids intercalated into phospholipid multilayers oriented at various angles with respect to the external magnetic field is shown in Figure 9 (Seelig and Niederberger, 1974). A quadrupolar splitting is indeed observed, but is significantly smaller than the value of 170 kHz expected for motionless chains. The origin of the reduced splitting is, of course, motional averaging. Note that one frequently deals with unoriented multilayers in membrane studies and only a powder spectrum is observed (Fig. 10) (Seelig and Seelig, 1974).

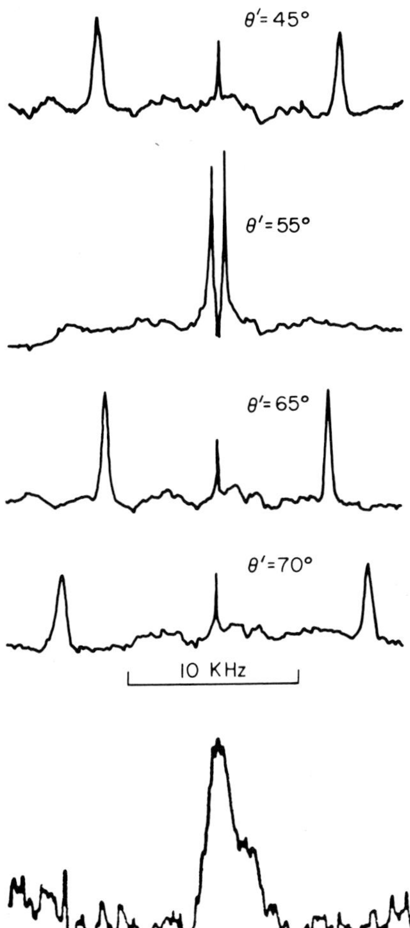

Fig. 9. The ^{2}H NMR spectra (13.8 MHz) of specifically deuterated fatty acids intercalated into phospholipid multilayers oriented at various angles with respect to the magnetic field. The angle θ' is between the magnetic field and the bilayer normal. The sharp signal in the center of the spectra arises from a small fraction of molecules which form an isotropic phase when the oriented multilayers are prepared (Seelig and Niederberger, 1974)

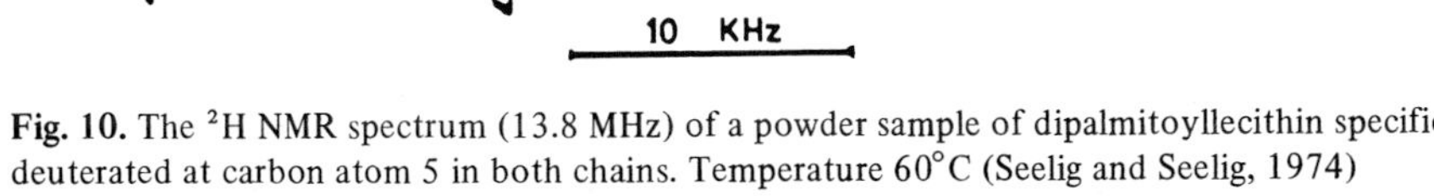

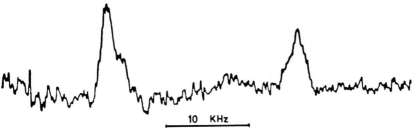

Fig. 10. The ^{2}H NMR spectrum (13.8 MHz) of a powder sample of dipalmitoyllecithin specifically deuterated at carbon atom 5 in both chains. Temperature 60°C (Seelig and Seelig, 1974)

The effect of motional averaging on electric quadrupolar interaction can be treated in exactly the same fashion as was done earlier for magnetic shielding anisotropy. For quadrupolar interactions, the timescale of the observation is $((\frac{3}{4})e^2qQ/h)^{-1}$, so that any processes which modulate the angular function in the Hamiltonian at a rate much faster than this timescale provide an effective averaging mechanism. In an analogous manner, we may rewrite the quadrupolar splitting, $\Delta\nu_Q$, in terms of the order parameters of the principal axes of the electric field gradient tensor relative to the magnetic field. For the case of I =1,

$$\Delta\nu_Q = \tfrac{3}{4}\frac{e^2qQ}{h}\left[2S_{cc}(1+\eta/3) + 4/3\,\eta S_{bb}\right] \tag{17}$$

and if $\eta = 0$, this simplifies further to

$$\Delta\nu_Q = \tfrac{3}{2}\frac{e^2qQ}{h}S_{cc}\,. \tag{18}$$

Thus, as in the case of an axially symmetric chemical shift tensor, only one order parameter is involved and can be measured for axially symmetric electric quadrupolar interaction.

Finally, we consider the case where the motion takes a molecular axis symmetrically about a director. Here the result for I = 1, $\eta = 0$ is

$$\Delta\nu_Q = \tfrac{3}{4}\frac{e^2qQ}{h}S_{\gamma\gamma}(3\cos^2\theta' - 1) \tag{19}$$

where θ' is again the angle between the applied field and the director and $S_{\gamma\gamma}$ for this experiment is the order parameter of the symmetry axis of the electric field gradient (along the C-D bond) relative to the director. It is important to realize that the principal axes of the magnetic shielding tensor do not necessarily coincide with those of the electric field gradient tensor. When these interactions can be independently measured in two separate experiments, care must be exercised in comparing the determined order parameters. We shall discuss this point in greater detail later in this chapter.

C. Dipolar Interactions

Magnetic nuclei interact through space. This dipolar interaction for a pair of spins is given by the following Hamiltonian

$$\mathcal{H}_D = \frac{\vec{\mu}_1\cdot\vec{\mu}_2}{r^3} - \frac{3(\vec{\mu}_1\cdot\vec{r})(\vec{\mu}_2\cdot\vec{r})}{r^5} \tag{20}$$

where $\vec{\mu}_1$ and $\vec{\mu}_2$ denote the magnetic moments of the two nuclei and $\vec{r}$ is the internuclear vector (Fig. 11). In terms of the principal axis system of the dipolar interaction tensor, this Hamiltonian can be written as

$$\mathcal{H}_D = \frac{\gamma_1\gamma_2\hbar^2}{r^3}(3I_{1c}I_{2c} - \vec{I}_1\cdot\vec{I}_2)\,. \tag{21}$$

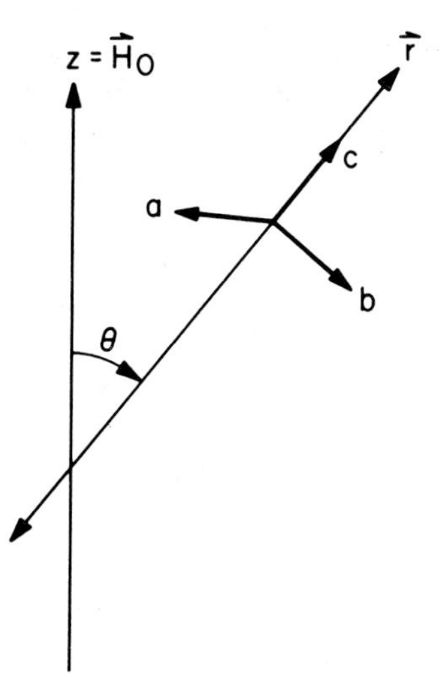

Fig. 11. The orientation of the dipolar interaction vector $\vec{r}$ relative to the molecule-fixed and laboratory coordinate systems (cf. Fig. 3)

When the dipolar interaction is observed in the presence of a strong magnetic field, the energy levels of the system are most conveniently obtained when the dipolar Hamiltonian is expressed to first order with the spin angular momentum quantized along the Zeeman axis, that is

$$\mathcal{H}_D = \frac{\gamma_1 \gamma_2 \hbar^2}{2r^3} (1 - 3\cos^2\theta)(3I_{1z}I_{2z} - \vec{I}_1 \cdot \vec{I}_2) . \qquad (22)$$

For membrane systems, we will be concerned primarily with $I = \frac{1}{2}$ nuclei. For the moment, we shall restrict our discussion to two spins, which may be identical, as in the case of two protons, or nonidentical, as in the case of a proton and a ^{13}C nucleus. In the former case, the nuclear transitions occur at

$$h\nu = \gamma_1 \hbar H_0 \pm \frac{3\gamma_1^2 \hbar^2}{r^3} (1 - 3\cos^2\theta) \qquad (23)$$

while in the latter they occur at

$$h\nu = \gamma_1 \hbar H_0 \pm \frac{\hbar^2 \gamma_1 \gamma_2}{2r^3} (1 - 3\cos^2\theta) \qquad (24)$$

and

$$h\nu = \gamma_2 \hbar H_0 \pm \frac{\hbar^2 \gamma_1 \gamma_2}{2r^3} (1 - 3\cos^2\theta) . \qquad (25)$$

In both cases, the spectrum is a doublet symmetrically displaced about the appropriate NMR frequency. Note that when the angle $\theta = 54°55'$, the dipolar splitting vanishes. This is the so-called "magic angle".

It is clear from Eqs. (23)-(25) that the magnitude of the dipolar interaction depends strongly on the separation between the magnetic nuclei. Because of the r^{-3} dependence, dipolar interactions are significant only when an abundant spin is observed, or when a dilute spin is monitored in the presence of an abundant magnetic nucleus. Also, the larger the magnetic moment, the larger is the dipolar interaction. For these reasons, dipolar interactions among ^1H, ^{19}F, and ^{31}P spins are particularly important. Typical dipolar interaction constants for molecular segments and geometries pertinent to membranes are given in Table 1.

Table 1. Static dipolar couplings for isolated $I = \frac{1}{2}$ spin pairs relevant for membrane systems

Segment [a]	Spin Pair	Dipolar Coupling (kHz)
CH_2	^1H, ^1H	31.8
CF_2	^{19}F, ^{19}F	13.6
CHF	^{19}F, ^1H	13.3
CH_2	^{13}C, ^1H	23.2
CF_2	^{13}C, ^{19}F	10.5
P. . . . H [b]	^{31}P, ^1H	6.05

[a] Standard bond lengths and tetrahedral geometry are assumed for the calculations.
[b] Since there are no $^{31}P - {}^1H$ bonds in phospholipid systems, an internuclear distance of 2Å was chosen for the calculation.

On the basis of the magnitude of the magnetic shielding anisotropies and quadrupolar interactions observed in oriented multilayer membranes (Seelig and Niederberger, 1974; Seelig and Gally, 1976), we expect the dipolar splitting for a methylene geminal pair of a specifically protonated perdeuterated phospholipid to be smaller because of motional averaging than predicted by Eq. (23). The powder spectrum for an assembly of noninteracting methylene pairs undergoing motions which take the interproton vector symmetrically about a director and with off-axis excursions over the range $-\Delta\beta \rightarrow +\Delta\beta$ is shown in Figure 12 for varying degrees of off-axis motion. These hypothetical powder spectra were

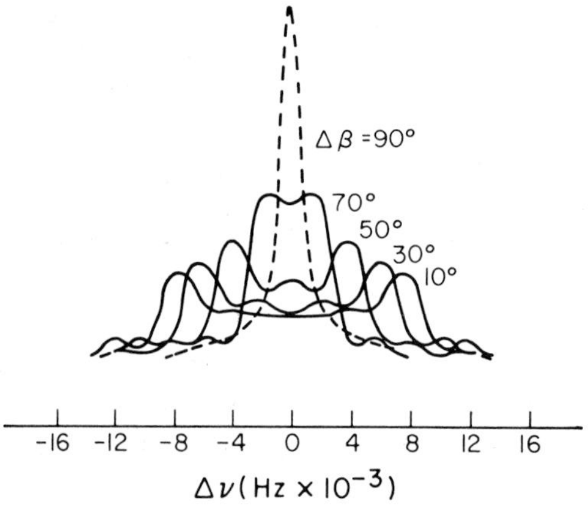

Fig. 12. The calculated powder spectra for an assembly of noninteracting methylene pairs undergoing motions which both take the interproton vector symmetrically about a director and have different amounts of off-axis excursion, $\Delta\beta$ (Seiter and Chan, 1973)

simulated by Seiter and Chan (1973), and can be understood in terms of the following relationship for the dipolar splittings when this type of motion occurs:

$$\Delta\nu_D = \frac{3\gamma_I^2\hbar}{4\pi r^3} S_{\beta\beta} (1 - 3\cos^2\theta') . \tag{26}$$

Thus, a comparison of the powder spectrum of a system where motional averaging occurs with that of a rigid system allows the order parameter, $S_{\beta\beta}$, to be determined.

The motionally averaged powder spectrum of the methyl rotors in phospholipid multilayers has also been simulated by Seiter and Chan (1973). For a methyl top which is rapidly reorienting about the top axis, the NMR transitions occur at

$$\nu - \nu_0 = 0;\ \pm \frac{3\gamma_I^2\hbar}{8\pi r^3} (1 - 3\cos^2\theta') \tag{27}$$

where θ' is now the angle between the rotor axis and the applied magnetic field. It is important to note that 50% of the intensity occurs at ν_0. When motional averaging takes the top axis symmetrically about a director, the results of Eq. (27) can be reexpressed in terms of the order parameter of the top axis relative to the director, $S_{\gamma\gamma}$, namely

$$\nu - \nu_0 = 0; \pm \tfrac{3}{8} \frac{\gamma_I^2\hbar}{\pi r^3} S_{\gamma\gamma} (1 - 3\cos^2\theta'') \tag{28}$$

where θ'' is the angle between the top axis and the director. The powder spectrum resulting from this set of transitions is shown in Figure 13 and consists of two principal features: a sharp central spike containing 50% or more of the intensity and broad dipolar wings containing the remaining intensity.

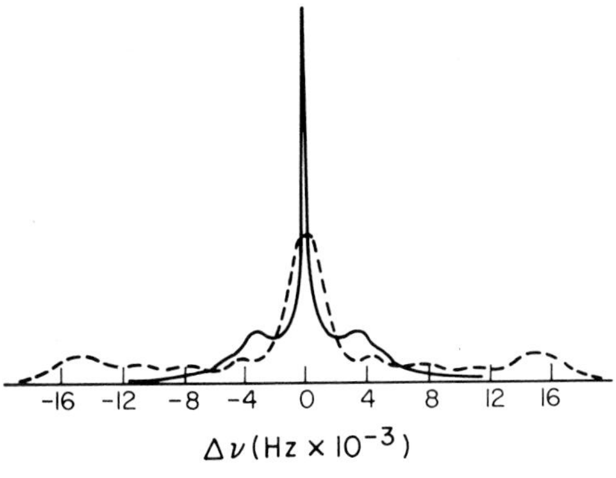

Fig. 13. The calculated powder spectra for methyl rotors rapidly reorienting about the top axis, including effects of nearest neighbor protons to the three-spin group. *Dashed line:* calculated spectrum for the reorienting methyl rotors including an off-axis motion of 10°; *solid line:* spectrum when the off-axis motion is increased to 70° (Seiter and Chan, 1973)

The determination of order parameters for proton NMR spectra is best done in the time domain, where one measures the free induction decay (FID) of the system. The FID, of course, is simply the Fourier transform of the lineshape function (Farrar and Becker, 1971), $F(\omega)$, and is given by

$$S(t) = \frac{1}{\sqrt{2\pi}} \int_{-\infty}^{\infty} F(\omega)\, e^{i\omega t} d\omega \,. \tag{29}$$

In the case of a motionally restricted system, the FID signal can be approximated by (Anderson, 1954)

$$S(t) = \langle e^{i\int_0^t \omega(t')dt'} \rangle . \tag{30}$$

Since the lineshape is symmetric about the Larmor precession frequency, ω_0, Eq. (30) can be rewritten as

$$\begin{aligned} S(t) &= e^{i\omega_0 t} \langle \cos\left(\int_0^t \omega(t')dt'\right) \rangle \\ &= e^{i\omega_0 t} s(t) \end{aligned} \tag{31}$$

In a time domain NMR experiment, the signal is generally phase-detected relative to the Larmor frequency; thus, s(t) is the quantity actually measured.

A typical FID for egg lecithin multilayers in the liquid crystalline phase is shown in Figure 14 (Chan et al., 1971). The FID contains contributions from all of the geminal

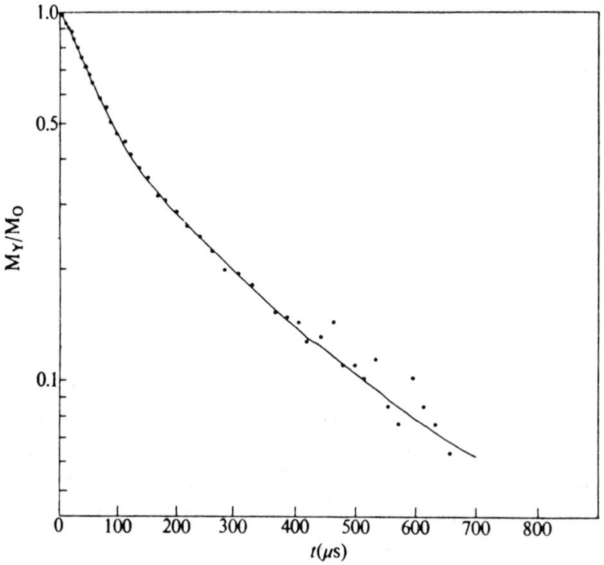

Fig. 14. The ^{1}H free induction decay of a lecithin bilayer sample at 30°C in a magnetic field of 14.1 kG (Chan et al., 1971)

methylene pairs, as well as the choline and terminal methyl protons. In the limit where the terminal methyl protons do not interact with the geminal proton pairs and there are no geminal interpair interactions, the time domain NMR spectrum should be a superposition of the FID signals of the individual segments. Accordingly, the frequency domain NMR spectrum is a superposition of the powder spectra of the individual segments, where the dipolar interaction for each segment is scaled by the order parameter for that segment. In actuality, however, the interpair dipolar interactions are not negligible and nearest neighbor interactions may be as much as 30% of the intrapair interaction.

Because of the complex character of the proton NMR lineshapes for phospholipid multilayers, it has been difficult to determine theoretical expressions which accurately describe the interactions present in these systems. All of the theoretical treatments argue that, in addition to incomplete averaging of static dipolar interactions, it is necessary to have effective axial symmetry in order to successfully predict the lineshape observed for phospholipid multilayers in the liquid crystalline phase (Chan et al., 1972; Wennerström, 1973; Seiter and Chan, 1973; Bloom et al., 1975, 1977). For this to be true, there must be rapid lateral diffusion of the phospholipid molecules in the plane of the bilayer, so that the intermolecular dipolar interactions become negligible[1]. This apparently is the case. In addition, the motions of the lipid molecules must take them symmetrically about a director. If effective axial symmetry is maintained, then the Hamiltonian for the multispin system can be written formally as

$$\langle \mathcal{H}_D \rangle (\theta') = \langle \mathcal{H}_D \rangle (0) \tfrac{1}{2} (3 \cos^2 \theta' - 1). \tag{32}$$

Because of the above symmetry property, the FID for a randomly oriented set of domains can be expressed as follows

$$s(t) = s(0) \int_0^{\pi/2} G_{\theta'}(t) \sin\theta' d\theta' \tag{33}$$

where

$$G_{\theta'}(t) = \sum_{n=0}^{\infty} \frac{(-1)^n}{(2n)!} M_{2n}(\theta') t^2 \tag{34}$$

In Eq. (34), the observed moments

$$M_{2n}(\theta') = M_{2n}(0) [P_2 (\cos \theta')]^{2n}$$

are the residual moments of the absorption line whose shape is determined for the time-averaged dipolar Hamiltonian. Note that for short times, the logarithm of the signal decays linearly with t^{-2}, from the slope of which the residual second moment can be determined. Bloom et al. (1977) have shown one should expect a t^{-1} dependence for the FID at long times. This t^{-1} dependence appears to be confirmed experimentally, but the above analysis neglects the homogeneous linewidths of the individual transitions, which contribute to the FID at long times. Recently, Hagan and co-workers (P.S. Hagan and S.I. Chan, unpublished) simulated the FID for the upper part of the hydrocarbon chain, where the order parameter has been shown to be nearly constant, by treating exactly the dipolar interactions between five geminal pairs of protons. The spectrum correspond-

1 Note that the intermolecular interactions strictly cannot vanish if lateral diffusion is confined rigorously to two dimensions.

ing to the Fourier transform of the calculated FID is shown in Figure 15. These results show that the early part of the FID is primarily sensitive to the wings of the bandshape

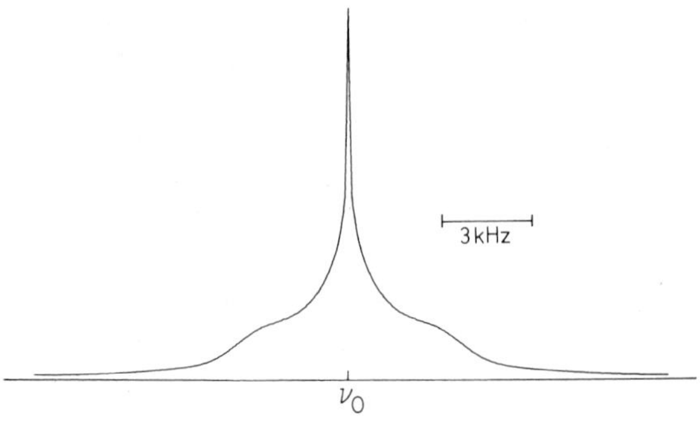

Fig. 15. The calculated powder spectrum expected for the five geminal proton pairs in the upper part of the phospholipid hydrocarbon chain where the order parameter is nearly constant

while the central component contributes to the FID at long times. The experimentally observed t^{-1} dependence of the FID at long times is also confirmed by the calculations, but the calculated magnitude of the t^{-1} component is smaller than the experimental value. This indicates that a sizeable contribution to the experimentally observed intensity in the central component of the bandshape comes from the methyl protons and from methylene protons with considerably smaller order parameters.

All of the above discussion has focused on proton-proton dipolar interactions. In view of the large amount of ^{13}C NMR work on membrane systems (Lee et al., 1974; James, 1975; Tiddy, 1975, 1977; Wennerström and Lindblom, 1977; Bocian and Chan, 1978), it is appropriate to consider $^{13}C - {}^{1}H$ dipolar interactions as well. Carbon-13 NMR studies of lipid systems are usually carried out in natural abundance or with enrichment at specific positions. Therefore, one usually deals with essentially isolated $^{13}CH_2$ geminal pair or $^{13}CH_3$ top. For the $^{13}CH_2$ geminal pair, there are dipolar interactions between the two protons and between the ^{13}C nucleus and each of the protons. Since the former interaction is stronger, one can approximate the system as a ^{13}C nucleus coupled to a singlet and triplet proton pair. Within this approximation the ^{13}C NMR transitions are expected at the following frequencies. For the triplet proton manifold,

$$h\nu = \gamma_1 \hbar H_0; \;\; \gamma_1 \hbar H_0 \pm \tfrac{1}{2} \frac{\gamma_1 \gamma_2 \hbar^2}{r^3} \left[(1 - 3\cos^2\theta_1) + (1 - 3\cos^2\theta_2) \right]. \qquad (35a)$$

For the singlet proton manifold,

$$h\nu = \gamma_1 \hbar H_0 \qquad (35b)$$

Here θ_1 and θ_2 denote the angles between the two different $^{13}C - {}^{1}H$ vectors and the magnetic field. These calculations show that the ^{13}C transitions associated with the singlet proton manifold and the $m_H = 0$ component of the triplet proton manifold is unshifted by $^{13}C - {}^{1}H$ dipolar interactions and hence not heterogeneously broadened. However, when the motion of the segment takes the two different $^{13}C - {}^{1}H$ interaction vectors over equivalent angular distributions, as has been shown in ^{2}H NMR measurements of membrane systems,

$$<1 - 3\cos^2\theta_1> = <1 - 3\cos^2\theta_2> \quad (36)$$

and the ^{13}C transitions associated with the $m_H = \pm 1$ components of the triplet proton manifold may be used to infer the motional averaging. These quantum effects are important even though ^{13}C NMR studies of membrane systems are usually carried out under conditions of proton noise decoupling, where the rate of proton spin flips is presumed to be rapid compared to the magnitude of the $^{13}C - {}^{1}H$ dipolar interactions. When the decoupling power is insufficient to completely decouple the protons, the observed lineshape can be artifactual, and care should be exercised in the interpretation of the signal intensities and spectral widths.

IV. Isotropic Motion in the Presence of Restricted Local Motion: Membrane Vesicles

In the previous section, we directed our attention toward the effect of restricted local molecular motions on the NMR spectra of condensed systems. The treatments and results are applicable to multilamellar bilayers and cell membranes, where the dimensions of the bilayer unit are microns or larger. Both the tumbling rate of these large bilayer units and the time necessary for individual molecules to diffuse laterally through one radian on the surface of the membrane are seconds or longer. These slow motions clearly cannot result in the further averaging of $\overline{\mathcal{H}}$ beyond that which occurs as a result of local segmental molecular motions.

Single-walled bilayer vesicles (Fig. 16), with or without membrane proteins, can be prepared by a variety of methods (Razin, 1972). Sonication of multilamellar dispersions of phospholipids yields these single-walled vesicles in varying sizes, with diameters ranging from 250Å-1000Å. Membrane protein reconstitution by the detergent dialysis method (Racker, 1972) often results in single-walled bilayer vesicles with intercalated protein, having diameters of 500Å or greater. These preparations are not generally homogeneous in particle size, a difficulty not often considered. While the smallest bilayer vesicles are expected to tumble isotropically, this is not necessarily the case for the larger membrane vesicles.

The extent to which collective motions of the bilayer unit and lateral diffusion of the individual lipid molecules are effective in removing the dispersion in the NMR spectrum produced by the angular dependence of $\overline{\mathcal{H}}$ depends, of course, on the rates of these motions compared with the magnitude of the dispersion which must be removed. On the basis of the expected orders of magnitude for the various interactions, chemical shift anisotropy is more readily averaged than dipolar or quadrupolar interactions. For the smallest membrane vesicles which can be prepared (diameter $\approx$ 250Å) the tumbling times, $\tau_V = \dfrac{4\pi R^3 \eta}{3kT} \approx 10^{-6}$ s, and the time-scale for a lipid molecule to diffuse one radian

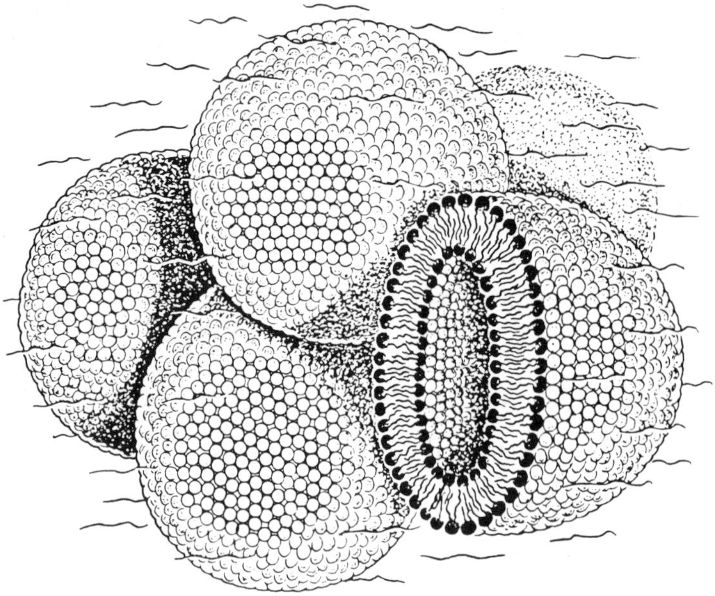

Fig. 16. A representation of single-walled bilayer vesicles. Each vesicle contains a volume of water which is separated from the extravesicular water (Khetrapal et al., 1975)

laterally across the surface, $\tau_{\ell d} = R^2/4D_{\ell d} \approx 4 \times 10^{-5}$ s, yield an effective averaging time-scale of 10^{-6} s. In the above relationships, R is the radius of the vesicle, η is the viscosity of the solution, and $D_{\ell d}$ is the lateral diffusion coefficient for phospholipid molecules in liquid crystalline phase bilayer membranes ($D_{\ell d} \approx 10^{-8}$ cm^2/s). Clearly, for very small vesicles the residual chemical shift anisotropy as well as dipolar and quadrupolar interactions will be averaged by these motions. The result is an NMR spectrum with more liquid-like, high resolution features. The relative importance of vesicle tumbling and lateral diffusion depends on the diameter of the bilayer unit. These two times become comparable for bilayer units with diameters of 1000Å but are then both much slower ($\tau_v \sim 10^{-4}$ s). For larger vesicles, whatever motional averaging occurs must necessarily be controlled by lateral diffusion of the molecules on the bilayer surface, but this rate becomes much too slow to average effectively all the residual first-order interactions, particularly the larger quadrupolar interactions.

There has been some controversy over whether tumbling of the bilayer unit, together with lateral diffusion of the lipid molecules on the bilayer surface, is sufficiently rapid to give rise to the high resolution spectra actually observed for smaller bilayer vesicles (Finer et al., 1972a, b; Horwitz et al., 1973; Seiter and Chan, 1973; Finer, 1974; Bloom et al., 1975; Lichtenberg et al., 1975; Stockton et al., 1976; Wennerström and Ulmius, 1976; Petersen and Chan, 1977). When the modulation of $\overline{\mathcal{H}}$ and the time variation of $\mathcal{H}(t) - \overline{\mathcal{H}}$ are taken together to estimate the resonance widths expected for a small bilayer vesicle, the calculated widths are approximately five to ten times larger than actually observed. It has been suggested that, due to the pronounced surface curvature of bilayer units with diameters of the order of 250Å, the packing of the lipid molecules is more disordered than in flat bilayer membranes (Horwitz et al., 1973; Seiter and Chan, 1973; Lichtenberg et al., 1975; Petersen and Chan, 1977). The expected result of this additional disorder is increased molecular and segmental motion, which may possibly occur at a more rapid rate as well. If this is the case, $\overline{\mathcal{H}}$ of the stationary bilayer unit would be decreased.

There is also the question of whether the rate of lateral diffusion of lipid molecules on the highly curved bilayer surface is more rapid. There is little evidence for this but, if true, $\overline{\mathcal{H}}$ would be modulated more effectively and would contribute to a lesser extent to the motionally averaged NMR lineshape. In principle, this point could be settled by studying the viscosity dependence of the observed vesicle NMR linewidths. The ^{31}P NMR spectra of vesicles have been shown to be strongly viscosity-dependent (Cullis, 1976). Although a corresponding viscosity dependence has not been observed for the proton NMR linewidths, the ^{31}P NMR results do indicate that $\tau_v < \tau_{\ell d}$ for small bilayer vesicles. Taken together, these NMR results indicate there is increased molecular motion when the bilayer membrane is highly curved. Whether the timescales of the local motions are also altered by curvature cannot be ascertained at this point, since both $\overline{\mathcal{H}}$ and $\mathcal{H}(t) - \overline{\mathcal{H}}$ contribute to the total linewidth. In the next section we shall show how some of these timescales can be inferred from relaxation time measurements.

The above difficulty with regard to the exact origin of the motional averaging in vesicles does not detract from the usefulness of these vesicles for membrane research. The motional information which can be obtained from these vesicles might not be representative of flat cell membranes, but should approximate the situation in intracellular storage organelles, such as synaptic vesicles and chromaffin granules, as well as highly curved regions of convoluted membrane systems, such as mitochondrial cristae. Because of their high resolution NMR spectra, they are also useful for the investigation of a number of membrane processes by NMR methods, for example, the inside-outside distribution of lipid molecules, phase separations, and ion permeabilities. These applications of bilayer vesicles to membrane studies will be discussed in more detail later.

V. Membrane Dynamics from NMR Relaxation Measurements

We have shown how anisotropic terms in $\overline{\mathcal{H}}$ can affect the NMR spectrum of membrane systems. We now turn to the so-called second order effects[2] arising from $(\mathcal{H}(t) - \overline{\mathcal{H}})$. These are manifested in the relaxation behavior of the nuclear spins. There are many interactions which can influence the relaxation behavior of the nuclear spins but for membrane systems the important ones are the same as those we discussed earlier, namely chemical shift anisotropy, nuclear dipolar and quadrupolar interactions. Scalar interactions arising from indirect (electron-coupled) spin-spin interaction can also contribute to relaxation in membrane systems, especially in small bilayer vesicles when the nuclei under study are coupled to quadrupolar nuclei (for example, ^{14}N).

2 The term $(\mathcal{H}(t) - \overline{\mathcal{H}})$ in the Hamiltonian can only contribute in second order since $\overline{\mathcal{H}(t) - \overline{\mathcal{H}})}$ is equal to zero.

There are two relaxation times in NMR spectroscopy, the longitudinal relaxation time, T_1, and the transverse relaxation time, T_2. The magnetic moments of the nuclei in the system together form the magnetization of the sample. For a weakly interacting spin system at equilibrium, this magnetization is directed along the applied Zeeman field (Fig. 17).

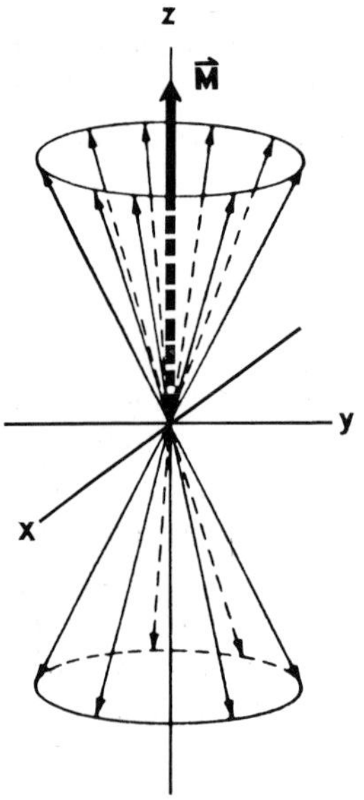

Fig. 17. The precession of an ensemble of identical magnetic moments of nuclei with $I = 1/2$. The net macroscopic magnetization $\vec{M}$ is directed along the external magnetic field (the z axis) (Farrar and Becker, 1971)

There is no net magnetization in the xy plane because there is a uniform azimuthal distribution of the spin magnetic moments of the individual nuclei; that is, there is no phase coherence. The relaxation time T_1 defines the timescale for the return to equilibrium of a nonequilibrium magnetization in the z-direction. Inasmuch as the magnetization in the z-direction is a measure of the distribution of the spins among the various Zeeman states, T_1 is a measure of the rate of return of a nonequilibrium spin population to equilibrium. Since there must be energy exchanged between the spin system and other degrees of freedom, frequently called the lattice, in order to reestablish the equilibrium, T_1 is frequently called the spin-lattice relaxation time. In a similar manner, the relaxation time T_2 defines the timescale for the return to equilibrium of a nonequilibrium magnetization in the xy plane. Since the magnetization in the xy plane is a measure of the phase coherence of the individual spin magnetic moments, the return to equilibrium does not require energy exchange. Two contributions to T_2 can be distinguished: (a) decay of a finite xy magnetization to equilibrium due to a dispersion of the NMR precessional frequencies, arising from first order interactions and static field inhomogeneities, and (b) dephasing of an otherwise homogeneous set of spin vectors due to second order fluctuations of the fields along the z-direction. The latter is the homogeneous contribution to T_2, and it is this on which we focus our attention in this section.

The relaxation time T_2 is typically measured by experimentally tilting the equilibrium magnetization by 90° and observing the decay of the magnetization in the xy plane

(Farrar and Becker, 1971). This dephasing signal is called the free induction decay (FID) and contains both homogeneous and inhomogeneous contributions. The spin-lattice relaxation time, T_1, can be measured experimentally by inverting the equilibrium magnetization to the negative z-direction and sampling its value at various intervals (Farrar and Becker, 1971) τ. The residual magnetization is sampled by quickly rotating it into the xy plane at a time $t = \tau$ and recording the FID. The latter can only be used when $T_1 \gg T_2$, which is the case for motionally restricted systems.

A detailed discussion of T_1 and T_2 is beyond the scope of this chapter. However, in view of the complexity of the subject matter, it is useful to outline certain elements of relaxation theory, in order to emphasize those features of the theory which pertain to membrane systems.

A. T_1

For simplicity, we consider two levels of a spin manifold of weakly interacting $I = \frac{1}{2}$ nuclei. A nonequilibrium spin population among these energy states can only return to equilibrium through transitions between the two spin levels. These transitions can be induced by interactions with fluctuating magnetic fields. Since there must be energy exchange between the spin and the lattice systems, and this energy exchange could only occur in quantized units of the spacing $\hbar\omega_0$ between the two spin levels, only the Fourier components of the fluctuating field which are near the Larmor frequency of the nuclei can contribute to this process. More specifically, if we write the fluctuating field as

$$\vec{H}(t) = \frac{1}{\sqrt{2\pi}} \int_{-\infty}^{\infty} g(\omega)\, e^{i\omega t} d\omega \tag{37}$$

then

$$1/T_1 \;\alpha\; |g(\omega_0)|^2 \; . \tag{38}$$

The frequency distribution $g(\omega)$ depends on the timescale of modulation of the interactions producing the fluctuating magnetic fields. For interactions which are modulated at a characteristic time, τ_c, the spectral distribution function maximizes at a frequency equal to τ_c^{-1}. For this reason the spin-lattice relaxation rate $1/T_1$ exhibits the correlation-time dependence shown in Figure 18. This figure shows that the spin-lattice relaxation rate is a maximum when the modulations of the interactions which couple the spins to the lattice occur near the Larmor frequency. In a given situation, the most efficient spin-lattice relaxation can be achieved by varying the factors affecting τ_c, such as temperature or viscosity, or by varying the NMR frequency.

The variation in the spin-lattice relaxation rate depicted in Figure 17 is appropriate for a molecule which reorients isotropically. In the case where the molecular motion is anisotropic, a plot of $1/T_1$ versus the τ_cs will yield a surface exhibiting multiple maxima, provided that the various correlation times are sufficiently temporally distinct. Thus, a rigid molecule undergoing axially symmetric anisotropic motion, where $\tau_\parallel \ll \tau_\perp$, may exhibit two maxima. For motionally restricted molecules, such as those found in mem-

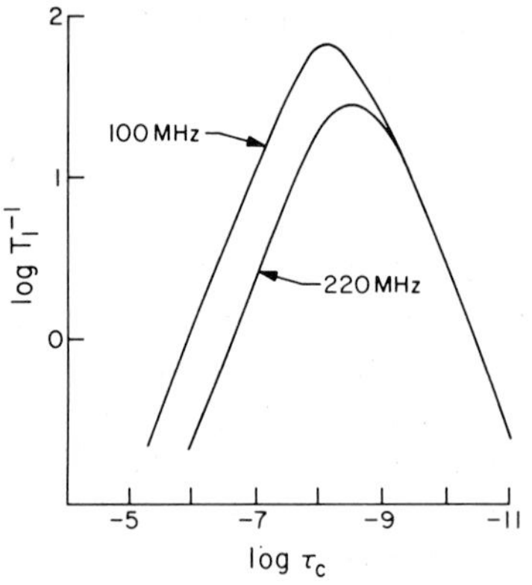

Fig. 18. Plot of the logarithm of the spin-lattice relaxation rate, T_1^{-1}, versus the logarithm of the correlation time, τ_c, for an isolated I = 1/2 spin pair undergoing isotropic orientation, where the NMR resonance frequencies are 100 MHz and 220 MHz

brane systems, similar behavior is expected, except that the spin-lattice relaxation rate may also depend on the average orientation of the molecule relative to the magnetic field. For example, for an assembly of methyl rotors where reorientation about the rotor axis is rapid and reorientation of the rotor axis is restricted to the range $-\Delta\theta' \leqslant \theta' \leqslant \Delta\theta'$, the dipole-dipole contribution to $1/T_1$ is given by the following expression (Feigenson and Chan, 1974)

$$\frac{1}{T_1} = \frac{9}{4}\frac{\gamma^4\hbar^2}{r^6}\left\{\tfrac{1}{4}\,\overline{(\sin\theta'\cos\theta')^2}\,\frac{2\tau_\perp}{1+\omega_0{}^2\tau_\perp{}^2} + \tfrac{1}{8}\,\overline{\sin^2\theta'(1+\cos^2\theta')}\,\frac{2\tau_c}{1+\omega_0{}^2\tau_c{}^2}\right.$$
$$\left. + \tfrac{1}{4}\overline{(\sin^2\theta')^2}\,\frac{2\tau_\perp}{1+4\omega_0{}^2\tau_\perp{}^2} + \tfrac{1}{8}\overline{(1+6\cos^2\theta'+\cos^4\theta')}\,\frac{2\tau_c}{1+4\omega_0{}^2\tau_c}\right\}.. \qquad (39)$$

Here, $1/\tau_c = 1/\tau_\perp + 1/\tau_\parallel$, where $\tau_\parallel$ and $\tau_\perp$ are the correlation times for reorientation about the rotor axis and for off-axis excursions respectively; θ' is the orientation of the methyl rotor axis in the magnetic field, and the bars denote averaging of the angular functions over the appropriate range of $\Delta\theta'$. Thus, for restricted motion, $1/T_1$ may depend on θ', depending on the amplitude of the off-axis excursion. For unrestricted anisotropic motion, the angular averages in the above expression are readily obtained, and Eq. (39) reduces to the well-known Woessner expression (Woessner, 1962) for $1/T_1$ of a methyl rotor undergoing rapid reorientation about the rotor axis, which in turn reorients isotropically in space.

$$\frac{1}{T_1} = \frac{9}{4}\frac{\gamma^4\hbar^2}{r^6}\left\{\left(\frac{1}{30}\right)\frac{2\tau_\perp}{1+\omega_0{}^2\tau_\perp{}^2} + \left(\frac{1}{10}\right)\frac{2\tau_c}{1+\omega_0{}^2\tau_c{}^2}\right.$$
$$\left. + \left(\frac{2}{15}\right)\frac{2\tau_\perp}{1+4\omega_0{}^2\tau_\perp{}^2} + \left(\frac{2}{5}\right)\frac{2\tau_c}{1+4\omega_0{}^2\tau_c{}^2}\right\} \qquad (40)$$

The above expressions for the dipolar contribution to the spin-lattice relaxation rate predict a complex frequency and temperature dependence. In the special case where $(\tau_\parallel\,\omega_0)^2 \ll 1$, the so-called extreme narrowing limit, and $(\omega_0\tau_\perp)^2 \gg 1$, the expression for $1/T_1$ is particularly simple and can be approximated by (Kroon et al., 1976)

$$1/T_1 \sim A\tau_{/\!/} + B\ \frac{1}{\omega_0{}^2\tau_\perp}\ . \tag{41}$$

This result predicts that the spin-lattice relaxation rate should approach $A\tau_{/\!/}$ with increasing frequency of observation, and that the frequency dependence of T_1 is only observed at sufficiently low frequency, with the frequency dispersion being more pronounced the lower the frequency of observation. This expected frequency dependence of T_1 is illustrated for the methylene protons in small sonicated vesicles in Figure 19. Note that Eq. (41) also predicts that a plot of $(1/T_1)\,\omega_0{}^2$ versus $\omega_0{}^2$ should be a straight line, which is apparently the case for lecithin multilayers.

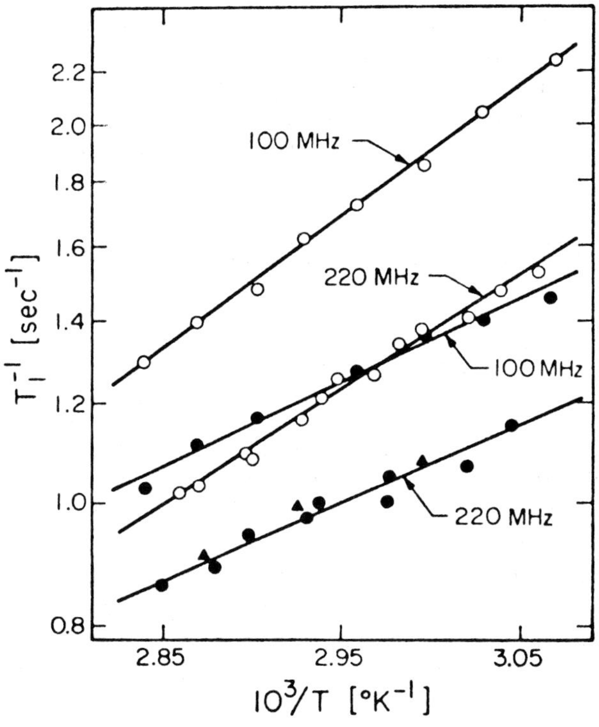

Fig. 19. The spin-lattice relaxation rates, $T_1{}^{-1}$, of the hydrocarbon chain methylene protons for small sonicated phospholipid bilayer vesicles as a function of reciprocal temperature and at NMR frequencies of 100 MHz and 220 MHz. ○, normal dipalmitoyllecithin; ●, 10% dipalmitoyllecithin dispersed in 90% dipalmitoyllecithin-d_{62}; ▲, 20% dipalmitoyllecithin-d_{31} dispersed in 80% dipalmitoyllecithin-d_{62} (Kroon et al., 1976)

The relative importance of the fast and slow motions to the spin-lattice relaxation rate can be inferred from the above frequency dependence. In general, it is more straightforward to ascertain this on the basis of the temperature dependence (Fig. 19). Inasmuch as both $\tau_{\|}$ and $\tau_\perp$ should increase with increasing temperature, the two terms should exhibit the opposite frequency dependence. In the case of lecithin bilayers, it has been argued on the basis of this temperature dependence that the spin-lattice relaxation rate is dominated by fast motion, namely $\tau_{//}$.

We have confined our remarks thus far to intramolecular dipolar interactions and their contribution to the spin-lattice relaxation rate. There are other contributions as well. For example, the intermolecular dipolar contribution has been measured by diluting normal lipid chains in a host system of perdeuterated lipids, and shown to be approximately 20% of the spin-lattice relaxation rate in vesicles (Kroon et al., 1976). These intermolecular dipolar interactions are modulated by lateral diffusion of the lipid molecules on the bilayer surface, as well as segmental motions within the molecule. The latter motions cannot be ignored, since they modulate the distance and the orientation of the internuclear vectors between molecules on a more rapid timescale than lateral diffusion. The problem is therefore very complicated and as yet the details have not been worked out. Nevertheless, the situation is not unlike that of anisotropic motion described earlier, and a similar temperature and frequency dependence is expected, and indeed has been observed experimentally. The only other mechanism which might be expected to contribute to the spin-lattice relaxation process is modulation of the chemical shift anisotropy. For ^{1}H and ^{13}C this contribution is probably not important, but for ^{19}F and ^{31}P the contribution of chemical shift anisotropy to $1/T_1$, can be as important as the dipolar contribution at sufficiently high magnetic fields.

For nuclei with $I \geqslant 1$, the dominant relaxation mechanism should be modulation of the nuclear quadrupolar interaction. The quadrupolar contribution to the spin-lattice relaxation rate for nuclei with $I = 1$ in a molecule undergoing isotropic reorientation is given by (Abragam, 1961)

$$1/T_1 = \frac{3}{40}\left(1+\frac{\eta^2}{3}\right)\left(\frac{e^2qQ}{\hbar}\right)^2\left[\frac{\tau_c}{1+\omega_0{}^2\tau_c{}^2}+\frac{4\tau_c}{1+4\omega_0{}^2\tau_c{}^2}\right]. \quad (42)$$

As is the case for nuclear dipolar interactions, the nuclear quadrupolar contribution to $1/T_1$ of nuclei in molecules undergoing restricted anisotropic motion will depend on the orientation of the molecules relative to the magnetic field. As yet, however, an explicit expression for $1/T_1$ has not been derived.

The molecular motions which can occur for a membrane system are expected to be extremely complex. There are motions which are principally molecular in nature and motions which involve a collection of molecules. For this reason, it may be an oversimplification to interpret T_1 data in terms of only two distinct correlation times. In doing so, we are not proposing that the system is totally defined by these two correlation times, but only that T_1 is sensitive to the faster motions, which tend to be molecular in origin rather than collective. In all likelihood, $\tau_{//}$ reflects segmental motions of the individual molecules, whereas $\tau_{\perp}$ represents only the most *probable* time for off-axis motions (Petersen and Chan, 1977). Inasmuch as the amplitude for the latter motions must necessarily be larger, a collection of molecules must be involved, and associated with these collective motions must be a range of frequencies. Consequently, a detailed characterization of this spectral distribution can only be determined by measuring T_1 over a wide range of frequencies.

Typical spin-lattice relaxation times for various nuclei in lecithin bilayers are summarized in Table 2. Inasmuch as these T_1s are temperature and frequency dependent,

Table 2. Spin-lattice relaxation times for lecithin in sonicated vesicles and multilamellar dispersions

Segment	T_1 (s)					
	Vesicles			Multilayers		
	^{13}C [a]	^{1}H [b]	^{2}H [c]	^{13}C [d]	^{1}H [e]	^{2}H [f]
$-CH_3$	2.8	0.76		1.9	0.39	
$-CH_2CH_3$	1.4					
$-CH_2CH_2CH_3$	0.64			0.43		
$-(CH_2)n$	0.40	0.47		0.50		
$-CH_2CO-$	0.26	0.34		0.39		
Carbonyl	1.8					
$-PO-CH_2$	0.41					0.028
$-CH_2-N-$	0.30					
$N(CH_3)_3$	0.62	0.41	0.046	0.39	0.40	0.080

[a] Godici and Lansberger, 1974. Temperature 34°C. NMR frequency 15.08 MHz.
[b] Horwitz et al., 1972. Temperature 40°C. NMR frequency 220 MHz.
[c] Stockton et al., 1974. Temperature 31°C. NMR frequency 15.4 MHz.
[d] Gent and Prestegard, 1974. Temperature 50°C. NMR frequency 20 MHz.
[e] Feigenson and Chan, 1974. Temperature 39°C. NMR frequency 220 MHz.
[f] Gally et al., 1975. Temperature 52°C. NMR frequency 13.8 MHz.

the values listed pertain only to the conditions specified. Note that the T_1s are not particularly sensitive to the state of aggregation of the lipids, that is, multilayers vis a vis vesicles. This result is to be expected since it is obvious from the above discussion that T_1 is determined principally by the high frequency motions in these systems.

B. Homogeneous T_2

As we mentioned earlier, T_2 is the characteristic time which describes the decay of a nonequilibrium magnetization in the xy plane. In general, and particularly for membrane systems, this decay is nonexponential. A reason for this is that both first order dispersions in the NMR spectrum and fluctuations in the magnetic field contribute to the decay. The latter is the so-called homogeneous contribution, which can be measured by spin-echo NMR techniques (Farrar and Becker, 1971). In such an experiment, the magnetization is rotated into the xy plane by an rf pulse and allowed to dephase for a time τ, at which point it is inverted by a second rf pulse. After a time 2τ the dephasing spin moments must necessarily refocus, so the variation with τ of the amplitude of the echo signal is a measure of the rate of homogeneous decay. The T_2s of selected protons in phospholipids in small bilayer vesicles have been measured by a modification of the basic spin-echo method, and are shown in Table 3.

The homogeneous transverse relaxation rate is the dephasing rate of an otherwise homogeneous set of spin vectors due to random fluctuation in the local fields seen by the individual spin moments. Although the various spins see the same average field, the instantaneous local fields vary from spin to spin, and accordingly there is a cumulative spreading of the spin vectors with time. For membrane systems where the homogeneous $T_2 \ll T_1$, this dephasing arises principally from fluctuation in the local fields along the

Table 3. Proton transverse relaxation times in sonicated vesicles of egg lecithin [a]

Segment	T_2 *(s)*
$-CH_3$	0.036
$-HC{=}CH-$	0.020
$CH_2-\overset{H}{\overset{\vert}{C}}{=}\overset{H}{\overset{\vert}{C}}-$	0.015
$-CH_2-$	0.056 (20%)
	<0.02 (80%)
$-CH_2-\overset{O}{\overset{\Vert}{C}}-O$	0.008
$-N(CH_3)_3$	0.075

[a] Horwitz et al., 1972. Temperature 40°C. NMR frequency 220 MHz.

Zeeman axis. Accordingly, the homogeneous T_2 depends essentially on the spectral density of these fluctuating fields at zero frequency. Hence, slow processes are expected to dominate this relaxation. Note that the same secular terms in the Hamiltonian which contribute to the first order dispersion, and hence the inhomogeneous T_2, also contribute to the homogeneous T_2. For the latter, however, it is the variance of these terms which is important rather than the mean. This is to be contrasted with the situation for the spin-lattice relaxation rate, which depends on those terms in the Hamiltonian which connect the various Zeeman states.

In a phospholipid multilamellar dispersion, only the local segmental motions and collective motions which tilt the chain axis can contribute to the homogeneous T_2. If the distribution of correlation times associated with these motions is large, which is likely to be the case for condensed systems such as membranes, then the homogeneous T_2 should be dominated by the slower of these motions. Moreover, the homogeneous T_2 is expected to be much less than T_1. At present, however, there are no accurate measurements of the homogeneous T_2 for a multilamellar dispersion. There has been only one attempt (Seiter and Chan, 1973) to average the first order dipolar interaction in the proton NMR spectrum by coherent averaging in spin space using a four-pulse sequence (WAHUHA; Waugh et al., 1968). The result of this spin averaging for a lecithin multilayer is shown in Figure 20. The proton NMR spectrum of the bilayer membrane is sharpened from several kHz to a width of ~ 100 Hz ($T_2 \cong 0.003$ s).

The situation in small bilayer vesicles is more difficult to treat. Here the residual first order interactions which are not modulated by the local segmental motions and collective motions of the molecules are averaged by a slower isotropic tumbling. As long as the rate of tumbling is sufficiently rapid compared with the frequency dispersion which must be averaged, there will be an additional homogeneous contribution to T_2. The additional dipolar contribution for an isolated two-spin system is given by (Seiter and Chan, 1973)

$$\left(\frac{1}{T_2}\right)_{\text{additional}} = \frac{4}{5}\left(\frac{3\gamma^2\hbar}{4r^3}\right)^2 S^2_{\gamma\gamma}\,\tau_v \tag{43}$$

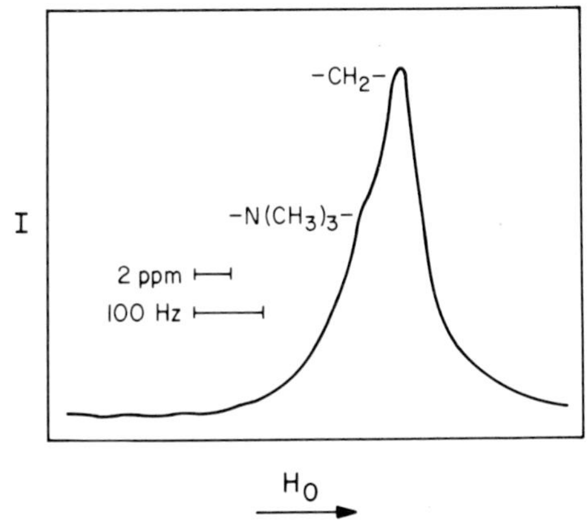

Fig. 20. Lecithin ^{1}H NMR spectrum at 55 MHz of dimyristoyllecithin bilayers, narrowed from an original width of ~ 3000 Hz by a Waugh multipulse sequence, with 60 μs cycle time for a complete eight-pulse cycle and intrumental resolution ~ 10 Hz. The chamical shift reference has been corrected for the scaling effect of the pulse sequence. The choline methyl and chain methylene protons (2.1-ppm separation) can be distinguished (Seiter and Chan, 1973)

while the analogous contribution in the case of symmetrical quadrupolar interaction is (Stockton et al., 1976)

$$\left(\frac{1}{T_2}\right)_{\text{additional}} = \frac{9}{80}\left(\frac{e^2qQ}{\hbar}\right)^2 S_{\gamma\gamma}^2 \tau_V \ . \tag{44}$$

In the above expressions τ_V is the correlation time for vesicle tumbling and $S_{\gamma\gamma}$ is the appropriate order parameter for the interaction being modulated in the otherwise stationary bilayer unit. It is important to recognize that $S_{\gamma\gamma}$ may be different for highly curved vesicles than for multilamellar systems. The reason for this is that the high surface curvature can alter the packing of the lipid molecules, introducing increased disorder with concomitant additional motions, which can further reduce the value of $S_{\gamma\gamma}$ in vesicles compared to that in multilayers (Petersen and Chan, 1977).

For the situation where $T_2 \ll T_1$, the homogeneous T_2 of the isolated two-spin system is directly related to the NMR linewidth of the nuclei being observed by the following expression

$$\Delta\nu = \frac{1}{\pi T_2} \ . \tag{45}$$

In the case of a multiple-spin system, however, such as the hydrocarbon chain in phospholipid molecules, the situation is much more complicated, in particular when the $S_{\gamma\gamma}$s are nonzero. Here the vesicle lineshape is a superposition of the Lorentzian lines of various widths if the rate of motional averaging is sufficiently rapid. A method of calculating this composite lineshape has been worked out by Wennerström and co-workers (Wennerström and Ulmius, 1976).

C. $T_{1\rho}$

The relaxation measurements described in the previous sections allow one to determine directly the characteristic times of motions which occur on the timescale of the inverse

Larmor frequency (MHz). However, collective motions of the lipid molecules in the bilayer unit, such as off-axis tilting of the molecules or the hydrocarbon chains, most likely occur over a wide range of frequencies. Therefore, in order to characterize adequately these motions, relaxation measurements should be made over as wide a frequency range as possible. A method which allows one to sample very slow motions directly, in the kHz regime, is the measurement of T_1 in the rotating frame, yielding what is usually referred to as $T_{1\rho}$.

$T_{1\rho}$ can be measured with the aid of a technique called forced transitory precession, or "spin locking" (Farrar and Becker, 1971). In this method the magnetization, $\vec{M}$, is first rotated to the y axis (in the rotating frame) by an rf pulse along the x axis. As soon as the pulse is completed, the phase of the rf is changed by $\pi/2$, which orients the perturbing field, $\vec{H}_1$, along the y axis. Since $\vec{H}_1$ and $\vec{M}$ are now colinear, no torque is exerted on $\vec{M}$, and $\vec{M}$ remains along the y axis. However, locking the spins along the y axis does not eliminate signal decay due to relaxation processes. The spin system will reestablish equilibrium along the y axis, where the equilibrium population distribution is now proportional to $\vec{H}_1$, which, in the rotating frame, plays the same role as the static field $\vec{H}_0$. The characteristic time for this relaxation is $T_{1\rho}$.

The dipolar contribution to the spin-lattice relaxation rate in the rotating frame can be shown to be, provided $H_1 \geqslant H_D$, the dipolar field

$$\frac{1}{T_{1\rho}} \quad \alpha \quad H_D{}^2 \; \frac{\tau_c^2}{1+4\omega_1{}^2\tau_c{}^2} \quad . \tag{46}$$

This expression is similar to that for $1/T_1$ in the laboratory frame, except that the precessional frequency, ω_1, of the spins in the field $\vec{H}_1$ has replaced the conventional Larmor frequency ω_0. Since $\omega_1 \ll \omega_0$, the maximum value of $1/T_{1\rho}$ occurs at a much lower frequency than $1/T_1$. It is this feature which permits one to sample motions at lower frequencies. Recently, $T_{1\rho}$ has been measured for a phospholipid bilayer dispersion as a function of the temperature and the magnitude of $\vec{H}_1$ (Fisher and James, 1978).

VI. Merits of Various Spin Probes in Membrane Studies

Earlier we outlined the effects of restricted motion in partially oriented systems on a number of NMR parameters. In order to measure and interpret these effects it is important that one chooses nuclear spin probes with the appropriate properties (Table 4). For example, only nuclei with spin $I \geqslant 1$ can exhibit quadrupolar interactions, and for dipolar interactions to be significant the interspin distance cannot be too great.

Beyond the purely spectroscopic or technical reasons for studying a particular spin probe, there are chemical and physical factors which may also influence the choice of the experimental probe. Thus chemical modifications or isotope substitutions may be necessary to optimize the experiments. In this section we outline specific advantages and limitations of a number of nuclei employed in studies of membrane structure and dynamics.

Table 4. Properties of nuclei relevant for membrane systems

Nucleus	Spin I (units of $h/2\pi$	Moment μ (units of β_N)	Q (e x 10^{-24} cm^2)	Resonance frequency (MHz/Tesla)	Chemical shift range (ppm)	Relative sensitivity	Natural abundance (%)	Relative detectability at natural abundance	Membrane location
^{1}H	1/2	2.79		42.577	10- 15	1	99.99	1	Ubiquitous
^{2}H	1	0.857	$2.77x10^{-3}$	6.536	10- 15	$9.6x10^{-3}$	$1.56x10^{-2}$	$1.5x10^{-6}$	Enrichment possible in many locations
^{13}C	1/2	0.702		10.705	200-300	$1.6x10^{-2}$	1.1	$1.7x10^{-4}$	Ubiquitous
^{14}N	1	0.403	$2x10^{-2}$	3.076	~ 600	$1x10^{-3}$	99.6	10^{-3}	Headgroup-interface
^{15}N	1/2	−0.283		4.315	~ 600	$1x10^{-3}$	0.3	$3x10^{-6}$	Headgroup-interface
^{17}O	5/2	−1.89	$-4x10^{-3}$	5.772	~ 700	$2.9x10^{-2}$	$3.7x10^{-2}$	10^{-5}	Headgroup-interface
^{19}F	1/2	2.63		40.055	~ 700	0.83	100	0.83	Substitution possible in many locations
^{23}Na	3/2	2.22	0.1	11.262	– –	$9.3x10^{-2}$	100	$9x10^{-2}$	In aqueous phase at surface
^{31}P	1/2	1.13		17.235	~ 700	$6.6x10^{-2}$	100	$6x10^{-2}$	Headgroup-interface

A. Proton NMR

Proton NMR measurements are technically the easiest to make, principally because of the high sensitivity and the large natural abundance of this nucleus. Since $I = \frac{1}{2}$, the nucleus has no quadrupole moment. It also exhibits very little chemical shift anisotropy, as expected on the basis of the small chemical shift range in ^1H NMR. As a consequence, practically all first and second order effects are dominated by dipolar interactions among the spins. This greatly simplifies the theoretical interpretation of the spectra.

Another important advantage of the proton probe is that it occurs naturally. Inasmuch as it is found in all parts of the lipid molecule, it can sample all regions of the bilayer membrane. However, the numerous spins cause complications, since there are multiple intra- and intermolecular dipolar interactions. Fortunately, it appears from a number of experiments that the geminal interproton interaction is dominant in the liquid crystalline phase (Seiter and Chan, 1973; Bloom et al., 1977). The remaining dipolar effects contribute no more than 30% of the second moment and are manifested in terms of additional broadenings and smoothing of the spectral features.

In multilamellar dispersions of phospholipids, the dipolar interactions give rise to inhomogeneously broadened lines with typical linewidths exceeding 3000 Hz (Lee et al., 1974; James, 1975; Tiddy, 1975, 1977; Wennerström and Lindblom, 1977; Bocian and Chan, 1978). Consequently, individual chemical shifts are not well resolved and much of the potential information pertaining to the separate regions of the bilayer is masked. It is well known that the proton NMR spectrum of phospholipid multilayers exhibits a super-Lorentzian lineshape. Within the framework of this interpretation, there has been a tendency to attribute the sharper features observed in the high resolution spectrum to the methylene protons, while the methyl protons are ignored (Wennerström and Ulmius, 1976; Bloom et al., 1977). Recent evidence indicates, however, that the methyl protons do contribute significantly to the super-Lorentzian features of the lineshape, as was originally proposed by Feigenson and Chan (1974). To support this latter interpretation, it has been observed that the sharper features become more intense as the number of methyl groups in the molecule is increased (Lindsey et al., 1979). Thus, the composite lineshape is a superposition of distinct sharp and broad signals. Because of this interference and instrumental limitations, accurate measurement of the width and intensity of the broad components is difficult, since correct baselines are not easy to ascertain. This problem can be solved, in principle, by measuring the early part of the FID (Chan et al., 1971), but if the decay is too rapid there are difficulties in this measurement as well. These can be alleviated in part by employing pulsed NMR echo techniques, and analyzing the shape of the resulting echo (Bloom et al., 1977).

With small vesicles, isotropic tumbling is sufficiently rapid to remove any inhomogeneity arising from first order dipolar interactions. This modulation of $\overline{\mathcal{H}}$ by vesicle tumbling, together with the modulation of $\mathcal{H}(t) - \overline{\mathcal{H}}$ by local motions, leads to linewidths which are about 20 Hz for these systems. Accordingly, all proton types are resolved in the spectrum, and it has been possible to measure their T_1 and T_2 by conventional techniques (see Tables 2 and 3). For those protons which are part of the central portion of the acyl chains, chemical shift dispersion is not sufficiently large to observe distinct resonances. The methylene signal is, therefore, inhomogeneous. In fact, at sufficiently high magnetic fields, the methylene signal is observed to be asymmetric (Kroon et al., 1976).

Because of the high resolution nature of the vesicle proton NMR spectrum, these systems are ideal for NMR studies and, in fact, have been utilized extensively for NMR investigation of membrane systems. Some of these typical applications will be discussed later in this chapter.

B. Deuteron NMR

Deuterium has in recent years played a major role in bilayer research and a separate chapter has been devoted to this subject. The principal advantage of ^{2}H NMR is that quadrupolar interactions dominate completely, making spectral interpretation relatively straightforward. This is particularly the case for multilamellar samples, both oriented and powder, where first order effects dominate. In fact, information about molecular order has come principally from these studies (Seelig, 1977). For vesicles, $\overline{\mathcal{H}}_Q$ is averaged and, hence, the information on orientational order is lost. Moreover, the quadrupole interaction is large compared to the chemical shift dispersion in the ^{2}H spectrum, so that only limited information can be gained from vesicle studies unless the lipid molecule is deuterated at a specific location. Of course, the greatest disadvantage of ^{2}H studies is that the necessary chemical modifications are often tedious and laborious.

C. Carbon-13 NMR

Although a large number of ^{13}C NMR experiments have been performed on lipid bilayer systems (Lee et al., 1974; James, 1975; Tiddy, 1977; Wennerström and Lindblom, 1977; Bocian and Chan, 1978), the full potential of this probe has not yet been realized. In principle, ^{13}C NMR can be used to sample all portions of the bilayer, since carbon is located throughout the molecule. Also the ^{13}C chemical shift range is larger than 100 ppm, and carbon resonances from the different environments are generally resolved. This is particularly the case when the ^{13}C spectra are obtained with proton noise decoupling. Unfortunately, under these conditions the high power proton irradiation removes all first order dipolar and scalar interactions between the ^{13}C and ^{1}H spins. Accordingly, a vast amount of information, especially concerning molecular order, is lost. There is the further complication that the ^{1}H decoupling power applied is insufficient to remove adequately *all* the dipolar interactions. The resulting partially decoupled spectra are then not meaningful. Thus the failure of several workers to obtain ^{13}C spectra of multilamellar bilayer systems or of vesicles in the gel phase is merely a consequence of insufficient decoupling. When adequate decoupling power is used, only chemical shift anisotropy can contribute in first order to the ^{13}C lineshape. This anisotropy has been shown to be approximately 25 ppm below the transition temperature and 10 ppm above, but as yet these results have not been interpreted (Urbina and Waugh, 1974).

In view of the above problems, the majority of the ^{13}C NMR studies have been performed on vesicles in the liquid crystalline phase, where local motions and vesicle tumbling average all the first order effects. Even for these studies, broad band ^{1}H decoupling is used to remove the ^{1}H-^{13}C scalar couplings, and to obtain nuclear Overhauser enhancements (NOE) for the ^{13}C signals. Again, when these spectra are obtained with adequate decoupling power, the linewidths of the ^{13}C resonances can be used to obtain information about those processes which modulate the ^{1}H-^{13}C dipolar interactions and ^{13}C chemical shift anisotropy.

Partially relaxed Fourier transform spectroscopy has been widely used to obtain T_1s of the individual carbons in these vesicle systems (see Table 2). These ^{13}C T_1 data have been instrumental in introducing the concept of a *motional* gradient along the hydrocarbon chain. Unfortunately, the interpretation of ^{13}C T_1 data has been greatly oversimplified. The tendency has been to interpret the data in terms of an isotropic motional model rather than in terms of restricted anisotropic motion, which is known to exist in these systems. It is clear that ^{13}C NMR can contribute to the understanding of the motional state of the lipid bilayer when additional measurements are performed over a range of frequencies in order to map the dispersion of correlation times, and when all the available data are interpreted in terms of a more realistic model.

One of the limitations of ^{13}C NMR for membrane research is the low natural abundance of these nuclei. However, it is possible to enhance selectively specific signals above the background by chemical enrichment with the ^{13}C isotope, either by direct chemical synthesis or by biosynthetic incorporation. Selective enrichment helps overcome the low sensitivity, but with the increased use of Fourier transform NMR spectroscopy, low sensitivity is becoming less of an obstacle, although experiments remain time-consuming.

D. Fluorine-19 NMR

Fluorine is not a natural constituent of membrane lipids, but may under favorable conditions be incorporated synthetically or biosynthetically into phospholipid components of the membrane. Because of the large electronegativity of fluorine atoms, some perturbations are introduced. For example, bilayers composed of 1,2-(7,7-difluoropalmitoyl) phosphatidylcholine exhibit a phase transition temperature approximately 10°C lower than that of the unsubstituted lipid (Longmuir et al., 1977). This may be compared with a 5°C lowering of the transition temperature upon perdeuteration of the fatty acid chains in the phosphatidylcholines (Petersen et al., 1975). Similarly, difluoro-substituted fatty acids incorporated biosynthetically into *E. coli* phospholipids cause some alteration in membrane structure and function (Chien-Ho, personal communication).

In spite of the perturbations introduced by fluorine substitutions, ^{19}F NMR may well become a useful tool in membrane research. The reason for this optimism is that ^{19}F chemical shifts are extremely sensitive to the environment, and the ^{19}F spectra provide good monitors of subtle changes in bilayer structure and dynamics. This is perhaps best illustrated by the finding that two well-resolved ^{19}F resonances are obtained for vesicles prepared from 1,2-(difluoropalmitoyl) phosphatidylcholines (Longmuir and Dahlquist, 1976). These two resonances arise from the difference in the structural and motional properties of the chains in the two monolayers of the highly curved vesicle bilayer.

Interpretation of ^{19}F spectra is more difficult than that of ^{1}H spectra simply because both chemical shift anisotropy and dipolar interactions contribute to the spectrum. The heteronuclear (^{19}F-^{1}H) and homonuclear (^{19}F-^{19}F) dipolar interactions are comparable and even though first order heteronuclear effects can be decoupled, the second order contribution remains. At this time, the theoretical progress and experimental applications have been too limited to provide a basis for predicting the success of this technique. Clearly the use of ^{19}F NMR merits more attention in future work.

E. Phosphorus-31 NMR

The potential of ^{31}P NMR as a probe of the lipid headgroup and the lipid-water interface in general was recognized early. It was not, however, until recent years, when Fourier transform ^{31}P NMR became routine, that detailed information about the structure and dynamics of the headgroup region emerged from these studies. The principal advantages of the ^{31}P probe are the high natural abundance and the sensitivity in chemical shift to environment and chemical structure. Moreover, the chemical shift tensor is highly anisotropic for the phosphate diester group (Griffin, 1976; Kohler and Klein, 1976).

As in the case of ^{13}C NMR there is some loss of information as a result of the ^{1}H decoupling typically used in ^{31}P NMR experiments, but in the present case the ^{1}H-^{31}P heteronuclear dipolar interactions are overshadowed by the large chemical shift anisotropy. Careful measurements of the chemical shift tensor values and identification of the principal axis system for a number of phospholipids have allowed interpretation of the ^{31}P NMR spectra in terms of headgroup orientation and motion (Kohler and Klein, 1976).

The above chemical shift anisotropy is modulated by vesicle tumbling and lateral diffusion of the lipid molecules in bilayer vesicles. This chemical shift anisotropy is apparently not significantly reduced upon sonication of the layer unit, and in view of its magnitude, its modulation contributes significantly to the ^{31}P resonance widths of vesicle systems. This contribution to the linewidth is proportional to H_0^2, so that it becomes increasingly important at higher magnetic fields. Cullis (1976) has in fact examined the viscosity dependence of the ^{31}P linewidth in the high field limit, in order to separate the contributions to the resonance width of vesicle tumbling and lateral diffusion of lipid molecules on the bilayer surface. In this manner, a lateral diffusion coefficient of $D \simeq 2 \times 10^{-8}$ cm^2/s was determined for lipid molecules in bilayer vesicles.

Phosphorus-31 chemical shifts are large, and accordingly, ^{31}P NMR affords a method for distinguishing the NMR signals arising from different lipids or from lipid molecules in different environments. Through exploitation of this sensitivity, ^{31}P chemical shifts have been used to monitor such processes as thermal phase transitions, phase separation, surface curvature effects, and the distribution of lipid molecules in the two monolayers of the bilayer in both single-component and mixed lipid vesicle systems (Michaelson et al., 1973; Berden et al., 1975; James, 1975; Yeagle et al., 1976).

In order to perform the above studies, it is necessary that the separation of resonances is not masked by broadening due to chemical shift anisotropy. Since the resolution of two chemically shifted resonances improves linearly with an increase in the applied magnetic field while the chemical shift anisotropy is proportional to H_0^2, the increased linewidths at high fields can mask the differences in chemical shifts. Thus, depending on what type of processes are being investigated with the ^{31}P probe, there is an optimum field for performing the ^{31}P NMR experiment. Based on the observed magnitudes of the ^{31}P chemical shift anisotropy and the ^{31}P chemical shift range, this field is approximately 6 tesla.

VII. Selected Applications of NMR to Membrane Studies

We will close this chapter by illustrating a number of selected applications of NMR to the study of membrane systems. Because of the limited space available, the illustrations are in no way intended to be exhaustive. These topics and others are discussed in more detail in a number of recent excellent reviews of NMR studies of membrane systems. These range from comprehensive descriptions of the current literature (Tiddy, 1975, 1977) to more specialized critical reviews of certain topics within this very large field (Lee et al., 1974; James, 1975; Seelig, 1977; Wennerström and Lindblom, 1977; Bocian and Chan, 1978).

A. The Motional State of the Hydrocarbon Region

Perhaps the most singularly important contribution of the NMR method to membrane research is the detailed elucidation of the motional state of the phospholipid bilayer. Two types of information have emerged from these studies and it is almost certain that a more complete picture will emerge as the field is further developed.

Deuterium NMR studies of multilamellar phospholipid systems have contributed detailed information about molecular order in different regions of the membrane (Seelig, 1977). This method has permitted order parameters to be determined for essentially all segments of the phospholipid molecule. From these results have emerged the *flexibility* gradient of the hydrocarbon chain, and the average orientation of the headgroup. Both of these topics will be discussed in further detail in the next chapter.

Despite the inherent difficulties in determining molecular order from ^1H NMR measurements, important information has nonetheless been obtained concerning the order in certain regions of the phospholipid molecules in bilayer membranes. Seiter and Chan (1973) have determined the ^1H order parameter for the upper portion of the hydrocarbon chain, where the order parameter is nearly constant. This orientational order is similar to that measured by ^2H NMR for this region of the molecule. Since the molecular interaction vectors sampled by the two different NMR methods are averaged over different angular regions of the distribution function for the spatial orientation of a methylene segment, it is possible to define more precisely those motions contributing to the motional averaging. Through comparison of the ^1H and ^2H order parameters, Petersen and Chan (1977) have suggested that the NMR data are best interpreted in terms of two distinct types of motion, one fast and the other slow.

Petersen and Chan have related the faster of the two motions proposed to chain isomerization and the slower to chain reorientation. In contrast to an EPR experiment, in an NMR experiment both of these motions are sampled. The timescale of observation in an EPR experiment is very short ($< 10^{-8}$ s), so that motions which occur on a timescale longer than this observation time appear to be static. For this reason, chain fluctuations are manifested in terms of a distribution of chain orientations even in EPR studies of oriented bilayers, and this observation has led to the concept of bent or tilted chains in these systems (McFarland and McConnell, 1971). Two distribution functions which explain either or both the EPR and NMR data are shown in Figure 21. In an NMR experiment the timescale of observation is long, so that the experiment samples over the distri-

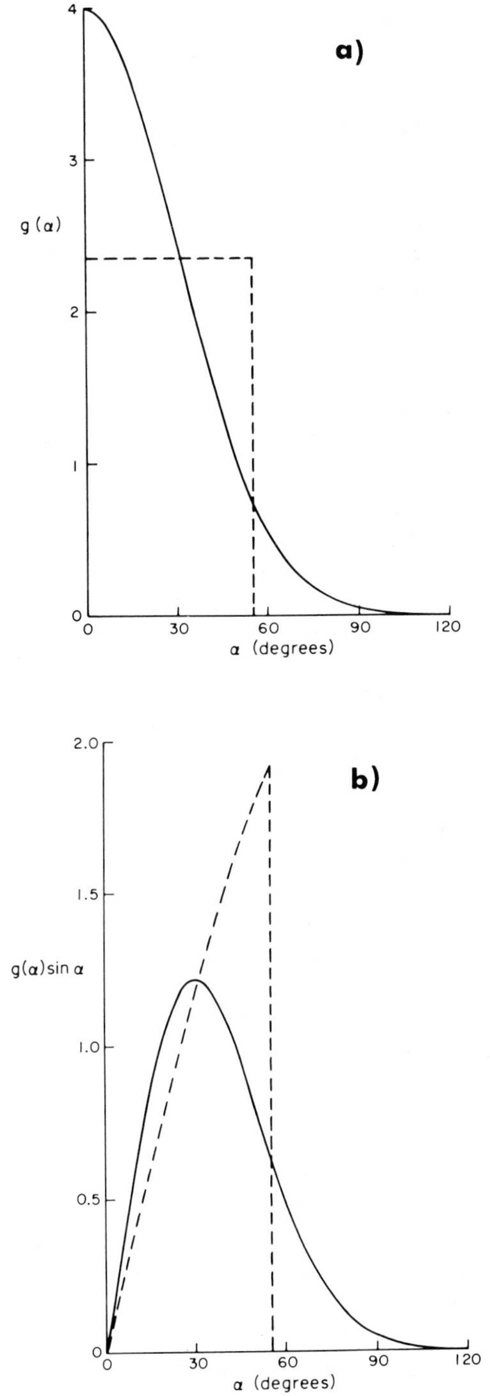

Fig. 21 a and b. Two distribution functions, g(α), for chain reorientation in phospholipid molecules which explain the EPR and NMR data. (a) Plots of g(α) versus α for a constant distribution within the angular range $0 \leqslant \alpha \leqslant 55°$ (- - - -) and for a normal distribution with a standard deviation $\alpha_0 = 30°$ (——). (b) The normalized weighted distributions g(α) sin α for a constant distribution function (- - - -) and a normal distribution function (——) with the same angular range and standard deviation as in (a)

bution function, and the order parameter must reflect this averaging. In fact, Petersen and Chan (1977) showed that under these conditions

$$S_{measured} = S_\gamma \cdot S_\alpha \tag{47}$$

where S_α is the order parameter associated with chain orientation and S_γ is the order parameter associated with chain isomerization. The idea is that only S_γ is measured in an EPR experiment, whereas the product, $S_\alpha \cdot S_\gamma$ is sampled in the NMR experiments. Both the value of S_α measured for the upper part of the hydrocarbon chain and the EPR results indicate that the maximum angular deflection of the chain from the bilayer normal is around 60°, while the most probable deflection is around 30°.

Measurement of the order parameter only allows the limits of the timescales for the motional processes to be ascertained. The actual timescales of the motions are better determined by NMR T_1 measurements. On the basis of ^{1}H NMR T_1 measurements, Feigenson and Chan (1974) estimated the following ranges of correlation times for the two types of motion in the liquid crystalline phase of phospholipid systems

$$\begin{aligned}\tau_\parallel &\approx 10^{-9} - 2 \times 10^{-10}\ \mathrm{s}\\ \tau_\perp &\geqslant 10^{-7}\mathrm{s}\end{aligned} \tag{48}$$

Later, Petersen and Chan (1977) offered a detailed interpretation of the motions associated with these correlation times. They suggest that $\tau_\parallel$ is associated with propagation of kinks along the chain via a δ-coupled trans-gauche isomerization (kink diffusion; Fig. 22).

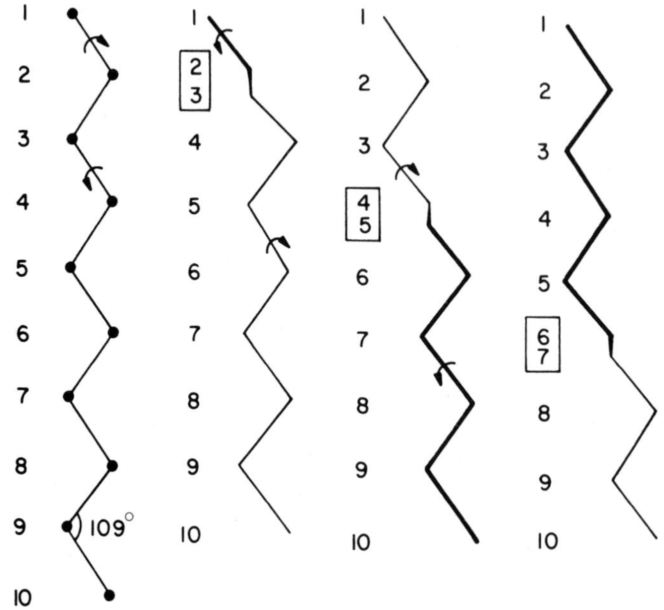

Fig. 22. δ-coupled trans-gauche ± rotations around a kink leading to kink diffusion. *Arrows:* direction of the rotations keeping the higher numbered carbon fixed; the sequence illustrated corresponds to, from left to right: all-trans→k_1^+→k_3^-→k_5^+. The carbon atoms are numbered, and the pair of methylene segments which are oriented off-axis with respect to the chain are indicated by the rectangle around the carbon numbers. This kink diffusion sequence will effect every methylene segment along the chain (Petersen and Chan, 1977)

More specifically, $\tau_{\parallel}$ is the characteristic time for a kink to diffuse along the lipid chain and subsequently return to its point of origin on the acyl chain. Petersen and Chan associated the slower correlation time $\tau_{\perp}$ with chain reorientations, and suggested that these motions reflect both motions of individual molecules as well as collective motions of domains of molecules in the bilayer unit. Since these motions must necessarily occur over a wide range of frequencies, the measured correlation time really then reflects the most effective combination of timescales and amplitudes of chain reorientation for the spin-lattice relaxation process.

The effect of extreme surface curvature on the structural and dynamical properties of bilayer membranes has also emerged from NMR studies. The bulk of the conclusion has been derived from detailed analysis of the factors affecting the vesicle linewidths. As pointed out earlier in this chapter, the linewidth for a pair of protons which are undergoing slow isotropic motion in the presence of rapid restricted anisotropic local motion can be approximated by

$$\Delta\nu = \frac{1}{\pi T_2} = <\mathcal{H}>^2\tau_e + <(\mathcal{H}(t) - \bar{\mathcal{H}})^2>\tau_{1\sigma c} \quad (49)$$

where τ_e is the characteristic time for the slow isotropic motion and $\tau_{1\sigma c}$ is the characteristic time pertinent to the faster local motions. For such a geminal pair in a bilayer vesicle, the relevant motions for the former are isotropic tumbling of the vesicle and lateral diffusion of the lipid molecules on the bilayer surface, so that $1/\tau_e = 1/\tau_v + 1/\tau_{\ell d}$. For the contribution from local motion, the important timescale is the slower of the two motions affecting the spin-lattice relaxation rate, that is, $\tau_{\perp}$. On the basis of the order parameters which have been determined for the methylene protons of the acyl chain, one can show that the first term in Eq. (49) should contribute significantly to the linewidth in vesicle systems. If this is true, then one expects a pronounced dependence of the observed linewidths on the solution viscosity since, $\tau_v \ll \tau_{\ell d}$. However, no such pronounced voscosity dependence of the methylene linewidth has been observed. This important observation has led two groups of workers (Sheetz and Chan, 1972; Horwitz et al., 1973) to conclude that the order parameter in stationary highly curved vesicles must be reduced from that in multilamellar systems. This conclusion is actually reasonable, since extreme curvature should introduce disorder in the packing of the lipid molecules, which in turn will introduce additional local motions which can further average the dipolar interactions. Hagan et al. (P.S. Hagan and S.I. Chan, unpublished) have obtained a more exact expression for the linewidth in vesicle systems within the motional model defined by Petersen and Chan (1977). On the basis of detailed analysis these workers concluded that upon sonication (i) the order parameter is reduced by a factor of 2-3, primarily through an increase in the average angle of deflection of the lipid chains from the bilayer normal, and (ii) the correlation time for chain reorientation $\tau_{\perp}$ is reduced by about a factor of 10, from 10^{-7} s to 10^{-8}-10^{-9}s.

B. Headgroup Orientation and Dynamics

Much of the motional information about the headgroup has come from ^{31}P and ^{2}H NMR experiments (Gally et al., 1975; Seelig and Gally, 1976; Seelig et al., 1977). Some information has emerged from ^{1}H NMR studies, but since the choline methyl proton signals are the only ones resolved in the ^{1}H spectrum, this information is necessarily confined to these methyl tops (Feigenson and Chan, 1974). However, improved ^{31}P and ^{2}H NMR measurements have allowed a more detailed interpretation of the headgroup orientation, conformation, and dynamics.

The determination of the values of the ^{31}P chemical shift tensor components and the orientation of the principal axes system make it possible to express the residual chemical shift anisotropy, σ_a, in terms of the order parameters of the principal axes connecting the two esterified and two nonesterified oxygens, S_{aa} and S_{cc}, respectively (Niederberger and Seelig, 1976). As discussed earlier two order parameters are necessary because the static shielding tensor is not axially symmetric. Since the ^{31}P NMR cannot yield both order parameter values, ^{2}H NMR measurements have been used in an attempt to provide a more complete description of the headgroup conformation (Gally et al., 1975; Seelig, 1977). These studies suggest a boomerang-shaped conformation for the headgroup (Fig. 23; Seelig et al., 1977), which is in accord with X-ray and neutron diffraction data (Franks, 1976; Worcester and Franks, 1976).

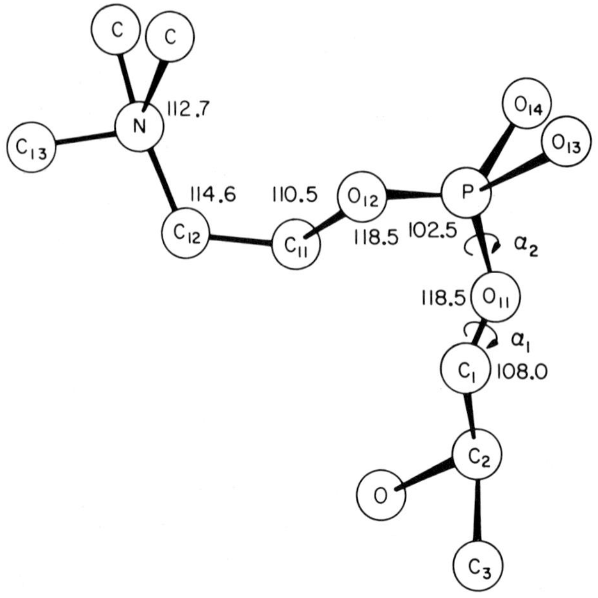

Fig. 23. Conformation of the headgroup in glycerylphosphorylcholine (Seelig et al., 1977)

In addition to the conformational model, there are currently two descriptions of the restricted motion of the headgroup. The data presently available cannot distinguish between these motional models. Kohler and Klein (1977) have found that two schemes of simple uncoupled rotations can successfully simulate the observed reduction of the ^{31}P shielding tensor. The first combination of headgroup motions (A) consists of a fast rotation about the P-O_{11} bond followed by a slower overall rotation about the glycerol C_1-C_2 bond, followed by a still slower overall rotation about the long molecular axis. The second possible combination of motions (B) includes only two rotations, a fast rotation about P-O_{11}, followed by a slower rotation about the long molecular axis. Kohler and

Klein also found that the temperature dependence of the residual chemical shift anisotropy is reproduced only when a wobble of the headgroup about the long axis is included. On the other hand, Seelig and Gally (1976) assume a fixed conformation about the bonds C_1-O_{11} and P-O_{11} and allow torsional angles α_1 and α_2 (see Fig. 23) to jump between specific values. Using these angles, the shielding tensor is transformed to a reference frame rotating about C_1-C_2 and averaged. Further averaging is performed by employing the ^{2}H NMR order parameters for the headgroup as a measure of the motion about C_1-C_2. It is not possible to distinguish between the above descriptions of the headgroup motion on the basis of the data available.

C. Lateral Diffusion

Lateral diffusion of the lipid molecule has been measured directly by both ESR and NMR methods. The ESR spin-label studies provided a value for $D \sim 10^{-8}$ cm^2/s (Devaux and McConnell, 1972; Träuble and Sackman, 1972). In practice, direct lateral diffusion measurements by NMR have been difficult. The conventional NMR technique for measuring diffusion is that of the pulsed magnetic field gradient spin-echo method (Farrar and Becker, 1971). In this experiment a spin-echo is generated at time 2τ by a ($\tau/2 - \tau - \pi$) pulse sequence in the traditional way, but a large field gradient is imposed on the sample twice, just before and just after the π-pulse. The echo amplitudes measured at different delay times, τ, depend on the diffusion of the molecules through the field gradient, and the diffusion coefficient can be extracted in a straightforward manner. However, in systems where large static dipolar interactions are present and T_2 is very short, the sample must be oriented and aligned at the "magic angle" to suppress the manifestations of these inhomogeneous contributions to T_2. A further complication initially encountered in the measurement of lateral diffusion by NMR methods was the interference from the proton signals of residual water in the lipids, even when measurements were made in D_2O. The latter problem has recently been resolved however, and a lateral diffusion coefficient of $\sim 5 \times 10^{-8}$ cm^2/s has been reported for dilauroyllecithin (Lindblom et al., 1976).

Recently Cullis (1976) also measured the lateral diffusion coefficient for lipids in sonicated vesicles by studying the ^{31}P NMR linewidths of the choline headgroups as a function of solution viscosity. As was pointed out earlier in this chapter, these ^{31}P linewidths at high magnetic fields of observation are determined primarily by modulation of the chemical shift anisotropy. If we include, in addition, modulation of the residual static dipolar interactions by these same motions, as well as contributions to the linewidth due to the local motions of the lipid molecules, then the linewidth can be approximated by

$$\Delta\nu - B \cdot \tau_\perp = A \cdot \left(\frac{3kT}{4\pi\eta R^3} + \frac{4D}{R^2}\right)^{-1} . \tag{50}$$

By varying the solution viscosity of the vesicle preparation over a large range, Cullis was able to attain a condition where the vesicle tumbling contribution to the linewidth became comparable to that arising from lateral diffusion. In this manner, Cullis obtained values of 2 x 10^{-8} cm^2/s and 10^{-9} cm^2/s for D in the liquid crystalline and gel phase, respectively, of phospholipid bilayer vesicles. Although the value determined for the gel phase is clearly an upper limit, the value obtained for the liquid crystalline phase is similar to that determined for multilamellar systems (Wu et al., 1977).

D. Studies of Phase Transitions and Lipid Mixtures

Of the numerous techniques employed to study the thermal phase transition in bilayer membrane systems, NMR has played a relatively minor role. The principal reason for this is that the low temperature gel phase is motionally so restricted that NMR experiments are difficult. Moreover, the spins located on different molecules are strongly coupled and the interpretation of the data cannot be made in simple molecular terms. The exception to this, of course, is when the experiments are performed under conditions where all the inter- and intramolecular static dipolar interactions are removed by coherent averaging of these interactions in spin space using multiple-pulse sequence techniques, by high-speed spinning of the sample at the magic angle, or by high-power proton decoupling in the case of ^{13}C and ^{31}P NMR experiments. A good example of the latter is provided by recent ^{31}P NMR studies of phospholipid systems. These studies permitted the ^{31}P chemical shift anisotropy to be investigated as a function of hydration and temperature, and have revealed a large amount of information regarding motions of the headgroup as the thermal phase transition is traversed (Griffin, 1976).

Because of the extreme motional restriction in the gel phase, the NMR lines are broadened to a considerable extent, even for bilayer vesicles. This is so because the tumbling of the bilayer unit and lateral diffusion are too slow to average the static dipolar interactions. For this reason, magic angle effects can be observed in the partially averaged NMR spectrum, provided that surface curvature sufficiently disrupts the packing of the hydrocarbon chains, which is apparently the case for highly curved vesicles (Lichtenberg et al., 1975). In addition, the extent of broadening of the NMR lines observed for the gel phase renders detection of the full signal intensities difficult with conventional high resolution NMR spectrometers. This *apparent* loss of intensity of the NMR signals has been used as an empirical indicator of the transition from the gel to the liquid crystalline phase in phospholipids systems. This method was first used to compare the thermal melting of vesicles of different sizes (Sheetz and Chan, 1972). The approach has been extended by von Dreele and Chan (P.H. von Dreele and S.I. Chan, unpublished) to investigate the miscibility of lipid components in vesicles. These workers measured the phase transition curves for equimolar mixtures of phospholipids of different chainlengths in which the acyl chains of the higher temperature melting lipid within the mixture were perdeuterated. In this manner, they were able to measure specifically the melting or freezing of one lipid component in another. The melting behavior of dimyristoyllecithin, and dimyristoyllecithin in 1:1 dimyristoyllecithin-dipalmitoyllecithin and 1:1 dimyristoyllecithin-distearoyllecithin vesicles is shown in Figure 24.

The application of perdeuterated phospholipids to allow observation of minor membrane components with ^1H NMR has been successful in several other cases. The first demonstration of this technique was the study of differential mobility of the cholesterol ring system and the isoprenoidal side chain when cholesterol is incorporated into phos-

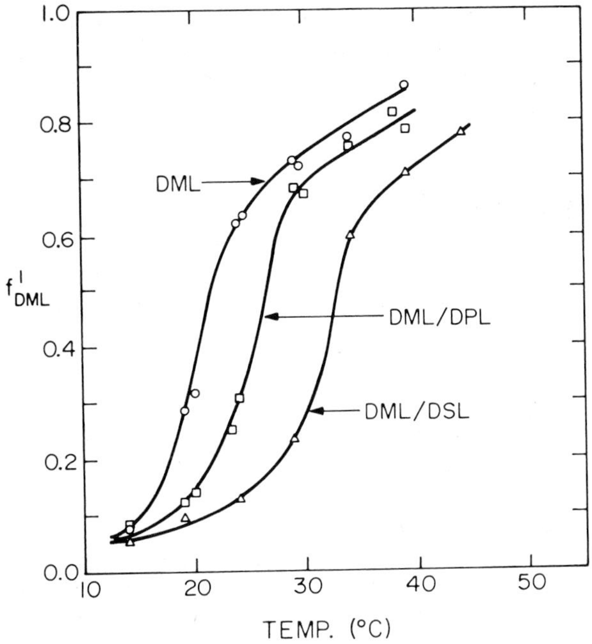

Fig. 24. A plot of the fraction of total dimyristoyllecithin (DML) melted into the liquid crystalline phase, $f^1{}_{DML}$, versus temperature for single bilayer vesicles of dimyristoyllecithin, 1:1 dimyristoyl-lecithin-dipalmitoyllecithin-d_{62} (DPL), and 1:1 dimyristoyllecithin-distearoyllecithin-d_{70} (DSL) (P.H. von Dreele and S.I. Chan, unpublished)

pholipid bilayers (Kroon et al., 1975). More recently the synthesis of phospholipids with all the hydrogens in both the headgroup and the side chains deuterated have allowed the extension of the technique to the studies of other lipid types as well as some protein units imbedded in the bilayer. Figure 25, for example, shows the spectrum of gramicidin A in perdeutero-DPL vesicles (G.W. Feigenson, personal communication).

There are studies which illustrate the usefulness of ^{31}P NMR examining lipid mixtures in bilayer vesicles. The chemical shift difference of the phosphate moiety in different phospholipid types such as phosphatidylcholines (PC), phosphatidylethanolamines (PE) and phosphatidylglycerols (PG) is large and these resonances are generally resolved at high magnetic fields. Upon the addition of membrane-impermeable shift or broadening reagents such as lanthanide and transition metal salts to the extravesicular compartment of vesicle preparations, it is also possible to distinguish those signals arising from phospholipids in the inner and outer monolayers of the bilayer vesicle. In this manner, the distribution of lipid species between the two vesicle monolayers can be determined (Michaelson et al., 1973; Berden et al., 1975; Yeagle et al., 1976). These studies show that in most cases there is asymmetric distribution of the lipids across the bilayer in highly curved membrane systems. For example, the concentration of PE is greater in the inner monolayer than in the outer when comixed with PC, and the PG concentration is greater in the outer monolayer in mixtures with PC. The implication is that in these lipid mixtures the extreme bilayer curvature imposes certain restrictions on the headgroup in-

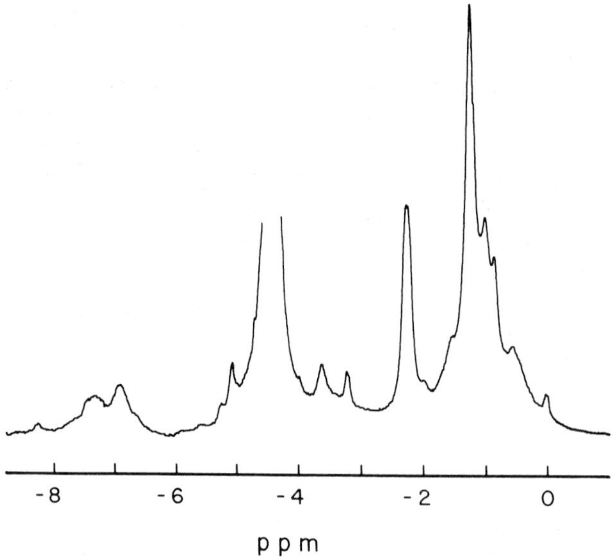

Fig. 25. The ^{1}H NMR spectrum (360 MHz) of 4 mole % gramicidin A in perdeuterated dimyristoyl-phosphatidylcholine (DMPC-d_{72}) vesicles at 50°C. The well-resolved gramicidin resonances are those from formyl protons at −8.3 ppm, tryptophan indole protons at −6.9 to −7.4 ppm, and upfield ring-current shifted leucine and valine methyl protons at −0.6 ppm. Courtesy of G.W. Feigenson

teractions which render the lipid with the bulkier headgroup more favorable thermodynamically in the outer monolayer. It should be noted that the bilayer vesicle is intrinsically asymmetric in that the two halves of the bilayer exhibit radii of curvature different in both magnitude and sign. This structural differences is manifested in the chemical shift difference of the phosphate moieties in the two halves of the bilayer vesicle.

VIII. Future Prospects

We have attempted to delineate those aspects of NMR spectroscopy which are relevant to its application to membrane systems, and have tried to illustrate these principles by a number of selected applications. This area of research is clearly still a virgin one despite the impressive number of NMR studies already performed on phospholipid bilayer systems. There is still valuable information to be gained with magnetic resonance techniques, as recent ^{2}H and ^{31}P studies of membrane systems dramatically illustrate. The development of new experimental techniques will undoubtedly also extend the range of possible investigations. Cross-polarization and multiple-pulse techniques have already been successfully applied to the investigation of phospholipid systems (Seiter and Chan, 1973; Urbina and Waugh, 1974; Kohler and Klein, 1977) and newer methods such as interferometric NMR spectroscopy (Stoll et al., 1977a, b) may also have application for the study of membranes. In addition, more work is necessary to confirm the validity of the various motional models for phospholipids in bilayers and, more importantly, to assess the importance of the motions themselves for the structure and function of natural membrane systems.

We expect the trends for future significant work to lie toward more complex model systems, which more closely simulate biomembranes, or the real membranes themselves. There is every reason to expect that sustained efforts in these directions, augmented by the detailed knowledge of membrane structure and dynamics already acquired through studies of simple model systems, will ultimately result in a more complete understanding of the complex processes involved in membrane-associated phenomena.

Acknowledgments. This work was supported by Grant GM-22432 from the National Institute of General Medical Sciences, U.S. Public Health Service.

References

Abragam, A.: The Principles of Nuclear Magnetism. London: Oxford University Press 1961.

Anderson, P.W.: A mathematical model for the narrowing of spectral lines by exchange or motion. J. Phys. Soc. Jpn. *9,* 316-339 (1954).

Berden, J.A., Barker, R.W., Radda, G.K.: NMR studies on phospholipid bilayers. Some factors affecting lipid distribution. Biochim. Biophys. Acta *375,* 186-298 (1975).

Bloom, M., Burnell, E.E., Valic, M.I., Weeks, G.: Nuclear magnetic resonance line shapes in lipid bilayer model membranes. Chem. Phys. Lipids *14,* 107-112 (1975).

Bloom, M., Burnell, E.E., Roeder, S.B.W., Valic, M.I.: Nuclear magnetic resonance line shapes in lyotropic liquid crystals and related systems. J. Chem. Phys. *66,* 3012-3020 (1977).

Bocian, D.F., Chan, S.I.: NMR studies of membrane structure and dynamics. Annu. Rev. Phys. Chem. *29,* 307-335 (1978).

Burnett, L.J., Muller, B.H.: Deuteron quadrupole coupling constants in three solid deuterated paraffin hydrocarbons: Ethane-d_6, butane-d_{10}, and hexane-d_{14}. J. Chem. Phys. *55,* 5829-5831 (1971).

Chan, S.I., Feigenson, G.W., Seiter, C.H.A.: Nuclear relaxation studies of lecithin bilayers. Nature *231,* 110-112 (1971).

Chan, S.I., Seiter, C.H.A., Feigenson, G.W.: Anisotropic and restricted molecular motion in lecithin bilayers. Biochem. Biophys. Res. Commun. *46,* 1488-1492 (1972).

Cullis, R.P.: Lateral diffusion rates of phosphatidylcholine in vesicle membranes: Effects of cholesterol and hydrocarbon phase transitions. FEBS Lett. *70,* 223-228 (1976).

Devaux, P., McConnell, H.M.: Lateral diffusion in spin-labeled phosphatidylcholine multilayers. J. Am. Chem. Soc. *94,* 4475-4481 (1972).

Edidin, M.: Rotational and translational diffusion in membranes. Annu. Rev. Biophys. Bioeng. *3,* 179-201 (1974).

Farrar, T.C., Becker, E.D.: Pulse and Fourier Transform NMR. New York: Academic Press 1971.

Feigenson, G.W., Chan, S.I.: Nuclear magnetic resonance relaxation behavior of lecithin multilayers. J. Am. Chem. Soc. *96,* 1312-1319 (1974).

Feigenson, G.W., Meers, P.R., Kingsley, P.B.: NMR observation of gramicidin A in phosphatidylcholine vesicles. Biochim. Biophys. Acta *471,* 487-491 (1977).

Finer, E.G.: Calculation of molecular motional correlation times from linewidths in nuclear magnetic resonance spectra of aggregated systems. Effects of particle size on spectra of phospholipid dispersions. J. Magn. Reson. *13,* 76-86 (1974).

Finer, E.G., Flook, A.G., Hauser, H.: Mechanism of sonication of aqueous egg yolk lecithin dispersions and nature of the resultant particles. Biochim. Biophys. Acta *260,* 49-58 (1972a).

Finer, E.G., Flook, A.G., Hauser, H.: The nature and origin of the NMR spectrum of unsonicated and sonicated aqueous egg yolk lecithin dispersions. Biochim. Biophys. Acta *260,* 59-69 (1972b).

Fisher, R.W., James, T.L.: Lateral diffusion of the phospholipid molecules in dipalmitoylphosphatidylcholine bilayers. An Investigation using nuclear spin relaxation in the rotating frame. Biochemistry *17,* 1177-1183 (1978).

Franks, N.P.: Structural analysis of hydrated egg lecithin and cholesterol bilayers. I. X-ray diffraction. J. Mol. Biol. *100,* 345-358 (1976).

Gally, H.-U., Niederberger, W., Seelig, J.: Conformation of the choline head group in bilayers of dipalmitoyl-3-sn-phosphatidylcholine. Biochemistry *14,* 3647-3652 (1975).

Gent, M.P.N., Prestegard, J.H.: Comparison of ^{13}C spin lattice relaxation times in phospholipid vesicles and multilayers. Biochem. Biophys. Res. Commun. *58,* 549-555 (1974).

Godici, P.E., Landsberger, F.R.: The dynamic structure of lipid membranes. A ^{13}C nuclear magnetic resonance study using spin labels. Biochemistry *13,* 362-368 (1974).

Griffin, R.G.: Observation of the effect of water on ^{31}P nuclear magnetic resonance spectra of dipalmitoyllecithin. J. Am. Chem. Soc. *98,* 851-853 (1976).

Habercorn, R.A., Herzfeld, J., Griffin, R.G.: High resolution ^{31}P and ^{13}C nuclear magnetic resonance spectra of unsonicated model membranes. J. Am. Chem. Soc. *100,* 1296-1298 (1978).

Horwitz, A.F., Horsley, W.J., Klein, M.P.: Magnetic resonance studies on membrane and model membrane systems: Proton magnetic relaxation rates in sonicated lecithin dispersions. Proc. Natl. Acad. Sci. USA *69,* 590-593 (1972).

Horwitz, A.F., Michaelson, D., Klein, M.P.: Magnetic resonance studies on membrane and model membrane systems. III. Fatty acid motions in aqueous lecithin dispersions. Biochim. Biophys. Acta *298,* 1-7 (1973).

Hubbell, W.L., McConnell, H.M.: Molecular motion in spin-labeled phospholipids and membranes. J. Am. Chem. Soc. *93,* 314-326 (1971).

James, T.L.: Nuclear magnetic Resonance in Biochemistry, Chap. 4. New York: Academic Press 1975.

Jost, P., Waggoner, A.S., Griffin, O.H.: Spin labeling and membrane structure. Structure and function of biological membranes (ed. Rothfield, L.), pp. 83-144. New York: Academic Press 1971.

Khetrapal, C.L., Kunwar, A.C., Tracey, A.S., Diehl, P.: Lyotropic Liquid Crystals. NMR Basic Principles and Progress (eds. Diehl, P., Fluck, E., Kosfeld, R.), Vol. 9, p. 5. Berlin-Heidelberg-New York: Springer 1975.

Kohler, S.J., Klein, M.P.: ^{31}P Nuclear magnetic resonance. Chemical shielding tensors of phosphorylethanolamine, lecithin, and related compounds: Applications to head-group motion in model membranes. Biochemistry *15,* 967-973 (1976).

Kohler, S.J., Klein, M.P.: Orientation of phospholipid head groups in bilayers and membranes determined from ^{31}P nuclear magnetic resonance shielding tensors. Biochemistry *16,* 519-526 (1977).

Kroon, P.A., Kainosho, M., Chan, S.I.: The state of molecular motion of cholesterol in lecithin bilayers. Nature *256,* 582-584 (1975).

Kroon, P.A., Kainosho, M., Chan, S.I.: Proton magnetic resonance studies of lipid bilayer membranes. Experimental determination of inter- and intramolecular relaxation rates in sonicated phosphatidylcholine bilayer vesicles. Biochim. Biophys. Acta *433,* 282-293 (1976).

Lee, A.G., Birdsall, N.J.M., Metcalf, J.C.: Nuclear magnetic relaxation and the biological membrane. Methods Membr. Biol. *2,* 1-156 (1974).

Lichtenberg, D., Petersen, N.O., Girardet, J.-L., Kainosho, M., Kroon, P.A., Seiter, C.H.A., Feigenson, G.W., Chan, S.I.: The interpretation of proton magnetic resonance linewidths for lecithin dispersions. Effect of particle size and chain packing. Biochim. Biophys. Acta *382,* 10-21 (1975).

Lindblom, G., Wennerström, H., Arvidson, G., Lindman, B.: Lecithin translational diffusion studied by pulsed nuclear magnetic resonance. Biophys. J. *16,* 1287-1295 (1976).

Lindsey, H., Petersen, N.O., Chan, S.I.: Physicochemical characterization of 1,2-di-phytanoyl-sn-glycerol-3-phosphorylcholine in model membrane systems. Biochim. Biophys. Acta *555,* 147-167 (1979).

Longmuir, K.J., Dahlquist, F.W.: Direct spectroscopic observation of inner and outer hydrocarbon chains of lipid bilayer vesicles. Proc. Natl. Acad. Sci. USA *73,* 2716-2719 (1976).

Longmuir, K.J., Capaldi, R.A., Dahlquist, F.W.: Nuclear magnetic resonance studies of lipid-protein interactions. A model of the dynamics and energetics of phosphatidylcholine bilayers that contain cytochrome *c* oxidase. Biochemistry *16,* 5746-5755 (1977).

McFarland, B.G., McConnell, H.M.: Bent fatty acid chains in lecithin bilayers. Proc. Natl. Acad. Sci. USA *68,* 1274-1278 (1971).

Michaelson, D.M., Horwitz, A.F., Klein, M.P.: Transbilayer asymmetry and surface homogeneity of mixed phospholipids in cosonicated vesicles. Biochemistry *12,* 2637-2645 (1973).

Niederberger, W., Seelig, J.: Phosphorus-31 chemical shift anisotropy in unsonicated phospholipid bilayers. J. Am. Chem. Soc. *98,* 3704-3706 (1976).

Petersen, N.O., Chan, S.I.: More on the motional state of lipid bilayer membranes: Interpretation of order parameters obtained from nuclear magnetic resonance. Biochemistry *16,* 2657-2667 (1977).

Petersen, N.O., Kroon, P.A., Kainosho, M., Chan, S.I.: Thermal phase transitions in deuterated lecithin bilayers. Chem. Phys. Lipids *14,* 343-349 (1975).

Pirs, J., Ukleja, P., Doane, J.W.: NMR in rapidly rotating samples or nematic liquid crystals. Phys. Rev. A *14,* 414-423 (1976).

Racker, E.: Reconstitution of cytochrome oxidase vesicles and conferral of sensitivity of energy transfer inhibitors. J. Membr. Biol. *10,* 221-235 (1972).

Razin, Shmuel: Reconstitution of biological membranes. Biochim. Biophys. Acta *265,* 241-296 (1972).

Saupe, A.: Das Protonenresonanzspektrum von orientiertem Benzol in nematisch-kristallinflüssiger Lösung. Z. Naturforsch. *20a,* 572-580 (1965).

Schempp, E., Bray, P.J.: Nuclear quadrupole Resonance spectroscopy. In: Physical Chemistry an Advanced Treatise (eds. Eyring, H., Henderson, D., Jost, W.), Vol. 4, Chap. II. New York-San Francisco: Academic Press 1970.

Schindler, H., Seelig, J.: Deuterium-magnetische Resonanzspektroskopie an spezifisch deuterierten flüssigen Kristallen. Ber. Bunsenges. Phys. Chem. *78,* 947-949 (1974).

Seelig, J.: Deuterium magnetic resonance: Theory and application to lipid membranes. Q. Rev. Biophys. *10,* 353-418 (1977).

Seelig, J., Gally, H.-U.: Investigation of phosphatidylethanolamine bilayers by deuterium and phosphorus-31 nuclear magnetic resonance. Biochemistry *15,* 5199-5204 (1976).

Seelig, J., Niederberger, W.: Deuterium-labeled lipids as structural probes in liquid crystalline bilayers. A Deuterium magnetic resonance study. J. Am. Chem. Soc. *96,* 2069-2072 (1974).

Seelig, J., Seelig, A.: Deuterium magnetic resonance studies of phospholipid bilayers. Biochem. Biophys. Res. Commun. *57,* 406-411 (1974).

Seelig, J., Gally, H.-U., Wohlgemuth, R.: Orientation and flexibility of the choline head group in phosphatidylcholine bilayers. Biochim. Biophys. Acta *467,* 109-119 (1977).

Seiter, C.H.A., Chan, S.I.: Molecular motion in lipid bilayers. A nuclear magnetic resonance study. J. Am. Chem. Soc. *95,* 7541-7553 (1973).

Sheetz, M.P., Chan, S.I.: Effect of sonication on the structure of lecithin bilayers. Biochemistry *11,* 4573-4581 (1972).

Singer, S.J., Nicolson, G.L.: The fluid mosaic model of cell membranes. Science *175,* 720-731 (1972).

Stockton, G.W., Polnaszek, C.F., Leitch, L.G., Tulloch, A.P., Smith, I.P.C.: A study of mobility and order in model membranes using ^{2}H NMR relaxation rates and quadrupole splittings of specifically deuterated lipids. Biochem. Biophys. Res. Commun. *60,* 844-850 (1974).

Stockton, G.W., Polnaszek, C.F., Tulloch, A.P., Hasan, F., Smith, I.P.C.: Molecular motion in single-bilayer vesicles and multilamellar dispersions of egg lecithin and lecithin-cholesterol mixtures. A deuterium magnetic resonance study of specifically labeled lipids. Biochemistry *15,* 954-966 (1976).

Stoll, M.E., Vega, H.J., Vaughn, R.W.: Explicit Demonstration of spinor character for a spin-1/2 nucleus via NMR interferometry. Phys. Rev. A *16,* 1521-1524 (1977a).

Stoll, M.E., Vega, A.J., Vaughn, R.W.: Double resonance interferometry: Relaxation times for dipolar forbidden transitions and off resonance effects in an AX spin system. J. Chem. Phys. *67,* 2029-2038 (1977b).

Tiddy, G.J.T.: N.M.R. of liquid crystals and micellular solution. Nucl. Magn. Reson. (Specialist Periodical Reports) *4*, 233-252 (1975).

Tiddy, G.J.T.: N.M.R. of liquid crystals and micellular solutions. Nucl. Magn. Reson. (Specialist Periodical Reports) *6*, 207-232 (1977).

Träuble, H., Sackman, E.: Studies of the crystalline-liquid crystalline phase transition of lipid model membranes. III. Structure of a steroid-lecithin system below and above the lipid-phase transition. J. Am. Chem. Soc. *94*, 4499-4510 (1972).

Urbina, J., Waugh, J.S.: Proton-enhanced ^{13}C nuclear magnetic resonance of lipids and biomembranes. Proc. Natl. Acad. Sci. USA *71*, 5062-5067 (1974).

Waugh, J.S., Huber, L.M., Haeberlen, U.: Approach to high-resolution NMR in solids. Phys. Rev. Lett. *20*, 180-182 (1968).

Wennerström, H.: Proton nuclear magnetic resonance lineshapes in lamellar liquid crystals. Chem. Phys. Lett. *18*, 41-44 (1973).

Wennerström, H., Lindblom, G.: Biological and model membranes studied by nuclear magnetic resonance of spin one half nuclei. Q. Rev. Biophys. *10*, 67-96 (1977).

Wennerström, H., Ulmius, J.: Proton NMR bandshapes in phospholipid bilayer vesicles. J. Magn. Reson. *23*, 431-435 (1976).

Woessner, D.E.: Spin relaxation in a two spin system undergoing anisotropic reorientation. J. Chem. Phys. *36*, 1-4 (1962).

Worcester, D.L., Franks, N.P.: Structural analysis of hydrated egg lecithin and cholesterol bilayers. II. Neutron diffraction. J. Mol. Biol. *100*, 359-378 (1976).

Wu, E.-S., Jacobson, K., Papahadjopoulos, D.: Lateral diffusion in phospholipid multilayers measured by fluorescence recovery after photobleaching. Biochemistry *16*, 3936-3941 (1977).

Wu, S.H., McConnell, H.M.: Phase separations in phospholipid membranes. Biochemistry *14*, 847-854 (1975).

Yeagle, P.L., Hutton, W.C., Huang, C., Martin, R.B.: Structure in the polar head group region of phospholipid bilayers: A ^{31}P $\{^{1}H\}$ nuclear Overhauser effect study. Biochemistry *15*, 2121-2124 (1976).

Electron Spin Resonance: Spin Labels

D. Marsh

I. Introduction

In general biological membranes possess no intrinsic paramagnetism and hence in the unlabelled state do not give rise to an electron spin resonance (ESR) spectrum. The introduction of a stable free radical ("spin label") thus enables one to use ESR spectroscopy to study specific environments within the membrane. The spin label which is invariably used is the nitroxide radical, which has a three-line nitrogen hyperfine structure whose splitting varies with the orientation of the magnetic field relative to the nitroxide axes. It is this spectral anisotropy which has made spin label ESR such a powerful tool in the study of the molecular motions which are the characteristic feature of the highly dynamic structure of biological membranes. The nitroxide hyperfine splittings are partially averaged by the anisotropic motion, which gives a measure of the motional *amplitude,* and the linewidths are differentially broadened by an extent which depends on the *rate* of molecular motion. Other important features of the spin label spectra are the broadening by intermolecular label-label interactions and the ability to detect compartmentation of the label by quantitating spectral lineheights after selectively removing accessible spin-label signals by chemical reducing agents. Label-label interactions arise from two sources: the Heisenberg exchange interaction which is essentially a contact interaction and is therefore capable of measuring translational diffusion; and the dipole-dipole interaction which depends on the distance apart of the labels and is therefore capable of measuring intermolecular separations. Quantitation after treatment with reducing agents is chiefly concerned with the measurement of transport properties: of spin-label substrates or reducing agents, or the translocation of labelled lipid molecules.

Since relatively versatile schemes are available for spin-label attachment, it is possible both to design environmental probes which will bind to membranes in a structurally sensitive manner, and to synthesize labelled membrane components whose spectra will reflect the motion of the labelled molecule in the membrane. In the latter case, the nitroxide "doxyl" group is particularly useful since this may be attached in a rigid stereo-specific manner either to steroid molecules or to the hydrocarbon chains of phospholipid molecules. The motion of the nitroxide will then directly reflect the motion of the labelled part of the molecule. The spin-label group must of course introduce some steric perturbation, but it is relatively small compared with the size of a phospholipid molecule, and comparison with non-perturbing techniques suggests that the size of this perturbation is tolerable at the label concentrations normally used. This certainly reflects in part

Max-Planck-Institut für biophysikalische Chemie, Abteilung Spektroskopie. D-3400 Göttingen, FRG the extremely cooperative nature of the lipid molecular motions in biological membranes, and makes one confident that spin-label methods can be reliably used to detect structural changes in membranes.

Several other spectroscopic methods are capable of giving similar structural information to that obtainable from spin labels, as can be seen from the other chapters of this book. All have their advantages and disadvantages and should be regarded as complementary in their application, especially when it comes to the extrapolation from model systems to real biological membranes. Thus ^{2}H NMR and ^{13}C NMR (see Chap. 1) can be used to measure motional amplitudes and rates without the disadvantage of the perturbing effect of the spin-label group. However, the problem with these techniques is the low sensitivity which means that they have so far only been applied to a very limited range of membranes. Raman and infrared spectroscopy also provide non-perturbing methods of determining lipid chain conformations. These methods are complementary to spin labelling in the sense that the information relates more to the ordered lipid chains, whereas the spin-label results relate more to the conformation of the fluid chains. Fluorescence spectroscopy can be used to gain information on motional rates, environmental polarity and translational diffusion of membrane-bound probes with considerably greater sensitivity than ESR. The disadvantage compared with ESR is that the probes are relatively more bulky and less flexible in their design than nitroxide spin labels. The main advantages of spin-label ESR thus lie in the relative sensitivity of the technique, the comparative ease and simplicity of the measurements, and the versatility in the range of different experiments which can be performed in both biological and model membranes. It is these features which have led to the considerable advances which spin-label spectroscopy has made in the field of membrane structure. (For a review of the achievements of spin-label ESR and those of other spectroscopic studies of membrane structure, see Marsh, 1975).

In the following the basic practical techniques and problems associated with spin-label ESR in membrane systems are discussed. After this comes a resumé of the theoretical considerations necessary for interpretation of the different types of measurements outlined above. This then leads on to a discussion of the basic features of the dynamic structure of membranes which are amenable to measurement by spin-label spectroscopy. The aim here is to illustrate the use of methods which are generally applicable to most membrane systems and to indicate the different sorts of information which can be obtained. Finally there is a discussion of new developments which are likely to extend the range of spin-label measurements possible in membrane systems.

Spin-label spectroscopy is a technique which has found extensive applications in molecular biology and consequently has already been very well served by a number of excellent reviews. These include: Hamilton and McConnell (1968); McConnell and McFarland (1970); Jost et al. (1971b); Jost and Griffith (1972); Melhorn and Keith (1972); Smith (1972); Keith et al. (1973); Gaffney and McNamee (1974); Smith et al. (1976); Jost and Griffith (1978); The different viewpoints of these various references give some idea of the breadth of application of spin-label studies, and of the range of spectral features which may be encountered. Special mention must be given to two volumes on "Spin Labeling" edited by Berliner (1976, 1979) which give a more extensive coverage of the subject, and to the review by Schreier et al. (1978) which pays special attention to the problems and artefacts which can arise in spin-label studies of membranes. A general view of the application of ESR to biological systems, including experimental aspects, is given in the book by Knowles et al. (1976). Where appropriate, individual reviews are referred to specifically in the various sections of this chapter.

II. Experimental Methods

A. Instrumentation

(Jost and Griffith, 1972; Knowles et al., 1976)
The basic requirements for the ESR experiment are a microwave frequency source, a steady applied magnetic field and a means of detecting the microwave absorption at resonance. Most spectrometers used in spin-label work operate at a frequency of 9 GHz which requires a DC magnetic field of 3.3 kG to observe the spin-label spectrum. Such magnetic fields are readily obtainable with laboratory electromagnets. Spectrometers are also designed which operate at 35 GHz, but the small microwave cavities required at this frequency pose problems in the handling of aqueous samples, and the larger g-value anisotropy, though sometimes advantageous, often complicates spectral interpretation at this frequency. In recording a spectrum it is practically more convenient to scan the magnetic field rather than the microwave frequency and this is the method of recording which is invariably used. A scan width of 100 gauss is the standard sweep normally used since this will accommodate the whole of the spin-label spectrum, under all conditions. In order to improve the detection sensitivity it is usual to modulate the magnetic field at a frequency of 100 kHz, via coils attached to the side of the microwave cavity, and to use signal amplification and phase-sensitive detection at this frequency. The phase-sensitive detection results in a signal which is the first derivative of the absorption spectrum, and this is the normal method of recording ESR spectra. Lower modulation frequencies can be used, though with reduction in detected signal-to-noise, and in particular double modulation can be employed, which results in a second derivative presentation which has improved resolution at the expense of signal-to-noise ratio.

A block diagram of a conventional form of spectrometer is given in Figure 1, with the operational units numbered in order. The sample is placed in the microwave cavity which is positioned between the poles of the electromagnet. A spectrum is then recorded according to the following scheme of adjustments:

1. At low power the microwave frequency is tuned to the resonant frequency of the microwave cavity with the sample in position (the tuning is dependent on the type of sample and the sample holder, dewar, etc.). The automatic frequency control is then brought into play, which locks the frequency of the microwave oscillator to that of the cavity.
2. The microwave power is then increased to the operating value. A value of 50-60 mW is normally suitable if one wants to obtain maximum signal-to-noise, since at this power the spin-label resonance is only moderately saturated and the absorption spectra have approximately their maximum intensity. If it is required to eliminate all effects of saturation then a power level of 1-10 mW is appropriate.
3. During the tuning and power adjustment the coupling iris to the microwave cavity is also adjusted. This adjustment is made so that all the microwave power incident

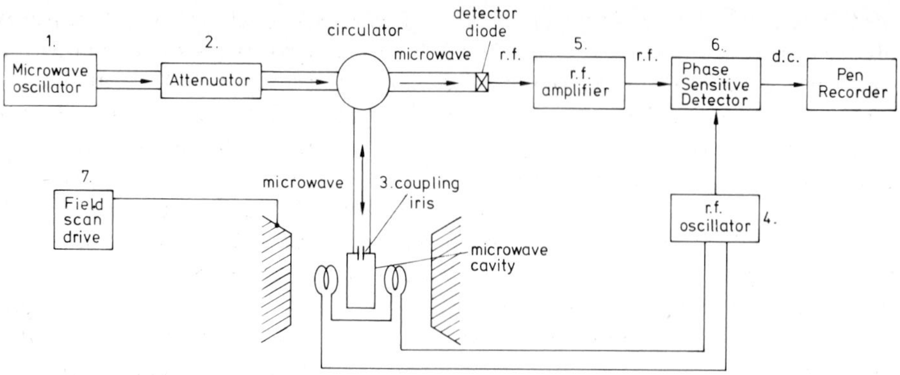

Fig. 1. Block diagram of a conventional ESR spectrometer. *Numbered units:* sequence of operations in obtaining a spectrum: (1) microwave frequency tuning; (2) microwave power adjustment; (3) cavity coupling; (4) rf field modulation amplitude; (5) signal gain; (6) filter time constant; (7) magnetic field centre, scan width and scan time

on the cavity is absorbed by it (zero detector current). Finally the cavity is slightly undercoupled to provide a working bias current for the detector diode.

4. The amplitude of the magnetic field modulation is then set. This is chosen as high as possible (to gain maximum signal strength) without causing broadening of the spectral lines by overmodulation. For no broadening or distortion at all, the modulation amplitude should be not greater than 0.2 x the peak-to-peak linewidth of the narrowest line in the spectrum. For maximum signal height, though with considerable broadening, the modulation amplitude required is 1-2 times the linewidth. A reasonable compromise would be to take 0.5 times the linewidth. In practice 1 gauss is a normal modulation amplitude for spin labels in membranes, though higher amplitudes can be used for very broad spectra and smaller values are required for spin labels in aqueous solution.
5. The amplifier gain can then be set to produce a reasonable signal height.
6. The filter time constant of the phase-sensitive detector is then chosen to reduce the noise to an acceptable level. The longer the filter time constant, the longer is the scan time required to record a spectrum which is not distorted by the filter response. Distortion will occur if the time taken to scan through the linewidth of the narrowest line in the spectrum is less than 10 times the time constant. In practice for spin-label spectra in membranes a value of 4 min scan for a 0.3 s time constant or an 8 min scan for a 1 s time constant is normally appropriate. (The phase of the rf modulation reference signal must also be adjusted at the phase-sensitive detector, but once set this is normally not changed.)
7. The magnet scan sweep time is set as indicated above, and the scan width is set: at 100 gauss for spin-label spectra. Clearly if a different scan width is employed then the scan time should be adjusted accordingly to avoid distortion by the response of the phase-sensitive detector filter.

Sensitivity is normally not a great problem in spin-label experiments with phospholipid bilayer model membranes, but can easily become so with biological membrane samples, especially if the lines are broad. In these latter cases it is necessary to optimize the spectrometer settings mentioned above to obtain the best signal-to-noise for the particular sample under study. For first derivative spectra the signal height *decreases* as the square of the linewidth, so care should be taken to optimize the modulation amplitude for spectra with broad lines. This is because, in this case, an increase in modulation amplitude is more likely to increase lineheight without an increase in linebroadening. In the case of composite spectra, e.g. from spin labels in the membrane and in the aqueous phase, it might be found that the narrow (aqueous) spectrum is over-modulated when the other, membrane-bound spectrum is still not broadened by the modulation. This can be an advantage in partially removing a troublesome aqueous spectrum, if only the spectrum for the spin label in the membrane is of interest. The second way to improve signal-to-noise is to increase the filter time constant and correspondingly the scan time. Values of time constant in the region of 2-16 s with scan times of 0.5-1 hour, or even longer, may be used for samples with weak signals. However, long scans require long-term stability of both spectrometer and sample; spectrometer drift or sample decomposition will mean that the end of the spectrum will correspond to different conditions from the beginning. One way of circumventing this problem is to use signal averaging in which the spectrum is scanned much more rapidly with faster time constant, and repeated scans are summed. In this case the spectral signal sums directly as the number of scans, but for the noise there is some statistical cancellation and this only accumulates proportional to the square root of the number of scans. Theoretically the increase in signal-to-noise is no better than by filtering (since this improves as the square root of the filter time constant; Feher, 1970), however the problems of long-term drifts and instabilities are largely eliminated. Signal averaging is carried out either by a CAT (*C*omputer of *A*verage *T*ransients) or by means of a digital computer directly interfaced to the spectrometer. The latter system has the advantage that many other spectral manipulations can be carried out (see Klopfenstein et al., 1972). For further details on instrumental aspects the reader is referred to Poole (1967); Wilmshurst (1967); Alger (1968).

B. Measurements

(Randolph, 1972, Knowles et al., 1976)
The principal measurements required for interpretation of spin-label experiments are of hyperfine splittings and to a lesser extent linewidths. Both of these quantities are most frequently quoted in gauss and require only a knowledge of the magnitude of the magnetic field scan width (assuming it to be linear). For the electromagnets of most commercial spectrometers, the magnetic field scan calibration can be relied upon over a sweep width of 100 gauss and the hyperfine splittings and linewidths can be read directly from the recorded spectrum. More exactly, the absolute magnetic field can be measured using a proton NMR magnetometer and the scan width calibrated from various fixed-field measurements. The field measurement is made by determining the rf frequency at which the proton NMR is observed for an aqueous sample placed in the magnetic field adjacent to the microwave cavity. The appropriate relationship is: $H(kG) = \nu_{proton}(MHz) \times 0.23487$. Less accurately, the field scan may be calibrated using a sample containing many narrow lines of known position, e.g. Wurster's Blue (Bolton et al., 1972). Measurements of g-values are less frequently required in membrane spin-label spectroscopy, and

these require measurement of both the absolute field and the microwave frequency, since $h\nu=g\beta H$. The field is best measured by proton NMR as explained above, and the microwave frequency is best measured digitally using a transfer oscillator with which a known harmonic of a variable fundamental frequency (~ 300 MHz) is brought into a fixed beat frequency (50 MHz) with the microwave frequency to be measured. The fundamental frequency is measured with a conventional frequency counter, and the harmonic number is obtained by determining both upper and lower beat positions. Less accurately, the microwave frequency can be measured with a cavity wavemeter, a tunable resonant cavity which is frequency-calibrated. A simpler alternative is to use a standard of known g-value, e.g. solid DPPH (diphenyl picryl hydrazyl): $g_S = 2.0037$, to determine either the absolute field or the microwave frequency, assuming the other is known. A useful approximation using a g-value standard is $g \simeq g_S(1+\Delta H/H_S)$, where only the field separation, ΔH, of the standard and unknown resonance is needed accurately, and the absolute field position of the standard, H_S, need only be approximate. A knowledge of the microwave frequency (or absolute field) is required for simulation of spin-label spectra, since this determines the size of the effect of the g-value anisotropies. However, at 9 GHz these effects are not so large, and so a high accuracy is not required, but the effects will be much more important at 35 GHz.

Lineheight measurements are also quite frequently applied in the analysis of membrane spin-label spectra. This can be either as a more sensitive alternative to linewidth measurements (since the lineheight varies as the inverse *square* of the linewidth for a fixed spin concentration) or to quantitate relative amounts of spin label either in composite spectra, or in spin-label reduction experiments. In this type of experiment it is essential to ensure that the spectra, particularly the narrower lines, are not disturbed by the recording conditions. Special care must thus be taken to eliminate modulation broadening, distortion by too fast a scan rate and differential power saturation of the spectra. Absolute determinations of spin concentrations or comparision of different samples pose additional problems caused by the difficulty in reproduceability of the microwave coupling, tuning and power setting. Another problem is that the microwave field pattern in the sample cavity is affected by the sample shape, size, orientation and composition, which can cause changes in all the above parameters; these effects are particularly large in the case of aqueous samples. Methods of overcoming these difficulties are described in the following section on samples. In comparing spectra of two samples recorded under different instrumental conditions (which cause no spectral distortion), it is of course necessary to scale the lineheights by all the factors which influence signal height. That is by dividing by the gain, modulation amplitude and square root of the microwave power. Absolute determinations of spin concentration are best made by comparison with standard nitroxide spin-label samples of known concentration. If lineheights are to be compared then it is essential that the standard and unknown spectra should be of identical lineshape and linewidth, otherwise only a relative calibration will be obtained. For normal, derivative-shaped lines, as is the case for spin labels in low viscosity solvents, the standard and unknown may be normalized by multiplying the lineheight by the square of the linewidth, but in general it is necessary to determine the second integral of the spectrum, since this is directly proportional to the area under the absorption curve. This latter operation is most easily accomplished by computer methods as described in Klopfenstein et al. (1972). Once a conversion factor has been established by double integration, then all determinations can be most conveniently determined by scaling lineheights relative to a narrow-line standard such as a small spin-label molecule in a non-viscous solvent, e.g. hydroxy TEMPO (4-hydroxy-2,2,6,6-tetramethylpiperidine-N-oxyl). For experiments comparing the partitioning of a spin label between an aqueous and a membrane phase, an absolute conversion factor can be established by difference, knowing the total concentration of spin label and the height of the same concentration of spin label in water alone. In practice, it is frequently not necessary to have an absolute calibration in these cases. For further details on operational techniques the reader is referred to the following: Bolton et al., (1972); Jost and Griffith (1972); Randolph (1972); Knowles et al. (1976).

C. Sample and Labelling Techniques

(Knowles et al., 1976)

Spectrometer sensitivity is frequently quoted as the minimum detectable number of spins in a standard sample recorded under standard conditions, and is a useful means of comparing spectrometer performance. For modern commercial spectrometers this quantity is in the region of 5 x 10^{10} spins per 1 gauss linewidth. However, this figure is not of much relevance in discussing minimum detectable spin-label concentrations in membrane systems, since it refers to a sample with no dielectric loss, whereas aqueous samples are extremely lossy at microwave frequencies. This dielectric loss arises from interaction of the water dipoles with the microwave electric field, and severely restricts the shape and size of aqueous samples that can be introduced into the microwave cavity. The optimum configuration for a normal rectangular microwave cavity, such as that indicated in Figure 2, is the aqueous "flat cell" in which the sample is contained in the volume enclosed be-

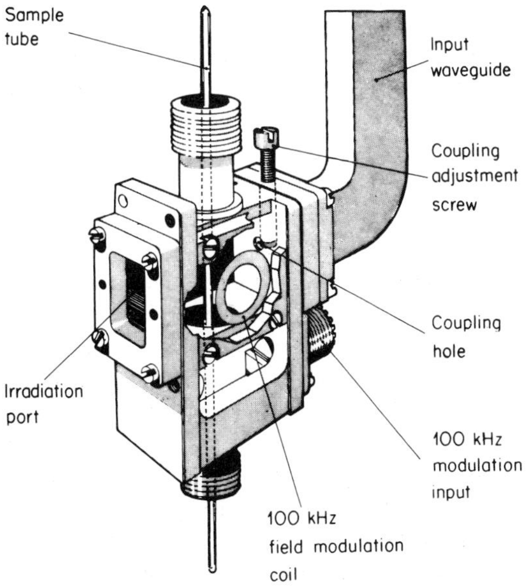

Fig. 2. Rectangular ESR microwave cavity resonating in the TE_{102} mode (Varian E-231). The quartz sample tube is either cylindrical and positioned along a central axis, by collets or within a cylindrical double-wall quartz dewar, or is a thin flat cell positioned along a central axis, transverse to the length of the cavity (Courtesy Varian Associates)

tween two quartz plates of dimensions approximately 6 x 1 cm and separated by 0.3 mm. By positioning the quartz flat cell transverse to the length of the microwave cavity, the thin flat lamella of aqueous sample can be located in the position of minimum microwave electric field, which also coincides with the position of maximum microwave magnetic field, in the centre of the cavity. Thus dielectric loss is minimized and magnetic absorption on resonance is maximized. A somewhat less optimal method is to use thin-walled capillaries of approximately 0.5 mm internal diameter. If the walls are sufficiently thin, $\leqslant$ 0.2 mm, problems with dielectric absorption and contaminating signals are not great, and glass instead of quartz can be used. These are conveniently made out of sealed-off 50 μl capillary pipettes, haematocrit or melting-point tubes, or even sealed-off dropping pipettes. The capillary pipettes or heamatocrit tubes are particularly useful in quantitating signal heights because of their reproduceable internal diameters. The sealed-off capillaries are then contained in standard, 4-mm diameter, quartz ESR tubes held in the microwave cavity. Capillaries have an advantage over flat cells when only small quantities of sample are available, since they require smaller volumes and have essentially no dead space. Use of capillaries is also virtually obligatory in the case of variable temperature measurements, since the flat cells are too large to be accommodated in the normal variable temperature dewars.

The working height of a 9 GHz cavity is 3 cm, so the minimum sample volume required is 30 μl in the case of capillaries and about 200 μl in the case of flat cells, because of dead volume effects. (It should be emphasized that all the above remarks and those to follow refer solely to 9 GHz spectrometers; at 35 GHz the microwave cavities are much smaller and the problems of dielectric loss are correspondingly much more acute, which is one of the reasons why 35 GHz studies are much less used.) The minimum working sample concentrations depend strongly on the linewidth of the spin-label spectra, for small spin-label nitroxides in non-viscous solvents a spin-label concentration of 5-10 μM should be sufficient, whereas for moderately immobilized spectra concentrations of 50 μM are required, and for strongly immobilized spectra concentrations of 300 μM. It must be stressed that, for normal measurements on membrane samples with lipid spin labels, the spin label must be magnetically dilute to avoid the effects of spin label-spin label interactions, i.e. less than or equal to 1 mol % of the total lipid concentration. This means that total lipid concentrations of the order of 0.5-1 mM, 5 mM and 30 mM are required for freely tumbling, moderately immobilized and strongly immobilized spectra respectively. For spin-labelled proteins the labels normally are automatically magnetically dilute, since there are only relatively few labelling sites per protein molecule, and so in this case the total membrane concentration required is approximately the same as for lipid labels.

Variable temperature measurements are normally performed using a nitrogen gas flow system with a double-wall, quartz, dewar insert within the microwave cavity. The dewar insert will accommodate the standard 4-mm quartz ESR tubes, which contain the membrane sample in a sealed capillary, and are filled with low viscosity silicon oil for thermal stability, to provide good thermal contact and also to minimize temperature gradients. The temperature is regulated by a heater and sensor in the gas flow below the cavity, and the sample temperature is measured by a thermocouple dipping into the silicon oil, just above the top of the microwave cavity. Good thermal stability and reproduceability can be obtained with such a system. An alternative arrangement which places the capillary directly and reproduceably in the gas flow has been described by Gaffney and McNamee (1974). A method of temperature variation not involving a quartz dewar insert has been described by Thomas et al. (1976). This involves blowing thermostatted nitrogen through the whole of the cavity, via the irradiation grid at the front, and has the advantage that it is then possible to use the large flat cells. The possible disadvantages of this system are that the temperature cannot be controlled so accurately, and that the frequency stability of the microwave cavity will not be so good.

The problems of accurately comparing sample intensities were discussed in the previous section. The problem of the resetting errors in microwave power, tuning and coupling can be overcome by using a dual cavity composed of two TE_{102} rectangular cavities placed back-to-back so as to resonate in a single TE_{104} mode. The samples in the two cavities have independent field modulation, and thus the spectra from the standard and unknown can be recorded separately under conditions of identical coupling, tuning and incident microwave power. However, the power at the two samples may still differ, because the two sample cells may concentrate the microwave field at the sample in different ways. Care thus must be taken to have identical sample cells, sample geometries and dewars, etc., and to interchange sample and standard positions, to check that this is the case. A simpler alternative is to attach a second capillary to the sample cell, be it either flat cell or identical capillary. This sealed capillary contains a reference compound, e.g. a small nitroxide spin label in water, or a speck of solid DPPH. Variations in microwave and sample settings can then be compensated by normalizing relative to the lineheight of this reference compound. A third possibility (Smith, 1972) is to use a flat cell with a reservoir syringe attached to its lower limb. The cell remains rigidly fixed in the cavity and microwave power, tuning and coupling are set with the cell filled with sample and then remain unchanged. The sample is changed by introduction from the top and removal by the syringe(s) at the bottom. Such an arrangement is particularly useful when performing titrations by adding small amounts of reagent from above and mixing by agitating the plunger of the syringe below, with no change in sample cell or tuning etc.

There is clearly no problem in labelling a membrane with a water-soluble spin label such as TEMPO which partitions between the aqueous and membrane phase. The spin label is simply added to the aqueous phase or the membrane dispersed in the spin-label solution, care only being taken that the spin-label concentration is not so high as to cause interaction broadening when taken up into the membrane. Labelling phospholipid bilayer model membranes or reconstituted lipid-protein complexes with spin labels which reside solely in the membrane phase again pose no problems. The spin label is mixed at 1 mol % with the lipids together in chloroform or other organic solvent. The organic solvent is then evaporated off in a stream of nitrogen and the residual solvent removed by placing under vacuum for at least 4 hours. The anhydrous lipid mixture is then dispersed in the appropriate amount of aqueous phase, either by vortex mixing which produces large multilamellar liposomes of approximately 0.5-5 μm in diameter, or by sonicating under nitrogen to produce small, single-bilayer vesicles of approximately 200-300 Å diameter. In the case of saturated phospholipids the dispersing should be carried out at a temperature above the lipid bilayer phase transition. Oriented bilayer model membranes can be produced by a similar method, except that the lipid solution is evaporated down on to a flat glass cover-slip or on to the internal surfaces of a quartz flat cell. After treatment under vacuum, the oriented lipid layers are hydrated by immersing in the aqueous phase, the excess of which is then drained off before measuring in the spectrometer. The flat cells can be measured directly, but the cover-slips must be examined in a quartz tissue cell or some other holder similar in geometry to the aqueous flat cell. Well-oriented lipid bilayers can also be produced, at hydrations less than maximum, by equilibrating the anhydrous bilayers over constant humidity solutions (see e.g. Jost and Griffith, 1976, for a simple constant-humidity sample cell). Alternatively the anhydrous lipid-spin label mixture can be equilibrated over a constant humidity atmosphere and then oriented by pressing between glass cover-slips. Very high degrees of orientation can be obtained in this way, but only for relatively low humidities.

The problems of labelling biological membranes with lipid spin labels, such as fatty acids, steroids or phospholipids, are somewhat greater. One of the simplest effective methods is to add the spin label to the membrane suspension as a small quantity of concentrated ethanol solution ($< 1\%$ ethanol). The free label which is not taken up by the membrane, and the ethanol, can then be removed by centrifugation and resuspending of the membrane pellet in fresh aqueous phase. An alternative method which avoids the presence of ethanol is to coat small glass beads with spin label by evaporating a chloroform solution as described above. The membrane suspension is then labelled from the large surface area of the beads which can be subsequently separated off. A method which achieves even closer contact with the membrane is to use spin-label-loaded bovine serum albumin (BSA); (see e.g. Gaffney and McNamee, 1974). The BSA is first labelled in a flask whose interior is coated with spin label, as for the glass beads described above; any excess solid particles are removed by centrifugation. The membrane suspension is then incubated with the BSA solution, and the BSA finally removed by centrifugation and resuspension of the labelled membrane pellet. Several phospholipid spin labels have been found to form reasonably stable sonicated dispersions from which membranes may be labelled (Scandella et al., 1972; McNamee and McConnell, 1973). The membrane suspension is incubated with the sonicated spin-label dispersion and the membrane then resolved from the unincorporated membrane vesicles by centrifugation. The small, single-bilayer, spin-labelled lipid vesicles do not sediment appreciably in normal centrifugation, but instabilities of the vesicle suspension may lead to larger aggregates which are more diffucult to separate. In these cases sucrose density gradient centrifugation may be required.

A problem sometimes encountered in spin-label studies with biological membranes is that of chemical reduction of the spin label by the membrane preparation. Two of the common causes of signal loss are reduction either by sulphydryl groups (Morrisett and Drott, 1969) or by the action of redox enzyme systems (Stier and Sackmann, 1973) such as respiratory chains. Reduction by protein sulphydryl groups can be effectively reduced by alkylating with such standard reagents as n-ethyl maleimide or iodoacetamide (Giotta and Wang, 1972; Baldassare et al., 1974), although this may impair biological activity. Reduction by electron transport chains and other redox enzyme systems may be stopped by the use of specific inhibitors or depletion of electron donor substrates by starving and extensive washing (Baldassare et al., 1974). Aeration is effective in some cases (Nakamura and Ohnishi, 1972) both in protecting against reduction and in restoring the reduced signal. The oxygenation procedure is also known to restore the nitroxide from the reduced hydroxylamine. Ferricyanide is generally useful in limiting the rate of reduction, or in reoxiding reduced label (Kaplan et al., 1973); however, high concentrations will broaden the lines of narrow spin-label signals, and liberated ferrous ions may catalyse spin-label reduction by sulphydryls. In the latter case it is better to work in an ion-free buffer containing EDTA (ethylene diaminotetra-acetic acid) to prevent catalysis of sulphydryl-mediated reduction.

For spin labelling membrane proteins, several nitroxide analogues of the standard protein modifying reagents are available for covalent labelling, e.g. the maleimide, iodacetamide and isothiocyanate derivatives. These reagents are capable of labelling nucleophilic groups by aqueous reaction, particularly sulphydryls or amino groups, with the specificity to some extent depending on the reaction conditions: amino groups e.g. are not labelled very extensively at pHs below their pKa and thus sulphydryl group labelling is normally favoured at neutral pH. Due to problems of solubility and aqueous hydrolysis of the labels, labelling is most conveniently carried out by adding a small volume of a concentrated stock solution of spin label in ethanol to the aqueous membrane suspension. The time-course of the reaction can be followed by monitoring the disappearance of the sharp ESR signal from the label in the aqueous phase, or by back-titration of the unlabelled groups with a standard colorimetric reagent such as DTNB (5,5-dithio bis(2-nitrobenzoic acid); (Jones et al., 1972). The excess unreacted label can then be removed by centrifugation and washing.

D. Spin Label Synthesis and Stability

(Hamilton and McConnell, 1968; McConnell and McFarland, 1970; Keith et al., 1973; Gaffney, 1976)

The two classes of nitroxide spin labels which are used in membrane studies are those derived from the oxazolidine and the piperidine or pyrrolidine rings. In both cases the nitroxide group is flanked by quaternary carbon atoms, protecting the radical from disproportionation reactions, which accounts for the high stability of the label.

The oxazolidine ring derivatives are prepared by the general procedure (Keana et al., 1967) for converting a ketone to an oxazolidine and hence to the corresponding nitroxide:

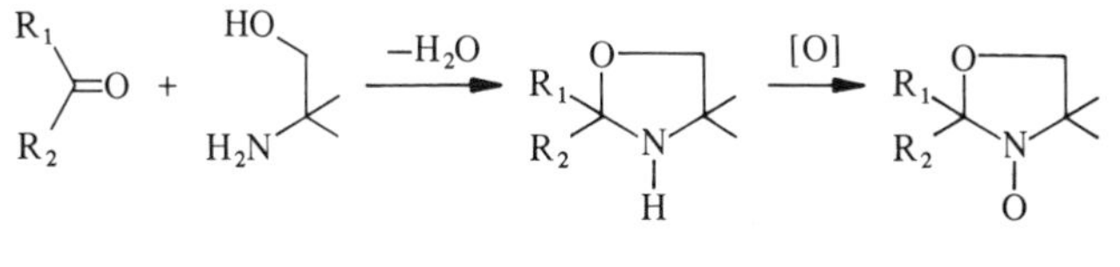

Here the first step is accomplished by heating under reflux with water removal by a Dean Stark trap, and the oxidation to the nitroxide performed with chlorobenzoic acid in organic solvent, or hydrogen peroxide-sodium tungstate if the compound is water soluble. These spin labels are especially useful in the analysis of molecular motion in membrane systems, because the spin-label group is rigidly and stereospecifically attached to the parent lipid molecule via the double bond derived from the ketone. Spin-labelled steroids of the type I have been synthesized by this procedure (Keana et al., 1967: I (a); Hubbell and McConnell, 1969: I (b), (c)):

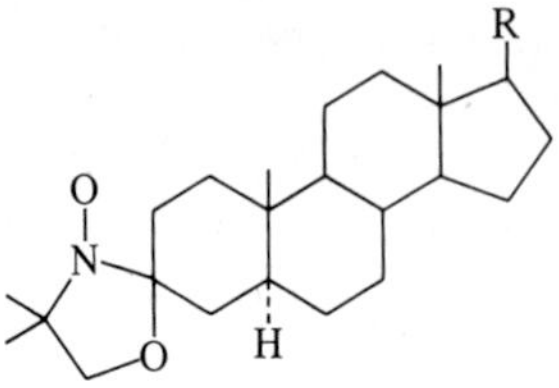

I

a) R = 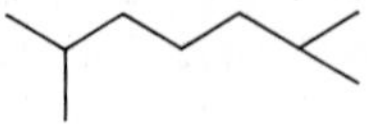 (Cholestane derivative)

b) R = –OH (Androstanol derivative)

c) R = –H (Androstane derivative)

The cholestane spin label I (a) can be considered a good analogue for studying the role of cholesterol in membranes, and the other two derivatives serve as general rigid probes of molecular motion in membranes. The androstanol label I (b) is particularly interesting for comparison purposes since it orients in the opposite direction to the other two. The spin label fatty acid drivatives II (m,n) have been prepared by a similar method:

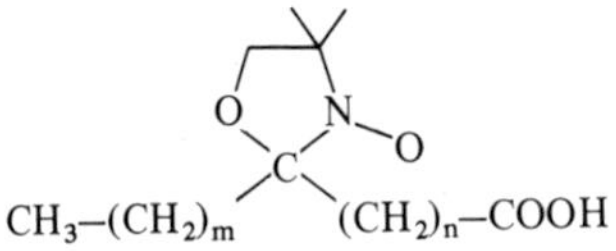

II (m, n)

where the II (5,10) derivative is prepared from the ketoester derived from the commercially available 12-hydroxystearic acid (Waggoner et al., 1969) and all other ketoester precursors are prepared according to the scheme:

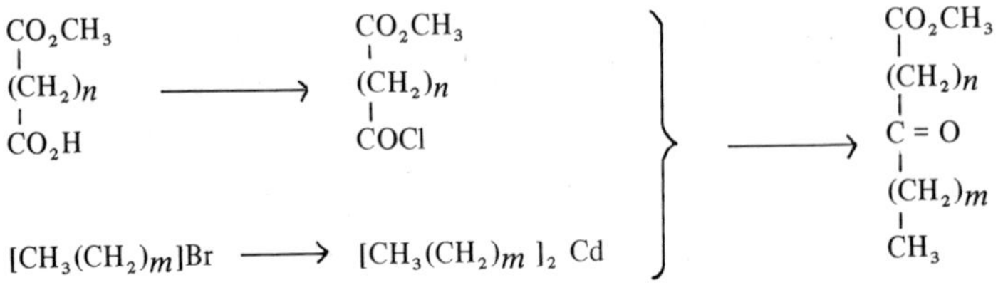

(Hubbell and McConnell, 1971; Gaffney, 1976). The phosphatidylcholine phospholipids, spin labelled on a single fatty acid chain, III (m,n), can be prepared by acylation of lysophosphatidylcholine with the acid anhydride derived from the corresponding fatty acid spin label II (m,n); (same references)

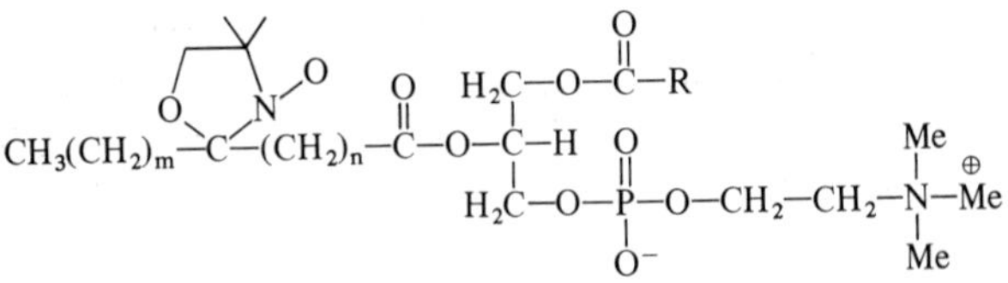

III (m, n)

Although the fatty acid spin labels II (m,n) have proved very useful probes for studying the motion of the lipid chains in membranes, it is clear that the phosphatidylcholine probes will have the further advantage that they will intercalate in the membranes almost exactly in register with the phospholipids and that they possess a well-defined phospholipid headgroup specificity. Spin-labelled phospholipids with other headgroups have been prepared starting from the corresponding phosphatidylcholine spin label III (m,n). Thus the phosphatidic acid spin label IV (m,n) has been prepared by enzymatic cleavage of III (m,n) with phospholipase D (Ito and Ohnishi, 1974), and hence the phosphatidylethanolamine V (m,n) and phosphatidylglycerol VI (m,n) spin labels (Tanaka and Ohnishi, 1976). The phosphatidylethanolamine spin label has also been prepared in a manner analogous to the phosphatidylcholine spin labels, with the amino group of the lysophosphatidylethanolamine blocked by the O-nitrophenylsulfenyl group (Devaux and McConnell, 1974). The phosphatidylserine spin label VII (m,n) has been prepared by the reaction of spin-labelled CDP-diglyceride (again derived from III (m,n) via IV (m,n)), with L-serine catalysed by phosphatidylserine synthetase (Ito et al., 1975):

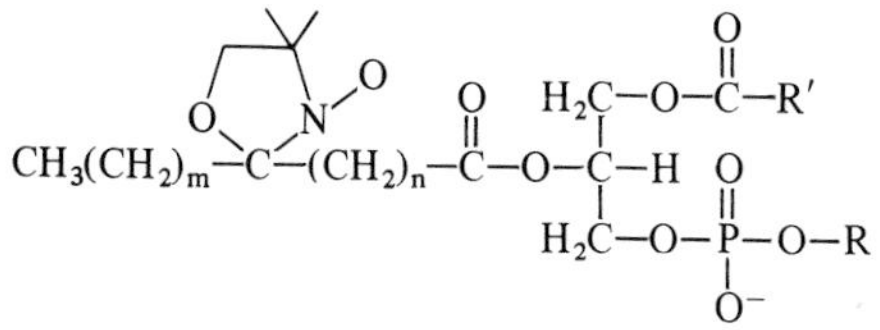

IV (m,n) R = H
V (m,n) $R = (CH_2)_2NH_3^+$
VI (m,n) $R = CH_2CH(OH)CH_2OH$
VII (m,n) $R = CH_2CH(\overset{+}{N}H_3)COO^-$

An alternative method of producing the phospholipid spin labels V-VII (m,n) is by headgroup exchange of the phosphatidylcholine spin label III (m,n), catalysed by phospholipase D in the presence of the corresponding base (Comfurius and Zwaal, 1977). Although the yields are less than for the unlabelled phospholipids, this is a viable general method (Watts et al., 1979). The different headgroup specificities of these phospholipid labels have been demonstrated by Tanaka and Ohnishi (1976) in a study of the differential fluidity and asymmetric distributions of the different lipid classes in the erythrocyte membrane.

The 2,2,6,6, tetramethylpiperidine and 2,2,5,5, tetramethylpyrrolidine spin-label derivatives are prepared by oxidation of the secondary amine with hydrogen peroxide-sodium tungstate in water or by chlorobenzoic acid in organic solvents:

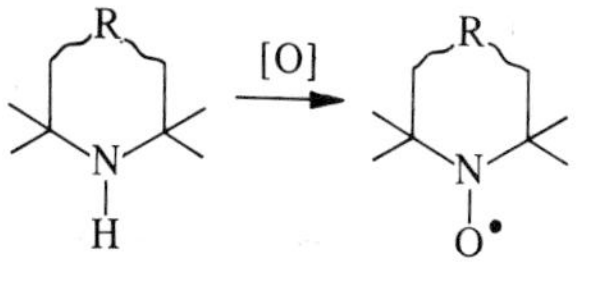

One of the simplest spin labels of this type is TEMPO VIII which is derived directly from tetramethylpiperidine (Rozantzev and Neiman, 1964; Rozentzev, 1970) and has proved to be an extremely valuable spin label for analysing membrane fluidity, phase transitions and lateral phase separations by means of the partitioning of the label between the aqueous and membrane phases (McConnell et al., 1972; Shimshick and McConnell, 1973).

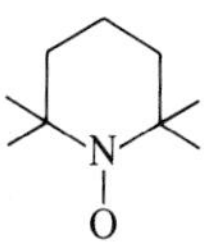

VIII

In general the tetramethylpiperidine and -pyrrolidine labels are less useful than the oxazolidine derivatives in directly studying molecular motion, since the ring is not normally attached by a rigid linkage, but via one or more single bonds about which rotations may take place, and in addition one has the possibility of conformational flexing of the piperidine ring. However, this type of label has proved extremely useful because of the versatility in the chemistry of the attaching group. This is aided by the high stability of the nitroxide, or corresponding secondary amine, during reactions which modify the R-group, resulting from the protecting effect of the methyl groups on the quaternary carbons. Most of these types of spin labels have been derived from a few precursors, notably the amino and also the hydroxy or carboxy derivatives which are either commercially available or themselves derived from the keto derivative. For a review of these schemes and details of precursor syntheses see Rozantzev (1970) and Gaffney (1976). Amongst the small molecules useful in membrane studies synthesized in this way can be mentioned the spin-label ions TEMPOcholine IX (Kornberg and McConnell, 1971) and TEMPO-phosphate X (Weiner, 1969). These molecules have proved useful both in transport studies and for the measurement of the internal volume of lipid vesicles.

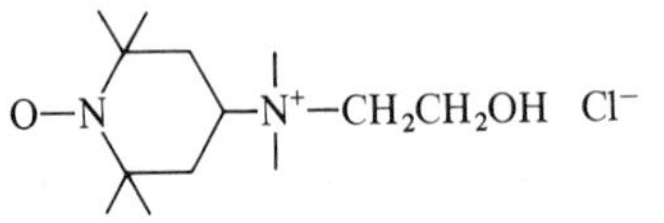

IX

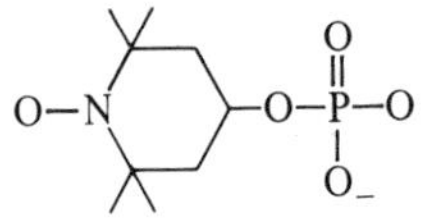

X

Piperidine or pyrrolidine derivatives have been used to prepare phospholipids spin-labelled in the headgroup region. Thus the headgroup labelled phosphatidylcholine XI has been prepared by condensation of TEMPOcholine IX with phosphatidic acid in the presence of TPS (2,4,6-triisopropylbenzensulfonylchloride) in pyridine (Kornberg and McConnell, 1971) and labelled phosphatidic acid XII has been prepared in a similar way from the corresponding spin-labelled alcohol (Aneja and Davies, 1970). Spin-labelled phosphatidylserine XIII has been prepared as an amide formed by reaction of the amino group with the carboxylic acid spin label (Tanaka and Ohnishi, 1976). Presumably this method will also be applicable to phosphatidylethanolamines.

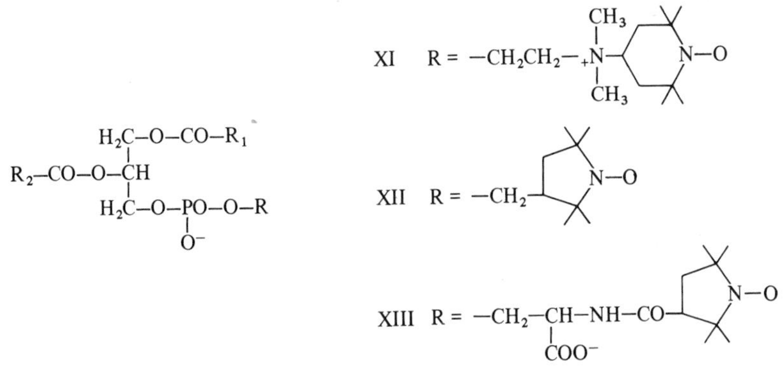

Other membrane probes have also been synthesized which locate the nitroxide in the phospholipid headgroup region of the membrane, e.g. the fatty acid amide probes XIV,

(Hsia et al., 1969), the fatty acid esters XV (Waggoner et al., 1968) and the long-chain quaternary ammonium derivative XVI (Hubbell et al., 1970)

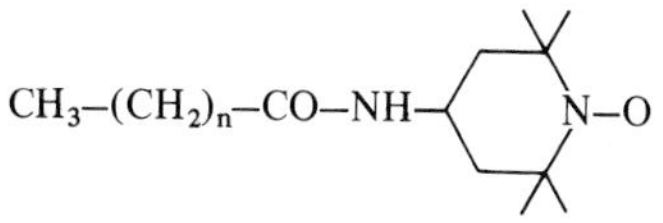

XIV

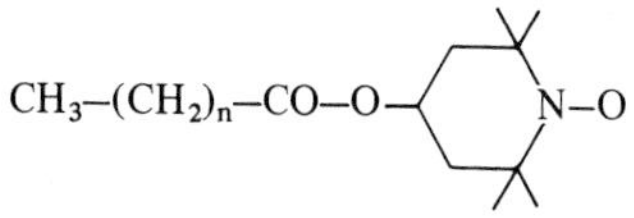

XV

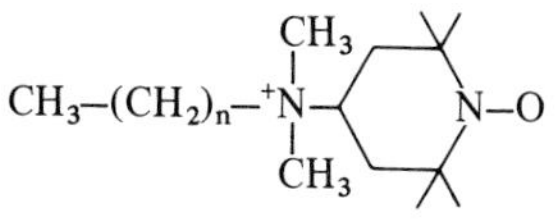

XVI

These headgroup labels are particularly useful in studying transbilayer flip-flop and asymmetric distribution of lipid molecules, lateral diffusion rates, responses to perturbations in the headgroup region and possibly also lateral phase separations and preferential segregation of different lipid types.

The final important application of labels based on the piperidine and pyrrolidine rings is in the synthesis of labels based on the standard covalent protein modifying reagents. The synthesis of these labels is frequently based on the 4-amino tetramethylpiperidine or 3-amino tetramethylpyrrolidine precursors: full details of these preparations may be found in Gaffney (1976). Particularly useful are the maleimide XVII and iodoacetamide XVIII analogues which are capable of alkylating both sulphydryl and amino groups:

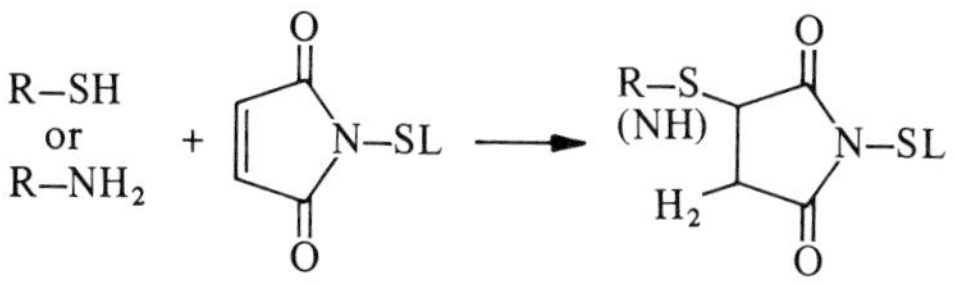

XVII

a) SL = 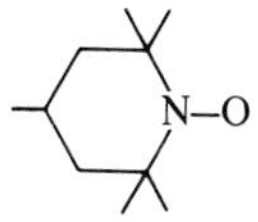

b) SL = 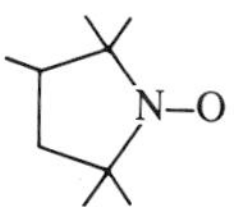

R-SH or R-NH_2 + I-CH_2-CO-NH-SL → R-S-CH_2-CO-NH-SL (NH)

XVIII

In general the maleimide derivatives are more reactive than the iodoacetamides and both derivatives are more reactive towards sulphydryls than amino groups. The pattern of labelling specificity can be changed somewhat by varying the reagent concentration and reaction times, and also by pre-blocking the more reactive groups. For a given spin-label moiety, SL, the maleimide derivatives in general give rise to a more strongly immobilized spectrum than the iodoacetamide derivatives, presumably because of the different flexibilities of the attaching linkage. Since the most strongly immobilized label is frequently not optimally sensitive to changes in the protein conformation, it is of advantage to have labelling reagents with varying distance between the nitroxide and attaching group. Such a series of maleimides XIX and iodoacetamides XX, which should also prove useful in studying the immediate environment of the protein and the geometry of the point of attachment, are commercially available from Syva Associates (1970).

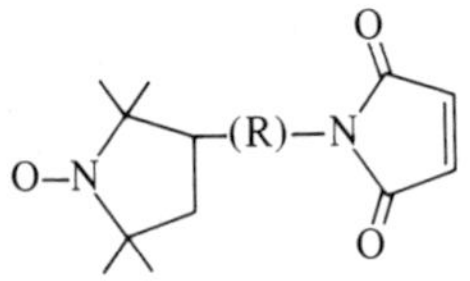

XIX

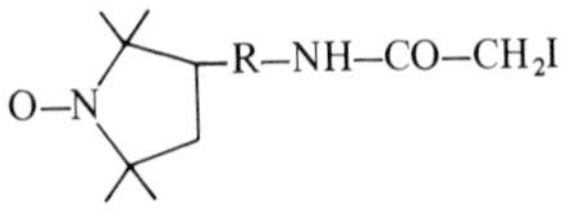

XX

a) R = –
b) R = $-CH_2-$
c) R = $-CO-NH-(CH_2)_2-$
d) R = $-CO-NH-(CH_2)_3-$
e) R = $-CO-NH-(CH_2)_2-O-(CH_2)_2-$

a) R = –
b) R = $-CH_2-$
c) R = $-CO-NH-CH_2-$
d) R = $-CO-NH-(CH_2)_3-$
e) R = $-CO-NH-(CH_2)_2-O-(CH_2)_2-$

Labels have also been synthesized with some specificity to amino groups. These include the isothiocyanate XXI (Gaffney, 1976), the mixed carboxylic-carbonic anhydrides XXII (Griffith et al., 1967) and the hydroxysuccinimide ester XXIII (Hoffman et al., 1970).

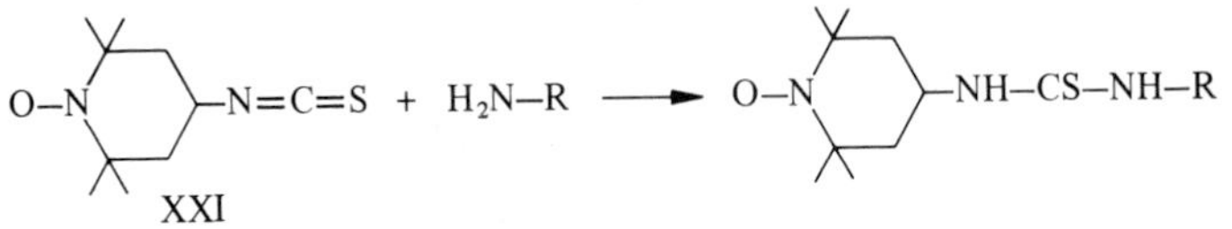

XXI

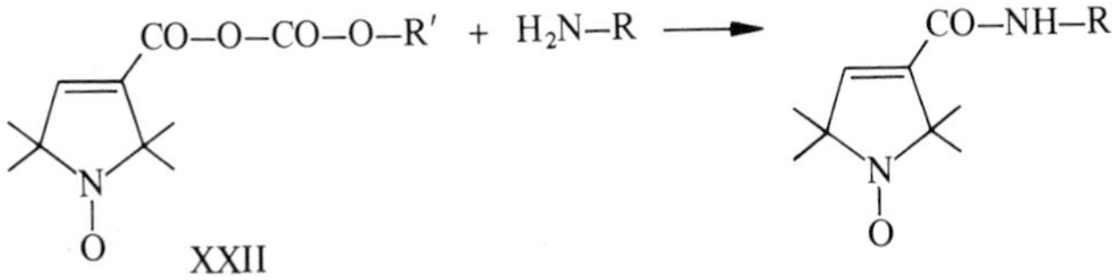

XXII

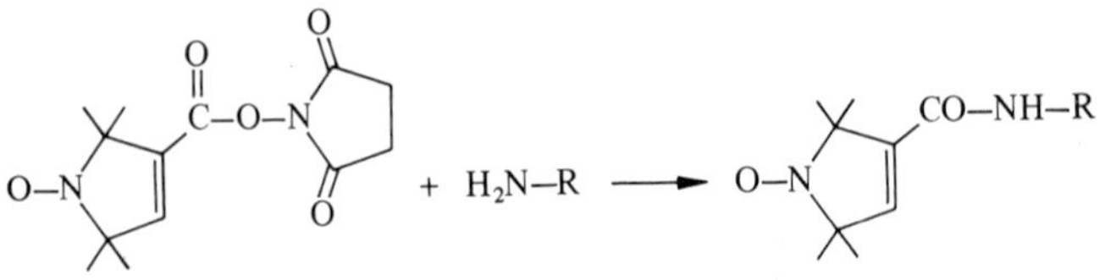

XXIII

Labelling by these reagents frequently gives rise to a rather mobile spectrum, possibly from the labelling of lysine groups resulting in a rather long, flexible linkage of the nitroxide to the polypeptide backbone. The pattern of specificity of labelling is, however, not well established.

As mentioned above, the protected nitroxide is an extremely stable free radical, the two main exceptions to this being its instability in strong acid and its susceptibility to reduction. In both cases the paramagnetism of the nitroxide may sometimes be restored by exposure to air. The reduction potential for the nitroxide group is about –150 mV (Gaffney, 1976) in the pH range 6-8, with a strong dependence on pH outside this range. Thus clearly a considerable range of reducing agents are capable of giving rise to loss of spin-label paramagnetism. Amongst these are: ascorbate, dithionite, hydroxylamine, phenylhydrazine, reduced glutathione, mercaptoethanol, cysteine, dithiothreitol and titanium III (Smith, 1972). The components of biological membranes which are capable of giving rise to spin-label reduction, and precautions and preventative measures have been discussed above. One of the exceptions of the susceptibility to reduction is that the nitroxide appears to be relatively resistent to reduction by sodium borohydride or lithium aluminium hydride, a fact which is used in synthetic work involving the reduction of the R-group attached to the nitroxide ring. The ready reduction of nitroxides by ascorbate which takes place according to the scheme:

$$>\text{N–O}^{\cdot} + \text{ascorbate} \rightarrow >\text{N–OH} + \text{dehydroascorbate}$$

has found wide application in the design of experiments to probe the accessibility of the spin-label group, or to selectively remove the ESR signal of the population of nitroxides which is readily accessible to the exterior aqueous phase. These applications are treated in detail in the appropriate sections below.

III. Theoretical Basis

A. Spectral Anisotropy – Angular Variation and Powder Spectra

(Knowles et al., 1976; Smith et al., 1976)

The anisotropy of the spectrum of a nitroxide spin label oriented in a single-crystal host is shown in Figure 3. It is seen that both the three-line nitrogen hyperfine splitting and the position about which the spectrum is centred depends upon which of the x, y, z molecular axes the magnetic field is oriented along. Formally the spectrum is described by the spin Hamiltonian for the magnetic energy of the nitroxide unpaired electron spin, $\underline{S}$:

$$\mathcal{H}_S = \beta \tilde{\underline{H}} \cdot \underset{\sim}{g} \cdot \underline{S} + \tilde{\underline{I}} \cdot \underset{\sim}{A} \cdot \underline{S} \qquad (1)$$

where β is the Bohr magneton, $\underline{I}$ is the ^{14}N nuclear spin and $\underline{H}$ is the laboratory magnetic field vector. The g-tensor, $\underset{\sim}{g}$, and the hyperfine tensor, $\underset{\sim}{A}$, are to a good approximation both diagonal in the molecular axis system indicated in Figure 3, thus:

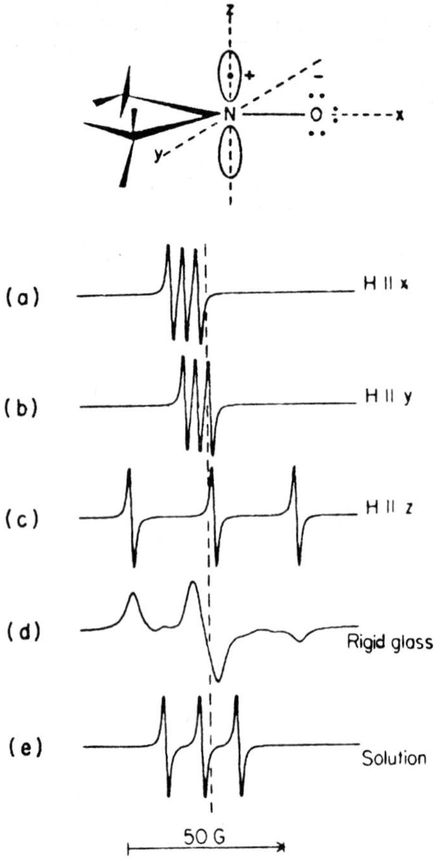

Fig. 3 a-e. Simulated nitroxide spectra. (a)-(c) Single crystal spectra with the magnetic field oriented along each of the principal axes, x, y, z. (d) Powder spectrum from a randomly oriented collection of rigidly immobilized nitroxides. (e) Isotropic spectrum from rapid and randomly tumbling nitroxides. (Adapted from Griffith and Jost, 1976)

$$\mathcal{H}_s = \beta (g_{xx} H_x S_x + g_{yy} H_y S_y + g_{zz} H_z S_z) + A_{xx} I_x S_x + A_{yy} I_y S_y + A_{zz} I_z S_z \quad (2)$$

The ESR spectroscopic transitions are found from the energy levels obtained by diagonalizing the spin Hamiltonian of Eq. (2). For the simple case of the magnetic field oriented along one of the molecular axes, say z, the spectrum is given by (Knowles et al., 1976):

$$h\nu = g_{zz} \beta H_{m_I} + A_{zz} m_I \quad (3)$$

where h is Planck's constant, ν the microwave frequency and $m_I = 0, \pm 1$ are the magnetic quantum numbers of the nitrogen nuclear spin. Equation (3) depicts the three transitions which are centred about $H_o = h\nu/g_{zz} \beta H$ and have equal hyperfine splittings: $H_o - H_{\pm 1} = \pm A_{zz}/g_{zz} \beta$, as shown in Figure 3. The principal values of the g- and A-tensor elements obtained in this way by single-crystal studies of the nitroxide groups most commonly used in spin-label studies are given in Table 1. A useful generalization from these results is that the hyperfine tensor is approximately axial: $6\ G \simeq A_{xx} \simeq A_{yy} \ll A_{zz} \simeq 32\ G$, whereas the g-value is non-axial, but with a smaller anisotropy (at 9 GHz): $g_{xx} > g_{yy} > g_{zz}$.

Table 1. Principal values of hyperfine and g-tensors of nitroxides oriented in single crystal hosts

	A_{xx}(G)	A_{yy}(G)	A_{zz}(G)	g_{xx}	g_{yy}	g_{zz}
DOXYL-cholestane [a]	5.8	5.8	30.8	2.0089	2.0058	2.0021
DOXYL-aliphatic [b] chain	5.9	5.4	32.9	2.0088	2.0058	2.0021
TEMPO/one [c]	5.2	5.2	31.0	2.0104	2.0074	2.0026
Pyrrolidine [d]	4.7	4.7	31.0	2.0101	2.0068	2.0028
DTBN [e]	7.6	6.0	31.8	2.0088	2.0062	2.0027

[a] Label Ia), Hubbell and McConnell (1971).

b 2-doxyl propane 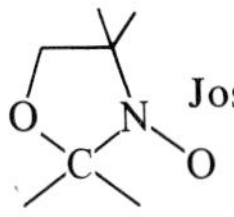 Jost et al. (1971a).

[c] Tempone: 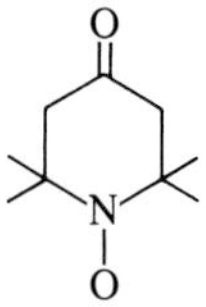 Griffith et al. (1965).

[d] Label 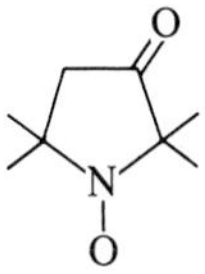 (A-tensor) and 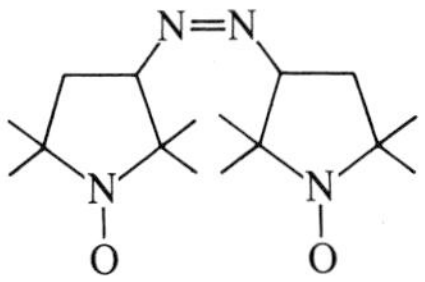 (g-tensor) see Ref. c.

[e] Di-t-butyl nitroxide: 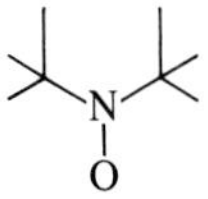 Libertini and Griffith (1970).

For a general orientation of the magnetic field (specified by polar coordinates θ, ϕ relative to the nitroxide z-axis) the analogue of Eq. (3) for the ESR transitions is:

$$h\nu = g(\theta,\phi)\,\beta\, H_{m_I}^{(\theta,\phi)} + A(\theta,\phi)\, m_I \tag{4}$$

where in the intermediate field approximation (Libertini and Griffith, 1970)

$$g(\theta,\phi) = (g_{xx}\cos^2\phi + g_{yy}\sin^2\phi)\sin^2\theta + g_{zz}\cos^2\theta \tag{5}$$

$$A(\theta,\phi) = [(A_{xx}{}^2\cos^2\phi + A_{yy}{}^2\sin^2\phi)\sin^2\theta + A_{zz}{}^2\cos^2\theta]^{1/2}. \tag{6}$$

This immediately shows that it is possible to determine the angular orientation of the lipid molecules (e.g. relative to the membrane surface) from the spin-label ESR spectrum. A simpler form of these equations, corresponding to axial symmetry, is normally used in such orientation studies, since the lipid spin-label tensors are nearly always reduced to axial symmetry by rapid long axis rotation:

$$g(\theta) = g_{\perp}\sin^2\theta + g_{\parallel}\cos^2\theta \tag{7}$$

$$A(\theta) = [A_{\perp}{}^2\sin^2\theta + A_{\parallel}{}^2\cos^2\theta]^{1/2}. \tag{8}$$

Here $A_{\parallel}$, $g_{\parallel}$ are the tensor components directed along the long axis of the molecule; $A_{\perp}$, $g_{\perp}$ are the axial components in the perpendicular plane, and θ is the angle between the molecular long axis and the magnetic field.

For samples which are not oriented single crystals, but consist of a random distribution of orientations of molecular axes, a spectrum which consists of a super-position of spectra from the three principal axes and all intermediate orientations is obtained. In this case outer extrema can be resolved in the so-called *powder spectrum* which correspond to the A_{zz} maximum hyperfine splitting as shown in Figure 3. However, inner extrema cannot be resolved corresponding to A_{xx} and A_{yy} because both the g-value anisotropy and the smallness of the splitting cause these lines to merge with the central hyperfine line (see Knowles et al., 1976; Smith et al., 1976). Thus A_{zz} and g_{zz} can be measured directly from the powder spectrum, whereas A_{xx}, A_{yy}, g_{xx} and g_{yy} can only be determined to considerably lower accuracy by simulation of the total spectral lineshape. In addition Hubbell and McConnell (1971) have shown that the outer extrema of the first derivative powder spectra correspond almost exactly to the single radical *absorption* lines for the magnetic field directed along the z molecular axis. This is because the Eqs. (4) and (6) for the line positions in the θ=0 region can be approximated by:

$$H_{m_I}(\theta) = \frac{h\nu}{g\beta} - \frac{A_{zz}\, m_I}{g\beta} \cdot \cos\theta \tag{9}$$

where the g-value anisotropy has been ignored. The first derivative powder lineshape is then given by:

$$\frac{\delta A_{m_I}(H)}{\partial H} = \int_0^{\frac{\pi}{2}} \frac{\partial a_{m_I}(\xi)}{\partial \xi} \cdot \sin\theta \cdot d\theta \tag{10}$$

where $\xi = H - H_{m_I}(\theta)$, and $a_{m_I}(\xi)$ is the lineshape function of the individual resonance absorptions. The integration of Eq. (10) using Eq. (9) is then straightforward:

$$\frac{\partial A_{m_I}(H)}{\partial H} = \frac{g\beta}{A_{zz} m_I} \left[a_{m_I}(\theta = 0) - a_{m_I}(\theta = \pi/2)\right] \tag{11}$$

showing that the powder lineshape consists essentially of two components, one arising from the $\theta = 0$ (z-axis) region and the other from the $\theta = \pi/2$ (x-, y-axis) region. The outer extrema ($\theta = 0$) are given exactly by the absorption lineshape corresponding to the z-axis orientation, but the same is not true for the inner ($\theta = \pi/2$) extrema because Eq. (9) does not hold for this region.

At the opposite extreme to the powder spectrum, arising from rigidly-immobilized, randomly-oriented radicals, is the isotropic spectrum, also illustrated in Figure 3, which arises from radicals rapidly and randomly tumbling in a non-viscous solvent. The result of this fast, isotropic motion is to completely average the g-value and A-tensor anisotropy, giving rise to a narrow, three-line spectrum which is independent of the magnetic field orientation. The hyperfine splitting and g-value are then given by the isotropic components:

$$a_o = (1/3)(A_{xx} + A_{yy} + A_{zz}) \tag{12}$$

$$g_o = (1/3)(g_{xx} + g_{yy} + g_{zz}). \tag{13}$$

The condition that this averaging of the spectral anisotropy should take place is that the motion should be fast compared with the frequency equivalent of the spectral anisotropies, i.e. $(A_{zz} - A_{xx}) \sim 0.73 \times 10^{-8}\ s^{-1}$ and $(g_{xx} - g_{zz}) \cdot \beta H \sim 0.29 \times 10^{-8}\ s^{-1}$. If the motion is very fast on the ESR time scale, i.e. with correlation time $\sim 10^{-11}$ s, then the spectrum will consist of three sharp lines of equal height, as illustrated in Figure 3. As the motion progressively slows there will be first a differential broadening of the spectrum, with the line positions remaining constant, and then a distortion of the line positions and lineshapes, the final limit being the rigidly immobilized powder spectrum. In addition, if the motion is fast but *an*isotropic, then there will only be partial averaging of the spectral anisotropy and the degree of motional averaging will be a measure of the angular amplitude of the anisotropic motion. The measurement of the amplitude and rate of motion from the spin-label spectrum are dealt with in detail in the next two sections.

B. Angular Amplitudes of Motion – Order Parameters

(Griffith and Jost, 1976; Seelig, 1976; Smith et al., 1976)
When the motion is fast on the ESR timescale, i.e. with correlation times faster than $\sim 3 \times 10^{-9}$ s, the line positions are determined by a time-averaged version of the spin Hamiltonian of Eq. (2), where the time-averaged hyperfine and g-tensor components are independent of the rate of motion and are determined solely by geometrical considerations as indicated below.

Consider the anisotropic motion of a spin-labelled lipid molecule oriented in a bilayer lipid membrane. Let the instantaneous orientation of the nitroxide x-, y-, z-axes relative to the bilayer normal be given by $\theta_1, \theta_2, \theta_3$ respectively, as seen in Figure 4. Transformation of the g- and A-tensor from the nitroxide axes to the bilayer axis system can be performed by the rotation matrix, $\underset{\sim}{R}$: $\underset{\sim}{A}' = \underset{\sim}{R}\,\underset{\sim}{A} \cdot \underset{\sim}{R}^{-1}$, $g' = \underset{\sim}{R} \cdot \underset{\sim}{g} \cdot \underset{\sim}{R}^{-1}$. Performing the transformation and the motional averaging relative to the bilayer normal gives rise to the following axial tensor components in the bilayer axis system (Seelig, 1970):

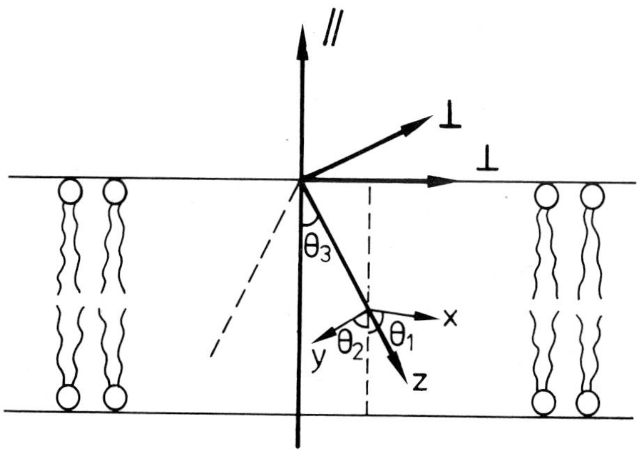

Fig. 4. Instantaneous orientation of the x, y, z nitroxide axes of a spin-labelled lipid, relative to the symmetry axes, ‖, ⊥, of the phospholipid bilayer. The lipid motion is such as to produce a time-average axial symmetry relative to the bilayer normal (‖)

$$A_{\parallel} = (A_{zz} - A_{xx}).<\cos^2\theta_3> + A_{xx} \quad (14)$$

$$A_{\perp} = \tfrac{1}{2}(A_{zz} - A_{xx}).(1-<\cos^2\theta_3>) + A_{xx} \quad (15)$$

and

$$g_{\parallel} = (g_{xx} - g_{yy}).<\cos^2\theta_1> + (g_{zz} - g_{yy}).<\cos^2\theta_3> + g_{yy} \quad (16)$$

$$g_{\perp} = \tfrac{1}{2}(g_{xx} - g_{yy}).(1-<\cos^2\theta_1>) + \tfrac{1}{2}(g_{zz} - g_{yy}).(1-<\cos^2\theta_3>) + g_{yy}. \quad (17)$$

These components involve only the time average of the angular orientation of the nitroxide axes relative to the bilayer normal (||), since complete averaging was performed about the other angles, corresponding to a motion which gives rise to axial symmetry in the plane of the membrane. The approximation $A_{yy} \simeq A_{xx}$ was made in Eqs. (14) and (15), which means that it is possible to obtain the time-average orientation of the nitroxide z-axis relative to the bilayer normal from the hyperfine splittings alone. The line positions in the motionally averaged spectrum can then be simply found by substituting the tensor components of Eqs. (14)-(17) into the expressions (7) and (8) above. In particular, for the magnetic field oriented along the bilayer normal the hyperfine splitting and g-value are simply $A_{\parallel}$ and $g_{\parallel}$ respectively, and similarly $A_{\perp}$ and $g_{\perp}$ are obtained with the magnetic field oriented in the plane of the bilayer. A further feature of the transformation Eqs. (14)-(17) is that the average values are unaffected by the motional amplitudes and remain equal to the isotropic values of Eqs. (12) and (13):

$$a_o = \tfrac{1}{3}(A_{\parallel} + 2A_{\perp}) \quad (18)$$

$$g_o = \tfrac{1}{3}(g_{\parallel} + 2g_{\perp}). \quad (19)$$

This is an important result since, as will be seen below, the isotropic g-values and hyperfine splittings are dependent on the polarity of the environment and thus Eqs. (18) and (19) can be used to probe the polarity of the region of the membrane which is sampled by the spin probe.

The angular amplitudes of motion are normally expressed as order parameters, S_{ii}, since these are conveniently related to the experimentally measured quantities and summarize all the tensor transformation properties arising from the motional averaging:

$$S_{ii} = \tfrac{1}{2}(3<\cos^2\theta_i> - 1) \qquad i = 1,2,3. \quad (20)$$

Here i = 1,2,3 corresponds to the order parameters of the x, y, z nitroxide axes respectively, relative to the bilayer normal as defined in Figure 4.

From Eqs. (14)-(17) it is clear that the order parameters are obtained from the experimental spectrum as follows:

$$S_{33} = \frac{A_{\parallel} - A_{\perp}}{A_{zz} - A_{xx}} \quad (21)$$

$$S_{11} = \frac{[3g_{\parallel} - (g_{xx} + g_{yy} + g_{zz}) - 2\,S_{33}(g_{zz} - g_{yy})]}{2\,(g_{xx} - g_{yy})} \quad (22)$$

$$S_{11} + S_{22} + S_{33} = 1. \tag{23}$$

The latter condition arises from the orthogonality property of the direction cosines and means that only two of the order parameters are independent. Normally one is most interested in the S_{33} order parameter, since this is sufficient to define the amplitude of motion of the long molecular axis relative to the bilayer normal. The second order parameter simply gives a measure of the non-axiality of the motion in the lateral plane; for axial motion: $S_{11} = S_{22} = -\frac{1}{2} S_{33}$.

The S_{33} order parameter is directly obtained from the hyperfine splittings, and in fact is simply the anisotropy of the splittings, $(A_{\parallel} - A_{\perp})$, normalized relative to that which would be obtained if the molecules were rigidly oriented with the nitroxide z-axis along the $\parallel$ direction: $(A_{zz} - A_{xx})$, e.g. in a single crystal. A complication arises in this calculation since the principal hyperfine tensor components of Table 1 are measured in single crystals in which the polarity of the environment (and hence the absolute values of the splittings) may be different from that of the bilayer membrane. For this reason a polarity correction is made to the order parameter (Hubbell and McConnell, 1971) by normalizing both hyperfine splitting anisotropies relative to the corresponding isotropic splitting constant, thus:

$$S_{33} = \frac{(A_{\parallel} - A_{\perp})}{(A_{zz} - A_{xx})} \cdot \frac{a_o \text{ (crystal)}}{a_o \text{ (bilayer)}} \tag{24}$$

where a_o (crystal) is given by Eq. (12) and a_o (bilayer) by Eq. (18). The molecular parameter one requires, of course, is the order parameter of the long axis of the lipid molecule relative to the bilayer normal, S_{mol}. In the case of the fatty acid spin labels II (m,n) and phospholipid spin labels III (m,n)-VII (m,n) this is identical with the order parameter S_{33}, since the nitroxide z-axis is parallel to the long molecular axis of the lipid chains. For cases in which this is not so, S_{mol} can be obtained from the measured S_{33} by the following transformation:

$$S_{mol} = S_{33} / \tfrac{1}{2} \, (3 \cos^2 \alpha - 1) \tag{25}$$

where α is the angle between the nitroxide z-axis and the long molecular axis. This formula is only valid for fast rotation about the long molecular axis, when $S_{11} \simeq S_{22}$. If one of the other two nitroxide axes coincides with the long molecular axis, then S_{mol} can be obtained simply by a redefinition of parameters and permutation of axes. An important example is the steroid spin labels I (a)-I (c) in which the nitroxide y-axis coincides with the long molecular axis. Here fast rotation about the long axis averages the A_{zz} and A_{xx} hyperfine constants to give an effective axial splitting of $\frac{1}{2}(A_{zz} + A_{xx})$ in the plane perpendicular to the long axis. Then for the steroid labels:

$$S_{mol} = \frac{A_{\parallel} - A_{\perp}}{A_{yy} - \frac{1}{2}(A_{zz} + A_{xx})} \cdot \frac{a_o \text{ (crystal)}}{a_o \text{ (bilayer)}} . \tag{26}$$

One other possible case of motion should be mentioned, since this cannot be described by this simplified order parameter formalism, that is the case of only *partial* rotation or oscillation about a given axis. Under these conditions the motionally averaged hyperfine tensor does not have axial symmetry. Consider for example the steroid label I performing limited oscillations in the angle ψ around the long molecular axis, i.e. about the nitroxide y-axis.

The observed hyperfine elements are then (Van et al., 1974):

$$A_{xx}^{osc} = A_{xx} + (A_{zz}\text{-}A_{xx}) < \cos^2 \psi > \quad (27)$$

$$A_{yy}^{osc} = A_{zz} - (A_{zz}\text{-}A_{xx}) < \cos^2 \psi > \quad (28)$$

$$A_{\parallel}^{osc} = A_{yy}. \quad (29)$$

In summary, it can be said that (except in cases where the long molecular axis does not coincide with the nitroxide z-axis and in addition there is not complete fast rotation about the long axis) the motionally averaged spectrum can be characterized by the two axial hyperfine splittings $A_{\parallel}$ and $A_{\perp}$. These together can be used to define both the polarity of the spin-label environment, via the isotropic splitting factor of Eq. (18), and the angular amplitude of motion of the molecular long axis relative to the bilayer normal, via the order parameters of Eqs. (24) and (25). Two practically useful expressions for the order parameter are obtained by using the single-crystal hyperfine tensor elements given in Table 1:

For fatty acid II (m,n) and phospholipid labels III (m,n)-VII (m,n):

$$S_{mol} = \frac{A_{\parallel} - A_{\perp}}{a_o\,(\text{bilayer})} \times 0.5407 \quad (30)$$

For steroid labels I:

$$S_{mol} = \frac{A_{\parallel} - A_{\perp}}{a_o\,(\text{bilayer})} \times (-1.131). \quad (31)$$

Finally some remarks must be made about the measurement of the $A_{\parallel}$ and $A_{\perp}$ parameters from the experimental spectra. The angular variation of the spectra is given by Eqs. (4), (7) and (8) and thus if one has an oriented bilayer sample, $A_{\parallel}$ and $A_{\perp}$ are simply obtained from the spectra obtained with the magnetic field oriented along the bilayer normal and in the bilayer plane respectively. When, as is more often the case, one has a random dispersion of bilayers or membranes, then one obtains the membrane analogue of the polycrystalline powder spectrum discussed in the previous Section III.A. Just as in that case, one can normally resolve the maximum, outer hyperfine splitting (which is $A_{\parallel}$ in the case of the fatty acid labels II (m,n) and phospholipid labels III (m,n)-VII (m,n)). However, unlike the polycrystalline powder spectra, it is also often possible to resolve the splittings of the inner extrema (which correspond to $A_{\perp}$ for the fatty acid II (m,n) and phospholipid III (m,n)-IV (m,n) labels). This is because the effect of motional averaging is to increase the inner splitting, so that the inner extrema move out away from the central (m=0) hyperfine line and thus can be resolved. Clearly, as the inner splitting increases with increasing amplitude of motion the outer splitting also decreases, until finally both coalesce to the isotropic value. There is thus a limited range of order parameters for which the outer and inner hyperfine splittings are both resolvable. At high order parameters the inner splitting is too small to be resolved and at low order parameters the two splittings overlap and are thus not resolvable. This is illustrated in Figure 5, which represents simulated spectra of the fatty acid II (m,n) or phospholipid

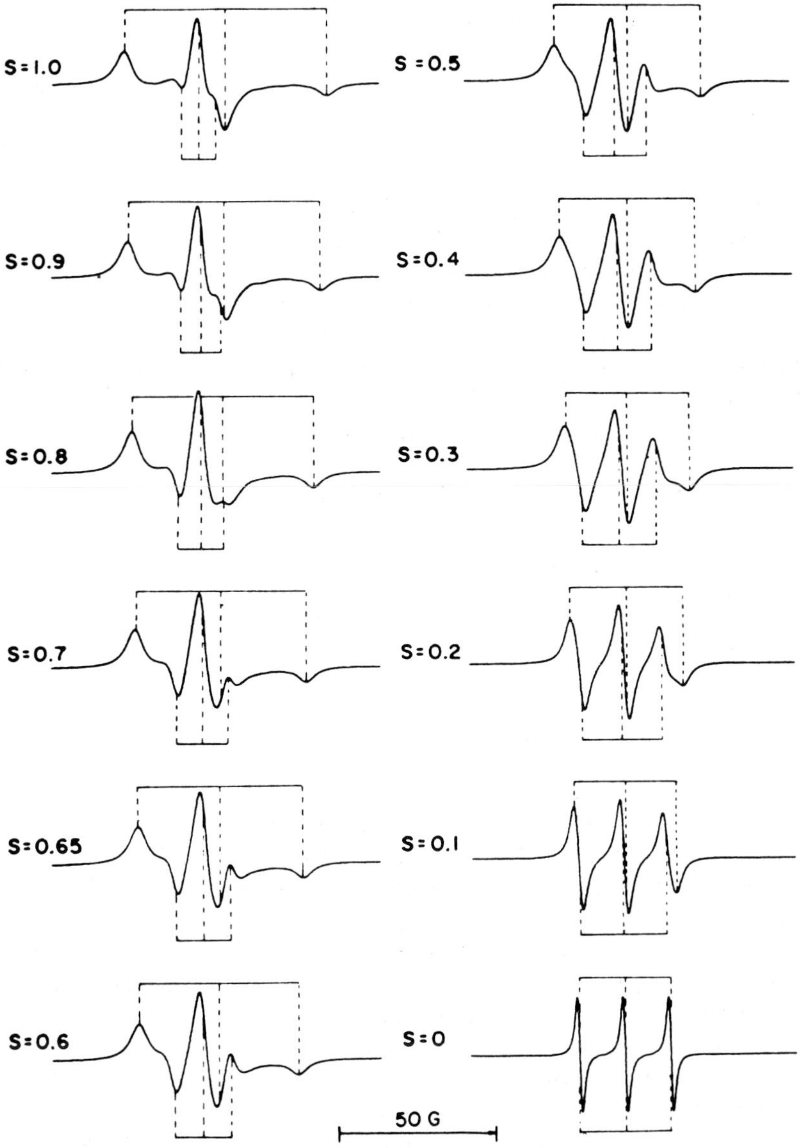

Fig. 5. Simulated ESR lineshapes of randomly oriented samples for lipid spin labels with different values of the order parameter, S_{mol}. *Dotted lines:* exact positions of the ESR absorptions defined by $A_{\parallel}$ and $g_{\parallel}$ (large splitting) and by $A_{\perp}$ and $g_{\perp}$ (smaller splittings). The intrinsic linewidths used were 4.0 G for S = 1 to S = 0.3; 3.0 G for S = 0.2; 2.5 G for S = 0.1; 1.5 G for S = 0.0. (Griffith and Jost, 1976)

III (m,n)-IV (m,n) labels in randomly oriented membranes and undergoing anisotropic motions characterized by various values of the order parameter, S_{mol}. The simulations correspond to a super-position of spectra from all different angular orientations calculated from Eqs. (4), (7) and (8) and weighted by the sinusoidal distribution function as given in Eq. (10). The effective outer and inner hyperfine splittings $A'_{\parallel}$, $A'_{\perp}$ are compared with the $A_{\parallel}$, $A_{\perp}$ hyperfine parameters used as input for the simulations. Clearly the outer splitting is a good measure for the $A_{\parallel}$ hyperfine parameter down to order parameters of 0.2. This is to be expected from the analysis given above in Eq. (10). However, the inner splitting differs significantly from $A_{\perp}$ even at high order parameters, and the discrepancy becomes larger as S decreases. Previously Hubbell and McConnell (1971) had suggested that a correction of 0.8 G be added to $A'_{\perp}$ to allow for this difference. This is clearly only a first approximation and a closer approximation is given by (Gaffney, 1976; Griffith and Jost, 1976):

$$A_{\perp}\,(\text{gauss}) \simeq A'_{\perp} + 1.4\left(1 - \frac{A'_{\parallel} - A'_{\perp}}{A_{zz}\text{-}A_{xx}}\right) \tag{32}$$

$$A_{\parallel}\,(\text{gauss}) \simeq A'_{\parallel}. \tag{33}$$

Thus order parameters in the range 0.2-0.8 can be estimated reasonably well from measurement of splittings in the lipid or membrane dispersion spectra by use of Eqs. (32), (33). More reliable values, especially outside this range, can be obtained by spectral simulation. A computer programme for doing such simulations is listed by Libertini et al. (1974). Even in the latter case, to obtain reliable values it is necessary to take account of the differential line broadening in the spectrum due to relaxation effects as described in the next section. In the limit of high order parameters when it is not possible to measure an inner splitting, the outer splitting $A'_{\parallel}$ is frequently used as an empirical parameter. This is sensitive to changes both in amplitude of motion and polarity, but if the polarity remains constant then changes in $A_{\parallel}$ are directly proportional to changes in the order parameter, since it can be shown that:

$$S = \frac{3(A_{\parallel} - a_o)}{2(A_{zz}\text{-}A_{xx})} \cdot \frac{a_o\,(\text{crystal})}{a_o\,(\text{bilayer})}\,. \tag{34}$$

Thus if a value for a_o can be assumed then S_{mol} can be calculated, alternatively $A_{\parallel}$ alone can be used. A word of caution is perhaps necessary here, since high order parameters are sometimes associated with slow rates of motion, although this is not always necessarily the case. If the spin-label motion is approaching the slow regime on the ESR time scale, then the $A'_{\parallel}$ splitting will depend not only on the amplitude of motion but also on the rate of motion, as is discussed in the next section.

C. Rates of Motion – Correlation Times

(Schindler and Seelig, 1973, 1974; Freed, 1976; Seelig, 1976; Polnaszek, 1979)
If the spin label motion is extremely fast on the ESR time scale, i.e. with correlation times $\tau \leqslant 10^{-11}$ s, then the spectra are essentially insensitive to the rate of molecular mo-

tion; in oriented samples three sharp lines of equal heights would be obtained, the degree of anisotropy of which would depend on the amplitude of motion. For rates of motion in the range $10^{-11} \leqslant \tau \leqslant 3.10^{-9}$ s, the motionally averaged line positions described in the previous section are obtained, but the lines are differentially broadened by an amount depending on both the rate and amplitude of motion. For rates slower than $\tau = 3.10^{-9}$ s both the line positions, lineshapes and linewidths depend on the rate of motion until eventually a totally immobilized powder spectrum is obtained.

In the fast motional averaging regime (10^{-11} s $\leqslant \tau \leqslant 3.10^{-9}$ s) the transverse relaxation times $T_2(m)$, and hence the differential linebroadening, may be treated by time-dependent perturbation theory. It is found that the linewidth depends on nuclear quantum number according to:

$$1/T_2(m) = a + bm + cm^2 \tag{35}$$

where the peak-to-peak derivative Lorentzian linewidth (in frequency units) is given by:

$$\Delta\nu(m) = 1/[\sqrt{3}\,\pi\, T_2(m)]. \tag{36}$$

Hence the a-term in Eq. (35) causes a uniform broadening of all three hyperfine lines, and in practice will be indistinguishable from static contributions to the linebroadening from unresolved proton hyperfine structure, instrumental broadening, spin-spin broadening, etc. The b-term causes a differential broadening of all three lines and the c-term causes a symmetrical broadening about the centre line.

The dominant contributions to the transverse relaxation are the motional modulations of the hyperfine and g-value anisotropies, and it is for this reason that the linewidths depend on the amplitude and rate of motion. For isotropic motion the dependence on the motional correlation time, τ_R, is given explicitly by the following expressions for the b and c parameters of Eq. (35); (Stone et al., 1965):

$$b = \tfrac{8}{45} \cdot \frac{\beta}{\hbar} [A_{zz} - \tfrac{1}{2}(A_{xx}+A_{yy})]\ [g_{zz} - \tfrac{1}{2}(g_{xx}+g_{yy})]\, H_o \cdot \tau_R \tag{37}$$

$$c = \tfrac{4}{72}\ [A_{zz} - \tfrac{1}{2}(A_{xx}+A_{yy})]^2 \cdot \tau_R. \tag{38}$$

For practical purposes it is more convenient to work in terms of linewidths in magnetic field units (gauss). The analogue to Eq. (35) is:

$$\Delta H(m) = A + Bm + Cm^2 \tag{39}$$

where $A = \Delta H(0) = \Delta\nu(0)h/g\beta$ is the central linewidth, and the differential broadening parameters B and C can be expressed in terms of the experimental lineheights, h_m, by:

$$B = \tfrac{1}{2}\, \Delta H(0)\, [\sqrt{h_o/h_{+1}} - \sqrt{h_o/h_{-1}}] \tag{40}$$

$$C = \tfrac{1}{2}\, \Delta H(0)\, [\sqrt{h_o/h_{+1}} + \sqrt{h_o/h_{-1}} - 2]. \tag{41}$$

Using the values for doxyl propane from Table 1, the numerical equivalents of Eqs. (37) and (38) are:

$$\tau_R = -1.22 \times 10^{-9}.B \tag{42}$$

$$\tau_R = 1.19 \times 10^{-9}.C \tag{43}$$

where τ_R is in seconds, $\Delta H(0)$ is in gauss and the calibration constant in Eq. (42) is directly proportional to H_o, here $H_o = 3300$ G corresponding to 9 GHz. Equations (40)-(43) are valid for the correlation time range 5.10^{-11} to 10^{-9} s and allow two independent determinations of the correlation time for isotropic motion from the experimental spectra. Differences between values of τ_R calculated from Eq. (42) and (43) can arise from anisotropy of the motion, molecular ordering effects, or motion in the slow regime ($\tau \geqslant 3.10^{-9}$ s) rather than in the motional narrowing region.

The usual situation in the membrane context is that of anisotropic rather than isotropic motion. For a cylindrically symmetrical molecule one will require at least two correlation times to describe its motion: $\tau_\parallel$ for rotation about the symmetry axis and $\tau_\perp$ for rotation perpendicular to this axis. Polnaszek (1979) has considered the case of anisotropic motion in an isotropic medium, for which there is anisotropy in the rate but not in the amplitude of motion (i.e. the order parameter is still zero). The $\tau_\parallel$ and the $\tau_\perp$ correlation times can then be expressed simply in terms of the B and C linewidth coefficients. However, the more relevant case for membrane systems is that of anisotropic motion in an anisotropic orientational potential, such as is the case for lipid molecules in lipid bilayers and in membranes. The linewidth coefficients of Eqs. (35) and (39) then no longer depend only on the correlation times $\tau_\parallel$, $\tau_\perp$, but also on rather complicated functions of the angular amplitude of motion. In this case the only possibility is to simulate the complete spectral lineshape (Schindler and Seelig, 1973, 1974; Hemminga, 1974, 1975). In order to limit the number of parameters which have to be fitted, an orientational potential of the Maier-Saupe (1959) type is assumed:

$$U(\theta_3) = \lambda.\cos^2\theta_3 \tag{44}$$

and the angular probability distribution is then given by:

$$P(\theta_3) = \exp(-\lambda \cos^2\theta_3/kT). \tag{45}$$

The strength of the pseudopotential, λ, is determined from the order parameter, since the required angular average is:

$$<\cos^2\theta_3> = \frac{\int_o^\pi \cos^2\theta_3.P(\theta_3).\sin\theta_3.d\theta_3}{\int_o^\pi P(\theta_3).\sin\theta_3.d\theta_3} \tag{46}$$

Having determined a self-consistent value of λ for a given value of S_{33} from Eq. (46), all the other angular averages required for the relaxation calculation can be determined. Thus the important parameters to be fitted in the simulation are the correlation times $\tau_\parallel$, $\tau_\perp$ and the order parameter, S_{33}. The reader is referred to the work by Schindler and Seelig (1974, 1975) and Hemminga (1974, 1975) for further details. Several important points emerge: the first is that accurate values of the polarity-corrected hyperfine and g-tensors are required for good simulations and that the pseudosecular terms and the asymmetry of the order parameter tensor must be taken into account; the second is that for long, rigid molecules such as the steroid molecules I, the correlation time $\tau_{\parallel}$ is much smaller than $\tau_{\perp}$; and finally for the chain-labelled lipids VI (m,n), III-VII (m,n): $\tau_{\parallel} = \tau_{\perp}$ as might be expected if the motion is dominated by rotational isomerism of the chain.

For spin labels in the slow motion regime ($\tau > 3.10^{-9}$ s) perturbation theory may no longer be applied and the relaxation calculations must be performed by the stochastic Liouville method (Freed, 1976; Polnaszek, 1979) or by means of the Bloch equations modified by adding a diffusion term (McCalley et al., 1972). In this case, even for isotropic motion, it is necessary to simulate the spectra, since both the positions, linewidths and lineshapes depend on the rate of motion. The simulations are also lengthier and more involved than those in the motional narrowing region, and again in the case of anisotropic motion an orientational pseudopotential of the form given by Eq. (44) is assumed. The reader is referred to the reviews by Freed (1976) and Polnaszek (1979) for further details of these simulations and to Cannon et al. (1975), Smith et al. (1976) and Schreier et al. (1978) for a discussion of the conditions under which the slow motion analysis (rather than the motional narrowing, perturbation analysis) is required.

In the case of isotropic slow motion, empirical calibrations in terms of the splitting and linewidths of the outer extrema of the spectra (see Fig. 6) have been established from

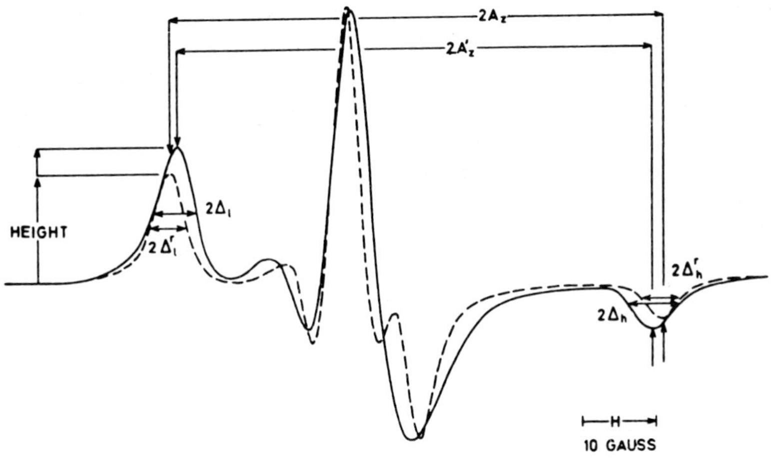

Fig. 6. Rigid limit nitroxide spectrum (*broken line*) and slow tumbling spectrum (*full line*), illustrating the measurement of the parameters A'_{zz} (= A'_z), A^R_{zz} (= A_z), ΔH_m (= Δ_m) and ΔH^R_m (= Δ^r_m), where m = h or l. (From simulations by Mason and Freed, 1974)

the simulations (Freed, 1976). By analogy with the well-known case of two-site exchange in magnetic resonance, it is seen that as the motional rate of the spin label increases from zero, the outer extrema of the rigid-limit powder spectrum will first broaden and then shift closer together. Thus the correlation time can be estimated from the linewidth of the high field outermost peak, ΔH_h, or of the low field outermost peak, ΔH_l, or from their separation, A'_{zz}, as compared to the corresponding values ΔH^R_h, ΔH^R_l, A^R_{zz} at the rigid limit (see Fig. 6). The empirical equations are:

$$\tau_R = a\,(1 - A'_{zz}/A^R_{zz})^b \tag{47}$$

$$\tau_R = a'_m\,(\Delta H_m/\Delta H^R_m - 1)^{b'_m} \tag{48}$$

where m = h or l, and the values for the calibration constants, which are obtained from simulations assuming Brownian motion, are (Freed, 1976):

$a = 5.4 \times 10^{-10}$ s, $b = -1.36$

$a'_h = 2.12 \times 10^{-8}$ s, $b'_h = -0.778$

$a'_l = 1.15 \times 10^{-8}$ s, $b'_l = -0.943$

For calibration constants corresponding to different intrinsic linewidths (here $\delta = 3.0$ G) or to different motional models, for example strong or intermediate collisions, the reader is referred to the article by Freed (1976). Care is required in choosing the appropriate set of rigid-limit parameters in these empirical equations, especially the value of A^R_{zz}, since this is also sensitive to the polarity of the environment, as will be seen in the following section. Since the nitroxide hyperfine splitting is approximately axial about the z-axis, the calibration Eqs. (47), (48) also apply in the presence of *fast* anisotropic motion about the z-axis. The correlation time measured in this case is that for *slow* rotation of the z-axis itself, i.e. $\tau_\perp$.

For slow anisotropic motion about the nitroxide y-axis, and very much slower motion perpendicular to this axis, Polnaszek (1977) has given a somewhat different calibration. For the situation in which the axial rotation is so slow that the spectra appear similar to the rigid limit spectra (cf. Fig. 6), an empirical calibration in terms of A'_{zz} is again possible. In this motional regime (2.10^{-9} s $\leqslant \tau_\parallel \leqslant 7.5.10^{-8}$ s), Eq. (47) can be used to estimate $\tau_\parallel$ with the following values for the calibration constants:

$a = 2.596 \times 10^{-10}$ s, $b = -1.396$

For the motional narrowing region ($7.5.10^{-11}$ s $\leqslant \tau_\parallel \leqslant 10^{-9}$ s) the following linear calibration is obtained:

$$\tau_\parallel\,(s) = (323 \times A'_{zz}/A^R_{zz} - 195) \times 10^{-10}. \tag{49}$$

Again in this region the motion is sufficiently fast that the spectra approximate to three derivative-like lines (although with considerably asymmetrical lineshapes). Effective values (B' and C') of the lineshape coefficients B and C can then be defined by Eqs. (40), (41) and the following empirical relations for the correlation time about the y-axis are obtained:

$$\tau_\parallel\,(\text{sec}) = (-1.650 - 4.438B' - 0.150B'^2) \times 10^{-10} \tag{50}$$

$$\tau_\parallel\,(\text{sec}) = (-7.782 + 3.262C' - 0.044C'^2) \times 10^{-10} \tag{51}$$

the range of validity again being $7.5.10^{-11}$ s $\leqslant \tau_\parallel \leqslant 10^{-9}$ s.

Thus in certain circumstances, very useful empirical calibrations are available for calculating correlation times, including the slow motional region. Eq. (47) also illustrates the very important point that if the spin-label motion is in the slow motion region, then the linesplittings will be dependent on the rate of motion and cannot be used directly to determine the motional amplitude: an overestimate of the order parameter would result.

D. Polarity Effects

(Seelig et al., 1972; Griffith et al., 1974)
The overall size of the hyperfine interaction, as measured by the isotropic hyperfine constant a_o (Eqs. (12), (18)), depends on the net unpaired spin density on the N-atom of the nitroxide. Since a more polar environment will favour the canonical structure B of the neutral free radical:

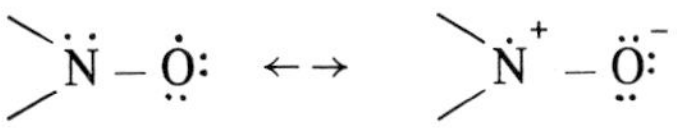

A B

which has greater spin density on the nitrogen atom, an increase in polarity will be accompanied by an increase in hyperfine splitting. Hydrogen bonding will also favour structure B and hence increase the isotropic hyperfine splitting constant.

The isotropic hyperfine splitting constants and g-values of the small spin-label molecule di-t-butyl nitroxide (DTBN), measured in various isotropic, non-viscous solvents of differing polarity, are given in Table 2. Increases in a_o of 2 gauss or more are observed on going from pure hydrocarbon solutions to aqueous solutions. Figure 7 gives the polar-

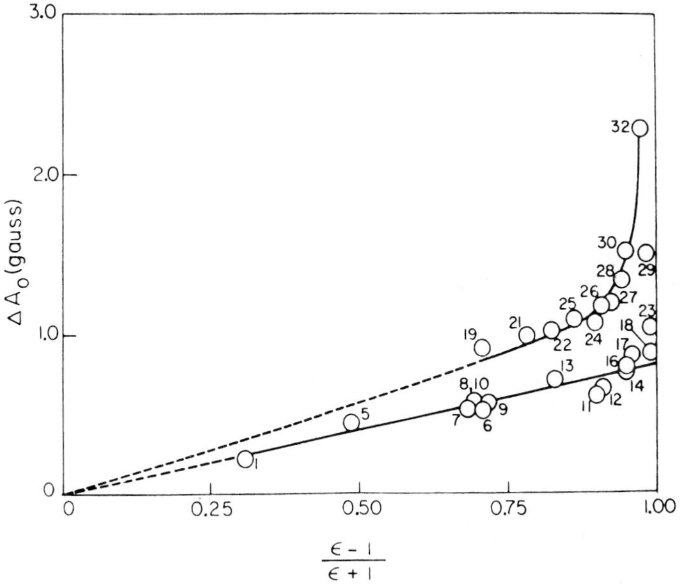

Fig. 7. Polarity dependence of the isotropic hyperfine constant, a_o, of di-t-butyl nitroxide. $\Delta A_o = a_o - 14.85$ G, ϵ is the dielectric constant of the solvent, and the numbers next to the experimental points refer to the solvent systems in Table 2 (Griffith et al., 1974)

Table 2. Isotropic hyperfine splitting constants and g-values of di-t-butyl nitroxide in solvents of various polarities, at 23°C (Griffith et al., 1974)

No.	Solvent	A_o	g_o
1.	Hexane	15.10	2.0061
2.	Heptane-pentane (1:1)[a]	15.13	2.0061
3.	2-Hexene	15.17	2.0061
4.	1,5-Hexadiene	15.30	2.0061
5.	Di-*n*-propylamine	15.32	2.0061
6.	Piperidine	15.40	2.0061
7.	*n*-Butylamine	15.41	2.0060
8.	Methyl propionate	15.45	2.0061
9.	Ethylacetate	15.45	2.0061
10.	Isopropylamine	15.45	2.0060
11.	2-Butanone	15.49	2.0060
12.	Acetone	15.52	2.0061
13.	Ethylacetate saturated with water	15.59	2.0060
14.	N,N-Dimethyl formamide	15.63	2.0060
15.	EPA[b] (5:5:2)[a]	15.63	2.0060
16.	Acetonitrile	15.68	2.0060
17.	Dimethyl sulfoxide	15.74	2.0059
18.	N-Methyl propionamide	15.76	2.0059
19.	2-Methyl-2-butanol	15.78	2.0059
20.	EPA[b] (5:5:10)[a]	15.87	2.0060
21.	1-Decanol	15.87	2.0059
22.	1-Octanol	15.89	2.0059
23.	N-Methyl formamide	15.91	2.0059
24.	2-Propanol	15.94	2.0059
25.	1-Hexanol	15.97	2.0059
26.	1-Propanol	16.05	2.0059
27.	Ethanol	16.06	2.0059
28.	Methanol	16.21	2.0058
29.	Formamide	16.33	2.0058
30.	1,2-Ethanediol	16.40	2.0058
31.	Ethanol/water (1:1)[a]	16.69	2.0057
32.	Water	17.16	2.0056
33.	10 M LiCl aqueous solution	17.52	2.0056

[a] By volume.
[b] EPA designates a mixture of *e*thyl ether (diethyl ether), iso*p*entane (2-methylbutane) and *a*lcohol (ethanol).

ity dependence of a_o in terms of the dielectric constant of the solvent. The values fall into two classes: the aprotic solvents which give rise to a polarity dependence explicable in terms of the polarizability of the solvent, and the H-bond donors (alcohols, primary and secondary amines, and aqueous solutions) which have high values of a_o.

From the slope of the lower curve in Figure 7, it is found that for aprotic solvents

$$a_o = K_v \cdot (\epsilon - 1) / (\epsilon + 1) + a_o^{\epsilon=1} \tag{52}$$

where for DTBN: $K_v = 0.8$ G, $a_o^{\epsilon=1} = 14.85$ G. For the rather more useful oxazolidine labels II (m,n)-VI (m,n): $K_v = 0.64$ G and $a_o^{\epsilon=1} = 13.85 \pm 0.09$ G. The form of Eq. (52)

arises from the polarization of the solvent by the dipole, μ, of the nitroxide, which gives rise to a reaction field at the nitroxide, given by (Onsager, 1936):

$$E_R = \frac{2(\epsilon-1)(n_D^2+2)}{3(2\epsilon+n_D^2)} \cdot \frac{\mu}{r^3} \tag{53}$$

where n_D is the refractive index of the pure nitroxide ($n_D^2 \simeq 2$) and r is its molecular radius. From Eq. (53) and the value of K_v for DTBN, Griffith et al. (1974) calculated that an electric field of 10^7 V/cm will produce a shift in a_o of 0.4 G. By similar methods, Seelig et al. (1972) calculated that a field of 10^6 V/cm is required to produce a shift of 0.4 G. The discrepancy of a factor of 10 arises from the different approaches used in the calculation: it is seen from Eq. (53) that the results will be very sensitive to the value assumed for r, the effective radius of the nitroxide. The results are nonetheless of considerable importance, since they give a method for estimating the electric field at the site of the nitroxide, and emphasize the fact that the presence of neighbouring charged groups can affect the value of a_o.

The hydrogen-bonding contribution to a_o has been investigated for the small spin-label molecule, 2,2,6,6-tetra-methyl-1N-oxyl-4-piperidinol (TEMPOL) by Gagua et al. (1978). They find the following relation for mixtures of aprotic and H-bonding solvents:

$$a_o = K_v(\epsilon-1)/(\epsilon+1) + a_o^{\epsilon=1} + K_h \cdot P \tag{54}$$

where $K_v = 0.35$ G, $a_o^{\epsilon=1} = 15.30$ G and the last term on the right is the contribution from hydrogen bonding, P being the proton donor concentration. The following values were obtained for mixtures with dioxane: $K_h = 0.30, 0.45, 0.50$ and 0.60 G for water, methanol, ethanol and isopropanol respectively. The dependence on proton donor concentration and the variation in K_h with donor were interpreted as indicating that the lifetime of the hydrogen bond is determined by the rotational correlation times of the solvent and spin-label molecules, and thus that the probability of H-bond formation is proportional to the collision frequency.

Not only the isotropic hyperfine splitting factor, a_o, but also the absolute value of the anisotropy of the A_{xx}, A_{yy} and A_{zz} hyperfine components is dependent on polarity. This is because both are determined by the amount of unpaired electron spin density on the nitrogen atom: a_o is determined by the contact interaction of the s-electron density, and the anisotropy by the dipolar interaction of the p-electron density. It is for this reason that polarity corrections have to be made in the calculation of the order parameter, as in Eq. (26), Section IIIB. The value of A_{zz}^R measured from the outer splitting of a rigidly-immobilized powder spectrum (cf. Fig. 6) is thus also a useful index of the polarity of the environment. Griffith et al. (1974) observed a linear relation between A_{zz}^R and a_o for the fatty acid labels II (m,n), which is given by:

$$A_{zz}^R(G) = 2.35 \times a_o(G) - 0.84 \text{ G}. \tag{55}$$

This was determined by choosing solvent mixtures which formed rigid glasses on cooling to −196°C, hence avoiding interaction broadening and giving rise to a clear powder spectrum from which A_{zz}^R could be determined. This calibration could prove useful in determining the appropriate values of A_{zz}^R for use in slow tumbling calculations such as Eqs. (47) and (49), if a_o can be obtained, e.g. from Eq. (18), from spectra of the probe in exactly the same environment, but under conditions of fast motion.

Finally, Table 3 gives an illustration of the use of the polarity sensitivity of the spin label in a membrane system. A difference in polarity is detected by the spin labels situated at different positions down the fatty acid chain. It can be seen that a_o, calculated from the measured $A_\parallel$ and $A_\perp$ splittings, decreases the further the nitroxide group is situated down the chain, corresponding to deeper penetration into the hydrophobic interior of the phospholipid bilayer membrane. The higher values at the top of the chain are caused by proximity to the aqueous membrane interface.

Table 3. Isotropic hyperfine splitting constants for stearic acid spin labels II (m,n) in chromaffin granule membranes and extracted lipid bilayers (Fretten et al., 1980)

Nitroxide	$a_o(G) = \frac{1}{3}(A_\parallel + 2A_\perp)$		
	Membrane	Lipid	Aqueous[a]
II (13,2)	15.2	15.2	15.6
(12,3)	15.1	15.1	15.7
(9,6)	15.0	15.1	15.7
(7,8)	14.4	14.4	–
(5,10)	14.0	14.1	–
(1,14)	14.0	14.1	15.7

[a] Label alone dispersed in aqueous buffer.

E. Label-Label Interactions

(Sackmann and Träuble, 1972; Devaux et al., 1973; Marsh and Smith, 1973)
There are two types of magnetic interaction which take place between labels at high concentration. The first is the classical magnetic dipole-dipole interaction which has the form:

$$\mathcal{H}_{d\text{-}d} = (1/r^3) \cdot [\underline{\mu}_1 \cdot \underline{\mu}_2 - 3(\underline{\mu}_1 \cdot \underline{r})(\underline{\mu}_2 \cdot \underline{r}) / r^2] \tag{56}$$

where $\underline{\mu}_1$ and $\underline{\mu}_2$ are the magnetic dipole moments of the two interacting spins and $\underline{r}$ is the vector joining them. In terms of the unpaired electron spins $\underline{S}_1, \underline{S}_2$:

$$\mathcal{H}_{d\text{-}d} = (g^2\beta^2/r^3) \cdot (3\cos^2\theta - 1) \cdot (\underline{S}_1 \cdot \underline{S}_2 - 3S_{1z}S_{2z}) \tag{57}$$

where θ is the angle between the vector $\underline{r}$ and the magnetic field direction. This shows that the interaction depends on the reciprocal of the spin-label separation cubed, and that it is completely anisotropic, thus it will be completely averaged out by isotropic motion: $<3\cos^2\theta - 1> = 0$. For incomplete motional averaging, the dipolar interaction will give rise to a linebroadening (Van Vleck, 1937):

$$\Delta H_d = \sqrt{<H_d^2>} \tag{58}$$

where $<H_d^2>$ is the mean-square dipolar field, given by:

$$<H_d^2> = (9/16) \cdot g^2\beta^2 \sum_i |r_{1i}|^{-6} <3\cos^2\theta_{1i} - 1>^2. \quad (59)$$

For a statistical distribution of label molecules within a triangular lipid lattice (Sackmann and Träuble, 1972):

$$\Delta H_d(G) = 3 \times 10^4 / d_{1a}(\text{Å})^3 \quad (60)$$

where d_{1a} is the mean label-label distance in Å units.

The second type of magnetic interaction is the exchange interaction; this is a quantum mechanical effect arising from the exclusion principle and has no classical analogue. It has a very short range, requiring that the spin labels should be virtually in contact, and takes the form of a spin-spin interaction:

$$\mathcal{H}_{ex} = -2J\, \underline{S}_1 \cdot \underline{S}_2 \quad (61)$$

where J is known as the exchange integral, which can be alternatively defined in terms of an exchange frequency, W_{ex}:

$$W_{ex} = J/h \quad (62)$$

where h is Planck's constant. The effect of the exchange interaction on the spin-label spectra can be analysed in terms of the usual model for two-site exchange in magnetic resonance. In the case of weak exchange where the exchange interaction is much smaller than the hyperfine interaction ($W_{ex} \ll a_N$), the line positions are unchanged but a line-broadening is observed, given by:

$$\Delta\nu = 2W_{ex}. \quad (63)$$

In the case of strong exchange ($W_{ex} \gg a_N$), the nitrogen hyperfine triplet collapses to a single, exchange-narrowed Lorentzian line, whose width is given by:

$$\Delta\nu = a_N^2/W_{ex}. \quad (64)$$

Thus in these two extreme cases the exchange frequency can be measured from the line-widths, if the contribution from dipolar broadening can be allowed for or eliminated. In the case of intermediate exchange, or as a help in distinguishing between exchange and dipolar broadening, it is necessary to simulate the whole lineshape. This is done by solving the Bloch equations modified for two-site exchange. The steady-state equations for the out-of-phase (real) and in-phase (imaginary) electron spin magnetizations, u_m and v_m, associated with the nitrogen nuclear spin quantum number, m, are given by (Sackmann and Träuble, 1972; Devaux et al., 1973):

$$u_m[1/T_{2m} + 3W_{ex}] - W_{ex} \cdot \sum_{m'=-1}^{+1} u_{m'} + (\omega_m - \omega)v_m = 0 \quad (65)$$

$$v_m[1/T_{2m}+3W_{ex}] - W_{ex} \cdot \sum_{m'=-1}^{+1} v_{m'} - (\omega_m\text{-}\omega)u_m = g\beta H_1 M_{om} \tag{66}$$

where ω_m is the frequency of the $m_I = m$ hyperfine line, M_{om} is the value of the static magnetization in the direction of the applied magnetic field, and H_1 is the microwave magnetic field. T_{2m} is the transverse relaxation time and is related to the linewidth by Eq. (36). The dipolar broadening can be allowed for by putting:

$$1/T_{2m} = 1/T_{2m}^{0} + 1/T_2^d \tag{67}$$

where $1/T_2^d$ describes the dipolar broadening. The Eqs. (65), (66) must be solved to yield the components v_m as a function of ω, by eliminating u_m. The total lineshape of the ESR absorption spectrum is then given by:

$$V(\omega) = v_{+1}(\omega) + v_o(\omega) + v_{-1}(\omega). \tag{68}$$

It is difficult to distinguish between exchange and dipolar contributions in the region of either weak or strong exchange, since then both interactions have similar effects on the spectra. However, in the region of intermediate exchange the two interactions have opposite effects: exchange shifts the two outer lines into the centre of the spectrum, whereas dipolar broadening shifts the sidebands outwards. Thus it is possible to distinguish unambiguously between dipolar broadening and exchange in this case, and the val-

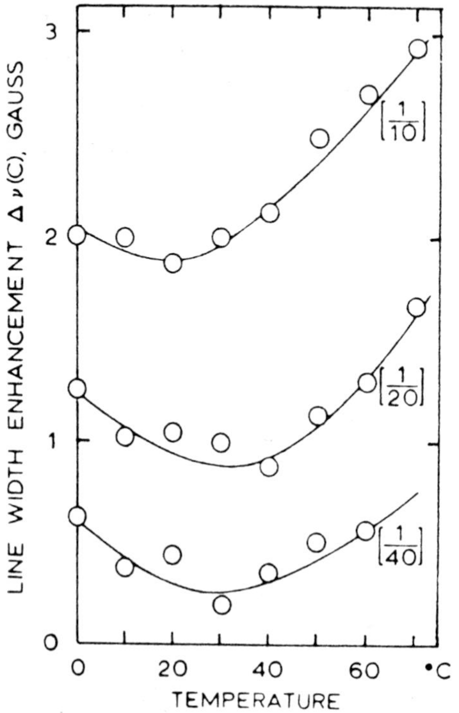

Fig. 8. Enhancement of the central linewidth as a function of temperature for various concentrations of phosphatidylcholine spin label III (1,14) in egg phosphatidylcholine/cholesterol (4:1) bilayers. Linewidth enhancement is defined relative to a very dilute concentration of spin label; numbers in brackets refer to the mole fraction of spin label (Devaux et al., 1973)

ues of ΔH_d (or $1/T_2^d$) can then be extrapolated to the two other regions, by assuming that ΔH_d is proportional to radical concentration (Galla and Sackmann, 1975).

Alternatively it is possible to choose the label position or the temperature such that the amplitude of motion of the nitroxide is large, and thus the dipolar interaction is almost completely averaged out. This is illustrated in Figure 8 for the spin label III (1,14), where the nitroxide is situated close to the terminal methyl end of the lipid chain and hence experiences a large amplitude motion. At the lower temperatures the interaction broadening decreases with increasing temperature, as a result of the increased motional averaging of the dipolar interaction. At the higher temperatures the interaction broadening increases with increasing temperature, corresponding to the increased exchange interaction, as a result of the increase in label collision frequency with temperature. The turnover point comes at lower temperature with increasing label concentration, as might be expected from the increased collision probability. It is important to be able to distinguish between the two contributions, since if the exchange interaction is determined by the collision frequency, it may be used to estimate the lateral diffusion rates of lipid molecules in membranes. This aspect is dealt with in more detail in Section IV.D below.

IV. Applications to Membrane Studies

A. Lipid Fluidity and Lateral Phase Separation – TEMPO Partitioning

(Shimshick and McConnell, 1973)
One of the simplest assays for membrane fluidity is provided by the partitioning of the small spin-label molecule, TEMPO VIII, between the aqueous phase and the fluid lipid phase of the membrane. The greater the fluidity of the membrane lipid, or the greater the proportion of fluid lipid, the greater will be the partitioning of TEMPO into the membrane. The spectra of TEMPO in the lipid phase and in water are easily resolved in the high field lines (see Fig. 9), because of the combined effect of the polarity differences

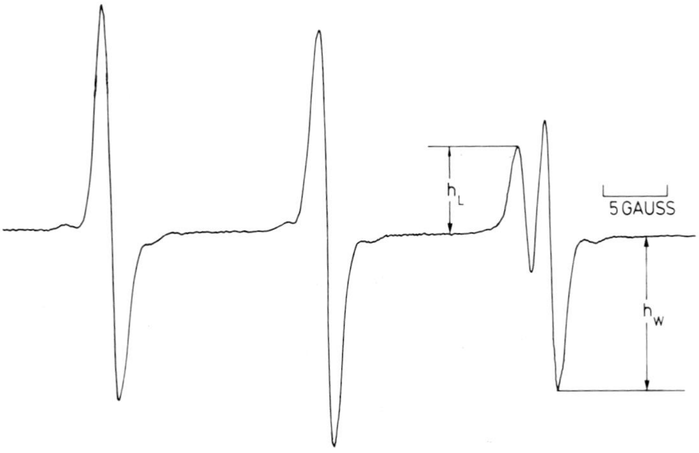

Fig. 9. ESR spectrum of the TEMPO spin label VIII in an aqueous dispersion of dimyristoylphosphatidylcholine bilayers at 30°C. h_L is the peak height of the spectrum of the label dissolved in the lipid phase, h_W that in the aqueous phase

from both the a_o splittings and the g-values. The degree of partitioning may thus be measured from the lineheight ratios in the high-field region of the spectrum; either of the following two parameters are commonly used:

$$\alpha = h_L/h_W \tag{69}$$

$$f = h_L/(h_L + h_W). \tag{70}$$

Strictly speaking, allowance should be made for the different linewidths of TEMPO in the water and in the lipid. This can be done by replacing h_L by $h_L \cdot \Delta H_L^2$ and h_W by $h_W \cdot \Delta H_W^2$ in Eqs. (69), (70), where ΔH_L and ΔH_W are the high-field linewidths of TEMPO in lipid and water respectively. A further correction, which is most necessary at low partitioning, is for the contribution of the ^{13}C satellite of the aqueous peak, which lies almost exactly under the high-field lipid peak in Figure 9. The ^{13}C satellite correction can be made by replacing h_L by $h_L - k.h_w$ in Eqs. (69), (70), where $k \simeq 0.04$ is the ratio of the satellite to the main aqueous peak.

The sensitivity of the TEMPO partitioning to the lipid fluidity is illustrated in Table 4, which gives the values of f for a series of egg phosphatidylcholine-cholesterol mixtures.

Table 4. TEMPO partition ratio α as a function of cholesterol content in egg phosphatidylcholine bilayers (PC). Normalized to [egg PC] = 25 mM (Marsh, unpublished)

Chol (mole %)	$\alpha = h_L/h_W$
0	0.39
10	0.31
20	0.25
30	0.21
40	0.16
50	0.14

These values are normalized to a constant amount of egg phosphatidylcholine, thus from the decreasing amount of TEMPO partitioning into the lipid it can be seen that there is a steady decrease in fluidity of the lipid chains due to interaction with the rigid cholesterol molecule. The values in Table 4 were not corrected for changes in linewidth of the lipid peak. Thus the decrease in f is a combined result of both decrease in partitioning and increase in linewidth, both of which are indicative of a decrease in lipid fluidity.

The TEMPO partition method is extremely useful for detecting lipid bilayer phase transitions, since essentially all the TEMPO is in the aqueous phase below the transition temperature and only partitions into the lipid on transition to the fluid phase. This is illustrated in Figure 10, which shows the phase transition of dimyristoylphosphatidylglycerol bilayers. In this case the transition has been recorded by continuously monitoring the height of the high-field aqueous peak (h_W in Fig. 9) as the temperature is scanned. Apart from the higher temperature resolution afforded by continuous scanning, this method has the advantage that the width of the aqueous peak will change only gradually with temperature, and thus decreases in this peak height will be a direct measure of the amount of spin-label partitioning into the lipid. The disadvantage is that, unlike

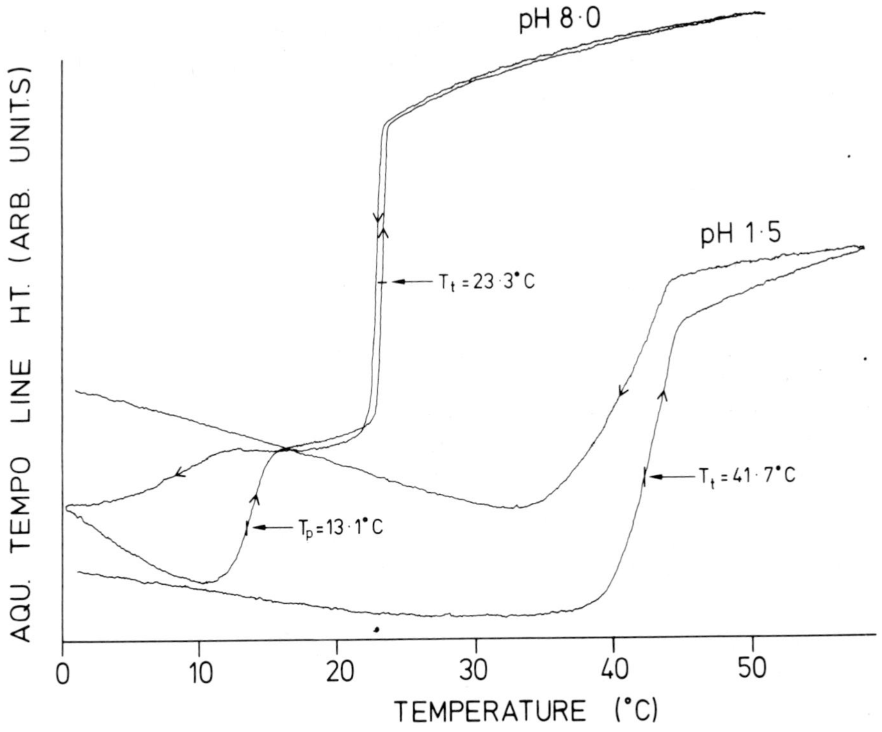

Fig. 10. Bilayer phase transition curves for dimyristoylphosphatidylglycerol bilayers (lipid concentration = 66 mM) in the fully charged (pH 8.0) and protonated state (pH 1.5). Recorded by continuously monitoring the height of the high-field aqueous peak (h_W in Fig. 9) of 10^{-4} M TEMPO VIII as the temperature was scanned (Watts et al., 1978)

the ratio methods, the peak height is also affected by spin label decomposition or reduction, and is sensitive to the settling of the sample or any other factors which affect spectrometer sensitivity. Figure 10 demonstrates that relatively sharp transitions can be detected by this method. At pH 8 a clear pretransition is observed, similar to that in phosphatidylcholines, and this displays a pronounced hysteresis, as the cooling transition is much broader and takes place at lower temperature than in the heating scan. At pH 1.5, when the phosphate of the headgroup is protonated, the transition takes place at a higher temperature, is considerably broader, and no pretransition is observed. The lack of reversibility of the scan in this case is due to chemical decomposition of the spin label at these high temperatures and low pHs. As well as the difference in transition temperatures of the two charge states of the bilayers, a difference in fluidity in the states above the phase transition is also found by measuring absolute values of the partition, f, as was done for the cholesterol case above (Watts et al., 1978).

The TEMPO method can be extended from studying simple bilayer phase transitions to the investigation of lateral phase separation in bilayers composed of mixtures of lipids with different phase transition temperatures. Plots of the temperature dependence of the TEMPO partitioning exhibit abrupt changes in slope at characteristic temperatures, corresponding to the onset and completion of lateral phase separation. In this way equilibrium phase diagrams can be constructed, indicating the regions of temperature and lipid composition which correspond to an all-solid phase, to an all-fluid phase, and to the coexistence of solid and fluid phases in equilibrium.

Binary mixtures of chemically similar phospholipids whose phase transitions are not very widely separated, e.g. dimyristoyl- and dipalmitoylphosphatidylcholine, often display solid phase miscibility as well as fluid phase miscibility. In this case the temperature dependence of the TEMPO partitioning has the appearance of a "broadened phase transition". The upper and lower boundaries of the region of lateral phase separation, T_f and T_s respectively, are then determined by the sharp discontinuities in the temperature dependence, which occur at the "ends" of the transition.

A less straightforward, and more interesting, situation occurs for lipid mixtures which exhibit solid phase immiscibility, as is illustrated for dimyristoyl- and distearoylphosphatidylcholine in Figure 11. The solid phase immiscibility is indicated by the nearly horizontal solidus line separating the solid phase region(s) from the region of coexist-

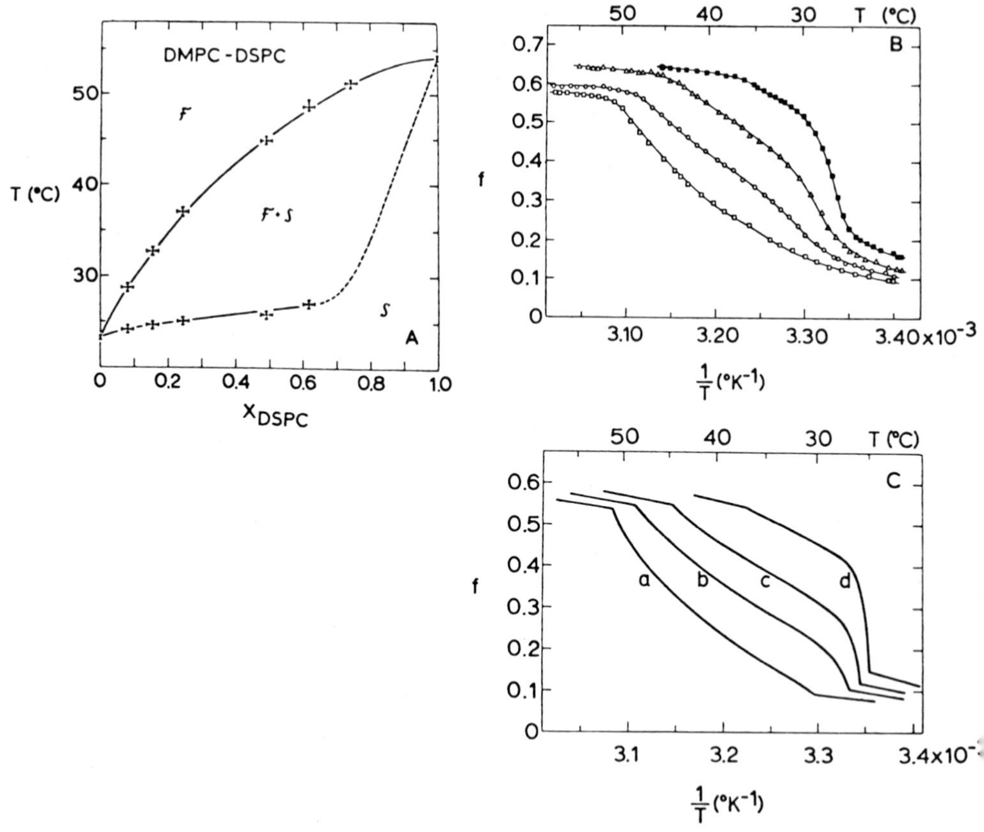

Fig. 11. (A) Phase diagram for aqueous dispersions of the dimyristoylphosphatidylcholine (DMPC)-distearoylphosphatidylcholine (DSPC) binary system. (B) Experimental TEMPO spectral parameter, f, as a function of 1/T for: (□) 74 mole % DSPC; (○) 62 mole %; (△) 49 mole %; (■) 24 mole %. (C) Calculated TEMPO spectral parameter, f, as a function of 1/T for: (a) 75 mole % DSPC; (b) 63 mole %; (c) 50 mole %; (d) 25 mole % (Shimshick and McConnell, 1973)

ence of fluid and solid phase (F + S) in Figure 11A. In this case there are found to be more than two discontinuities in the slope of the temperature dependence of f (here plotted against 1/T, since this is found empirically to have a linear variation in the fluid phase). It is only the two outer extremes which mark the boundaries of the region of lateral phase separation; the temperature dependence in the region in between is determined by the exact shape of the solidus and fluidus curve of the phase diagram. At high distearoylphosphatidylcholine concentrations a distinct low-temperature break is not observed, which means that the solidus curve is indeterminate at $X_{DSPC} > 0.62$. To overcome this difficulty, and also to check that the temperature dependence of the TEMPO partitioning does agree with the phase diagram deduced from the high and low temperature breaks, it is necessary to calculate f as a function of temperature from the phase diagram.

The TEMPO parameter is approximately related to the partition coefficient, K, by:

$$f = Kv_L / (Kv_L + v_W) \tag{71}$$

where v_L and v_W are the volumes of lipid and water respectively. This equation should hold exactly if corrections have been made for the linewidths and ^{13}C satellites, as mentioned above. The partition coefficient can be assumed to be a linear combination of the partition coefficients K_F, K_S of fluid and solid phase lipid respectively:

$$K = K_F \cdot F_F + K_S \cdot (1-F_F) \tag{72}$$

where F_F is the fraction of lipids in the fluid phase. For a binary mixture, F_F can be determined directly from the phase diagram; by material balance the following expression holds in the lateral phase separation region:

$$X^o = F_F \cdot X_F + (1-F_F) \cdot X_S \tag{73}$$

where X_F is the mole fraction of the higher-melting component in the F-phase and X_S that in the S-phase, as defined in Figure 12. X^o is the (total) mole fraction of the higher-melting component. The partition coefficients K_F and K_S can be assumed to be a linear combination of the partition coefficients: K_{AF}, K_{BF} and K_{AS}, K_{BS} of the pure melting components A and B respectively. Assuming that B is the higher-melting component:

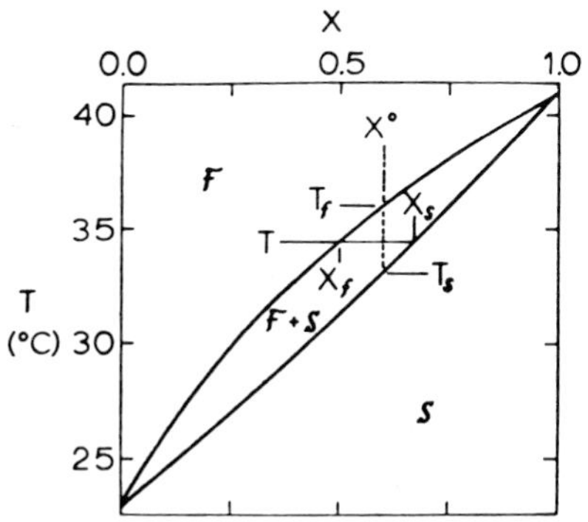

Fig. 12. Schematic phase diagram for two components which are completely miscible in the S-phase. (F + S) indicates the region of lateral phase separation. For a constant temperature, T, the compositions of the F- and S-phases co-existing in equilibrium are X_f and X_s as indicated. For a constant composition, X^o, the temperatures corresponding to onset and completion of phase separation are T_f and T_s, as indicated (Shimshick and McConnell, 1973)

$$K_F = (1-X_F) \cdot K_{FA} + X_F \cdot K_{FB} \quad (74)$$

$$K_S = (1-X_S) \cdot K_{SA} + X_S \cdot K_{SB} \quad (75)$$

where again X_S and X_F are obtained from the phase diagram, as in Figure 12. Eqs. (71)-(75) allow one to calculate f in terms of K_{FA}, K_{FB}, K_{SA} and K_{SB} for any given value of X^o, by using the corresponding values of X_F and X_S deduced from the phase diagram. Experimentally the partition coefficients of pure components in the F- and S-phase are found to have a temperature dependence of the form: $K = c - d/T$, where values of the c and d coefficients for various phospholipids are given by Shimshick and McConnell (1973). The f versus 1/T curves calculated in this way from the phase diagram of Figure 11A are given in Figure 11C. Reasonably good agreement is obtained with the experimental curves in Figure 11B, which both confirms the validity of the phase diagram established from the upper and lower break points and also serves to define the broken region of the phase diagram for which a distinct break in the TEMPO data was not observed.

Possible artefacts which can arise from measurements with TEMPO are discussed by Polnaszek et al. (1978). They originate mainly from changes in linewidth and from the ^{13}C satellites, as discussed above, and also from the ordering of TEMPO in lipid bilayers.

B. Angular Amplitudes and Rates of Motion – Rigid Steroids

(Schreier-Muccillo et al., 1973; Schindler and Seelig, 1974; Hemminga, 1974, 1975) Another method of characterizing the lipid fluidity in membranes is from the ESR spectra of rigid steroid spin labels of the type Ia), b). The order parameters give a measure of the angular amplitude of motion of steroids, such as cholesterol, in the lipid bilayer. The spectra of the cholestane spin label Ia) in oriented bilayers of egg phosphatidylcholine-cholesterol are given in Figure 13a[1]. This spin label has the nitroxide y-axis oriented along the long molecular axis, thus the direction of maximum hyperfine splitting (the nitroxide z-axis) is perpendicular to the long axis. The fact that the larger hyperfine splitting is observed with the magnetic field in the plane of the bilayer immediately suggests that the steroid molecule is oriented with its long axis preferentially perpendicular to the plane of the membrane, as would be expected in a bilayer structure. This is confirmed by the fact that the spectra with the magnetic field in the bilayer plane indicate that the spin label is rotating rapidly about its long axis, since otherwise the spectrum would be a two-dimensional powder spectrum rather than the normal three-line derivative spectrum obtained from the rapid motional averaging of the A_{zz} and A_{xx} splittings.

Under these conditions of axial averaging by long axis rotation, the order parameter for the motion of the long molecular axis can be calculated from the observed anisotropy of the spectral splittings using Eq. (26) or (31). It is found that S_{mol} increases from 0.55 to 0.80 as the cholesterol content of the bilayer increases from 0 to 50 mole %. This or-

1 The steroid labels, I, are best studied in oriented bilayers because the maximum hyperfine anisotropy of these labels is considerably reduced by the rapid rotation about the long molecular axis. Thus, in random bilayer dispersions these labels give rise to pseudo-three-line spectra, which are relatively insensitive to motion of the long molecular axis. To determine order parameters easily, oriented bilayers are required.

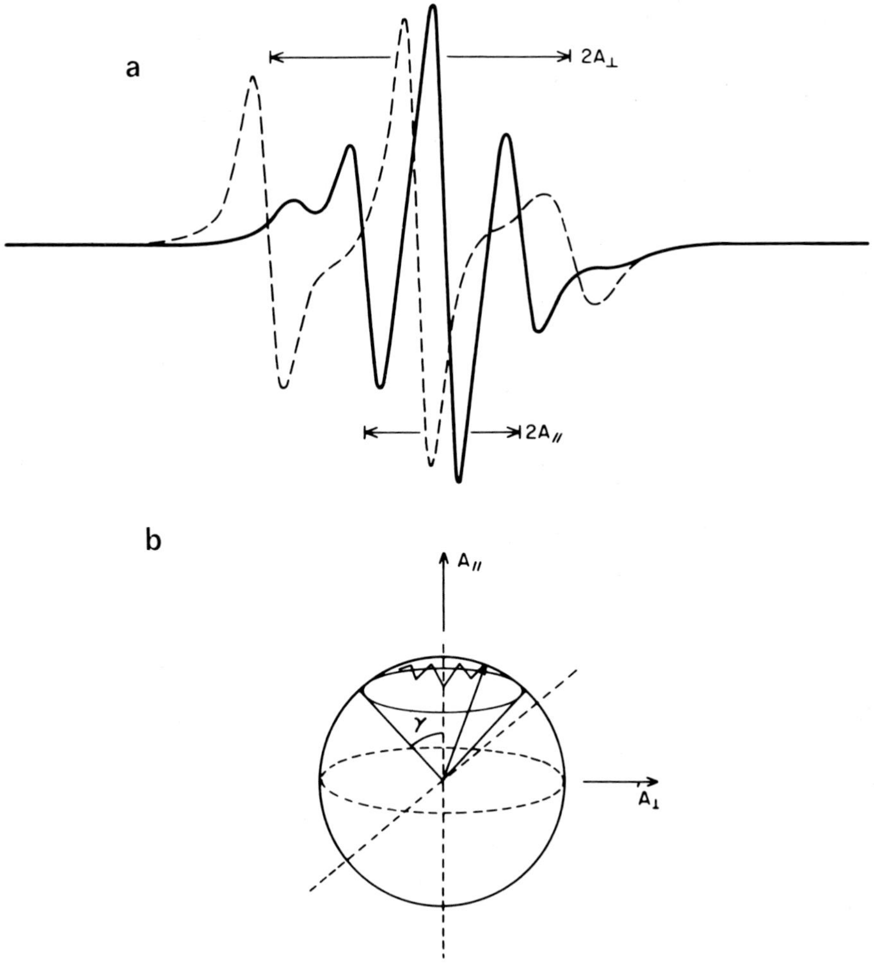

Fig. 13. (a) ESR spectra of the cholestane spin label Ia) in oriented bilayers of egg phosphatidylcholine + 30 mole % cholesterol with magnetic field parallel (||) to the bilayer normal (*full line*) and with the magnetic field in the plane (⊥) of the bilayer (*broken line*). (b) Restricted random walk model: the end of the spin label traces out a random walk on the surface of the sphere, but restricted within a cone of angle γ (Marsh, 1975)

dering effect corresponds to the well-known condensing effect and physiological stabilizing effect of cholesterol in membranes. The increase in order parameter can be related to the decrease in angular amplitude of motion, via Eq. (20), if a model is assumed for the angular motion. That is, if the angular probability distribution, $P(\theta_3)$, over which the motional averaging in Eq. (46) takes place, is known. A simple model is the restricted random walk shown in Figure 13b, in which the spin label performs a random motion restricted within a cone of fixed angle, γ. For this the angular distribution is: $P(\theta_3) = 1$, $0 \leqslant \theta_3 \leqslant \gamma$; $P(\theta_3) = 0$, $\theta_3 > \gamma$ and the order parameter is determined simply by the average value of $\cos^2\theta$ between $\theta = 0$ and γ, hence:

$$S_{mol} = (1/2)(\cos\gamma + \cos^2\gamma) \quad (76)$$

Calculated values of γ for egg phosphatidylcholine-cholesterol bilayers are given in Figure 14, which indicates a decrease in amplitude of motion from 45° to less than 20° with increasing cholesterol.

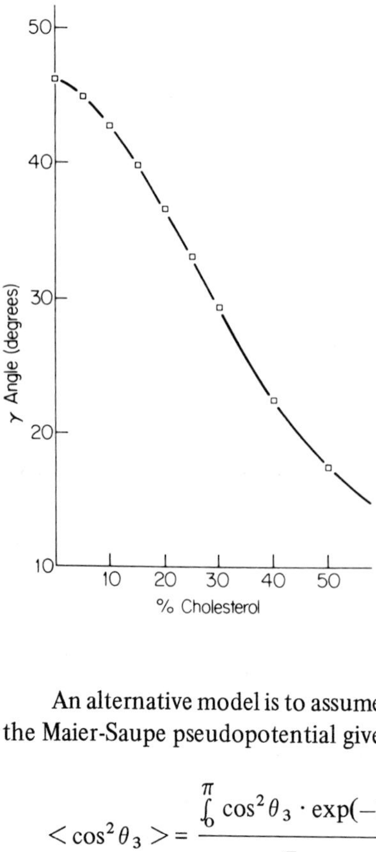

Fig. 14. Variation of the angular amplitude of restricted random walk with cholesterol concentration in egg phosphatidylcholine/cholesterol bilayers (Schreier-Muccillo et al., 1973)

An alternative model is to assume that the angular amplitude of motion is governed by the Maier-Saupe pseudopotential given in Eq. (44). Eq. (46) then becomes:

$$\langle \cos^2\theta_3 \rangle = \frac{\int_0^\pi \cos^2\theta_3 \cdot \exp(-\lambda\cos^2\theta_3/kT) \cdot \sin\theta_3 \cdot d\theta_3}{\int_0^\pi \exp(-\lambda\cos^2\theta_3/kT) \cdot \sin\theta_3 \cdot d\theta_3} \quad (77)$$

This equation can be integrated numerically for a given value of λ/kT, and the resulting values of the order parameter are given as a function of the strength of the pseudopotential in Figure 15. It is found that this model reasonably well fits the temperature dependence of the order parameters of the cholestane spin label Ia) in dipalmitoylphosphatidylcholine-cholesterol bilayers, where λ is given by the values in Table 5 (Marsh, unpublished). The ordering effect of cholesterol is clearly seen by the increase in the strength of the orientational pseudopotential with increasing cholesterol content. As already mentioned, the values of λ in Table 5 can then be used to calculate the mean value of any function of the amplitude of motion, using equations analogous to Eq. (46).

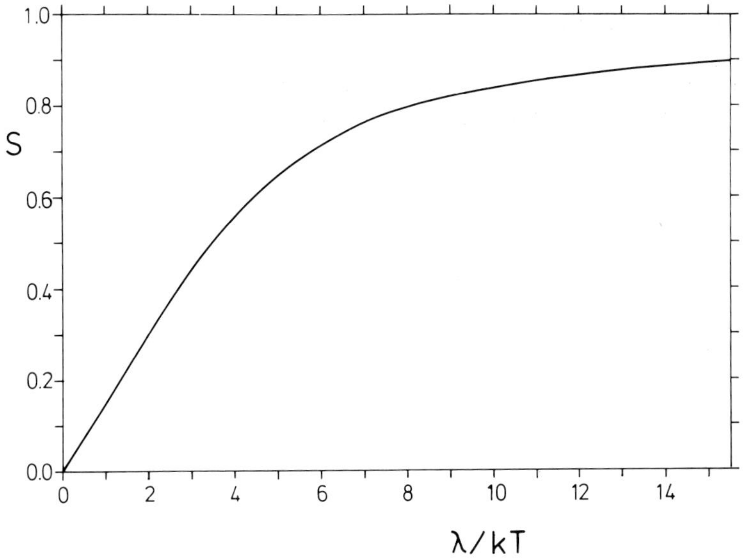

Fig. 15. Variation in order parameter (S) with strength of the Maier-Saupe pseudopotential. Calculated from Eqs. (20), (77)

Table 5. Strength of the Maier-Saupe pseudopotential, λ, for the cholestane spin probe Ia, in bilayers of dipalmitoylphosphatidylcholine-cholesterol (chol). Limiting values in the fluid phase, T = 70°C (Marsh, unpublished)

Chol (mole %)	λ (cal/mol)
0	2.000
5	3.200
10	4.200
20	5.000
30	6.500
50	15.000

The fluidity of the bilayer is characterized not only by the angular amplitude of motion, but also by the rate of motion of the spin label. This latter information is obtained from the linewidths, which requires simulation of the lineshape in terms of two rotational diffusion coefficients or correlation times, $\tau_{\parallel}$ and $\tau_{\perp}$, for rotation around and perpendicular to the long molecular axis, also taking into account the effect of the order parameters, as discussed in Section III-C. In general it is found that the rate of rotational motion around the long axis is at least ten times faster than the angular motion of the long axis itself (Hemminga, 1974, 1975; Schindler and Seelig, 1974). A value of $\tau_{\parallel} \sim 3.10^{-10}$ s was found for the cholestane label Ia) in a sodium decanoate-decanol liquid crystalline bilayer (Schindler and Seelig, 1974), and $\tau_{\parallel} \sim 2.10^{-9}$ s in dipalmitoylphosphatidylcholine bilayers at less than limiting hydration (Hemminga, 1974). For the latter system it was found that $\tau_{\parallel}$ was relatively insensitive to cholesterol content, but $\tau_{\perp}$ increased two- to fourfold with increasing cholesterol content. In this case the decrease in motional rate

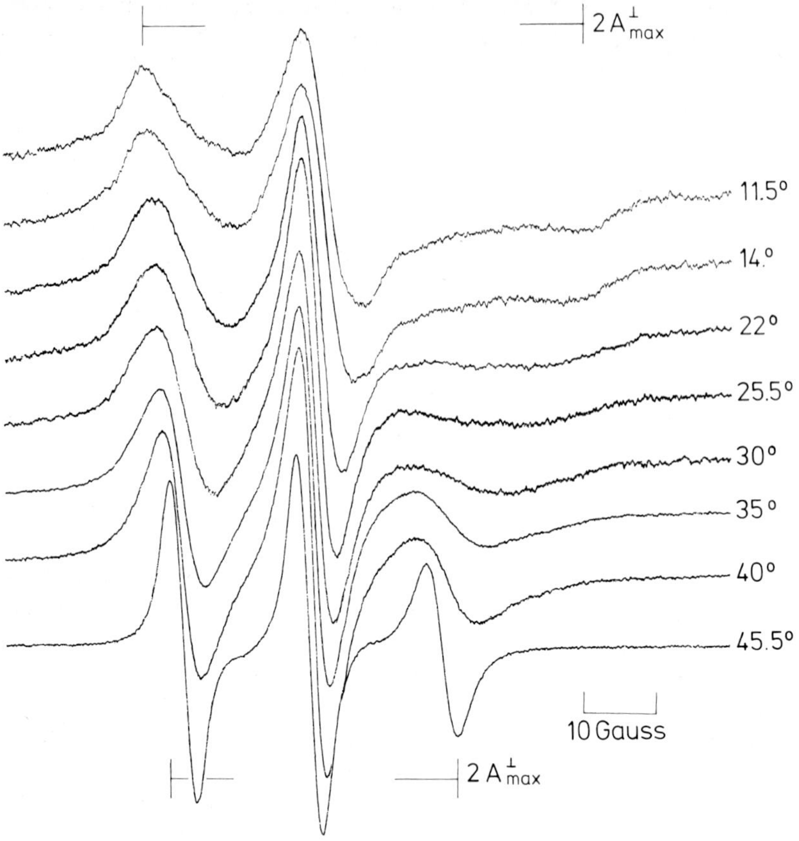

Fig. 16. ESR spectra, as a function of temperature, of the cholestane spin label Ia) in oriented bilayers of dipalmitoylphosphatidylcholine, with the magnetic field oriented in the plane of the bilayer (Marsh, 1980)

Table 6. Rotational correlation times of the cholestane spin label Ia in gel-phase bilayers of dipalmitoylphosphatidylcholine (T_t = 41°C) (Marsh, 1980)

$\tau_{\parallel}^{R}$ (s)	4°C	14°C	24°C	34°C	44°C
A'_{zz}/A^{R}_{zz} [a]	$> 10^{-8}$	4.10^{-9}	$1.9.10^{-9}$	$1.2.10^{-9}$	$\sim 10^{-10}$
B′ [b]	$> 2.10^{-9}$	$> 2.10^{-9}$	2.10^{-9}	$0.4.10^{-9}$	$\sim 10^{-10}$
C′ [c]	$> 2.10^{-9}$	$> 2.10^{-9}$	$1.5.10^{-9}$	$0.2.10^{-9}$	$\sim 10^{-10}$

[a] Deduced from Eq. (49) and calibrations given by Polnaszek (1977).
[b] Deduced from Eq. (50).
[c] Deduced from Eq. (51).

parallels the decrease in motional amplitude with increasing cholesterol content. However, the molecular order and rate of molecular motion must not necessarily be correlated, particularly when comparing absolute values from different systems. For instance, the liquid crystalline soap systems mentioned above are more ordered, i.e. have smaller amplitude of motion, than phospholipid bilayers, although their rate of molecular motion is faster (Schreier et al., 1978).

For situations of high order, the linewidths of oriented bilayer spectra with the magnetic field in the plane of the bilayer can be considered to be dominated by rotation around the long molecular axis. For the steroid labels Ia), b), Schreier et al. (1978) give the following approximations for the correlation time in the motional narrowing region:

$$\tau_{\|} \text{ (sec)} = -2.82 \times 10^{-10} \cdot B \tag{78}$$

$$\tau_{\|} \text{ (sec)} = 2.25 \times 10^{-10} \cdot C \tag{79}$$

where B and C are the linewidth coefficients defined previously in Eqs. (40), (41). Values for $\tau_{\|}$ in various systems, obtained using these equations, are given by Schreier et al. (1978).

Another situation in which rotation about the lipid long molecular axis is detected, essentially in the absence of motion of the long axis itself, is in the ordered or gel phase of phosphatidylcholine bilayers. In this phase the lipid chains are in their rigid, all-trans conformation and the molecular packing density is relatively high. The effects of onset of long axis rotation of the cholestane Ia) label are seen in Figure 16, which clearly shows the transition from a two-dimensional powder spectrum to a sharp three-line spectrum, on passing through the pretransition region of dipalmitoylphosphatidylcholine bilayers. Rapid rotation about the long molecular axis is detected below the main phase transition; the latter takes place at 41°C and is accompanied by a further narrowing of the three-line spectrum, arising from both an increase in the rate of rotation around the long axis and also the onset of motion of the long axis itself. The spectra in Figure 16 are from oriented bilayers and are complicated by the fact that, below the main transition, the lipid molecules are tilted relative to the bilayer normal. The spectra of lipid dispersions are not complicated by molecular tilt and, in the region below the phase transition, the rotational correlation time for motion about the long molecular axis can be determined from Eqs. (49)-(51) and from the slow motion calibrations given by Polnaszek (1977). Typical values are given in Table 6, which demonstrates the onset of a rotational motion in the slow regime ($\tau_{\|} > 10^{-9}$ s), whose rate increases with temperature to a value in the fast motional regime ($\tau_{\|} \leqslant 10^{-9}$ s), at a temperature which is still below that of the main phase transition.

C. Angular Amplitudes and Rates of Molecular Motion – Flexible Chains

(Seelig, 1970, 1971; Hubbell and McConnell, 1971)
Similar experiments to those with the steroid labels can be performed with the fatty acid labels II (m,n) or the phospholipid labels III-VII (m,n). The two main differences are that the nitroxide z-axis is directed along the long molecular axis for these labels,

and that the nitroxide group may be situated at various different positions down the flexible chain. Typical phospholipid bilayer spectra from the various positional isomers of the phosphatidylcholine spin label III (m,n) are given in Figure 17. These spectra illustrate

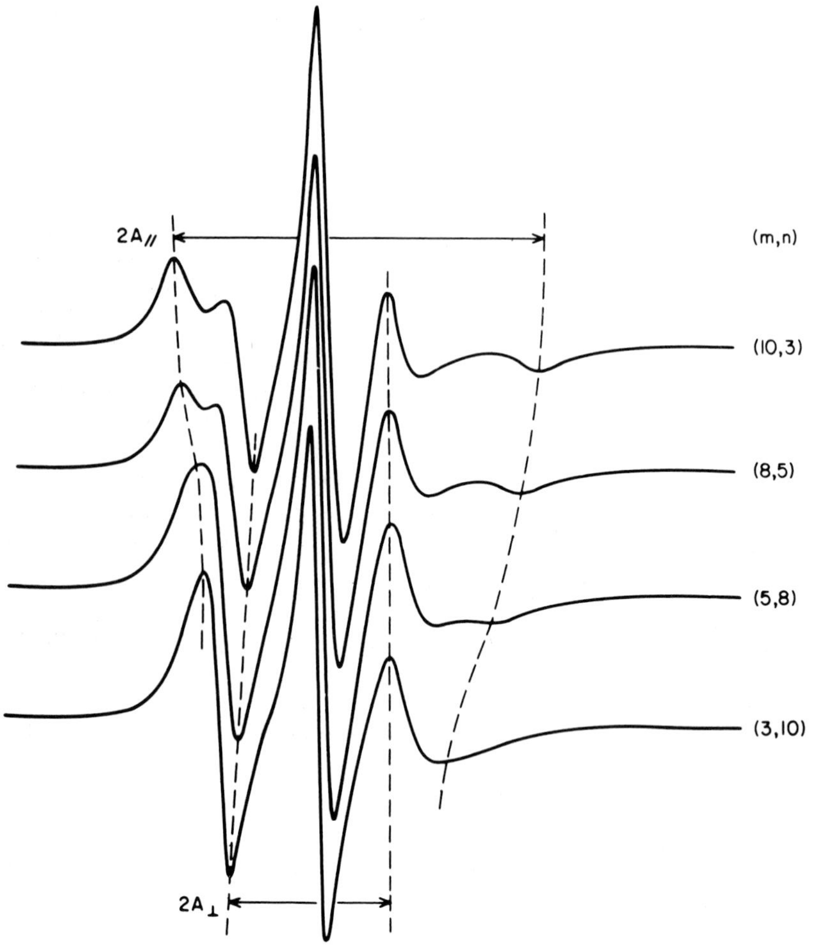

Fig. 17. ESR spectra of the phosphatidylcholine spin labels III (m,n) in random dispersions of egg phosphatidylcholine bilayers (Knowles et al., 1976)

the "flexibility gradient" which is characteristic of these types of labels in a lipid bilayer environment. As the spin label is situated further down the chain, away from the polar headgroup, the anisotropy of the spectrum ($A_{\parallel} - A_{\perp}$), and hence the order parameter, decreases. (The spectrum of label III (3,10) in Fig. 17 is an example of a situation in which the order parameter cannot be measured directly from the spectrum, but requires simulation as pointed out in Sects. III.B and C)

The flexibility gradient arises because the lipid chains are anchored at the glycerol backbone region of the phospholipid molecule. Thus as the spin-label position is stepped down the chain towards the terminal methyl end, there are increasingly more carbon-carbon single bonds about which rotation can take place, giving rise to greater angular displacements of the spin-label group. Since the chains are not free, but packed into a bilayer arrangement, there are additional intermolecular restrictions which modify the fluidity gradient. This is illustrated in Figure 18: single *gauche* conformations give rise

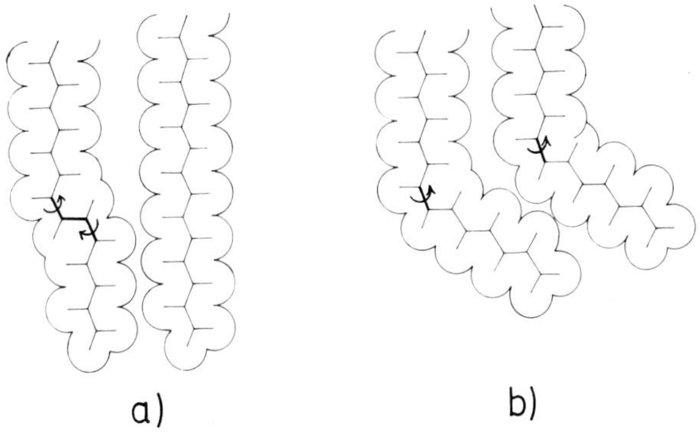

Fig. 18 a and b. Rotational isomerism of the hydrocarbon chains in lipid bilayers. *Gauche* isomers, $g^{\pm}$, are created by ± 120° rotations about C-C single bonds. (a) The all-*trans* chain and a g^+tg^- kink conformation. (b) A single *gauche* (g^+) conformation (Marsh, 1974)

to a 60° deviation of the all-*trans* chain and thus become more likely towards the terminal methyl end, where the net lateral displacement of the chain will be smaller. Further up the chain, the more compact conformations such as the g^+tg^- kinks (Fig. 18a) will be favoured, since they give rise to smaller lateral displacements. Thus the fluidity gradient also arises as a result of the chain packing: a wider range of *gauche* conformations is allowed towards the end of the chain, and in addition the intrinsic probability of kink formation also increases.

If it is assumed that the rotations about the individual C-C bonds of the chain are independent, then it is a general result[2] that the order parameter resulting from the rota-

2 This can easily be seen from a double application of Eqs. (14) and (15). Consider a motion which is characterized by an angle θ relative to the instantaneous director (or molecular axis). The time-averaged hyperfine splittings $A_{\parallel}(\theta)$, $A_{\perp}(\theta)$ for the magnetic field directed along or perpendicular to this instantaneous director are given by Eqs. (14), (15). If the second motion gives rise to angular displacements, β, of the instantaneous director relative to the bilayer normal, then the net time-averaged hyperfine splittings, relative to the bilayer normal, are:

$$A_{\parallel}(\beta,\theta) = A_{\perp}(\theta) + [A_{\parallel}(\theta) - A_{\perp}(\theta)] < \cos^2\beta >$$

$$A_{\perp}(\beta,\theta) = 1/2\ [A_{\parallel}(\theta) + A_{\perp}(\theta)] - 1/2\ [A_{\parallel}(\theta) - A_{\perp}(\theta)] < \cos^2\beta >$$

Hence the resultant order parameter, defined by:

$$S(\beta,\theta) = \frac{[A_{\parallel}(\beta,\theta) - A_{\perp}(\beta,\theta)]}{(A_{ZZ} - A_{XX})},$$

becomes:

$$S(\beta,\theta) = \frac{A_{\parallel}(\theta) - A_{\perp}(\theta)}{A_{ZZ} - A_{XX}} \cdot \tfrac{1}{2}\ (3 < \cos^2\beta > - 1)$$

$$= S(\theta) \cdot S(\beta)$$

Thus the order parameter for the combined motion is simply the product of the order parameters for the two individual motions.

tions about two bonds is the product of the order parameters for rotation about the individual bonds: $S = S_1 \times S_2$. Generalizing for the whole chain we get (see e.g. Seelig, 1970):

$$S_n = S_\sigma^n \times S_o \tag{80}$$

where S_σ is the order parameter for rotation about a single bond, S_o is the order parameter for motion of the long axis of the lipid molecule as a whole, and n is the number of C-C single bonds about which the rotation can take place (i.e. between the nitroxide and the carboxyl group). Thus by using two different spin-label positional isomers, e.g. III (m_1,n_1) and III (m_2,n_2), the intrinsic order parameter for rotation about individual C-C single bonds, S_σ (and also S_o), can be determined. In general the value of S_σ will vary down the chain, because of the interchain packing restrictions discussed above, i.e. $S_n = S_o \times \prod_i^n S_{\sigma i}$, and Eq. (80) will thus only apply to restricted sections of the chain, over which S_σ remains approximately constant. An approximately exponential dependence of S_n on n, over extended regions of the chain, has been found, however, in some cases (Seelig, 1970. Hubbell and McConnell, 1971).

Hubbell and McConnell (1971) have treated the flexibility gradient analysis explicitly in terms of the probabilities, p_t, p_g, of *trans* and *gauche* conformations about a given C-C bond. If p_g is small compared with p_t (*gauche* is the energetically unfavourable conformation), and both are constant along the chain, then the order parameter can be approximated by a rapidly converging series:

$$S_n = p_t^n \cdot \eta_o + n \cdot p_t^{n-1} \cdot p_g \cdot \eta_1 + \frac{n(n-1)}{2!} \cdot p_t^{n-2} \cdot p_g^2 \cdot \eta_2 + \ldots \tag{81}$$

where $p_g + p_t = 1$, and it is assumed that rotations about individual bonds are statistically independent. Thus η_m is the average value of $\frac{1}{2}(3\cos^2\theta - 1)$ when the chain has m *gauche* conformations, i.e. $\eta_o = 1$, $\eta_2 = 1/16$, $\eta_3 = -1/32$. The rapid convergence of Eq. (81) again gives rise to an approximate exponential dependence on n, as in Eq. (80), thus:

$$\log S_n = n \cdot \log p_t + C_n + \log S_o \tag{82}$$

where the correction term C_n depends only weakly on n, and is given by:

$$C_n = \log \left\{\eta_o + n \cdot (p_g/p_t) \cdot \eta_1 + \frac{n(n-1)}{2!} \cdot (p_g/p_t)^2 \cdot \eta_2 + \ldots \right\}. \tag{83}$$

For the phosphatidylcholine spin labels III (m,n) in egg phosphatidylcholine-cholesterol bilayers, Hubbell and McConnell (1971) found that to explain the data it was necessary to assume that p_g varied down the chain. Generally it is found that $p_g \sim 0.1$ for fluid lipid bilayers, which corresponds to approximately 2 *gauche* conformations per chain, and that p_g decreases to $p_g \sim 0.02$ on the addition of 50 mole % cholesterol (Schreier-

Muccillo et al., 1973). Within the approximation of independent rotations about C-C bonds, the temperature dependence of p_g, and hence of S_n, is given by:

$$p_g = 2\sigma/(1 + 2\sigma) \tag{84}$$

$$\sigma = \exp(-E_g/kT) \tag{85}$$

where E_g is the effective energy difference between *trans* and *gauche* conformations. E_g contains contributions from the intermolecular, steric interactions as well as the normal, bond rotational potential, which is known to be $\sim$ 500-800 cal/mol per CH_2 group from paraffins in dilute solution.

Seelig (1971) has presented a more detailed analysis which takes into account the interdependence of rotations about adjacent C-C single bonds ($g^{\pm}g^{\mp}$ combinations are strongly disfavoured on intramolecular, steric grounds). Two-bond conformational probabilities have then to be taken into account. Seelig (1971) finds that the order parameter corresponding to the relative orientation between two adjacent bonds, S_β, is related to the single-bond order parameter, S_σ, by: $S_\beta \simeq S_\sigma^2$. Hence this model also predicts an exponential dependence of order parameter on bond position, as is given by Eq. (80). The mean value of $\langle \cos^2\beta \rangle$ corresponding to the order parameter, S_β, is then given by:

$$\langle \cos^2\beta \rangle = (1/3)(2S_\sigma^2 + 1) = p_{tt} + p_{tg} \tag{86}$$

where p_{tt} is the probability that two adjacent bonds are in the *trans* conformation, and p_{tg} is the probability that one bond is in the *trans* conformation and the following bond is in the *gauche*$^{\pm}$ conformation. In Eq. (86) it has been explicitly assumed that $p_{g^{\pm}g^{\mp}} = 0$. Evaluating the two-bond probabilities, the following expression is obtained for the temperature dependence of S_σ:

$$(1/3)(2S_\sigma^2 + 1) = 1/[(1+\sigma)^2 + 4\sigma]^{1/2} \tag{87}$$

and in this case the probability of *gauche* conformation ($p_g = p_g^+ + p_g^- = 2p_g^+$) is given by:

$$p_g = (1/2)\left\{1 - \frac{1-\sigma}{[(1+\sigma)^2+4\sigma]^{1/2}}\right\} \tag{88}$$

Within the same framework, Marsh (1974) has attempted to extend this analysis to take some account of the intermolecular steric restrictions on the chain conformation, produced by the bilayer packing. This is done by excluding $g^{\pm}g^{\pm}$ conformations which, although allowed on intramolecular grounds, are strongly disfavoured by intermolecular packing, since they produce a right-angle bend in the chains. Using Eq. (86) the corresponding expressions to Eqs. (87), (88) are:

$$(2/3)(1 - S_\sigma^2) = 6\sigma/[1+8\sigma+(1+8\sigma)^{1/2}] \tag{89}$$

$$p_g = (1/2)\left[1 - \frac{1}{(1+8\sigma)^{1/2}}\right]. \tag{90}$$

Either the Eqs. (82)-(85) or the Eqs. (87), (89) together with Eqs. (80), (85) can be used to characterize the temperature dependence of the order parameters of the II (m,n) or the III-VII (m,n) labels in terms of E_g, the energy-of-*gauche*-conformation parameter. The Eq. (89) yields the smallest value of E_g, since this makes some explicit allowance for the intermolecular interaction. Nevertheless, the values obtained are all considerably larger than the 500-800 cal/mol value for the rotational potential in dilute paraffins, indicating that E_g must still be considered a pseudopotential, containing contributions from intermolecular effects as well as the intrinsic, intramolecular rotational potential.

Typical values for E_g, obtained in lipid bilayers and in a natural membrane, are given in Table 7. From this table it can be seen that this is an adequate way of describing the

Table 7. Values of the Effective Energy of Gauche Conformations, E_g, of the Lipid Chains in Dimyristoylphosphatidylcholine, DMPC, and Dipalmitoylphosphatidylcholine, DPPC, bilayers (Seelig et al., 1973) and in Chromaffin Granule Membranes, CGM (Marsh et al., 1976a). Deduced from the ESR Spectra of the Fatty Acid Spin Labels, II (m,n)

E_g (kcal/mole)[a]	15°	20°	25°	30°	40°	50°	60°
DMPC	2.13	2.00	1.93	1.83	1.80	1.80	1.80
DPPC	–	2.29	2.20	2.08	1.86	1.80	1.80
CGM	2.40	2.40	2.40	2.40	2.25	2.20	2.20

[a] Deduced from Eq. (87)

temperature dependence, and also is sensitive to structural changes taking place in the lipid part of the membrane. Above the ordered-fluid phase transitions of dimyristoyl- and dipalmitoylphosphatidylcholine bilayers, the lipid chain fluidity is characterized by a single, temperature-independent value of E_g = 1800 cal/mol for both systems. As the bilayer goes down through the ordered-fluid phase transition, at 41°C for dipalmitoyl- and 23°C for dimyristoylphosphatidylcholine, E_g increases, indicating the onset of rigidifying of the chains. Below the phase transition the model is inapplicable, since the chains are then in their ordered, all-*trans* configuration; (and the chain motion is also too slow for the spin label spectra to be analysed to yield order parameters, as was discussed in Sects. III.B and C). The chromaffin granule membrane is in the fluid state over the whole of the temperature range studied. Below 35°C the chain fluidity is characterized by a single, temperature-independent value of E_g = 2400 cal/mol. At ∿ 35°C a structural change takes place in the membrane (Marsh et al., 1976a) to a high-temperature state which is characterized by a lower value of E_g = 2200 cal/mol. In both states the value of E_g is considerably higher than that for the pure, fluid lipid bilayers, as a result of the high cholesterol content of these membranes.

Although the above statistical mechanical models meet with considerable success in analysing the order parameters of the chain-labelled lipids II-VII (m,n), they are deficient in the sense that they refer essentially to single, isolated chains and the interchain interactions are only taken into account in an empirical way. Thus the models refer only to the fluid phase, and are incapable of predicting the existence of the ordered-fluid phase transition, and the increase in the intrinsic probability of *gauche* conformations towards the centre of the bilayer can only be introduced in an empirical manner. A statistical model which takes into account the intermolecular interactions has been given by

Marčelja (1974), and this has been used by Schindler and Seelig (1975) in interpreting the order parameters obtained from ^{2}H-NMR of specifically chain-deuterated phospholipids. In this model, the chain statistics are treated in a similar manner to that of Seelig (1971) and Marsh (1974). The intermolecular dispersion interactions are accounted for by a self-consistent orientational pseudopotential of the Maier-Saupe type (cf. Eq. 44). The pseudopotential is modified to apply to flexible chains, giving the following expression for the dispersion energy of a chain in configuration (i):

$$E_{disp}^{(i)} = -\phi(n_{tr}^{(i)}/n) \sum_{j=1}^{n} (1/2)(3\cos^2\beta_j^{(i)} - 1) \tag{91}$$

where β_j is the angle between the bilayer normal and the jth chain segment, and $(n_{tr}^{(i)}/n)$ is the fraction of C-C bonds in the trans conformation. The strength of the orientational molecular field, ϕ, in turn depends on the average order in the bilayer:

$$\phi = (V_o/n) \langle (n_{tr}^{(i)}/n) \sum_{j=1}^{n} (1/2)(3\cos^2\beta_j^{(i)} - 1) \rangle \tag{92}$$

where the average is taken over all configurations (i). The potential must be determined self-consistently because the average on the right-hand side of Eq. (92) is dependent on ϕ. V_o is the so-called coupling constant and is a parameter to be fitted. The hydrophobic forces, together with steric and electrostatic repulsions, were accounted for by an interfacial energy term of the form $\gamma A^{(i)}$, where $A^{(i)}$ is the effective cross-sectional area of the chain in conformation (i), and γ is a constant to be fitted. In this way it was possible to fit the ^{2}H NMR order parameter profile of dipalmitoylphosphatidylcholine with values of $V_o = 590$ cal/mol and $\gamma = 18.5$ dyn/cm, and with these values it was also possible to account for several of the structural and thermodynamic properties of the bilayer phase transition.

The results from spin-label ESR and ^{2}H NMR on chain-labelled lipids are in basic agreement as to the existence of the fluidity gradient, and in the fact that the intrinsic probability of *gauche* conformations becomes considerably greater close to the terminal methyl end of the chain. However, the methods differ considerably in the details of the shape of the order parameter profile. In particular, the deuterium results indicate that the order parameter is approximately constant in the region at the top of the chain, indicating that kink conformations of the type shown in Figure 18a are the only rotational isomers that are allowed in this region of the chain (Schindler and Seelig, 1975). In contrast, the ESR spin-label results indicate a steady decrease in order parameter in this region. There are at least three reasons why the spin-label and ^{2}H NMR results might differ. The first is the perturbing effect of the nitroxide label group, both on the labelled chain and on its immediate surrounding chains. The second is the time scale of the two methods: conventional spin-label ESR is only sensitive to motions faster than $\sim 10^8$ s^{-1}, whereas ^{2}H NMR is sensitive to motions down to as slow as $\sim 2.10^5$ s^{-1}. The third possible source of discrepancy is that, if the motion is in the slow motion regime for ESR (but fast for ^{2}H NMR), the spin label order parameters may be artefactually high because the line positions are sensitive to the rates of motion, as discussed in Sections III.B and C.

Because of the difference in time scale, an orientational distribution which appears as a static tilt in spin-label ESR may be rapidly motionally averaged in ^{2}H NMR. Gaffney and McConnell (1974) have demonstrated a static tilt of the section of the lipid chains close to the polar headgroup, from the ESR spectra of phosphatidylcholine spin labels III (m,n) in oriented bilayers of egg phosphatidylcholine. The tilt distribution, obtained by spectral simulation, was:

$$P(\theta) \cdot \sin\theta = \exp[-(\theta-\overline{\theta})^2/2\theta_o^2] \cdot \sin\theta \tag{93}$$

where $\overline{\theta}$ is the tilt parameter and θ_o is the angular spread of the distribution. The most probable angle of tilt is given by:

$$\tan\theta_{mp} = \frac{\theta_o^2}{(\theta_{mp}-\overline{\theta})} \, . \tag{94}$$

Using the measured tilt values of $\overline{\theta} = 30°$ and $\theta_o = 5°$, Gaffney and McConnell (1974) then recalculated the order parameter profile by assuming that the tilt distribution was averaged by rapid angular motion. This average is simply performed by using the measured tilt distribution of Eq. (93) in the integral average given in Eq. (46). In this way an order parameter profile was obtained which very closely resembled those found by ^{2}H NMR (see Fig. 19). Whether or not this is a general explanation for the order parameter

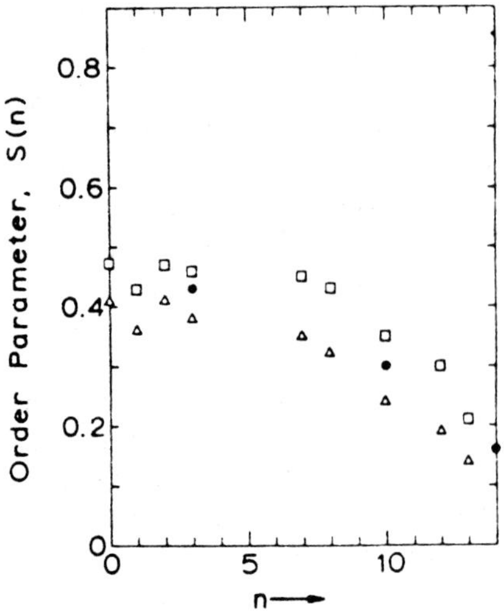

Fig. 19. Comparison of the order parameter obtained directly from ^{2}H NMR of selectively deuterated dipalmitoylphosphatidylcholine (△: 41°C; □: 57°C; Seelig and Seelig, 1974) with that obtained from the spin labels III (m,n) in egg phosphatidylcholine bilayers at 25°C, by a calculation which motionally averages the static tilt observed by ESR (•; Gaffney and McConnell, 1974; McConnell, 1976)

differences between spin-label and ^{2}H NMR measurements, the fact still remains that the flexibility gradient of the form depicted in Figure 17 is a general characteristic of fluid phospholipid bilayer membranes, and as such may be used as a diagnostic feature to identify lipid bilayer, or other similar liquid crystalline structures, in biological membranes.

The rates of motion of lipid labels of the types II-VII (m,n) have been far less studied than the amplitudes of motion. Scandella et al. (1974) have investigated the acyl chain motion in the PM2 virus membrane, using the simulations for anisotropic motion developed by Schindler and Seelig (1973, 1974) (see Sect. III.C). They found relatively little change in motional correlation time with position of the label on the acyl chain.

Values of $\tau \sim 3.10^{-9}$ s (which is at the limit of the motional narrowing region) were obtained at 30°C, decreasing to values of $\tau \sim 1.10^{-9}$ s at 60°C. Cannon et al. (1975) have simulated the spectra of the stearic acid label II (12,3) at 2°C in intact mitochondria from the brown adipose tissue of cold-adapted hamsters, and in the mitochondrial lipids of warm-adapted hamsters. The simulations were performed by the stochastic method (see Sect. III.C) and gave correlation times of the order of $\tau \sim 1.10^{-9}$ s in the former, and $\tau \sim 2.10^{-10}$ s in the latter case. A correlation time for overall motion of the label of 5.10^{-8} s was also included in the simulation, and it was pointed out that the spectrum for intact mitochondria at 2°C ($\tau \sim 10^{-9}$ s) probably lay outside the motional narrowing limit. Thus it is possible that the differences in correlation times observed by the two sets of authors may arise not only from the different membranes studied, but also from the different methods used for the simulations.

D. Lateral Diffusion and Lipid Segregation

(Träuble and Sackmann, 1972; Devaux et al., 1973)
Another important feature of the fluidity of the lipid phase of membranes is the rapid lateral diffusion of the lipid molecules within the plane of the membrane. Since the spin label-spin label exchange interaction requires that the nitroxide groups be virtually in van der Waals contact, it can be used to measure the spin-label collision frequency, and hence the lateral diffusion rate in these fluid membranes (see Sect. III.E). Somewhat higher spin label concentrations (> 2 mole %) than are normally used are required to obtain appreciable interaction broadening. The dipolar and exchange contributions to the interaction broadening are distinguished by the temperature dependence and line-shape simulation, as was explained in section III.E. Diffusion-controlled interaction is distinguished from static broadening due to label segregation by the different concentration dependence of these two mechanisms.

The spin-label exchange frequency, defined in Section III.E, is directly related to the collision frequency, ν_{coll}. The two are not identical, however, since even for strong collisions only those labels differing in both their electron and nuclear spin quantum numbers will give rise to spin exchange. Of the nine possible combinations of the nitrogen nuclear spin quantum numbers, only two-thirds have $m_{I_1} \neq m_{I_2}$. Only half of the collisions with these combinations will be between radicals with different spin orientations ($+^1/_2, -^1/_2$ or $-^1/_2, +^1/_2$) and it is only between these that spin exchange can occur. Thus for strong collisions the collision frequency is given by[3]:

$$\nu_{coll} = 3x\, W_{ex} \tag{95}$$

which allows the determination of ν_{coll} from the spectra, as explained in Section III.E.

There are two methods for determining the lateral diffusion constant from the collision frequency. The first method (Träuble and Sackmann, 1972) considers the process as a two-dimensional, homogeneous diffusion. If d_c is the effective interaction distance between nitroxides, then ν_{coll} is the product of the area swept out by the interaction diameter ($2d_c$) per second, times the density of labels, i.e.:

---

3 Träuble and Sackmann (1972) did not include this factor of 3 in their original analysis.

$$\nu_{coll} = 2d_c \cdot 1 \cdot \frac{c}{F} \tag{96}$$

where 1 is the integrated distance of travel per second, c is the mole fraction of the spin label and F is the area per lipid molecule. The Einstein expression for the two-dimensional diffusion coefficient is[4]:

$$D_{diff} = (1/4) \cdot \nu \cdot \lambda^2 \tag{97}$$

where ν is the label hopping frequency and λ is the length of one jump, determined by the lattice constant. Since $1 = \nu \cdot \lambda = 4D_{diff}/\lambda$, Eq. (96) becomes:

$$\nu_{coll} = \frac{8d_c}{F \cdot \lambda} \times D_{diff} \times c \tag{98}$$

Thus for the diffusion-controlled situation, the collision frequency, and hence the exchange frequency, are linearly related to the spin-label concentration, c. The diffusion coefficient can be determined from the concentration dependence via Eqs. (95), (98), using the methods of spectral analysis given in Section III.E. The first term on the right-hand side of Eq. (98) involves only geometrical factors. For the steroid spin label Ib) in fluid bilayers of dipalmitoylphosphatidylcholine, Träuble and Sackmann (1972) assumed: $F = 58\ \text{Å}^2$, $\lambda = 8\ \text{Å}$, $d_c = 20\ \text{Å}$, and obtained a lateral diffusion coefficient of $D_{diff} \simeq 10^{-8}\ \text{cm}^2/\text{s}$.

The second method (Devaux et al., 1973) considers the diffusion process as a nearest-neighbour exchange on a lattice in which each lipid molecule is situated at the centre of a hexagon, surrounded by six nearest-neighbours. In a single nearest-neighbour exchange, each exchanging molecule thus encounters three new nearest neighbours (and still remains in contact with three of its previous neighbours). Thus the spin label-spin label collision frequency is given by:

$$\nu_{coll} = 3 \cdot \nu \cdot c \tag{99}$$

where again ν is the hopping frequency and c the mole fraction of spin label. Hence using Eq. (97) to eliminate ν:

$$\nu_{coll} = \frac{12}{\lambda^2} \times D_{diff} \times c \tag{100}$$

which has the same concentration dependence as Eq. (98). Here $\lambda = a$, the lattice constant, and the area/molecule is given by $F = \sqrt{3}/2 \cdot a^2$, and an alternative form for the geometrical constant in Eq. (100) is $12\sqrt{3}/(2F)$, which is the same order of magnitude as the constant in Eq. (98). By this method Scandella et al. (1972) obtained a lateral diffusion coefficient $D_{diff} \simeq 6.10^{-8}\ \text{cm}^2/\text{s}$ for the phosphatidylcholine spin label III (1,14) in sarcoplasmic reticulum membrane vesicles.

---

4 Träuble and Sackmann (1972) used the expression for one-dimensional diffusion in their original analysis.

A general feature of these lateral diffusion measurements is that they require higher concentrations of spin label than for a normal experiment. It is thus desirable to use spin labels which as closely resemble the endogeneous lipids as possible, and to check the membrane viability in the presence of the spin label. It is also necessary to check that a linear dependence of collision frequency on label concentration is obtained (cf. Eqs. (98), (100)), indicating that the diffusion rate is independent of spin-label concentration. When labelling biological membranes the label concentration in the membrane must be determined by analysis of the extracted lipid, since it cannot be assumed that all the added spin label has been incorporated. In addition, the unincorporated label must be separated from the membrane by centrifugation, before measurements can be made. This is particularly important for phospholipid labels which tend to form their own bilayers, and may even be added as spin-label vesicles. In this case resolution by density gradient centrifugation will be required (cf. Sect. II.B).

At the opposite extreme to the diffusion-controlled exchange is the situation in which the spin label is not homogeneously mixed, but forms clusters within the host lipid. In this case the measured exchange interaction can be used to analyse the lipid segregation in terms of the cluster size and density (Träuble and Sackmann, 1972; Galla and Sackmann, 1975). Each circular cluster of radius r is assumed to be divided into a core of radius $(r-d_c)$ in which all the radicals have the same, maximum exchange frequency $\hat{W}_{ex}$, and an outer belt of thickness d_c in which $W_{ex} \simeq (1/2)\hat{W}_{ex}$. In the cluster growth phase, it is only because of this reduction in the exchange interaction at the perimeter of the cluster that the exchange frequency displays a dependence on label concentration. The reduction in exchange frequency, below the maximal value $\hat{W}_{ex}$ is proportional to the fractional area of the outer belt. For $d_c \ll r$ the reduction in area is $\delta A/A = 2\delta r/r = 2d_c/r$. This fraction of the area has exchange frequency $W_{ex} = (1/2)\hat{W}_{ex}$, thus the reduction in exchange frequency is $\delta W_{ex} = -(d_c/r) \cdot W_{ex}$, and the measured exchange frequency is:

$$W_{ex} = \hat{W}_{ex} \cdot [1 - d_c/r]. \tag{101}$$

The area of a cluster, πr^2, is related to the total fractional area occupied by the labels, i.e. c, by multiplying by the density of clusters, n/cm^2:

$$c = n \times \pi r^2 \tag{102}$$

where it is assumed that the clusters are composed solely of spin label. Thus the concentration dependence of the exchange frequency is:

$$W_{ex} = \hat{W}_{ex} \cdot [1 - \sqrt{\pi n} \cdot d_c/\sqrt{c}]. \tag{103}$$

Hence, in the cluster growth phase where n remains constant, W_{ex} should have a linear dependence on $1/\sqrt{c}$. The maximal exchange frequency in the cluster, $\hat{W}_{ex}$, is determined by the intercept on the W_{ex} axis, and from both the intercept and the slope the density of clusters, n, can be determined, assuming a value for d_c. The cluster size depends on c and is obtained from n by Eq. (102). Equation (103) is valid in the growth phase of non-overlapping clusters with radii greater than d_c, i.e. $d_c \leqslant r \leqslant (1/2)(1/\sqrt{n} - d_c)$.

For the steroid label Ib) in dipalmitoylphosphatidylcholine bilayers below the phase transition, Träuble and Sackmann (1972)[5] found that Eq. (103) was obeyed, the size of the clusters increasing with increasing label concentration, whereas the density n of clusters was independent of c. At 19°C they found a cluster density of $3.4.10^{11}$ per cm^2, corresponding to an inter-cluster distance of $D \sim 1/\sqrt{n} = 170$ Å; and the cluster size increased from r = 18 Å at c = 0.036 to r = 44 Å at c = 0.213, corresponding to an increase in number of molecules per cluster from 25 to 150 (assuming $d_c = 15$ Å and F = 40 Å^2). The value of $\hat{W}_{ex}$ at 19°C was 13.4 MHz and decreased with temperature ($d\hat{W}_{ex}/dT = -0.1$ MHz/°C), corresponding to a lattice constant expansion coefficient of $da/dT = -(a/2\hat{W}_{ex}) \cdot d\hat{W}_{ex}/dT \simeq 0.02$ Å/°C.

Galla and Sackmann (1975)[5] also found that Eq. (103) was obeyed for mixtures of the phosphatidylcholine spin label III (10,5) with dipalmitoyl phosphatidic acid in the presence of Ca^{2+}. In this way it was found that a lateral phase separation of the two species was induced by binding Ca^{2+} to the phosphatidic acid. At a mole ratio of Ca^{2+} to phosphatidic acid of 1:1 (pH 9, 59°C), the extrapolated value of $\hat{W}_{ex} = 27$ MHz was very close to that found for bilayers of the pure spin label: $\hat{W}^{\circ}_{ex} = 30$ MHz, indicating that the clusters in the mixed system are composed solely of spin label, i.e. there is a complete phase separation. The density of clusters was $n = 3.2.10^{11}$ per cm^2 and at c = 0.2 the cluster radius was r = 45 Å. At lower mole ratios of Ca^{2+}, for which not all the phosphatidic acid was Ca^{2+}-bound, the values of $\hat{W}_{ex}$ were lower than $\hat{W}^{\circ}_{ex}$, indicating that that the label clusters contained some (unlabelled) phosphatidic acid, i.e. that the phase separation was not total. In this case the above analysis has to be modified somewhat. Both if the spin labels are statistically distributed, or if they are undergoing rapid lateral diffusion (cf. Eqs. (95), (98)) within the cluster, the exchange frequency will depend linearly on the label concentration:

$$\hat{W}_{ex} = \hat{W}^{\circ}_{ex} \cdot c_{cl} \tag{104}$$

where c_{cl} is the mole fraction of the label in the cluster. Equation (101) is unaffected, since it does not make the assumption that the clusters are composed solely of spin label, merely that they can be characterized by a single, homogeneous value of W_{ex}. However, Eq. (102) must be modified, the appropriate expression being:

$$c = n \times \pi r^2 \times c_{cl}. \tag{105}$$

The corresponding version of Eq. (103) is then:

$$W_{ex} = \hat{W}_{ex} \cdot [1 - \sqrt{\pi \cdot n \cdot c_{cl}} \cdot d_c/\sqrt{c}] \tag{106}$$

Hence, in the case of clusters which are not composed of pure spin label: $\hat{W}_{ex}$ is first obtained from the W_{ex}-axis intercept in the plot against $1/\sqrt{c}$; c_{cl} is then obtained from Eq. (104), assuming one knows the value of W°_{ex}; the cluster density n can then be obtained from the slope in Eq. (106) and hence the cluster dimensions from Eq. (105). (If

5 The exact numerical values for cluster sizes and densities in the work of Träuble and Sackmann (1972) and Galla and Sackmann (1975) may require slight change, since they determined values of $\hat{W}_{ex}$ from extrapolation to $1/\sqrt{c} = 1$ rather than $1/\sqrt{c} = 0$.

W°_{ex} is not known, the cluster size, but not the cluster density, can be obtained.) In this way Galla and Sackmann (1975) were able to analyse the Ca^{2+} concentration dependence of the chemically induced lateral phase separation.

E. Phospholipid Flip-Flop and Lipid Asymmetry

(Kornberg and McConnell, 1971)
Although the lipid molecules in phospholipid bilayers can translocate laterally at very high rates, it appears that the transverse bilayer motion, "flip-flop" between the opposite sides of the bilayer, is very slow. This illustrates the inherent structural stability of the phospholipid bilayer as a permeation barrier, and indicates that functional transbilayer asymmetries of lipid distribution can persist in biological membranes.

Kornberg and McConnell (1971) have measured the rate of transbilayer migration of the phosphatidylcholine spin label XI in egg phosphatidylcholine vesicles. The spin label XI is labelled in the phospholipid headgroup region and an asymmetric label distribution can be set up by reducing the spin label in the outer half of the bilayer using ascorbate at 0°C, at which temperature the ascorbate does not penetrate the vesicle (see Fig. 20). The ratio of the spin label intensities (assayed at 0°C) after and before the reduction then immediately gives the fraction of lipids in the inner half of the bilayer:

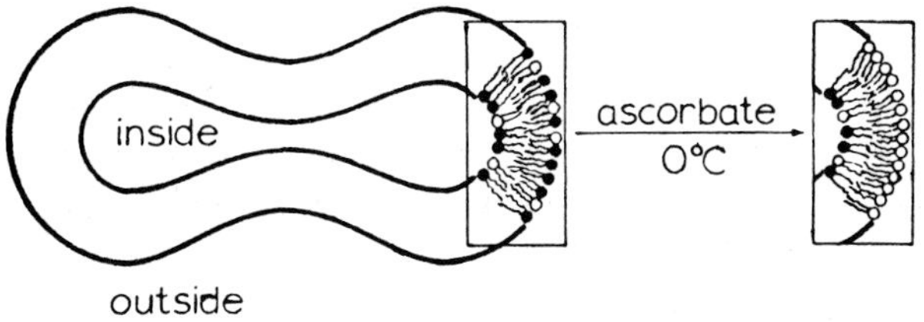

Fig. 20. Schematic representation of the effect of sodium ascorbate at 0°C on the distribution of paramagnetic molecules in a spin-labelled phosphatidylcholine vesicle. *Filled circles* and *open circles:* polar headgroups of the labelled and unlabelled phospholipid molecules (the proportion of spin-labelled molecules is exaggerated for illustration purposes; actual mole ratio of spin label = 0.03 – 0.05; Kornberg and McConnell, 1971)

$$\gamma = N_i / (N_i + N_o) \tag{107}$$

where N_i and N_o are the numbers of spin-label molecules in the inner and outer monolayer respectively. By using labels with different headgroups, e.g. XI-XIII, or chain compositions, this method can be used to investigate the asymmetrical distributions of different phospholipid types. For small, sonicated phospholipid vesicles $\gamma \sim 0.25$-0.35, because in these highly curved structures the area of the inner monolayer is smaller than the outer monolayer. For large, single-bilayer vesicles (in the absence of any intrinsic lipid asymmetry), one would expect $\gamma = 0.5$, and for large multibilayer vesicles $\gamma = 1 - 0.5/m$, where m is the number of bilayers. Double integration of the spectra is necessary to determine the spin-label intensities since, particularly in the case of small, sonicated vesicles, the spectral lineshapes can be considerably different in the outer and inner monolayer.

After the first reduction the ascorbate is immediately removed by column chromatography at 0°C. Then the rate of phospholipid flip-flop between the outer and inner halves of the bilayer can be followed by incubating for various times at the desired temperature. At the given time, the remaining spin-label asymmetry is measured by means of a further ascorbate reduction at 0°C. (The stability of the vesicles to chromatography may be checked by assaying the retention of a water-soluble spin label such as TEMPO-choline IX, again by ascorbate reduction at 0°C, as described in the next section IV.F.)

The phospholipid flip-flop can be considered as a reversible, pseudo-unimolecular process:

$$(\text{lipid spin label})_i \underset{k_i}{\overset{k_o}{\rightleftharpoons}} (\text{lipid spin label})_o \, .$$

In general the rate constants k_o, k_i for outward and inward transfer will not be equal, because the outer and inner monolayers have unequal numbers of lipid sites to which lipid molecules can be transferred. This is simply related to the bilayer asymmetry, γ, thus:

$$k_i = \gamma \cdot k \tag{108}$$

$$k_o = (1-\gamma) \cdot k \tag{109}$$

where $k = (k_o + k_i)$ is the true first order rate constant. The kinetic equation for the change of the internal spin-label signal is given by:

$$\frac{dN_i}{dt} = k_i \cdot N_o - k_o \cdot N_i. \tag{110}$$

The condition for the constancy of the total number of spin labels is:

$$N_i + N_o = N_i^o \tag{111}$$

where N_i^o is the initial value for the number of spin labels in the inner monolayer, immediately after the first ascorbate reduction. Using Eqs. (108), (109), (111), the kinetic Eq. (110) may be simply transformed to:

$$\frac{d(N_i - \gamma N_i^o)}{dt} = -k \cdot (N_i - \gamma N_i^o). \tag{112}$$

Hence, integrating and writing N_i = constant x A(t), where A(t) is the amplitude of the spin-label ESR signal from the inner half of the bilayer (as measured by ascorbate reduction of the outer layer), we get:

$$\frac{A(t) - \gamma A(O)}{A(O) \cdot (1-\gamma)} = e^{-kt}. \tag{113}$$

The equilibrium value of A(t) is $\gamma A(O)$, where A(O) is the initial spin-label amplitude immediately after the original reduction, and γ is the equilibrium inner-outer asymmetry defined by Eq. (107). Equation (113) states that the distance from equilibrium of the paramagnetic spin-label distribution decays exponentially from its initial value. Both A(O) and γ are determined on the first reduction; hence the rate constant k may be determined from a logarithmic plot of the left-hand side of Eq. (113) against time. For the measurement of A(t) it is sufficient simply to take the peak height of the spin-label signal, since firstly these quantities are simply involved as a ratio in Eq. (113), and secondly they refer always to the internal signal, hence the lineshape will be the same in all cases.

Kornberg and McConnell (1971) concluded that at 30°C the asymmetry in egg phosphatidylcholine vesicles decayed with a half-time of 6.5 h, and that the probabilities of outward and inward transitions were 0.07/h and 0.04/h respectively. On the other hand McNamee and McConnell (1973) found much faster rates with excitable membrane vesicles from the electroplax of *Electrophorus electricus*. A half-time for phosphatidylcholine flip-flop of 4-7 min at 15°C was estimated both by the above method and by the direct rate of reduction of the "internal" spin label at 15°C. Clearly the method involving column chromatography is not very suitable for membranes with fast flip-flop, and in this case direct reduction may be used provided that the flip-flop rate is greater than the rate of translocation of the ascorbate into the vesicles.

F. Membrane Permeability, Vesicle Internal Volumes, and Transmembrane Potentials

(Marsh et al., 1976b; McNamee and McConnell, 1973)
The ascorbate reduction method can also be applied to measuring the rate of transfer of water-soluble spin-label molecules across membranes. If the permeability to the spin label is low at 0°C, ascorbate reduction at this temperature may be used to assay the amount of spin label contained within the membrane vesicles after incubation for various times at given temperatures. Two types of experiment are possible, either uptake in which the membrane vesicles are incubated with spin label added on the outside, or release experiments in which the vesicles are loaded with spin label and the external label removed by chromatography or dialysis at low temperature.

Marsh et al. (1976b) have measured the permeability of small, sonicated, single-bilayer vesicles of dimyristoylphosphatidylcholine, to the TEMPOcholine spin-label cation IX. The method of the spin-label assay is indicated in Figure 21. Since the spectra both before and after reduction were recorded at 0°C, the lineshapes are identical and quantitation could be performed simply in terms of h/h_0, the ratio of the low-field line-heights (the central line is sometimes contaminated by the ascorbate free radical signal), without double integration. Typical uptake curves are shown in Figure 22, which plots the fractional spin-label signal remaining after ascorbate reduction at 0°C for samples which were incubated at different temperatures for various times, t, after addition of TEMPOcholine to the preformed vesicle suspension at time t = 0. The uptake rises steeply initially, giving a measure of the permeability to TEMPOcholine, and then flattens off to a limiting value which is a measure of the final, limiting uptake by the vesicles. This latter value is a direct measure of the internal volume of the vesicles, and gives a method

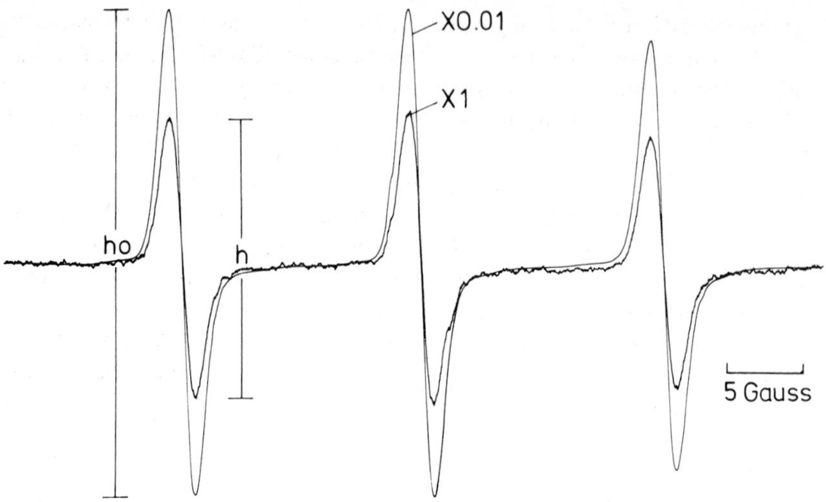

Fig. 21. ESR spectra at 0°C of the TEMPOcholine spin label IX. *Heavy line* (h): spectrum of TEMPO-choline taken up into sonicated vesicles of dimyristoylphosphatidylcholine. Vesicles incubated for 20 min at 29°C with 3 mM TEMPOcholine added to the outside, then treated with ascorbate at 0°C prior to taking the spectrum. *Fine line* (h_o): spectrum of control sample, similarly incubated, but without added ascorbate – note decreased gain (Marsh et al., 1976b)

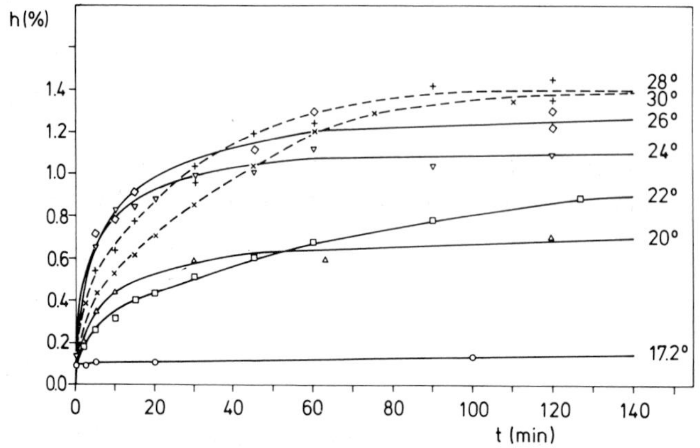

Fig. 22. Time course of uptake of TEMPOcholine IX into dimyristoylphosphatidylcholine vesicles, on incubation at various temperatures. Plotted is the height (h) of the TEMPOcholine ESR spectrum at 0°C (expressed as percentage of the 3 mM TEMPOcholine originally outside the vesicles, obtained from the control sample) remaining after treatment of the vesicles with ascorbate at 0°C, against time of incubation of the vesicles at various incubation temperatures (Marsh et al., 1976b)

for determining the variation in internal volume with temperature (Watts et al., 1978). The vesicle trapped volume per g of lipid is simply given by:

$$v_c = h_{max}/(h_o \cdot c_{lipid}) \tag{114}$$

where h_o is the lineheight of the untreated sample, h_{max} is the limiting lineheight of the ascorbate-treated sample, obtained after long incubation with TEMPOcholine, and c_{lipid} (g/ml) is the final lipid concentration of the sample. An alternative method of obtaining the internal volume of phospholipid vesicles is to prepare the vesicles in the presence of TEMPOcholine.

The TEMPOcholine permeability was also determined by assaying the release of TEMPOcholine from vesicles which had been sonicated in the presence of TEMPOcholine and then dialysed at 4°C to remove the TEMPOcholine on the outside (Marsh et al., 1976b). In this case a plot of the logarithm of the ESR lineheight of the TEMPOcholine remaining inside the vesicles, against time of incubation, gives the first order rate constant for the release directly from the slope of the linear plot. The vesicle permeability coefficient to TEMPOcholine, P, can then be determined from either the uptake or the release experiments respectively:

$$P = R_{init} \cdot v_o/A \tag{115}$$

$$P = k_{release} \cdot v_i/A \tag{116}$$

where R_{init} is the initial, *fractional* rate of uptake obtained from Figure 22, $k_{release}$ is the first order rate constant for the release process, v_o is the total sample volume and $v_i = \phi \times v_o$ is the total internal volume of the vesicles, ϕ being the fractional occluded volume. (From the limiting TEMPOcholine uptake in Fig. 22: $\phi = 0.014$.) The total vesicle surface area is given by:

$$A = a_{ves} \cdot n_{ves} \tag{117}$$

where the area/vesicle, a_{ves}, and the total number of vesicles, n_{ves}, can be calculated if the mean vesicle radius, r_m, and the vesicle molecular weight, M, are known from independent measurements:

$$a_{ves} = 4\pi r_m^2 \tag{118}$$

$$n_{ves} = (c_{lipid}/M)N_{av} \cdot v_o \tag{119}$$

where $N_{av} = 6.023 . 10^{23}$. In this way, maximum permeability coefficients $P \sim 7\text{-}30 . 10^{-11}$ cm/s were obtained at the centre of the phase transition, corresponding to diffusion coefficients: $D = P \times d \sim 2\text{-}9 . 10^{-17}$ cm^2/s (Marsh et al., 1976b).

A different application of the ascorbate method is to use the equilibrium distribution of charged spin labels across the membrane to measure the transmembrane potential. The internal volume of the vesicles is first measured as described above, in the situation in which there is no transmembrane potential. The equilibrium distribution is then determined again, in exactly the same manner, in the presence of the membrane potential. The inside-outside distribution is then simply related to the transmembrane potential, V, by the Nernst equation:

$$[SL^{\pm}]_i / [SL^{\pm}]_o = \exp(\mp eV/kT) \tag{120}$$

where $[SL^{\pm}]_i$, $[SL^{\pm}]_o$ are the spin-label concentrations inside and outside the vesicle respectively, e is the electron charge and k Boltzmann's constant. Since the internal volume of the vesicles is small compared with the total volume, $[SL^{\pm}]_o$ can be taken as the (final) concentration of spin label added. $[SL^{\pm}]_i$ is determined from the ESR lineheights h_o, h before and after ascorbate reduction (cf. Fig. 21); in the presence of the membrane potential:

$$[SL^{\pm}]_i = h/h_o \times [SL^{\pm}]_o / \phi \tag{121}$$

where $\phi = v_i/v_o$ is determined from the ascorbate reduction experiment in the absence of the transmembrane potential, as described above. In this way McNamee and McConnell (1973) measured an induced Na^+ diffusion potential in excitable membrane vesicles, using the TEMPOcholine spin-label cation.

The method does not apply if the membrane is impermeable to the charged spin label. However, Kornberg et al. (1972) found that in these cases spin-label analogues of weak acids may be used, if the vesicles are permeable to H^+ ions and to the uncharged form of the spin label. Kornberg et al. (1972) showed that the inside-outside distribution of TEMPOtartrate XXIV

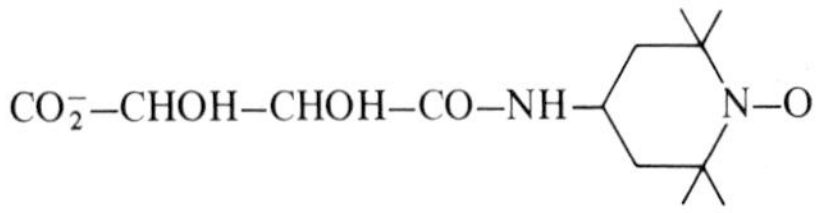

XXIV

is coupled to the inside-outside distribution of H^+ in egg phosphatidylcholine vesicles, and thus could be used to measure the transmembrane potential. The vesicle membrane is permeable to the uncharged form of TEMPOtartrate (HT) and thus at equilibrium the concentrations of the uncharged form on either side of the membrane must be equal ($[HT]_i = [HT]_o$). Further, taking into account the acid-base equilibrium of TEMPOtartrate ($[H^+]_i [T^-]_i / [HT]_i = K_D = [H^+]_o [T^-]_o / [HT]_o$), one finds:

$$[T^-]_i / [T^-]_o = [H^+]_o / [H^+]_i = \exp(-eV/kT) \tag{122}$$

where $[H^+]_o$, $[H^+]_i$ are the H^+ concentrations outside and inside the vesicles, and these are assumed to be at equilibrium across the membrane, under the influence of the transmembrane potential, V. $[T^-]_i$ and $[T^-]_o$ are the inside and outside concentrations of the charged form of TEMPOtartrate, which at the pH used (6.8-7.0) can be approximated by the total TEMPOtartrate concentrations (charged plus uncharged). Note, however, that the rate of approach to the equilibrium distribution is pH-dependent (typically 4 h at pH 7.0), since this is determined by the concentration of the uncharged species [HT]. For this reason, high pHs are to be avoided. Thus the membrane potential may be obtained from an ascorbate reduction experiment with TEMPOtartrate, in exactly the same way as outlined above in Eqs. (120), (121). Kornberg et al. (1972) were able to use the distribution of TEMPOtartrate to measure the diffusion potentials induced by K^+-valinomycin, in this way.

An alternative approach to the application of the ascorbate reduction technique in studying membrane permeability is to follow the kinetics of spin-label reduction directly by monitoring the decrease in ESR lineheight with time. The reduction rate is then determined by the rate of translocation both of the spin label and of the ascorbate. In the case where the former is slow, it is possible to study the ascorbate permeability. Schreier-Muccillo et al. (1976) have measured the reduction rate of intercalated, spin-labelled lipid molecules, to determine the permeation profile of ascorbate into egg phosphatidylcholine bilayers. Figure 23 shows a typical exponential decay of the ESR spectrum of

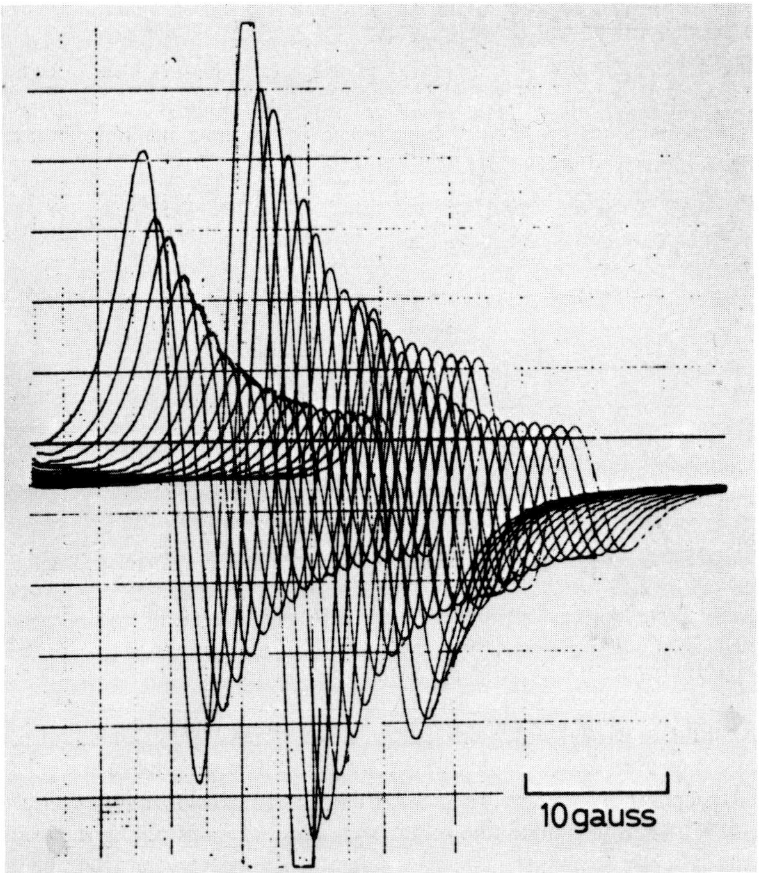

Fig. 23. ESR spectra of the cholestane spin label Ia) in oriented multibilayers of egg phosphatidylcholine (PC = 50 mole %) + cholesterol (chol = 35 mole %) + hexadecyltetramethylammonium (HDTMA = 15 mole %) as a function of time after hydration with 10^{-3} M sodium ascorbate at pH 6.5, 6°C. First spectrum (*left*) was taken immediately after hydration (< 1 min), the others at 2-min intervals. For each measurement the starting point of the magnetic field was stepped by 1 gauss. Spectra were taken with the applied field perpendicular to the plane of the multibilayer film (Schreier-Muccillo et al., 1976)

the steroid label Ia) intercalated in oriented lipid bilayers which were hydrated with ascorbate in the aqueous phase. Plots of the logarithm of the spin-label low-field lineheight against time indicated that, for all the labels studied, the reduction follows pseudo-first-order kinetics, as expected for a fixed excess of ascorbate:

$$d[SL]/dT = -k \cdot [SL] \tag{123}$$

$$k = k_o \cdot [asc] \tag{124}$$

The half-times of reduction, $t_{1/2} = 0.69315/k$, of the various spin probes, whose nitroxide groups are situated at different depths, d, into the bilayer, are given in Table 8. The half-time of reduction very clearly increases the deeper the nitroxide group is situated into the bilayer; the differences in reduction rates with label position define the permeability profile. Addition of cholesterol decreases the rate of permeation, as would be expected, and also changes the shape of the profile somewhat. Incorporation of charged lipids into the bilayer changes the rate of reduction of all the labels uniformly. Positively charged hexadecyltetramethyl ammonium increases the penetration rate, and negatively charged phosphatidylserine decreases the penetration rate. The effect arises from the change in the surface potential, ψ, of the bilayer. The ascorbate concentration at the bilayer surface is given by a Boltzmann factor:

Table 8. Half-times ($t_{1/2}$) at 19°C for the reaction between ascorbate (10^{-2} M, pH 6.2-6.8) and spin probes in multibilayers of various compositions (Schreier-Muccillo et al., 1976)

Spin probe	$t_{1/2}$ (min) [a]				d (Å) [b]
	PC	PC (65%) + chol (35%)	PSer⁻ (30%) + PC (35%) + chol (35%)	HDTMA⁺ (15%) + PC (50%) + chol (35%)	
XIV	⩽ 1	⩽ 1	⩽ 1	⩽ 1	0
II	–	3.2	12	5.9 [c]	1
III	9.5	11.5	51	–	6.5
IV	16	29	85	16	10.5
V	30	60	136	–	15
VI	32	–	166	34	20

[a] PC, egg phosphatidylcholine; PSer⁻, phosphatidylserine; HDTMA⁺, hexadecyltetramethyl ammonium; chol, cholesterol.
[b] Maximum extended length from aqueous interface to label position.
[c] [Ascorbate] = 10^{-3} M.

$$[asc] = [asc]_o \cdot \exp(-ze\psi/kT) \tag{125}$$

where $[asc]_o$ is the ascorbate concentration in the bulk aqueous phase and ze is the charge on the ascorbate ion (z = –1 at pH 6.4-6.8). Thus the presence of the charged phospholipids directly affects the rate constant, or half-time, for ascorbate reaction, and this can be used to determine the change in surface potential. From Eqs. (124), (125) it is calculated that addition of 15 mole % HDTMA⁺ gives rise to a change in surface potential: $\Delta\psi = +15$ mV, and addition of 30 mole % phosphatidylserine gives $\Delta\psi = -30$ mV.

Thus the method can be used to investigate both changes within the hydrocarbon interior of the membrane and changes at the membrane surface.

G. Lipid-Protein Interactions

(McConnell et al., 1972; Jost et al., 1973a; Knowles et al., 1979)
One of the important determinants of lipid-protein interactions in membranes is the extent to which the protein changes the fluid bilayer properties of the membrane lipids. McConnell et al. (1972) have devised a simple spin-label assay for the fraction of lipid in a biological membrane which is in a fluid-bilayer state. This is based on the extent of partitioning of the TEMPO spin label VIII into the membrane (cf. Sect. IV.A). Since the partition coefficient depends on the intrinsic fluidity of the lipid, it is necessary to correct for fluidity differences. This is done by means of an empirical calibration which relates the specific TEMPO partition parameter, ρ, to the fluidity characterized by the order parameter of the II (5,10) stearic acid spin label. The specific TEMPO parameter is defined by:

$$\rho = ((A/B) \times 10^2) / [\text{Fluid lipid}] \qquad (126)$$

where $A = h_L$ as defined in Figure 9, and B is the peak-to-peak lineheight of the high-field aqueous line (designated 'W' in Fig. 9). The (A/B) ratio is corrected for the underlying ^{13}C hyperfine structure, as described in Section IV.A. The calibration curve is determined from measurements in lipid bilayers, for which [Fluid lipid] = [Total lipid]. The fraction of fluid lipid, F, in a membrane is then determined from the ratio of the specific TEMPO parameter calculated from the measured (A/B) partitioning ratio, to that predicted for 100% fluid lipid from the calibration curve, using the measured value of the II (5,10) order parameter in the membrane:

$$F = (A/B \times 10^2) / ([\text{Total lipid}] \times \rho) \qquad (127)$$

where ρ is deduced from the measured value of S. In this way McConnell et al. (1972) found that approximately 75% of the lipid in sarcoplasmic reticulum membranes were in a fluid state. An alternative way to correct for the dependence of the partitioning on intrinsic fluidity is to compare the TEMPO partition coefficient of the membrane with that of bilayers of the extracted lipids. Using this method, Metcalfe et al. (1972) estimated that nearly all the lipids in Acholeplasma membranes were in a fluid state. This latter type of comparison does not allow for changes of the bilayer fluidity by the presence of the protein, whereas the first method relies on the exact applicability of the empirical calibration.

More detailed information regarding the nature of the lipid interacting directly with the protein can be obtained using the spin-labelled lipid analogues II-VII (m,n). Since the spin label spectra are only sensitive to motions faster than $\sim 10^8$ s^{-1}, any lipid-protein association with a lifetime longer than this might in principle give rise to a separate ESR spectrum, distinct from the bilayer spectrum. Cytochrome oxidase is an example of a large integral membrane protein (mol. wt. 200,000), and this represents an extensive hydrophobic surface in the membrane interior, with which the lipid chains may interact. Jost et al. (1973a) have investigated cytochrome oxidase membranes of varying lipid content, using the stearic acid spin label II (1,14). This label gives rise to a narrow, motionally averaged spectrum in fluid lipid bilayers (cf. Sect. IV.C), giving optimum conditions for the detection of any specific immobilization of the lipid by the protein. The ESR spectra are given in Figure 24. At lipid/protein ratios below 0.2 mg lipid/mg protein the spectrum

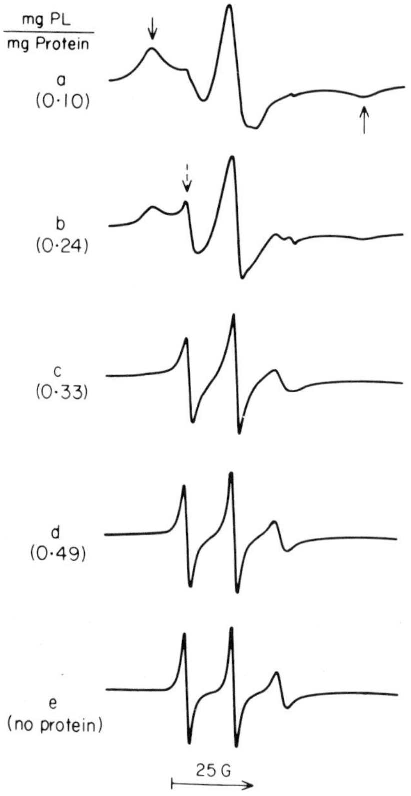

Fig. 24 a-e. ESR spectra of the stearic acid label II (1,14) in progressively lipid-depleted samples of beef heart cytochrome oxidase. Outer *full line arrows:* spectrum from the immobilized spin label; inner *broken line arrow:* spectrum from the spin labels in the fluid lipid bilayer. Proportions of the immobilized and fluid components obtained by spectral synthesis are: (a) 100:1; (b) 95:5; (c) 69:31; (d) 34:66; (e) 0:100 (Jost et al., 1973a)

is composed of a single, strongly immobilized component; whereas at lipid/protein ratios higher than this, the spectra are a super-position of both an immobilized and a fluid bilayer component, the proportional of the latter increasing with increasing lipid/protein ratio. The immobilized component arises from interaction of the lipid chains with the protein, since it is not present in bilayers of the lipid alone. Jost et al. (1973a) suggested that the 0.2 mg of immobilized lipid per mg protein is just sufficient to form a single boundary layer around the protein, which is then surrounded by fluid bilayer lipid. The dimensions of the protein used in this calculation were obtained from a rather low resolution electron microscopy study; thus it is possible that the actual protein perimeter could be considerably larger than this. However, there is one important feature of the spectra in Figure 24 which suggests that the immobilized lipid, if not forming a complete boundary, is in some way directly associated with the structure of the individual protein units; this is that a constant amount of immobilized lipid is associated with the protein, independent of the total lipid/protein ratio.

This fixed stoichiometry of the immobilized lipid is illustrated in Figure 25, in this case for a reconstituted cytochrome oxidase system. For this lipid-protein titration the

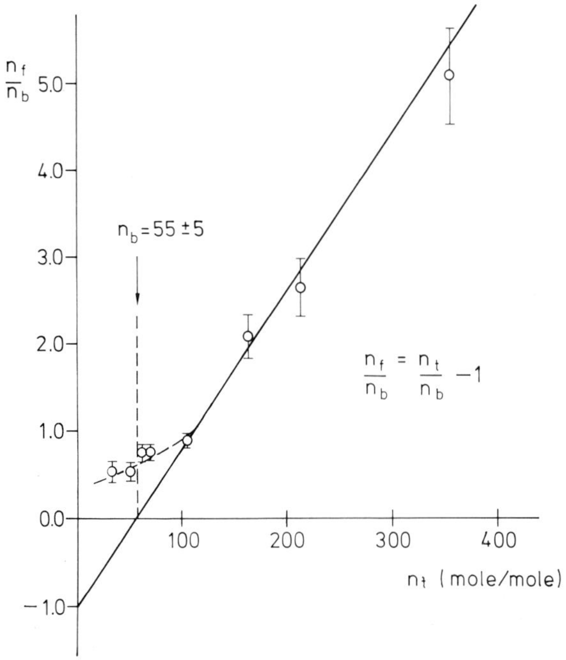

Fig. 25. Lipid-protein titration of yeast cytochrome oxidase-dimyristoylphosphatidylcholine reconstituted complexes, obtained from ESR difference spectra of the III (3,12) phosphatidylcholine spin label. n_t is the total lipid/protein ratio in the complex; n_f and n_b are the number of molecules per protein in the fluid and immobilized lipid components respectively. n_f/n_b is determined from the fraction of the total integrated intensity which is removed from spectra of the complexes on subtracting the immobilized component (Knowles et al., 1979)

relative proportions of the two components in the spectra were quantitated by subtracting the immobilized component, given by a lipid-depleted sample such as in Figure 24a, from the spectra of the complexes to yield the fluid component.[6] In this way the double-integrated intensity of the individual components relative to the total complex could be obtained (Knowles et al., 1979). Since the total lipid/protein ratio is simply the sum of the fluid and immobilized components: $n_t = n_f + n_b$, where n_f (n_b) is the number of fluid (immobilized) lipid molecules per protein, the lipid-protein titration can be expressed as:

$$n_f/n_b = n_t/n_b - 1 \tag{128}$$

6 For technical details of such spectral subtractions see Jost and Griffith (1978) and Klopfenstein et al. (1972).

Thus if the number of immobilized lipids per protein, n_b, remains constant, a plot of n_f/n_b (obtained from the spectral subtraction) against n_t (obtained from chemical analysis) should be linear with intercept −1 on the n_f/n_b axis. This is the case in Figure 25, except at the very low lipid/protein ratios, and a constant number, $n_b = 55 \pm 5$, of immobilized lipid molecules is associated with each protein, independent of the total lipid/protein ratio. In Eq. (128) it has been tacitly assumed that the III (3,12) phosphatidylcholine spin-label distribution directly reflects the distribution of the unlabelled lipid in an exact 1:1 fashion. The fact that the functional form of Eq. (128) is obeyed justifies this assumption.

An alternative analysis has been developed by Griffith and Jost (1978), who modelled the immobilization of the lipid by the protein as a multiple-site exchange equilibrium between labelled (L*) and unlabelled (L) lipid:

$$L^* + L_{n-m}L^*_m P \rightleftharpoons L + L_{n-m-1}L^*_{m+1}P$$

With the assumption that all n sites on the protein are always occupied, and that the total label concentration is much smaller than the total lipid concentration, the effective binding equation is:

$$n_f/n_b = n_t / (n \cdot K_r) - 1/K_r \tag{129}$$

where K_r is the relative binding constant of labelled versus unlabelled lipid. Clearly from Figure 25: $K_r = 1$, again indicating that there is no selectivity between the labelled and unlabelled lipid.

The example of cytochrome oxidase given above (Fig. 24) is a case in which the protein induces immobilization of the lipid chains. In principle the opposite might be possible and a protein might induce a fluidization of the chains. In this case also, separate spectra should be observable if the lifetime of the interaction is greater than 10^{-8} s, and if the spectra of the bilayer and protein-associated lipid are sufficiently different to be resolvable. The spectra would be best resolved for labels with the nitroxide group close to the polar headgroup of the molecule, e.g. II-VII (12,3), since in this case the bilayer spectrum has a relatively large anisotropy (cf. Sect. IV.C), making it easier to distinguish a more isotropic component. In general it is not easy to distinguish the spectra of two superimposed components with similar degrees of anisotropy, especially if the proportion of one component is small. Thus a superimposition of two components may be mistaken for a general broadening of the bilayer spectrum by the protein. An example is given by the spectra of the II (12,3) label in cytochrome oxidase preparations (Jost et al., 1973b); an immobilized component can only be distinguished in these spectra by means of computer subtraction. A further example is afforded by the analysis of the effect of the protein on the shells of lipid beyond the first boundary layer – the fluid component in Figure 24. In cytochrome oxidase-dimyristoylphosphatidylcholine complexes, Knowles et al. (1979) observed progressive broadening of the fluid component difference spectrum with increasing protein content of the complexes. It could not be distinguished whether the fluid difference spectra consisted of a single component or of a super-position of unresolved components from the various lipid shells, thus the data had to be analysed in terms of the two possibilities of either fast or slow exchange between the lipid shells.

Finally we consider the interaction of extrinsic proteins with lipid membranes; this interaction is probably primarily electrostatic in origin. Galla and Sackmann (1975) have demonstrated a lateral phase separation of negatively charged phosphatidic acid on binding the basic polypeptide, polylysine to mixed bilayers with phosphatidylcholine spin label III (10,5). Somewhat similar effects have been observed by Birrell and Griffith (1976) with the androstanol label, Ib), on binding cytochrome c to cardiolipin-containing bilayers. These experiments are analogous to the Ca^{2+}-induced lipid segregation discussed in Section IV.D and can be analysed in a similar manner. Mateu et al. (1978) have studied the binding of basic proteins to negatively charged phospholipid bilayers using the stearic acid spin label II (5,10). From X-ray studies they found that interactions between lysozyme and phosphatidic acid at pH 6 were essentially electrostatic, whereas at pH 4 the protein-lipid interactions were of a hydrophobic type. The spin-label spectra of the "electrostatic phase " were single-component above the phase transition, and similar to those from the lipid alone. The spectra from the "hydrophobic phase", however, consisted of two components: one similar to the lipid alone and the other strongly immobilized (not unlike the spectra of Fig. 24). This illustrates the difference between the two types of interaction: the electrostatic interaction produces very little perturbation of the lipid chains, as was also found for cytochrome c binding (Van and Griffith, 1975), whereas the hydrophobic interaction gives rise to a specific immobilization of the lipid chains. Although the latter interaction might be considerably different from the cytochrome oxidase case, similar methods of spectral analysis should apply.

H. Membrane Proteins

Far less detailed information has been obtained from the spectra of spin-labelled membrane proteins than that obtained from spin-labelled lipids, which has been discussed above. The reason lies in the far greater size and complexity of the protein molecules: it is not possible to synthesize stereospecifically labelled molecules, and the pattern of labelling of reactive amino-acid residues is often complicated. The labelling reagents most commonly used are the maleimide XVII and the iodoacetamide XVIII nitroxide derivatives which, at neutral pH, principally alkylate sulphydryl groups. Integral membrane proteins often contain many -SH groups whose extent of labelling differs markedly, depending on the reaction conditions. For example, the ATPase of sarcoplasmic reticulum has approximately 10 -SH groups, some of which are kinetically distinguishable, and one of which is protected by the presence of ATP and, when labelled, inactivates the enzyme. In principle the sensitivity of labelling to reaction conditions can be used to achieve some degree of specificity, but it also means that the reaction conditions must be carefully controlled to obtain reproduceable labelling. The aim in labelling is normally to alkylate a group which gives rise to a conformationally sensitive ESR spectrum, but which does not inactivate the protein.

Typical ESR spectra obtained from the non-specific labelling of chromaffin granule membranes with the maleimide XVIIa and iodoacetamide XVIIIa labels are given in Figure 26. These can be taken as being representative of the general labelling of membrane proteins with these reagents. The maleimide XVIIa label spectrum consists of a strongly immobilized component ($A_{z'z'}$ = 33.5 gauss, corresponding to an $a_o \simeq$ 14.5-

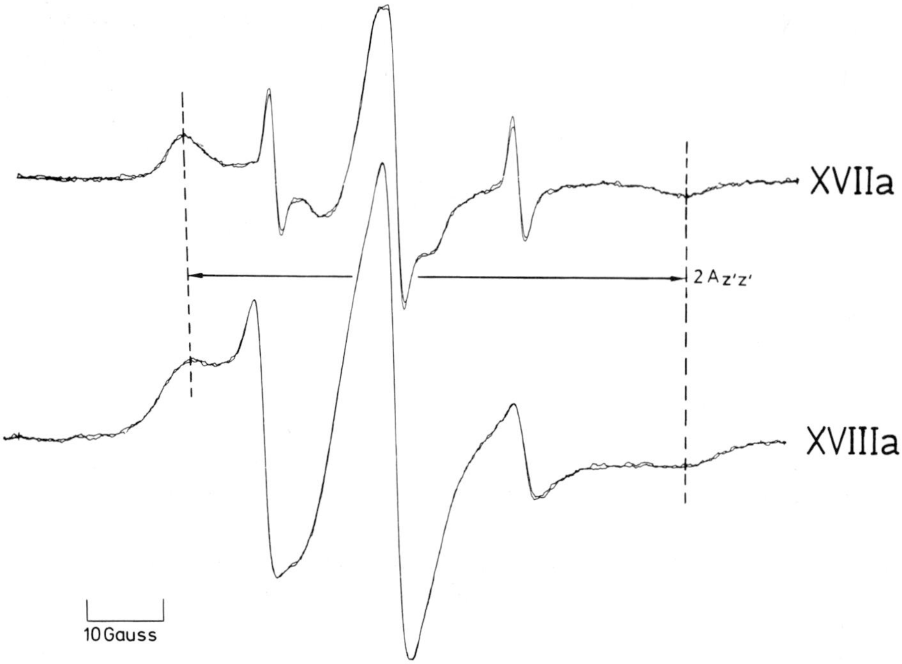

Fig. 26. ESR spectra of the maleimide XVIIa (*upper*) and iodoacetamide XVIIIa (*lower*) spin labels covalently bound to protein in the chromaffin granule membrane at 20°C (Marsh et al., 1976a)

14.9 gauss), with a small mobile component superimposed. The iodoacetamide XVIIIa label gives a similar pattern, except that the immobilized component has a smaller splitting and is somewhat broader. The very mobile components occur in the same region of the spectrum as would unreacted spin label, but these components could not be removed by repeated washing or dialysis. They possibly arise from a partial labelling of lysine groups with low pKs (the isothiocyanate label XXI, which has some specificity for amino groups, often gives rise to a single, very mobile component), or from very mobile -SH groups. In any case the double-integrated intensity in this mobile component is much lower than that in the immobilized component.

The smaller splitting and broader lines in the immobilized component of the iodoacetamide label suggest that it has considerably higher mobility than the maleimide label, and thus is more likely to prove to be a conformationally sensitive probe. (Both the immobilized component and the very mobile component of the maleimide label lie at the extremes of the range of motional sensitivity of the nitroxide spectrum and thus will give only very small spectral changes in response to changes in motion.) This is demonstrated by the relative temperature dependences of the spectra in Figure 27. The outer splitting $A_{z'z'}$ of the maleimide label remains almost constant with temperature, whereas that of the iodoacetamide label begins to decrease steeply with temperature at ∿ 32°C, corresponding to the temperature at which a structural transition is detected in the lipid phase of the membrane (Marsh et al., 1976a). It is not known whether this transition directly reflects the transition in the lipid or whether it is a result of a conformational

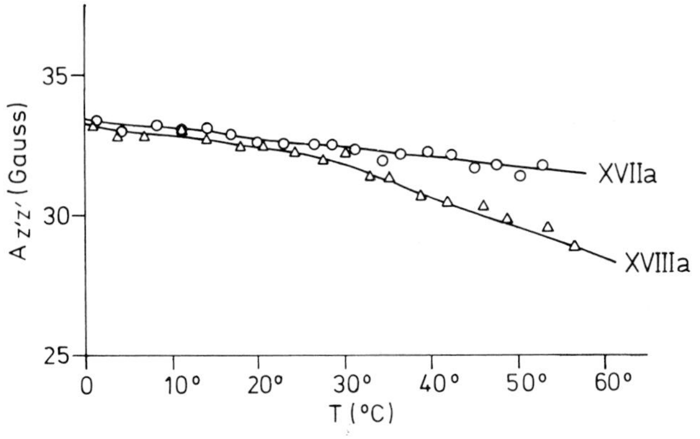

Fig. 27. Temperature variation of the maximum splitting, $A_{z'z'}$, of the strongly immobilized spin label covalently bound to the chromaffin granule membrane. (○): Maleimide spin label XVIIIa. (△): Iodoacetamide spin label XVIIIa (Marsh et al., 1976a)

change taking place in the protein driven by the lipid transition. However, it does correlate well with the breaks observed in Arrhenius plots of the ATPase and NADH oxidase activities in the membrane.

One reason for the different behaviour of the maleimide and iodoacetamide labels is probably the small number of flexible bonds joining the nitroxide moiety to the point of covalent attachment in the maleimide derivative, compared with the considerable number of single bonds in the iodoacetamide linkage (cf. Sect. II.C). Another factor may also lie in the different patterns of labelling: in general the maleimide derivatives are considerably more reactive than the iodoacetamide derivatives and also have an approximately twofold greater lipid partition coefficient. Whatever the origin of the differences, it seems likely that the short iodoacetamide labels XVIII will prove more useful in the detection of conformational changes in membrane proteins, whereas the short maleimide labels XVII will be more rigidly bound and thus can be applied to detection of motion of the protein as a whole, by the saturation transfer ESR technique discussed in one of the following sections. The situation regarding the longer maleimide XIX and iodoacetamide XX labels is not so clear, but it is clear that these labels add to one's flexibility in attempting to find a conformationally sensitive probe.

One of the very important aspects of the application of covalent protein labels is the detection of ligand-induced conformation changes in membrane-bound enzymes, receptors and transport proteins. An extensively studied system is the Ca^{2+}-transporting ATPase of sarcoplasmic reticulum. Pang and Briggs (1974) measured the ratio of the low-field peak height of the weakly immobilized component to that of the strongly immobilized component in the spectrum of the maleimide label XVIIa (a commonly used empirical index), and showed that there were small changes on adding ATP. Such changes could arise from changes in the strongly immobilized component, or in the weakly immobilized component, or both, but as noted above, both components are likely to show only small spectral changes in response to changes in the mobility of the label. In this connection it is interesting to note that the weakly immobilized component was much smaller in preparations of purified ATPase than in sarcoplasmic reticulum vesicles, and in a more recent study it was found to be completely absent in the purified ATPase (Hidalgo and Thomas, 1977). In contrast Tonomura and Morales (1974) found no change in the peak height ratio of maleimide-(XVIIa)-labelled sarcoplasmic reticulum on addition of ATP. This apparent contradiction probably results from the difference in labelling conditions, which is also reflected in the variability of the peak height ratio between preparations, and also from the intrinsic insensitivity of the maleimide label.

Clearer spectral changes have been obtained with the iodoacetamide XVIIIa label. Coan and Inesi (1977) labelled sarcoplasmic reticulum vesicles using conditions under which approximately three -SH groups are labelled, none of which are protected by ATP. The kinetics and stoichiometry of labelling were followed by monitoring the disappearance of the sharp ESR signal arising from the unreacted label in aqueous solution, and the stoichiometry was checked by colorimetric back-titration with DTNB (5,5′-dithiobis (2-nitrobenzoic acid)). The ESR spectra are given in Figure 28. In the absence of ATP

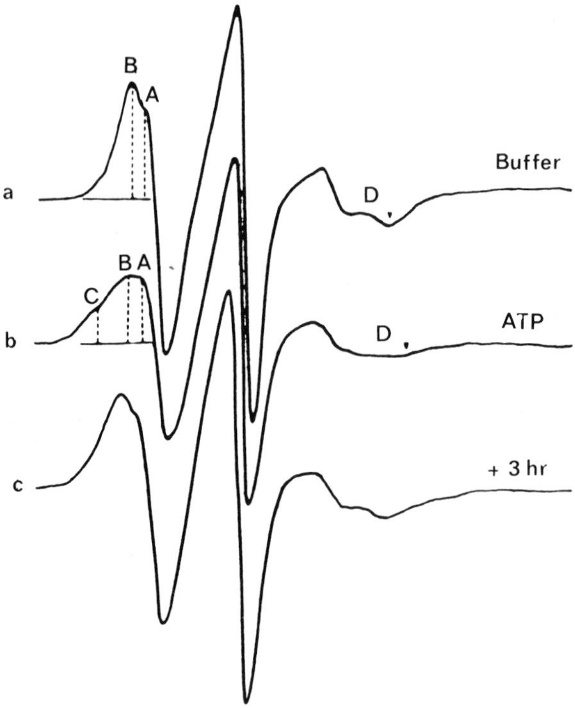

Fig. 28 a-c. ESR spectra of the iodoacetamide spin label XVIIIa covalently bound to sarcoplasmic reticulum membrane proteins (a) in buffer (20 mM MOPS, 80 mM KCl, pH 6.8) with $CaCl_2$ (10 mM), $MgCl_2$ (25 mM); (b) immediately after the addition of ATP (100 mM); (c) suspension 3 h after addition of ATP. (A)-(D): see text. Scan range = 100 gauss (Coan and Inesi, 1977)

the spectra are composed of a mobile and a more immobilized component, A and B respectively. The outer splitting of the "immobilized" component, B, is considerably less than that of a strongly immobilized label, suggesting that it may be conformationally sensitive. The proportion of the mobile component, A, varied somewhat from preparation to preparation, and was attributed to more slowly reacting residues. Addition of ATP (in the presence of Ca^{2+} and Mg^{2+}) gave rise to a third component, C, which increased in intensity at the expense of peak B, and was more strongly immobilized than either of the other two components. This was taken as direct evidence for a conformational change induced by ATP binding, and since the A component remained approximately constant, the ratio of the peak heights C/A or A/B was used as a measure of the degree of conformational conversion. In this way it was possible to follow the relaxation of the conformational change as the ATP was used up by enzymatic hydrolysis. Figure 28c shows that the spectrum relaxes back to its original state after 3 h. It was shown that ADP binding produces the same spectral changes as ATP, but addition of AMP produced no spectral change. Very little change was observed in the absence of Ca^{2+} and Mg^{2+}. Ca^{2+} titrations were performed in the presence of Mg^{2+} and ATP, using the C/A peak height ratio, and after correcting for the competitive effect of the EGTA present in the mixture, a binding constant of 4.8 μM was calculated. This identified the spectral change with binding to the high-affinity Ca^{2+} site on the enzyme. From these studies it was concluded that the observed spectral change corresponded to a conformational change which takes place only on binding of both the substrate (Mg-ATP) and the activator (Ca^{2+}) to the enzyme:

$$2Ca^{2+} + (\text{Mg-ATP}) + E \longleftrightarrow Ca_2 \cdot E \cdot (\text{Mg-ATP})$$

The conformational change takes place before formation of the phosphorylated enzyme intermediate, $Ca_2 \cdot E \sim P$, since the spectral changes take place on binding ADP, which is not hydrolysed. It is possible that different labelling conditions might give rise to the labelling of other groups which may be sensitive to possible conformational changes taking place at other stages in the enzymatic and transport cycle.

A different approach to studying the conformational changes in membrane-bound enzymes is to determine the accessibility of covalently bound label groups to ascorbate. Tonomura and Morales (1974) found that the reduction of maleimide-(XVIIa)-labelled sarcoplasmic reticulum vesicles by ascorbate was biphasic, corresponding to two distinct spin-label populations. The ascorbate concentration was sufficiently high that the reduction of each population corresponded to the pseudo-first-order kinetics given by Eqs. (123), (124) and the decay of the total spin-label concentration is thus given by:

$$[SL(t)]_{tot} = [SL(O)]_s \cdot \exp(-k_s t) + [SL(O)]_f \cdot \exp(-k_f t) \tag{130}$$

where the subscripts s and f refer to the slow and fast reducing populations respectively. By varying the ligands present in the suspending medium, it was possible to study the rate of spin-label reduction when the ATPase was poised in different states of the enzymatic cycle. The results are summarized in Table 9; the rate constants for the two populations are the same in each of the various enzymatic states, but the distribution of labels between the two populations is sensitively dependent on the enzymatic state (and on the degree of labelling). These results can be most readily explained in terms of conformational changes taking place in the ATPase, which change the accessibility of certain of the spin-label groups to ascorbate. Clearly there is no such change on the binding of Ca^{2+} alone, but a conformational change, or shift in conformational equilibrium, takes place on binding MgATP alone, with a further change taking place on binding MgATP plus Ca^{2+}, going through to the phosphorylated enzyme intermediate, $Ca \cdot E \sim P$. If the two spin-label populations were rigorously homogeneous, the constancy of the rate con-

Table 9. Kinetic constants for the reduction of maleimide (XVIIa)-labelled sarcoplasmic reticulum vesicles by ascorbate (Tonomura and Morales, 1974)

Medium	Intended state	$(k_o)_f$ [a] $(mM^{-1}\ min^{-1})$	$(k_o)_s$ $(mM^{-1}\ min^{-1})$	$[SL(O)]_s / [SL(O)]_f$	
				Observed [b]	Corrected [c]
2 mM EGTA, 15 mM $MgCl_2$	E	0.48	0.086	32	32
5 mM $CaCl_2$, 30 μM $MgCl_2$, creatine kinase	Ca · E	0.48	0.086	32	32
2 mM EGTA, 15 mM $MgCl_2$, 4 mM ATP	E · (Mg-ATP)	0.48	0.086	57	77
5 mM $CaCl_2$, 30 μM $MgCl_2$, 4 mM ATP, creatine kinase	Ca · E ∿ P	0.48	0.086	46	57

[a] Second order rate constant: $k_o = k/[asc]$.
[b] Directly observed fraction of spin labels in the slowly reduced phase.
[c] "Corrected for the fact that the ATPase constitutes only ∿ 60% of the total, labelable membrane protein.

stants would make it necessary to assume a single conformational equilibrium which is displaced to different extents in the different enzymatic states. However, it is more likely that the populations are heterogeneous and only approximate to two distinct groups. (The rotating carrier model proposed by the original authors, if taken literally, is unlikely on energetic grounds in view of the amphipathic nature of integral membrane proteins.)

A quite different method of labelling membrane proteins is to utilize specific protein ligand molecules to which the spin-label group is covalently attached. Such labels can be used for investigating ligand binding equilibria and, if the binding is sufficiently tight, also for studying details of the ligand binding site (Dwek et al., 1975; Sutton et al., 1977). A spin-labelled substrate of the β-galactoside transport system in the erythrocyte membrane has been synthesized by Struve and McConnell (1972), and several spin-labelled analogues of cholinergic agonists have been synthesized, some of which bind to the acetylcholine receptor or acetylcholine esterase in postsynaptic membranes (see e.g. Weiland et al., 1976).

An elegant adaptation of the above method is the synthesis of specific protein ligands to which are acylated spin-labelled fatty acid chains of the type II (m,n). These bifunctional probes allow one selectively to study the hydrophobic surfaces of a particular membrane protein in situ. Devaux et al. (1975) have used spin-labelled acyl-Coenzyme A's XXV (m,n) to investigate the ADP-ATP carrier in the inner mitochondrial membrane.

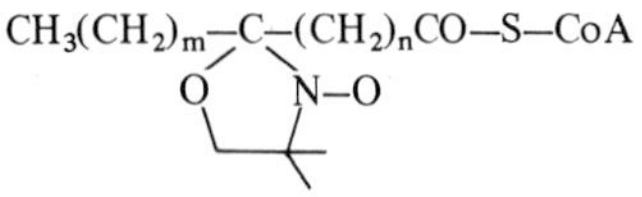

XXV (m, n)

The natural acyl-CoA's have long fatty acid chains and are known to bind in an inhibitory way to the ADP carrier, and thus the spin-labelled analogues should be a good probe for this interaction. The spin-labelled acyl-CoA's were shown to inhibit ADP transport in a strictly competitive manner, with an inhibition constant $K_i \sim 10^{-7}$ M, similar to that of the unlabelled CoA's. Direct evidence for this interaction with the carrier protein was obtained from the ESR spectra. At low concentrations the (10,3)acyl-CoA in pigeon heart mitochondria gave rise to a strongly immobilized spectrum at 20°C, with outer splitting $A_{\parallel}$ = 31.0 G, which is slightly less than the maximum obtainable ($A_{\parallel}$ = 33.0 G at −18°C). In contrast, the corresponding fatty acid probe II (10,3) in the same preparation gave rise to a typical bilayer spectrum with outer splitting $A_{\parallel}$ = 26.8 G. This spectral difference was reversed on adding specific ligands of the ADP carrier: ADP, ATP or carboxyatractyloside. The (10,3)acyl-CoA then gave a spectrum similar to that of the II (10,3) fatty acid, indicating that the acyl-CoA spin label had been displaced from the protein into the fluid lipid bilayer. A similar immobilization was observed with the (7,6)acyl-CoA, but the (5,10) and (1,14)acyl-CoA's gave spectra which were rather similar to those of the II (5,10) and II (1,14) fatty acids respectively. Thus, since the interaction with the protein could not be detected in the ESR spectra of acyl-CoA's with the nitroxide situated closer to the terminal methyl end of the fatty acid chain, it was concluded that the ADP carrier (or the acyl-CoA binding subunit) was not a trans-membrane protein, but extended only part of the way through the lipid bilayer. Further, because the (10,3) and (7,6) acyl-CoA isomers showed less than total immobilization and the (5,10) and (1,14) isomers registered essentially the same fluidity as the lipid bilayer, it was concluded that the carrier protein was situated in a fluid lipid bilayer environment.

These studies were extended using spin-labelled acyl-atractylosides, XXVI (m,n):

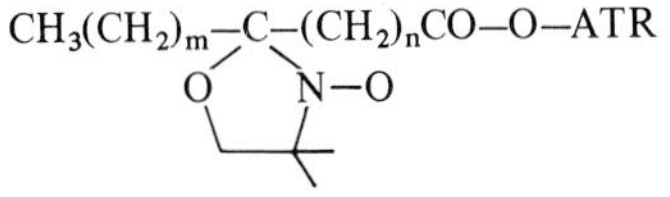

XXVI (m, n)

(Lauquin et al., 1977). In this case, the unlabelled parent atractyloside does not bear an acyl chain. Nevertheless the spin-labelled acyl-ATR's were found to inhibit ADP transport and were partially displaced from the carrier by addition of atractyloside or other specific ligands. The ESR spectra of the (10,3)-, (7,6)- and (7,8)acyl-ATR's were strongly immobilized; the (5,10)acyl-ATR was somewhat less strongly immobilized, and the (1,14)-acyl-ATR was only weakly immobilized, as shown by the values of the outer hyperfine splitting given in Table 10. These results are consistent with those obtained with the

Table 10. Outer hyperfine splitting 2 $A_{\parallel}$ (gauss) for spin-labelled acyl-ATR's and fatty acids II (m,n) in beef heart mitochondria at 25°C (Lauquin et al., 1977)

(m,n)	(m,n)acyl-ATR		II (m,n)
	(Alone)	(+ATR) [a]	
10,3	66.5	61	54
7,6	65	59	51
7,8	63	56.5	–
5,10	60.5	53	41.5
1,14	36	32	32

[a] In the presence of excess unlabelled atractyloside.

(m,n)acyl-CoA's, indicating that the ADP carrier only penetrates part of the way through the membrane. Differences from the acyl-CoA's are observed in the details of the immobilization profile, particularly with regard to the (5,10) isomer, and also in the effects of displacement with atractyloside. These effects can probably be attributed to the different influence of the anchoring ligand "headgroup", which has always to be taken into account with this type of label.

V. New Developments and Future Trends

A. New Spin Labels

(Keana, 1978, 1979)
Further developments of the method can always be expected in the field of spin-label synthesis. Of particular importance in membrane applications should be mentioned the recent introduction of the more stable "proxyl" labels and of the new "minimum perturbation" probes, both of which have been reviewed by Keana (1978, 1979).

The "proxyl" labels (Keana et al., 1976) are intended to replace the oxazolidine labels II-VI (m,n), having the advantage over the latter of a higher chemical stability and a lower polarity, whilst retaining the same spectral characteristics. The oxazolidine ("doxyl") ring is replaced by the pyrrolidine ring in these labels:

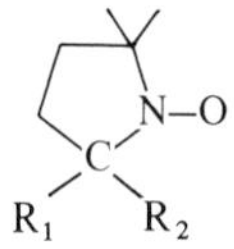

The absence of the oxygen atom decreases the polarity of the ring and reduces its susceptibility to chemical modification or decomposition. Thus the proxyl nitroxides partition more readily into hydrophobic environments and show an increased resistance to loss of the ESR signal by chemical reduction. Such labels should be particularly useful in preparations involving lengthy biochemical manipulations, for instance in reconstitution experiments with labelled membrane proteins.

The "minimum perturbation" probes attempt to integrate the nitroxide ring directly into the molecular structure, hence minimizing the intermolecular perturbation caused by the label. The first labels of this type incorporated the oxazolidine ring directly into the lipid hydrocarbon chain (Williams et al., 1971):

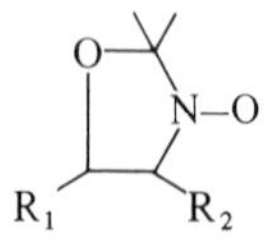

These labels, however, introduce a double-bond-like, steric restriction in the chain, and in addition the nitroxide x-axis, rather than the z-axis, lies along the long molecular axis, giving them a different, and in many ways less useful, motional sensitivity than the oxazolidine labels. More recently introduced are the azethoxyl nitroxides (Lee et al., 1978), which incorporate the pyrrolidine ring into the hydrocarbon chain. In this case the nitroxide group is part of the chain and *cis* or *trans* isomers are possible:

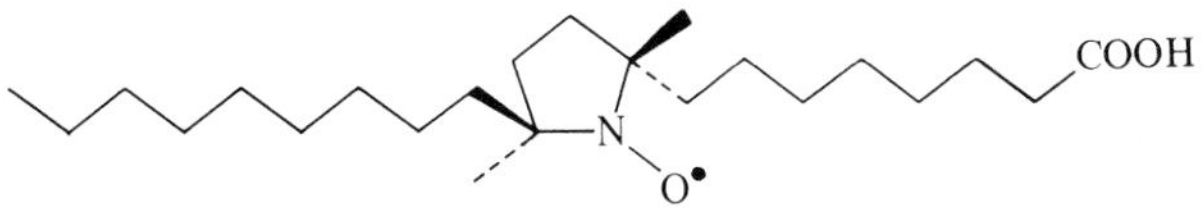

trans-10-azethoxyl eicosanoic acid

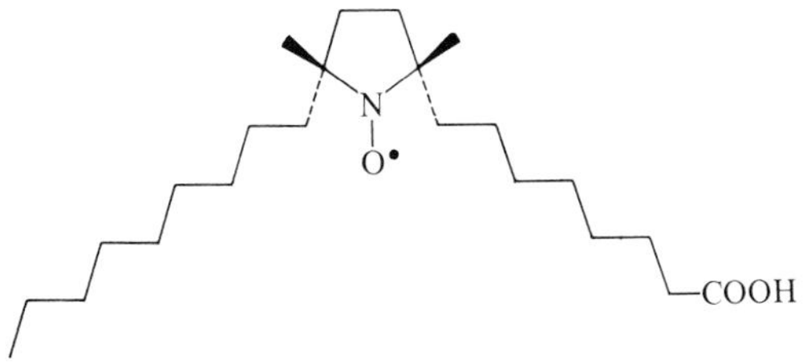

cis-10-azethoxyl eicosanoic acid

These labels also somewhat modify the rotational isomerism in the chains. The *cis* isomers resemble the *cis* double bond in unsaturated lipids and the *trans* isomers resemble more closely the saturated lipids. Again the z-axis of the nitroxide is not directed along the long molecular axis, but makes an angle of 50°-60° with it. In other respects, however, these labels will introduce the minimum perturbation of the lipid membrane and thus be a better reporter of the membrane environment.

B. Saturation Transfer ESR

(Thomas et al., 1976; Hyde and Dalton, 1979)
Undoubtedly the most important recent technical development, and the one which is likely to give rise to significant advances in membrane research in the future, is the technique of saturation transfer ESR. This method extends the motional sensitivity of the spin-label technique down to slower correlation times, almost to within the millisecond time range.

Conventional spin-label ESR which relies on the motional averaging of the spectral anisotropies (a T_2 mechanism) has a motional sensitivity to correlation times in the range 10^{-11}-10^{-7} s (see Sects. III.B and C). Saturation transfer ESR (STESR), which relies on the anisotropy of the saturation characteristics, extends the motional sensitivity to the range 10^{-7}-10^{-3} s. The reason for this is that motional averaging requires significant motion within a time T_2 which specifies the inhomogeneous linewidth (Eq. 36), whereas to affect saturation, significant motion need only take place within a time T_1, the spin-lattice relaxation time. For slow-moving nitroxides $T_2 \simeq 2.4.10^{-8}$ s, as determined from the outer linewidths of the conventional spectrum, and $T_1 \simeq 6.6.10^{-6}$ s (Huisjen and Hyde, 1974). Thus the saturation transfer method is expected to be sensitive to motions $\sim$ 300x slower than is conventional ESR. More precisely (McConnell, 1976), the nitroxide is required to move through an angle $\Delta\theta$ such that the resonance position shifts by one inhomogeneous linewidth:

$$\Delta\theta \,/\, (\pi/2) \simeq (1/\pi T_2) \,/\, (A_{zz} - A_{xx}) \tag{131}$$

For rotational diffusion, the time taken to move through this angle is:

$$t = 3/2 \cdot \tau_c \cdot \overline{(\Delta\theta)^2} \tag{132}$$

where τ_c is the rotational correlation time. For conventional motional averaging: $t \leqslant T_2$, and for the motion to affect saturation: $t \leqslant T_1$. Hence the motionally sensitive ranges of the two methods are:

$$\tau_c(\text{conv}) \leqslant (8/3) \cdot T_2^3 \cdot (A_{zz} - A_{xx})^2 \tag{133}$$

$$\tau_c(\text{ST}) \leqslant (8/3) \cdot T_1 \cdot T_2^2 \cdot (A_{zz} - A_{xx})^2 \tag{134}$$

Hence conventional ESR is sensitive to correlation times down to $\tau_c \sim 2.10^{-7}$ s and STESR down to $\tau_c \sim 10^{-4}$ s.

In practice the saturation transfer spectra are observed in the normal way with high frequency modulation of the magnetic field, but by detecting the signal which is *90°-out-of-phase* with respect to the field modulation, rather than detecting the in-phase signal as is done in conventional ESR spectroscopy. In the absence of saturation this signal should rigorously be zero, but in the presence of saturation the response is no longer linear and a non-zero out-of-phase signal is detected. Angular molecular motion, if it is fast enough, will alleviate saturation by saturation transfer, and hence decrease the 90°-out-of-phase signal. Since the spectral anisotropy is such that rotational motion affects different parts of the spectrum to different extents (cf. Sect. III.A) – in particular the extreme turning points are totally unaffected – the overall shape of the spectrum will be sensitive to the extent of saturation transfer, i.e. to the rate of molecular motion. Empirically it was found that the second harmonic, 90°-out-of-phase, absorption, or the first harmonic, 90°-out-of-phase, dispersion, lineshapes are most sensitive to the rate of motion. Since these are passage experiments, it is clear that the effects will be dependent on the modulation frequency, ω_m. In fact the optimal response is obtained for: $\omega_m^{-1} \sim \tau_c \sim T_1$, and since $T_1 \sim 10^{-5}$ s, it is clear that the method will be optimally sensitive to correlation times $\sim 10^{-5}$ s, using a modulation frequency of approximately 100 kHz.

The phase setting is a crucial part of the STESR experiment, since small mis-settings will crucially affect the size of the out-of-phase signal. The experimental protocol for phase-setting and recording of an STESR spectrum is given in Figure 29. The 90° phase setting is obtained by locating the null in the signal at low power, typically $\leqslant 1$ mW, for which there should be no saturation. The origin of possible instrumental phase inhomogeneities and phase shifts has been discussed in detail by Thomas et al. (1976), but typically out-of-phase nulls of $\leqslant 1\%$ should be obtainable (cf. Fig. 29). Since the sample itself and also sample conditions such as temperature can affect the phase of the modulation, the phase should be set by the "self-null" method indicated in Figure 29, before recording each spectrum. After obtaining the null the spectrum is recorded simply by turning up the microwave power, without changing any of the other instrumental settings. In this way any possible phase shifts introduced by the modulation or detection settings are avoided. A power of ~ 63 mW, corresponding to a microwave magnetic field

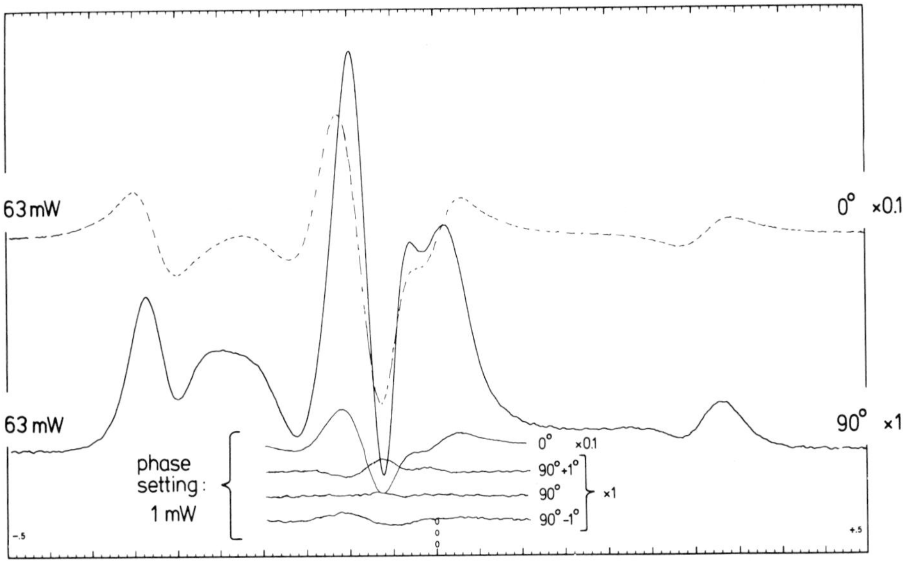

Fig. 29. Second harmonic, 90°-out-of-phase, absorption STESR experiment: setting the phase and recording the spectrum. For details see text. At 63 mW, out-of-phase/in-phase signal ($\propto \omega_m \cdot T_1$) = 14.7%. ω_m = 50 kHz, H_m = 5 G

at the sample of H_1 = 0.25 G, is normally used for recording the spectrum, since the nitroxide is moderately saturated at this power and the ESR signal has approximately its maximum height. A further reason is that most of the theoretical simulations and experimental standard calibrations have been performed at this H_1 microwave field. For similar reasons, a modulation amplitude of 5.0 G is normally taken as standard, and this also produces a good signal strength without undue modulation broadening. At 63 mW the out-of-phase signal is typically ∿ 10% of the in-phase signal (cf. Fig. 29). In fact the out-of-phase/in-phase ratio is directly proportional to $\omega_m T_1$ (Thomas and McConnell, 1974) and this provides a useful experimental parameter with which to check for changes in T_1, since this as well as molecular motion will also give rise to saturation transfer (cf. Marsh, 1980). From simulations Thomas and McConnell (1974) concluded that for ω_m = 50 kHz, an out-of-phase/in-phase ratio of 10% corresponds to a $T_1 \simeq 1$ μs. For further details of the experimental aspects of STESR the reader is referred to the paper by Thomas et al. (1976).

The sensitivity of STESR spectra to slow molecular motion is illustrated in Figure 30, which gives calibration spectra of covalently spin-labelled haemoglobin in solutions of various viscosities. The maleimide spin label XVIIa was found to be rigidly bound to the haemoglobin (cf. Sect. IV.H) and thus the rotational correlation times were calculated from the Debye equation for Brownian motion:

$$\tau_2 = 4\pi\eta R^3/3kT \tag{135}$$

where R = 29 Å for haemoglobin. Clearly both types of saturation transfer spectra in Figure 30 are sensitive to motions in the correlation time range: $\tau \sim 10^{-7}$-10^{-3} s. In

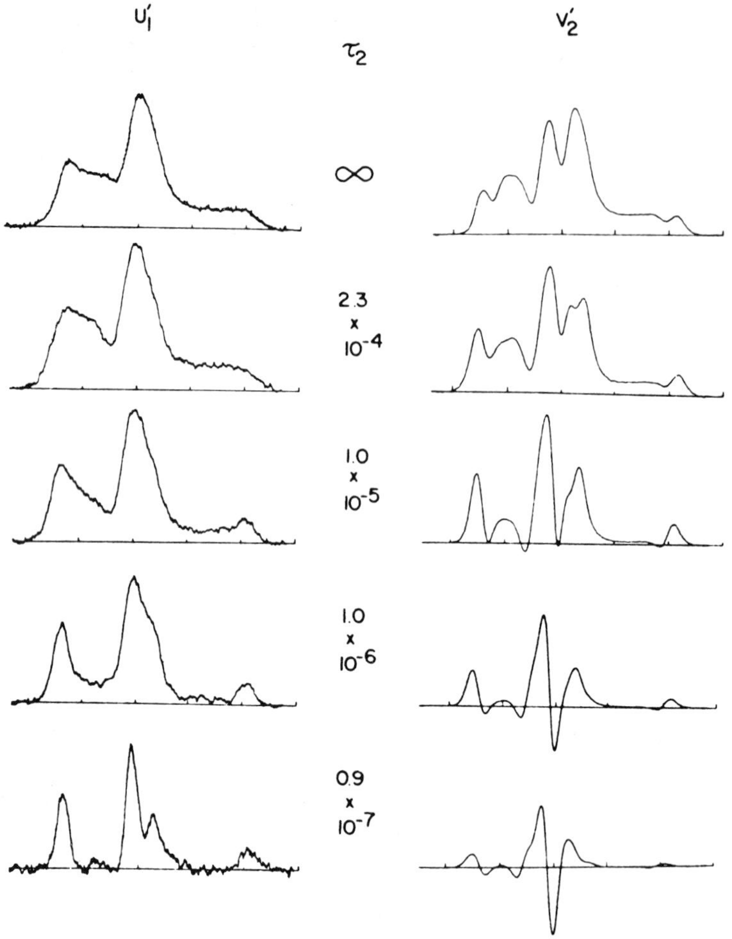

Fig. 30. Saturation transfer ESR spectra from maleimide (XVIIa) spin-labelled haemoglobin. ***Right:*** second harmonic, 90°-out-of-phase absorption (ω_m = 50 kHz); ***left:*** first harmonic, 90°-out-of-phase dispersion (ω_m = 100 kHz). From the top: precipitated haemoglobin; 90% glycerol, −12°C; 80% glycerol, 5°C; 60% glycerol, 5°C; 40% glycerol, 5°C. H_1 = 0.25 G, H_m = 5 G (Thomas et al., 1976)

practice, the second harmonic absorption mode of detection is preferred in membrane applications because of signal-to-noise considerations. In the second harmonic spectra in Figure 30 it is seen that the peaks corresponding to the stationary points at the extreme ends and in the centre of the spectrum remain fairly constant, whereas the regions in between change quite dramatically with motional rate. Thus, by taking the ratio of the lineheight in the rapidly changing region to that at the turning point, a measure of the rate of molecular motion can be obtained. Three independent measurements are possible: in the low-field, high-field and central regions of the spectrum. The lineheight ratio calibration in the low-field region of the haemoglobin reference spectra is given in Figure 31.

The spin-labelled haemoglobin reference spectra refer, of course, to isotropic motion. An important feature in membrane systems is the anisotropy of the motion, as has

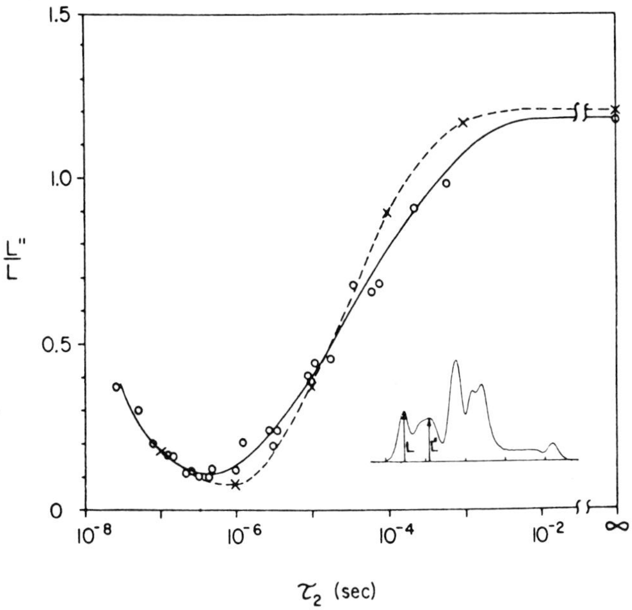

Fig. 31. Calibration curve deduced from the low-field region of the spin-labelled haemoglobin reference spectra (cf. Fig. 30). Circles (*solid curve*) are from the experimental spectra; crosses (*broken curve*) are from computer simulation (Thomas et al., 1976)

been seen above, and some progress has been made with STESR already. Figure 32 gives the STESR spectra of the phosphatidylcholine spin label III (12,3) in the ordered (gel) phase of dipalmitoylphosphatidylcholine (DPPC) and of dimyristoylphosphatidylethanolamine (DMPE) bilayers. Since the transition temperatures of these two lipids are quite close (41°C for DPPC and 48°C for DMPE), it is valid to compare spectra from the two systems at the same temperature in the gel phase. Of particular interest is the sharp change taking place in the central region of the DPPC spectra at around 25°C, which is not seen in the spectra from DMPE. The temperature dependence of the spectra is quantitated in terms of the diagnostic peak height ratios in Figure 33. It is clear that there is an abrupt change in C′/C for DPPC at 25°C, whereas the change in H″/H is only small at this temperature. This change is most probably associated with the phosphatidylcholine pretransition, since it is not observed in DMPE (Fig. 33b), which has no pretransition. (Both DPPC and DMPE show changes in all the peak height ratios at the main transition, of course.) The transition in C′/C at 25°C for DPPC is interpreted as the onset of fast, anisotropic rotation around the long axis of the phospholipid molecule, with relatively little motion of the long axis itself. Rotation about the long molecular axis would modulate the anisotropy of the g-tensor in the x-y plane, leaving all the other tensor values unchanged, and hence only giving rise to saturation transfer in the central part of the spectrum. In contrast, motion of the long axis itself would modulate all the spectral anisotropies, giving rise to saturation transfer throughout the entire spectrum. These conclusions are borne out when the effective rotational correlation times for isotropic motion are calculated from the lineheight ratios, using the haemoglobin reference spectra as in Figure 31 – see Table 11. It is implicit in these calibrations that if the motional rates about all axes are the same, i.e. if the motion is isotropic, then all three peak height

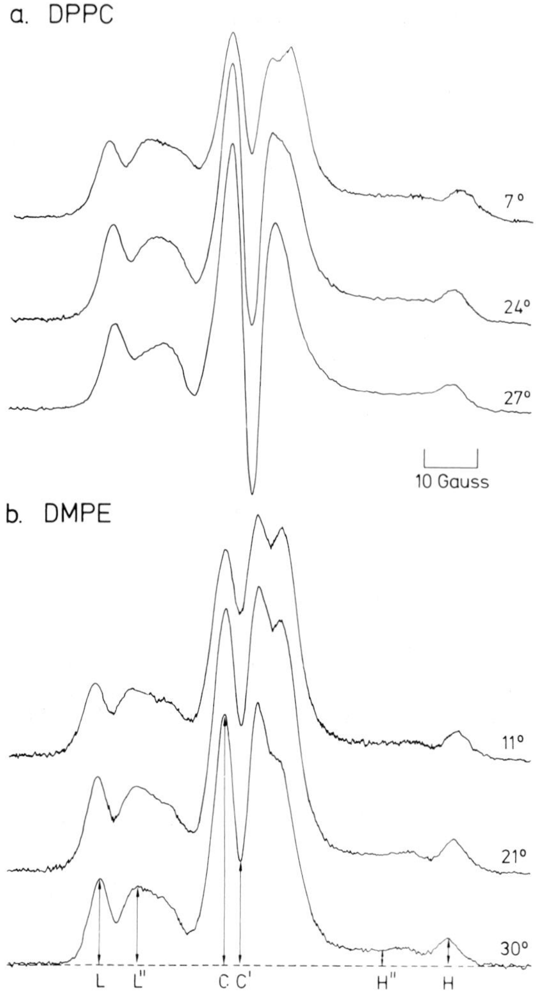

Fig. 32 a and b. Second harmonic, 90°-out-of-phase, absorption STESR spectra of the phosphatidylcholine spin label III (12,3) in aqueous bilayer dispersions of (a) dipalmitoylphosphatidylcholine (DPPC), and (b) dimyristoylphosphatidylethanolamine (DMPE), as a function of temperature (Marsh, 1980)

ratios should give the same correlation time. This is the case for DMPE, all correlation times are $\sim 10^{-4}$ s in the gel phase. For DPPC, however, the onset of a strongly anisotropic motion is detected. Whereas all the peak height ratios give a correlation time of $\sim 10^{-4}$ s below 25°C, and the L''/L and H''/H ratios give correlation times close to this throughout the gel phase, the C'/C ratio undergoes a sharp transition from $\tau \sim 10^{-4}$ s to $\tau \sim 10^{-6}$ s at 25°C. This is interpreted as the onset of long axis rotation in the region of the phosphatidylcholine pretransition.

An extremely interesting and important application of saturation transfer ESR is in the detection of rotational motion of membrane proteins. It is known from transient photodichroism measurements that rhodopsin rotates rapidly about an axis perpendicular to the plane of the rod outer segment disc membrane, with a rotational relaxation time of ~ 20 μs. This lies in the centre of the range of motional sensitivity of saturation transfer ESR. Baroin et al. (1977) have investigated bovine rod outer segment mem-

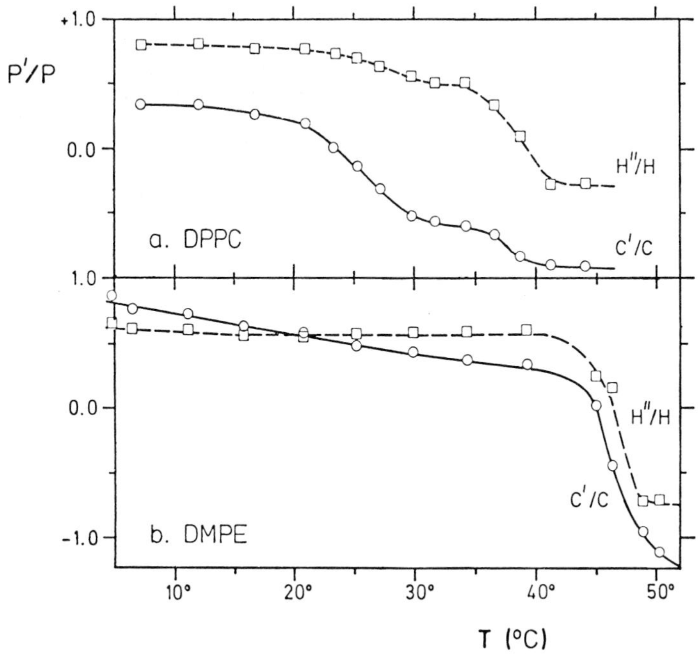

Fig. 33 a and b. Central, C'/C, and high-field, H''/H, peak height ratios (see Fig. 32) in the second harmonic STESR spectra of (a) DPPC, and (b) DMPE, as a function of temperature (Marsh, 1980)

Table 11. Effective rotational correlation times in gel-phase dipalmitoylphosphatidylcholine (DPPC) and dimyristoylphosphatidylethanolamine (DMPE), deduced from calibration of the peak height ratios (see Figs. 31, 32) (Marsh, 1980)

τ_2 (s) [a]	12°C	25°C	30°C	45°C	50°C
DPPC:					
L''/L	6.10^{-4}	1.10^{-4}	$0.8.10^{-4}$	$\sim 10^{-9}$	–
H''/H	4.10^{-4}	2.10^{-4}	1.10^{-4}	$\sim 10^{-9}$	–
C'/C	$0.8.10^{-4}$	5.10^{-6}	$0.9.10^{-6}$	$\sim 10^{-9}$	–
DMPE:					
L''/L	2.10^{-4}	2.10^{-4}	2.10^{-4}	8.10^{-6}	$\sim 10^{-9}$
H''/H	1.10^{-4}	1.10^{-4}	1.10^{-4}	1.10^{-5}	$\sim 10^{-9}$
C'/C	$\leqslant 10^{-3}$	10^{-4}	10^{-4}	6.10^{-6}	$\sim 10^{-9}$

[a] Calibration from haemoglobin reference spectra of Thomas et al. (1976).

branes, covalently labelled with the maleimide spin label XVIIb, by using STESR. The conventional ESR spectra were strongly immobilized with a small contamination by a weakly immobilized component (cf. Sect. IV.H). Motion was detected on the saturation transfer timescale, and this motion was abolished by cross-linking the protein with glutaraldehyde. The cross-linking experiment indicates that the spin label was rigidly attached to the protein, and that the motion detected by STESR was the rotation of the protein as a whole. Effective correlation times for isotropic motion, deduced from the lineheight ratios H″/H and L″/L, are given in Table 12. The two lineheight ratios give somewhat different correlation times, as expected for anisotropic motion in view of the discussion of the lipid example above. Nevertheless, values of $\tau_2 \sim 20$ μs were obtained at 20°C, which correlates with the optical measurements of the rotation of the protein about an axis perpendicular to the plane of the membrane.

Table 12. Rotational correlation times of spin-labelled rhodopsin in bovine rod outer segment disc membranes. Deduced from calibration of the STESR peak height ratios (Baroin et al., 1977)

τ_2 (μs) [a]	−4°C	+4°C	+20°C	+37°C	5% Gluta. +4°C
H″/H	80-100	50-80	20	7-10	⩾ 1000
L″/L	60	30	15-20	15-20	⩾ 1000

[a] Calibration from isotropic haemoglobin reference spectra of Thomas et al. (1976). The uncertainties in τ_2 are due to: (a) the difficulty in measuring L″ because of overlap with the weakly immobilized component; (b) the difficulty in measuring H″ because of its small value; (c) the discrepancy between theoretical and experimental values of the calibration spectra.

Similar experiments by Kusumi et al. (1978) on frog rod outer segment disc membranes yielded effective correlation times, at 20°C, of 20, 3.6 and 13 μs from the peak height ratios in the low-field, central and high-field parts of the spectrum respectively. These correlation times were only increased to 45, 9.6 and 45 μs respectively, in this case, on glutaraldehyde fixation. Again the results are indicative of a relatively rapid anisotropic motion of the protein, with possibly some motion of the label relative to the protein in this case. A small, but definite change was observed on bleaching the rhodopsin.

Two sets of measurements of the rotational motion of the Ca^{2+}-ATPase in sarcoplasmic reticulum membranes have been made by saturation transfer ESR (Kirino et al., 1978; Thomas and Hidalgo, 1978). In this case the protein rotation has been first detected by STESR before any other method. Thomas and Hidalgo (1978) demonstrated that the covalent maleimide label XVIIa could be rigidly attached to the Ca^{2+}-ATPase, and measured an effective correlation time of 60 μs for the protein rotation at 4°C, using the low-field lineheight ratio. Kirino et al. (1978) measured effective rotational correlation times of 800 μs at 4°C and 200 μs at 20°C, using the same label. The difference between the two results might be attributed to different labelling conditions, in particular the contaminating effect of the weakly immobilized component. The latter was practically absent in the spectra of Thomas and Hidalgo (1978), whereas Kirino et al. (1978) corrected for this peak by extrapolating the STESR lineheight ratios to zero value of W/S, the ratio of the weakly to the strongly immobilized component. In this latter way, similar correlation times were obtained from both the low-field and high-field lineheight ratios.

Rousselet and Devaux (1977) have applied STESR to the study of the spin-labelled acetylcholine receptor in membrane fragments from the electroplax of *Torpedo marmorata*. They were unable to detect any motion of the receptor protein on the saturation transfer ESR timescale, concluding that the receptor was immobilized in the electroplax membrane, with any rotational motion having a correlation time of longer than 1 ms.

These studies all illustrate the considerable potential of STESR in studying the rotational motion of proteins in membranes. Clearly more work is required to clarify the anisotropic nature of these motions: both on the theoretical side in spectral simulations, and on the experimental side in measurements on oriented membrane preparations.

References

Alger, R.S.: Electron Paramagnetic Resonance: Techniques and Applications. New York: Interscience 1968.

Aneja, R., Davies, A.P.: The synthesis of a spin-labelled glycero-phospholipid. Chem. Phys. Lipids *4*, 60-71 (1970).

Baldassare, J.J., Robertson, D.E., McAfee, A.G., Ho, C.: A spin-label study of energy-coupled active transport in Escherichia coli membrane vesicles. Biochemistry *13*, 5210-5214 (1974).

Baroin, A., Thomas, D.D., Osborne, B., Devaux, P.F.: Saturation transfer paramagnetic resonance on membrane-bound proteins. I. Rotational diffusion of rhodopsin in the visual receptor membrane. Biochem. Biophys. Res. Commun. *78*, 442-447 (1977).

Berliner, L.J. (ed.): Spin-Labelling. Theory and Applications. New York: Academic Press Vol. I, 1976; Vol. II, 1979.

Birrell, G.B., Griffith, O.H.: Cytochrome *c* induced lateral phase separation in a diphosphatidylglycerol-steroid spin label model membrane. Biochemistry *15*, 2925-2929 (1976).

Bolton, J.R., Borg, D.C., Swartz, H.M.: Experimental aspects of biological electron spin resonance studies. In: Biological Applications of Electron Spin Resonance (eds. Swartz, H.M., Bolton, J.R., Borg, D.C.) pp. 63-118. New York-London: Wiley 1972.

Cannon, B., Polnaszek, C.F., Butler, K.W., Eriksson, L.G., Smith, I.C.P.: The fluidity and organization of mitochondrial membrane lipids of the brown adipose tissue of cold-adapted rats and hamsters as determined by nitroxide spin probes. Arch. Biochem. Biophys. *167*, 505-518 (1975).

Capiomont, A., Chion, B., Lajzerowicz, J., Lemaire, H.: Interpretation and utilization for crystal structure determination of nitroxide free radicals in single crystals. J. Chem. Phys. *60*, 2530-2535 (1974).

Coan, C.R., Inesi, G.: Ca^{2+}-dependent effect of ATP on spin-labelled sarcoplasmic reticulum. J. Biol. Chem. *252*, 3044-3049 (1977).

Comfurius, P., Zwaal, R.F.A.: The enzymatic synthesis of phosphatidylserine and purification by CM-cellulose column chromatography. Biochim. Biophys. Acta *488*, 36-42 (1977).

Devaux, P., McConnell, H.M.: Equality of the rates of lateral diffusion of phosphatidylethanolamine and phosphatidylcholine spin labels in rabbit sarcoplasmic reticulum. Ann. N.Y. Acad. Sci. *222*, 489-496 (1974).

Devaux, P., Scandella, C.J., McConnell, H.M.: Spin-spin interactions between spin-labelled phospholipids incorporated into membranes. J. Mag. Res. *9*, 474-485 (1973).

Devaux, P.F., Bienvenüe, A., Lauquin, G., Brisson, A.D., Vignais, P.M., Vignais, P.V.: Interaction between spin-labelled acyl-coenzyme A and the mitochondrial adenosine diphosphate carrier. Biochemistry *14*, 1272-1280 (1975).

Dwek, R.A., Jones, R., Marsh, D., McLaughlin, A.C., Press, E.M., Price, N.C., White, A.I.: Antibody-hapten interactions in solution. Phil. Trans. R. Soc. Lond. Ser. B *272*, 53-74 (1975).

Feher, G.: Electron Paramagnetic Resonance with Application to Selected Problems in Biology. New York: Gordon and Breach 1970.

Freed, J.H.: Theory of slow tumbling ESR spectra for nitroxides. In: Spin-Labelling. Theory and Applications (ed. Berliner, L.J.) Vol. I, pp. 53-132. New York: Academic Press 1976.

Fretten, P., Morris, S.J., Watts, A., Marsh, D.: Lipid-Lipid and lipid-protein interactions in chromaffin granule membranes. A spin label ESR study. Biochim. Biophys. Acta *598*, 247-259 (1980).

Gaffney, B.J.: Practical considerations for the calculation of order parameters for fatty acid or phospholipid spin labels in membranes. In: Spin Labelling. Theory and Applications (ed. Berliner, L.J.) Vol. I, pp. 567-571. New York: Academic Press 1976.

Gaffney, B.J., McConnell, H.M.: The paramagnetic resonance spectra of spin labels in phospholipid membranes. J. Magn. Reson. *16*, 1-28 (1974).

Gaffney, B.G., McNamee, C.M.: Spin label measurements in membranes. Methods Enzymol. *32 B*, 161-198 (1974).

Gagua, A.V., Malenkov, G.G., Timofeev, V.P.: Hydrogen bond contribution to the isotropic hyperfine splitting constant of a nitroxide free radical. Chem. Phys. Lett. *56*, 470-473 (1978).

Galla, H.-J., Sackmann, E.: Chemically induced lipid phase separation in model membranes containing charged lipids: A spin label study. Biochim. Biophys. Acta *401*, 509-529 (1975).

Giotta, G.J., Wang, H.H.: Reduction of nitroxide free radicals by biological materials. Biochem. Biophys. Res. Commun. *46*, 1576-1580 (1972).

Griffith, O.H., Jost, P.C.: Lipid spin labels in biological membranes. In: Spin Labelling. Theory and Applications (ed. Berliner, L.J.) Vol. I, pp. 453-523. New York: Academic Press 1976.

Griffith, O.H., Jost, P.C.: The lipid-protein interface in cytochrome oxidase. In: Proceedings of the Japanese-American Seminar on Cytochrome Oxidase. (eds. Chance, B., King, T.E., Okunuki, K., Orii, Y.). Amsterdam: Elsevier 1978.

Griffith, O.H., Cornell, D.W., McConnell, H.M.: Nitrogen hyperfine tensor and g-tensor of nitroxide radicals. J. Chem. Phys. *43*, 2909-2910 (1965).

Griffith, O.H., Keana, J.F.W., Noall, D.L., Ivey, J.L.: Nitroxide mixed carboxylic-carbonic anhydrides. A new class of versatile spin labels. Biochim. Biophys. Acta *148*, 583-585 (1967).

Griffith, O.H., Dehlinger, P.J., Van, S.P.: Shape of the hydrophobic barrier of phospholipid bilayers. Evidence for water penetration in biological membranes. J. Membr. Biol. *15*, 159-192 (1974).

Hamilton, C.L., McConnell, H.M.: Spin labels. In: Structural Chemistry and Molecular Biology (eds. Rich, A., Davidson, N.) pp. 115-149. San Francisco-London: Freeman 1968.

Hemminga, M.A.: Angular dependent linewidths of ESR spin probes in oriented smectic systems. An Application to oriented lecithin-cholesterol multibilayers. Chem. Phys. *6*, 87-99 (1974).

Hemminga, M.A.: An ESR study of the mobility of the cholestane spin label in oriented lecithin-cholesterol multibilayers. Chem. Phys. Lipids *14*, 141-150 (1975).

Hemminga, M.A., Berendsen, H.J.C.: Magnetic resonance of ordered lecithin-cholesterol multibilayers. J. Magn. Reson. *8*, 133-143 (1972).

Hidalgo, C., Thomas, D.D.: Heterogeneity of SH groups in sarcoplasmic reticulum. Biochem. Biophys. Res. Commun. *78*, 1175-1182 (1977).

Hoffman, B.M., Schofield, P., Rich, A.: Spin-labelled transfer RNA. Proc. Natl. Acad. Sci. USA *62*, 1195-1202 (1969).

Hoffman, B.M., Schofield, P., Rich, A.: Spin labelling studies of aminoacyl transfer ribonucleic acid. Biochemistry *9*, 2525-2532 (1970).

Hsia, J.C., Piette, L.H., Noyes, R.W.: Binding of spermatozoa with surface active spin labels. J. Reprod. Fertil. *20*, 147-149 (1969).

Hubbell, W.L., McConnell, H.M.: Motion of steroid spin labels in membranes. Proc. Natl. Acad. Sci. USA *63*, 18-22 (1969).

Hubbell, W.L., McConnell, H.M.: Molecular motion in spin-labeled phospholipids and membranes. J. Am. Chem. Soc. *93*, 314-326 (1971).

Hubbell, W.L., Metcalfe, J.C., Metcalfe, S.M., McConnell, H.M.: The interaction of small molecules with spin-labelled erythrocyte membranes. Biochim. Biophys. Acta *219*, 415-427 (1970).

Huisjen, M., Hyde, J.S.: A pulsed EPR spectrometer. Rev. Sci. Instrum. *45*, 669-675 (1974).

Hyde, J.S., Dalton, L.A.: Saturation transfer spectroscopy. In: Spin Labelling. Theory and Applications (ed. Berliner, L.J.), Vol. II, pp. 1-70. New York: Academic Press 1979.

Ito, T., Ohnishi, S.-I.: Ca^{2+}-Induced lateral phase separations in phosphatidic acid-phosphatidylcholine membranes. Biochim. Biophys. Acta *352*, 29-37 (1974).

Ito, T., Ohnishi, S., Ishinaga, M., Kito, M.: Synthesis of a new phosphatidylserine spin-label and calcium-induced lateral phase separation in phosphatidylserine-phosphatidylcholine membranes. Biochemistry *14*, 3064-3069 (1975).

Jost, P.C., Griffith, O.H.: Electron spin resonance and the spin labelling method. In: Methods in Pharmacology (ed. Chignell, C.), Vol. II, pp. 223-276. New York: Appleton-Century-Crofts 1972.

Jost, P.C., Griffith, O.H.: The spin-labelling technique. In: Methods in Enzymology, Vol. XLIX (eds. Hirs, C.G.W., Timasheff, S.N.). New York: Academic Press 1978.

Jost, P.C., Libertini, L.J., Hebert, V.C., Griffith, O.H.: Lipid spin labels in lecithin multilayers. A study of motion along fatty acid chains. J. Mol. Biol. *59*, 77-98 (1971a).

Jost, P.C., Waggoner, A.S., Griffith, O.H.: Spin labelling and membrane structure. In: Structure and Function of Biological Membranes (ed. Rothfield, L.), pp. 83-144. New York: Academic Press 1971b.

Jost, P.C., Griffith, O.H., Capaldi, R.A., Vanderkooi, G.: Evidence for boundary lipid in membranes. Proc. Natl. Acad. Sci. USA *70*, 480-484 (1973a).

Jost, P.C., Griffith, O.H., Capaldi, R.A., Vanderkooi, G.: Identification and extent of the fluid bilayer regions in membranous cytochrome oxidase. Biochim. Biophys. Acta *311*, 141-152 (1973b)

Keana, J.F.W.: New aspects of nitroxide chemistry. In: Spin Labelling. Theory and Applications (ed. Berliner, L.J.), Vol. II, pp. 115-172. New York: Academic Press 1979.

Jones, R., Dwek, R.A., Walker, I.O.: Conformational states of rabbit muscle phosphofructokinase investigated by a spin label probe. FEBS Lett. *26*, 92-96 (1972).

Kaplan, J., Canonico, P.G., Caspary, W.J.: Electron spin resonance studies of spin-labeled mammalian cells by detection of surface membrane signals. Proc. Natl. Acad. Sci. USA *70*, 66-70 (1973).

Keana, J.F.W.: Newer aspects of synthesis and chemistry of nitroxide spin labels. Chem. Rev. *78*, 37-64 (1978).

Keana, J.F.W., Keana, J.B., Beetham, D.: A new versatile ketone spin label. J. Am. Chem. Soc. *89*, 3055-3056 (1967).

Keana, J.F.W., Lee, T.D., Bernard, E.M.: Side-chain substituted 2,2,5,5-Tetramethyl pyrrolidine-N-oxyl (Proxyl) nitroxides. A new series of lipid spin labels showing improved properties for the study of biological membranes. J. Am. Chem. Soc. *98*, 3052-3053 (1976).

Keith, A.D., Sharnoff, M., Cohn, G.E.: A summary and evaluation of spin labels used as probes for biological membrane structure. Biochim. Biophys. Acta *300*, 379-419 (1973).

Kirino, Y., Ohkuma, T., Shimizu, H.: Saturation transfer electron spin resonance study on the rotational diffusion of calcium- and magnesium-dependent adenosine triphosphatase in sarcoplasmic reticulum membranes. J. Biochem. *84*, 111-115 (1978).

Klopfenstein, C., Jost, P., Griffith, O.H.: The dedicated computer in electron spin resonance. In: Computers in Chemical and Biochemical Research (eds. Klopfenstein, C., Wilkins, C.), Vol I, pp. 175-221. New York: Academic Press 1972.

Knowles, P.F., Marsh, D., Rattle, H.W.E.: Magnetic Resonance of Biomolecules. London-New York: Wiley 1976.

Knowles, P.F., Watts, A., Marsh, D.: Spin label studies of lipid immobilization in dimyristoylphosphatidylcholine-substituted cytochrome oxidase. Biochemistry *18*, 4480-4487 (1979).

Kornberg, R.D., McConnell, H.M.: Inside-outside transitions of phospholipids in vesicle membranes. Biochemistry *10*, 1111-1120 (1971).

Kornberg, R.D., McNamee, M.G., McConnell, H.M.: Measurement of transmembrane potentials in phospholipid vesicles. Proc. Natl. Acad. Sci. USA *69*, 1508-1513 (1972).

Kusumi, A., Ohnishi, S., Ito, T., Yoshizawa, T.: Rotational motion of rhodopsin in the visual receptor membrane as studied by saturation transfer spectroscopy. Biochim. Biophys. Acta *507*, 539-543 (1978).

Lauquin, G.J.M., Devaux, P.F., Bienvenüe, A., Villiers, C., Vignais, P.V.: Spin-labelled acyl atractyloside as a probe of the mitochondrial adenosine diphosphate carrier. Asymmetry of the carrier and direct lipid environment. Biochemistry *16*, 1202-1208 (1977).

Lee, T.D., Birrell, G.B., Keana, J.F.W.: A new series of minimum steric perturbation nitroxide spin labels. J. Am. Chem. Soc. *100*, 1618-1619 (1978).

Libertini, L.J., Griffith, O.H.: Orientation dependence of the electron spin resonance spectrum of di-t-butyl nitroxide. J. Chem. Phys. *53*, 1359-1367 (1970).

Libertini, L.J., Burke, C.A., Jost, P.C., Griffith, O.H.: An orientation distribution model for interpreting ESR line shapes of ordered spin labels. J. Magn. Reson. *15*, 460-476 (1974).

Maier, W., Saupe, A.: Eine einfache molekular-statistische Theorie der nematischen kristallinflüssigen Phase. Teil I. Z. Naturforsch. Teil A *14*, 882-889 (1959).

Marčelja, S.: Chain ordering in liquid crystals. II. Structure of bilayer membranes. Biochim, Biophys. Acta *367*, 165-176 (1974).

Marsh, D.: Statistical mechanics of the fluidity of phospholipid bilayers and membranes. J. Membr. Biol. *18*, 145-162 (1974).
Marsh, D.: Spectroscopic studies of membrane structure. In: Essays in Biochemistry (eds. Campbell, P.N., Aldridge, W.N.), Vol. 11, pp. 139-180. London: Academic Press 1975.
Marsh, D.: Molecular motion in phospholipid bilayers in the gel phase: Long axis rotation. Biochemistry *19*, 1632-1637 (1980).
Marsh, D., Smith, I.C.P.: An interacting spin label study of the fluidizing and condensing effects of cholesterol on lecithin bilayers. Biochim. Biophys. Acta *298*, 133-144 (1973)
Marsh, D., Radda, G.K., Ritchie, G.A.: A spin-label study of the chromaffin granule membrane. Eur. J. Biochem. *71*, 53-61 (1976a).
Marsh, D., Watts, A., Knowles, P.F.: Evidence for phase boundary lipid. Permeability of tempo-choline into dimyristoylphosphatidylcholine vesicles at the phase transition. Biochemistry *15*, 3570-3578 (1976b).
Mason, R., Freed, J.H.: Estimating microsecond rotational correlation times from lifetime broadening of nitroxide ESR spectra near the rigid limit. J. Phys. Chem. *78*, 1321-1323 (1974).
Mateu, L., Caron, F., Luzzati, V., Billecocq, A.: The influence of protein-lipid interactions on the order-disorder conformational transitions of the hydrocarbon chain. Biochim. Biophys. Acta *508*, 109-121 (1978).
McCalley, R.C., Shimshick, E.J., McConnell, H.M.: The effect of slow rotation on magnetic resonance spectra. Chem. Phys. Lett. *13*, 115-119 (1972).
McConnell, H.M.: Molecular motion in biological membranes. In: Spin Labelling. Theory and Applications (ed. Berliner, L.J.), Vol. I, pp. 525-560. New York: Academic Press 1976.
McConnell, H.M., McFarland, B.G.: Physics and chemistry of spin labels. Q. Rev. Biophys. *3*, 91-136 (1970).
McConnell, H.M., Wright, K.L., McFarland, B.G.: The fraction of the lipid in a biological membrane that is in a fluid state: A spin label assay. Biochem. Biophys. Res. Commun. *47*, 273-281 (1972).
McNamee, M.G., McConnell, H.M.: Transmembrane potentials and phospholipid flip-flop in excitable membrane vesicles. Biochemistry *12*, 2951-2958 (1973).
Melhorn, R.J., Keith, A.D.: Spin-labelling biological membranes. In: Membrane Molecular Biology (eds. Fox, C.F., Keith, A.D.), pp. 192-225. Stanford: Sinauer 1972.
Metcalfe, J.C., Birdsall, N.J.M., Lee, A.G.: ^{13}C NMR spectra of acholeplasma membranes containing ^{13}C-labelled phospholipids. FEBS Lett. *21*, 335-340 (1972).
Morrisett, J.D., Drott, H.R.: Oxidation of the sulfhydryl and disulfide groups by the nitroxyl radical. J. Biol. Chem. *244*, 5083-5087 (1969).
Nakamura, M., Ohnishi, S.: Spin-labeled yeast cells. Biochem. Biophys. Res. Commun. *46*, 926-932 (1972).
Onsager, L.: Electric moments of molecules in liquids. J. Am. Chem. Soc. *58*, 1486-1493 (1936).
Pang, D.C., Briggs, F.N.: Analysis of the ATP-induced conformational changes in sarcoplasmic reticulum. Arch. Biochem. Biophys. *164*, 332-340 (1974).
Polnaszek, C.F.: 6th International Symposium on Magnetic Resonance, Banff, Abstr, p. 282 (1977).
Polnaszek, C.F.: The analysis and interpretation of electron spin resonance spectra of nitroxide spin probes in terms of correlation times for rotational reorientation and of order parameters. Unpublished preprint (1979).
Polnaszek, C.F., Schreier, S., Butler, K.W., Smith, I.C.P.: Analysis of the factors determining the ESR spectra of spin probes that partition between aqueous and lipid phases. J. Am. Chem. Soc. *100*, 8223-8232 (1978).
Poole, C.P.: Electron Spin Resonance. A Comprehensive Treatise on Experimental Techniques. New York-London: Wiley 1967.
Randolph, M.R.: Quantitative considerations in electron spin resonance studies of biological materials. In: Biological Applications of Electron Spin Resonance (eds. Swartz, H.M., Bolton, J.R., Borg, D.C.), pp. 119-153. New York-London: Wiley 1972.
Rousselet, A., Devaux, P.F.: Saturation transfer electron paramagnetic resonance on membrane-bound proteins. II. Absence of rotational diffusion of the cholinergic receptor protein in *Torpedo marmorata* membrane fragments. Biochem. Biophys. Res. Commun. *78*, 448-454 (1977).

Rozantzev, E.G.: Free Nitroxyl Radicals. New York-London: Plenum 1970.

Rozantzev, E.G., Neiman, M.B.: Organic radical reactions involving no free valence. Tetrahedron *20*, 131-137 (1964).

Sackmann, E., Träuble, H.: Studies of the crystalline-liquid crystalline phase transition of lipid model membranes. II. Analysis of electron spin resonance spectra of steroid labels incorporated into lipid membranes. J. Am. Chem. Soc. *94*, 4492-4498 (1972).

Scandella, C.J., Devaux, P., McConnell, H.M.: Rapid lateral diffusion of phospholipids in rabbit sarcoplasmic reticulum. Proc. Natl. Acad. Sci. USA *69*, 2056-2060 (1972).

Scandella, C.J., Schindler, H., Franklin, R.M., Seelig, J.: Structure and synthesis of a lipid-containing bacteriophage. Acyl-chain motion in the PM2 virus membrane. Eur. J. Biochem. *50*, 29-32 (1974).

Schindler, H., Seelig, J.: EPR spectra of spin labels in lipid bilayers. J. Chem. Phys. *59*, 1841-1850 (1973).

Schindler, H., Seelig, J.: EPR spectra of spin labels in lipid bilayers. II. Rotation of steroid spin probes. J. Chem. Phys. *61*, 2946-2949 (1974).

Schindler, H., Seelig, J.: Deuterium order parameters in relation to thermodynamic properties of a phospholipid bilayer. A statistical mechanical interpretation. Biochemistry *14*, 2283-2287 (1975).

Schreier, S., Polnaszek, C.F., Smith, I.C.P.: Spin labels in membranes: Problems in practice. Biochim. Biophys. Acta *515*, 375-436 (1978).

Schreier-Muccillo, S., Marsh, D., Dugas, H., Schneider, H., Smith, I.C.P.: A spin probe study of the influence of cholesterol on motion and orientation of phospholipids in oriented multibilayers and vesicles. Chem. Phys. Lipids *10*, 11-27 (1973).

Schreier-Muccillo, S., Marsh, D., Smith, I.C.P.: Monitoring the permeability profile of lipid membranes with spin probes. Arch. Biochem. Biophys. *172*, 1-11 (1976).

Seelig, J.: Spin label studies of oriented liquid crystals (a model system for bilayer membranes). J. Am. Chem. Soc. *92*, 3881-3887 (1970).

Seelig, J.: On the flexibility of hydrocarbon chains in lipid bilayers. J. Am. Chem. Soc. *93*, 5017-5022 (1971).

Seelig, J.: Anisotropic motion in liquid crystalline structures. In: Spin Labelling. Theory and Applications (ed. Berliner, L.J.), Vol. I, pp. 373-409. New York: Academic Press 1976.

Seelig, J., Hasselbach, W.: A spin label study of sarcoplasmic vesicles. Eur. J. Biochem. *21*, 17-21 (1971).

Seelig, A., Seelig, J.: The dynamic structure of fatty acyl chains in a phospholipid bilayer measured by deuterium magnetic resonance. Biochemistry *13*, 4839-4845 (1974).

Seelig, J., Limacher, H., Bader, P.: Molecular architecture of liquid crystalline bilayers. J. Am. Chem. Soc. *94*, 6364-6371 (1972).

Seelig, J., Axel, F., Limacher, H.: Molecular architecture of bilayer membranes. Ann. N.Y. Acad. Sci. *222*, 588-596 (1973).

Shimshick, E.J., McConnell, H.M.: Lateral phase separations in phospholipid membranes. Biochemistry *12*, 2351-2360 (1973).

Smith, I.C.P.: The spin label method. In: Biological Applications of Electron Spin Resonance (eds. Swartz, H.M., Bolton, J.R., Borg, D.C.), pp. 483-539. New York-London: Wiley 1972.

Smith, I.C.P., Schreier-Muccillo, S., Marsh, D.: Spin labelling. In: Free Radicals in Biology (ed. Pryor, W.A.), Vol. I, pp. 149-197. New York: Academic Press 1976.

Stier, A., Sackmann, E.: Spin labels as enzyme substrates. Heterogeneous lipid distribution in liver microsomal membranes. Biochim. Biophys. Acta *311*, 400-408 (1973).

Stone, T.J., Buckman, T., Nordio, P.L., McConnell, H.M.: Spin-labeled biomolecules. Proc. Natl. Acad. Sci. USA *54*, 1010-1017 (1965).

Struve, W.G., McConnell, H.M.: A new spin-labelled substrate for β-galactosidase and β-galactoside permease. Biochem. Biophys. Res. Commun. *49*, 1631-1637 (1972).

Sutton, B.J., Gettins, P., Givol, D., Marsh, D., Wain-Hobson, S., Willan, K.J., Dwek, R.A.: The gross architecture of an antibody combining site as determined by spin-label mapping. Biochem. J. *165*, 177-197 (1977).

Tanaka, K.-I., Ohnishi, S.-I.: Heterogeneity in the fluidity of intact erythrocyte membrane and its homogenization upon haemolysis. Biochim. Biophys. Acta *426*, 218-231 (1976).

Thomas, D.D., Hidalgo, C.: Rotational motion of the sarcoplasmic reticulum Ca^{2+}-ATPase. Proc. Natl. Acad. Sci. USA *75*, 5488-5492 (1978).

Thomas, D.D., McConnell, H.M.: Calculation of paramagnetic resonance spectra sensitive to very slow rotational motion. Chem. Phys. Lett. *25*, 470-475 (1974).

Thomas, D.D., Dalton, L.R., Hyde, J.S.: Rotational diffusion studied by passage saturation transfer electron paramagnetic resonance. J. Chem. Phys. *65*, 3006-3024 (1976).

Tonomura, Y., Morales, M.F.: Change in state of spin labels bound to sarcoplasmic reticulum with change in enzymic state, as deduced from ascorbate-quenching studies. Proc. Natl. Acad. Sci. USA *71*, 3687-3691 (1974).

Träuble, H., Sackmann, E.: Studies of the crystalline-liquid crystalline phase transition of lipid model membranes. III. Structure of a steroid-lecithin system below and above the lipid-phase transition. J. Am. Chem. Soc. *94*, 4499-4510 (1972).

Van, S.P., Griffith, O.H.: Bilayer structure in phospholipid-cytochrome *c* model membranes. J. Membr. Biol. *20*, 155-170 (1975).

Van, S.P., Birrell, G.B., Griffith, O.H.: Rapid anisotropic motions of spin labels. Models for averaging of the ESR parameters. J. Magn. Reson. *15*, 444-459 (1974).

Van Vleck, J.H.: On the anisotropy of cubic ferromagnetic crystals. Phys. Rev. *52*, 1178-1198 (1937).

Waggoner, A.S., Keith, A.D., Griffith, O.H.: Electron spin resonance of solubilized long-chain nitroxides. J. Phys. Chem. *72*, 4129-4132 (1968).

Waggoner, A.S., Kingzett, T.J., Rottshaeffer, S., Griffith, O.H.: A spin-labelled lipid for probing biological membranes. Chem. Phys. Lipids *3*, 245-253 (1969).

Watts, A., Harlos, K., Maschke, W., Marsh, D.: Control of the structure and fluidity of phosphatidylglycerol bilayers by pH titration. Biochim. Biophys. Acta *510*, 63-74 (1978).

Watts, A., Volotovski, I.D., Marsh, D.: Rhodopsin-lipid associations in bovine rod outer segment membranes. Identification of immobilized lipid by spin labels. Biochemistry *18*, 5006-5013 (1979).

Weiland, G., Georgia, B., Wee, V.T., Chignell, C.F., Taylor, P.: Ligand interactions with cholinergic receptor-enriched membranes from Torpedo: Influence of agonist exposure on receptor properties. Mol. Pharmacol. *12*, 1091-1105 (1976).

Weiner, H.: Interaction of a spin-labeled analog of nicotinamide-adenine dinucleotide with alcohol dehydrogenase. I. Synthesis, kinetics and electron paramagnetic resonance studies. Biochemistry *8*, 526-533 (1969).

Williams, J.C., Melhorn, R., Keith, A.D.: Syntheses and novel uses of nitroxide motion probes. Chem. Phys. Lipids *7*, 207-230 (1971).

Wilmshurst, T.H.: Electron Spin Resonance Spectrometers. London: Hilger 1967.

Absorption and Circular Dichroism Spectroscopies

M.M. Long and D.W. Urry

I. Introduction

The obtaining of absorption and circular dichroism spectra of a suspension of biological membranes is a relatively simple and quick procedure. Extracting correct and meaningful information from these spectra is not so simple and quick. In this chapter an effort will be made stepwise to present the problem of the application to the study of biomembranes of the physical methods of ultraviolet absorption and circular dichroism spectroscopies. This will be approached by first considering solutions, the phenomena that occur and type of information that is available in solution. Next will be considered suspensions of homogeneous particles and how to improve their distorted spectra, and finally the added complication of suspensions of heterogeneous particles (biomembranes) will be noted by presenting data on the three major sources of membrane absorption (lipids, protein side chains, and protein backbone) in several biomembranes and carrying out approximate corrections on experimental ellipticities.

II. Solution Studies

A. Absorption

Light is described as an oscillating electromagnetic vector of amplitude E_0 with electric and magnetic components arranged at right angles, as depicted in Figure 1. The expres-

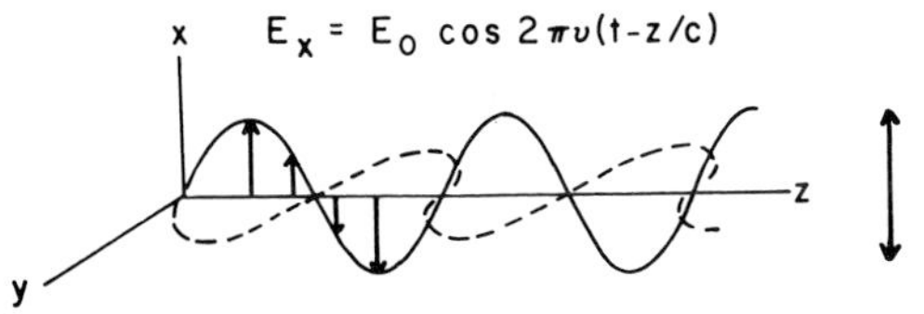

Fig. 1. Plane (linearly) polarized electromagnetic radiation showing both the electric (——) and magnetic (- - -) components

sion as a function of time, t, for a beam of light polarized in the x-z plane and traveling in the z-direction is

Laboratory of Molecular Biophysics and the Cardiovascular Research and Training Center University of Alabama Medical Center, Birmingham, Alabama 35294, USA

$$E_x = E_o \, \mathrm{Cos}\, 2\pi\nu(t\text{-}z/c) \tag{1}$$

where ν is the frequency and c is the velocity of light. The intensity of a beam, I, is proportional to E_o^2. The absorption process is observed as the loss of intensity of a beam of light as it passes through an absorbing medium. The loss in intensity, -dI, is proportional, k, to the intensity, I, the concentration of absorbers, C, and the increment of path, dl.

$$-dI = kI\,Cdl. \tag{2}$$

On integrating intensity over the total path length of a uniformly absorbing, homogeneous solution, this becomes

$$-\ln \frac{I}{I_o} = k\,Cl. \tag{3}$$

Changing from natural logarithms to log to the base 10 gives

$$\log \frac{I}{I_o} = -\epsilon\,Cl = -A \tag{4}$$

where I_o is the initial intensity of the beam; I is the intensity of the emergent beam; ϵ is the molar extinction coefficient, and A is the absorbance. The concentration C is in moles/liter.

The absorbing moiety, the chromophore, may be defined as the grouping of atoms over which the distribution of electrons is significantly altered on absorption of a photon with a wavelength corresponding to an absorption band for the molecule. Atoms in a heterogeneous grouping tend to carry a net charge. Some atoms will have more electrons in their immediate vicinity, in their orbitals, than the number of positive charges in their nuclei. These will have a net negative charge, e.g., the oxygen and nitrogen atoms in a peptide chromophore (see Fig. 2). Other atoms have fewer surrounding electrons than

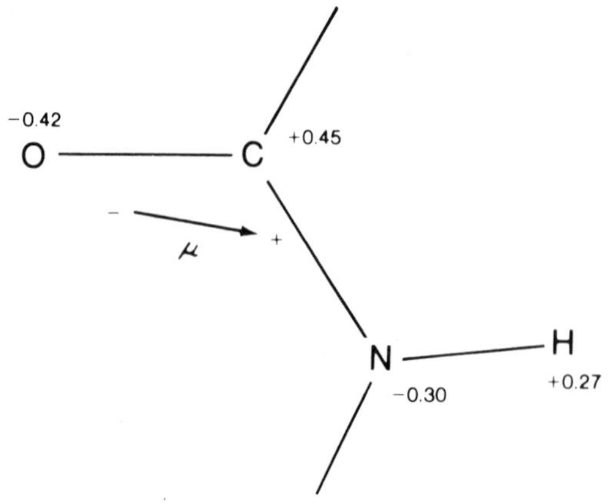

Fig. 2. The peptide chromophore, comparising four different atoms with different net charges (see Ooi et al., 1967). The charges sum to give the permanent (ground state) dipole moment, μ, indicated by the vector with positive and negative ends

the number of positive charges in their nuclei and have a net positive charge, e.g., the carbon and hydrogen atoms of a peptide chromophore (see Fig. 2). The sum of net charges in a chromophoric unit is zero unless it is an ionized species where the value will be an integer, e.g., –1 for a carboxylate anion or +1 for a pyridinium moiety. The distribution of charges in a chromophore may be represented by a dipole moment vector, μ, as shown in Figure 2.

On absorbing a photon the distribution of electrons in a chromophore changes, resulting in a dipole moment for the excited state which is different from that of the ground state. The dipole moment, which when added to the ground state dipole moment of the excited state, is called the transition dipole moment, μ_i. It has both magnitude and orientation. The orientation of the vector defines the plane of polarization of the absorption, and the square of the magnitude of the transition dipole moment (called the dipole strength, D_i) is proportional to the area of the absorption divided by the wavelength of the absorption maximum (Moscowitz, 1960).

$$D_i = |\mu_i|^2 = 1.63 \times 10^{-38} \frac{\epsilon_i^o \Delta_i}{\lambda_i} \quad (5)$$

where ϵ_i^o is the molar extinction coefficient for the ith absorption band corresponding to the ith electronic transition, Δ_i is half-band width at ϵ_i^o/e, and λ_i is the wavelength of the absorption maximum.

Changes in the absorption maximum and/or the dipole strength of an absorption curve can provide information on the relative orientation of transition dipole moments of juxtaposed chromophores, that is, on the relative orientation of the chromophoric groups or, in more general terms, on the conformation of a molecule. When two identical chromophores are juxtaposed, during the absorption of light, an excitation resonance interaction occurs which results in a shifting to shorter wavelengths of the absorption maximum when the relative orientation of the transition dipole moments is one of stacking, i.e., $\overset{\mu_i}{\underset{\mu_i}{\rightrightarrows}}$; or a shifting to longer wavelengths when the relative orientation is head-to-tail, i.e., $\mu_{i\rightarrow}\ \mu_{i\rightarrow}$; or a splitting of the band when the relative orientation is oblique, e.g., $\mu_i\nearrow\nwarrow\mu_i$ (Kasha et al., 1961; Davydov, 1962; Kasha, 1963). The more intense transition dipole moments are in the plane of the usually aromatic or unsaturated chromophore, so that one is actually discussing the relative orientation of the planes of the chromophores.

In addition to the excitation resonance interaction there occurs, simultaneously, an interaction of the ith transition dipole moment in one chromophore with the transition dipole moments of other absorption bands, the jth bands, in the adjacent chromophore. These are called dispersion force interactions. When the jth bands are at shorter wavelengths, a stacked orientation of transition dipole moments, e.g., $\overset{\mu_i}{\underset{\mu_j}{\rightrightarrows}}$, results in a decrease in absorption – a hypochromism; a head-to-tail or coplanar orientation of chromophores, e.g., $\mu_{i\rightarrow}\ \mu_{j\rightarrow}$, results in an increase in the intensity of the absorption band – a hyperchromism (Rhodes, 1961; Tinoco, 1961). When applied to polypeptides, a helically arranged set of peptide moieties, as in the α-helix, results in a splitting and a hypochromism of the major accessible peptide absorption band with $\epsilon^{190} \simeq 4000$; a sheet-like arrangement of peptide moieties, as in the β-pleated sheet conformations, leads to a hyperchromism with $\epsilon^{190} \simeq 8000$; and a disordered arrangement of peptide moieties results in the major accessible absorption band having an intermediate value $\epsilon^{190} \simeq 7000$. Thus one can qualitatively relate shape and intensity of peptide absorption bands to polypeptide conformation.

B. Circular Dichroism (CD)

Light may be made to be circularly polarized by passing plane-polarized light through a birefringent plate which splits the light into two plane-polarized beams, one oscillating along the fast axis and the other along the slow axis. If the plane of linearly polarized light bisects the slow and fast axes, the two beams will be of equal intensity. When the birefringent plate is made to be a quarter wave retarder, then on emerging the two beams will be out of phase by 90°, as depicted in Figure 3B for the component emergent beams. The two beams, when added together as in Figure 3C, result in light in which the electric

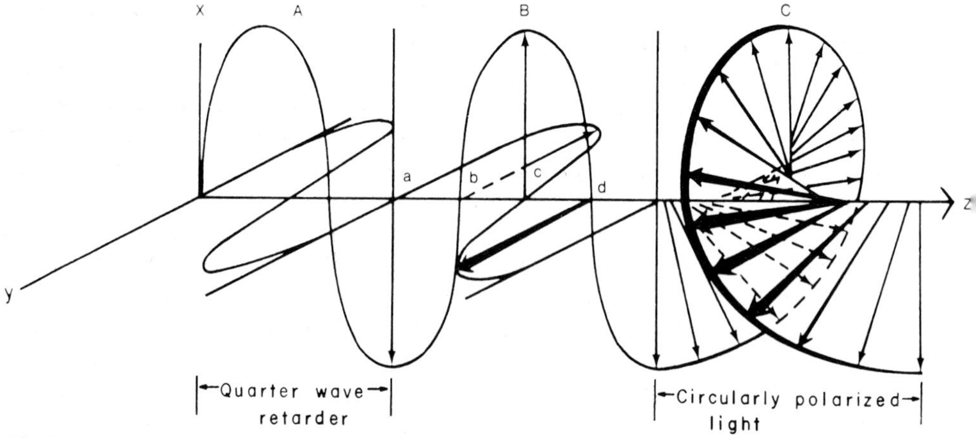

Fig. 3 a-c. Generation of circularly polarized light by means of a quarter wave retarder, (A). Two beams with electric vectors polarized at right angles begin in phase on entering the retarder but on emerging they are 90° out of phase as shown in (B). Addition of the two beams 90° out of phase in (C) gives circularly polarized light (Urry, 1974)

vector moves in a circular manner as it travels in the z-direction. The emergent beam drawn in Figure 3C is right circularly polarized. By inverting the fast and slow axes with an alternating electric field, as in the electro-optic effect, the emergent beam can be made to alternate between left and right circularly polarized light. This is one of several methods of obtaining circularly polarized light (Velluz et al., 1965).

The circular dichroism experiment is simply a difference absorbance experiment but instead of using two beams, i.e., two different light trains with different samples, one commonly uses one beam and alternately passes left and right circularly polarized light through the same sample to the phototube and electronically determines the difference.

The difference absorbance of left and right circularly polarized light is reported as a molar ellipticity, $[\theta]$, where

$$[\theta] = 3300\,(\epsilon_L - \epsilon_R) = 3300\,(A_L - A_R)\,/\,C_o l. \tag{6}$$

C_o is the analytical concentration in moles/liter and l is in centimeters. Since both positive and negative differences are obtained, there are positive and negative bands. The area of the band normalized to wavelength is called the rotational strength, R_i, and may be calculated as (Moscowitz, 1960)

$$R_i = 1.23 \times 10^{-42}\,\frac{[\theta_i^o]\,\Delta_i}{\lambda_i} \tag{7}$$

where $[\theta_i^o]$ is the molar ellipticity of the extremum. Δ_i and λ_i are again the half-band width at $[\theta_i^o]/e$ and the wavelength of the extremum.

The sources of rotational strength may be divided into three major categories depending on the nature of the transition and of the chromophoric group: (1) strong absorption bands with large electric transition dipole moments; (2) weak absorption bands with small electric transition dipole moments but with large magnetic moments; and (3) inherently dissymmetric chromophores (Tinoco, 1962).

Strong electric transitions derive their rotational strengths primarily by coupling with strong electric transition dipole moments in adjacent chromophores. This gives rise to an interesting effect called reciprocal relations (Urry, 1965), where bands close to the same wavelength and with favorable rotations can develop large and oppositely signed circular dichroism bands (Miles and Urry, 1967). Such an effect may be used to demonstrate the proximity of two chromophores in a protein. Weak absorptions with large magnetic transition dipole moments derive rotational strength from the dissymmetric distribution of charges and polarizable groups in their vicinity. This is the type of transition, e.g., the n-π^* transition of carbonyl groups, which obeys a quadrant rule. An approach called partial molar rotatory powers has been undertaken to solve for the contribution of the vicinal group, e.g., a methyl moiety, such that given a position relative to the peptide carbonyl its contribution to the rotational strength can be calculated (Urry, 1968). In proteins disulfide bridges can be dissymmetric in either a left- or right-handed screw orientation and the signs of the rotation can be used to deduce handedness or chirality (Urry, 1970).

Characteristic Circular Dichroism Patterns. There are well-recognized patterns for proteins. These are those of the α-helix (see Fig. 4B), of the β-pleated sheet structure (see Fig. 5B), and of disordered or random structures. This last has little or no ellipticity at 220 nm and a large negative band near 200 nm. In addition, there is recurrent conformational feature in proteins and polypeptides known as the β-turn (Urry and Ohnishi, 1970). It is a ten-atom hydrogen-bonded ring. The resolved absorption and CD spectra for a polymer comprised almost exclusively of a repeating β-turn are given in Figure 6A and B (Urry et al., 1974). This is a distinct pattern with, however, a much reduced ellipticity when compared to that of α-helices and β-pleated sheet structures. Accordingly, when the β-turn is present with α-helix and β-pleated sheet structures the latter two will dominate the CD pattern. In the case of membranes where specialized structures for

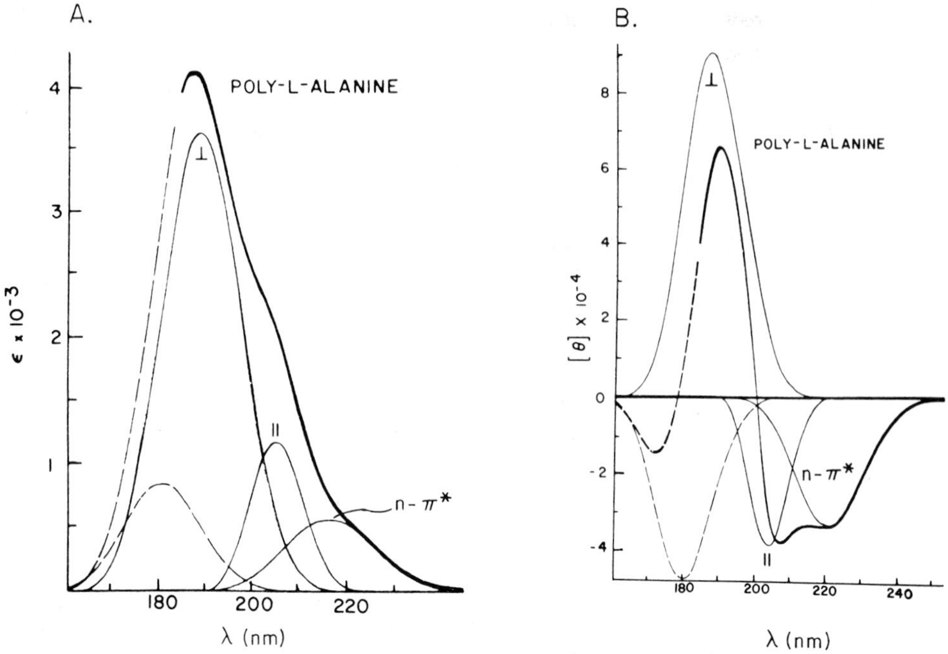

Fig. 4 a and b. Resolved absorption (A) and dircular dichroism (B) curves of poly-L-alanine, showing the n-π* band near 220 nm, the parallel polarized band resolved near 205 nm, and the perpendicularly polarized band resolved near 190 nm. This is the classical α-helical structure without complicating side chain chromophores (Quadrifoglio and Urry, 1968a)

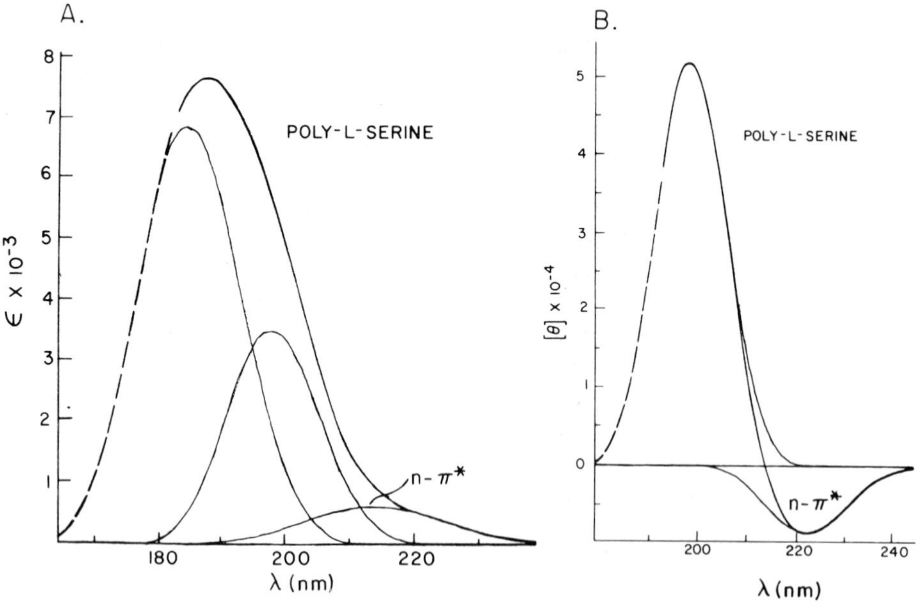

Fig. 5 a and b. Resolved absorption (A) and circular dichroism (B) curves of poly-L-serine in the β-pleated sheet conformation. The hydroxyl in the side chain may contribute to the resolved absorption band at shortest wavelengths but is not expected to near 195 nm, which is responsible for the strong positive CD band (Quadrifoglio and Urry, 1968b)

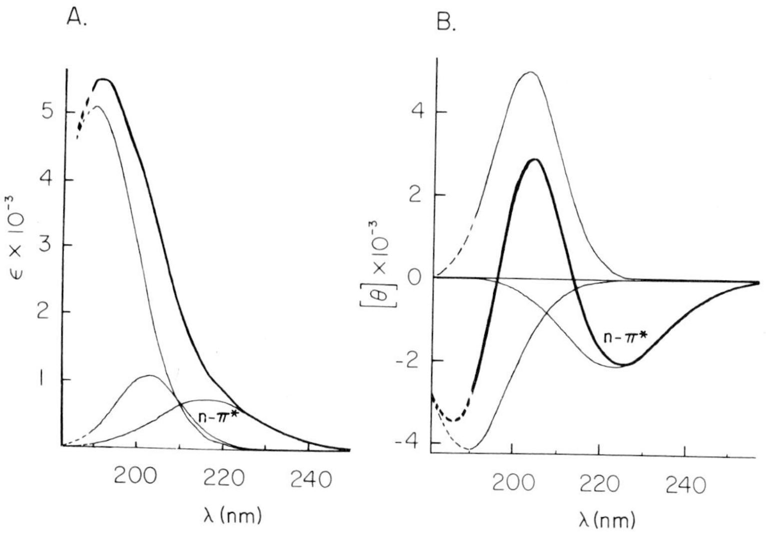

Fig. 6 a and b. Resolved absorption (A) and circular dichroism (B) spectra of the polytetrapeptide, (Val-Pro-Gly-Gly)$_n$, which is a polymer model for the β-turn. *Bold-faced curves:* experimental curves. Note that the ellipticities are an order of magnitude smaller than for the α-helix in Fig. 4 and the β-pleated sheet in Fig. 5 (Urry et al., 1974)

specialized functions can be expected, e.g., selective and voltage-dependent ion transport, then additional structures such as β-helices (Urry et al., 1976) and β-spirals (Long et al., 1974; Urry and Long, 1976) could also be present.

An analysis of the absorption and CD data of the α-helix and β-pleated sheet conformations can, with least ambiguity, be carried out with model systems which do not contain complicating absorptions in the side chains. Accordingly, perhaps the most favorable model would be poly-L-alanine. Since poly-L-alanine occurs under usual circumstances in the right-handed α-helical conformation, it provides for a good characterization of the backbone absorption and ellipticity bands of the α-helix. The analysis can be carried out by means of appropriate simultaneous resolution of the absorption and CD data of poly-L-alanine in a trifluoroethanol solution, as shown in Figure 4A and B (Quadrifoglio and Urry, 1968a). The calculated dipole strengths, rotational strengths, etc., are given in Table 1. There are three major bands – the n-π* band near 220 nm and the parallel polarized (204 nm) and perpendicularly polarized (189 nm) components deriving from excitation resonance interactions of the strong electric transition (Moffitt, 1956a, b; Gratzer et al., 1961; Holzwarth and Doty, 1965). There is a marked hypochromism of the absorption band near 190 nm on going from a disordered structure to an α-helix. The absorption intensity of the n-π* transition, which is dominantly a magnetic transition originating on a nonbonding orbital on the oxygen and becoming delocalized over the peptide chromophore, shows very little change with conformation.

For the antiparallel β-pleated sheet conformation, poly-L-serine serves as an adequate model because the OH moiety absorption is at much shorter wavelengths. Again simultaneous resolution allows characterization of the absorption and ellipticity bands, as shown in Figure 5A and B (Quadrifoglio and Urry, 1968b). The critical values are given in Table 2.

Table 1. Critical values for resolved Gaussian curves for poly-L-alanine (Quadrifoglio and Urry, 1968a)

Wavelength of extremum	Molar extinction coefficient	Oscillator strength	Dipole strength	Molar ellipticity	Rotational strength	Anisotropy | Ri/dI |
	x 10^{-3}		x 10^{36}	x 10^{-3}	x 10^{40}	x 10^{4}
216 (221)[a]	0.55	0.013	0.58	−33	−24	40
204	1.16	0.017	0.71	−38	−18	24
189	3.62	0.065	2.71	91	71	32
180[b]	0.84	0.018	0.78	−47	−33	40

[a] The wavelength 216 nm is the position of the longest wavelength resolved band in absorption. This represents the only relaxation of the constraints that the same Gaussian bands must simultaneously fit absorption and CD data by changing only the sign and magnitude of the resolved bands

[b] The existence of a 180-nm band is only inferred, in that it markedly facilitates and improves the resolution

Table 2. Critical values for resolved Gaussian curves for poly-L-serine (Quadrifoglio and Urry, 1968b)

Wavelength of extremum	Molar extinction coefficient	Oscillator strength	Dipole strength	Molar ellipticity	Rotational strength	Anisotropy | Ri/Di |
	x 10^{-3}		x 10^{36}	x 10^{-3}	x 10^{40}	x 10^{4}
214 (222)[a]	0.6	0.019	0.9	−9.9	−5.6	6.2
197	3.5	0.065	2.85	52	32	11.2
184[b]	6.8	0.17	6.8	Small	Small	Small

[a] See Table 1, footnote (a)

[b] The absence of ellipticity corresponding to the 184 nm band is inferred, in that no optical activity is required to improve the resolution

For our purposes here, what may be noted is the hyperchromism of the absorption band with an extinction coefficient of approximately 8000 near 190 nm and the absence of any significant differences in the intensity of the n-π^* band near 220 nm. The molar extinction coefficient for the n-π^* transition in the α-helix is 550, whereas it is 600 for the β-pleated sheet conformation. This difference is within the experimental error for this low-intensity band. This information will be useful when considering particulate systems and making appropriate corrections by means of finding a suitable solubilized state.

III. Suspensions of Homogeneous Particles

When compared to solution spectra, the absorption spectra of suspensions are flattened and the background absorption is raised due to light scattering. The CD spectra are even more severely distorted, with red-shifted extrema and dramatic and progressive dampening of extrema as shorter wavelengths are reached. Given these distortions, our objective is to calculate the CD and absorption curves that would be expected if the protein and polypeptide molecules of a membrane or other particulate system were made molecularly dispersed in solution, with retention of their membrane or particulate cpnformation. With such corrected data it becomes possible to utilize in a comparative manner the wealth of information already available for polypeptides and proteins molecularly dispersed in solution.

A. Absorption by Suspension

One of the fundamental problems of absorption by suspensions was treated by Duysens (1956); it derives from the fact that the loss of intensity is not uniform throughout the light path in a suspension but is discontinuous as light enters and leaves a particle. Because of the intense absorption within a particle, the precipitous drop in light intensity on passing through a particle results in a significantly less intense beam leaving the particle. Since the probability of absorption is proportional to the beam intensity (see Eq. (2)), the effect of the particle is one of casting a shadow obscuring chromophores behind the particle. To account for this effect Duysens (1956) defined an absorption flattening quotient, Q_A, at a given wavelength, λ, as

$$Q_A^\lambda = \frac{A_{susp}^\lambda}{A_{soln}^\lambda} . \tag{8}$$

We prefer to define a flattened absorbance

$$A_F^\lambda = Q_A^\lambda \, A_{soln} \tag{9}$$

where $A_{susp} \geqslant A_F$.

In suspensions concern is with all of the factors which alter the intensity of light entering the phototube. Another factor, of course, is light scattered in directions other than the phototube. This scattered light is referred to as A_S and is defined in analogy to Eq. (4) as

$$A_S = -\log \frac{I_o - I_S}{I_o} \tag{10}$$

$$A_S = -\log\ 1 - X_S \tag{11}$$

where

$$X_S = \frac{I_S}{I_o} \tag{12}$$

that is, X_S is the fraction of light that is scattered. Similar expressions may be written for the absorbed light, i.e.,

$$A = -\log \frac{I_o - I_A}{I_o} = -\log (1 - X_A) \tag{13}$$

where I, the emergent beam intensity of Eq. (4), is the initial beam intensity, I_o, minus the intensity loss due to absorption, I_A. X_A is the fraction of absorbed intensity, or in other words, the probability of absorption.

It is useful to consider another aspect of the scattered light. When a photon strikes the solvent-particle interface and is scattered out of the sample in a direction other than the phototube, the photon no longer has a probability of being absorbed as it would have had at the same position in traversing a solution. Because light was scattered, some absorption did not occur. Since the probability of absorption is X_A and the probability of scatter is X_S, the product X_A times X_S gives the fraction of the beam that would have been absorbed had the light scattering not occurred. This is referred to as an obscured absorption, A_{OBSC} (Urry, 1972; Urry and Long, 1974), which is written in analogy to Eqs. (11) and (13) as

$$A_{OBSC} = -\log(1 - X_A X_S) = -\log(1 - X_O). \tag{14}$$

We then write for the absorbance of a suspension three components, a flattened absorbance, A_F, an obscured absorbance, $-A_O$, and an absorbance due to scatter, A_S, i.e.,

$$A_{susp} = A_F - A_O + A_S. \tag{15}$$

Using Eq. (14) with a given value for A_S and A_F a value can be obtained for A_O as plotted in Figure 7. This becomes particularly significant when making corrections for distortions in the CD spectra for suspensions.

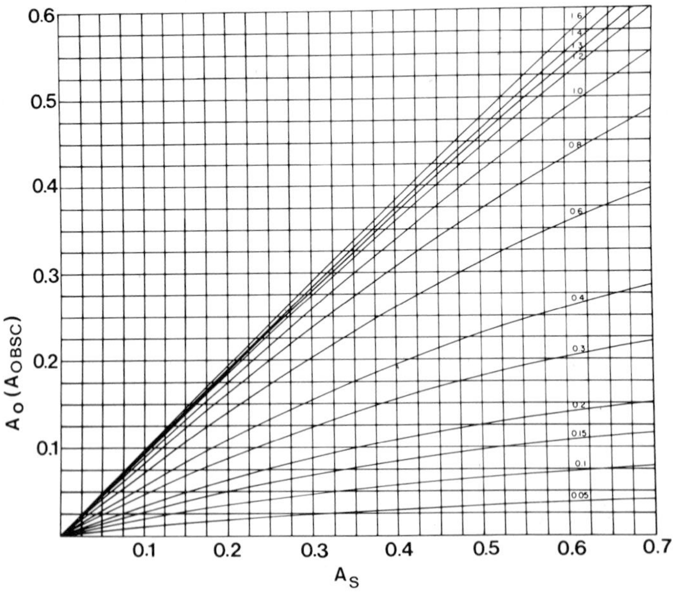

Fig. 7. Relationship between A_S and A_O for different values of A_{soln}. Note that when true absorbance becomes greater than 1.5, A_S is the same magnitude as A_O. For example, when A_{soln} is 1.6 (*top curve*), choosing a value of 0.3 for A_S the value of A_O is about 0.29. This means that at the absorption maxima distortion is almost totally due to differential absorption flattening (Urry, 1972)

B. Circular Dichroism of Suspensions: Differential Absorption Effects

By analogy to Eq. (6) the molar ellipticity for a suspension is written as

$$[\theta]_{susp} = \frac{3300}{C_o l} (A_L - A_R)_{susp}. \tag{16}$$

An ellipticity distortion quotient, Q_E, is defined as

$$Q_E = \frac{[\theta]_{susp}}{[\theta]_{soln}} = \frac{A_L - A_R \ susp}{A_L - A_R}. \tag{17}$$

With Eqs. (15) and (9) we write

$$Q_E = \frac{(Q_{AL} A_L - Q_{AR} A_R) - (A_{OL} - A_{OR}) + (A_{SL} - A_{SR})}{A_L - A_R}. \tag{18}$$

The first term in the numerator is the differential absorption flattening quotient; the second is the concentration obscuring effect, and the third is the differential light-scattering effect. The first and second terms have been evaluated (Urry, 1972) as

$$\frac{Q_{AL} A_L - Q_{AR} A_R}{A_L - A_R} \simeq Q_A^2 \tag{19}$$

and

$$\frac{A_{OL} - A_{OR}}{A_L - A_R} \simeq Q_A^2 (1 - e^{-\sigma}) \tag{20}$$

where

$$\sigma = \frac{A_O}{A_{soln}}. \tag{21}$$

When neglecting differential scatter of the left and right circularly polarized beams, the ellipticity distortion quotient becomes

$$Q_E^{A_{SL} = A_{SR}} = Q_A^2 e^{-\sigma} \tag{22}$$

and the corrected ellipticity becomes

$$[\theta]_{corr}^{A_{SL} = A_{SR}} = \frac{[\theta]_{susp}}{Q_A^2 e^{-\sigma}} \tag{23}$$

Equation (23) holds at points where the optical rotatory dispersion curve is zero. For α-helical conformations this is near 220 nm and 190 nm, which are two wavelengths of great interest in the CD data. The differential scatter distortion can be subtracted to give

$$[\theta]_{corr} \simeq \frac{[\theta]_{susp} - (A_{SL} - A_{SR})\, 10^{-A_F} \times \frac{3300}{C_o\, l}}{Q_2^A\, e^{-\sigma}} \tag{24}$$

Practical means of evaluating the quantities in Eq. (24) are outlined below following a short section which shows the adequacy of the approach in calculating the distorted spectra of suspensions of α-helical poly-L-glutamic acid.

C. Calculation of Distorted Spectra of Particulate Poly-L-Glutamic Acid

Poly-L-glutamic acid (PGA) undergoes a pH-elicited disorder → α-helix transition on lowering the pH to 4, where the degree of ionization is reduced to about 0.05. At this pH there is some aggregation, estimated to be about 10 molecules, i.e., an aggregate of about 10 helical rods (Tomimatsu et al., 1966). This aggregation is not sufficient to cause significant particulate distortions in the absorption and CD data; it is sufficient, however, to have already introduced the majority of any effects arising from intermolecular excitation resonance and dispersion force interactions. Also, absorption and optical rotation effects resulting from protonating the remaining approximately 5% carboxylate side chains are negligible. With PGA, association to form larger aggregates can be controlled by dropping the pH further, to 2.4 for example, and by sonication at that pH which promotes association. Infrared studies have been used in D_2O to demonstrate that the α-helical conformation remains as aggregation progresses (Urry et al., 1970). Accordingly, aggregation of PGA provides a suitable system for studying the particulate distortions.

Figure 8A shows a set of absorption curves a to c of increasing average particle size, and in Figure 8B are the corresponding CD data. Both the absorption and CD data were obtained simultaneously on the same phototube. This is a critical consideration in initial efforts to correlate absorption and CD data and to use the absorption data to correct the CD data. Once an approach is substantiated this criterion can be relaxed. Using the equations outlined above, calculated curves b and c are given in Figure 9A. The details and numerical values obtained for each term are given elsewhere (Urry, 1972). In Figure 9B are values obtained using the data in Figure 8 and applying the Mie scattering theory (Gordon, 1972). While to our knowledge the ad hoc phenomenological approach outlined here provides the most accurate calculation of distorted spectra, the purpose of making the comparison in Figure 9 is to demonstrate that the relatively simple approach used in the following section on several membranous systems can be expected to improve the CD and absorption data for the purpose of qualitative discussion of conformation and differences in conformation.

IV. Suspensions of Biomembranes (Heterogeneous Particles)

Absorption and CD spectra of biomembrane suspensions exhibit the same distortions as do suspensions of the poly-L-glutamic acid particles described above. Biomembranes have

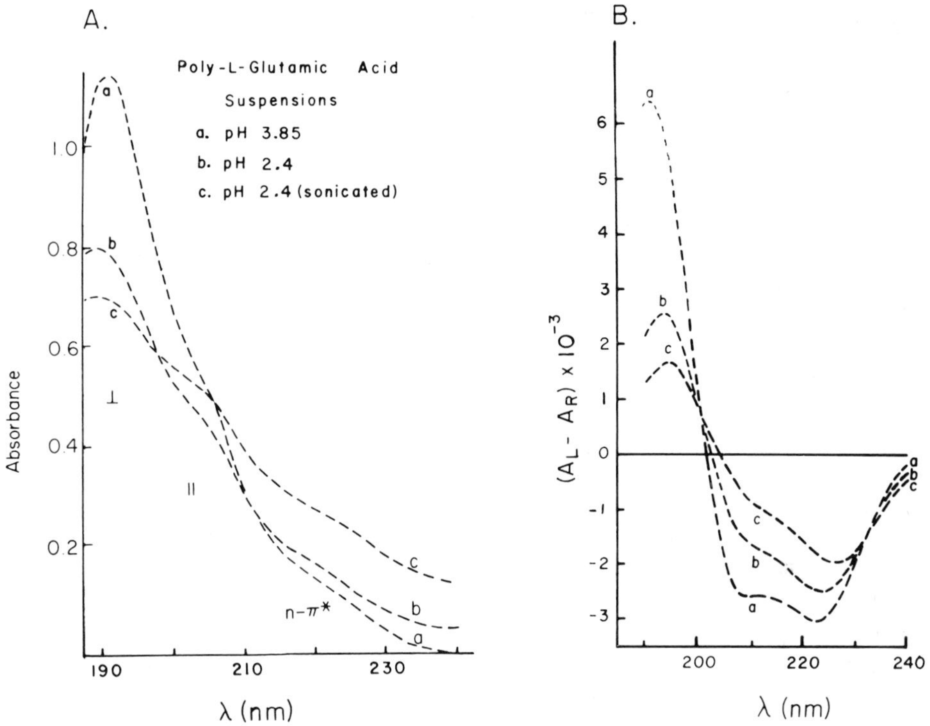

Fig. 8 a and b. Effect of increasing particle size on the absorption (A) and circular dichroism (B) spectra of poly-L-glutamic acid. Curve (a): a molecularly dispersed state with little or no particulate distortions. As the particles increase in size the spectra become more and more distorted with red-shifting and progressively greater dampening as shorter wavelengths are approached. Using the absorption data in (A), curves (b) and (c) of (B) are calculated (see Fig. 9A) by means of the approach outlined in the text (Urry et al., 1970)

the added complication of being heterogeneous. The poly-L-glutamic acid particles contain only the peptide chromophore and the carboxyl chromophore as a single repeating unit. Biomembranes, on the other hand, contain much lipid and side chain absorbance in addition to the polypeptide or protein backbone absorbance. The CD spectra, however, are dominantly due to the peptide backbone chromophore arranged in various conformations.

This difference between biomembranes and PGA is significant, as an analysis of the absorbances of biomembranes demonstrates. As will be shown in some detail below, the lipid and side chain absorbances account for about 60% of the total absorbance at 190 nm, whereas the peptide backbone contributes about 40%. The absorbance of the carboxyl side chain of PGA contributes very little to the absorbance at 190 nm. The peptide chromophore of PGA is the source of almost all of the absorbance at 190 nm. This is the limitation in applying the same approach, which was shown to be so adequate for PGA, to the heterogeneous membrane particles. The more intricately involved the lipid is with the protein, the better the approach will work. If there were large expanses of lipid devoid of protein and dense areas of protein, the approach would be more limited.

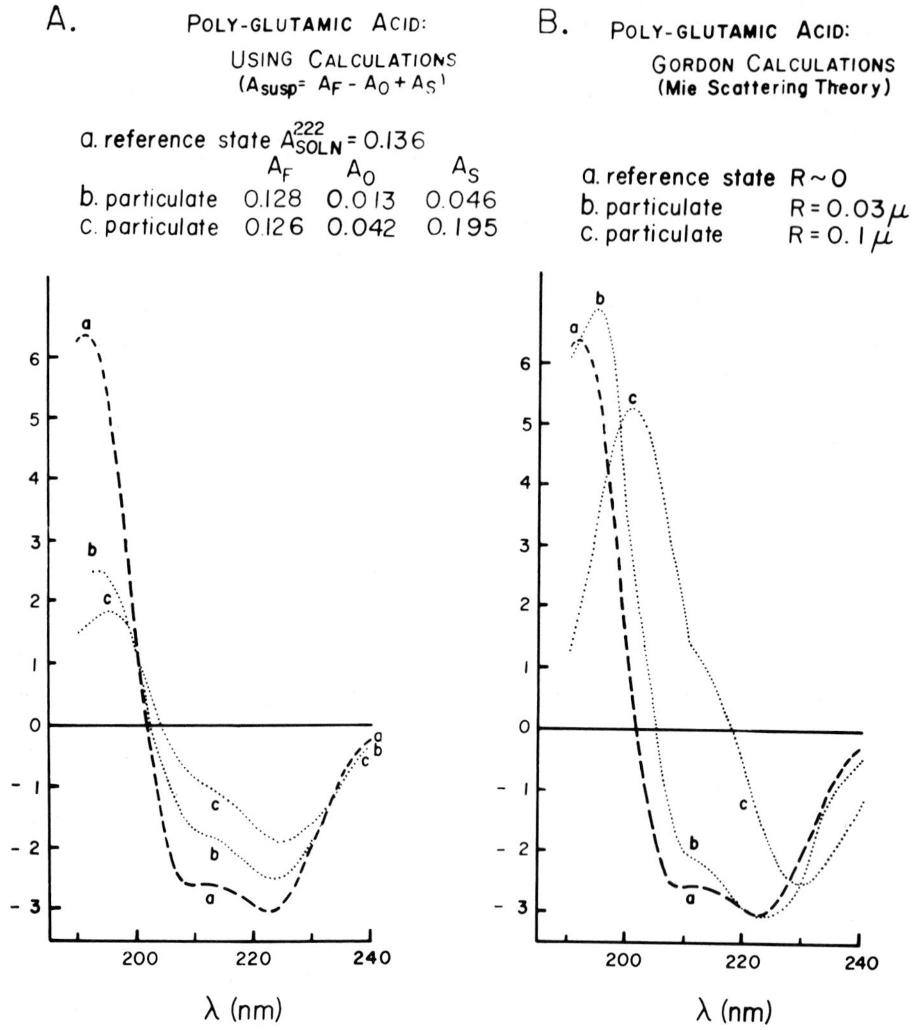

Fig. 9 a and b. Calculations of the distorted spectra of Fig. 8B by means of the approach outlined in the text (A) and the plotted values obtained by Gordon (1972) using the Mie scattering theory (B). For detailed calculations for curves (b) and (c) of (A) see Urry (1972). A comparison of the curves in (A) with the experimental data in Fig. 8B shows how satisfactorily the approach works (Urry and Long, 1974)

Another operational difference is important. In the PGA system there is a well-defined reference state, i.e., pH 4, where the particle size, about 10^6 daltons, provides a good solution spectrum from which to derive the factor Q_A and $e^{-\sigma}$. In correcting the spectra of biomembranes it becomes necessary to find a membrane-dissolved state to serve as the reference state for the absorption data. This is called a pseudo-reference state and severe restrictions are placed on the absorption and CD data of this dissolved state, which often makes it difficult to obtain. With some effort a suitable pseudo-reference state can be found and the CD suspension data can be corrected.

A. Absorption of Biomembranes

Analysis of absorption data will be carried out for three membrane systems – red blood cell ghosts, sarcotubular vasicles, and brain microsomes. The red blood cell ghosts were prepared by the method of Dodge et al. (1963) from venous human blood at pH 7.4 with hemolysis on ice in 20 mM phosphate buffer. Spectroscopic examination of the ghosts in the visible and near ultraviolet spectral regions indicated that the ghosts were free of hemoglobin. Both fresh and frozen samples were studied. The procedure of Seraydarian and Mommaertz (1965) was used for the preparation of rabbit sarcotubular vesicles. The appropriate centrifugation band was found to contain calcium-stimulated ATPase activity and was used in the spectroscopic studies without freezing. Liberated phosphate was determined by the method of Fiske and Subbarow (1925). Calcium-stimulated ATPase activity of the sarcotubular vesicles by a factor of 4.1. Porcine brain microsomes were isolated by differential centrifugation. The microsomal material was subsequently treated with NaI according to the method of Nakao et al. (1965). The total ATPase activity of the NaI-treated microsomes varied from 14.1 to 16.9 μmol P_i/mg h with 82% to 97% inhibition by ouabain. Protein concentrations were determined by amino acid analyses and by the method of Lowry et al. (1951). The buret method yielded values that differed significantly from those calculated by the aforementioned techniques.

1. Amino Acid Analyses and Mean Residue Side Chain Extinction Coefficients. The membrane samples were hydrolyzed in evacuated sealed tubes at 110°C for 24 h in 6 M hydrochloric acid and analyzed on a Beckman Model 116 amino acid analyzer (Starcher, 1968). Tryptophan was analyzed separately by the procedure of Barman and Koshland (1967) using 2-hydroxy-5-nitrobenzyl bromide (HNB) as the tryptophan-specific reagent and exhaustive dialysis against 0.18 M acetic acid at pH 2.7. When the tryptophan content of α-chymotrypsin was determined with this procedure, the ratio of moles of tryptophan to moles of α-chymotrypsin was found to be 7.93 ± 0.11. This compared favorably with Barman and Koshland's value of 8.1. Tryptophan analyses of the membrane samples necessitated the addition of 0.1% (w/v) sodium-dodecyl sulfate to the urea incubation medium during the unfolding and solubilization step. The protein concentration was determined by amino acid analyses of acid hydrolysates of known aliquots of the HNB-membrane protein complex.

Table 3 summarizes the results of amino acid analysis of four membrane preparations. As a check of the amino acid analysis procedure, the composition of α-chymotrypsin was determined and the results were within experimental error both for the amino acids quantitated on the amino acid analyzer and for the HNB determination of tryptophan. The composition of the red blood cells agrees well with that determined by Rosenberg and Guidotti (1968), while the amino acid composition of the sarcotubular vesicles is in accord with that determined by Martonosi and Halpin (1971). The brain microsomes and NaI-treated microsomes have an amino acid composition similar to that of the sarcotubular vesicles and red blood cells, being relatively high in amino acids with nonpolar side chains.

The data in Table 3 were used to calculate the contribution of the amino acid side chains to the total ultraviolet absorbance at λ_{max}. The percentage of the total amino acid that each amino acid represented was multiplied by that amino acid extinction co-

Table 3. Amino acid composition of the four membrane preparations residues/1000

Amino acid residue	Membrane							
	NaI Microsomes		Microsomes		Sarcotubular vesicles		Red blood cells	
	Mean	σ	Mean	σ	Mean	σ	Mean	σ
Lys	71.1	2.7	62.4	2.5	63.4	5.8	55.6	3.8
His	27.8	2.0	25.6	2.3	18.1	1.4	25.8	2.7
Arg	53.1	4.0	53.4	1.0	47.5	4.7	57.7	4.2
Asp	91.0	4.6	95.1	5.1	100.1	4.0	81.3	1.3
Thr	55.8	2.8	56.6	2.9	62.0	3.3	54.7	2.9
Ser	75.1	2.7	72.4	1.9	63.7	1.3	72.3	3.0
Glu	126.2	4.5	129.5	5.8	116.2	4.4	131.2	5.8
Pro	51.2	2.2	50.4	5.2	50.2	4.3	49.6	2.3
Gly	77.1	2.7	76.6	1.3	74.2	2.3	69.0	3.9
Ala	80.9	2.2	81.8	2.5	87.8	3.7	80.7	3.6
Cys	10.5	1.7	14.0	1.7	13.5	1.5	2.9	2.1
Val	56.7	2.1	58.0	3.1	69.4	1.3	62.4	3.3
Met	15.4	2.0	13.7	4.2	13.4	8.2	15.8	2.4
Ileu	43.2	2.0	43.5	0.8	55.2	1.5	47.4	1.7
Leu	94.5	2.0	94.8	1.8	101.1	2.8	121.2	4.9
Tyr	29.8	1.9	29.6	0.5	25.0	1.6	27.0	2.1
Trp	24.6	3.4	20.4	4.5	28.2	1.3	18.4	3.3
Phe	39.3	2.6	40.4	1.6	40.6	1.3	49.3	2.9

efficient. For example, in the case of the red blood cells, 4.93% of all the amino acid residues were phenylalanyl residues whose extinction coefficient at 190 nm is 54,500 (Sober, 1970). The contribution of the phenylalanyl side chain to the total absorbance would be the percentage of phenylalanine multiplied by its extinction coefficient, i.e., 4.93% × 54,500 = 2686. Similarly, tyrosine constitutes 2.70% of the amino acids in red blood cells and the extinction coefficient is 42,800, resulting in a 2.70% × 42,800 or 1155 value for tyrosine's contribution to the total absorbance value of the membrane. In the same way, the contribution of all the other amino acid residues was calculated and summed to give the values in Table 4. As the intention is to sum only side chain absorbance, it is necessary to substract the carboxyl and amino contributions. These may be estimated by noting the molar extinction coefficients for those amino acids which contain no chromophoric side chains, e.g., Gly, Ala, Val, Ileu, Leu, and Pro. The average molar extinction coefficient at 190 nm for these six amino acids is 598. Accordingly 600 is subtracted as the contribution of the carboxyl and amino groups which become the peptide chromophore in the protein. This gives mean residue side chain extinction coefficients of approximately 4900 for the microsomes, 4700 for the sarcotubular vesicles, and 5200 for the red blood cell ghosts. When these values are subsequently used to obtain the mean residue protein backbone extinction coefficients, it will be assumed that all hyper- and hypochromism effects, within the diverse protein side chain interactions in a complex membrane, cancel out.

2. Lipid Extraction and Equivalent Mean Peptide Residue Extinction Coefficients. Lipid extraction of the different membranes used chloroform-methanol as follows: 2 ml

Table 4. Calculated mean residue side chain extinction coefficients for four membrane preparations

Amino acid	Extinction[a] coefficient	NaI Microsomes		Microsomes		Sarcotubular vesicles		Red blood cells	
		Residues/1000	$\Delta\epsilon$	Residues/1000	$\Delta\epsilon$	Residues/1000	$\Delta\epsilon$	Residues/1000	$\Delta\epsilon$
Lys	0.9×10^3	71.1	64	62.4	56	63.4	57	55.6	50
His	5.57×10^3	27.8	155	25.6	143	18.1	101	25.8	144
Arg	13.1×10^3	53.1	696	53.4	699	47.5	622	57.7	756
Asp	1.61	91.0	146	95.1	153	100.1	161	81.3	131
Thr	0.75	55.8	42	56.6	42	62	47	54.7	41
Ser	0.61	75.1	46	72.4	44	63.7	39	72.3	44
Glu	1.61	126.2	203	129.5	208	116.2	187	131.2	211
Pro	0.54	51.2	28	50.4	27	50.2	27	49.6	27
Gly	0.57	77.1	44	76.6	44	74.2	42	69.0	39
Ala	0.57	80.9	46	81.8	47	87.8	50	80.7	46
Cys	2.82	10.5	29	14.0	39	13.5	38	2.9	8
Val	0.57	56.7	32	58.0	33	69.4	39	62.4	35
Met	2.67	15.4	41	13.7	36	13.4	36	15.8	42
Ileu	0.67	43.2	29	43.5	29	55.2	37	47.4	32
Leu	0.67	94.5	63	94.8	63	101.1	68	121.2	81
Tyr	42.8	29.8	1275	29.6	1267	25.0	1070	27.0	1155
Phe	54.5	39.3	2142	40.4	2201	40.6	2213	49.3	2686
Trp	17.6	24.6	433	20.4	359	28.2	496	15.9	280
			5514		5490		5330		5808
Subtract carboxyl and amino $\Delta\epsilon$/mean residue			−600		−600		−600		−600
Side chain mean residue extinction coefficients (ϵ_{SC})			~ 4900		~ 4900		~ 4700		~ 5200

[a] From *Handbook of Biochemistry*, ed. H.A. Sober, The Chemical Rubber Company, Cleveland, Ohio, 1970, p. B-73; data compiled by R.S. McDiarmid

of the membrane suspension in 200 mM Tris-HCl pH 7.4, with a known protein concentration, were added to 50 ml of chloroform-methanol (2:1, v/v) and incubated overnight at 4°C. This mixture was filtered into a separatory funnel and washed four times with chloroform-methanol. The chloroform-methanol solutions were pooled and added to 50 ml of saline. The organic and water phases were allowed to separate for three hours, at which time the bottom layer was withdrawn. The top layer was washed three times with chloroform-methanol and the bottom layers combined with the first bottom layer. To this pool was added granular anhydrous sodium sulfate. After filtration, the extract was evaporated to dryness under vacuum and then brought to a known volume with chloroform and/or cyclohexane. The lipid phosphorus of the known volume was determined by the method of Bartlett (1959) and with the membrane protein concentration known, the ratio of μgm lipid phosphorus to mg of membrane protein was calculated. It was assumed that the lipid extraction was quantitative.

The lipid absorbance of the samples is quite significant, although less than the absorbance of the amino acid side chains. Of the total membrane OD lipid contributed about 20% (Table 5), as calculated in the following manner. The optical density of a total lipid extract from each membrane type was measured, along with the inorganic phosphorus of each lipid solution. The absorbance was then normalized relative to a standard concentration of lipid phosphorus and to a standard pathlength. Because the ratio of micrograms lipid phosphorus to milligrams membrane protein (i.e., μgm Pi/mg protein) had been determined, as described above, the absorbance normalized to a standard phosphorus concentration was related further to a standard protein concentration. The conversions were necessary because it is the contribution of the lipid equivalent mean peptide residue extinction coefficient that is to be determined. Once extracted, the lipid is examined spectroscopically, ideally at the concentration in which it occurs in the membrane, and then compared to the membrane's total absorbance. Experimentally, however, the lipid concentration of the spectroscopic sample is different from its membrane concentration. This difference can be corrected by expressing the lipid phosphorus concentration in terms of membrane protein concentration, simply by multiplying the former by the ratio of microgram lipid phosphorus to milligram protein. An example of this calculation is in Table 5. When the total membrane peptide residue extinction coefficient is corrected for the contributions of the amino acid side chains and the lipids, it is reduced by an average of 60% (see below). These corrected mean residue extinction coefficients at 190 nm represent an approximate value for the peptide moiety in juxtaposition with other carbonyl moieties along the peptide backbone. In other words, the corrected extinction coefficients at 190 nm provide information about the peptide's secondary structure. The corrected suspension absorbances and mean peptide residue extinction coefficients for the four membrane preparations are obtained in combination with the corrected CD data by means of the pseudo-reference state approach. This is outlined below.

B. Circular Dichroism of Biomembranes

1. Pseudo-Reference State Approach to Improving Ellipticity Data. The differential light-scattering term is zero when the optical rotatory dispersion curve is zero. For mem-

Table 5. Lipid equivalent mean peptide residue extinction coefficients

Membrane	Total μg lipid Pi	Total μg membrane protein	μg Lipid Pi/ mg membrane protein	Lipid [Pi]	[Pi] in terms of [Protein]	190 nm OD/ 1.0 mm cell	190 nm[a] ε(lipid)
NaI Microsomes	79	6.33	12.5	3.17	0.254	0.57	2490
Microsomes	138	9.32	14.9	5.54	0.372	0.81	2417
Sarcotubular vesicles	106	5.66	18.7	4.24	0.227	0.51	2494
Red blood cells	82	4.02	20.4	3.29	0.161	0.55	3792

[a] Calculation of lipid extinction coefficient in terms of its contribution to the total membrane's extinction coefficient:

$OD^{190} = 0.57$

Pathlength = 0.1 cm

[Pi] = 3.17 μg lipid Pi/ml

But there are 12.5 μg lipid Pi/mg membrane protein.

Therefore this sample represents the concentration of lipid found in a membrane sample whose protein concentration is (3.17) (1/12.5) or 0.253 mg/ml

$\epsilon = OD/Cl$

$$\epsilon = \frac{0.57}{\left(\frac{0.254\ \text{mg/ml}}{111\ \text{mg/mmol}}\right)\left(1 \times 10^{-1}\ \text{cm}\right)} = \frac{0.57}{2.288 \times 10^{-4}} = .2490 \times 10^{4}$$

$\epsilon = 2490$

branes containing sufficient α-helix, which is most membranes, this is near 222 nm and 192 nm. (These wavelengths are also not far from those wavelengths where $A_{SL} = A_{SR}$ for β-pleated sheet conformations.) Because of this we can neglect the differential light-scattering term and write

$$[\theta]_{corr}^{222} = \frac{[\theta]_{susp}^{222}}{\left(Q_A^{222}\right)^2 \; Q_\sigma{}^{222}} \tag{25}$$

and

$$[\theta]_{corr}^{192} = \frac{[\theta]_{susp}^{192}}{\left(Q_A^{192}\right)^2 \; Q_\sigma^{192}} \tag{26}$$

where $e^{-\sigma}$ has been defined as Q_σ for ease of superscripting. At 222 nm the absorbance is low for membranous particles whose size is small enough to correct, i.e., for those vesicles which do not become black to 192 nm light. This is because the absorbance at 192 nm is usually some four to five times larger than at 222 nm for membrane proteins and lipids. Because of this Q_A^{222} is close to unity, usually about 0.9 or larger. Accordingly

$$\left| [\theta]_{corr}^{222} \right| \gtrsim \left| \frac{[\theta]_{susp}^{222}}{Q_\sigma^{222}} \right| . \tag{27}$$

As $[\theta]_{corr}^{222}$ is negative neglecting Q_A^{222} results in a calculated extremum which is too small. Also at 192 nm where the suspension absorbance is usually greater than 1, (see Fig. 7), A_s approaches A_o in magnitude such that $Q_\sigma \to 1$ and $A_{susp} \to A_F$, giving

$$[\theta]_{corr}^{192} \lesssim \frac{[\theta]_{susp}^{192}}{\left(Q_A^{192}\right)^2} . \tag{28}$$

These inequalities serve as the basis for finding a suitable pseudo-reference state (prs).

Another feature can be utilized: the absorption of a polypeptide at 222 nm is essentially independent of conformation, that is, the n-π* band near 222 nm exhibits little or no change in intensity on changing conformation. This means that a dissolved state of the membrane, a molecularly dispersed state (mds) in which the dispersing solvent system does not add special effects such as hyper- and hypochromism or n-π* band shifting on interacting with the solute, can be used to obtain an approximately correct absorbance at 222 nm. The solution absorbance at 222 nm is subtracted from the suspension absorbance, i.e.,

$$A_{susp}^{222} - A_{mds}^{222} = {}^{o}A_S^{222} . \tag{29}$$

This value of ${}^{o}A_S^{222}$ is used in Figure 10 to find ${}^{o}Q_\sigma^{222}$, i.e., ($e^{-\sigma}$), by choosing the curve with a value closest to A_{mds}^{222}. The value of ${}^{o}Q_\sigma^{222}$ is then used in Eq. (27) to obtain a calculated ${}^{o}[\theta]^{222}$. If the cross-over between the suspension absorption curve and the mds absorption curve is too near 222 nm, then ${}^{o}Q_A^{192}$ is calculated from the ratio

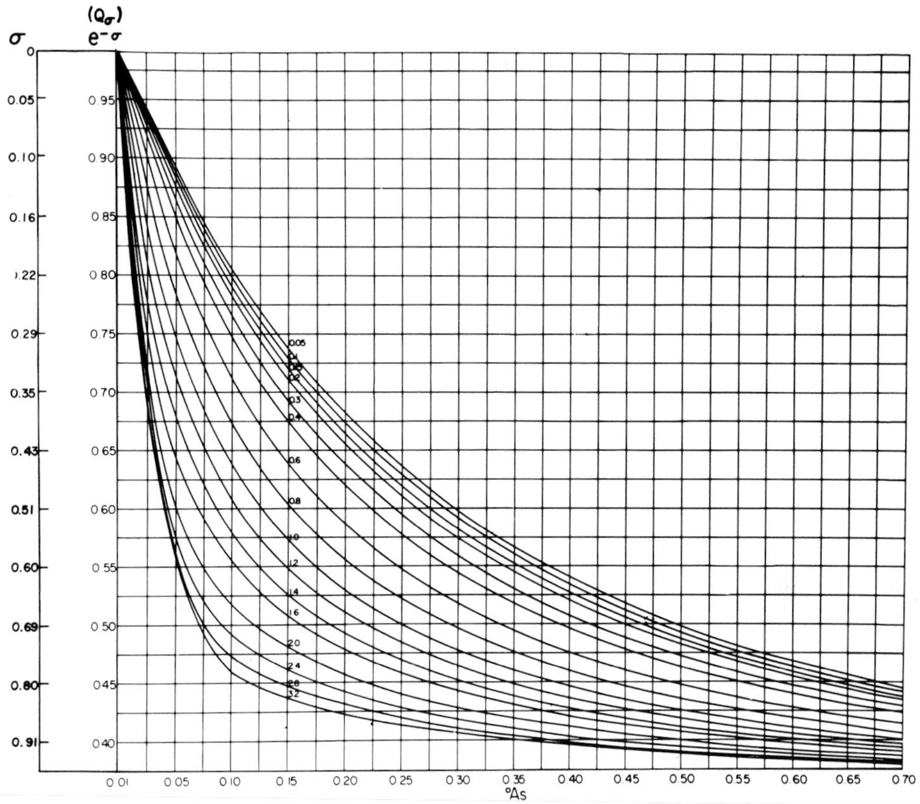

Fig. 10. Plot of $e^{-\sigma}$ against $^{o}A_S$. The latter is defined in Eqs. (30) and (32). With a given value of $^{o}A_S$ the appropriate curve for the A_{soln} or A_{mds}, at the wavelength of interest, is chosen and $e^{-\sigma}$ is obtained from the ordinate (Urry and Long, 1974)

$A^{192}_{susp}/A^{192}_{mds}$; this value of $^{o}Q^{192}_{A}$ is then used in Table 6 to get A_p (192) which can be used to obtain A_p (222) by the equation

$$A_p(\lambda_1) = \frac{A_{soln}(\lambda_1)}{A_{soln}(\lambda_2)} A_p(\lambda_2). \tag{30}$$

The value for A_p (222) is then used to approximate $^{o}Q^{222}_{A}$ by use of Table 6 and a new $^{o}A_s$ is obtained:

$$A_{susp} - {}^{o}Q^{222}_{A} A_{mds} = {}^{o}A_s. \tag{31}$$

The $^{o}A_s$ by Figure 10 leads to $^{o}Q^{222}_{\sigma}$ and a calculated value for $^{o}[\theta]^{222}$. (The NaI microsomal membrane preparation is an example of when the latter procedure was necessary because $A^{222}_{mds} = A^{222}_{susp}$.)

In Table 7 are three molecularly dispersed states of NaI microsomes. Only one, where solubilization is with 0.5% sodium dodecyl sulfate (SDS) plus 20% trifluoroethanol

Table 6. Flattening quotients and associated particle absorbances

A_p	Q_A(spheres)[a]	Q_A(vesicles)[a]	A_p	Q_A(spheres)[a]	Q_A(vesicles)[a]
0.02	0.99	0.97	0.86	0.73	0.67
0.04	0.98	0.95	0.88	0.73	0.66
0.06	0.97	0.94	0.90	0.73	0.66
0.08	0.97	0.92	0.92	0.72	0.65
0.10	0.96	0.91	0.94	0.72	0.65
0.12	0.95	0.90	0.96	0.71	0.65
0.14	0.94	0.89	0.98	0.71	0.64
0.16	0.94	0.88	1.00	0.70	0.64
0.18	0.93	0.87	1.02	0.70	0.63
0.20	0.92	0.86	1.04	0.69	0.63
0.22	0.92	0.85	1.06	0.69	0.62
0.24	0.91	0.85	1.08	0.68	0.62
0.26	0.90	0.84	1.10	0.68	0.62
0.28	0.90	0.83	1.12	0.68	0.61
0.30	0.89	0.82	1.14	0.67	0.61
0.32	0.88	0.81	1.16	0.67	0.61
0.34	0.88	0.81	1.18	0.66	0.60
0.36	0.87	0.80	1.20	0.66	0.60
0.38	0.87	0.79	1.22	0.66	0.59
0.40	0.86	0.79	1.24	0.65	0.59
0.42	0.85	0.78	1.26	0.65	0.59
0.44	0.85	0.77	1.28	0.64	0.58
0.46	0.84	0.77	1.30	0.64	0.58
0.48	0.84	0.76	1.32	0.64	0.58
0.50	0.83	0.76	1.34	0.63	0.57
0.52	0.82	0.75	1.36	0.63	0.57
0.54	0.82	0.75	1.38	0.62	0.57
0.56	0.81	0.74	1.40	0.62	0.56
0.58	0.81	0.73	1.42	0.62	0.56
0.60	0.80	0.73	1.44	0.61	0.56
0.62	0.80	0.72	1.46	0.61	0.55
0.64	0.79	0.72	1.48	0.61	0.55
0.66	0.79	0.71	1.50	0.60	0.55
0.68	0.78	0.71	1.52	0.60	0.54
0.70	0.78	0.70	1.54	0.59	0.54
0.72	0.77	0.70	1.56	0.59	0.54
0.74	0.76	0.69	1.58	0.59	0.53
0.76	0.76	0.69	1.60	0.58	0.53
0.78	0.75	0.68	1.62	0.58	0.53
0.80	0.75	0.68	1.64	0.58	0.52
0.82	0.74	0.68	1.66	0.57	0.52
0.84	0.74	0.67	1.68	0.57	0.52
1.70	0.57	0.51	1.86	0.54	0.49
1.72	0.56	0.51	1.88	0.54	0.49
1.74	0.56	0.51	1.90	0.54	0.48
1.76	0.56	0.50	1.92	0.53	0.48
1.78	0.56	0.50	1.94	0.53	0.48
1.80	0.55	0.50	1.96	0.53	0.48
1.82	0.55	0.50	1.98	0.53	0.47
1.84	0.55	0.49	2.00	0.52	0.47

[a] Using the formalism of Duysens (1956) for relating A_p and Q_A for spherical particles and of Gordon and Holzwarth (1971) for vesicles to relate Q_A (vesicles) with an associated particle absorbance. From Urry and Long (1974)

Table 7. NaI microsomes (for details see text)

Molecularly dispersed state	Molar ellipticity x 10^{-4}					
	222 nm		208 nm		190 nm	
	$^{o}[\theta]_{calc}$	$[\theta]_{prs}$	$^{o}[\theta]_{calc}$	$[\theta]_{prs}$	$^{o}[\theta]_{calc}$	$[\theta]_{prs}$
0.5% SDS, 20% TFE	−1.2	−1.5	−1.2	−1.7	2.5	2.5
0.5% SDS	−1.3	−1.3	−0.8	−0.9	3.3	2.3
0.5% TFA, 80% TFA	−1.1	−2.9	−0.9	−3.2	1.8	4.8

(TFE), satisfies the requirements of a pseudo-reference state. The other two solvent systems give unsatisfactory ellipticities at either (190 nm in the case of 0.5% SDS) or both (222 nm and 190 nm for the third solvent system) of the wavelengths of interest. As differential scatter is not introduced, the values at 208 nm are expected to differ. By varying the amount of trifluoroethanol the ellipticity can be steadily enhanced as shown in Fig. 11B, such that a suitable dissolved state can often be found in this way. Note that

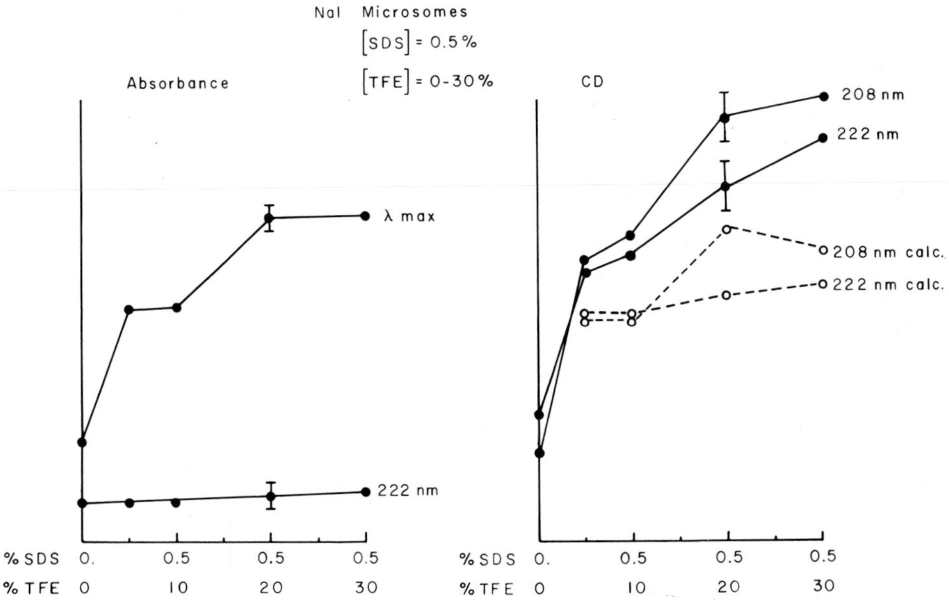

Fig. 11 A and B. Changes in absorption intensity (A) and (B) difference absoebance (circular dichroism) as a function of solvent

while the absorbance at the maximum changes dramatically as does the ellipticity both at 222 and 208 nm, the absorbance at 222 nm changes very little. This makes it possible to use this molecularly dispersed state to obtain $^{o}A_S^{222}$ and with the use of Fig. 10 to obtain Q_σ^{222}, which is then used in Eq. (28). If the calculated value of $[\theta]^{222}$ is about

20% below $[\theta]_{m\,ds}^{222}$, then the first criterion for a pseudo-reference state is met. The second criterion is set by Eq. (29). It is assumed that the absorption curve of the pseudo-reference state can be used in combination with the absorption curve of the membrane suspension to improve the membrane ellipticity data by providing values for Q_A and Q_σ at each wavelength.

The differential scatter of left and right circularly polarized light can be subtracted empirically by obtaining the optical rotatory dispersion curve of the prs and correcting the $[\theta]_{susp}^{201}$ to zero, because this is the wavelength where the ellipticity is near zero for the α-helical conformation. A parameter κ is calculated

$$\kappa \simeq \frac{[\theta]_{susp}^{201}}{[m]_{prs}^{201} \quad 10^{-A_F^{201}}} \qquad (32)$$

where $[m]_{prs}^{201}$ is the mean residue rotation for the pseudo-reference state. Thus the improved value for each wavelength is

$$[\theta]_{corr} = \frac{[\theta]_{susp} - \kappa[m]_{prs} \; 10^{-A_F}}{Q_A^2 \cdot Q_\sigma} \; . \qquad (33)$$

At this point one may ask whether it would not be reasonable directly to consider the pseudo-reference state CD pattern itself as an estimate for the correct ellipticities, since it satisfied the stringent requirements of Eqs. (27) and (28). The answer is a prs with ellipticities which are 10% to 15% greater in magnitude than $^{o}[\theta]^{222}$, and $^{o}[\theta]^{192}$ is probably a reasonable representation of the membrane conformation. The set of values for the corrected ellipticities are given in Table 8 together with the equivalent mean residue exctinction coefficient and the values for the backbone contribution (Table 9).

Table 8. Membrane mean residue ellipticities and absorption maximum extinction coefficients

Membrane	$[\theta]_{susp}^{224} \times 10^{-4}$	$[\theta]_{corr}^{224} \times 10^{-4}$	$[\theta]_{corr}^{192} \times 10^{-4}$	190 nm $\epsilon_T \times 10^{-4}$	$\epsilon_{BB} \times 10^{-4}$
NaI Microsomes	−0.7	−1.2	+2.5	1.4	0.66
Microsomes	−0.8	−1.2	1.8	1.2	0.47
Sarcotubular vesicles	−0.6	−1.1	2.8	1.3	0.58
Red blood cells	−1.5	−2.2	3.4	1.6	0.70

ϵ_T is the corrected membrane absorbance given in terms of a mean residue extinction coefficient.
ϵ_{BB} is the calculated mean residue backbone extinction coefficient obtained by subtraction of the mean residue side chain and equivalent mean residue lipid contribution to ϵ_T, i.e., $\epsilon_{BB} = \epsilon_T - \epsilon_{SC} - \epsilon(\text{lipid})$

2. Application of Pseudo-Reference State Approach to the Purple Membrane of Halobacterium halobium. This membrane is 25% lipid and 75% protein – with the protein being comprised of a single protein type. In 1975, Henderson and others used electron microscopy and X-ray analysis to determine the peptide backbone conformation of this single protein in the membrane (Blaurock, 1975; Henderson, 1975; Henderson and Un-

Table 9. Side chain, backbone, and lipid components of membrane absorbance

Membrane	Total membrane absorbance (190 nm)	Amino acid side chain absorbance (190 nm)	Lipid absorbance (190 nm)	Net absorbance (190 nm)
NaI microsomes	14,000	4900	2500	6600
Microsomes	12,000	4900	2400	4700
Sarcotubular vesicles	13,000	4700	2500	5800
Red blood cells	16,000	5200	3800	7000

win, 1975; Unwin and Henderson, 1975). Their results indicated that the structure was 70% to 80% α-helix, suggesting that this membrane would be a good test case to study the applicability of the pseudo-reference state approach to the correction of membrane CD spectra. Figure 12 shows the ultraviolet absorption spectra for the purple membrane

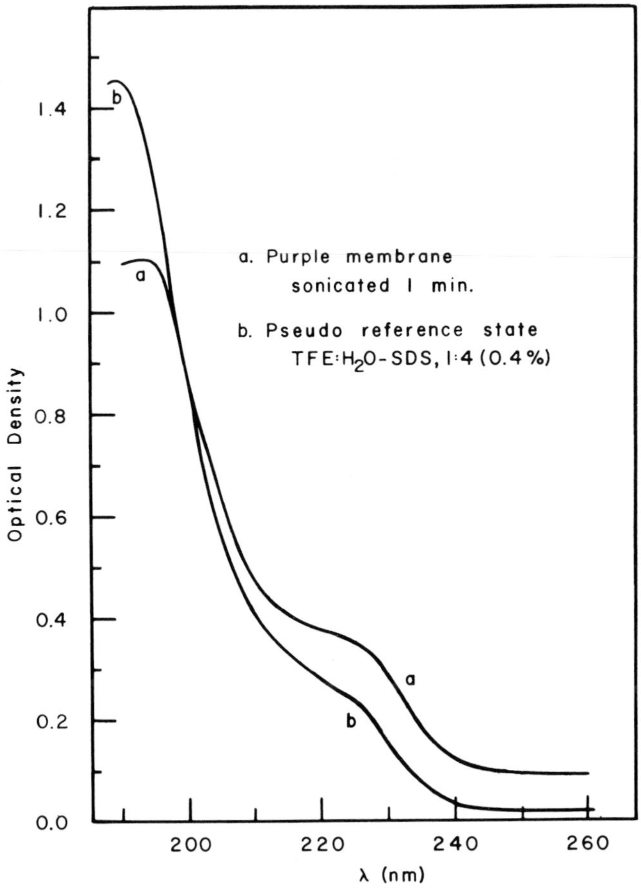

Fig. 12. Purple membrane suspension (**a**) and purple membrane solution (**b**) ultraviolet absorbance data corrected for protein concentration and pathlength to make the spectra directly comparable (Long et al., 1977)

(a) and its pseudo-reference state (b); (Long et al., 1977; Urry and Long, 1978). Characteristically curve (a) has enhanced optical density at 260 nm and dampened absorption at 190 nm relative to the solution state of curve (b). These data, together with the optical rotatory dispersion spectrum of the pseudo-reference state, were used to correct (as described above) the CD spectrum of the membrane suspension (Fig. 13, curve (b)). The

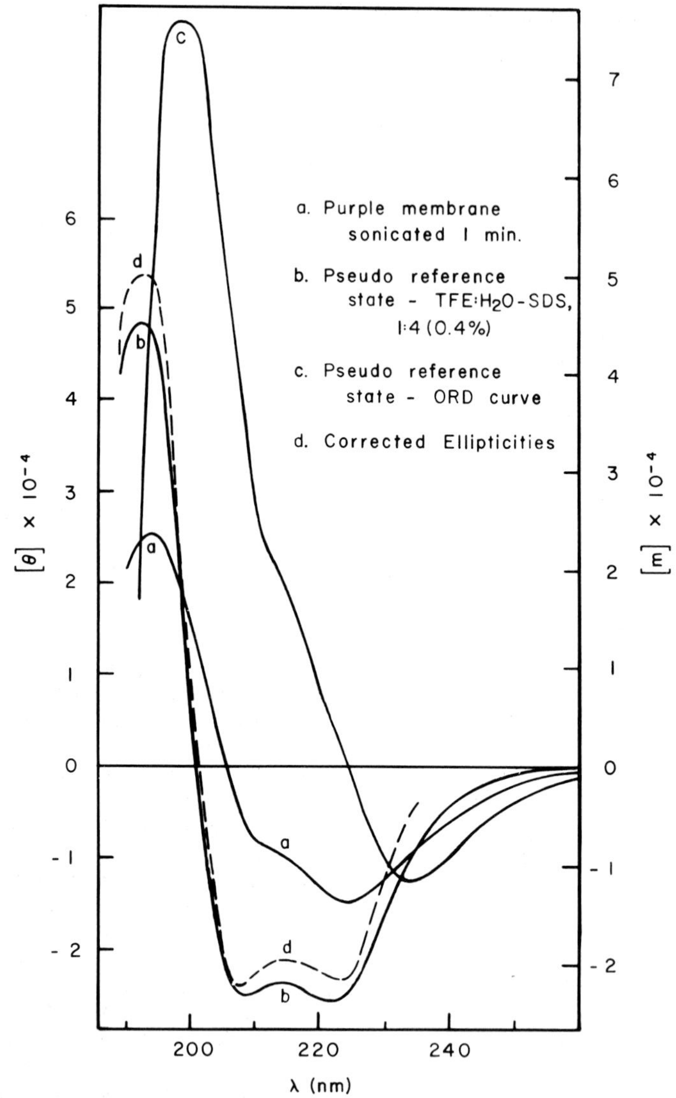

Fig. 13. Ultraviolet CD spectra of a sonicated suspension of the purple membrane (**a**), the purple membrane solution – pseudo-reference state (**b**), the pseudo-reference state optical rotatory dispersion (**c**), and the corrected curve (**d**); (Long et al., 1977)

corrected spectrum is curve (d), from which the proportion of α-helix was calculated to be 73% α-helix at 222 nm and 81% at 190 nm – in marked correspondence with the X-ray data. Not only do the absolute values parallel those found by Henderson and Unwin (1975), but the shape of the corrected curve closely follows that of metmyoglobin, which is 70% α-helix as determined by X-ray (Fig. 14, curve (e)). The striking parallel between the X-ray and corrected CD spectra estimates of percentage of α-helix with this membrane system emphasizes applicability of the pseudo-reference state approach.

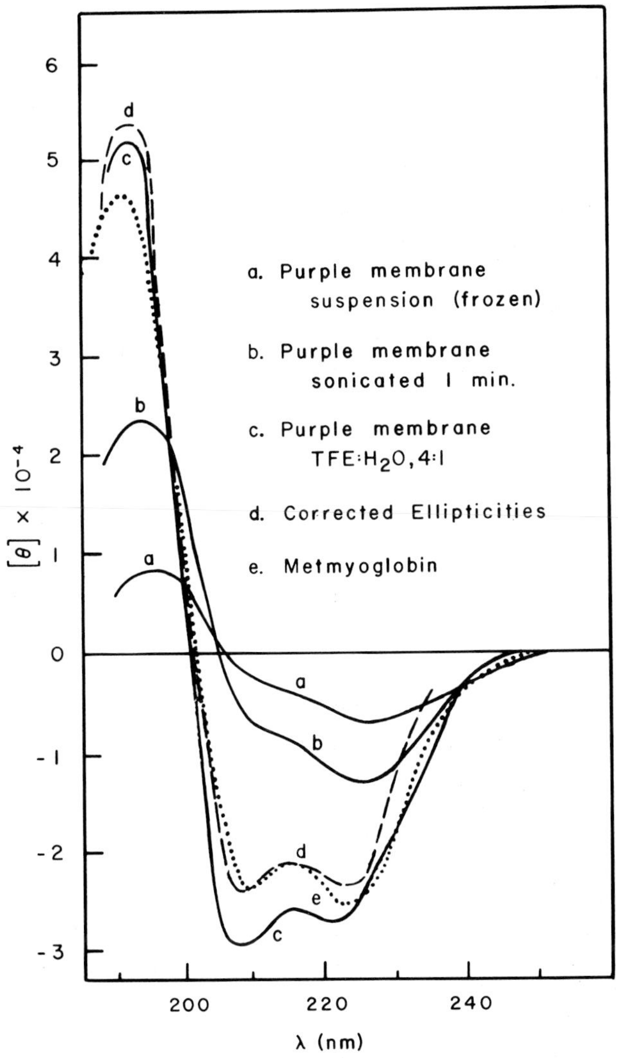

Fig. 14. Ultraviolet CD spectra of the purple membrane after freezing (**a**), after subsequent sonication (**b**), and after subsequent dissolution with trifluoroethanol (TFE); (**c**). The TFE spectrum is similar to the corrected spectrum (**d**), which is similar in shape and magnitude to (**e**), the spectrum of metmyoglobin which is 70% α-helix as shown by X-ray analysis (Long et al., 1977). Curve (**e**) is reprinted with permission from Quadrifoglio and Urry (1968a). Copyright by the American Chemical Society

Acknowledgment. This work was supported by the National Institutes of Health, Grant No. HL-11310. The authors wish to thank H. Spitzer for his quidance in the techniques of lipid extraction and quantitation.

References

Barman, T.E., Koshland, D.E.: A colorimetric procedure for the quantitative determination of tryptophan residues in proteins. J. Biol. Chem. *242*, 5771-5776 (1967).

Bartlett, G.R.: Phosphorus assay in column chromatography. J. Biol. Chem. *234*, 466-471 (1959).

Blaurock, A.E.: Bacteriorhodopsin: Transmembrane pump containing α-helix. J. Mol. Biol. *93*, 139-158 (1975).

Davydov, A.S.: Theory of Molecular Excitons (translated by Kasha, M., Oppenheimer, M., Jr.) New York: McGraw-Hill 1962.

Dodge, J.T., Mitchell, C., Hanahan, D.J.: The preparation and chemical characteristics of haemoglobin-free ghosts of human erythrocytes. Arch. Biochem. Biophys. *100*, 119-130 (1963).

Duysens, L.N.M.: The flattening of the absorption spectrum of suspensions, as compared to that of solutions. Biochim. Biophys. Acta *19*, 1-12 (1956).

Fiske, C.H., Subbarow, Y.: The colorimetric determination of phosphorus. J. Biol. Chem. *66*, 375-400 (1925).

Gordon, D.J.: Mie scattering by optically active particles. Biochemistry *11*, 413-420 (1972).

Gordon, D.J., Holzwarth, G.: Artifacts in the measured optical activity of membrane suspensions. Arch. Biochem. Biophys. *142*, 481-488 (1971).

Gratzer, W.B., Holzwarth, G.M., Doty, P.: Polarization of the ultraviolet absorption bands in α-helical polypeptides. Proc. Natl. Acad. Sci. USA *47*, 1785-1795 (1961).

Henderson, R.: The structure of the purple membrane from *Halobacterium* halobium: Analysis of the X-ray diffraction pattern. J. Mol. Biol. *93*, 123-138 (1975).

Henderson, R., Unwin, P.N.: Three-dimensional model of purple membrane obtained by electron microscopy. Nature *257*, 28-32 (1975).

Holzwarth, G., Doty, P.: The ultraviolet circular dichroism of polypeptides. J. Am. Chem. Soc. *87*, 218-228 (1965).

Kasha, M.: Energy transfer mechanisms and the molecular excitor model for molecular aggregates. Radiat. Res. *20*, 55-71 (1963).

Kasha, M., El-Bayoumi, A., Rhodes, W.: Excited states of nitrogen base-pairs and polynucleotides. J. Chem. Phys. *58*, 916-925 (1961).

Long, M.M., Urry, D.W., Ohnishi, T.: Circular dichroism of repeat pentapeptide of elastin. Int. Res. Commun. Syst. *2*, 1352 (1974).

Long, M.M., Urry, D.W., Stoeckenius, W.: Circular dichroism of biological membranes: purple membrane of *Halobacterium halobium*. Biochem. Biophys. Res. Commun. *75*, 725-731 (1977).

Lowry, O.H., Rosenbrough, N.L., Farr, A.L., Randall, R.J.: Protein measurement with the folin phenol reagent. J. Biol. Chem. *193*, 255-275 (1951).

Martonosi, A., Halpin, R.A.: X. The protein composition of sarcoplasmic reticulum membranes. Arch. Biochem. Biophys. *144*, 66-77 (1971).

Miles, D.W., Urry, D.W., Reciprocal relations in the circular dichroism of adenosine mononicotinate. J. Phys. Chem. *71*, 4448-4454 (1967).

Moffitt, W.: Optical Rotatory dispersion of helical polymers. J. Chem. Phys. *25*, 467-478 (1956a).

Moffitt, W.: The optical rotatory dispersion of simple polypeptides. Proc. Natl. Acad. Sci. USA *42*, 736-746 (1956b).

Moscowitz, A.: Optical Rotatory Dispersion. (ed. Djerassi, C.) pp. 150-177. New York: McGraw-Hill 1960.

Nakao, T., Tashima, Y., Nagano, K., Nakao, M.: Highly specific sodium-potassium-activated adenosine triphosphatase from various tissues of rabbit. Biochem. Biophys. Res. Commun. *19*, 775-788 (1965).

Ooi, T., Scott, R.A., Vanderkooi, G., Scheraga, H.A.: Conformational analysis of macromolecules. IV. Helical structures of poly-l-alanine, poly-l-valine, poly-β-methyl-l-aspartate, poly-γ-methyl-l-glutamate and poly-l-tyrosine. J. Chem. Phys. *46*, 4410-4416 (1967).

Quadrifoglio, F., Urry, D.W.: Ultraviolet rotatory properties of polypeptide in solution. I. Helical poly-L-alanine. J. Am. Chem. Soc. *90*, 2755-2760 (1968a).

Quadrifoglio, F., Urry, D.W.: Ultraviolet rotatory properties of polypeptides in solution. II. Poly-L-serine. J. Am. Chem. Soc. *90*, 2760-2765 (1968b).

Rhodes, W.: The hypochromism and other spectral properties of helical polynucleotides. J. Am. Chem. Soc. *83*, 3609-3617 (1961).

Rosenberg, S.A., Guidotti, G.: The protein of human erythrocyte membranes. I. Preparation, solubilization and practical characterization. J. Biol. Chem. *243*, 1985-1992 (1968).

Seraydarian, K., Mommaertz, W.F.H.M.: Density gradient separation of sarcotubular vesicles and other particulate constituents of rabbit muscle. J. Cell. Biol. *26*, 641-656 (1965).

Sober, H.A. (ed.), in: Handbook of Biochemistry B-73. Cleveland, Ohio: The Chemical Rubber Company 1970.

Starcher, B.C.: A modified two-column procedure for the analysis of the basic amino acids found in elastin, collagen and antibiotics. J. Chromatogr. *38*, 293-295 (1968).

Tinoco, I., Jr.: Hypochromism in polynucleotides. J. Am. Chem. Soc. *82*, 4785-4790 (1960) and Optical and other electronic properties of polymers. J. Chem. Phys. *34*, 1067 (1961).

Tinoco, I., Jr.: Theoretical aspects of optical activity. Part 2: Polymers. Adv. Chem. Phys. *4*, 113-160 (1962).

Tomimatsu, Y., Vitello, L., Gaffield, W.: Effect of aggregation on the optical rotatory dispersion of poly(α, L-Glutamic Acid). Biopolymers *4*, 653-662 (1966).

Unwin, P.N., Henderson, R.: Molecular structure determination by electron microscopy of unstained crystalline specimens. J. Mol. Biol. *94*, 425-440 (1975).

Urry, D.W.: Protein-heme interactions in heme proteins: Cytochrome C. Proc. Natl. Acad. Sci. USA *54*, 640-648 (1965).

Urry, D.W.: Optical rotation. Annu. Rev. Phys. Chem. *19*, 477-530 (1968).

Urry, D.W.: Spectroscopic Approaches to Biomolecular Conformation (ed. Urry, D.W.), pp. 33-121. Chicago, Ill.: American Medical Association Press 1970.

Urry, D.W.: Protein conformation in biomembranes: Optical rotation and absorption of membrane suspensions. Biochim. Biophys. Acta Biomembr. Rev. *265*, 115-168 (1972).

Urry, D.W.: Determining biomolecular conformations. IV. Circular dichroism. Res. Dev. *25*(1), 20-24 (1974).

Urry, D.W., Long, M.M.: Methods in Membrane Biology (ed. Korn, E.D.), Vol. 1, pp. 105-141. New York: Plenum Press 1974.

Urry, D.W., Long, M.M.: Conformations of the repeat peptides of elastin in solution. CRC Crit. Rev. Biochem. *4*, 1-45 (1976).

Urry, D.W., Long, M.M.: Ultraviolet absorption. Circular Dichroism, and optical toratory dispersion in biomembrane studies. In: Physiology of Membrane Disorders (eds. Andreoli, T.E., Hoffman, J.F., Fanestil, D.D.), pp. 107-124. New York: Plenum Press 1978.

Urry, D.W., Ohnishi, M.: Spectroscopic approaches to biomolecular conformation (ed. Urry, D.W.), pp. 263-300. Chicago, Ill.: American Medical Association Press 1970.

Urry, D.W., Hinners, T.A., Masotti, L.: Calculation of distorted circular dichroism curves for poly-L-glutamic acid suspensions. Arch. Biochem. Biophys. *137*, 214-221 (1970).

Urry, D.W., Long, M.M., Ohnishi, T., Jacobs, M.: Circular dichroism and absorption of the polytetrapeptide of elastin: A polymer model for the β-turn. Biochem. Biophys. Res. Commun. *61*, 1427-1433 (1974).

Urry, D.W., Long, M.M., Jacobs, M., Harris, R.D.: Conformation and molecular mechanisms of carriers and channels. Ann. N.Y. Acad. Sci. *264*, 203-220 (1976).

Velluz, L., Legrand, M., Grosjean, M.: Optical Circular Dichroism Principles, Measurements and Applications, p. 21. New York: Academic Press 1965.

Optical Spectroscopy of Monolayers, Multilayer Assemblies, and Single Model Membranes

J. Heesemann[1] and H.P. Zingsheim[2]

I. Introduction

During the past few years considerable advances in the knowledge of membrane structure have been made. Undoubtedly, spectroscopic probes – dyes in particular – have greatly contributed to that progress. The success of the probe approach depends on the ability to introduce the probe into a selected region of the membrane and to observe interpretable signals from it.

The response of spectroscopic probes to changes in environment (polarity, fluidity, electric field strength, charge density, ion concentrations, molecular conformations, etc.) will always be determined by the same physical principles, regardless of the origin of the system under investigation (see Chapter 3, this volume, and Radda, 1975). This is the fundamental reason why so much effort has been devoted to the study of dyes in model systems. The aim of such work is to understand the probes' behavior under well-characterized and controlled conditions. With model systems, one assumes that these conditions can be designed by the experimenter to represent situations found (or anticipated) in biological membranes. The biological relevance of work with model systems depends almost entirely on the validity of those assumptions which, in turn, depend on the existence of a priori knowledge of the properties of the model system.

In many cases, unfortunately, rather little attention was paid to these considerations and – equally important – to perturbations caused by the incorporation of the probes. Frequently, the net result of considerable experimental effort has been merely an increased understanding of a probe's behavior, rather than of membrane structure.

The growing awareness of these difficulties has led to numerous attempts to design lipid-like dye probes. It also seems clear now that it is not sufficient just to calibrate a spectroscopic probe by using a well-defined model system. Equally important would be experimental results, which may provide guidelines for anticipating the effects caused by the eventual incorporation of the probe into a biological membrane.

From this methodical point of view, spectroscopic studies of monolayers, multilayer lamellar systems, or single lipid bilayer membranes offer unique advantages. Apart from their geometric simplicity, such systems are often quite amenable to experimentation, particularly to precisely controlled manipulation of their composition, surface pressure, and surface charge; transmembrane concentration or potential gradients may

1 Institut für Med. Mikrobiologie und Immunologie, Universitäts-Krankenhaus Eppendorf, D-2000 Hamburg 26, FRG

2 Max-Planck-Institut für biophysikalische Chemie (Karl-Friedrich-Bonhoeffer-Institut) Abt. molekularer Systemaufbau, D-3400 Göttingen-Nikolausberg, FRG

be easily established; the study of transport processes may be carried out by well-established techniques. In principle, all these techniques may be combined with simultaneous spectroscopic investigations on incorporated dye probes, natural pigments, or chromophore-labeled functional membrane components (e.g., carriers, pores).

In this article we have concentrated on reviewing a few specific examples, which particularly clearly illustrate the points made above. Therefore we have deliberately sacrificed encyclopedical completeness for the sake of a critical evaluation biased by our own opinions and experience.

II. General Considerations and History

A. Monolayers and Multilayer Assemblies

Monolayers at the air-water interface provide a simple direct method for testing dye probes under membrane-like conditions because of their simple geometry and their controlled composition and surface pressure. Much of the relevance of monolayer work rests on the fact that a bilayer membrane – to a first approximation – may be visualized as two apposed monolayers (Fettiplace et al., 1975). Of course, apart from artificial probes the behavior of natural pigments (so-called intrinsic probes), e.g., chlorophyll, are studied spectroscopically in monolayers in order to facilitate the interpretation of absorption and fluorescence spectra obtained from the photosynthetic membrane.

Below we present a brief list of the information obtainable from studies on mono- and multilayer systems. It will be discussed in detail in the following sections.

1) The interactions of the dye with the lipid at the interface can be studied by measuring the surface pressure-surface area isotherms.
2) The orientation of transition moments can be determined from dichroic absorption spectra.
3) The existence of dye monomers, dimers, or aggregates can be verified via the absorption or fluorescence spectra.
4) The location of the chromophores may be estimated by energy transfer measurements.
5) Surface potentials and interfacial ion concentrations (interfacial pH) can be measured with lipid pH-indicators.
6) Electrochromic dyes can be calibrated for their eventual use in transmembrane potential measurements.

Originally, monolayer spectroscopy did not originate from a biologically motivated interest in testing dye probes but from the concept of constructing – by the monolayer assembly technique – functional units of molecular dimensions with surface-active dyes and fatty acids as structural elements. The refined spectroscopic techniques developed for this field entailed a deep insight into the behavior of dyes in monolayers, e.g., the equilibrium between monomers and dimers, the molecular arrangement of aggregates, the orientation of chromophores in monolayers and the quenching of fluorescence by energy transfer.

Monomolecular films at the air-water interface can be obtained in a simple manner as first published by Pockels (1891). Lipid-like molecules are dissolved in an organic solvent and dropped on to a clean water surface. After a few minutes, the organic solvent evaporates and the solute molecules remain at the water surface forming a monomolecular film. The surface pressure (π) associated with the packing pressure of the spread molecules can be measured as a function of the surface area per molecule (a). From the π-a isotherm one obtains first indications of the packing and conformation of the film molecules under membrane-like conditions. Compressed monolayers can be transferred on to solid supports and it is even possible to build up multilayer systems (Langmuir, 1920).

Several reviews of fluorescence and absorption spectroscopy of monolayers and multilayers have been published (Gaines, 1966; Kuhn et al., 1972; Möbius, 1978). In this contribution we will cover selected aspects of dye probes in monolayers and multilayers which may be relevant for biomembrane spectroscopy.

Spectrometers to measure the optical absorption, reflection, and fluorescence of monomolecular films have been described by Tweet (1963), Brody (1971), Kuhn et al. (1972), Fromherz (1973a), and Aoshima et al. (1973).

B. Black Lipid Membranes

The black lipid membrane (BLM) (Mueller et al., 1962) is one of the best characterized model membranes. Stable membranes may be formed from a wide range of lipids and the membranes may be made in a large variety of geometrical configurations (planar, hemispherical, spherical) and over a large range of sizes (from a few microns to a few millimeters) (Fettiplace et al., 1975). Their main advantage is the fact that these membranes separate two experimentally accessible aqueous compartments; this makes the system eminently suitable for the measurement of transport processes. The BLM is not exactly a bilayer, a fact that is often neglected. More precisely, the system must be viewed as a thin film of hydrocarbon solvent that is stabilized by two lipid monolayers. Thus, the nature of the system demands that it contains a certain fraction of hydrocarbon solvent because it is in thermodynamic equilibrium with an adjacent bulk solution (the so-called Plateau-Gibbs border). The thermodynamic theory of the BLM as well as contact angle measurements show that the tension of each of the lipid monolayers stabilizing the film is only very slightly ($< 1\%$) smaller than the tension of the respective monolayer at the interface between the bulk phases (cf. Fettiplace et al., 1975). This is of great significance because – for most practical purposes – one may apply to the BLM the data obtained from individual monolayers.

A slightly different type of bilayer model membrane are the so-called "solventless" bilayers (Montal and Mueller, 1972), which can be made by the mechanical apposition of two surface lipid monolayers. The difference between this system and the "classical" BLM is of little consequence in the context of this contribution; most spectroscopic measurements discussed here have been made with the black lipid films, which contained some solvent.

Spectroscopic investigations of lipid bilayer model membranes have developed independently from monolayer spectroscopy. This was due to the close relation of bilayer work to membrane biology, where spectroscopic studies of probe molecules had been carried out for many years.

The information gained by spectroscopic studies on these model membranes is in many respects similar to that listed above for monolayers:

1) The membrane concentration of the absorbing or fluorescent species can be accurately determined. This is particularly important in combination with kinetic studies of transport mechanisms.
2) The orientation and location of the dye molecules can be established by polarization measurements.
3) Dimerization and the formation of higher dye aggregates can be investigated.
4) Rotational and translational diffusion can be measured by fluorescence depolarization and concentration correlation spectroscopy.
5) The study of many of the above effects as a function of membrane potential may help in understanding the potential dependent effects observed with a number of dyes in biological membranes.

Spectroscopic measurements on BLM entail a number of technical problems, which are direct consequences of the nature of the system. They may be overcome by a careful design of the experiment. A virtually complete discussion of these aspects has been given by Fettiplace et al. (1975), so that the following brief list should suffice:

1) The Plateau-Gibbs border must not contribute to the measurements.
2) Bulk solvent trapped in the black film (lenses) may cause problems due to light scattering or due to contributions from chromophores present in the bulk solvent.
3) The solid film support must not contribute to scattered light.
4) Artifacts due to emulsion droplets created by rupture of films must be eliminated.

III. Design of Spectroscopic Probes

A monomolecular layer at an aqueous interface may be considered as a phase consisting of two regions: (a) the interfacial region consisting of the head groups of the constituent molecules, and (b) the nonaqueous region consisting of the hydrocarbon tails of the tensides.

A successful method of putting dye probes into specific regions within a monolayer is the covalent attachment of chromophores at certain points along a paraffin chain. Probes for the interfacial region are obtained in the form of ionic chromophores with long-chain paraffin substituents (Giles and Neustadter, 1952; Kuhn et al., 1972; Fukuda et al., 1976; Gruda and Leblanc, 1976). Labeling in the nonpolar region is achieved by hydrophobic chromophores, e.g., neutral aromatic compounds, which are attached to the upper part of the paraffin chain of a fatty acid (Waggoner and Stryer, 1970; Heesemann, 1976; Cadenhead et al., 1977). An alternative is the attachment of chromophores to the ω-positions of short-chain fatty acids and the linking of the obtained "chromophoric acids" via ester bonds to monoglycerides (Heesemann, 1976, see p. 191 dye **XIV**). The next step is to choose the appropriate chromophore. This depends on the response of the probe to its environments: pH-indicator dyes are of course sensitive to the interfacial pH; energy transfer experiments require strong fluorophores with a sufficient overlap of the fluorescence band of the sensitizer and the absorption band of the acceptor; chromophores used as electric field indicators are found in the group of solvatochromic dyes. In Table 1 we give some examples of surface-active dyes used in monolayer spectroscopy.

Table 1

STRUCTURE	APPLICATION, WAVELENGTH OF ABSORPTION AND FLUORESCENCE MAXIMA, λ^A AND λ^F, RESPECTIVELY IN MIXED FILMS.
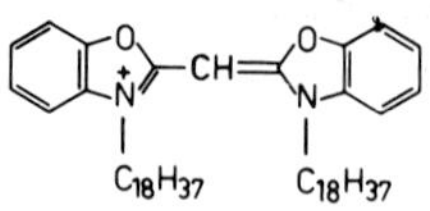 I	SENSITIZER FOR ENERGY TRANSFER MIXTURE: DYE/ARACHIDIC ACID = 1/50 λ^A = 384 [NM]; λ^F = 420 [NM].
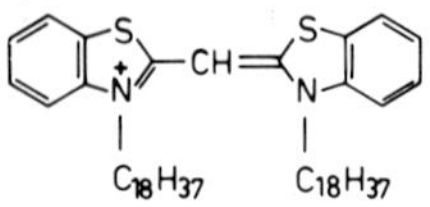 II	ACCEPTOR FOR ENERGY TRANSFER MIXTURE: DYE/ARACHIDIC ACID = 1/50 λ^A = 430 [NM]; λ^F = 490 [NM].
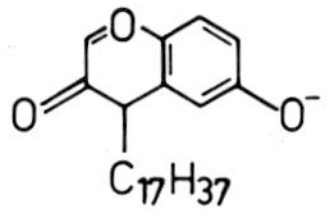 III	INTERFACIAL PH-INDICATOR AND INTERFACIAL ELECTRIC POTENTIAL PROBE MIXTURE: DYE/LIPID = 1 : 400 EXCITING WAVELENGTH λ^E = 366 [NM], λ^F = 440 [NM].
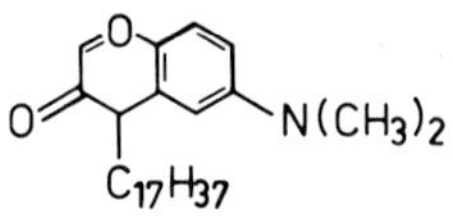 IV	EXCITING WAVELENGTH λ^E = 366 [NM], λ^F = 450 [NM].
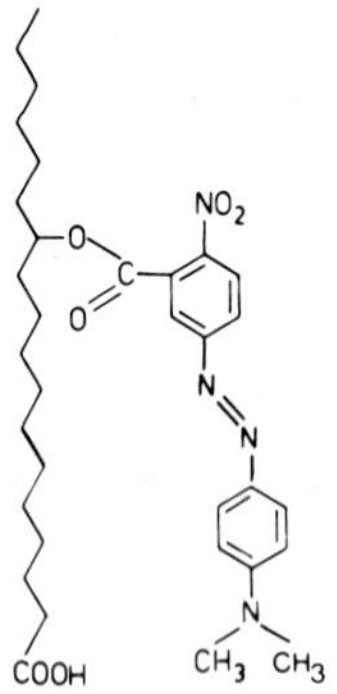 V	TRANSMEMBRANE ELECTRICAL POTENTIAL PROBE MIXTURE: DYE/STEARIC ACID/ STEARYL AMINE/ = 1/15/16 MONOMER BAND λ^A = 475 [NM], DIMER BAND λ^A = 425 [NM], ASSOCIATION CONSTANT FOR THE MONOMER-DIMER EQUILIBRIUM $4.4 \cdot 10^{-14}$ cm^2/DYE MOLECULE.

For historical reasons mainly, the dye probes used in bilayer work are different from those used in monolayer spectroscopy, although the design criteria are similar (Waggoner and Stryer, 1970). However, the effect of a surface-active dye on the stability of the bilayer membranes is an additional point of concern. Table 2 lists a few dyes frequently used in spectroscopic studies with bilayer lipid membranes. A new dimension has been introduced into this field by the use of dye-labeled functional membrane constituents, such as Gramicidin A analogs labeled with dansyl or azo dyes (Veatch et al., 1975; Veatch and Stryer, 1977).

Table 2

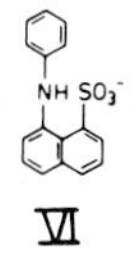 VI	1-ANILINONAPHTALENE-8-SULFONATE (ANS)
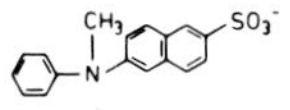 VII	2-(N-METHYLANILINO)NAPHTALENE-6-SULFONATE (2,6-MANS)
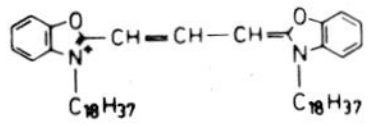 VIII	N,N'-DI(OCTADECYL)OXACARBOCYANINE
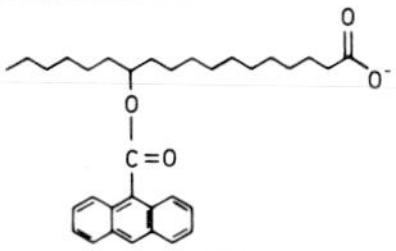 IX	12-(9-ANTHROYL)-STEARIC ACID
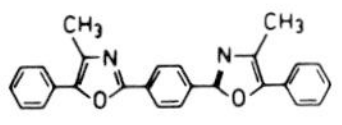 X	P-BIS[2-(4-METHYL-5-PHENYLOXAZOLYL)]BENZENE (DIMETHYL-POPOP)
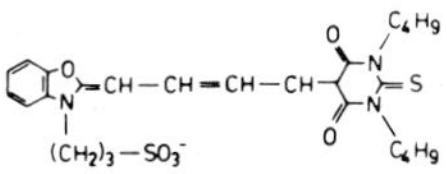 XI	MEROCYANINE 540
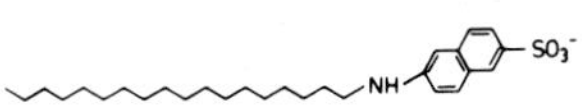 XII	2-(OCTADECYLAMINO)NAPHTALENESULFONATE
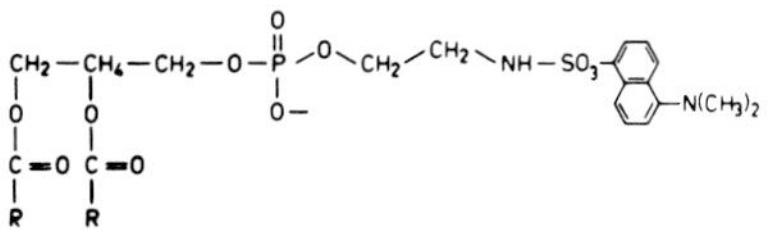 XIII	DANSYLATED PHOSPHATIDYLETHANOLAMINE

IV. Concentration, Orientation, and Location of Chromophores

A. Monolayers and Multilayer Assemblies

Monolayer spectroscopy with polarized light is a valuable tool for determining the orientation of chromophores in monolayers. The dye monolayer is traversed by a beam of polarized light at an oblique angle and the absorption (A) is measured for s-polarized light (electric vector of the light is normal to the plane of incidence, superscript s) and for p-polarized light (electric vector of the light is parallel to the plane of incidence, superscript p). Under the condition that the direction of the transition moment is known relative to the geometry of the dye molecule, the dichroic ratio $D = A^s/A^p$ provides information about the orientation of the chromophores in the monolayer. For chromophores with transition moments normal to the layer plane the expected value is $D = 0$ (as $A^s = 0$ and $A^p \neq 0$). This value is experimentally obtained with closely packed monolayers of dye V and some others (Heesemann, 1980a,b). With transition moments statistically oriented in the layer plane the dichroic ratio is calculated as $D = 1.4$ for monolayers at the air-water interface (angle of incidence $\alpha = 45°$). The same value of D was found for the expanded monolayer of dye V at $a = 115$ Å^2/molecule, where the chromophores are lying flat on the water surface, as concluded from the π-a curve (Heesemann, 1980a). From polarized spectra of monolayers of cyanine dyes such as dye I and II deposited on glass slides a statistical orientation of the transition moments in the layer plane was inferred (Bücher et al., 1967; Kuhn et al., 1972). From the angular distribution of the fluorescence of mixed monomolecular films containing oleyl alcohol and a few percent of chlorophyll a, the direction of the transition moments was determined (Tweet et al., 1964). The angles of the transition moments in the red gel and in the blue were estimated as 20° and 28° respectively.

B. Black Lipid Membranes

The most fundamental application of optical spectroscopy to black lipid membranes is the measurement of the membrane concentration of chromophores.

The first such work was reported by Alamuti and Läuger (1970), who estimated, by fluorescence measurements, the proportion of chlorophyll a in BLMs formed from dioleoyl-lecithin in n-decane. Steinemann et al. (1971), in an investigation of the same system, also determined fluorescence polarization and optical absorption. They concluded that, depending on its bulk concentration in the membrane-forming solution, up to $n = 3 \times 10^{13}$ molecules/cm^2 of chlorophyll a were incorporated into the membranes; this is equivalent to a mean distance of 20 Å between porphyrin rings. For distances below 30 Å ($n > 10^{13}$/cm^2) the fluorescence intensity was quenched. This was attributed to energy transfer (Förster, 1959). These results agree well with results from spread monolayers (Trosper et al., 1968) at the air-water interface (see above) and with estimates based on adsorption data from hydrocarbon-water interfaces (Trosper, 1972). There is also good agreement between these data and the interpretation of polarized absorption spectra of chlorophyll a in BLMs by Cherry et al. (1971a,b).

A problem common to all these measurements was their calibration. Regardless of how the sensitivity of the apparatus is determined (this is generally a straightforward exercise), assumptions had to be made as to the extinction coefficient or the quantum yield of chlorophyll a. Generally, data from bulk solutions have been applied to black lipid membranes (see also Trissl, 1974). The justification of this procedure has already been discussed by Steinemann et al. (1971).

The composition of black lipid membranes containing chlorophyll a seems to agree well with estimates based on adsorption data from interfacial tension measurements (Trosper, 1972). Nevertheless, these data are not sufficiently accurate and complete to allow the conclusion that calculations based on interfacial tension measurements and the application of the Gibbs adsorption equation (cf. Aveyard and Haydon, 1973; Fettiplace et al., 1975) yield correct results. Experimental proof of this contention is by no means trivial; in fact, there are only very few experimental verifications of the Gibbs adsorption equation in the literature (cf. Aveyard and Haydon, 1973).

A suitable system for this purpose is represented by BLMs onto which a surface-active, water-soluble fluorescent dye is adsorbed from the aqueous phase. If both lipid and dye are in adsorption equilibrium in such a system, the Gibbs equation should be applicable. Zingsheim and Haydon (1973) reported on an investigation in which they determined the adsorption of 1-anilinonaphtalene-8-sulfonate(ANS) (see Table 2) to BLMs made from glyceroyl-monooleate (GMO) in n-decane. The system is relatively complicated because ANS contributes (with different quantum yields) to the fluorescence of the membrane and the adjacent aqueous phase. This makes the above-mentioned technical prerequisites particularly stringent. In order to obtain the amount of ANS adsorbed to the membrane, use was made of the fact that one side of a black film does not differ significantly from the interface between the equilibrium bulk lipid solution and the aqueous phase, as outlined above. The authors determined the interfacial tension between a bulk aqueous phase containing dissolved ANS and a bulk phase of n-decane containing GMO at a concentration above the critical micelle concentration. Application of the Gibbs equation yielded the amount of adsorbed ANS. This analysis was further simplified by the fact that ANS and GMO are, for all practical purposes, mutually insoluble in their opposite phases. The corrected fluorescence spectra (λ_{max} = 505 nm) and the quantum yield (ϕ = 0.2) of ANS in GMO black films agree well with the respective values on sonicated aqueous dispersions of the lipid. This work is of great practical significance, because it provides an easily accessible membrane fluorescence standard, which allows the calibration of other experiments. The confirmation of the validity of the Gibbs adsorption equation for black lipid films is of considerable fundamental significance.

Spectroscopy has been used successfully in order to determine the concentrations of active species in black lipid films. Veatch et al. (1975) have carried out simultaneous measurements of the fluorescence and conductance of dansyl Gramicidin C in BLMs. Dansyl Gramicidin C is a fluorescent, active analog of Gramicidin A, which forms trans-membrane channels for cations. At sufficiently low membrane concentrations, discrete conductance fluctuations can be observed (Hladky and Haydon, 1970), which arise from the formation and breakdown of individual conducting channels. A number of observations (Tosteson et al., 1968; Goodall, 1970) suggested that the conducting species is a dimer, in equilibrium with the nonconducting monomer in the membrane. This hypothesis is consistent with the results of a kinetic analysis of channel formation (Bamberg and Läuger, 1973; Zingsheim and Neher, 1974; Kolb et al., 1975).

Veatch et al. (1975) succeeded in determining the membrane concentration of dansyl Gramicidin C in BLMs via fluorescence measurements. Their instrument was calibrated with the aid of a GMO-ANS membrane fluorescence standard (see above, Zingsheim and Haydon, 1973). (The lowest density of dansyl Gramicidin C that could be detected corresponds to 60 fluorescent molecules per μm^2.) The density of conducting channels was ascertained from simultaneous membrane conductance measurements. It was concluded that the active channel is a dimer formed from the nonconducting species and that almost all dimers are active channels.

Information on the orientation of chromophores in lipid bilayer membranes can be obtained from polarization measurements in a manner analogous to that outlined for monolayers. Such measurements are relevant to at least three questions:

a) Comparison of the model situation with the situation found in vivo (e.g., with chlorophyll a)
b) Orientation of transition moments, wich must be known for the interpretation of energy transfer experiments (see below)
c) Orientation of transition moments as an important parameter in models explaining the behavior of potential sensitive dyes in excitable membranes and in lipid bilayer model membranes.

Chlorophyll a was among the first molecules studied by polarization measurements on black lipid films. The independent investigations of Steinemann et al. (1971) and Cherry et al. (1971a,b) yielded results which were comparable to those obtained from dichroism measurements in spread monolayers (Tweet et al., 1964).

Yguerabide and Stryer (1971) have reported on measurements of fluorescence excitation, emission, and polarization spectra of dyes VIII to X (Table 2) in spherical black films (Pagano and Thompson, 1967). These membranes can be freely suspended in a density gradient and offer the advantage that all possible orientations of the membrane with respect to the light path are available at the same time. Yguerabide and Stryer (1971) concluded that the transition moment of dye VIII was oriented parallel to the plane of the membrane, as one would expect for this type of dye (Bücher et al., 1967). The transition moment of dye IX was also preferentially oriented in the membrane plane, whereas dye X showed a preferential orientation of its transition moment perpendicular to the membrane plane.

A detailed analysis of the orientation of two water-soluble fluorescence probes, 2,6-MANS (dye VII, Table 2) and 1.8-ANS (dye VI, Table 2) has been presented by Carbone et al. (1976). This work is relevant to the interpretation of potential induced effects (see below).

V. Monomers, Dimers and Aggregates of Dye Probes

Dyes tend to form associates which can be directly observed by spectroscopic methods. Compared to dyes in solution, the chromophore in a monolayer is essentially locally fixed and a much smaller average distance between two chromophores can be realized. Therefore it is not surprising that in dye monolayers aggregate bands can be observed, which cannot be found in dye solutions. A good example is given by dye V. In solution one observes exclusively a monomer band (ethanol: λ_{max} = 470 nm). On the other hand, in monolayers one finds monomers and dimers (monomer: λ_{max} = 475 nm; dimer: λ_{max} = 430 nm), where the dimer is thought to be arranged with the transition moments (which are also the molecular long axis) side by side. With mixed films of dye I and octadecane (ratio 1:1), a narrow absorption and an almost coinciding fluorescence band is observed which is attributed to J-aggregates (Bücher and Kuhn, 1970a). A brick-stone-work-like arrangement of the chromophores was inferred from molecular models and corresponding to this model the monomer-to-aggregate band shift was calculated. It was found to be in good agreement with the experimental shift (Czikkely et al., 1970). H-aggregates (transition moments are arranged in a pin-cushion planar lattice) were found with 4-nitro-4'amino azobenzene chromophores (Heesemann, 1980a,b). With the aid of a single-beam, single-reflectance photometer, Fromherz (1973a) investigated the monomer-dimer equilibrium of dye II at the air-water interface as a function of the surface area.

VI. Energy Transfer

The electronic excitation energy of a chromophore S (sensitizer) can be transferred across a certain distance to another chromophore A (acceptor) via dipole-dipole resonance interaction which is known as the Förster type of energy transfer (Förster, 1948, 1959). The quenched fluorescence of S can be demonstrated by a sandwich system of monolayers of dye S and dye A which are separated by arachidic acid monolayers acting as spacers (Zwick and Kuhn, 1962; Kuhn et al., 1972). The fluorescence intensity I_d of S as a function of the distance d between S and A is predicted by the expression

$$\frac{I_d}{I_\infty} = \left(1 + \left(\frac{d_o}{d}\right)^4\right)^{-1} \qquad (1)$$

$$d_o = \alpha \frac{\lambda_s}{n} (q_s A_{AS})^{\frac{1}{4}} \qquad (2)$$

where I_∞ is the fluorescence intensity of S in the absence of A (Drexhage et al., 1963; Kuhn, 1970). The critical distance d_o depends on q_s (the quantum yield of S in the absence of A), on A_{AS} (the absorption of the acceptor for the vacuum wavelength λ_s of the fluorescence radiation), on the refractive index n of the medium and on the orientation factor α which is a function of the orientation of the transition moment of S relative to that of A. As d_o is a function of many variables, it is practically impossible to predict theoretically exact values of d_o for complex systems.

With the monolayer assembly technique d_o can be measured; with dye I as sensitizer, dye II as acceptor, and Cd-arachidate interlayers a value of 73 Å for d_o was found, in agreement with the theoretical prediction. The sensitizer and acceptor layers can also be separated by a biological membrane, so that the membrane thickness can be determined by energy transfer measurements. This was done with air-dried human erythrocyte ghost membranes which had been embedded between a monolayer of a fluorescent dye (sensitizer) and a thin gold layer (acceptor) (Peters, 1973a,b). The thickness of the ghost membrane was derived by comparing the fluorescence intensity on ghosts with the intensity from a reference system consisting of Cd-arachidate interlayers (thickness of one layer: 28 Å). The thickness of a single membrane was found to be 65 Å.

Fluorescence quenching in mixed monolayers of chlorophyll a (sensitizer) and copper pheophytin a (nonfluorescent acceptor) was observed at the air-water interface (Tweet et al., 1964). The quenched fluorescence of chlorophyll was measured as a function of the quencher concentration, and interpreted in terms of a Förster energy transfer mechanism with a critical transfer distance of 40 Å between chlorophyll and quencher. The fluorescence depolarization from chlorophyll in monolayers was observed to increase with increasing surface density of the pigment. This can be explained by a Förster energy transfer mechanism (Trosper et al., 1968).

There have been only a few investigations of energy transfer in black lipid films. In their work on chlorophyll a, Alamuti and Läuger (1970) and Steinemann et al. (1971) presented evidence for the existence of energy transfer between chlorophyll a molecules. Pohl (1972) reported a detailed investigation with black lipid membranes of the energy transfer (including orientation effects) between chlorophyll b and chlorophyll a, between dye X (Table 2) and chlorophyll a and between 1.6-diphenyl-1,3,5-hexatriene and chlorophyll b. It should be noted that such measurements in black lipid films relate to energy transfer within, not across, the plane of the membrane.

Measurements of energy transfer across black lipid films have not yet been made. In this respect it is interesting to refer to some earlier work, not on black films in water, but on soap films in air, by Mormann and Kuhn (1969). Soap films in air, being thin films of water stabilized by detergent, are in many ways analogous to black lipid films, which are thin films of hydrocarbon solvent, stabilized in water by lipids. Mormann and Kuhn (1969) demonstrated energy transfer between two soap films which had been brought into molecular contact and could be made to fuse. Although similar manipulations can be performed with black lipid films (Neher, 1974), analogous spectroscopic measurements have not been reported.

VII. Lipoid pH-Indicators in Monolayers

A charged interface can interact with an electrolyte solution. As a result, a diffuse electrical double layer associated with an electrical potential is formed near the interface. As a consequence of the surface potential the ion concentration at the surface differs from that of the water solution. For example, the proton concentration at the interface $[H^+]_i$ is related to the bulk concentration $[H^+]_w$ by a Boltzmann factor

$$[H^+]_i = [H^+]_w \exp(e\psi^i/kT) \qquad (3)$$

where ψ^i is the electrical potential in the plane of the fixed charges and e, k, and T have their usual meaning. According to the Gouy-Chapman theory the electrical potential ψ^i for a 1:1 electrolyte is given by

$$\psi^i = \frac{2kT}{e} \operatorname{arsinh} (\sigma^2 e^2 / 8\epsilon\epsilon_o ckT)^{\frac{1}{2}} \tag{4}$$

where σ is the effective charge density of the surface, c the concentration of the 1:1 electrolyte, ϵ and ϵ_o the relative and the vacuum permittivities respectively.

The validity of Eqs. (3) and (4) for lipid monolayers of various compositions has been tested experimentally via the determination of the "interfacial" pH_i with fluorescent pH-indicators. Fromherz (1973b) and Fromherz and Masters (1974) doped monolayers of various charge densities (obtained by mixing neutral, positively, and negatively charged tensides) with the proton-sensitive dye III. The fraction of dissociated dye was measured directly by the relative fluorescence intensity as a function of the bulk pH_w. From titration curves of different monolayers the $pK_{m\,w}$ of the membrane-bound indicator was obtained and related to the interfacial pH_i and to the environment polarity of the membrane-bound dye. A comparison of the experimental ψ^i with the theoretical ψ^i [according to Eq. (4)] shows agreement for neutral and positively charged films. However, with negatively charged films significant deviations were found, indicating strong interactions between cations in the bulk phase and the negatively charged headgroups of the tensides.

The theoretical basis for the application of lipoid pH-indicators as probes of interfacial potentials in membrane systems was pointed out by Fernandez and Fromherz (1977) and was tested with micelles by the application of dyes III and IV. In Fig. 1, the two acid-base equilibria of the indicator, dissolved in water (w) and membrane-bound (m), are depicted. The experimentally determined $pK_{m\,w}$ is related to the interfacial pK_i and the interfacial pH_i by

$$pK_{m\,w} = pK_i - (pH_i - pH_w). \tag{5}$$

There are two approaches for calculating pH_i from the experimentally measured values of $pK_{m\,w}$, pK_w, and pH_w.

1. According to Eq. (4) $\psi^i = O$ for an electrically neutral surface and consequently $pH_i = pH_w$. Therefore [Eq. (5)] the observed $pK_{m\,w}$ of an electrically neutral membrane is equal to pK_i. The difference $\Delta pK_i = pK^o_{m\,w} - pK_w$ is then attributed to a polarity shift. Under the condition that the pK_i of the indicator bound to a charged membrane remains unchanged ($pK^o_{m\,w} = pK_i$), the observed pK-shift is then due to the electrical potential difference $\Delta\psi^i$ as shown by equation

$$pK_{m\,w} = pK_i = -(pH_i - pH_w) = -\frac{1}{2.3\,RT} \cdot F\,\Delta\psi^i. \tag{6}$$

2. It is assumed that the work required to bring the indicator from the water solution to the nonaqueous medium of the membrane is due to the standard chemical potential difference $\Delta\mu^{om}$ and may be characterized by $\Delta pK_i = pK_i - pK_w$. For dye III (neutral acid, ionic base) and dye IV (ionic acid, neutral base) it is therefore expected that the values of ΔpK_i are of the same magnitude but of opposite sign, $\Delta pK_i^{III} \simeq -\Delta pK_i^{IV}$. This is supported by experiments with micelles of Triton X-100 and sodium dodecyl-

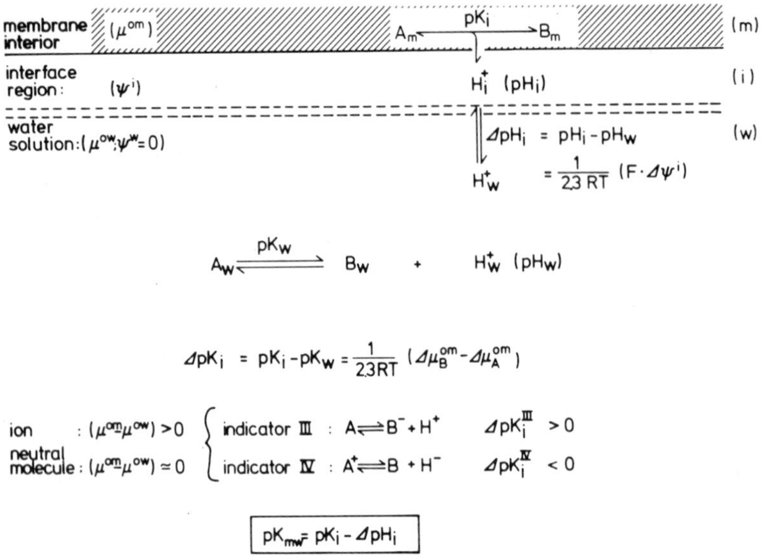

Fig. 1. The membrane-water system is separated schematically into 4 phase compartments: the nonaqueous medium of the membrane interior (m); the interface region (i) which may correspond to the plane of the headgroups of the lipids; the diffuse electrical double layer which is represented here by the two dotted lines; the bulk water (w). The pKi characterizes the two-phase reaction of the membrane-bound acid (Am) and the base (Bm) in the nonaqueous medium, and of the proton (H_i) in the aqueous medium of the interphase. The pK_W characterizes the equilibrium of the dissolved species Aw, Bw, and Hw in bulk water. μ^{om} and μ^{ow} are the single standard chemical potentials of the membrane-bound indicator and the dissolved indicator in water, respectively, and may be related to the environmental polarity of the dye. It will be assumed that the proton equilibrium between the interface and the bulk is entirely due to the interfacial electrical potential, ψ^1. From titration curves (relative fluorescence intensity of the indicator as a function of pH_W) one obtains pK_W and $pK_{mw} = pK_i - \Delta pH_i$. The pK-shift $\Delta pK_{mw} = \Delta pK_i - \Delta pH_i$ is composed of a chemical potential contribution and an electrical potential contribution

sulfate. $|\Delta pK_i| = 1.1$ was found for both indicators by Fernandez and Fromherz (1977). This fact makes it possible to calculate the interfacial ψ^i by adding the pK differences $\Delta pK_{mw} = pK_{mw} - pK_w$ of dye III and IV.

$$\frac{1}{2}\left(\Delta pK_{mw}{}^{III} + \Delta pK_{mw}{}^{IV}\right) = -\frac{1}{2.3\,RT} \cdot F \cdot \Delta\psi^i. \tag{7}$$

By subtracting the ΔpK_{mw} s one obtains the absolute value of ΔpK_i

$$\frac{1}{2}\left(\Delta pK_{mw}{}^{III} - \Delta pK_{mw}{}^{IV}\right) = |\Delta pK_i|. \tag{8}$$

This value was related to the observed pK-shift of the indicators dissolved in a dioxane-water mixture of a dielectric constant of 32. In biomembrane research the method should find a successful application for the determination of the polarity of the interfacial region and the interfacial potential of cell membranes.

VIII. Lateral Diffusion in Model Membranes

Under physiological conditions, most cell membranes are in a fluid state, so that lipids and proteins may move in lateral directions (Edidin, 1974). There have been many observations suggesting that cytoplasmic structures (the "cytoskeleton") are actively involved in the control of membrane movement (Bretscher and Raff, 1975; Edelman, 1976; Nicolson, 1976). On the other hand, these observations have also led to a number of studies which were aimed at an understanding of pure passive diffusion in lipid model membranes.

To this end two techniques have been used. The first, fluorescence photobleaching recovery (FPR), is based on bleaching a sharply bounded area on a membrane with an intensive light beam. Subsequently, at a much reduced light intensity, the reappearance of fluorescence in this area (by diffusion) is monitored as a function of time, so that the diffusion constant of the fluorescent species may be estimated. This is usually done in a fluorescence microscope (Peters et al., 1974). The second method, fluorescence correlation spectroscopy (FCS), measures the stochastic fluctuations around equilibrium of the number of fluorescent molecules within a small membrane area, which is usually defined by a focussed laser beam (Elson and Webb, 1975). The time dependence of the fluctuations is most conveniently determined by calculating their correlation function, from which the diffusion coefficient may be obtained (Madge et al., 1972, 1974; Elson and Madge, 1974).

Fluorescence photobleaching recovery has been used by Wu et al. (1977) to determine the diffusion of dye-labeled lipids in multibilayers made by pressing hydrated lipids between two flat surfaces. The measured diffusion constant values ranged between 4×10^{-8} and 1.5×10^{-7} cm^2/s, with a twofold change at a temperature increase of 10°C. Equimolar cholesterol in the lipid multibilayers reduced the diffusion constants by a factor of 2. An approximately 100-fold reduction was observed when the temperature was lowered below the gel-to-liquid-crystalline transition temperature. Also by FPR, Wu et al. (1978) determined the diffusion of a fluorescence-labeled Gramicidin S derivative in phospholipid multibilayers. Gramicidin S is a hydrophobic, cyclic decapeptide (mol.wt. 1141). Its diffusion coefficient was similar to that of dye-labeled lipids (3.5×10^{-8} cm^2/s). Again, equimolar cholesterol in the multilayer structure lowered this value (to 1.4×10^{-8} cm^2/s) and a reduction of $\simeq$ 100-fold was observed below the phase transition temperature.

Fluorescence correlation spectroscopy was used by Herbert et al. (1974) and by Fahey et al. (1977) to determine the diffusion of dye VIII (Table 2) (Herbert et al., 1974), and of 3,3'-dioctadecylindocarbocyanine iodide (di I) and rhodamine-labeled dipalmitoyl phosphatidylethanolamine in planar black lipid membranes and in the "solvent-free' bilayer membranes (see Sect. I). Diffusion constants of $1\text{-}2 \times 10^{-7}$ cm^2/s were measured. The values turned out to be remarkably independent of membrane composition and temperature. The values for (di I) are approximately tenfold higher than the diffusion coefficients of the probe in rat myotube membranes (Schlessinger et al., 1976, 1977).

An interesting application of the FPR technique was reported by Wolf et al. (1977), who investigated diffusion and patching of marcomolecules in black lipid membranes. The authors were interested in devising a model system in which cell surface phenomena (e.g., "capping", Bretscher and Raff, 1975; Edelman, 1976; Nicolson, 1976) could be studied under conditions where cell metabolism or the cytoskeleton would not be effective. Dextran derivatives (mol.wt. 82,000), to which fatty acid, rhodamine, and an antigenic hapten (2,4,6-trinitrophenyl [TNP]) had been covalently bound have been absorbed to the black film from the aqueous phase and have a diffusion constant of $\simeq 2 \times 10^{-8}$ cm^2/s in black fims of oxydized cholesterol and of $\simeq 5 \times 10^{-8}$ cm^2/s in black films of egg lecithin. Clumps ("patches") are formed after the addition of anti-TNP. Antigen diffusion in the membrane was significantly reduced by cross-linking with anti-TNP. Further cross-linking by anti-Ig completely immobilized the antigen. The diffusion of a fluorescent lipid probe was unaffected by cross-linking. This work has demonstrated that patching can be a purely passive phenomenon induced by cross-linking ligands, that receptor diffusion is retarded by cross-linking, and that cross-linking can be sufficient for receptor immobilization.

IX. Electric Field-Induced Effects

A. General

The effects of electric fields on spectra in the optical region can be observed with dissolved dyes which interact either with the Onsager reaction field of the solvent (solvatochromism) or with an applied external field (electrochromism). The theory of electrochromism has been given by several authors (Lippert, 1955; Liptay and Czekalla, 1960; Labhart, 1961; Platt, 1961; detailed reviews by Liptay, 1969, 1974). The electric-field-induced changes of the absorption spectrum are essentially attributed to three sources:

1. Field-induced dichroism (orientational effect)

   In fluid phases dye molecules are oriented by the external electric field. The resulting absorption change depends on the square of the external electric field and is proportional to the absorption spectrum. In rigid systems, e.g., in solid monolayers, the chromophores remain in a fixed orientation and the orientational effects can be neglected.

2. Field-induced changes of the transition dipole moment

   The wavefunctions of the dye molecules for the ground state and the excited state are altered by an electric field F_e, and accordingly the transition moment is changed:

$$\mu^F_{ge} = \mu_{ge} + \alpha_{ge} \cdot F_e \tag{9}$$

   where μ^F_{ge} and μ_{ge} are the transition moments for $F_e \neq 0$ and $F_e = 0$ respectively, and α_{ge} is the transition polarizability tensor (Liptay, 1969). For $|\mu_{ge}| \gg |\alpha_{ge} F_e|$ the absorption band is practically unaffected by F_e, as observed for the strong π-π^+ transitions of azo- and stilbene dyes.

3. Field-induced band shifts

Consider a molecule in a uniform electric field, F_e. The work done on the dipole of the molecule by changing the field from zero to F_e is

$$E(F_e) = -\mu \cdot F_e \cdot \cos\varphi - \frac{1}{2}\alpha \cdot F_e{}^2 \tag{10}$$

where μ is the permanent dipole moment, φ its angle with the electric field lines, and α is the polarizability in the direction of F_e.

In general, the dipole moment and the polarizability are different in the ground and in the excited state. This difference in potential energy of the dipole causes the energy levels to be shifted, as illustrated schematically in Fig. 2.

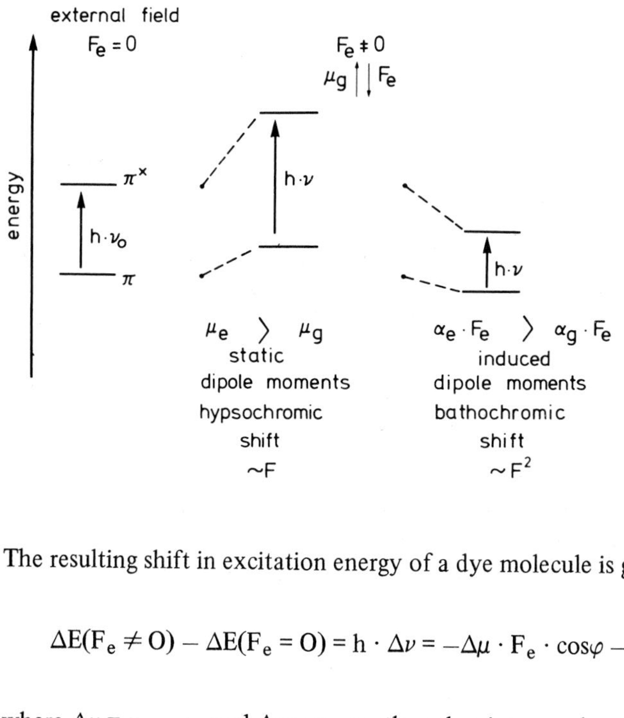

Fig. 2. Energy level diagram for a dye whose dipole moment is antiparallel to the external electric field. The dipole moment change ($\Delta\mu > O$) produces a blue-shift of the absorption band proportional to F_e and the polarizability change ($\Delta\alpha > O$) a red-shift proportional to F_e^2

The resulting shift in excitation energy of a dye molecule is given by

$$\Delta E(F_e \neq O) - \Delta E(F_e = O) = h \cdot \Delta\nu = -\Delta\mu \cdot F_e \cdot \cos\varphi - \frac{1}{2}\Delta\alpha \cdot F_e^2 \tag{11}$$

where $\Delta\mu = \mu_e - \mu_g$ and $\Delta\alpha = \alpha_e\text{-}\alpha_g$, the subscripts e and g standing for excited and ground state respectively; $\Delta\nu$ is the electrochromic frequency shift of the absorption band. For small values of $\Delta\nu$ we may write for a bathochromically shifted absorption band

$$A(\nu\text{-}\Delta\nu) = A(\nu) - \Delta\nu \cdot \frac{dA}{d\nu} + \frac{\Delta\nu^2}{2} \cdot \frac{d^2A}{d\nu^2} \tag{12}$$

where A (ν) is the absorption as a function of the frequency ν for $F_e = O$. The measured absorption of the monolayer is given by $A = 1\text{-}I/I_o$ where I_o and I are the transmitted light intensities of the substrate and substrate-plus-dye monolayer. As $A \ll 1$ we may write $\ln I/I_o = \ln(1\text{-}A) \simeq -A$, which is related to the molar extinction by $E = \lg I_o/I = 2.3 \cdot A$. The electric-field-induced absorption change ΔA is defined by $\Delta A = 1 - \frac{I(F_e \neq O)}{I(F_e = O)}$ where I is the transmitted light intensity of the substrate-plus-dye monolayer modulated by the electric field F_e.

From Eqs. (11) and (12) the field-induced absorption change is given by (only linear and quadratic terms are considered):

$$A(\nu\text{-}\Delta\nu) - A(\nu) = \Delta A(\nu) = \Delta A_L(\nu) + \Delta A_q(\nu) \tag{13}$$

with

$$\Delta A_L(\nu) = \frac{\Delta\mu \cos\varphi}{h} \cdot \frac{dA}{d\nu} \cdot F_e$$

and

$$\Delta A_q(\nu) = \left(\frac{\Delta\alpha}{2h} \cdot \frac{dA}{d\nu} + \frac{\Delta\mu^2 \cos\varphi}{2h^2} \cdot \frac{d^2A}{d\nu^2}\right) \cdot F_e^2 .$$

The absorption change ΔA_L is proportional to the first derivative of the absorption with respect to the frequency, $dA/d\nu$, and to the applied electric field strength F_e (linear electrochromism). It is obvious that $\Delta A_L = O$ in case of $\cos\varphi = O$ and reaches its extreme value in case of $\cos\varphi = \pm 1$.

The quadratic electrochromism ΔA_q is composed of two contributions, one of which is proportional to $dA/d\nu$ and the other proportional to $d^2A/d\nu^2$. When $dA/d\nu$ reaches an extremum, $d^2A/d\nu^2$ vanishes. This relation allows for the separate calculation of $\Delta\alpha$ and $\Delta\mu^2 \cdot \cos\varphi$ from the quadratic effect.

B. Monolayer Assemblies

The monolayer assembly technique is particularly useful for studying electrochromism, because it allows the application of a strong electric field ($F_e = 5 \cdot 10^6$ V/cm) and enables one to assemble the dye molecules into a fixed orientation. By this method thin dye-capacitors are prepared in the following manner (Bücher et al., 1969): a semitransparent aluminium film is evaporated onto a glass slide; this is followed by the deposition of fatty acid- and dye monolayers. Finally, a transparent aluminium electrode is deposited. Such a dye capacitor is shown schematically in Fig. 3. With several dyes, electric-field-induced linear and quadratic changes of the absorption of monolayer condensors were detected, using a lock-in technique (a detailed description of this method is given by Kuhn et al., 1972).

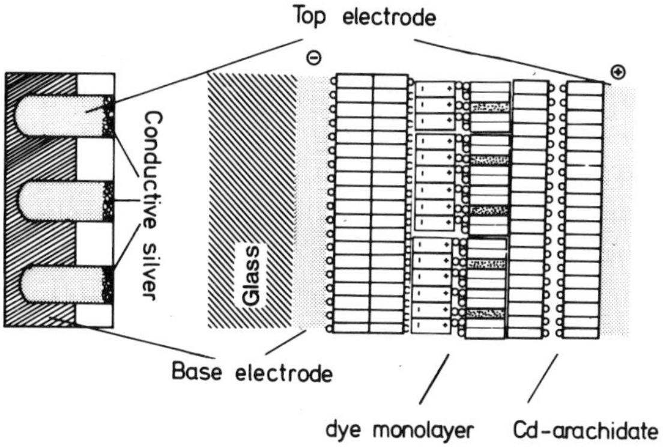

Fig. 3. Schematic representation of a thin-layer capacitor used for measuring electrochromism of organic dyes (the shown sample is associated with the dyes in Fig. 4-6)

Measurements were also performed with merocyanines by Bücher et al (1969); with cyanines by Bücher and Kuhn (1970b); with chlorophyll by Kleuser and Bücher (1969); Schmidt et al. (1969); Reich and Scheerer (1976); with rhodamin B by Schmidt and Reich (1972a); with lutein by Schmidt and Reich (1972b); with all-trans-β-apo-8'-carotenoic acid by Reich and Sewe (1977); with lutein-chlorophyll complexes by Sewe and Reich (1977), and with stilbene- and azo-dyes by Heesemann (1976, 1978). Electric-field-induced fluorescence changes of dye monolayers have not yet been studies.

A quantitative interpretation of the electrochromic spectra is of interest in relation to the use of dyes as electric field indicators in biological membranes, but also for an interpretation of flashlight-induced absorption changes of photosynthetic membranes, and of course for quantum chemical calculations. Three factors complicating the interpretation of electrochromic spectra should be mentioned.

1. Interference effects. The incident light beam is multireflected at the aluminium electrodes. Standing light waves formed in the multilayer system by interference affect the absorption of the dye layer. Presumably, the measured absorption of such a system differs from that without aluminium electrodes. However, absorption measurements of dye capacitors failed because ist was impossible to produce reference and sample electrodes of identical optical properties. (This problem does not arise for electrochromic measurements because one electrode alternately acts as reference and sample.)

For an evaluation of electrochromic spectra most workers used uncorrected absorption spectra of an analogous multilayer system without aluminium electrodes. This method is incorrect as investigated theoretically and experimentally for cyanine and merocyanine dye layers, the transition moments of which are oriented statistically in the layer plane (Bücher and Kuhn, 1970; Kuhn et al., 1972). Based on the interference filter properties of the dye capacitor, the absorption $A(\nu)$ and the absorption change $\Delta A(\nu)$ were predicted to increase by a factor of $\frac{1+r}{1-r} = 2.6$ for a zero order interference filter (the reflection coefficient of the aluminium electrodes for normal incidence of light is $r = 0.45$). At smaller electrode distances, drastic deformations for the electrochromic spectra were predicted. These theoretical predictions were essentially confirmed by experiments.

With azo- and stilbene dyes (transition moment normal to the layer plane) only slight changes of the electrochromic spectra are caused by varying electrode distances (Heesemann, 1976). However, a decrease of the transmission of the second electrode from 30% to 15% causes a shift of the ΔA_L versus ν curve of about $\Delta\nu = 0.25^{-1}$, which may be attributed to a bathochromic shift of the absorption band due to the electrodes. A theoretical treatment of such an optical system has not yet been presented.

2. Orientation of the chromophore. If the exact orientation of the chromophores in the monolayer is known, the dipole and polarizability differences obtained from electrochromic spectra can be related to the geometry of the molecule. A simple situation is realized by monolayers of cyanines and merocyanines whose molecular long axes are statistically oriented in the layer plane, and by azo- and stilbene dyes whose long molecular axis is normal to the layer plane, as indicated by optical dichroism (see Sect. II). The long molecular axis is also the direction of the transition moment. With monolayers of lutein, chlorophyll, carotenoic acid, and rhodamine rather bold assumptions are necessary in order to infer the molecular orientation from polarized absorption spectra. Therefore the reported values for $\Delta\mu$ and $\Delta\alpha$ obtained from the electrochromic spectra have to be treated with caution.

3. Permanent oriented electric fields within the monolayers. A permanent inner electric field, F_p, which can be built up by the dipole molecules of the monolayer, produces an additional linear electrochomism caused by the induced dipole moment difference, $\Delta\mu_i = \Delta\alpha \cdot F_p$ (Reich and Schmidt, 1972). Taking this into account, the linear electrochromism effect is given by

$$\Delta A = \frac{(\Delta\mu \cdot \cos\varphi + \Delta\alpha \cdot F_p \cdot \cos\theta)}{h} \cdot F_e \tag{14}$$

where θ is the angle between the vectors of F_p and F_e. A strong influence of the permanent electric field on the linear electrochromism is observed with 4-nitro-4'-amino-azobenzene chromophores as reported by Heesemann (1978). In a mixed monolayer of dyes XIV and XV at a molar mixed ratio of 1:2 a narrow and strong absorption band for p-polarized light, typical for H-aggregates, appears (see Fig. 4-5). H-aggregates are assumed to consist of planar pin-cushion-like arrangements of the molecular long axis of the chromophores. For such a molecular arrangement the inner electric field, F_p, produced by the dipole moment of the chromophores may be estimated using a capacitor model:

$$F_p = \frac{4 \cdot \pi \cdot \mu_g}{\epsilon \cdot L \cdot a} . \tag{15}$$

With the values of $\mu_g = 8$ debye, a length of the dipole of $1 = 16 \cdot 10^{-8}$ cm, a dielectric constant of $\epsilon = 4$, and a surface area per chromophore of $a = 28 \cdot 10^{-16}$ cm^2, the electric field is calculated as $F_p = 2.5 \cdot 10^7$ V/cm. As $\cos\varphi$ and $\cos\theta$ are of opposite sign for this arrangement, the induced dipole moment difference, $\Delta\mu_i$, compensates the static dipole moment difference $\Delta\mu$. In Fig. 6 the ΔA_L versus ν and ΔA_q versus ν curves are shown for a field strength of $F_e = 1.2 \cdot 10^6$ V/cm. The linear effect corresponds to a hypsochromic shift of the absorption band, as expected. With the aid of Eq. (14) the total dipole moment difference $\Delta\mu t = \Delta\mu(.-1) + \Delta\mu_i$ is calculated as 2 debye. From the quadratic

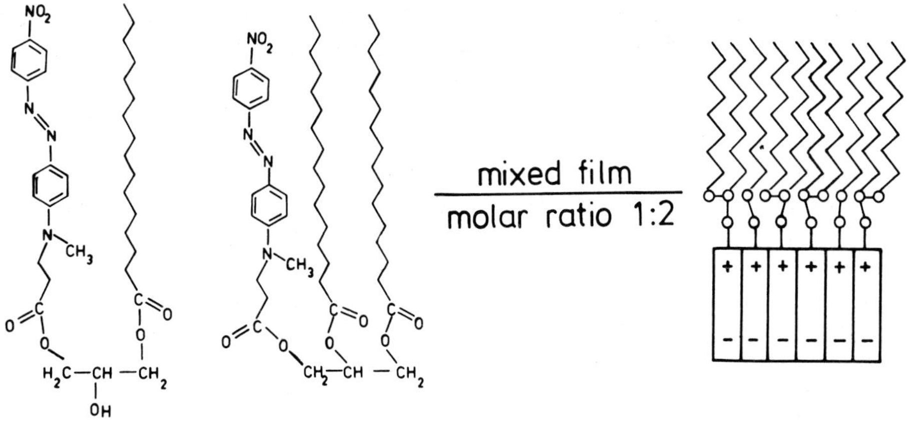

Fig. 4. A mixture of the chromophoric monoglyceride XIV and the chromophoric diglyceride XV (molar ratio 1:2) forms a closely packed monolayer, each azo dye chromophore (bar with indication of the molecular dipole) occupies 28 Å^2

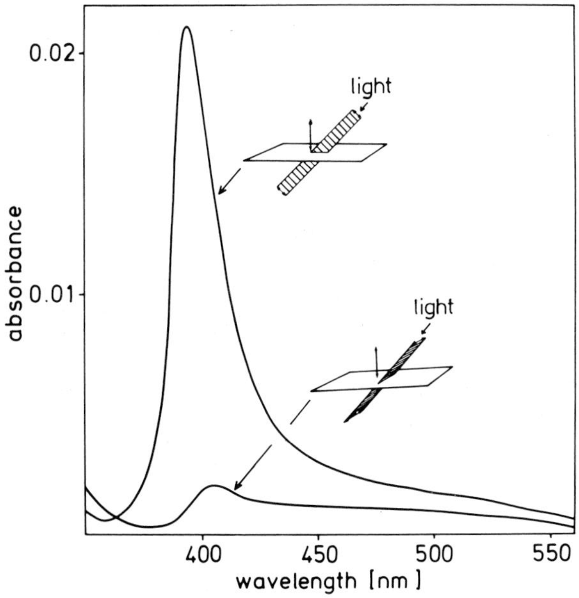

Fig. 5. Absorption spectra of the mixed monolayer of dye XIV and XV (molar ratio 1:2) measured with polarized light under a 45°-incidence. The multilayer assembly corresponds to that of Fig. 3, but without electrodes. The strong band observed with p-polarized light is assigned to H-aggregates

effect $\Delta\alpha$ is calculated as 220 Å^3; with an approximate value of $F_p = 2.5 \cdot 10^7$ V/cm this yields $\Delta\mu_i = 18.5$ debye. With these values one obtains $\Delta\mu = 20.5$ debye for the static dipole moment difference. This value agrees well with that found for 4-nitro-4'-(dimethylamino)-azobenzene in solution ($\Delta\mu = 17.0$ debye) by Liptay (1969). As the total dipole moment difference is relatively small, the quadratic effect is essentially attributed to $\Delta\alpha$, corresponding to a bathochromic shift of the absorption band.

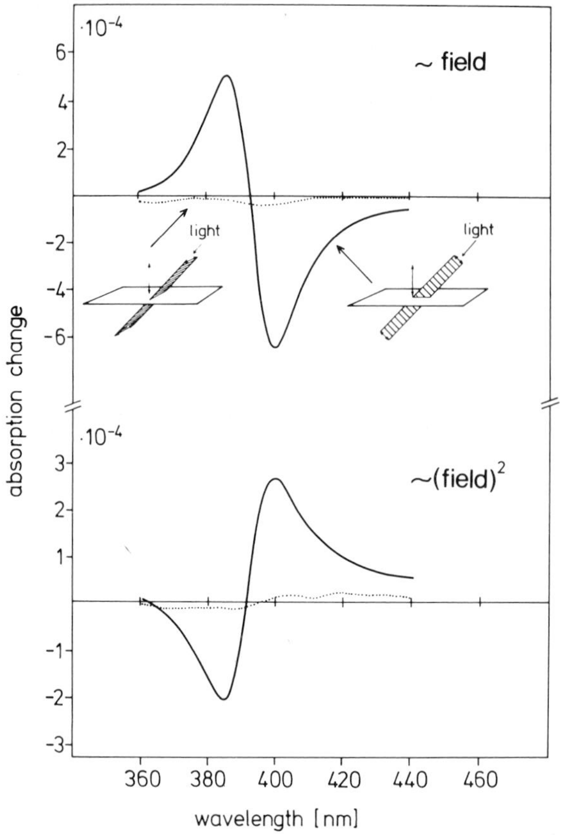

Fig. 6. Polarized linear (*upper part*) and quadratic electrochromism spectra (*lower part*) of the mixed monolayer of dye XIV and XV (molar ratio = 1:2; the multilayer assembly corresponds to that of Fig. 3). The applied electric field is $F_e = 1.2 \cdot 10^6$ V/cm, the linear contribution corresponds to a blue-shift of the absorption band ($\Delta\mu > 0$, $\cos\varphi \simeq -1$) and the quadratic contribution corresponds to a red-shift ($\Delta\alpha > 0$)

C. Black Lipid Membranes

During the last few years there have been a large number of reports on studies of potential-dependent changes in the spectral properties of dye probes in membranes. Studies with lipid bilayers have been performed mainly in order to understand the mechanism of similar effects observed in cell membranes (Tasaki et al., 1968; Hoffman and Laris, 1974; Sims et al., 1974; Conti, 1975; Ross et al., 1977; for an excellent recent review see Waggoner, 1979). With cell membranes, a rather empirical approach has been taken which involved the testing of approximately 300 dyes, more than half of which gave positive results. Quite small changes in absorbance ($\simeq 10^{-5}$ for cyanine dyes: Waggoner et al., 1977) and relatively large fluorescence changes (10^{-3} with a merocyanine dye: Cohen et al., 1974a) have been observed during typical action potentials. Frequently, fast spectral responses (of the order of milliseconds) to changes in membrane potential are superimposed on relatively slow responses (of the order of seconds). The following mechanisms have been considered in order to account for the observed effects:

1) A potential-dependent partition coefficient of the dye between the membrane and the aqueous phase, which may alter the membrane dye concentration and, as a consequence, the membrane fluorescence. Slow responses might be explained by such an effect.

2) A potential-dependent orientation of the dye, which may be responsible for fast responses.
3) The position of the dye in the membrane, which may be potential-dependent. This could be an important contribution to responses observed with solvent-sensitive chromophores.
4) Monomer-dimer equilibria for membrane-bound dyes (e.g., cyanine dyes), which may be disturbed by the membrane potential.
5) Electrochromic effects.

Of these effects, the first four have received special consideration. Recently Dragsten and Webb (1978) have reported a very complete and detailed investigation of the potential sensitivity of the fluorescent probe merocyanine 540 (dye XI, Table 2). They used spherical bilayers which are well adapted to polarization measurements. Dye concentrations in the membrane and diffusion coefficients of the dye in aqueous solutions were measured by fluorescence correlation spectroscopy (FCS, see above). The authors concluded that slow responses were due to voltage-induced changes in the membrane dye concentration. At saturating dye membrane concentrations this response decreased. These results are in agreement with an earlier study by Conti et al. (1974). The fast response (rise time $< 6\ \mu s$) is attributed to field-induced changes in the distribution of dye orientations. A fraction of dye monomers (oriented parallel to the membrane surface) is in equilibrium with nonfluorescent dimers in the membrane. This equilibrium is disturbed by the membrane potential.

Whereas investigations on potential-sensitive dyes in bilayers, such as reported by Dragsten and Webb (1978), are too numerous to be exhaustively reviewed here (see also Conti and Malerba, 1972; Conti et al., 1974; Conti, 1975; Waggoner and Grinvald, 1977), there has been no report of the measurement of a true electrochromic effect in single lipid bilayer membranes. Until recently, suitable dyes which can be incorporated into membranes with their transition moment parallel to the electric field, have not been available. Note that most dyes used as potential probes are oriented transverse to the electric field. A systematic attempt at the design and characterization of suitable electrochromic dyes has been made by Heesemann (1980a,b). Below we briefly outline a hypothetical application of an electrochromic dye as a transmembrane field indicator:

The surface-active dyes XIV and XV cannot be used as transmembrane potential probes because (a) the linear electrochromism near the interface has been analyzed only for H-aggregates, and (b) the chromophore which is located at the interface cannot interact with the electric field of the hydrophobic region of the membrane. This difficulty is overcome with dye V which, in mixed films with stearic acid and stearylamine, is in an equilibrium between monomers and dimers, the long axis of the chromophore being parallel to the hydrocarbon chains, as described by Heesemann (1980b). For a molar mixing ratio of dye V: stearic acid: stearylamine of 1:15:16 a total dipole moment difference of 10 debye for dimers as well as for monomers is calculated from the linear electrochromism by Heesemann (unpublished results). These results make dye V a candidate for transmembrane potential measurements.

1. A realistic value of the transmembrane electric field is about $3 \cdot 10^5$ V/cm. For a chromophore with $\Delta\mu_t = 10$ debye and $\Delta\alpha = 200$ Å^3 the linear electrochromism, ΔA_L, is 100 times the quadratic electrochromism, ΔA_q.

2. In order to obtain a strong field-induced absorption change, $\Delta\mu_t$ should be larger than 10 debye, and oriented parallel to the electric field lines. The molar extinction coefficient should be of the order of 10^7-10^8 cm^2/mol.
3. The dye probe must be asymmetrically distributed in the membrane, because otherwise the linear electrochromism would be averaged out.

Consider now a bilayer which is formed from two monolayers, e.g., by the method of Montal and Mueller (1972). One monolayer is assumed to contain dye V at a surface concentration of $4 \cdot 10^{12}$ dye/cm^2 (this value is obtained from the molar mixing ratio of dye V: glycerol monoleate = 1:100). Assuming an angle of the transmitted light beam (p-polarized) of 45°, a change of the electric field of $3 \cdot 10^5$ V/cm, the optical porperties of dye V ($\Delta\mu_t$ = 10 debye, molar extinction coefficient in chloroform $\epsilon = 3 \cdot 1 \cdot 10^7$ cm^2/mol, half-width of the monomer band $\Delta\nu_{1/2} = 0.481 \cdot 10^{14} s^{-1}$), one predicts a maximal absorption (λ = 475 nm) of $7 \cdot 10^{-4}$ and a linear absorption change (λ = 514 nm) of $1.4 \cdot 10^{-5}$ (at this point the absorption band has its maximal slope).

References

Alamuti, N., Läuger, P.: Fluorescence of thin chlorophyll membranes in aqueous phase. Biochim. Biophys. Acta *211*, 362-364 (1970).

Aoshima, R., Iriyama, K., Asai, H.: High sensitivity fluorophotometer using photon counting. Appl. Opt. *12*, 2748-2750 (1973).

Aveyard, R., Haydon, D.A.: An Introduction to the Principles of Surface Chemistry. Cambridge: University Press 1973.

Bamberg, E., Läuger, P.: Channel formation kinetics of gramicidin A in lipid bilayer membranes. J. Membr. Biol. *11*, 177-194 (1973).

Bretscher, M.S., Raff, M.C.: Mammalian plasma membranes. Nature *258*, 43-49 (1975).

Brody, S.S.: Interactions between ferredoxin and chlorophyll in a monolayer system. Z. Naturforsch. *26b*, 922-929 (1971).

Bücher, H., Kuhn, H.: Scheibe aggregate formation of cyanine dyes in monolayers. Chem. Phys. Lett. *6*, 183-185 (1970a).

Bücher, H., Kuhn, H.: Difference between ground and excited state dipole moments and polarizabilities as determined from electrochromism of Scheibe-aggregates in monolayer assemblies. Z. Naturforsch. *25b*, 1323-1327 (1970b).

Bücher, H., Drexhage, K.H., Fleck, M., Kuhn, H., Möbius, D., Schäfer, F.P., Sondermann, J., Sperling, W., Tillmann, P., Wiegand, J.: Controlled transfer of excitation energy through thin layers. Mol. Cryst. *2*, 199-230 (1967).

Bücher, H., Wiegand, J., Snavely, B.B., Beck, K.H., Kuhn, H.: Electric field induced changes in the optical absorption of a merocyanine dye. Chem. Phys. Lett. *3*, 508-510 (1969).

Cadenhead, D.A., Kellner, B.M.J., Jacobson, K., Papahadjopoulos, D.: Fluorescent probes in model membranes I: Anthroyl fatty acid derivatives in monolayers and liposomes of dipalmitoylphosphatidylcholine. Biochemistry *16*, 5386-5392 (1977).

Carbone, E., Malerba, F., Poli, M.: Orientation and rotational freedom of fluorescent probes in lecithin bilayers. Biophys. Struct. Mech. *2*, 251-266 (1976).

Cherry, R.J., Hsu, K., Chapman, D.: Absorption spectroscopy of chlorophyll in bimolecular lipid membranes. Biochem. Biophys. Res. Commun. *43*, 351-358 (1971a).

Cherry, R.J., Hsu, K., Chapman, D.: Polarized absorption spectroscopy of chlorophyll-lipid membranes. Biochim. Biophys. Acta *267*, 512-522 (1971b).

Cohen, L.B., Salzberg, B.M., Davila, H.V., Ross, W.N., Landowne, D., Waggoner, A.S., Wang, C.H.: Changes in axon fluorescence during activity: Molecular probes of membrane potential. J. Membr. Biol. *19*, 1-36 (1974).

Conti, F.: Fluorescent probes in nerve membranes. Annu. Rev. Biophys. Bioeng. *4*, 287-310 (1975).

Conti, F., Malerba, F.: Fluorescence signals in ANS-stained lipid bilayers under applied potentials. Biophysik *8*, 326-332 (1972).

Conti, F., Fioravanti, R., Malerba, F., Wanke, E.: A comparative analysis of extrinsic fluorescence in nerve membranes and lipid bilayers. Biophys. Struct. Mech. *1*, 27-45 (1974).

Czikkely, V., Försterling, H.D., Kuhn, H.: Extended dipole model for aggregates of dye molecules. Chem. Phys. Lett. *6*, 207-210 (1970).

Dragsten, P.R., Webb, W.W.: Mechanism of the membrane potential sensitivity of the fluorescent membrane probe merocyanine 540. Biochemistry *17*, 5228-5240 (1978).

Drexhage, K.H., Zwick, M.M., Kuhn, H.: Sensibilisierte Fluoreszenz nach strahlungslosem Energieübergang durch dünne Schichten. Ber. Bunsenges. Physik. Chem. *67*, 62-67 (1963).

Edelman, G.M.: Surface modulation in cell recognition and cell growth. Science *192*, 218-226 (1976).

Edidin, M.: Rotational and translational diffusion in membranes. Annu. Rev. Biophys. Bioeng. *3*, 179-201 (1974).

Elson, E.L., Madge, D.: Fluorescence correlation spectroscopy. I. Conceptual basis and theory. Biopolymers *13*, 1-27 (1974).

Elson, E.L., Webb, W.W.: Concentration correlation spectroscopy: A new biophysical probe based on occupation number fluctuations. Annu. Rev. Biophys. Bioeng. *4*, 311-334 (1975).

Fahey, P.F., Koppel, D.E., Barak, L.S., Wolf, D.E., Elson, E.L., Webb, W.W.: Lateral diffusion in planar lipid bilayers. Science *195*, 305-306 (1977).

Fernandez, M.S., Fromherz, P.: Lipoid pH indicators as probes of electrical potential and polarity in micelles. J. Phys. Chem. *81*, 1755-1761 (1977).

Fettiplace, R., Gordon, L.G.M., Hladky, S.B., Requena, J., Zingsheim, H.P., Haydon, D.A., in: Methods in Membrane Biology (ed. Korn, E.D.), Vol. IV, pp. 1-75. New York, London: Plenum Press 1975.

Förster, T.: Zwischenmolekulare Energiewanderung und Fluoreszenz. Ann. Phys. 6. Folge 55-84 (1948).

Förster, T.: Transfer mechanisms of electronic excitation. Discuss. Faraday Soc. *27*, 7-17 (1959).

Fromherz, P.: Monolayer states of a cyanine dye studied by a spectroscopic technique. Z. Naturforsch. *28c*, 144-148 (1973a).

Fromherz, P.: A new method for investigation of lipid assemblies with a lipoid pH indicator in monomolecular films. Biochim. Biophys. Acta *323*, 326-334 (1973b).

Fromherz, P., Masters, B.: Interfacial pH at electrically charged lipid monolayers investigated by the lipoid pH-indicator method. Biochim. Biophys. Acta *356*, 270-275 (1974).

Fukuda, K., Nakahara, H., Kato, T.: Monolayers and multilayers of anthraquinone derivatives containing long alkyl chains. J. Coll. Interface Sci. *54*, 430-438 (1976).

Gaines, G.L., Jr.: Insoluble Monolayers at Liquid-Gas Interfaces. New York-London: Interscience 1966.

Giles, C.H., Neustadter, E.L.: Researches on monolayers. Part I. Molecular areas and orientation at water surfaces of aromatic azo-compounds containing long alkyl chains. J. Chem. Soc., part I, 918-923 (1952).

Goodall, M.C.: Structural effect in the action of antibiotics on the ion permeability of lipid bilayers. III. Gramicidin "A" and "S" and lipid specificity. Biochem. Biophys. Acta *219*, 471-478 (1970).

Gruda, I., Leblanc, R.M.: Synthesis of some long-chain spiropyranindolines. Can. J. Chem. *54*, 576-580 (1976).

Heesemann, J.: Diglyceridchromophorsysteme als Bausteine für organisierte Schichtverbände. Ph.D. Thesis, University of Göttingen (1976).

Heesemann, J.: Elektrochromism of oriented dyes in monolayer capacitors. Ber. Bunsenges. Phys. Chem. *82*, 868 (1978).

Heesemann, J.: Studies on monolayers I: Surface tension and absorption spectroscopic measurements of monolayers of surface-active azo- and stilbene-dyes. J. Am. Chem. Soc. *102*, 2167-2176 (1980a).

Heesemann, J.: Studies on Monolayers. II: Designed monolayer assemblies of mixed films of surface-active azo dyes. J. Am. Chem. Soc. *102*, 2176-2181 (1980b).

Herbert, T.J., Elson, E., Webb, W.W.: Fluorescence correlation spectroscopy of lipid bilayer membranes. Fed. Proc. *33*, 1303 (1974).

Hladky, S.B., Haydon, D.A.: Discreteness of conductance change in bimolecular lipid membranes in the presence of certain antibiotics. Nature *225*, 451-453 (1970).

Hoffman, J.F., Laris, P.C., Determination of membrane potentials in human and *Amphiuma* red blood cells by means of a fluorescent probe. J. Physiol. *239*, 519-552 (1974).

Kolb, H.-A., Läuger, P., Bamberg, E.: Correlation analysis of electrical noise in lipid bilayer membranes: Kinetics of gramicidin a channels. Membr. Biol. *20*, 133-154 (1975).

Kleuser, D., Bücher, H.: Elektrochromie von Chlorophyll-a und Chlorophyll-b in monomolekularen Filmen. Z. Naturforsch. *24b*, 1371-1374 (1969).

Kuhn, H.: Classical aspects of energy transfer in molecular systems. J. Chem. Phys. *53*, 101-108 (1970).

Kuhn, H., Möbius, D., Bücher, H., in: Physical Methods of Chemistry, Part III (eds. Weissberger, A., Rossiter, B.W.), pp. 577-702. New York-London: Interscience 1972.

Labhart, H.: Beeinflussung der Lichtabsorption organischer Farbstoffe durch äussere elektrische Felder. I. Theoretische Betrachtung. Helv. Chim. Acta *44*, 447-456 (1961).

Langmuir, I.: The mechanism of the surface phenomena of flotation. Trans. Faraday Soc. *15*(3), 62-74 (1920).

Lippert, E.: Dipolmoment und Elektronenstruktur von angeregten Molekülen. Z. Naturforsch. *10a*, 541-545 (1955).

Liptay, W.: Elektrochemie – Solvatochromie. Angew. Chem. *81*, 195-232 (1969).

Liptay, W.: Excited states, vol. 1, 129-148. New York: Academic Press (1974).

Liptay, W., Czekalla, J.: Die Bestimmung von absoluten Übergangsmomentrichtungen und von Dipolmomenten angeregter Moleküle aus Messungen des elektrischen Dichroismus. I. Theorie. Z. Naturforsch. *15a*, 1072-1079 (1960).

Madge, D., Elson, E.L., Webb, W.W.: Thermodynamic fluctuations in a reacting system – measurement by fluorescence correlation spectroscopy. Phys. Rev. Lett. *29*, 705-711 (1972).

Madge, D., Elson, E., Webb, W.W.: Fluorescence correlation spectroscopy. II. An experimental realization. Biopolymers *13*, 29-61 (1974).

Möbius, D.: Designed monolayer assemblies. Ber. Bunsenges. Phys. Chem. *82*, 848-858 (1978).

Montal, M., Mueller, P.: Formation of bimolecular membranes from lipid monolayers, and a study of their electrical properties. Proc. Natl. Acad. Sci. USA *69*, 3561-3566 (1972).

Mormann, W., Kuhn, H.: Seifenlamellen mit Doppelsandwichstruktur. Z. Naturforsch. *24b/10*, 1340-1341 (1969).

Mueller, P., Rudin, D.O., Tien, H.T., Wescott, W.C.: Reconstitution of cell membrane structure in vitro and its transformation into an excitable system. Nature *194*, 979-980 (1962).

Neher, E.: Asymmetric membranes resulting from the fusion of two black lipid bilayers. Biochim. Biophys. Acta *373*, 327-336 (1974).

Nicolson, G.: Transmembrane control of the receptors on normal and tumor cells. I. cytoplasmic influence over cell surface components. Biochim. Biophys. Acta *457*, 57-108 (1976).

Pagano, R., Thompson, T.E.: Spherical lipid bilayer membranes. Biochim. Biophys. Acta *144*, 666-669 (1967).

Peters, R.: The effect of rehydration on the thickness of erythrocyte membranes as observed by energy transfer. Biochim. Biophys. Acta *318*, 469-473 (1973a).

Peters, R.: The thickness of air-dried human erythrocyte membranes as determined by energy transfer. Biochim. Biophys. Acta *330*, 53-60 (1973b).

Peters, R., Peters, J., Tews, K.H., Bähr, W.: A microfluorimetric study of translational diffusion in erythrocyte membranes. Biochim. Biophys. Acta *367*, 282-294 (1974).

Platt, J.R.: Electrochromism, a possible change of color producible in dyes by an electric field. J. Chem. Phys. *34*, 862-863 (1961).

Pockels, A.: Surface tension (Letters to the editor). Nature *43*, 437-439 (1891).

Pohl, G.W.: Energy transfer in black lipid membranes. Biochim. Biophys. Acta *288*, 248-253 (1972).
Radda, G.K.: Fluorescent probes in membrane studies. In: Methods in Membrane Biology (ed. Korn, E.D.), Vol. IV, pp. 97-188. New York: Plenum Press 1975.
Reich, R., Scheerer, R.: Effect of electric field on the absorption spectrum of dye molecules in lipid layers. IV. Electrochromism of oriented chlorophyll-b. Ber. Bunsenges. Phys. Chem. *80*, 542-547 (1976).
Reich, R., Schmidt, S.: Über den Einfluß elektrischer Felder auf das Absorptionsspektrum von Farbstoffmolekülen in Lipidschichten. I. Theorie. Ber. Bunsenges. Phys. Chem. *76*, 589-598 (1972).
Reich, R., Sewe, K.-U.: The effect of molecular polarization on the electrochromism of carotenoids. I. The influence of a carboxylic group. Photochem. Photobiol. *26*, 11-17 (1977).
Ross, W.N., Salzberg, B.M., Cohen, L.B., Grinvald, A., Davila, H.V., Waggoner, A.S., Wang, C.H.: Changes in absorption, fluorescence, dichroism and birefringence in stained giant axon: Optical measurement of membrane potential. J. Membr. Biol. *33*, 141-183 (1977).
Schlessinger, J., Koppel, D.E., Axelrod, D., Jacobson, K., Webb, W.W.: Lateral transport in cell membranes: Mobility of concanavalin A receptors on myoblasts. Proc. Natl. Acad. Sci. USA *73*, 2409-2413 (1976).
Schlessinger, J., Axelrod, D., Koppel, D.E., Webb, W.W.: Lateral transport of a lipid probe and labelled proteins on a cell membrane. Science *195*, 307-309 (1977).
Schmidt, S., Reich, R.: Über den Einfluß elektrischer Felder auf das Absorptionsspektrum von Farbstoffmolekülen in Lipidschichten. II. Messungen an Rhodamin B. Ber. Bunsenges. Phys. Chem. *76*, 599-602 (1972a).
Schmidt, S., Reich, R.: Über den Einfluß elektrischer Felder auf das Absorptionsspektrum von Farbstoffmolekülen in Lipidschichten. III. Elektrochemie eines Carotinoids (Lutein). Ber. Bunsenges. Phys. Chem. *76*, 1202-1208 (1972b).
Schmidt, S., Reich, R., Witt, H.T.: Absorptionsänderungen von Chlorophyll-b im elektrischen Feld. Z. Naturforsch. *24b*, 1428-1431 (1969).
Sewe, K.-U., Reich, R.: The effect of molecular polarization on the electrochromism of carotenoids. II. Lutein-chlorophyll complexes: The origin of the field-indicating absorption-change at 520 nm in the membrane of photosynthesis. Z. Naturforsch. *32c*, 161-171 (1977).
Sims, P.J., Waggoner, A.S., Wang, C.-H., Hoffman, J.F.: Studies on the mechanism by which cyanine dyes measure membrane potential in red blood cells and phosphatidylcholine vesicles. Biochemistry *13*, 3315-3330 (1974).
Steinemann, A., Läuger, P.: Interaction of cytochrome c with phospholipid monolayers and bilayer membranes. J. Membr. Biol. *4*, 74-86 (1971).
Steinemann, A., Alamuti, N., Brodmann, W., Marschall, O., Läuger, P.: Optical properties of artificial chlorophyll membranes. J. Membr. Biol. *4*, 284-294 (1971).
Tasaki, I., Watanabe, A., Sandlin, R., Carnay, L.: Changes in fluorescence, turbidity, and birefrigence associated with nerve excitation. Proc. Natl. Acad. Sci. USA *61*, 883-888 (1968).
Tosteson, D.C., Andreoli, T.E., Tiefenberg, M., Cook, P.: The effects of macrocyclic compounds on cation transport in sheep red cells and thin and thick lipid membranes. J. Gen. Physiol. *51*, 373S-384S (1968).
Trissl, H.W.: Studies on the incorporation of fluorescent pigments into bilayer membranes. Biochim. Biophys. Acta *367*, 326-337 (1974).
Trosper, T.L.: Some properties of chlorophyll a at hydrocarbon-water interfaces and in black lipid membranes. J. Membr. Biol. *8*, 133-148 (1972).
Trosper, T.L., Park, R.B., Sauer, K.: Excitation transfer by chlorophyll a in monolayers and the interaction with chloroplast glycolipids. Photochem. Photobiol. *7*, 451-469 (1968).
Tweet, A.G.: Spectrometer for optical studies of ultra-thin films. Rev. Sci. Instrum. *34*, 1412-1417 (1963).
Tweet, A.G., Bellamy, W.D., Gaines, G.L., Jr.: Fluorescence quenching and energy transfer in monomolecular films containing chlorophyll. J. Chem. Phys. *41*, 2068-2077 (1964).
Tweet, A.G., Gaines, G.L., Jr., Bellamy, W.D.: Angular dependence of fluorescence from chlorophyll a in monolayers. J. Chem. Phys. *41*, 1008-1010 (1969).
Veatch, W., Stryer, L.: The dimeric nature of the gramicidin A transmembrane channel: Conductance and fluorescence energy transfer studies of hybrid channels. J. Mol. Biol. *113*, 89-102 (1977).

Veatch, W.R., Mathies, R., Eisenberg, M., Stryer, L.: Simultaneous fluorescence and conductance studies of planar bilayer membranes containing a highly active and fluorescent analog of gramicidin A. J. Mol. Biol. *99*, 75-92 (1975).

Waggoner, A.S.: Dye indicators of membrane potential. Annu. Rev. Biophys. Bioeng. *8*, 47-68 (1979).

Waggoner, A.S., Grinvald, A.: Mechanisms of rapid optical changes of potential sensitive dyes. Ann. N.Y. Acad. Sci. *303*, 217-241 (1977).

Waggoner, A.S., Stryer, L.: Fluorescent probes of biological membranes. Proc. Natl. Acad. Sci. USA *67*, 579-589 (1970).

Waggoner, A.S., Wang, C.H., Tolles, R.L.: Mechanism of potential-dependent light absorption changes in lipid bilayer membranes in the presence of cyanine and oxonol dyes. J. Membr. Biol. *33*, 109-140 (1977).

Wolf, D.E., Schlessinger, J., Elson, E.L., Webb, W.W., Blumenthal, R., Henkart, P.: Diffusion and patching of macromolecules on planar lipid bilayer membranes. Biochemistry *16*, 3476-3483 (1977).

Wu, E.-S., Jacobson, K., Papahadjopoulos, D.: Lateral diffusion in phospholipid multibilayers measured by fluorescence recovery after photobleaching. Biochemistry *16*, 3936-3941 (1977).

Wu, E.-S., Jacobson, K., Szoka, F., Portis, A., Jr.: Lateral diffusion of a hydrophobic peptide, N-4-nitrobenz-2-oxa-1,3-diazole gramicidin S, in phospholipid multibilayers. Biochemistry *17*, 5543-5550 (1978).

Yguerabide, J., Stryer, L.: Fluorescence spectroscopy of an oriented model membrane. Proc. Natl. Acad. Sci. USA *68*, 1217-1221 (1971).

Zingsheim, H.P., Haydon, D.A.: Fluorescence spectroscopy of planar black lipid membranes. Probe adsorption and quantum yield determination. Biochim. Biophys. Acta *298*, 755-768 (1973).

Zingsheim, H.P., Neher, E.: The equivalence of fluctuation analysis and chemical relaxation measurements: A kinetic study of ion pore formation in thin lipid membranes. Biophys. Chem. *2*, 197-207 (1974).

Zwick, M.M., Kuhn, H.: Strahlungsloser Übergang von Elektronenanregungsenergie durch dünne Schichten. Z. Naturforsch. *17a*, 411-414 (1962).

Fluorescence Spectroscopy of Biological Membranes

J. Yguerabide and M.C. Foster*

Fluorescence spectroscopy is a highly sensitive technique with many applications in the study of membranes, as shown in Table 1. Several reviews have recently describing uses of the technique in several specialized areas of membrane research [1-5]. The object of this chapter is to describe the basic theory of fluorescence spectroscopy and present a quantitative description of some of the applications listed in Table 1.

Table 1. Membrane properties studied by fluorescence spectroscopy

Property	*Technique*
Orientation and rotational mobility of lipids	Polarized fluorescence
Phase transitions and lateral phase separations	Fluorescence intensity and polarized fluorescence
Microviscosity of internal regions of membranes	Polarized fluorescence
Organization of lipids in the immediate vicinity of proteins	Fluorescence intensity and polarized fluorescence
Lateral mobility of lipids and proteins	Photobleaching, fluorescence quenching, and fluorescence capping
Dimensions of membrane proteins	Excitation energy transfer
Proximity relations between functional sites of proteins	Excitation energy transfer
Interaction (attachment and detachment) of cytoplasmic proteins with membrane	Excitation energy transfer
Organization of cytoplasmic components with respect to cytoplasmic side of membrane	Fluorescence tracing
Electrical potentials across biological membranes	Voltage-sensitive fluorescent probes
Electrostatic surface potentials	Reversible binding of charged fluorescent probes
Organization of simple translocators in membranes	Excitation energy transfer
Aggregation (subunit interaction) of membrane proteins and its role in gated membrane transport	Excitation energy transfer
Membrane asymmetry	Membrane-impermeable fluorescent probes
Membrane fusion	Fluorescence intensity and polarized fluorescence

* Department of Biology, University of California at San Diego, La Jolla, California 92093, USA

I. Basic Theory

The use of fluorescence spectroscopy in the study of biological membranes is based on the observation that the emission properties of a fluorescent molecule depend on the molecular conformation and dynamics of its environment. Thus when such a molecule is inserted into a specific region of a membrane, the structure and dynamics of that region, as well as alterations induced by changes in pH, temperature, and bathing electrolyte, or by addition of drugs or hormones, can be monitored by measurement of emission parameters. The parameters most commonly used are the fluorescence spectrum, fluorescence efficiency, excitation spectrum, lifetime, and degree of polarization of emitted light. In addition, the absorption spectrum of the fluorophore is of interest in most applications. In the following paragraphs we present a brief description of these important spectroscopic quantities.

A. Energy Level Diagrams and Absorption Spectra

The emission and absorption properties of a *fluorophore* (fluorescent substance) can be best described in terms of an *energy level diagram*, as shown in Fig. 1 [6, 7]. The diagram

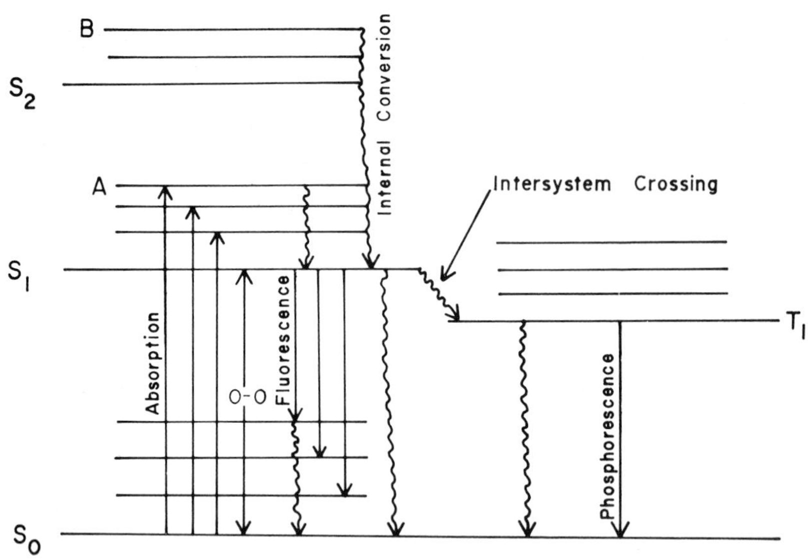

Fig. 1. Energy levels of a fluorophore. ***Straight arrows:*** light absorbing or emitting processes, ***wiggly arrows:*** nonradiative processes. S_0 is the ground singlet electronic state while S_1 and S_2, respectively, are the first and second electronically excited states. T_1 is the first triplet state

gives a schematic representation of the different vibrational and electronic energy levels of the fluorophore. The distance between two horizontal lines is proportional to the energy difference between the two levels represented by the lines. The long horizontal lines represent the different electronic energy levels which can be reached by *exciting* (elevating) an electron into high orbitals. Within each electronic state, several vibrational levels are possible and are represented in the diagram by the shorter horizontal lines. In addition, several rotational levels are possible for each vibrational level, but these are not shown in the diagram because they are very closely spaced. The energy separation between electronic levels is usually 0.4-3.0 eV, between vibrational levels 0.04-0.4 eV and between rotational levels 0.001-0.04 eV [8]. The singlet energy states (S_0, S_1. . .) arise from electronic configurations where all electron spins are paired so that the molecule has zero net spin. On the other hand, the triplet states (T_1, for example) arise from electronic configurations where two electron spins are unpaired and the molecule has a net spin number of 1. The lowest electronic energy level of most fluorophores is a singlet state.

At room temperature, the molecule of the fluorophore are essentially all in the lowest electronic and vibrational energy level, which is thus called the ground electronic-vibrational energy level or, simply, the ground level. More exactly, the fraction of molecules in any one level is proportional to $e^{-\epsilon/kT}$, where ϵ is the energy of the level (measured from ground level) and $kT = 0.026$ eV at 25°C. The energy of the first excited vibrational or electronic level is usually so much greater than kT that almost all molecules are in the ground level. When a solution of fluorophore molecules is illuminated with a beam of light of wavelength λ, the molecules can absorb energy from the beam if the photon energy $h\nu$ is equal to the energy difference between the ground energy level and one of the upper levels. Each molecule can only absorb one photon at a time, and during the absorption process a molecule undergoes a transition (is excited) from the ground to one of the upper states. If the wavelength of the exciting beam of light is scanned continuously, absorption of energy from the beam occurs each time the photon energy coincides with the energy between two levels. A plot of strength of absorption, usually expressed as optical density (OD) or extinction coefficient (ϵ), versus wavelength gives the absorption spectrum. The optical density and extinction coefficient at a given wavelength are defined by the expression

$$\begin{aligned} OD(\lambda) &= \log(I_o/I) \\ &= \epsilon(\lambda)\, cx \end{aligned} \qquad (1)$$

where I_o is the intensity of light incident on the solution of fluorophore molecules, I is the amount of light transmitted after passing through a distance x (in cm) of solution, and c is the concentration of fluorophore in moles per liter. Optical density is usually measured in a 1 cm pathlength cell.

In fluorescence spectroscopy, the most important transitions are those involving excitations from the singlet ground state to higher singlet electronic energy levels. (Optical transitions between singlet and triplet states are very weak and are usually not detected in an ordinary absorption spectrometer.) The absorption spectra resulting from transitions involving electronic excitation are called electronic absorption spectra. For fluorophores of interest in biological studies, these spectra are in the visible and ultraviolet regions of the spectrum (2500-6000 Å). Figure 2 shows the electronic absorption spectrum for anthracene in solution in the region 2900-3800 Å. This band is produced by transitions from the ground to the first excited electronic state. The narrower bands within the broad absorption band, often referred to as vibrational structure, correspond

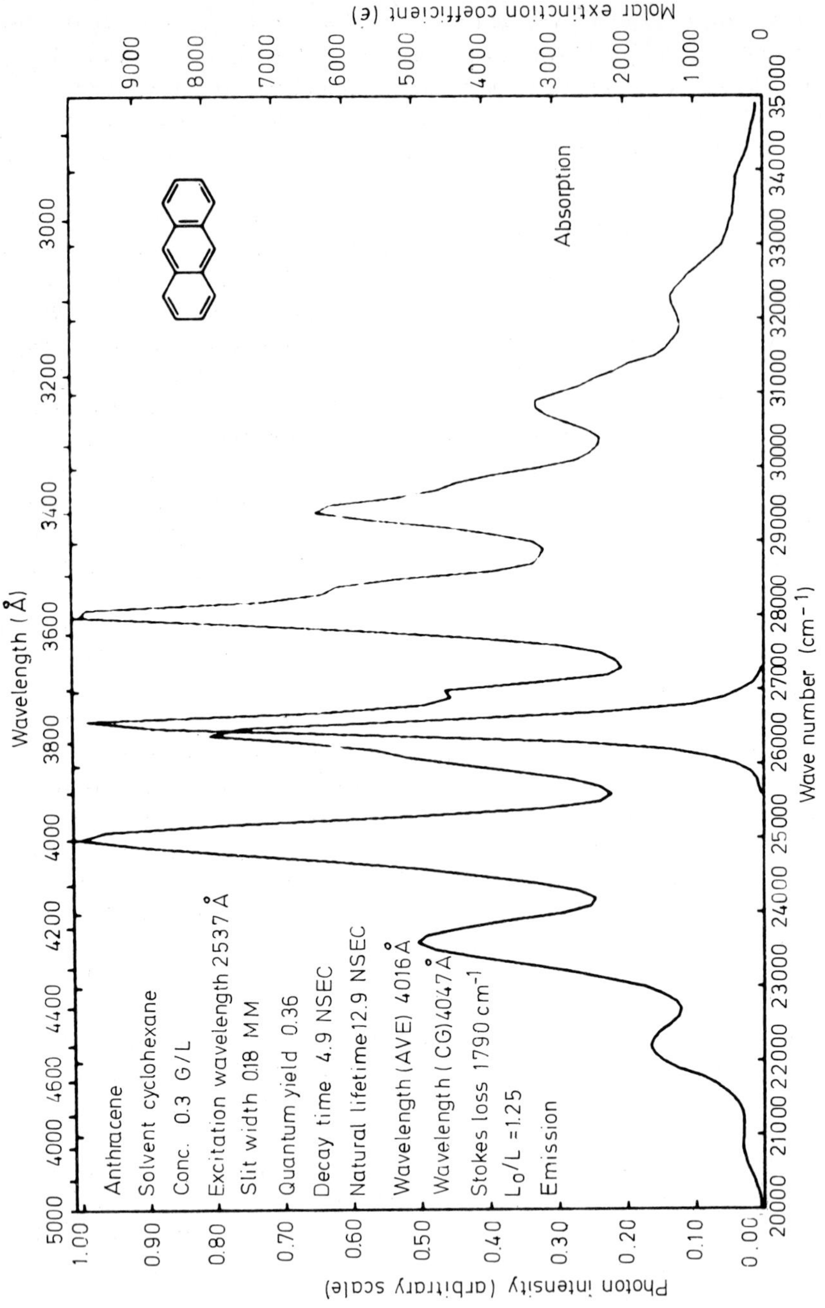

Fig. 2. Absorption and fluorescence spectra of anthracene in cyclohexane [70]

to transitions between the ground energy level and different vibrational levels of the first excited electronic state. The vibrational band centered at 3780 Å corresponds to the transition between the zero vibrational levels of the ground and upper electronic states and is called the 0-0 transition. Additional electronic absorption bands (not shown in Fig. 2) appear below 2900 Å corresponding to transitions between the ground level and higher electronic states.

It should be noted that many fluorophores have absorption spectra without vibrational structure, as shown in Fig. 3, and in some instances the spectrum of a given fluorophore may display structure in a nonpolar solvent but lose it in a polar solvent. Continuous spectra (spectra with little or no vibrational structure) result from closely spaced vibrational levels and solvent-solute interactions which alter the separation between energy levels in the energy level diagram. Statistically, the solvent environment around a fluorophore molecule fluctuates with time and is not exactly the same for all fluorophore molecules at any one time [9, 10]. Thus the energy levels of a fluorophore in solution are not sharp lines as shown in Fig. 1 but are in reality wide lines which have been broadened by solvent-solute interactions. Consequently, the vibrational bands in the absorption spectrum are broad (solvent broadening) and when the interactions are large (as in a polar solvent), the vibrational bands may broaden to the point where the spectrum becomes continuous.

B. Fluorescence Spectrum

When a fluorophore molecule is excited into an upper level such as A or B in Fig. 1, it rapidly cascades (in times $< 10^{-10}$ s) down to the zero vibrational energy level of the first excited state S_1, dissipating excess energy into vibrational modes of the solvent molecules, without emission of light. That is, no light emission occurs from the upper levels. Light emission, however, can occur from the zero vibrational level of S_1 and, in fact, the fluorescent light detected in a fluorescence experiment results entirely from radiative transitions between this level and vibrational levels of the ground electronic state as shown in Fig. 1. We will therefore refer to this level as the radiating emitting level.

The types of fluorescence spectra which result from radiative transitions are shown in Figs. 2 and 3. The fluorescence spectrum can display vibrational structure or be continuous. Important features of fluorescence spectra can be listed as follows [6, 7].

1. A fluorophore in a given homogeneous solvent normally displays only one fluorescence spectrum and the spectrum is independent of wavelength of exciting light. This is due to the fact, that, as explained above, a fluorophore has only one radiative level and emission occurs only from this level regardless of the energy level into which the molecule is initially excited.
2. The fluorescence spectrum is always present at longer wavelengths than the absorption spectrum. This can be seen in Fig. 1, which shows that the vertical lines representing the absorption processes are longer (higher energy, shorter wavelength) than the lines representing the emission processes, except for the 0-0 transition where the two transitions have the same energy. The 0-0 vibrational peak on the fluorescence

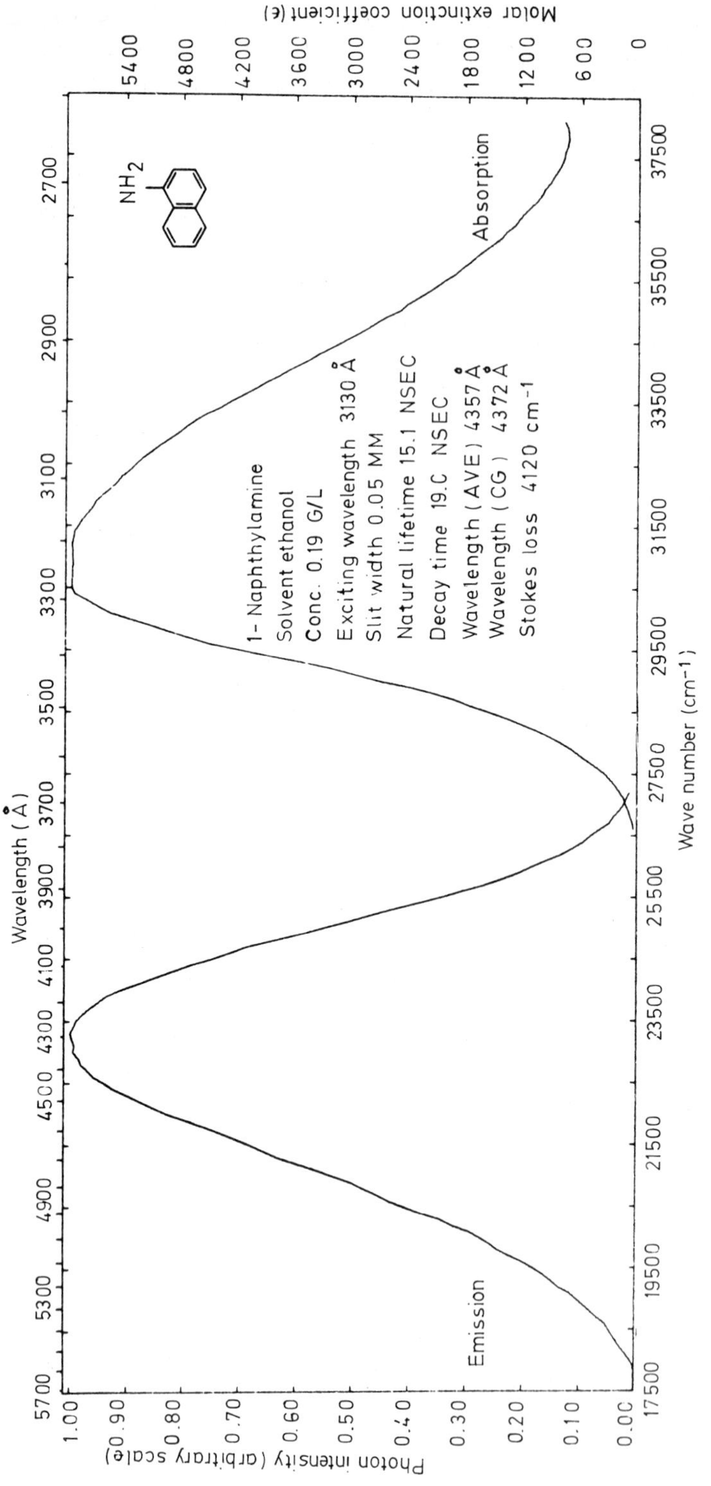

Fig. 3. Absorption and fluorescence spectra of 1-naphthylamine in ethanol [70]

spectrum usually overlaps the 0-0 transition peak of the long wavelength absorption band (see Fig. 1). This relation however may be altered by solvent-solute interaction, as explained above, in which case the absorption and emission spectra may not overlap at all.

3. The shape and wavelength position of the fluorescence spectrum are sensitive to the molecular properties of the environment. This is due to perturbation of the energy levels of the fluorophore by fluctuating dipole and electrostatic (in case of charged species) interactions between solvent and fluorophore molecules. In general, these interactions increase with solvent polarity, and they shift the absorption and fluorescence spectra toward longer wavelengths, and broaden the vibrational bands. The magnitude of the shift, however, is not necessarily the same for the absorption and fluorescence spectra. This is due to differences in the interactions of the solvent molecules with fluorophore molecules in the ground and excited states. In the ground state the solvent molecules have an average arrangement around the fluorophore molecules which is partly determined by the interaction between solvent and ground state molecules. This arrangement determines the effect of the solvent on the absorption spectrum. When a molecule is excited, the solvent arrangement in the ground state is maintained at the instant of excitation, but shortly after excitation the solvent molecules relax into a different arrangement dictated by new interactions with the excited molecule [9, 10]. Emission takes place from this relaxed state. The influence of solvent may thus differ for emission and absorption spectra. Solvent-solute interactions are usually greater in the excited state than in the ground state, so that the emission is usually broader and shifted more toward longer wavelengths than the absorption spectrum.

C. Fluorescence Efficiency

Because of competition between radiative and nonradiative processes not all excited molecules emit light. For any fluorophore in a given environment, the fraction of excited molecules which do emit light is a well-defined number and is called the fluorescence efficiency or quantum yield. Experimentally, the fluorescence efficiency is usually measured in a steady-state spectrofluorimeter. In this instrument the fluorophore solution is excited by a steady beam of light of wavelength λ_{ex}. The beam excites a number n_{ex} of fluorophore molecules per second, and a number n_{em} of excited fluorophore molecules each emits a photon per second. The fluorescence efficiency is then defined by the expression

$$\phi_F(\lambda_{ex}) = \frac{n_{em}}{n_{ex}}$$

The fluorescence efficiency may also be defined in terms of kinetic parameters, as shown in the following section [see Eq. (5)]. The fluorescence efficiency for different fluorophores may range from 1 for highly fluorescent substances to 0 for nonfluorescent compounds.

In general, the fluorescence efficiency may depend on the exciting wavelength λ_{ex}. That is, it may depend on the energy level to which the molecule is excited. The nature of this dependence can be seen more explicitly from the following expression:

$$\phi_F(\lambda_{ex}) = \phi_I(\lambda_{ex})\,\phi_0 \tag{2}$$

which states that the value of $\phi_F(\lambda_{ex})$ is determined by $\phi_I(\lambda_{ex})$, the fraction of molecules in the initially excited upper level which cascade down to the radiative level, and ϕ_0, the fraction of molecules at the radiative level which actually emit light. ϕ_0 is independent of wavelength and its value is determined by competition between radiative processes from the radiative level and such nonradiative processes as internal conversion (nonradiative cascade into the ground level), intersystem crossing (nonradiative conversion of the excited molecules into the nonradiative triplet state), and decomposition. $\phi_I(\lambda_{ex})$ is similarly determined by competition between the cascade processes and such processes as intersystem crossing, decomposition, and molecular rearrangement in the upper levels.

For many fluorophores of use in biological studies, $\phi_I(\lambda_{ex})$ is independent of wavelength and usually equal to 1. For these fluorophores, the fluorescence efficiency is equal to ϕ_0 and independent of exciting wavelength.

D. Fluorescence Lifetime

The fluorescence lifetime τ is defined by the expression

$$I(t) = I_0\, e^{-t/\tau} \tag{3}$$

which describes the decay of fluorescence intensity I(t) following excitation of a fluorophore solution by a very fast pulse of light. The lifetime is essentially defined as the time required for the intensity I(t) to fall to 0.37 (i.e., to e^{-1}) of its initial value I_0.

The lifetime is usually independent of the exciting wavelength. This is due to the fact that molecules excited into high energy levels internally convert to the emitting level in times which are too short (10^{-12} to 10^{-11} s) to be detected by a nanosecond spectrofluorimeter. The lifetime measured by the fluorimeter is thus essentially the lifetime of excited molecules in the emitting level. In addition to its experimental definition, the lifetime can also be defined in terms of kinetic parameters by the expression

$$\frac{1}{\tau} = k_e + k_i \tag{4}$$

where k_e is specific rate for light emission from the radiative level (i.e., it is the sum of the specific rates for all radiative processes from the emitting level to the ground electronic level) and k_i is the sum of the specific rates for all nonradiative processes from the emitting level. It should also be noted that the fluorescence efficiency Φ_F is related to the lifetime and kinetic parameters by the expressions [assuming that the fluorescence efficiency is independent of wavelength, see Eq. (2)]

$$\phi_F = \frac{k_e}{k_e + k_i} \tag{5}$$

$$= k_e \tau \tag{6}$$

E. Polarized Emission and Absorption Properties of a Fluorophore

Light emitted by a fluorescent molecule is polarized and is most intense along certain specific directions with respect to the molecular framework. In addition, the strength with which photons are absorbed from a polarized beam of light depends on the orientation of the molecule with respect to the direction of polarization of the light beam. These properties are of interest in biological studies because they allow the orientation and rotational mobility of membrane components to be determined by polarized emission measurements.

The laws describing the properties mentioned above can be most easily discussed in terms of absorption and emission transition dipole moments, which have specific directions within the molecular framework of the fluorophore [11-14]. A transition moment can be visualized classically as a dipole antenna which is responsible for absorption or emission of radiation and has its direction fixed along the direction of the transition moment [15]. Experiment and theory indicate that each electronic absorption band of a fluorophore has a transition dipole moment associated with it. Similarly, the emission band has an associated emission moment.

In the case of anthracene, for example, the long wavelength absorption band shown in Fig. 2 is associated with an absorption moment which lies along the short axis of the molecule shown in Fig. 4 [14]. The next absorption band at shorter wavelength (second

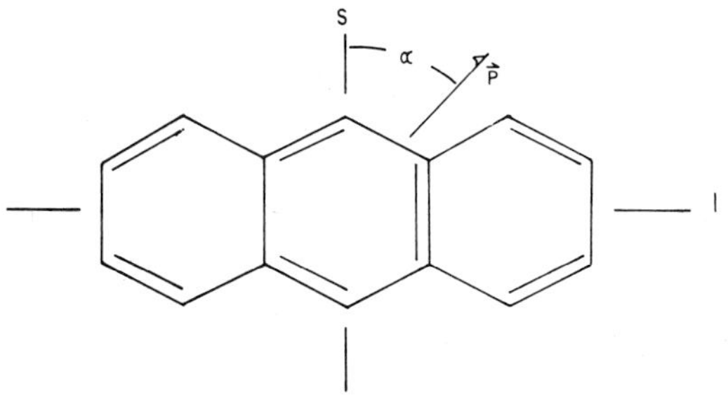

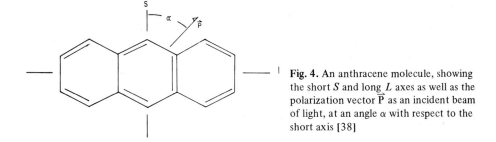

Fig. 4. An anthracene molecule, showing the short S and long L axes as well as the polarization vector $\vec{P}$ as an incident beam of light, at an angle α with respect to the short axis [38]

absorption band corresponding to transitions between ground and second electronic state S_2) in Fig. 1 has its transition moment along the long axis. For most fluorophores the emission moment lies close to the transition moment of the long wavelength absorption band, since this absorption band and the emission band usually involve the same electronic transition. The two moments, however, usually do not coincide but make an angle in the range 10°-40°. This angle is determined by internal molecular mechanisms discussed further below.

The laws describing the absorption and emission of polarized light, as well as the spatial distribution of fluorescence intensity around a molecule, can be described in terms of transition moments as follows:

1. A fluorophore molecule absorbs plane polarized light with maximum efficiency when the direction of polarization $\vec{E}$ of the light coincides with the direction of the absorption transition moment of the corresponding absorption band. Thus, when an anthracene molecule is illuminated with light whose wavelength coincides with the long wavelength absorption band, maximum absorption occurs when $\vec{E}$ is along the short axis of the molecule. When $\vec{E}$ makes an angle θ with the transition moment, the efficiency of absorption is proportional to $\cos^2\theta$.
2. Light is emitted with highest intensity along any direction perpendicular to the emission moment. For an anthracene molecule the emission moment is oriented along a direction close to the direction of the short axis. For any other direction $\vec{d}$ (see Fig. 4) which makes an angle ϕ with the emission moment, the intensity is proportioned to $\sin^2\phi$.
3. The light emitted by a fluorophore molecule along any direction $\vec{d}$ is plane-polarized. The direction of polarization is perpendicular to $\vec{d}$ and lies in the plane defined by $\vec{d}$ and the emission moment.

The manner in which polarized emission is used to study orientation and rotational mobility in membranes is discussed in Sect. III.A. It should be noted that the use of dipole transition moments to describe polarized emission and absorption of light applies only to fluorophores which emit and absorb light by means of a so-called dipole mechanism. In general the emitting and light absorbing properties of a molecule can be considered to arise from a superposition of several mechanism which can be classified as dipole, quadrupole, octapole, and higher multipole moments mechanisms. For electronic transitions which have a high extinction coefficient (e.g., $> 10^3$), the dipole mechanism usually dominates and can be used alone to describe emitting and absorbing properties. There are certain transitions, however, which cannot occur by a dipole mechanism (dipole-forbidden transitions) but instead take place as a result of higher multipole interactions. Such transitions are usually weak, having extinction coefficients $< 10^2$, and do not obey the laws listed above. Most of the fluorophores used in membrane studies, however, can be described in terms of dipole transitions. The following discussions assume such fluorophores.

F. Methods of Approach in the Application of Fluorescence Spectroscopy to the Study of Membranes

As previously mentioned, the application of fluoroscence spectroscopy to the study of membranes is based on the sensitivity of emission parameters to the molecular conformation and dynamics of the environment [16]. A quantitative understanding of the dependence of these parameters on environmental properties must ultimately involve complex quantum-mechanical considerations. Fortunately, however, most of the useful applications do not require a detailed quantum-mechanical understanding but can be developed instead on a semi-empirical basis. The usual approach in a particular application is first to select a fluorophore which has an emission property (or properties) that is particularly sensitive to the environmental parameter of interest and then to "calibrate" the fluorescent probe by studying the dependence of the emission parameter on the environmental parameter of interest, using model systems. The calibration may be guided by semi-theoretical expressions obtained by relatively simple classical or quantum-mechanical considerations. Thus, for example, the microviscosity η of a membrane can be studies by measuring the fluorescence anisotropy A of perylene inserted into the membrane [17]. Classical diffusion theory indicates that the fluorescence anisotropy is a function of the ratio $T\tau/\eta$, where T is temperature and τ is the lifetime of the fluorophore. The exact functional dependence between A and $T\tau/\eta$, however, cannot be established by theoretical calculations but can be determined experimentally by measuring A and τ for perylene in oils of known microviscosity η at different temperatures and making a plot of A versus $T\tau/\eta$. This plot then serves as a calibration graph for determining η of a membrane from the measurement of A and τ of perylene inserted into the membrane (see Sect. III.B.2 for details).

In the following section we discuss some of the more interesting applications of fluorescence spectroscopy in the study of membranes, describing in each case the emission parameter or parameters most appropriate for the application, the calibration procedure, and the information which is obtained from the fluorescence measurements.

II. Types of Fluorophores Used in the Study of Membranes

The fluorophores used in the study of biological membranes may be classified as intrinsic or extrinsic [16]. Intrinsic fluorophores are fluorescent substances or groups which are naturally present in membranes. An example is the amino acid tryptophan which is found in many membrane proteins. Intrinsic probes are not often used in fluorescence studies, because none may be located at a site of interest or the group may be present in too many different sites, such as tryptophan, to yield interpretable results. Extrinsic probes, on the other hand, can be designed to enter a specific site or region of a membrane and may be selected to have emission properties which are particularly sensitive to the membrane parameter of interest. Table 2 gives examples of extrinsic probes which have been used to study various aspects of membranes.

The amphipathic, noncovalent probes shown in Table 2 spontaneously enter lipid bilayer regions of membranes when added to a membrane suspension, and can therefore be used to study these regions of membranes. They have indeed been used in many investigations of pure, synthetic lipid bilayers as well as natural membranes. In the case of natural membranes, however, the probe may also be attached to membrane proteins. Some of the noncovalent amphipathic probes have been designed to monitor specific transverse regions of the lipid bilayer portion of a membrane. In a lipid bilayer the polar head of an amphipathic probe is expected to be located in the polar region of the bilayer while the nonpolar part can be assumed to project toward the membrane interior. Nonpolar probes such as perylene would presumably be located in the interior central region of the bilayer. The actual location of the probes can be determined or ingerred by a vari-

Table 2. Examples of fluorescent probes used in membrane studies

NONCOVALENT

NONPOLAR

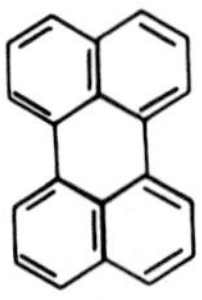

Perylene

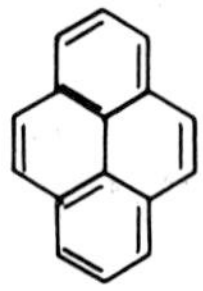

Pyrene

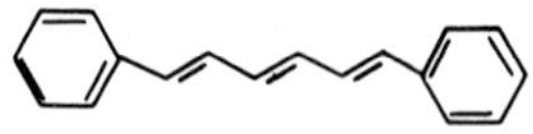

trans-Diphenylhexatriene (DPH)

POLAR

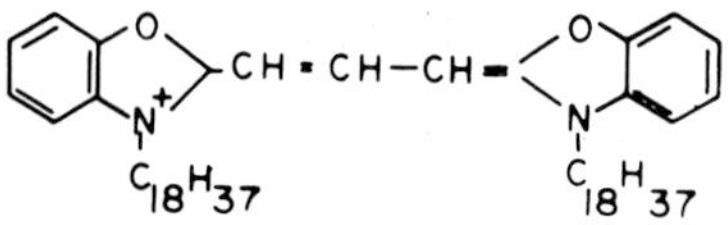

N,N'-Di(octadecyl)-oxacarbocyanine (K-I)

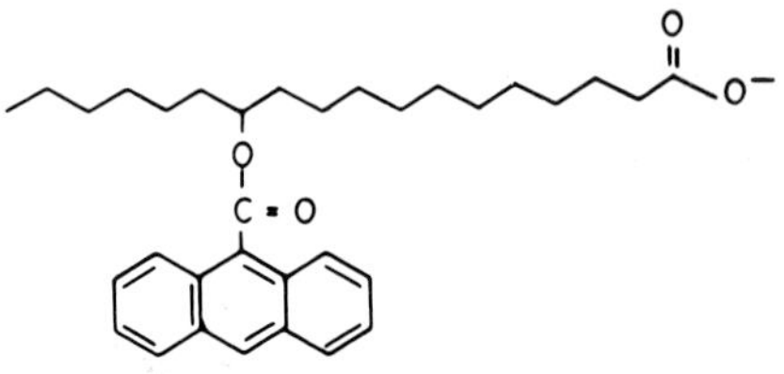

12-(9-Anthroyl)-stearic acid (AS)

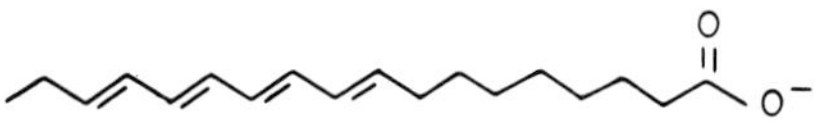

trans-Parinaric acid

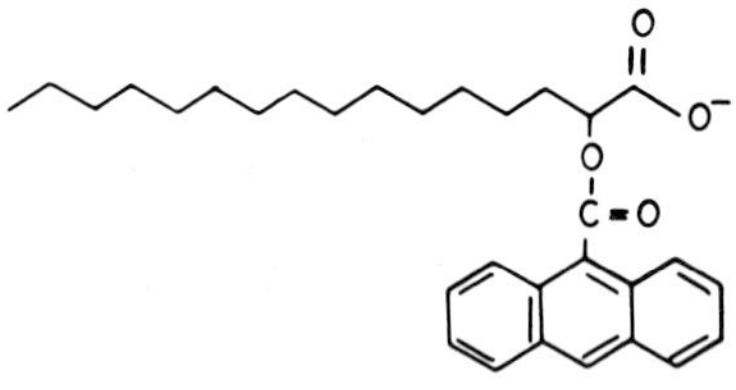

2-(9-Anthroyl)-palmitic acid (AP)

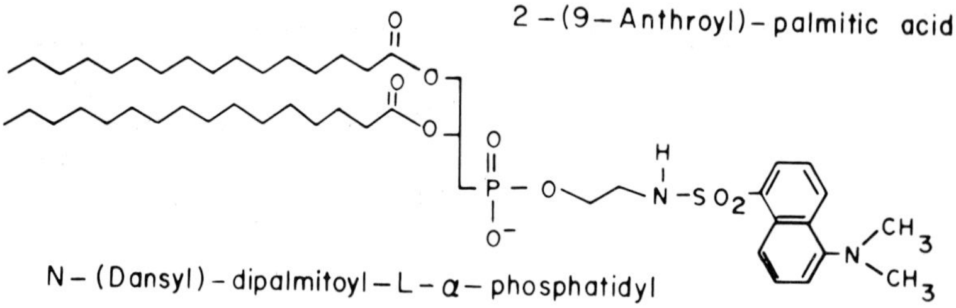

N-(Dansyl)-dipalmitoyl-L-α-phosphatidyl ethanolamine

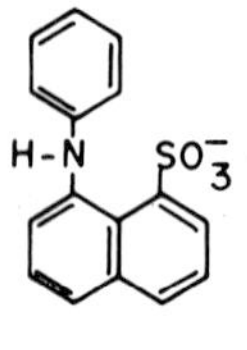

1-Anilino-8-naphthalene sulfonate (ANS)

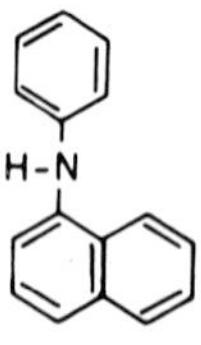

N-phenyl-1-naphthylamine (NPN)

Tabel 2 continued

HIGH AFFINITY

AChE (acetylcholinestrase) AChR (acetylcholine receptor)

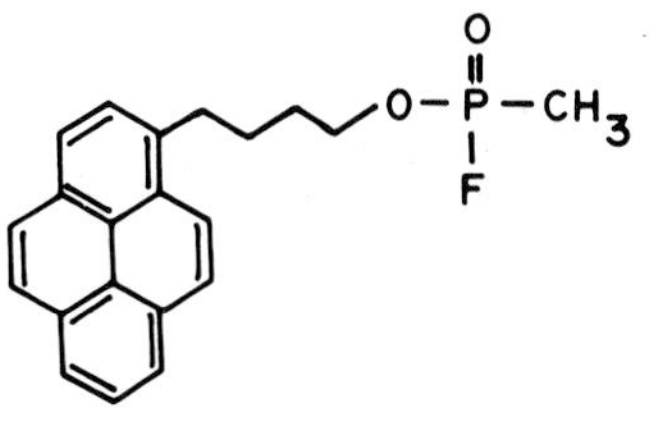

Pyrenebutylmethyl-phosphonofluoridate (AChE and serine hydrolases)

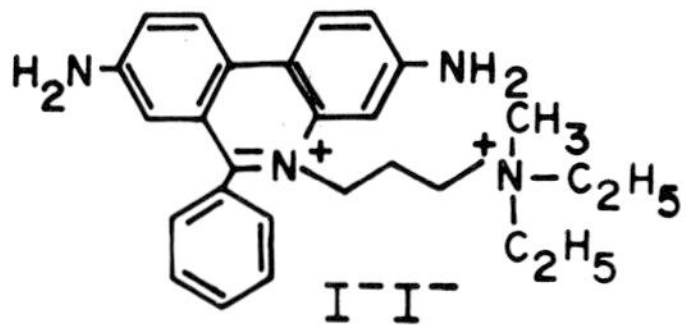

Propidium diiodide (AChE and AChR)

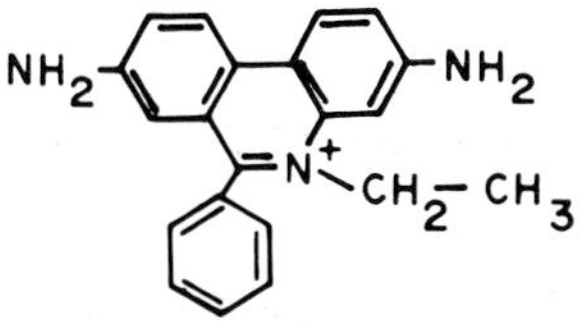

Ethidium bromide (AChE)

COVALENT

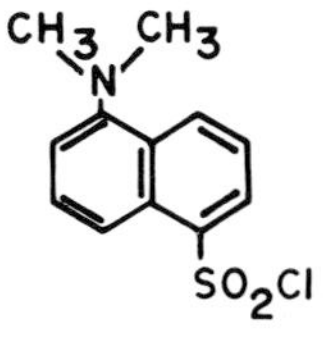

1-dimethylamino-naphthalene-5-sulfonyl chloride (dansyl chloride)

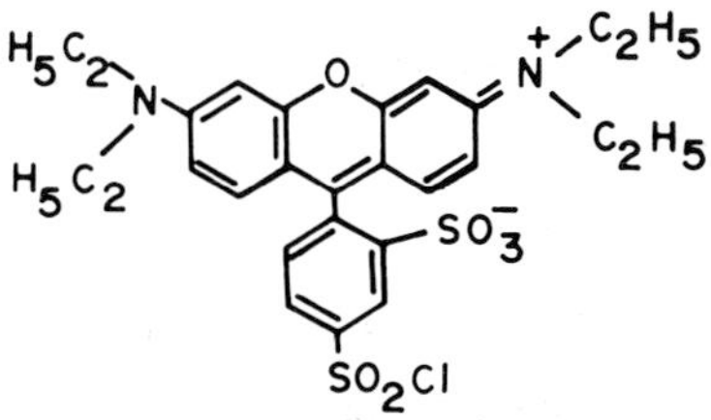

Lissamine rhodamine B sulfonyl chloride

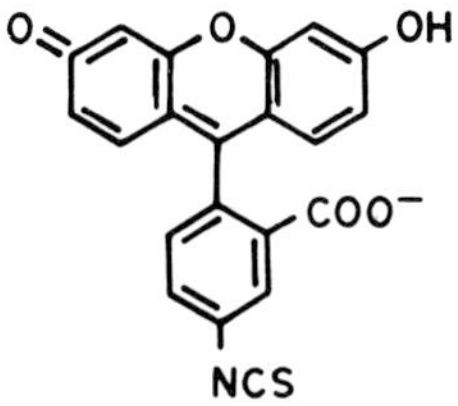

Fluorescein isothio-cyanate (FITC)

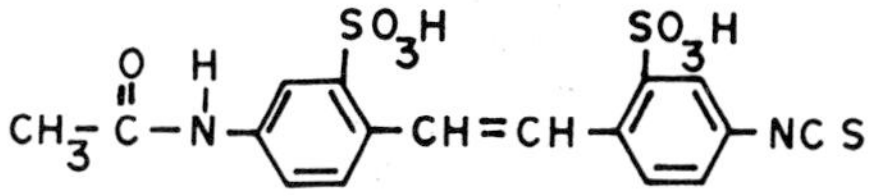

4-Acetamido, 4'-isothiocyanostilbene-2,2'-disulfonic acid (SITS)

ety of techniques. Lesslauer et al. [18] have used X-ray diffraction techniques to study the location of ANS, AS, and ONS in oriented lipid multibilayers. From electron density maps they conclude that the probes are located in a lipid bilayer, as expected on the basis of their amphipathic character. Others have inferred the location of specific probes by comparison of the values of their emission parameters in a bilayer with values obtained in solvents of different polarities [19]. Thus, the fluorescence lifetime of AS increases from 1.6 to 8.5 ns in going from methanol to hexane. In a phosphatidyl choline vesicle, the lifetime of AS is 12.4 ns, which indicates that the anthroyl group is located in a very nonpolar environment, as expected if the nonpolar chain of AS extends into the membrane interior.

Other results, however, suggest that the probe may not be located at a unique site in a bilayer and that it may not always be located where expected on the basis of its molecular structure. Thus, lifetime studies have shown that AS in vesicles of dipolmitoyl lecithin at 50°C displays at least two different lifetimes (about 5.5 and 10 ns), indicating that it senses at least two different environments [20]. Similarly, lifetime measurements of perylene in a bilayer indicate that this probe also senses more than one environment [21]. The probes, therefore, do not seem to sit at a unique site. It is, however, unclear at the present time whether the multiple lifetimes displayed by fluorophores in membranes are due to location of the probe in more than one region of a bilayer or whether a given bilayer region exists in several different conformational states, depending on the method used to prepare the bilayer. Evidence that a probe may not always be located where expected can be obtained, for example, from studies of the emission properties of perylene in the liquid and gel states, respectively, of a bilayer. On general grounds it is expected that the fluorescence efficiency and lifetime of a fluorophore will increase as the bilayer passes from the liquid to the gel phase. Experiments indicate that the fluorescence efficiency of perylene actually decreases. This result has been interpreted as indicating that the "crystallization process" forces the perylene molecules from the interior into the polar regions of the bilayer [21]. If this interpretation is correct, then perylene molecules in the gel phase do not reside in the central region of the bilayer, as expected from simple considerations. In addition, at high probe concentrations some probes may be forced out of the bilayer into clusters of pure probe surrounded by lipid molecules [22, 23]. Moreover, when probes such as perylene or AS are added to a natural membrane (in contrast to a synthetic pure lipid bilayer) the probe may become attached to membrane proteins as well as to bilayer regions of the membrane. Finally, it should be noted that in the case of intact cells the probe may not be exclusively situated on any one particular cellular membrane, for example, plasma, mitochondrial or endoplasmic reticulum membrane. Although the plasma membrane is exclusively labeled when a probe is first introduced into the bathing electrolyte, experiments indicate that at least in some cells the label quickly migrates and labels internal membranes and possibly other internal structures [24]. This multiple labeling of membranes and internal structures introduces serious complications of course, in the study of intact cells. One possible approach for circumventing this problem is to purify the various cell membranes and study each individually. There are, however, insufficient data at the present time with which to determine the extent to which membrane properties are altered during the purification steps. Moreover, the study of the role of membranes in specific physiological functions may require an intact cell in order for the functions to be expressed.

A variety of covalent fluorescent probes are also available for studying membrane components. Many of these probes are of the type which have been used to label water-soluble proteins. They consist of a fluorescent moiety with a reactive group that allows the fluorophore to be covalently attached for example to amino acid side groups of proteins. Examples of these probes are shown in Table 2. The use of these probes in the study of membrane proteins is usually restricted to the study of purified membrane proteins, since their indiscriminate labeling properties often lead to complicated and uninterpretable results when applied to natural membranes containing many different proteins. Some of the probes also react with specific lipids such as phosphatidylethanolamine. More recently, probes with more discriminating labeling properties have been developed by attaching fluorescent groups to substrates or substrate analogs of membrane receptors. Relatively specific labeling of protein components has also been achieved by the use of fluorescent labeled antibodies and lectins which bind to membrane glycoprotein and glycolipids.

The particular fluorophore to use in a specific application depends on factors which may vary from one application to the next, as discussed below in the the section on applications. However, a general consideration in the use of fluorescence techniques to study membranes in intact tissues or cells is that of background fluorescence due to natural fluorescence from the cell or tissue. This fluorescence becomes quite intense when a tissue or cell is exposed to light with a avelength below 4000 Å. Much of the background fluorescence excited by light with a wavelength around 3600 Å comes from $NADH_2$, which absorbs strongly in this region of the spectrum and emits light in the range 4200-5600 Å. Natural protein emission (from tryptophan) is excited at wavelengths below 2900 Å. Because of background emission, fluorophores such as AS and ANS, which have proved to be very useful in the study of vesicles and purified membranes, are sometimes diffucult to use in the study of intact tissue or cells since their exciting wavelength is below 4000 Å.

An important question which arises in the use of probes to study membranes is to what extent the probes disturb membrane properties. That is, does a probe give information on the inherent properties of a membrane or does it report on properties which have been altered by probe perturbations? Experience with water-soluble proteins, such as enzymes and antibodies, indicates that the activity and conformation of a protein is not significantly altered by many fluorescent probes at low labeling ratios (1 or 2 probe molecules per protein molecule) unless the probe is introduced directly into an active site or attached to some essential amino acid side group. This result is due to the large difference in sizes between probes and proteins. Thus, we expect that membrane proteins will also not be greatly affected by labeling. In any particular experiment, however, the effects of labeling should be tested by measuring activity, absorption spectrum, optical rotating dispersion, circular dichroism, and other indications of protein structure. In contrast to proteins, bilayer properties could be expected to be affected by fluorescent probes, since their sizes are a significant fraction of the thickness of a bilayer. However, a variety of experiments indicate that the perturbation of a bilayer is not significantly large at low labeling ratios (less than 1 probe molecule in 100 lipid molecules). Thus, measurements with fluorescent probes of phase-transition temperatures of synthetic lipid bilayer (see Sect. III.F) yield values which are very close to those obtained by calorimetry. Similarly, phase diagrams determined with fluorescent probes for two component lipid bilayers agree well with those obtained by calorimetric measurements. At high labeling ratios, however, the probes often alter phase-transition temperatures, lipid mobility, and other membrane properties [25]. In addition, probes which have a net charge may significantly change the electrostatic surface potential of a membrane and alter functions such as membrane permeability as well as distribution of lipid phases within a bilayer.

III. Applications

A. Determination of the Orientation of Membrane Components by Polarized Fluorescence Measurements on Oriented Membranes

When a fluorophore molecule is inserted into a membrane its molecular axes will assume a distribution of orientations which reflects the orientational distribution of membrane components. Thus if such a molecule is inserted into the lipid bilayer, its orientation will reflect that of the lipid molecules [1, 26, 27]. On the other hand, if the fluorophore is attached to a membrane protein or carbohydrate, its orientation will reflect that of these components. Information on the orientation of membrane components can therefore be obtained from studies of the orientation of a fluorophore inserted into a specific site in the membrane. In this section we describe how the orientational distribution of a fluorophore can be evaluated by polarized fluorescence intensity measurements. Before doing so, however, we should note that these measurements do not actually give direct information on the orientation of the molecular axes of a fluorophore but instead give information on the orientational distribution of the absorption and emission moments of the molecule. However, since these moments have fixed and known directions with respect to the molecular axes of the fluorophore molecule, some information concerning the orientation of the molecule can be obtained from a knowledge of the orientation of the emission and absorption moments. Thus, if an anthracene molecule is inserted into a membrane and excited with light having a wavelength coincident with the long-wavelength absorption band, then polarized fluorescence intensities give information on the orientation of the short axis of the molecule, since the absorption and emission moments lie along this axis. If the molecule is excited in the second absorption band, information is obtained about the orientational distribution of the long axis, since the absorption moment lies in this direction for excitation in the second absorption band. The following discussions are limited to excitation in the longest wavelength band.

The method used to determine orientation of a fluorophore in a membrane consists of exciting an oriented patch of membrane, such as shown in Fig. 5, with a beam of light polarized along a direction defined by the unit vector $\vec{P}$ and measuring the intensity of the emitted light through an analyzing polarizer oriented in the direction $\vec{A}$. The polarized fluorescence intensity I_{PA} measured in this manner is then related to the orientation of the absorption moments as follows. Let $\vec{a}(\theta,\phi)$ be a unit vector which describes the direction of the absorption moment of a given fluorophore molecule, let $\rho(\theta,\phi) \sin\theta d\theta d\phi$ be the fraction of fluorophore molecules which have their absorption moments within the solid angle $\sin\theta d\theta d\phi$ centered at θ,ϕ. That is, $\rho(\theta,\phi)$ describes the orien-

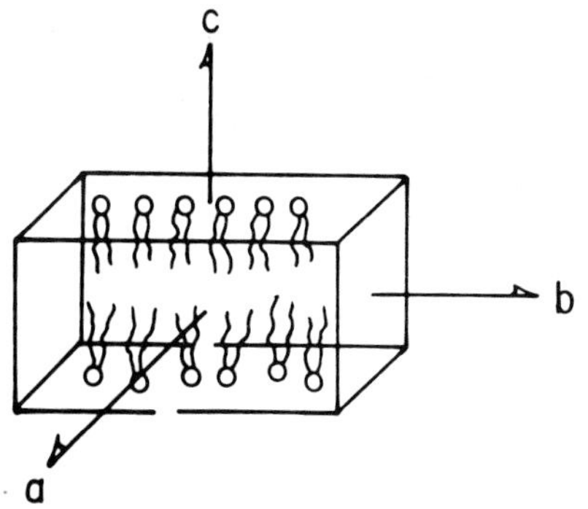

Fig. 5. A lipid bilayer and its principal *a*-, *b*-, and *c*-axes

tational distribution of the absorption moments. The polarized intensity I_{PA} is then related to $\rho(\theta,\phi)$ by the expression (see Fig. 6 for definition of angles)

$$I_{PA} = \int\int \rho(\theta,\phi)\ [\vec{a}(\theta,\phi) \cdot \vec{P}]^2\ [\vec{a}(\theta,\phi) \cdot \vec{A}]^2 \sin\theta d\theta d\phi \tag{7}$$

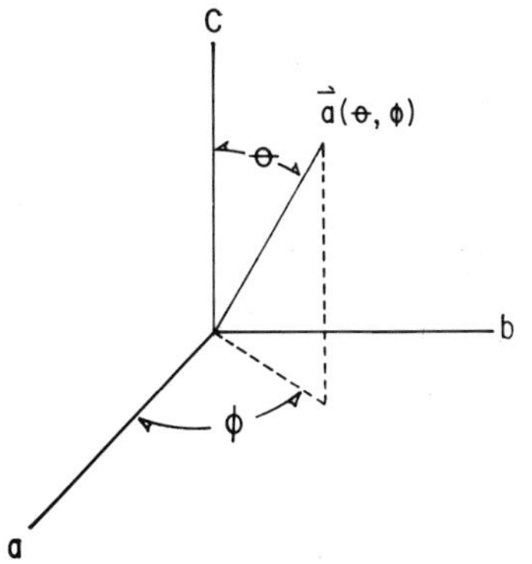

Fig. 6. System of coordinates used to express the orientation of the absorption moment vector $\vec{a}$

This equation is for the restricted case where the emission and absorption moments are in the same direction within the molecular axes and where the molecules do not rotate significantly during the lifetime τ of the fluorophore. The first restriction applies to a close approximation when a fluorophore is excited in its longest wavelength absorption band, but even in this case the emission moment does not in general coincide exactly with the absorption moment. Moreover, various measurements have shown that fluorophore molecules in a membrane undergo significant rotation during the time τ. Removal of these restrictions, however, highly complicates the relation between I_{PA} and $\rho(\theta,\phi)$. In this section we will limit our discussion to the relatively simple Eq. (7). This expression contains all the essential aspects necessary to demonstrate the use of polarized emission spectroscopy in the study of orientation in membranes.

The object of a polarized fluorescence experiment is to evaluate $\rho(\theta,\phi)$ with Eq. (7) from measurement of I_{PA} for various orientations of $\vec{P}$ and $\vec{A}$. Although $\vec{P}$ and $\vec{A}$ can have any orientation which the experimenter desires, there are certain fundamental orientations which are suggested by the structure of a membrane, namely those in which $\vec{P}$ and $\vec{A}$ are each oriented along one of the membrane axes a, b, or c. Thus if $\vec{P}$ and $\vec{A}$ are oriented alternatively along a, b, and c, then nine principle polarized fluorescence intensities are obtained and can be represented by the matrix

$$I = \begin{pmatrix} I_{aa} & I_{ab} & I_{ac} \\ I_{ba} & I_{bb} & I_{bc} \\ I_{ca} & I_{cb} & I_{cc} \end{pmatrix} \tag{8}$$

where the first and second subscripts refer to the axes along which $\vec{P}$ and $\vec{A}$ are oriented respectively. The nine intensities, however, are not independent. First, theoretical considerations based on the symmetry of the membrane (uniaxial symmetry) indicate that when the absorption and emission moments are in the same direction, the following relations apply [26]:

$$I_{aa} = I_{bb} = \alpha_0 \tag{9}$$

$$I_{ac} = I_{cb} = I_{ca} = I_{cb} = \gamma_0 \tag{10}$$

$$I_{ab} = I_{ba} = \beta_0 \tag{11}$$

$$I_{cc} = \delta_0 \tag{12}$$

More explicitly, from Eq. (7) we can write:

$$\alpha_0 = \frac{3\pi}{4} \int_0^{\pi/2} \rho(\theta) \sin^5\theta d\theta \tag{13}$$

$$\beta_0 = \frac{\pi}{4} \int_0^{\pi/2} \rho(\theta) \sin^5\theta d\theta \tag{14}$$

$$\gamma_0 = \pi \int_0^{\pi/2} \rho(\theta) \sin^3\theta \cos^2\theta d\theta \tag{15}$$

$$\zeta_0 = 2\pi \int_0^{\pi/2} \rho(\theta) \cos^4\theta \sin\theta d\theta \tag{16}$$

where $\rho(\theta)$ is normalized so that

$$\int_0^{\pi/2} \rho(\theta) \sin\theta d\theta = \frac{1}{2\pi}$$

The subscript zero emphasizes that these relations are for the case where the emission and absorption moments coincide in each fluorophore molecule and the molecules do not rotate. Equations (9)-(16) show that there are only four different values for the nine principle intensities. Second, additional theoretical considerations show that only two of the four intensities are really independent. More specifically, the following relations apply:

$$\alpha = 3\beta \tag{17}$$

$$\tfrac{3}{8}\alpha_0 + \delta_0 + 4\gamma_0 = 1 \tag{18}$$

Finally, I_{PA} for any arbitrary orientation of $\vec{P}$ and $\vec{A}$ can be evaluated from the principle intensities. In summary, we can therefore state that *two principle polarized fluorescence intensities contain all the information which can be obtained about orientation in a bilayer for the case where the emission and absorption moments coincide and the molecules do not rotate during their fluorescence lifetime.*

We now address the question, what precisely do the polarized fluorescence intensities tell us about $\rho(\theta,\phi)$? First, these intensities can tell us, by a simple comparison of the values of I_{aa} and I_{cc}, whether a fluorophore has a random or specific (nonrandom) distribution of orientations within a bilayer and, in case it has a specific distribution, what is the direction of preferred orientation. Thus if $I_{aa} \neq I_{cc}$ the distribution is nonrandom. Furthermore, if $I_{cc} > I_{aa}$, the emission moments are preferentially aligned along the c axis of the membrane, whereas if $I_{cc} < I_{aa}$ they are aligned preferentially along the ab plane. Second, the polarized intensities can also give information on how tell the moments are aligned, i.e., the degree of orientation along the preferred direction of orientation. This information can best be obtained by evaluating $\rho(\theta,\phi)$. Unfortunately, the polarized fluorescence intensities do not give sufficient information to define $\rho(\theta,\phi)$ completely. In fact, theory shows that the principle polarized fluorescence intensities give values for the fourth moments of the distribution $\rho(\theta,\phi)$ [28, 29], whereas a large range of moments beginning from 1 to a very high order (infinite in principle) is required to define $\rho(\theta,\phi)$ completely. In view of these limitations, the procedure normally used to obtain information on the degree of orientation is to assume a form, e.g., a gaussian form, with a variable width parameter for $\rho(\theta,\phi)$ and use the polarized fluorescence intensities to evaluate the width parameter, which then serves as measure of the degree of orientation. Thus if the moments are preferentially aligned along the c axis we can assume that $\rho(\theta,\phi)$ has the form

$$\rho(\theta,\phi) = \frac{e^{-\theta^2/\theta_g^2}}{N} \tag{19}$$

while for alignment along the ab plane we can assume

$$\rho(\theta,\phi) = \frac{e^{-(\pi-\theta)^2/\theta_g^2}}{N} \tag{20}$$

where N is a normalizing factor and θ_g is a measure of the width of the distribution. To evaluate θ_g from the polarized intensities, the preferred direction is determined first by comparing I_{aa} with I_{cc}. Then, depending on the preferred direction of orientation, Eq. (19) or (20) is inserted into Eqs. (13)-(16) and values for the principle polarized intensities are calculated for different values of θ_g. These values are then compared with the experimental values of θ_g giving the best agreement is chosen as the width of the distribution.

A method for measuring the principle polarized fluorescence intensities from a spherical bilayer of plasma membrane of a cell is demonstrated by reference to the spherical membrane in Fig. 7. In this figure it is assumed that the experimental arrangement is such

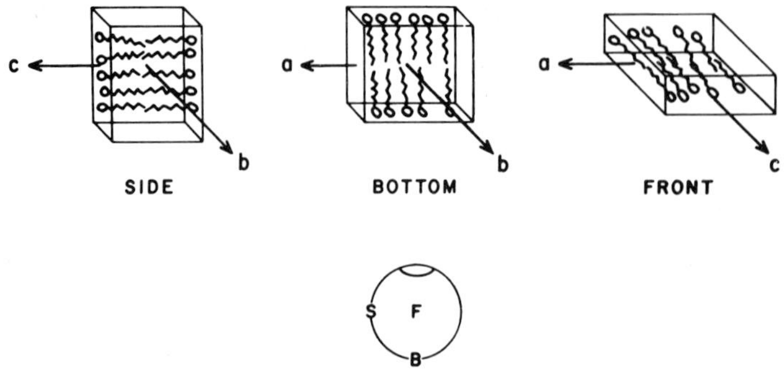

Fig. 7. Orientation of a patch of bilayer membrane at the side, bottom, and front of a spherical bilayer [26]

that a narrow exciting beam of polarized light travels from left to right in the plane of the figure. The emitted light is detected in a direction perpendicular to the plane of the figure (right angle excitation-detection arrangement). When the narrow beam of light falls on a small patch of membrane on the side of the spherical membrane, the bilayer being excited is oriented as shown in Fig. 7. Under these conditions, the exciting beam can be polarized along a or b while the analyzing polarizer can be oriented along a or c. Thus the intensities I_{aa}, I_{ac}, I_{ba}, and I_{bc} can be measured from a small patch on the side of the sphere. Other polarized fluorescence intensities can similarly be measured by illuminating small patches of membrane on the bottom or front of the sphere.

Table 3 gives the results of polarized fluorescence intensity measurements from a synthetic spherical bilayer membrane of about 4 mm diameter, made by a procedure described elsewhere. The polarized intensities were measured with a microspectrofluorimeter which uses a polarized beam of light of about 0.1 mm diameter to excite a patch of membrane. The emitted light is detected at right angles through an analyzing polarizer and a microscope which views only the small area on the membrane being illuminated. Features which should be noted in the data of Table 3 are as follows.

1. The probes are not randomly oriented in the bilayer, since for all three probes $I_{aa} \neq I_{cc}$.
2. The direction of preferred orientation for each probe, determined by comparing I_{aa} and I_{cc}, is the direction expected from the molecular structures of the probes. Thus for AS, the emission moment is perpendicular to the long axis of the fatty acid chain (i.e., the moment is along the short axis of the anthracene moiety) and in a bilayer one would expect the moment to be preferentially aligned along the ab plane, as is indeed indicated by the data of Table 3. For N,N′-di(octadecyl)-oxacarbocyanine (K-1) the emission moment is along the long axis of the cyanine moiety. The polarized intensities show that this moment is preferentially aligned along the ab

Table 3. Polarized fluorescence intensities and orientation and relaxation parameters for some fluorescent probes in an oriented spherical membrane

Probe	Membrane	ξ	I_{aa}	I_{ab}	I_{cb}	I_{bc}	I_{cc}	PO[a]	θ_G	θ_G	θ_H
AS	Egg lecithin	30°	1.0	0.89	0.68	0.79	0.62	a-b plane	78°	47°	0.8
AS	Oxidized cholesterol	30°	1.0	0.6	0.52	0.51	0.41	a-b plane	57°	38°	0.26
K-1	Oxidized cholesterol	20°	1.0	0.43	0.13	0.10	0.07	a-b plane	29°	20°	0.06
Dimethyl-popop	Oxidized cholesterol	20°	0.4	0.17	0.27	0.29	1.0	c-axis	73°	45°	0.12

Source: J. Yguerabide, unpublished results.
[a] Direction of preferred orientation

plane. This is consistent with what is expected from the molecular structure of K-1, namely that the cyanine moiety should lie on the surface of the bilayer with the hydrocarbon chains extended into the bilayer. For dimethylpopop the emission moment lies along the long axis of the molecule and the polarized intensities indicate that this axis is preferentially oriented along the c-axis of the membrane, as expected to facilitate packing.

3. The results indicate that I_{bc} is not equal to I_{cb}, which is contrary to the relation shown in Eq. (10). This discrepancy is due to the fact that for the probes in Table 3, the direction of the emission moment at the time of emission does not coincide with the direction which the absorption moment had at the time of excitation. Such coincidence was assumed in the derivation of Eq. (7). More general theoretical consideration indicates that when these two moments do not coincide, γ is split into two values, γ_1 and γ_2, which are related to the polarized intensities by the expressions

$$\gamma_1 = I_{ca} = I_{cb} \tag{21}$$

$$\gamma_2 = I_{ac} = I_{bc}. \tag{22}$$

The relations by Eqs. (9), (10), and (12) however are not altered. The direction of the absorption moment $\hat{a}$ at the time of excitation can differ from the direction of the emission moment $\hat{e}$ at the time of emission as a result of (1) internal motions within the molecule, (2) rotational motion of the fluorophore molecule before emission of light, or (3) a combination of these two processes. The extent to which internal motions contribute to the difference in the direction of $\hat{a}$ and $\hat{e}$ can be estimated in the following experiment. (For a quantum-mechanical discussion of internal factors which make $\hat{a}$ and $\hat{e}$ differ in direction, see [30].) The fluorophore is dissolved in a solvent which forms a glass at low temperatures and in which the fluorophore molecules have a random distribution of orientations. Next, the solution is frozen so as to immobilize the molecules and the system is then excited with a beam of light polarized in a vertical direction with respect to the laboratory axes. The emitted light intensity is measured through an analyzing polarizer oriented first in a vertical and then in a horizontal direction. Values for the two polarized fluorescence intensities, I_{VV} and I_{VH}, are thus obtained. The fluorescence anisotropy A is then calculated with the expression

$$A = \frac{I_{VV} - I_{VH}}{I_{VV} + 2\,I_{VH}} \tag{23}$$

which defines A. Theory indicates that for a system of immobilized fluorophore molecules, the fluorescence anisotropy is related to the angle ξ between the emission and absorption moment by the relation

$$A = \frac{0.4}{2}\,(3\cos^2\xi - 1) \tag{24}$$

This angle is a measure of the extent to which $\hat{e}$ and $\hat{a}$ differ in direction as a result of internal processes. Values of ξ calculated with Eq. (24) and experimental values of A are shown in Table 3.

4. Information about rotational motion (which is a measure of membrane fluidity) can be obtained from the ratio I_{aa}/I_{ab}, which is especially sensitive to this motion. If $\xi = 0$, the value of I_{aa}/I_{ab} ranges from 3 for no motion [see Eq. (17)] to 1 for complete rotational relaxation during the lifetime τ of the probe. The degree of relaxation in a given membrane can be stated in terms of a rotational relaxation factor R defined (for $\xi = 0$) by the expression

$$R = \frac{(I_{ab}/I_{aa})_{exp} - (I_{ab}/I_{aa})}{1 - (I_{ab}/I_{aa})_0} \tag{25}$$

where $(I_{ab}/I_{aa})_{exp}$ is the value measured experimentally for these ratios and $(I_{ab}/I_{aa})_0$ is the value for zero motion (which for $\xi = 0$ is equal to 3). The value of R ranges from 0 for no rotational motion to 1 for complete relaxation. When $\xi \neq 0$, Eq. (25) can still be used but $(I_{ab}/I_{aa})_0$ must now be corrected for changes introduced by the finite value of ξ. This can be done with expressions derived in our laboratory, but because of space limitations these are not presented here [31]. Table 3 shows relaxation values calculated for several probes in two different membranes. Caution must be exercised in comparing R values for different probes. It is necessary to keep in mind that R measures the degree of relaxation during the fluorescence lifetime of the fluorophore. Thus if two probes have very different lifetimes but the same rates of rotational motion, then the value of R will be higher for the probe with the longer lifetime. R values should therefore be compared only for probes which have the same lifetime. The R values are probably most useful in determining the relative "fluidity" of membranes with a given probe, but caution must again be exercised if the lifetime changes with type of membrane used.

A different method for computing rotational relaxation in an oriented bilayer has been presented by Bradley et al. [27]. Their approach is to calculate values of the principle polarized fluorescence intensities for different degrees of relaxation, where the degree of relaxation is defined as the average angle Ω through which an emission moment moves during the fluorescence lifetime. By comparing the experimental and calculated intensities, a value can be assigned to Ω. Here again, however, the value of Ω depends not only on the rates of rotational motion but also on lifetime, and caution must be exercised in comparing values of Ω for fluorophores with different lifetimes.

5. The evaluation of the width parameter by means of Eqs. (13)-(16) as outlined above is complicated by the fact that $\xi \neq 0$ and significant rotation occurs during the fluorescence lifetime for most fluorophores inserted into a lipid bilayer. The effects of ξ can be taken into consideration in a fairly rigorous fashion and, as mentioned above, we have derived generalized expressions for the principle polarized intensities which explicitly account for the finite value of ξ [31]. The effects of rotational motion, however, are more difficult to treat and no general formulation is available at the present time which accounts for both ξ and rotational motion, although we are working on such a formulation in our laboratory. One must therefore use an approximate approach in the analysis. Our experience indicates that for preferred orientation along the c-axis, the ratio I_{cc}/I_{aa} is more sensitive to the width of the distribution than to rotational motion and that a width parameter can be obtained by comparing this ratio to ratios calculated with expressions derived for $\xi \neq 0$ but zero rotational motion. For preferential orientation along the ab plane a similar analysis can be made using the ratio I_{ab}/I_{ca}. The values of θ_G resulting from such an analysis are shown in Table 3. In order to give some physical meaning to θ_G we have also calculated the angle θ_H, which is the half-angle of the cone encompassing 50% of the emission moments in the distribution.

B. Determination of the Orientation and Mobility of Membrane Components by Polarized Fluorescence Measurements of Membrane Dispersions

In the preceding section we discussed the kinds of information which can be obtained by steady-state polarized fluorescence measurements on oriented membranes. Although very informative, these systems are difficult to handle and require special instrumentation for their study. Most experimeters therefore prefer to work with membrane suspensions where membranes with a dimension of around one micron or less are simply dispersed in solution. Measurements can then be made with a standard spectrofluorimeter. In this instrument a broad beam of exciting light simultaneously illuminates many membranes instead of a small patch of a single membrane. Membranes with all orientations are thus excited at the same time. In fact, the principle axes of the membrane can be considered to have a random distribution of orientations with respect to the direction of polarization of the incident light.

The usual procedure for doing polarized fluorescence measurements with a membrane dispersion is to excite the sample with a steady beam of light polarized in a vertical direction with respect to the laboratory axes. The emitted light is then detected through an analyzing polarizer oriented first in a vertical direction and then in a horizontal direction. Two polarized intensities, I_{VV} and I_{VH}, are thus obtained. These are then used to calculate the fluorescence anisotropy A defined by Eq. (23).

Polarized fluorescence measurements can also be made with a pulsed beam of light. In this case the sample is excited with a very fast pulse of light polarized in a vertical direction and the time-dependent intensities $I_{VV}(t)$ and $I_{VH}(t)$ are recorded. The time-dependent anisotropy is then defined by the equation

$$A(t) = \frac{I_{VV}(t) - I_{VH}(t)}{I_{VV}(t) + 2\, I_{VH}(t)} \tag{26}$$

In the following section we will discuss the kinds of information which can be obtained about the conformation and dynamics of membranes from measurements of steady-state and time dependent fluorescence anisotropy on membrane dispersions. The discussion will be divided into two parts. The first part will be concerned with probes such as AS, which in a bilayer have a specific orientation with respect to the bilayer axes. The other part will be concerned with probes such as perylene, which have an approximate random distribution of orientations with respect to the bilayer axes.

1. Oriented Probes.

a) Time-Dependent Anisotropy Measurements. Time-dependent (nanosecond) anisotropy measurements are more informative than steady-state measurements [32]. However, until recently very few studies using nanosecond anisotropy had been reported because of difficulties encountered in removing artifacts (deconvolution) resulting from the finite duration of the exciting light pulse and the finite time resolution of present nanosecond spectrofluorimeters. During the past two years, methods for deconvoluting nanosecond anisotropy data have been developed and, although presently restricted to a few laboratories, their application will probably increase in the near future.

The time profile for A(t) typically displayed by a fluorescent labeled membrane dispersion is shown in Fig. 8. The striking features of the graph are that the fluorescence anisotropy initially decays very fast but levels off with time and reaches a steady value at longer times. This behavior is to be contrasted with a freely rotating probe in an iso-

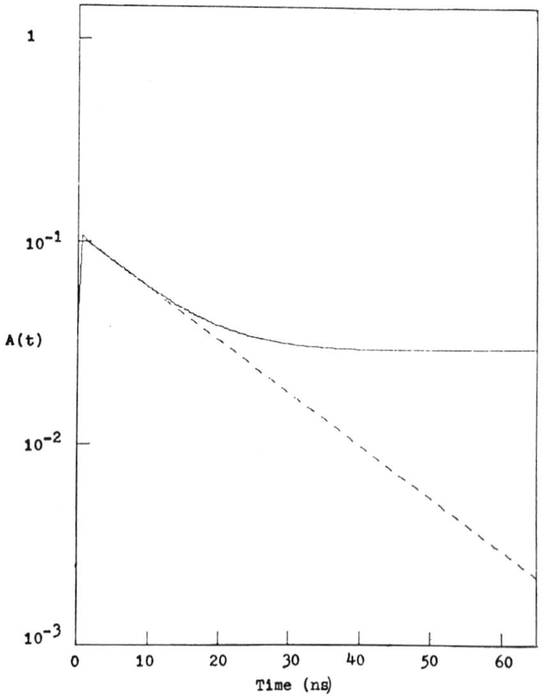

Fig. 8. Time-dependent anisotropy A(t) (*solid line*) for 12-anthroyl stearic acid in a dispersion of dipalmitoyl lecithin vesicles at 31°C [34]

tropic solution where A(t) decays continuously toward zero. The time profile for A(t) displayed by a membrane dispersion can be understood as follows: in a membrane dispersion the fluorophore molecules have a random distribution of orientations with respect to the direction of polarization of the exciting light. In this sense the fluorophore molecules have the same distribution as in an isotropic solution. In an isotropic solution, however, an excited fluorophore can move freely over all angles in space. When such a fluorophore molecule is excited at time zero, its emission moment will have a specific orientation with respect to the orientation of the analyzing polarizer, but as a result of random (free) rotational motion the emission moment will eventually become randomly oriented and its fluorescence anisotropy will decay to zero. For a freely rotating spherical molecule in an isotropic system, the decay of anisotropy follows the expression [32, 33]

$$A(t) = A_0 \, e^{-t/\phi} \tag{27}$$

where ϕ, the rotational correlation time, is related to the microviscosity η of the solution and volume of the molecule V by the expression

$$\phi = \frac{\eta V}{kT} \tag{28}$$

For a freely rotating, nonspherical molecule, the decay of anisotropy is given by

$$A(t) = A_0 \sum_{i=1}^{5} f_i \, e^{-t/\phi_i} \tag{29}$$

where ϕ_i are rotational correlation times that depend on the rotational diffusion coefficients D_1, D_2, and D_3 for rotational diffusion about each of the three principle axes of the molecule.

In a membrane, however, the fluorophore molecules are not free to move over all angles because of restrictions imposed by the bilayer structure. To a first approximation we can describe the rotational motion in terms of a model in which the emission transition moment of the fluorophore can move only within a cone having an average half-angle θ_H. The axis of the cone is along the direction in which the absorption transition moment is preferentially oriented in the membrane (see Sect. III.A). Because of the restricted motion, the emission moment of the probe excited at time zero cannot become randomly oriented. The anisotropy therefore does not decay to zero at long times but reaches a steady level determined by the value of θ_H.

It should be noted that A(t) can also decay as a result of rotational motions of the membranes themselves. These motions however are very slow for membranes having dimensions around 1μ and in the nanosecond time range the membranes can be considered stationary.

An expression for A(t) has not yet been rigorously derived for the model presented above, but a semi-empirical expression for anisotropy similar to that used to describe a system of macromolecules displaying segmental flexibility can be used [32]. Thus we may write for the decay of fluorescence anisotropy from a labeled membrane dispersion [34, 35, 36]

$$A(t) = A_0 \left(\sum_{i=1}^{n} f_i \, e^{-t/\phi_i} + c\right) \tag{30}$$

where A_0, the anisotropy at $t = 0$, is related to the angle ξ between the absorption and emission moments at $t = 0$ by the expression

$$A_0 = \frac{0.4\,(3\cos^2 \xi - 1)}{2} \;. \tag{31}$$

The ϕ_i are correlation times for rotations which describe the motion within the cone of allowed motion. A value of $n = 2$ usually suffices to describe the decay of anisotropy empirically. The f_i and c must satisfy the relation

$$c + \sum_{i=1}^{n} f_i = 1 \tag{32}$$

The limting anisotropy $A(\infty)$, that is the value of $A(t)$ at long times where $A(t)$ has reached a steady value, is according to Eq. (30) given by the expression

$$A(\infty) = A_0 \, c \tag{33}$$

Now according to elementary theory we can write

$$A(\infty) = \frac{0.4\,(3\cos^2 \xi - 1)}{2} \; \frac{(3\cos^2 \theta_H - 1)}{2} \tag{34}$$

or

$$c = \frac{3\cos^2 \theta_H - 1}{2} \;. \tag{35}$$

Furthermore, analysis of $A(t)$ data with Eq. (30) gives values for A_0, c, and f_i. The value of c can be used to calculate θ_H with Eq. (35), which gives a measure of how much order exists in a bilayer. The smaller the value of θ_H, the higher is the order for a given probe. The ϕ_i on the other hand give an indication of the fluidity of the bilayer. The smaller the values of ϕ_i, the higher is the fluidity for a given probe. The parameters θ_H and ϕ_i can then be used to determine the effects on the conformation and dynamics of a bilayer induced by changes in pH, lipid composition or temperature, or upon addition of, e.g., hormones or anesthetics. It should be noted that the value of θ_H ranges from 0 to 54.7°. The value of $\theta_H = 0$ corresponds to the case where the probe molecules are perfectly oriented in the bilayer ($c = 1$), whereas $\theta_H = 54.7°$ corresponds to the case where the probes have a random distribution of orientations in the bilayer ($c = 0$).

Very few experiments have so far been reported in which time-dependent anisotropy has been used to study membranes, but applications will probably increase as the technique becomes better established. Table 4 gives values obtained for AS in synthetic vesicles.

Table 4. Time-dependent anisotropy parameters for 12-anthroyl stearic acid incorporated in vesicles of dipalmitoyl lecithin

Temperature	20°C	31°C	40°C	51°C
A(0) [a]	0.11	0.11	0.11	0.11
A(∞) [b]	0.0355	0.0291	0.0020	0.0014
ϕ_1 [c] (ns)	24	17	7	5
ξ [d]	44°	44°	44°	44°
θ_H [e]	21°	28.4°	52.9°	53.4°

Source: D.H. Sherwood and J. Yguerabide, unpublished results; D.H. Sherwood, Ph.D. Thesis, University of California at San Diego, 1974 [34]

[a] Values of anisotropy at time t = 0.

[b] Steady values of anisotropy at very long times.

[c] Rotational relaxation time obtained by fitting Eq. (30) to experimental graphs of A(t) versus t with n = 1.

[d] Calculated from value of A(0) with Eq. (31).

[e] Calculated from value of A(∞) and A(0) with Eqs. (33) and (35).

b) Steady-State Anisotropy Measurements. These are relatively simple to perform and have been used to detect, for example, phase transitions in membranes. A detailed interpretation of steady-state anisotropy measurements in terms of membrane structure and conformation, however, has so far been limited by the lack of an expression which relates steady-state anisotropy to membrane parameters. A semi-empirical expression, however, can be derived from Eq. (30) as follows. For a given system, the steady-state anisotropy A is related to the time-dependent anisotropy A(t) by the expression

$$A = \frac{1}{\tau} \int_0^{\infty} A(t)\, e^{-t/\tau} dt \tag{36}$$

where τ is the fluorescence lifetime of the fluorophore. This expression assumes that the fluorophore has a single lifetime in the membrane. Introducing Eq. (30) into Eq. (36) gives

$$A = A_0 \left[c + \Sigma \left(\frac{f_i}{1 + \frac{\tau}{\phi_i}} \right) \right] \tag{37}$$

This equation relates the steady-state anisotropy to the dynamic parameters θ_i and the structural parameter c.

Equation (37) has several interesting properties in the limits where $\phi_i \ll \tau$ or θ_H approaches 57.4°. When $\phi_i \ll \tau$, Eq. (37) reduces to

$$A = A_0\, c = A(\infty) \tag{38}$$

This is the case where the probe molecules rotate very fast and completely relax to the limiting value of anisotropy $A(\infty)$ in times short compared to τ. The steady-state anisotropy is thus completely determined by $A(\infty)$. If the value A_0 is known, for example from a measurement of anisotropy of the probe in a viscous solvent at low temperature, then we can calculate c with Eq. (38) and θ_H with Eq. (35). It is now of interest to note that the *definition of c given by Eq. (35) is identical in form to the definition of the so-called structural parameter S used to describe the orientation of paramagnetic resonance probes in membranes* [37, 38]. *Thus, c can be considered as a structural parameter and we can state that the steady anisotropy is a direct measure of the structural parameter c when $\phi_i \ll \tau$.* The evaluation of S in paramagnetic resonance (PMR) spectroscopy also assumes that the PMR probe rotates very fast compared to its "lifetime". Unfortunately, the condition $\phi_i \ll \tau$ does not usually apply.

In the limit where θ_H approaches 57.4°, c becomes zero and Eq. (37) reduces to the form

$$A = A_0 \sum_{i=1}^{n} \frac{f_i}{1 + \frac{\tau}{\phi_i}} \tag{39}$$

The limit θ_H = 57.4° describes the case where the fluorophore has a random orientation and moves as if it were in an isotropic environment. Equation (39) indeed has the well-known form for freely rotating probes in an isotropic medium. This limiting case has been assumed to apply to such probes as perylene and DPH, and a method for evaluating microviscosity with these probes has been developed by Weber and his colleagues. The method is described below.

In most instances the limiting cases described above do not apply, so that the steady-state fluorescence anisotropy depends on structural as well as dynamic parameters. Unfortunately, no method for separating these two sets of parameters from steady-state measurements has so far been developed. Caution, however, must be exercised in the interpretation of steady-state fluorescence anisotropy solely in terms of dynamic parameters, as is presently being done. Differences in anisotropy between different membranes or in the same membrane under different conditions may be due to differences in both structural and dynamic parameters.

2. Microviscosity.

As mentioned above, Weber and colleagues have devised a technique in which steady anisotropy measurements are used to determine the microviscosity of membranes with nonpolar probes, such as perylene and DPH. The basis of the technique is as follows. It is assumed that the probe has a random distribution of orientations with respect to the bilayer axes, in which case Eq. (39) applies. Now, according to theory, the ϕ_i for a freely rotating molecule are functions of the quantify $\eta V/T$. The denominator Eq. (39) will thus consist of terms which are dependent on $\eta V/kT\tau$. Since V, the volume of the probe, is independent of environment, assuming solvation does not vary with solvent or temperature, then A is essentially a function of $T\tau/\eta$. For a spherical particle, Eq. (39) reduces to

$$\frac{A_0}{A} = 1 + \frac{T\tau}{\eta V} \tag{40}$$

and a plot of $\frac{1}{A}$ versus $\frac{T\tau}{\eta}$ should be a straight line. For the asymmetric probes used in membrane studies the more general Eq. (39) applies and a plot of $\frac{1}{A}$ versus $\frac{T\tau}{\eta}$ will be curved.

To evaluate the microviscosity of membranes with nonpolar probes, Weber and colleagues [17] have taken advantage of the fact that the value of A is essentially determined by the ratio $T\tau/\eta$ and have used a viscous oil at different temperatures to calibrate the relation between A and $T\tau/\eta$ for specific probes. Essentially, a calibration graph of A versus $T\tau/\eta$ is thus obtained. Such a graph of perylene is shown in Fig. 9. To determine

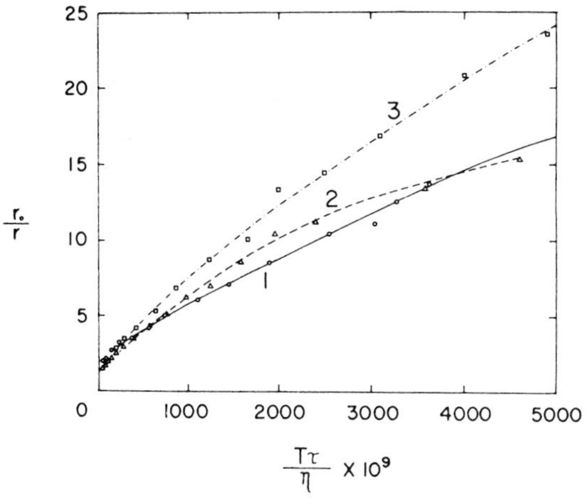

Fig. 9. Calibration graphs used to determine microviscosity η of membranes from anisotropy measurements. Graphs shown are for perylene (curve 1), 2-methylanthracene (curve 2), and 9-vinylanthracene (curve 3). Anisotropy (r) was determined in white oil USP 35 and the anisotropy r_0 was measured in propylene glycol at $-50°C$. r and r_0 correspond to A and A_0 in the text [71]

the microviscosity of a membrane, the probe is inserted into the membrane A and τ are then measured at a specific temperature T. Next, the value of $R = T\tau/\eta$ corresponding to the measured value of A is determined from the calibration graph and the value of η is finally calculated with the relation

$$\eta = \frac{T\tau}{R} \tag{41}$$

using the value of T and τ for the membrane. Values of η obtained in this manner for a variety of membranes are shown in Table 5. Table 6 lists fluorescence anisotropy measurements which have been used to study the role of microviscosity in the expression of physiological functions.

Caution must be exercised in the absolute interpretation of microviscosities determined by the method here described, because of assumptions involved in the technique.

Table 5. Microviscosities η of some bilayers as determined by polarized fluorescence measurements using the probe perylene

Bilayer	η in poise (25°C)
Egg lecithin	1.3
Egg lecithin-cholesterol (1.5:1)	12
Dipalmitoyl lecithin	9.4
Dipalmitoyl lecithin-cholesterol (6:1)	36
Dipalmitoyl lecithin-cholesterol (3:1)	53

Source: U. Cogan, M. Shinitzky, G. Weber, T. Nishida: Biochemistry *12*, 521 (1973) [71]

First, perylene and, especially, DPH are not actually randomly oriented in a membrane as assumed in the method [36]. Therefore, the value of anisotropy is determined by structural as well as dynamic parameters. Moreover, viscosity is a macroscopic property of a liquid and is difficult to define for a molecule as small as a fluorescent probe. In fact, molecular theories of rotational motion in solution indicate that the rotational motions of a molecule in a given solvent may depend on the detailed molecular structure of the solvent molecules and not simply on the macroscopic viscosity. Thus, the calibration graph of A versus $(T\tau/\eta)$ for a given probe may depend on the oil used for calibration. With the graph for DPH this has indeed been shown to be the case [39].

C. Lateral Mobility of Membrane Components

The ability of membrane components to move laterally in the plane of the membrane is a subject of great interest in membrane research [40, 41]. Several fluorescence techniques have recently been developed to measure this lateral motion. Here we discuss three of them.

1. Measurement of Lateral Mobility from Recovery of Fluorescence Intensity Following Bleaching of a Small Membrane Area.

In this technique the membrane component of interest is labeled with a fluorophore. A small area of the membrane is then illuminated by a weak beam of exciting light and fluorescence is recorded as a function of time. At a predetermined time the intensity of the exciting light beam is momentarily increased several orders of magnitude so as to irreversibly bleach a significant fraction of the fluorescent molecules in the illuminated area, i.e., to photochemically change them into a permanently nonfluorescent form. Following the bleaching pluse, the fluorescence intensity is low, but it increases with time as fluorescent labeled molecules move from unbleached areas into the bleached area. Lateral mobility can thus be evaluated from the rate of recovery of fluorescence intensity [42]. Figure 10 shows a typical fluorescence intensity versus time graph obtained in a fluorescence photobleaching experiment.

A lateral diffusion coefficient can be evaluated from the graph of Fig. 10 as follows. When the illuminated area is much smaller than the planar of the membrane, then theory

Table 6. Some studies of microviscosities of natural membranes

Probe	*Study*	*Result*	*Ref.*
Perylene	Microviscosity of red cell membrane	Microviscosity η is around 1.32 P for a red cell ghost membrane at 37°C. η is not much affected by ionic strength ortrypsin digestion but significantly reduced by phospholipid digestion	1
DPH	Effect of cholesterol on microviscosity of normal lymphocytes and malignant lymphoma cells	Microviscosity increases with cholesterol content. Lymphoma cells have lower cholesterol content and microviscosity than normal lamphocytes. It is suggested that malignancy may be controlled by changes in microviscosity through changes in amount of cholesterol	2
NPN	Effect of colicin E1 on *E. coli*	Colicin El increases microviscosity of the *E. coli* membrane	3
Perylene	Microviscosity and effect of local anesthetics on microviscosity of a variety of natural membranes	Fluidity of some natural membranes at 25°C: polymorphonuclear leucocytes, 3.35 P; bovine brain myelin, 2.7 P; human erythrocyte, 1.9 P; rat liver microsomes, 0.95 P; rat liver mitochondria, 0.10 P. In most cases a correlation between microviscosity and cholesterol/lipid ratio was found	4
DPH	Relation between microviscosity and protein mobility in normal and malignant fibroblasts and normal and malignant lymphocytes	Mobility of concanavalin A receptors increases with decrease in fluidity. It was postulated that decrease in fluidity causes the proteins to be less embedded in the membrane	5
DPH	Role of fluidity in the enhancement by tertiary amine local anesthetics of the agglutination of untransformed mouse and hamster cells by concanavalin A	Anesthetics increase membrane fluidity but not sufficiently to account for enhancement of agglutination	6
Perylene	Abnormalities in membrane microviscosity and ion transport in genetic muscular dystrophy	Membranes of muscle, liver, and erythrocytes of dystrophic chicks have a significantly higher microviscosity than normal cells. The higher microviscosity is correlated with higher cholesterol content	7
DPH	Microviscosity of membranes of normal, transformed, and revertant 3T3 cells	Transformation increases microviscosity by about 50%. The enhancement upon transformation of the agglutinability by lectins cannot be due to increase in fluidity	8
DPH	Microviscosity or retinal rod outer segment disk membrane	Microviscosity changes from 15 P to 1.4 P in the temperature range 0-40°C but no phase transitions were detected	9

Table 6 continued

Probe	*Study*	*Result*	*Ref.*
DPH	Changes in microviscosity of the polymorphonuclear leukocyte plasma membrane during phagocytosis	Phagocytosis induces a decrease in microviscosity which parallels the extent of phagocytosis	10
Perylene	Role of fluidity in the differences in mobility of antigens and receptors in neonatal and adult human erythrocytes	Fluidity is the same in neonatal and adult erythrocytes. Difference in mobility of antigens and receptors is not due to differences in fluidity	11
Perylene	Role of microviscosity in the increase in water permeability induced by vasopressin in toad urinary bladder mucosal cells	Changes in permeability induced by vasopressin are accompanied by increase in membrane fluidity	12

1 R. Rudy and C. Gitler: Biochim. Biophys. Acta *288*, 231 (1972).
2 M. Shinitzky and M. Inbar: J. Mol. Biol. *85*, 603 (1974).
3 S.H. Helgerson, W.A. Cramer, J.M. Harris, F.E. Lytle: Biochemistry *13*, 3057 (1974).
4 M.B. Feinstein, S.M. Fernandez, R.I. Sha'afi: Biochim. Biophys. Acta *413*, 354 (1970).
5 M. Shnitzky, M. Inbar: Biochim. Biophys. Acta *433*, 133 (1976).
6 G. Poste, D. Papahadjopoulos, K. Jacobson, W.J. Vail: Biochim. Biophys. Acta *394*, 520 (1975).
7 R.I. Sha'afi, S.B. Rodan, R.L. Hintz, S.M. Fernandez, G.A. Rodan: Nature *254*, 525 (1975).
8 P. Fuchs, A. Parola, P.W. Robbins, E.R. Blout: Proc. Natl. Acad. Sci. USA *72*, 3351 (1975).
9 F.W. Stubbs, B.J. Litman, Y. Barenholz: Biochemistry *15*, 2766 (1976).
10 R.D. Berlin, J.P. Fera: Proc. Natl. Acad. Sci. USA *74*, 1072 (1977).
11 M. Kehry, J. Yguerabide, S.J. Singer: Science *195*, 486 (1977).
12 B.R. Masters, J. Yguerabide, D.D. Fanestil: J. Membr. Biol. *40*, 179 (1978)

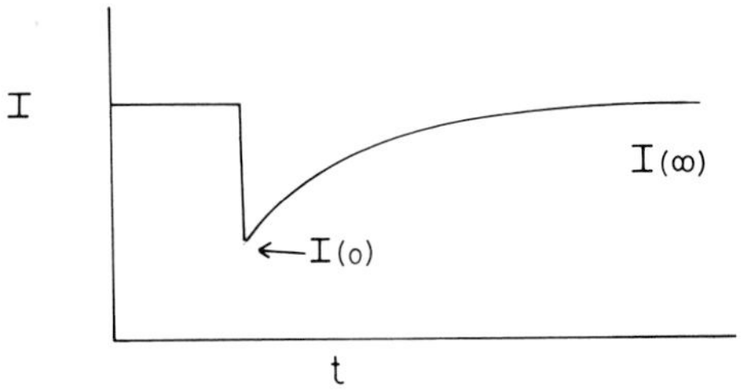

Fig. 10. Intensity, I, versus time, t, in a fluorescence photobleaching recovery experiment

based on the solution of the lateral diffusion equation shows that the fluorescence intensity I(t) following the bleaching pulse is given by the expression [43]

$$\bar{I} = \frac{I(\infty)-I(t)}{I(\infty)-I(O)} = \int_{o}^{1/2\tau} \frac{I_1(Z)\,dZ}{Z\,e^Z} \tag{42}$$

where $I(\infty)$ is the fluorescence intensity at very long times after bleaching, I(O) is the intensity right after the bleaching pulse, $\tau = Dt/r_o^2$, $I_1(z)$ is the modified Bessel function of the first kind of order zero, D is the lateral diffusion coefficient in cm^2/s, and r_0 is the radius of the bleached area on cm. Time zero in this equation is selected as immediately after the bleaching pulse. $\bar{I}$, defined by Eq. (42), will here be called the reduced intensity. Figure 11 shows a plot of $\bar{I}$ versus τ. If the experimental data I versus t is replotted

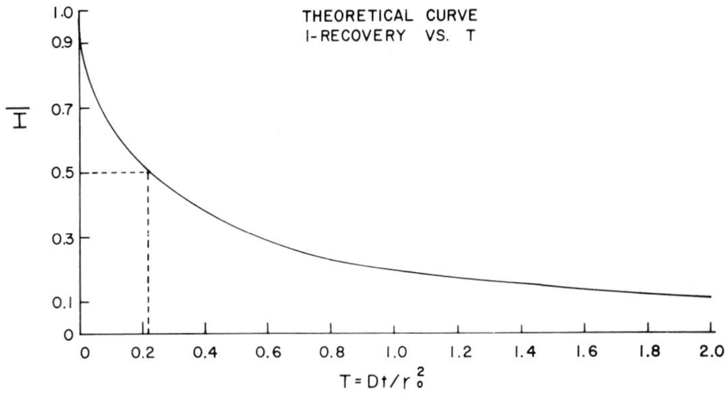

Fig. 11. $\bar{I}$ versus τ calculated with Eq. (42) [43]

as $\bar{I}$ versus τ, then according to Eq. (42), the resulting plot is independent of the values of D and r_0. That is, the graph $\bar{I}$ versus τ applies to all membranes and fluorophores, since the plot is the same for all values of D and r_0. We will refer to a plot of $\bar{I}$ versus τ as a normalized plot of reduced intensity.

The method for applying the theoretical graph of Fig. 11 to the analyses of experimental data is as follows. Inspection of this theoretical graph shows that the normalized intensity $\bar{I}$ decreases to the value 0.5 when $\tau = 0.22$ [44]. Thus we may write

$$t_{1/2} = \frac{0.22\, r_o^2}{D} \tag{43}$$

where $t_{1/2}$ is the time required for $\bar{I}$ versus t to decrease to 0.5. Now if the experimental graph of I versus t is plotted as $\bar{I}$ versus t, then the value of $t_{1/2}$ for the system under consideration can be read from the graph. This value can then be used to evaluate D with Eq. (43) and the value of r_0 used in the experiment. With these values of r_0 and D, the plot $\bar{I}$ versus t can be replotted as $\bar{I}$ versus τ. This reduced plot can then be compared with the theoretical plot of Fig. 11 to determine whether the experimental data do indeed conform to the graph expected for recovery of fluorescence intensity by a lateral diffusion mechanism. The procedure outlined in this paragraph thus gives a value of D and also allows one to check whether the recovery of fluorescence intensity is consistent with predictions of a diffusion mechanism.

The range of values of $t_{1/2}$ expected in a photobleaching experiment can be estimated with Eq. (43). For a lipid covalently labeled with a fluorophore, measurements indicate that the value of D is around 10^{-8} cm^2/s. The highest value of r_0 is limited by the size of the membrane under study, since r_0 has to be smaller than the planar dimension of the membrane. Thus for a cell of about 10 μ in diameter, r_0 should be approximately 1 μ. With $r_0 = 1$ μ and $D = 10^{-8}$ cm^2/s, Eq. (43) gives $t_{1/2} = 0.22$ s. For a protein component, $D \sim 10^{-10}$ cm^2/s, in which case $t_{1/2} \sim 22$ s. For a synthetic multibilayer membrane which has dimensions of the order of 1 cm, one is free to adjust r_0 to a value which is convenient for measurements with a conventional microscope and lamp using a manually operated shutter. Thus if r_0 is 40$_\mu$, $t_{1/2}$ for a lipid component will be in the range of 6 minutes.

Figure 12 is a plot of I versus t for AS in a hydrated multibilayer of soybean lecithin [44]. The multibilayer was formed by evaporating the solvent from a thin film of a solution of soybean lecthin and AS in chloroform that had been deposited on a microscope

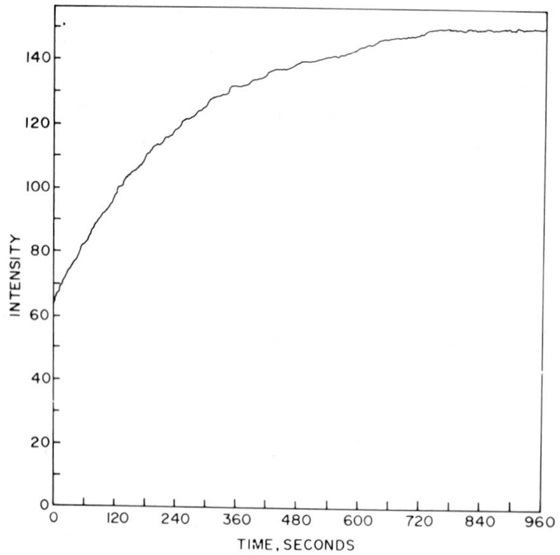

Fig. 12. Recovery of fluorescence intensity I versus t following a bleaching pulse for AS in a hydrated multibilayer of soybean lecithin at 25°C. r_0 is 37 μ, $t_{1/2} = 170$ s. (R. Walter and J. Yguerabide, unpubl. results)

slide. The lipid to AS ratio was 1000. The bleaching experiment was done on a conventional fluorescence microscope using a 200-watt mercury lamp. The illuminated area had a radius of 35 μ. A value of $t_{1/2} = 170$ s was determined by plotting the experimental data as $\bar{I}$ versus t which corresponds, according to Eq. (43), to $D = 1.6 \times 10^{-8}$ cm^2/s. These values were then used to make a plot if $\bar{I}$ versus τ, as shown in Fig. 13. The points

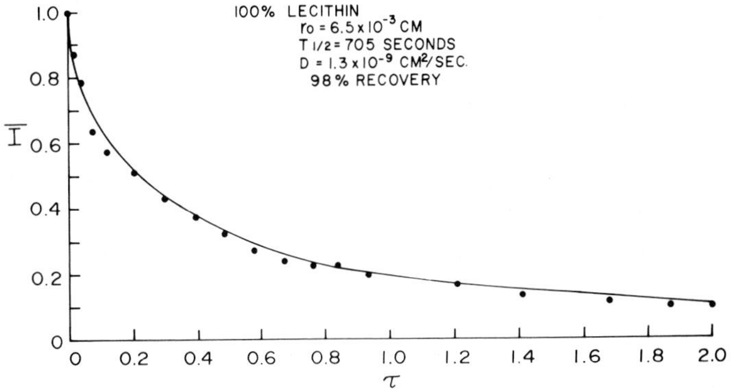

Fig. 13. $\bar{I}$ versus τ for AS in a hydrated multibilayer of soybean lecithin at 25°C. *Solid line* is the theoretical graph of Fig. 11. Points are experimental intensities. r_0 is 65 μ and $t_{1/2} = 170$ s (R. Walter and J. Yguerabide, unpubl. results)

correspond to the experimental data, while the solid line is the theoretical graph of Fig. 11. The agreement of the two graphs indicates that the recovery of fluorescence intensity following a bleaching pulse is well described by a diffusion mechanism. An additional test was made by measuring $t_{1/2}$ as a function of r_0. The results showed that $t_{1/2}$ is proportional to r_0^2, as expected from Eq. (43).

It should be noted that no recovery of fluorescence intensity over a period of one day was detected for unhydrated multibilayers of soybean lecithin, indicating that the apparent value of D is very large. However, we have not established whether the AS molecules in the unhydrated bilayer are individually dispersed or are in the form of small clusters of pure AS dispersed in the bilayer. The formation of AS clusters is suggested by the following observations. A variety of experiments indicate that the reaction responsible for the photobleaching of AS is (see Sect. III.C.3)

$$AS^* + AS \rightleftharpoons AS_2 \qquad \mathit{1}$$

in which an excited molecule of AS, i.e., AS*, forms a stable dimer when it encounters a ground state AS molecule. Since the rate of this reaction depends on the rate of encounter of AS* and AS molecules in the bilayer, it follows that the rate of photobleaching will depend on the lateral diffusion coefficient D, and if D was very small it would be difficult to photobleach. Experiments indicate that the unhydrated bilayer can be bleached quite easily, contrary to expectations if D were small. This result, however, is consistent with cluster formation, since the excimer reaction *1* could easily occur in such a cluster, but the overall value of D in the bilayer overall would be very small.

When using a new fluorophore to measure lateral mobility by the bleaching technique here described, it is necessary to show that the recovery of fluorescence is indeed due to lateral diffusion and not to spontaneous recovery of the bleached molecules. This possibility can be tested in several ways.

One is to measure $t_{1/2}$ for different values of r_0. If the recovery is due to a spontaneous mechanism, $t_{1/2}$ will be independent of r_0, whereas it will vary as r_0^2 for the diffusion mechanism. Another method is to bleach the whole membrane. Fluorescence intensity recovers in the case of a spontaneous recovery process but not in the case of a lateral diffusion mechanism. Table 7 lists some studies of lateral diffusion in membranes using the photobleaching technique.

Table 7. Mobility studies with fluorescence photobleaching/recovery experiments

System	*Results*	*Ref.*
1. Acetylcholine receptors in rat embryonic muscle culture, labeled with rhodamine bungarotoxin	Two popolations of acetylcholine receptors were found: (1) mobile, uniformly distributed, (2) nonmobile, in dense patches	1
2. Fluorescein con-A receptors in mouse fibroblasts	Binding of con-A causes restriction of mobility of surface receptors. The degree of restriction is a function of time after binding of the lectin	2
3. Membrane proteins in red blood cell ghosts labeled with fluorescein isothiocyanate	Proteins were immobile	3
4. Membrane proteins in rat myoblasts	A fraction of the labeled proteins were immobile and a fraction were mobile. The mobile fraction was slowed by cytochalasin B and by con-A but was not affected by colchicine or by azide	4
5. Lipid probe in rat myoblasts	The lipid probe diffuses about 50 times more rapidly than the mobile con-A receptors. The mobility of the lipid probe is *not* affected by colchicine, con-A, cytochalasin, or azide	4

1 D. Axelrod, R. Ravdin, D.E. Koppel, J. Schlessinger, W.W. Webb, E.L. Elson, T.R. Podleski: Proc. Natl. Acad. Sci. USA *73*, 4594-4598 (1976)
2 Y. Zagyanski, M. Edidin: Biochim, Biophys. Acta *433*, 209-214 (1976)
3 R. Peters, J. Peters, K.H. Tews, W. Bähr: Biochim. Biophys. Acta *367*, 282-294 (1974)
4 J. Schlessinger, D. Axelrod, D.E. Koppel, W.W. Webb, E.L. Elson: Science *195*, 307-309 (1977)

2. Measurement of Lateral Mobility by Fluorescence Quenching.

Fluorescence quenchers are molecules which quench the emission of a fluorescent molecule during an encounter in solution. That is, they decrease fluorescence efficiency. The quenching process can be represented by the bimolecular reaction

$$A^* + Q \rightarrow A + Q \qquad 2$$

where A* is the excited fluorophore and Q is quencher. Quenching may occur by any one of several mechanisms. Thus, quenchers with heavy atoms, such as bromobenzene, magnetically perturb excited molecules and quench by enhancing singlet-triplet crossings. Molecules whose first electronic excited state has a lower energy than that of the fluorophore can quench by an excitation transfer mechanism (see Sect. III.E). One of the effects of a quencher is to reduce the fluorescence lifetime of the fluorophore. In a regular solution of quencher and fluorophore molecules the effect on lifetime can be described by the equation

$$\frac{1}{\tau} = \frac{1}{\tau_0} + k_Q\, c_Q \tag{44}$$

where τ and τ_0, in seconds, are the lifetimes of the fluorophore in the presence and absence of quenchers, k_Q, in l/mole/s, is the specific rate constant for the quenching process, and c_Q, in mol/l, is the concentration of quencher. Moreover, when the quenching reaction is diffusion-controlled, that is, when quenching occurs on almost every collision so that the rate of quenching is determined by the rate at which molecules encounter each other as they diffuse through the solution, then we can write for k_Q (45)

$$k_Q = \frac{4\pi N D_{AQ} R_{AQ}}{1000} \tag{45}$$

where D_{AQ} (cm^2/s) is the sum of the translational diffusion coefficients of A and Q, R_{AQ} (cm) is the sum of the radii of A and Q, and N is Avogadro's number. k_Q as before has units of l/mol/s. This expression is most important in the present discussion for it indicates that the diffusion coefficient D_{AQ} can be evaluated if the value of k_Q is known. k_Q can indeed be evaluated experimentally from lifetime measurements, as discussed further below.

In a membrane where concentrations are expressed as surface densities we can write

$$\frac{1}{\tau} = \frac{1}{\tau_0} + k_Q\, \sigma_Q \tag{46}$$

and

$$k_Q = 2\,\pi\, D_{AQ} \tag{47}$$

where k_Q has units of cm^2/molecule/s and σ_Q has units of molecules/cm^2, and as before, D is in cm^2/s and τ is in seconds. It should be noted that expressions other than Eq. (47) have been derived in the literature for the relation between k_Q and D_{AQ} by authors using different theoretical models for diffusion. (For a discussion of diffusion models see ref. 45). Thus Gala and Sackmann [23], using a hopping model, get the expression

$$k_Q = \frac{8 R_{AQ} D_{AQ}}{\lambda} \tag{48}$$

where λ is the length of a diffusional jump. They assume that $\lambda \sim 2R_{AQ}$ so that

$$k_Q = 4\, D_{AQ} \tag{49}$$

This expression is about 1.6 times smaller than Eq. (47). The density of quencher molecules in a membrane can be calculated from the molar ratio M of lipid to quencher with the expression

$$\sigma_Q = \frac{1}{\alpha M} \tag{50}$$

where α is the area per lipid molecule in cm^2, or with the expression

$$\sigma_Q = \frac{c_Q}{\alpha\, c_L} \tag{51}$$

where c_Q and c_L are the overall concentrations of quencher and lipid in the membrane dispersion.

To evaluate k_Q experimentally, the quencher is introduced into the membrane at a suitable density σ_Q, and the lifetime in the presence, τ, and absence, τ_0, of quencher is measured using a membrane suspension. k_Q is then calculated from Eq. (46) and its value is finally used to calculate D with Eq. (47) or Eq. (48). Alternatively, k_Q can be evaluated from steady-state fluorescence measurements. Thus if I_0 and I are the steady-state fluorescence intensities in the absence and presence of quencher, then we have

$$\frac{I_O}{I} = 1 + k_Q\, \tau_0\, \sigma_Q \tag{52}$$

The slope of a plot of $\frac{I_O}{I}$ versus σ_Q gives $k_Q\, \tau_0$, from which k_Q can be evaluated after measuring τ_0 with a nanosecond spectrofluorimeter.

The density of quencher σ_Q necessary to get a measurable change in lifetime or fluorescence intensity in a quenching experiment can be estimated as follows. The magnitude of the effect on the lifetime and fluorescence intensity depends on the value of $k_Q\sigma_Q$ compared to $1/\tau_0$. Thus, for example, in order to get 50% quenching we must have $k_Q\, \sigma_Q = 1/\tau_0$; for 10% quenching $k_Q\, \sigma_Q = 0.11/\tau_0$. Now molecules of the size commonly used as quenchers and fluorophores in quenching experiments have diffusion coefficients D_{AQ} with values around 10^{-7} cm^2/s, which gives, from Eq. (47), $k_Q \simeq 6.3 \times 10^{-7}$. Using this value for k_Q and a value of $\tau_0 = 10$ ns, we get $\sigma_Q = 1.73 \times 10^{13}$ molecules/cm^2 for 50% quenching. This density corresponds to a lipid to quencher molar ratio of 10 [using $\alpha = 58$ Å^2 in Eq. (50) for the area of a phospholipid molecule]. For $\tau_0 = 100$ ns, $\sigma_Q = 1.73 \times 10^{12}$, which corresponds to a lipid to quencher molar ratio of 100. This calculation emphasizes an important aspect of the quencher experiment, namely that the lifetime of the fluorophore τ_0 must be large in order to keep the quencher to lipid ratios low and thus minimize perturbations of the bilayer by the quencher.

Galla and Sackman [23] have used a fluorescence quenching technique to measure the lateral motion of the fluorophore pyrene in bilayers of dipalmitoyl lecithin and mixtures of dipalmitoyl lecithin and cholesterol. Pyrene has a lifetime of around 100 ns and is therefore most suitable for quenching studies. In their experiments, Galla and Sackman made use of the fact that a pyrene molecule A in the ground state can interact with a pyrene molecule A* in the excited state to form an excimer as shown by the reaction

$$A + A^* \rightarrow A_2^* \qquad 3$$

Since this reaction removes A*, it results in quenching of monomer fluorescence. Thus, in this case A acts as its own quencher. The process is thus called self-quenching. The equations presented above apply in this case if we let $\sigma_Q = \sigma_A$. In addition, the pyrene excimer A_2^* is fluorescent, and one can write for the ratio of the fluorescence intensity of the monomer I_M and excimer I_E

$$\frac{I_E}{I_M} = \frac{k_{eE}}{k_{eM}} \tau_{0E}\, k_Q\, \sigma_A \qquad (53)$$

where k_{eM} and k_{eE} are the specific rates for emission of monomer and excimer respectively, and τ_{0E} is the lifetime of the excimer. For pyrene $k_{eE}/k_{eM} \simeq 10$. Table 8 gives

Table 8. Lateral diffusion coefficients D calculated from self-quenching fluorescence data for pyrene in dipalmitoyl lecithin bilayers containing 1, 2, 5, 10, 30, and 50 mole % cholesterol[a]

	D x 10^7 (cm²/s)			
% Cholesterol	35°C	45°C	50°C	60°C
0		1.0	1.40	1.88
1		1.0	1.35	1.85
2		0.91	1.25	1.80
5		0.81	1.12	1.53
10	0.38	0.71	0.96	1.37
30	0.23	0.40	0.64	1.12
50	0.18	0.34	0.46	0.78

Source: H.J. Galla, E. Sackman: Biochim, Biophys. Acta *339*, 103 (1974) [23]
[a] Values of D were calculated from experimental values of k_Q and Eq. (49).

values of k_Q evaluated with Eq. (53) from measurement of I_E/I_M and τ_0 at different temperatures for a range of values of σ_A corresponding to lipid to pyrene molar ratios of 133, 68, 45, 34, and 27. Values of D were calculated from k_Q with Eq. (49) for mixed lipid bilayers consisting of different ratios of dipalmitoyl lecithin (PC) and cholesterol and are shown in Table 8.

It is of interest to note that for a pure PC bilayer containing pyrene at a lipid to probe ratio greater than 1000, k_Q was found to increase below 41°C where the lipid bilayer begins to enter the gel phase. This result is contrary to predictions of the diffusion model, since k_Q is expected to decrease in the more viscous gel phase. The result, however, can be explained by assuming that for high densities of pyrene in a gel phas, the pyrene moleluces are forced out of the bilayer to form pyrene clusters. The apparent value of k_Q is consequently high because the density of pyrene molecules is much higher in the clusters than if the molecules were individually dispersed throughout the area of the bilayer.

In summary, pyrene self-quenching can be used as a convenient method for evaluating the fluidity, in terms of the lateral diffusion coefficient D, of membrane suspensions.

3. Measurement of Lateral Mobility from Decrease of Fluorescence Intensity During Photobleaching of a Membrane Suspension.

This technique is essentially restricted to the measurement of lateral mobility with fluorophores in the presence of a quencher which in the process of quenching forms a stable complex AQ that does not dissociate spontaneously. The reaction can be represented by the equation

$$A^* + Q \rightarrow AQ \qquad 4$$

This reaction decreases the concentration of A in solution and results in a decrease in fluorescence intensity with time. If the reaction is diffusion-controlled the rate of decrease of fluorescence intensity with time will depend on D_{AQ} and therefore can be used to evaluate this parameter which, as discussed above, is a measure of membrane fluidity. It should be noted that the decrease in fluorescence intensity will occur in a finite time only if the exciting beam is intense. In the quenching studies described in Sect. III.C.2, the intensity of the exciting beam is kept low so as to minimize, during the time requirement for measurement, changes in fluorescence intensity due to photobleaching. In the photobleaching experiment being considered here, the beam intensity is made intentionally high.

The photobleaching experiment can be done on membrane suspensions as well as on single membranes or regular solutions of fluorophores. The experimental data consists of a plot of fluorescence intensity versus time.

McGrath et al. [22] have used the photobleaching technique described here to measure the lateral mobility of AS in bilayers. In their experiments they made use of the self-quenching properties of AS. That is, AS plays the role of Q and A* in reaction (*4*). The experimental data, which in this case form a plot of decreasing fluorescence intensity versus time, can be analyzed in terms of the following mechanism

$$A + h\nu_0 \rightarrow A^* \qquad 5$$

$$A^* \rightarrow A + \text{heat} \qquad 6$$

$$A^* \rightarrow A + h\nu_F \qquad 7$$

$$A^* + A \rightarrow A_2 \qquad 8$$

A refers to fluorophore molecules in the ground state and A* in the excited state. Steps 5-8 represent, respectively, light absorption, internal quenching. light emission, and dimer formation. For the low densities of σ_A used in the photobleaching experiment, the rate of step 8 is low compared to steps 6 and 7 and we may write

$$\frac{d[A^*]}{dt} = N_p \epsilon [A] x - (k_e + k_i)[A^*] \tag{54}$$

where N_p is the exciting light intensity in moles of photons per second, ϵ is extinction coefficient, x is optical pathlength, and k_i and k_e are the specific rate constants for steps 6 and 7 respectively. Note that [A] is not the concentration of A within the membrane but the overall concentration of A in the system. Thus in a membrane dispersion [A] is moles of A per liter of dispersion. [A] is related to the density of fluorophore molecules in the membrane σ_A (molecules/cm^2) as shown in Eq. (51) with $c_Q = [A]$.

In the steady state where $d[A^*]/dt = 0$, Eq. (54) gives

$$[A^*] = \frac{Np\epsilon [A] x}{k_e + k_i} \tag{55}$$

Fluorescence intensity is then given by

$$I = \delta\, k_e [A^*] \tag{56}$$

$$= \frac{\delta\, k_e N_p\, \epsilon [A] x}{k_e + k_i} \tag{57}$$

where δ is an instrumental factor. The rate of bleaching $d\sigma_A/dt$ is equal to the rate of step 8, so that we can write

$$\frac{d\sigma_A}{dt} = -k_Q\, \sigma_{A^*}\, \sigma_A \tag{58}$$

where σ_{A^*} and σ_A are the molecular densities (molecules/cm^2) of A* and A in the bilayer and k_Q is the specific rate constant for step 8. As before, k_Q is given by Eq. (47). Equation (58) can be written in terms of [A] and [A*] as follows

$$\frac{d[A]}{dt} = -b\, k_Q [A^*][A] \tag{59}$$

where $b = 1/\sigma c_L$ [see Eq. (51)]. Finally from Eqs. (56), (57), and (58) we can write

$$\frac{dI}{dt} = \frac{-b\, k_Q I^2}{\delta\, k_e} \tag{60}$$

$$= -K I^2 \tag{61}$$

K will be called the second order bleaching constant. Integrating Eq. (61) and making use of Eq. (47) finally gives

$$\frac{1}{I} - \frac{1}{I_0} = K\,t \tag{62}$$

$$= \frac{2\pi\, bD_{AQ}t}{\delta\, k_e} \tag{63}$$

where I_0 is the intensity at t = o.

Equation (63) indicates that the slope K of a plot of $\frac{1}{I} - \frac{1}{I_0}$ versus t is proportional to D. To obtain an absolute value of D from K one must know the values of the other parameters appearing in Eq. (63). An alternate approach to the direct evaluation of these parameters is to measure under comparable conditions the bleaching rate K for a reference system in which D is known. The diffusion coefficient for any other sample is then given by

$$D_{sample} = \frac{K_{sample}\, D_{ref}}{K_{ref}} \tag{64}$$

The experiments of McGrath et al. [22] were done with a membrane suspension compressed into a thin film between two microscope slides. The photobleaching was done under a microscope. The reference system was a thin film of AS in butanol. In this system, $D_{ref.}$ was estimated by calculation to be approximately 0.11 x 10^{-5} cm^2/s. Values of K (in arbitrary units) measured for AS in butanol, dilauroyl phosphatidylcholine (DLP), and dielaidoyl phosphatidylcholine (DEP) were respectively 2.9 x 10^{-7}, 6.8 x 10^{-9}, and 1.25 x 10^{-9}. The temperature was 23°C, in which case DLP and DEP are in the fluid phase. Diffusion coefficients calculated with Eq. (47) were 2.6 x 10^{-9} for DLP and 1.25 x 10^{-8} for DEP.

For bilayers in the gel phase AS was found to bleach faster than in the fluid phase, contrary to expectations since D should be lower in the gel phase. This result indicates that in the gel phase AS is forced out of the bilayer into clusters where self-quenching is high. This result is similar to observations made with pyrene as described in the previous section.

D. Electrostatic Surface Membrane Potential

Electrostatic potentials are usually present on the surface of membranes as a result of ionization of membrane proteins, carbohydrates, and head groups of lipids. These potentials are of great interest in the study of physiological functions because they may be involved in the regulation of processes by which charged molecules bind to and are transported through membranes. Here we propose a method for measuring surface membrane potentials using charged, amphipathic fluorescent probes. We develop the theory using the negatively charged fluorophore ANS as an example, but the resultant equations apply equally well to any other charged fluorophores.

Experiments indicate that when membranes are suspended in an aqueous solution of ANS, the singly charged ANS ions partition between the membrane and the electrolyte

[46-51]. The bound ANS ions have their hydrophobic part extended toward the inner hydrocarbon region of the membrane, while the charged part is in the polar region of the lipid heads. The binding of ANS to the membrane is accompanied by a large increase in fluorescence intensity, indicating that the fluorescence efficiency is high in the membrane and low in the electrolyte. Because of its negative charge, the binding constant for binding of ANS to the membrane is expected to depend on the membrane electrostatic surface potential. Indeed, experiments have shown that ANS binding depends on the ionic strength of the electrolyte, is increased by addition of positively charged lipids to the membrane, and is decreased by negatively charged lipids [49]. This sensitivity of binding to the electrostatic surface potential suggests that the potential for a given membrane can be determined from binding studies of ANS to the membrane.

The method which we propose is based on the following additional experimental observations for the binding of ANS ions to a membrane. A plot of fluorescence intensity versus total ANS concentration, $[ANS]_0$, added to a membrane suspension has the form shown in Fig. 14. This is a typical binding graph. The steady value of intensity attained

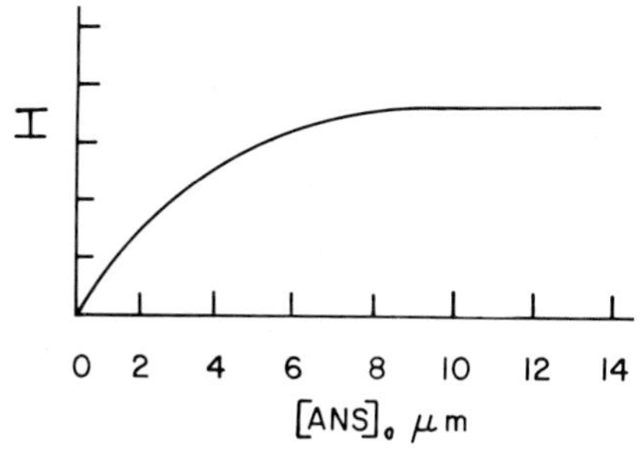

Fig. 14. Intensity I versus ANS concentration, $[ANS]_0$, for addition of ANS to a membrane dispersion

at high ANS concentrations may at first sight be interpreted as due to saturation of finite numbers of binding sites on the membrane. However, nanosecond measurements, as well as studies on the effect of ionic strength on binding (which show that the steady intensity at high ANS concentrations increases with ionic strength) indicate that the constant level is not die to saturation of binding sites but rather to the surface potential created by ANS ions as they bind to the membrane [50, 51]. They create a negative membrane surface potential which increases with density of ANS ions on the membrane surface. At high concentrations the surface potential reaches a value which prevents further binding of ANS ions and the I versus $[ANS]_0$ plot levels off to a steady value. Increasing the ionic strength reduces the electrostatic surface potential by a shielding effect and allows more ANS ions to bind to the membrane.

The binding of ANS to a membrane can be described by the equilibrium reaction

$$P + M \rightleftharpoons PM$$

where P refers to probe (e.g., ANS) and M refers to binding sites on the membrane. The binding constant for this reaction is defined by the expression

$$K = \frac{[PM]}{[P]\,[M]} \tag{65}$$

where concentrations are in moles per liter in the volume of the membrane suspension. The binding constant is given by

$$K = K_c\, e^{-h_B F\psi/RT} \tag{66}$$

where K_c is the part of the binding constant related to the chemical potential and ψ is the electrostatic surface potential produced by charges naturally present on the membrane and by probe ions bound to the membrane surface. ψ is positive when the net surface charge is positive and negative when the net charge is negative. The parameter h_B accounts for the sign of the charge on the probe ions. It has a value of +1 when the charge is positive and −1 when negative.

Equation (65) can be simplified by assuming, as discussed above, that the membrane is far from saturation, so that at the concentrations of ANS used in the binding experiments we can assume that $[M] \simeq [M]_0$ where $[M]_0$ is the total concentration of binding sites in the membrane suspension. Introducing this assumption into Eq. (65), we get

$$K = \frac{[PM]}{[P]\,[M]_0} \tag{67}$$

Now the concentration of binding sites is proportional to the amount of membrane in the suspension. Therefore, we can write

$$[M]_0 = \beta\,[L]_0 \tag{68}$$

where $[L]_0$ is the total concentration of membrane lipids and β is the number of moles of binding sites per mole of lipid. Introducing Eq. (68) into Eq. (67) we get

$$K = \frac{[PM]}{\beta[P]\,[L]_0} \tag{69}$$

Finally, from Eq. (69) we can write

$$k = \frac{[PM]}{[P]} \tag{70}$$

where

$$k = K\,\beta\,[L]_0 = \beta\,[L]_0 K_c\, e^{-h_B F\psi/RT} \tag{71}$$

It should be noted that Eq. (65) is basically the same as the high-concentration Langmuir isotherm, while Eq. (70) is similar to the low-concentration Langmuir isotherm [53].

The surface potential of the membrane is related to the surface charge density by the Gouy Chapman equation [52] which can be written as

$$\sinh\left[\frac{F\psi}{2RT}\right] = (\sigma_0' + \sigma_B')\sqrt{\frac{500\pi}{DRTC}} \tag{72}$$

where ψ is the potential in electrostatic units (note that 1 esu of potential = 300 V), σ_0' is the density of charge (in electrostatic units of charge/cm^2) normalls present on the membrane, σ_B' is the charge density of the probe ions bound to the membrane, D is the dielectric constant of water (80 at 20°C), T is temperature in K, C is concentration in moles/liter of electrolyte bathing the membrane (assuming a uni-univalent electrolyte), R is the universal gas constant (8.31 x 10^7 erg mole^{-1} K^{-1}, and F is the Faraday in esu (2.89 x 10^{14} esu mole^{-1}). We can rewrite Eq. (71) in terms of number charge densities σ_0' and σ_B' follows:

$$\sinh\left[\frac{F\psi}{2RT}\right] \doteq \left(\frac{h_0\sigma_0' + h_B\sigma_B'}{\sqrt{C}}\right)\gamma \tag{73}$$

where

$$h_0 = \begin{cases} +1 \text{ if } \sigma_0' > 0 \\ -1 \text{ if } \sigma_0' < 0 \end{cases} \tag{74}$$

$$h_B = \begin{cases} +1 \text{ if } \sigma_B' > 0 \\ -1 \text{ if } \sigma_B' < 0 \end{cases} \tag{75}$$

$$\gamma = e\sqrt{\frac{500\pi}{DRT}} \tag{76}$$

e is the absolute value of the elementary electronic charge in esu (4.77 x 10^{-10}), σ_0' and σ_B' have positive values in units of elementary charges/cm^2.

For small values of the electrostatic potential ψ, we can expand the left hand side of Eq. (73) into a power series and retain only first order terms. We then get

$$\psi = \frac{2RT}{F}\,\frac{(h_0\sigma_0' + h_B\sigma_B')\gamma}{\sqrt{C}} \tag{77}$$

This equation is a reasonable approximation for $\psi < 50$ mV at room temperature. The surface charge of which $\psi < 50$ mV depends on the concentration of electrolyte as shown in Eq. (73). For a 0.1 molar solution of univalent electrolyte, the total charge density for $\psi = 50$ mV is calculated to be 0.003 elementary charges/Å^2. Assuming that the area of a lipid molecule is 58 Å^2, this charge density is equivalent to one elementary charge for every 6 lipid molecules.

Equation (77) indicates that for $\psi < 50$ mV, the electrostatic potential can be considered as the sum of the potential ψ_0 due to the normal charge on the membrane and the potential ψ_B produced by the membrane-bound probe, that is

$$\psi = \psi_0 + \psi_B \tag{78}$$

where

$$\psi_0 = \frac{2RT\gamma h_0 \sigma_0'}{F\sqrt{C}} \tag{79}$$

and

$$\psi_B = \frac{2RT\gamma h_N \sigma_B'}{F\sqrt{C}} \tag{80}$$

We now show how the ratio [PM]/[P] in Eq. (70) can be evaluated from fluorescence intensity measurements and how these measurements can be used to evaluate ψ_0. The total fluorescence intensity I from a membrane suspension containing probe at a total concentration $[P]_0$ is related to the fluorescence intensities I_F and I_B of free and bound probe respectively by the expressions

$$I = I_B + I_F \tag{81}$$

where

$$I_B = a\,\phi_B[PM] \tag{82}$$

$$I_F = a\,\phi_F[P] \tag{83}$$

$$[P]_0 = [P] + [PM] \tag{84}$$

ϕ_B and ϕ_F are the fluorescence efficiencies of bound and free probe respectively. We assume in Eqs. (82) and (83) that the extinction coefficient is the same for bound and free probe. [This assumption can be removed if necessary by introducing the appropriate extinction coefficients in Eqs. (81) and (82)]. The maximum fluorescence intensity I_M which is produced when the total amount of probe $[P]_0$ is all bound to the membrane is given by

$$I_M = a\,\phi_B[P]_0 \tag{85}$$

I_M can be measured experimentally by increasing the membrane concentration until all probe is bound. From Eqs. (81)-(85) we can write

$$\frac{I - I_F}{I_M - I} = \frac{[PM]}{[P]\left(1 - \dfrac{\phi_F}{\phi_B}\right)} \tag{86}$$

Equation (86) can be simplified by taking into account the fact that for polarity-sensitive probes as ANS, $\phi_F \ll \phi_B$. Specifically for ANS, $\phi_F \simeq 10^{-2}\phi_B$. Thus we can ignore the ratio ϕ_F/ϕ_B in the denominator of Eq. (86) and write

$$\frac{I - I_F}{I_M - I} = \frac{[PM]}{[P]} \tag{87}$$

With this equation we can thus evaluate [PM]/[P] using experimentally measurable intensities. I is the intensity of the membrane suspension at a total probe concentration of $[P]_0$ (under conditions where not all probe is bound to the membrane). I_F is the fluorescence intensity of free probe in the suspension and can be determined by sedimenting the membranes through centrifugation and measuring the fluorescence of the supernatant which is I_F. If $I_F \ll I$, then I_F can be ignored in Eq. (87). I_M is obtained by adding membranes to a solution of probe at a concentration $[P]_0$ until the fluorescence reaches a steady value. The steady value is I_M.

Equation (87) can be used to evaluate ψ_0 as follows. From Eqs. (70), (71), and (87) we can write

$$\frac{I - I_F}{I_M - I} = \beta \, [L]_0 \, K_c \, e^{-h_B F \psi / RT} \tag{88}$$

and using this equation and Eq. (77) we can further write

$$\frac{I - I_F}{I_M - I} = \beta \, [L_0] \, K_c \, e^{-\frac{-2 h_B \gamma (h_0 \sigma_0 + h_B \sigma_B)}{\sqrt{C}}} \tag{89}$$

Taking the natural log, we get

$$\ln \frac{I - I_F}{I_M - I} = \ln (\beta \, [L]_0 K_c) - \frac{2 \, h_B h_0 \gamma \sigma_0'}{\sqrt{C}} - \frac{2 \, h_B^2 \, \sigma_B'}{\sqrt{C}} \tag{90}$$

The density σ_B' in this equation is related to [PM] by means of the equation

$$\sigma_B' = \frac{[PM]}{\alpha \, [L]_0} \tag{91}$$

where α is the area (in cm^2) occupied per lipid molecule in the membrane. Moreover, from Eqs. (81) and (82) we have

$$[PM] = \frac{(I - I_F)}{a \, \phi_B} \tag{92}$$

Finally, introducing Eqs. (91) and (92) into Eq. (90) we get

$$\ln \frac{I - I_F}{I_M - I} = \ln (\beta \, [L]_0 K_c) - \frac{2 h_B h_0 \gamma \sigma_0'}{\sqrt{C}} - \frac{2 \gamma \, h_B^2 \, (I - I_F)}{a \, \alpha \, [L]_0 \, \phi_B \sqrt{C}} \tag{93}$$

Equation (93) can be used to evaluate ψ_0 as follows. First the intensities I, I_F, and I_M are measured for a given concentration of electrolyte C. Equation (93) then indicates that a plot of ln $(I-I_F)/(I_M-I)$ versus $(I-I_F)$ is a straight line with an intercept J given by the expression

$$J = -\frac{2h_B h_0 \gamma \sigma_0'}{\sqrt{C}} + \ln(\beta [L]_0 K_c) \tag{94}$$

If J is now measured for different values of C, then a plot of J versus $1/\sqrt{C}$ has a slope S given by

$$S = -2h_B h_0 \gamma \sigma_0' \tag{95}$$

The value of h_B (±1) is known from the charge on a probe ion. Using the values of S and h_B we can calculate $h_0 \gamma \sigma_0'$ from Eq. (95). Finally, these values can be introduced into Eq. (79) to calculate ψ_0.

E. Proximity Relations in Membranes

An important aspect of membrane research is the study of proximity relations within and between membrane components. These studies include determination of the size of proteins, distance between functional sites within a protein (such as the distance between active and regulatory sites of a membrane enzyme), distance between membrane proteins, aggregation and disaggregation of proteins, and attachment and detachment of cytoplasmic components to the membrane. In this section we discuss the use of singlet-singlet excitation energy transfer to study proximity relations in membranes.

The methods of excitation transfer are based on the observation that an excited fluorophore molecule can transfer its excitation energy to a suitable acceptor molecule when the two are brought close to each other [16, 54]. The rate at which excitation energy is transferred from the donor D to the acceptor A depends on the distance between the two groups. Thus if D is positioned at a site S_1 in a membrane and A at a site S_2, the distance between S_1 and S_2 can be calculated from the rate at which energy is transferred from D to A using the theory presented below.

1. General Theory of Excitation Transfer

The specific rate k_{DA} (s^{-1}) for excitation transfer from A to D separated by a distance R (in Å) is given by the expression

$$k_{DA} = 8.7 \times 10^{23} J K^2 n^{-4} k_e R^{-6} \tag{96}$$

where n is the refractive index of the medium separating D and A, and k_e (in s^{-1}) is the rate of emission of donor. K is an orientation factor defined by the equation

$$K = \hat{a} \cdot \hat{d} - 3(\hat{a} \cdot \hat{r})(\hat{d} \cdot \hat{r}) \tag{97}$$

where $\hat{a}$ and $\hat{d}$ are, respectively, unit vectors along the absorption transition moment of A and emission moment of D, and $\hat{r}$ is a unit vector along the line joining the centers of A and D. J is the spectral overlap parameter defined in the equation

$$J = \frac{\int F_D(\lambda)\, \epsilon_A(\lambda)\lambda^4\, d\lambda}{\int F_A(\lambda)\, d\lambda} \qquad (98)$$

where $F_D(\lambda)$ (in photons/cm) is the fluorescence intensity of the fluorophore at the wavelength λ (in cm) and $\epsilon_A(\lambda)$ is the molar decadic absorption coefficient (in units of liter mole^{-1} cm^{-1}) of the acceptor. The integral is taken over by the region of wavelengths where the fluorescence spectrum of the donor overlaps the absorption spectrum of the acceptor. J has units of (cm^3 liter mole^{-1}).

In order to determine the distance between D and A it is necessary to determine the values of the parameters k_{DA}, k_e, n, J, and K. Methods used to evaluate these parameters are as follows (32):

a) k_{DA} can be evaluated by measuring the lifetime of the donor in the absence, τ_D°, and in the presence, τ_D, of the acceptor. k_{DA} is given by the expression

$$k_{DA} = \frac{1}{\tau_D} - \frac{1}{\tau_D^\circ} \qquad (99)$$

b) k_e is usually evaluated from τ_D° and the fluorescence efficiency ϕ_D° of the donor in the absence of the acceptor using the expression

$$k_e = \frac{\phi_D^\circ}{\tau_D^\circ}\,. \qquad (100)$$

Introducing this expression into Eq. (96) gives

$$k_{DA} = 8.7 \times 10^{23}\, J K^2 n^{-4} \phi_D^\circ\, (\tau_D^\circ)^{-1} R^{-6}. \qquad (101)$$

c) The value of n is usually estimated from average values of n given in the literature for specific components. Thus for D and A attached to two different sites on a protein, a value of 1.4 is usually used. Errors introduced in the calculation of R using estimated values of n are usually not greater than 10%.

d) The spectral overlap integral J is evaluated with Eq. (98) using the fluorescence spectrum (F_D versus λ) of the donor and the absorption spectrum (ϵ_A versus λ) of the acceptor.

e) The orientation factor K^2 cannot usually be evaluated experimentally and its estimation presents one of the major uncertainties in the use of excitation transfer to study proximity relations. Its value ranges from 0 to 4, and in the absence of any knowledge concerning the true value of K^2 only an upper limit, corresponding to $K^2 = 4$, can be calculated for R using Eq. (101). The method used in practice to get some estimate for the actual value of K^2 is based on the following observations. The transition moments $\hat{a}$ and $\hat{d}$ do not usually have a fixed orientation with respect to each other as we have assumed so far but instead can rapidly rotate through restricted but finite angles. These rotations involve rotational motions of the donor and acceptor within their sites of attachment as well as internal motions of the transition moments within the acceptor and donor molecules. (For a discussion of internal factors which rotate $\hat{a}$ and $\hat{d}$ see ref. 30.) Regardless of their origin, these rotations cause K^2 to assume a value which is an average of several orientations. When $\hat{a}$ and $\hat{d}$ rotate rapidly and randomly over a complete solid angle, K^2 has a value of 2/3.

For donors and acceptors attached to proteins, experiments indicate that $\hat{d}$ and $\hat{a}$ can rapidly rotate over solid angles with half-angles in the range of 10° to 30°. This angle can be experimentally determined for the donor, and also for the acceptor if it is fluorescent, by measurement of the fluorescence anisotropy, A_0, as described in Sect. III.A. It is however common practice simply to assume that $K^2 = 2/3$, the value for complete random rotations, in the evaluation of R with Eq. (101). The magnitude of the errors which may result in the value of R when it is assumed that $K^2 = 2/3$ can be calculated as follows. If we let R_t and K_t^2 be the true values of R and K^2 and we let R_c be the value of R calculated with $K^2 = 2/3$, then the error factor $\alpha = R_t R_c$ is given by this expression

$$\alpha = \sqrt[6]{\frac{3}{2} K_t^2} \tag{102}$$

A plot of α versus K_t^2 covering all possible values of K_t^2 is given in Fig. 15. The error in R is less than 10% for values of K_t^2 in the range of 0.4 to 1. For K_t^2 greater than 1, α in-

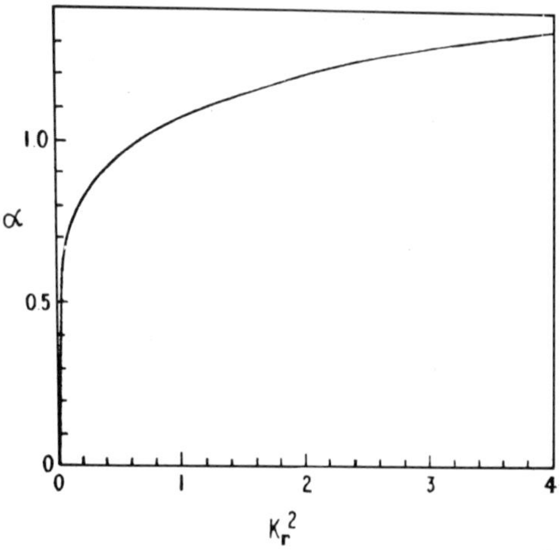

Fig. 15. Error factor α for the distance R_C calculated from excitation energy transfer measurements assuming $K^2 = 2/3$ versus the true orientation factor K_T^2 · α was calculated with Eq. (102) [51]

creases slowly with increasing K_t^2 and approaches a value of about 1.35 for $K_t^2 = 4$. For values of K_t^2 less than 0.4, α decreases very rapidly with decreasing K_t^2 and the error in R_c rapidly becomes much greater than 30% for $K_t^2 > 0.1$. It is evident from the graph that the greatest errors occur for very small values of K_t^2. These small values correspond to orientation where $\hat{a}$ and $\hat{d}$ are approximately perpendicular to each other as well as perpendicular to the intermolecular vector $\hat{r}$ [see Eq. (97)].

Dale and Eisinger [55] have calculated a complete set of values of K^2 for different degrees of rotational motions of $\hat{a}$ and $\hat{d}$ about specified directions. The tables and graphs which they present can be consulted for a more detailed inspection of the errors which can arise in the measurement of distances by excitation transfer. In practice, uncertainties in the values of R resulting from uncertainties in the value of K^2 can be reduced by using several acceptors with different orientations of $\hat{a}$ and $\hat{d}$. If all of the acceptors yield similar values of R (using the assumption $K^2 = 2/3$), then the assumed value of K^2 is not greatly in error.

It should be noted that the extent to which the donor fluorescence overlaps the acceptor absorption spectrum greatly influences the ability of a particular chromophore to act as an acceptor for a particular donor. If the overlap is zero, the two chromophores cannot be used as a donor-acceptor pair. The suitability of a donor-acceptor in a particular experiment can be expressed in terms of the parameter R_0, the distances between D and A where $k_{DA} = 1/\tau_D^0$. This condition corresponds to 50% excitation transfer. The value of R_0 can be calculated from Eq. (101) and $k_{DA} = 1/\tau_D^0$

$$R_0 = (8.7 \times 10^{23} J K^2 n^4 \phi_D^\circ)^{1/6} \tag{103}$$

Values of R_0 for commonly used donor-acceptor pairs are in the range of 15-100 Å. In terms of R_0, Eq. (101) can be written as

$$k_{DA} = \frac{1}{\tau_D^\circ} \left(\frac{R_0}{R}\right)^6 \tag{104}$$

Above we have discussed methods for evaluating R from a value of k_{DA} obtained by lifetime measurements. R can however also be evaluated from steady-state measurements. In fact, steady-state and lifetime measurements are related through the following expressions:

$$\frac{1}{\tau_D} = \frac{1}{\tau_D^\circ} + k_{DA} \tag{105}$$

$$\frac{I_0}{I} = \frac{\tau_D^\circ}{\tau_D} = 1 + k_{DA}\,\tau_D^\circ \tag{106}$$

$$= 1 + \left(\frac{R_0}{R}\right)^6 \tag{107}$$

where I_0 and I are the donor steady-state fluorescence intensities in the absence and presence of the acceptor. Steady-state measurements thus yield similar information about R to the lifetime measurements. Lifetime measurements, however, have an advantage in that steady-state measurements are subject to a variety of artifacts (such as inner filter effects and light scattering, which reduce fluorescence intensity but not the lifetime) that lead to erroneous conclusions if ignored.

The range of distances R which can be measured with a particular donor-acceptor pair having a specific value of R_0 can be estimated from Eq. (107) by assuming that precise measurements can be made only if I/I_0 or τ/τ_0 is in the range 0.9-0.1 (corresponding to 10%-90% fluorescence quenching). Introducing this range into Eq. (107), we find that a particular donor pair is useful for measuring values of R in the range 1.4 R_0-0.69 R_0. In practice, it is common to tabulate values of R_0 for different donor-acceptor pairs using $K^2 = 2/3$. Values of R_0 calculated with this assumed value of K^2 are usually written as prime R_0, i.e., R_0'.

2. Measurement of Distance Between Sites on a Specific Membrane Component

This usually requires a purified preparation in order to insert the donor and acceptor pairs at specific sites. This is particularly true when the donor or acceptor is a probe with low selectivity, such as dansyl chloride [56]. Some selectivity, however, may be obtained for example in the case of enzymes by attaching the probe to a natural substrate or inhibitor of the enzyme, in which case the system may not have to be as pure.

The use of excitation transfer to study dimensions of membrane proteins is exemplified by the work of Wu and Stryer [67] on rhodopsin. The object of their experiments was to determine whether the rhodopsin molecule is sufficiently long to span a bilayer and therefore act as a gated channel for translocation of ions across visual receptor membranes. Rhodopsin which had been purified and solubilized with detergents was labeled with the fluorophores shown in Table 9. The 11-cis retinal chromophore that is naturally present in rhodopsin served as acceptor for these fluorophores.

Table 9

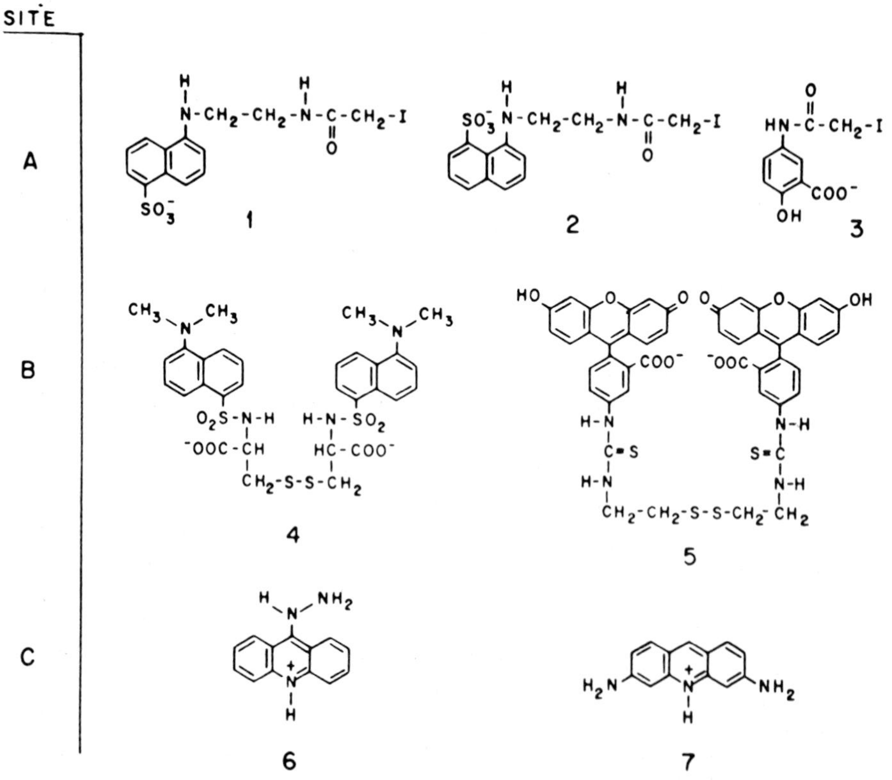

The iodoacetamide probes 1-3 and the disulfide probes 4 and 5 are reagents which form covalent bonds with sulfhydryl groups on proteins. The acridine probes 6 and 7 are non-reactive but spontaneously bind to the rhodopsin molecule through noncovalent interactions. Stoichiometric studies indicated that the probes were bound to the rhodopsin molecule in a 1:1 ratio and were located in one of three different binding sites designated A, B, and C. The site location of each probe is shown in Table 9. The results of excitation transfer measurements are shown in Table 10. The results indicate that the apparent dis-

Table 10. Excitation transfer parameters and distance r' between a given site and 11-cis retinal for rhodopsin

	Energy donor						
	1	2	3	4	5	6	7
Site	A	A	A	B	B	C	C
λ_e(nm)*	350	350	323	350	495	440	470
λ_f(nm)*	495	495	405	520	518	470	512
Q_d	0.68	0.28	0.44	0.61	0.12	0.04	0.14
Q_b†	0.75	0.30	0.45	0.95	0.15	0.05	0.18
Q_d/Q_b	0.91	0.94	0.97	0.64	0.80	0.80	0.79
τ_d(ns)	19.4	17.8	2.0	11.5	2.5	3.5	3.6
τ_b(ns)	21.3	18.5	2.1	17.9	3.2	4.0	4.7
τ_d/τ_b	0.91	0.96	0.96	0.64	0.78	0.88	0.77
J x 10^{13} (cm^3 M^{-1})	1.84	1.75	0.31	1.29	2.26	1.38	1.87
R_0' (Å)	51	45	41	52	42	33	41
E (%)	9	4	3	36	22	12	23
r' (Å)	75	77	73	57	52	46	49

* λ_e is the long-wavelength excitation maximum of the energy donor and λ_f is its fluorescence emission maximum.

† For energy donor 3, Q_b refers to the quantum yield of 3 attached to opsin.

λ_e is the long-wavelength excitation maximum of the donor, λ_f is the fluorescence emission maximum; Q_b and Q_d are, respectively, the fluorescence efficiencies of donor in the absence of (I_0) and presence of (I) of acceptor (Q_b was determined with bleached rhodopsin, in which case the bleached 11-cis retinal no longer acts as an acceptor); τ_b and τ_d are the donor lifetimes in the absence ($\mathring{\tau}_D$) and presence (τ_D) of acceptor; E (%) is percent of excitation transfer (I/I_0), r' is the distance between a given site and 11-cis retinal calculated from Eq. (103) assuming $K^2 = 2/3$.

tance of the acceptor to site A (based on $K^2 = 2/3$) is about 75 Å, to site B about 55 Å, and to site C about 48 Å. The conclusion drawn from these measurements is that the rhodopsin molecule is elongated. If it were spherical its diameter would be 45 Å. The results indicate that the molecule is at least 75 Å long and therefore sufficiently long to span the thickness of a lipid bilayer, which is usually around 60 Å. The experiments, however, do not give any information on how rhodopsin is oriented on the membrane, i.e., whether the long axis is in the plane of the bilayer or perpendicular to it. It has recently been suggested, on the basis of studies on the orientation of rhodopsin by hydrodynamic shear, that the rhodopsin molecule in its native membrane may have its long axis in the plane of the bilayer and not perpendicular as would be required in order to span the bilayer [58].

The apparent dimension of the rhodopsin molecule measured by the excitation transfer experiments is of course subject to the uncertainties which arise from the assumed value for K^2. The authors argue that the uncertainties are not very large since the apparent distances obtained with different probes at the same site are very similar.

3. Aggregation and Disaggregation (Subunit Interactions) of Membrane Proteins

The possibility that membrane proteins may aggregate and disaggregate in response to particular stimuli has recently received much attention and it has been postulated that such processes may be involved in the triggering of specific membrane functions by agents such as hormones or light [59]. More specifically it has been suggested that in the case of transport functions which can be gated on and off, the translocator proteins under the appropriate stimulus can aggregate into complexes that form channels through which specific solutes cross the membrane.

In the study by excitation energy transfer of aggregation-disaggregation processes in membranes, a fraction of the molecules of the protein of interest is labeled with donor groups (one donor group per molecule) and another fraction is labeled with acceptor groups. In the monomeric state, the protein molecules will usually be too far apart energy transfer to occur. If the proteins aggregate, however, donor and acceptor groups come sufficiently close for excitation transfer to occur and the fluorescence efficiency ϕ and lifetime τ of the fluorophore are decreased. Aggregation and disaggregation processes can therefore be detected through changes in Φ and τ. The use of excitation transfer to study protein aggregation will usually require a pure protein preparation in order selectively to label one fraction of the protein with donor groups and the other with acceptor groups. Vanderkooi et al. [60] have used the technique to study possible aggregation of the Mg^{++}-Ca^{++}-activated ATPase of the sarcoplasmic reticulum. It has been postulated on the basis of electron microscope studies that the ATPase molecules in the sarcoplasmic reticulum membrane exist as clusters of several ATPase molecules which form a hydrophilic channel for the passage of Ca^{++} ions. In their experiments, Vanderkooi et al. used a purified ATPase that had been reconstituted into lipid vesicles. The ATPase in a fraction of vesicles was labeled with N-iodoacetyl-N'-(5-sulfo-1-naphthyl) ethylenediamine (1,5-IAEDANS) which served as donor, while another fraction was labeled with iodoacetamidofluorescein (IAF) which served as the acceptor R_0 for this donor-acceptor pair is 48 Å. The acceptor IAF is fluorescent so that in this case energy transfer from donor to acceptor can be detected by increase in fluorescence intensity of acceptor as well as decrease in fluorescence intensity and lifetime of donor.

When vesicles containing IAF-ATPase were added to vesicles containing 1,5-IAEDANS-ATPase under conditions where the vesicles do not mix, the fluorescence intensity of the mixed system was the same as the sum of the fluorescence intensities of the separate vesicles. That is, there was no evidence of excitation transfer. When conditions were adjusted (by addition of detergent) so that the vesicles fuse and mix, the fluorescence intensity and lifetime of the donor fluorescence decreases while the fluorescence intensity of the acceptor increased, indicating the occurrence of excitation energy transfer. In general such transfer can occur as a result of random collisions between acceptor- and donor-labeled ATPase molecules or as a result of aggregation. To distinguish between these two possibilities the degree of excitation transfer was studied as a function of surface density of donor- and acceptor-labeled ATPase molecules as well as a function of temperature. The degree of excitation transfer was unaffected by decrease in temperature and surface density, which is inconsistent with a collisional mechanism. It was thus concluded that the excitation transfer experiment supports the hypothesis that the ATPase molecules are aggregated in the membrane.

Veatch and Stryer [61] have used the technique of excitation transfer to study the mechanism by which gramicidin A increases membrane permeability of alkali cations and protons. gramicidin is a linear polypeptide of 15 amino acids. On the basis of studies with black lipid membranes it has been suggested that the translocation of ions across membranes in the presence of gramicidin is facilitated by a dimer rather than monomer form of gramicidin A. To test this hypothesis, gramicidin C labeled with dansyl (dansyl-gramicidin C) was used as a donor and gramicidin C labeled with 4-(dithylamino)-phenylazobenzene-4-sulfonylchloride (DPBS derivative of gramicidin C) was used as acceptor. The R_0 for this system, assuming $K^2 = 2/3$, is 39 Å. It was shown that the labeling did not destroy the translocator properties of gramicidin C. In the excitation transfer experiments, the dansylgramicidin C was incorporated into lipid vesicles and the fluorescence intensity was measured as a function of the mole fraction a of the acceptor DPBS gramicidin C added to the vesicles. Addition of acceptor resulted in a decrease in fluorescence intensity indicating that the gramicidin molecules were aggregated, since the distance between monomer gramicidin molecules would be too large (∿ 250 A at the densities used) for excitation transfer to occur. To determine the number of molecules in the aggregate, a plot I/I_0 versus a was analyzed with equations for a dimer, trimer, and tetramer model. For the dimer model, the intensity expression is

$$\frac{I}{I_0} = 1 - E\,a \tag{108}$$

where E is the efficiency of energy transfer within the aggregated complex. For the simplest trimer mechanism the pertinent expression is

$$\frac{I}{I_0} = E + E\,(1\text{-}a)^2 \tag{109}$$

while for the simplest tetramer model we have

$$\frac{I}{I_0} = 1 - E + E\,(1\text{-}a)^3 \tag{110}$$

The results could best be fitted by the dimer model but not by the tetramer model, indicating that the aggregates are dimers. A trimer model, however, could not be eliminated with high certainty. E was found to have a value of 75%, indicating that the distance between the donor and acceptor groups in the complex is about 33 Å.

4. Attachment and Detachment of Proteins to Plasma Membrane

Attachment and detachment of cytoplasmic proteins have recently been shown to be involved in the regulation of mobility of membrane components, and it has been suggested these interactions may have an important role in such processes as cell motility, pinocytosis, phagocytosis, down regulation, and segregation of proteins into specialized areas of membranes, such as in the postsynoptic membrane of neural junctions [59, 62].

The attachment and detachment of proteins to membranes can be detected and characterized by excitation energy transfer. In this application, a donor fluorophore is anchored on the membrane and acceptor groups are attached to the protein. When the labeled protein attaches to the membrane, the acceptor and donor groups come sufficiently close for excitation transfer to occur. Attachment and detachment can therefore be detected through changes in ϕ and τ of the donor fluorophore. Indeed, we have recently developed a theory by which this change in ϕ and τ can be used to evaluate the binding constant and number of binding sites as well as the distance between protein binding sites in the membrane where the donor is located [63]. The theory has been used to study the interaction of hemoglobin with the cytoplasmic surface of the red cell membrane [63, 64]. This study is briefly reviewed below to demonstrate the basic aspects of the technique.

Red cell ghosts were labeled with the negatively charged fluorophore AS (which served as the donor) at a molar ratio of lipid to probe of 1000. The heme groups on hemoglobin served as acceptors. There is experimental evidence that negatively charged molecules added to the red cell membrane are chiefly located on the outer membrane surface [65]. It was thus assumed that AS is located on this outer surface of the membrane. Addition of hemoglobin at low ionic strength was found to reduce the steady-state fluorescence intensity of the membrane-bound AS. A decrease in fluorescence intensity can result from excitation transfer due to bonding of Hb molecules to the membrane, trivial inner filter effects, or simple collision of the Hb molecules with the membrane. These possibilities were distinguished by varying the ionic strength. Increase in ionic strength decreased the quenching effect and practically eliminated it at high ionic strength (~ 0.1 M NaCl). Since ionic strength is not expected to significantly change inner filter effects or collision frequencies, the result indicates that Hb binds to the membrane at low ionic strengths but its binding affinity decreases with increasing salt concentration.

To quantitate the quenching data (I versus [Hb]), we developed the following theory. The theory considers a donor fluorophore molecule located at a specific distance from one surface of the membrane shown in Fig. 16 and quencher molecules randomly distri-

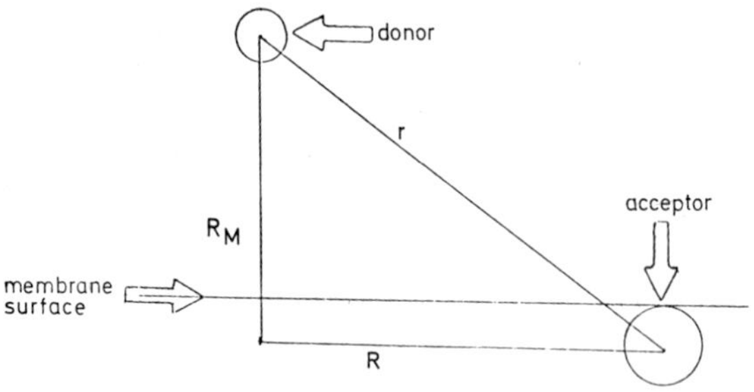

Fig. 16. Coordinate system used to develop theory of excitation energy, transfer between a donor probe located at a distance R_M from the surface of a membrane and an acceptor molecule located on that surface [63]

buted on that surface at a density of σ molecules/cm^2. The closest distance to which an acceptor group can approach the donor is designated R_M and is determined by the extent to which the quencher penetrates the membrane. The quencher molecule may have several acceptors attached to it. Thus, hemoglobin has four acceptor heme groups. The closest distance to which each group can approach the donor depends on the orientation and degree of penetration of Hb on the membrane. We will consider the general case where each quencher has j identical acceptor groups and designate by R_{Mj} the distance of closest approach for the jth acceptor group. Solution of the appropriate kinetic equations yields the following expression for the decay of fluorescence intensity following excitation by a very short pulse of light (for hemoglobin, n = 4).

$$I = I_0\, e^{-t/\tau} \prod_{j=1}^{n} e^{-\sigma L_j(t)} \tag{111}$$

where

$$L_j(t) = \int_{R_{Mj}}^{\infty} (1 - e^{-k(r)t}) 2\pi r \, dr \tag{112}$$

and

$$k(r) = \frac{1}{\tau_0} \left(\frac{R_0}{r} \right)^6 \tag{113}$$

where k(r) is the specific rate for energy transfer between the donor and an acceptor at a distance r from the donor. These expressions assume that K^2 is the same for all positions of the acceptor with respect to the donor, so that R_0 is the same for all acceptors on a given quencher. In order to use Eq. (111) to analyze experimental data it is necessary to evaluate the integral of Eq. (112). This integral cannot in general be evaluated analytically. However, an analytical integration can be performed for the case where R_{Mj} is much greater than R_0 (more specifically where $R_{Mj} > 1.7\, R_0$). For the large protein quencher molecules used as quenchers in the technique under discussion this approximation will often apply. (The conditioning $R_M > 1.7\, R_0$ may at first sight seem to be inconsistent with the calculation given in Sect. III.E.1 where we indicated that excitation transfer measurements were suitable for determining distances in the range 1.4 R_0 to 0.69 R. The calculation of that section however was made for one donor and one acceptor by a given distance. In the present case we are considering one donor and many acceptors. Thus for $R_M > 1.7\, R_0$, the value of τ/τ_0 can be brought into the range of 0.9 to 0.1 by adjusting the density σ of the quencher molecules.) For the case $R_{Mj} > 1.7\, R_0$ the integral can be expanded in a power series and if we retain only first terms, integration yields

$$I(t) = I_0 e^{-(\frac{1}{\tau_0} + k_Q \sigma)t} \tag{114}$$

where

$$k_Q = \frac{\Pi R_0^2}{2\,\tau_0} \left[\frac{R_0}{\overline{R}_M} \right]^4 \tag{115}$$

and

$$\left(\frac{1}{\overline{\overline{R}}_M}\right)^4 = \sum_{j=1}^{n} \left(\frac{1}{R_{M_j}}\right)^4. \tag{116}$$

k_Q has units of cm^2/molecule s, R_0 is in cm, τ_0 is in seconds and σ in molecules/cm^2. Equation (114) has the same form as the so-called Stern-Volmer equation for quenching of fluorescence in solution.

The steady-state intensity I is given by the expression

$$I = a \int_0^\infty I(t)dt \tag{117}$$

where a is a constant of proportionality. Introducing Eq. (114) into Eq. (117) yields

$$\frac{I_0}{I} = 1 + K_Q\,\sigma \tag{118}$$

where

$$K_Q = \tau_0\,k_Q$$

$$= \frac{\pi R_0^6}{2} \frac{1}{\overline{R}_M^4} \tag{119}$$

The quenching constant K_Q has units of cm^2/molecules. Equation (118) indicates that a plot of I_0/I versus σ should be linear with a slope equal to K_Q. In addition $\overline{R}_M$ can be evaluated from the value of K_Q with Eq. (119).

Figure 17 shows a plot of I_0/I versus total concentration of hemoglobin $[Hb]_0$ for red cell ghosts labeled with AS. At the cell concentration used in the experiment all of the added Hb is attached to the membrane for $[Hb]_0 < 6 \times 10^{13}$ hemoglobin molecules/ml. Under these conditions

$$\sigma = \frac{[Hb]_0}{\alpha\, c_L} \tag{120}$$

where $[Hb]_0$ and c_L are, respectively, the concentration of hemoglobin and membrane lipid in moles/liter, and α is the area in cm^2 per lipid molecule. With this expression we calculate that under conditions where all the added Hb is attached to the membrane (i.e., for $[Hb]_0 < 6 \times 10^{13}$ molecules/ml) each unit on the concentration axis of Fig. 17 is equal to a density of Hb on the membrane of 10^{11} molecules/cm^2. The plot indicates that I_0/I is a linear function of σ for $[Hb]_0 < 6 \times 10^{13}$ molecules/ml as predicted by Eq. (118). The deviation from linearity between $[Hb]_0$ equal to 6.3×10^{13} and 12×10^{13} Hb molecules/ml is due to the fact that all the added Hb is not attached to the membrane at this higher concentration of Hb. The leveling of the plot at still higher concentrations of Hb is due to saturation of binding sites. The fact that the plot has a small but finite slope at this higher concentration, instead of being constant as expected for complete saturation, indicates that at these higher concentrations, sites with low binding constants are being filled. Our main interest, however, was in the high-affinity binding

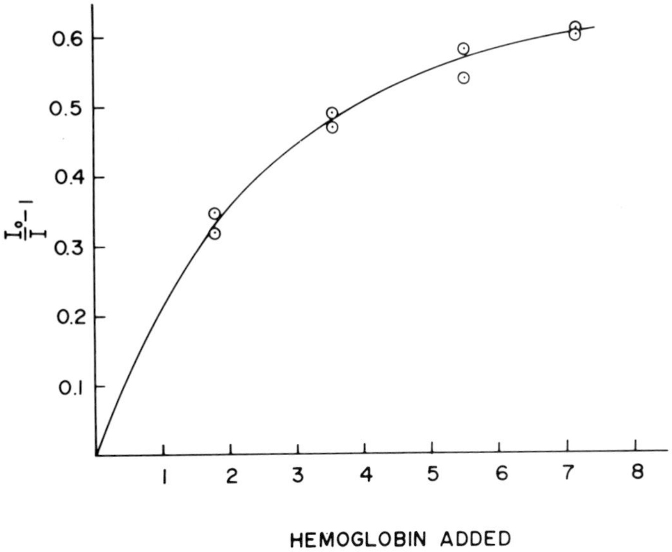

Fig. 17. ($I_0/I - 1$) versus total hemoglobin concentration added to a suspension of red cell ghosts at a cell concentration of 7.26 x 10^7 cells/ml in 5 mM phosphate buffer, pH = 6.0. Each unit on ordinate scale is equal to 1.05 x 10^{13} total hemoglobin molecules/ml. For total hemoglobin concentration < 6 x 10^{13} molecules/ml, all the hemoglobin is bound to the membrane. For those concentrations, each unit on the ordinate scale corresponds to 10^{11} hemoglobin molecules bound to membrane/cm^2 [63]

sites. The ordinate of the point where the two straight lines meet in the plot of Fig. 17 gives 1.4 x 10^6 for the number of higher affinity sites per ghost cell. The slope of I_0/I versus σ for [Hb] < 6 x 10^{13} gives K_Q = 4.9 x 10^{-13} cm^2/molecule, from which we can calculate, with Eq. (119), $\bar{R}_M$ = 42 Å (using the value R_0' = 46 Å). The value of R_{M_j} for the four heme groups can be calculated from Eq. (116) if we know how the Hb molecules are oriented on the membrane surface. If we assume for example that two of the subunits of Hb are in contact with the membrane surface while the other two are directed away from the surface, then the distant heme groups would be too far away from the surface, given the dimensions of the Hb molecules, to contribute to the excitation transfer. In this case, the two equivalent heme groups close to the surface are responsible for the excitation and we may write

$$R_{M_j} = \bar{R}_M \sqrt[4]{2} = 1.19\,\bar{R}_M \tag{121}$$

from which we get R_{M_j} = 50 Å. That is, with the indicated assumption, we calculate that the heme groups closest to the surface are about 50 Å from the region of the membrane where the donor group is located.

To evaluate the binding constant, we determined I_0/I versus $[Hb]_0$ at a low cell concentration where only a fraction of the added hemoglobin is bound to the membrane. The graph is shown in Fig. 18. The concentration $[Hb]_b$ of hemoglobin bound to the membrane at a total concentration $[Hb]_0$ can be calculated from the graphs of Figs. 17

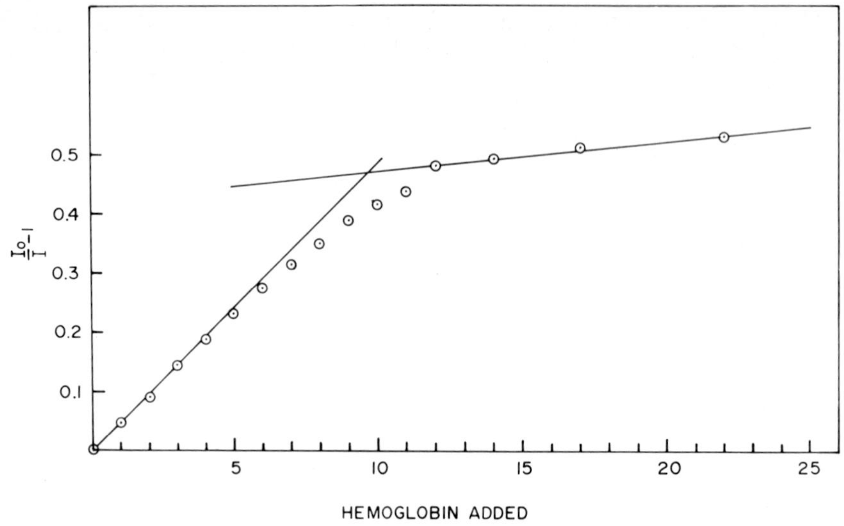

Fig. 18. ($I_0/I - 1$) versus total hemoglobin concentration added to a suspension of red cell ghosts at a cell concentration of 4.6 x 10^6 cells/ml in 5 mM phosphate buffer, pH = 6.0 [63]

and 18 as follows. In the region of Fig. 17 below $[Hb]_0 = 6 \times 10^{13}$ molecules/ml, where all the hemoglobin molecules are bound to the membrane, that plot I_0/I versus $[Hb]_0$ serves as a calibration graph giving a relation between I_0/I and density σ of Hb molecules bound to the membrane. Using this calibration graph we can therefore convert the values of I_0/I in Fig. 18 to values of σ, the density of molecules of Hb bound at total concentration $[Hb]_0$. Concentration of bound hemoglobin $[Hb]_b$ is then given by

$$[Hb]_b = \sigma\, c_L \alpha \tag{122}$$

where c_L is concentration of membrane lipid in moles/liter and α is area occupied by one lipid molecule. Concentration of free Hb is given by

$$[Hb]_f = [Hb]_0 - [Hb]_b. \tag{123}$$

According to the Hill equation we can write

$$\log \frac{\nu}{1-\nu} = C \log [Hb]_f + \log K \tag{124}$$

where K is the binding (association) constant, C is a measure of binding cooperatively, and

$$\nu = \frac{[Hb]_b}{[M]_0} = \frac{\sigma}{\sigma_0} \tag{125}$$

where $[M]_0$ is the total concentration of binding sites and σ_0 is total density of binding sites. A plot of log $\nu/(1-\nu)$ versus $[Hb]_f$ is shown in Fig. 19. This plot yielded C = 1.1 and K = 0.85 x 10^8 M^{-1}.

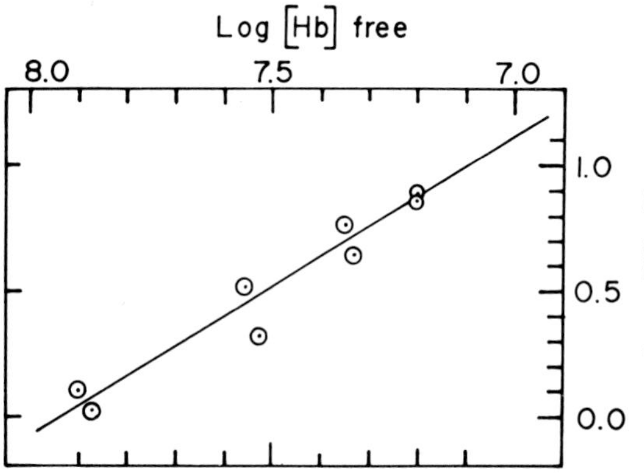

Fig. 19. Hill plot for the data of Fig. 18. See Eqs. (122)-(125) [63]

F. Phase Transitions

It has long been known that as the temperature of a lipid bilayer composed of a single lipid is decreased, the bilayer undergoes a transition from a liquid crystalline to a gel phase at a specific temperature T_c [66]. In the gel phase (which we will also call the solid phase), the hydrocarbon chains are relatively well oriented and have low rotational and lateral mobilities, while in the liquid or fluid phase their motions are enhanced and their organization is decreased. The phase transition temperature T_c of a bilayer composed of a single lipid component depends on the nature of the polar head and hydrocarbon chains. T_c increases with increase in chain length and decreases with increase in the number of double bonds. For a given chain length, phosphatidylcholine (PC) has a lower phase transition temperature than phosphatidylethanolamine (PE). A complex bilayer composed of a mixture of phospholipids also displays a phase transition but the width of the transition (temperature range over which the transition occurs) is broader than that of a single lipid bilayer where the transition usually occurs over a temperature range of about 2°C. In addition, the temperature region over which the transition occurs depends on the composition of the system, which is usually expressed in terms of the mole fraction X_i of the lipid components. Phase transitions in complex bilayers are often called lateral phase separations. At temperatures above the phase transition temperature range, a complex bilayer usually consists of a homogeneous fluid phase (assuming that the lipids are completely miscible in the fluid phase). Within the temperature region of the phase transition the bilayer consists of solid and liquid phases of different composition in equilibrium with each other. The composition of the solid phase depends on the miscibility of the lipids in this phase. For a given mixture of lipids, the number and composition of phases at any temperature T can be described by a phase diagram. Figure 20 shows two types of phase diagrams which are often displayed by two component lipid mixtures. Diagram **a** is for a system where the lipids are completely miscible in the solid phase (i.e., the lipids cocrystallize) whereas diagram **b** is for lipids which do not cocry-

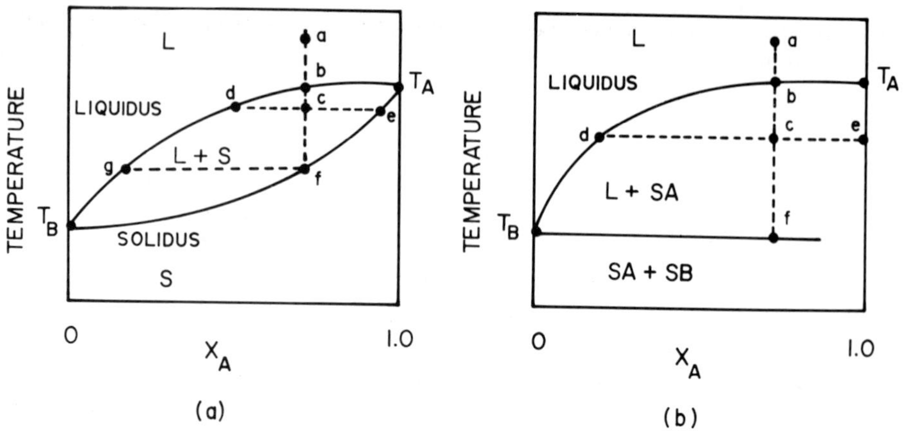

Fig. 20 a and b. Types of phase diagrams shown by two-component lipid bilayers. **a** is representative of a bilayer in which the two lipids in the bilayer cocrystallize. **b** is representative of a bilayer in which the two lipids do not cocrystallize

stallize. The ordinate in these diagrams expresses the composition of the system in terms of the mole fraction X_A of component A, that is

$$X_A = \frac{N_A}{N_A + N_B} \tag{126}$$

whereas N_A and N_B are the number of moles of components A and B, respectively, in the bilayer. The liquidus curve gives the temperature T_F at which crystallization begins, as the temperature is lowered for a bilayer with a specific composition X_A. Thus if a bilayer has the temperature and composition given by the ordinate and abscissa of point *a*, then as the temperature is lowered crystallization begins at the temperature of point *b*. If the temperature continues to decrease, the bilayer will eventually go completely into the solid phase. The temperature T_S at which this occurs marks the termination of the transition and is given by point *f* on the solidus curve. For a temperature T between T_F and T_S (e.g., point *c*), the system consists of both solid and fluid phases. For a specific value X_A, the composition of the fluid and solid phases can be found from the phase diagram as follows (using point *c* as an example). A horizontal line is drawn through the point *c*. The composition X_{AF} of the fluid phase is then given by the ordinate of the point where the horizontal line crosses the liquidus curve, which is point *d*. The composition of the solid phase X_{AS} is given by the ordinate of the point where the line crosses the solidus graph, which is point *e*. X_{AF} and X_{AS} are defined by the expression

$$X_{AF} = \frac{N_{AF}}{N_{AF} + N_{BF}} \tag{127}$$

$$X_{AS} = \frac{N_{AS}}{N_{AS} + N_{BS}} \tag{128}$$

where N_{AF} and N_{AS} are the number of moles of A in the fluid and solid phases respectively. Note that

$$N_A = N_{AF} + N_{AS} \tag{129}$$

The study of phase transitions and phase diagrams is important in membrane research because it gives information on the structural organization of the membrane and because such transitions may be involved in modulating membrane functions. Indeed, it has been shown that phase transitions can be induced in lipid bilayers not only by changes in temperature but also by changes in pH, ionic strength, bivalent ions, and charged polypeptides, and it has been suggested that changes in these parameters may be involved in modulating membrane functions through changes in phase equilibria in the bilayer [38].

The use of fluorescence spectroscopy in the study of phase transition is based on the observation that the emission properties, e.g., fluorescence efficiency and fluorescence anisotropy, of a fluorophore in a bilayer change during a phase transition. The appearance of a phase transition in a fluorescence intensity versus temperature diagram is shown in Fig. 21.

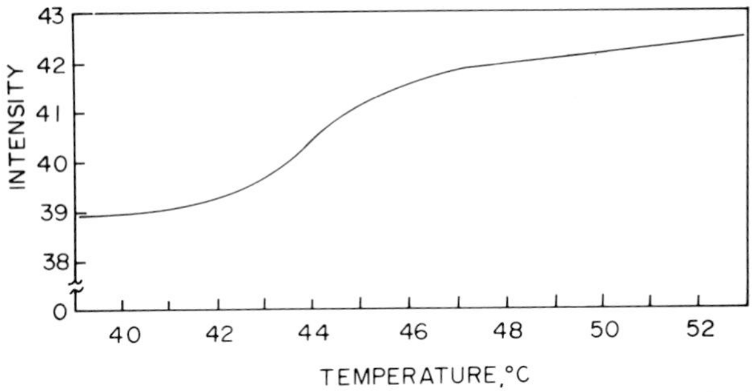

Fig. 21. Detection of a phase transition with a fluorescent probe. Intensity versus temperature for a suspension of bilayer lipid vesicles composed of dipalmitoyl phosphatidylcholine and distearoyl phosphatidylcholine (DSPC) labeled with perylene. Mole fraction of DSPC was 0.434 [67]

Graphs of the type in Fig. 21 have been used to construct phase diagrams and to study the effects of ionic strength, pH, etc. on phase distributions [38]. To construct a phase diagram for a two-lipid component system by fluorescence intensity measurements the experimerter measures I versus T for different values of X_A. It is then assumed that T_F and T_S for each X_A are given by the temperatures at which breaks appear in the I versus T graph. Such breaks are shown arrows in Fig. 21. In general, the validity of this assumption depends on the value of the partition coefficient K for the partitioning of the probe between fluid and solid phases, and also on the shape of the phase diagram. Below we present a theory which we recently developed [67] to gain a better understanding of the relation between the plot I versus T and phase diagrams. The theory also provides a basis for evaluating the partition coefficient K for a fluorophore in the fluid and solid phases of a two-component bilayer from measurements of I versus T for different values of X_A and with the aid of a phase diagram determined by an independent method such as differential thermal calorimetry [38]. Knowledge of partition coefficients is important in the interpretation of fluorescence spectroscopic studies of natural membranes.

The theory which we present here gives an expression for I versus T for a two-component system having a composition X_A. We assume that the bilayer is labeled with a probe that is insoluble in water or that has no cluorescence intensity in water. The lipid to probe molar ratio is $\geqslant 1000$. At any temperature T, the bilayer may be in the fluid or solid phase, or consist of a mixture of fluid and solid phases. The number and composition of phases at any temperature is determined by the phase diagram of the lipids in the bilayers. The intensity I is essentially a function of temperature and of the mole fraction X_A, i.e., $I = I(T,X_A)$.

In general we may write for the fluorescence intensity

$$I = a(\phi_S P_S + \phi_F P_F) \tag{130}$$

where ϕ_S and P_S are the fluorescence efficiency and number of moles of fluorophore in the solid phase, while ϕ_F and P_F are corresponding quantities in the fluid phase. The partition coefficient of the probe between the solid and liquid phases is defined by the expression

$$K = \frac{c_F}{c_S} \tag{131}$$

where c_F and c_S are the concentrations in the membrane of fluorophore in the solid and fluid phases respectively. If V_S and V_F are the volumes of the crystalline and fluid phases then we can write

$$K = \frac{P_F V_S}{P_S V_F} \tag{132}$$

Note that the total amount of probe P_0 in the bilayer is given by

$$P_0 = P_F + P_S \tag{133}$$

We will assume that K is independent of composition of fluid and solid phases. Introducing Eqs. (132) and (133) into Eq. (130) gives

$$I = a\,\phi_S P_0 + \frac{a(\phi_F - \phi_S) K P_0 V_F}{V_S + K V_F} \tag{134}$$

We now define the following intensities:

$$I_S = a\,\phi_S P_0 \tag{135}$$

$$I_F = a\,\phi_F P_0 \tag{136}$$

I_S is the intensity one would get if all the probe P_0 were in the solid phase, I_F if all the probe were in the fluid phase. I_S and I_F are essentially measures of ϕ_S and ϕ_F. Note that ϕ_S and ϕ_F may depend on temperature and composition as discussed in more detail below. Methods for evaluating I_S and I_F from an I versus T graph are also described below.

Introducing Eqs. (135) and (136) into Eq. (134) gives

$$\frac{I - I_S}{I_F - I_S} = \frac{KV_F}{V_S + KV_F} \tag{137}$$

Finally, if we let D_F and D_S be the molar densities of the lipid in the fluid and solid phases respectively (for example, $D_F = N_F/V_F$) we can write

$$\frac{I - I_S}{I_F - I_S} = \frac{K\beta N_F}{N_S + K\beta N_F} = \frac{K\beta N_F/N_S}{1 + K\beta N_F/N_S} \tag{138}$$

where $\beta = D_S/D_F$. N_F is the total number of moles of lipid in the fluid phase and N_S is the total number of moles of lipid in the solid phase.

Equation (138) allows us to predict I versus T for a lipid mixture whose phase diagram is known. Conversely, it can be used to analyze I versus T data so as to obtain information about phase diagrams. The manner in which the parameters of Eqs. (137) and (138) make their appearance in an I versus T plot is shown in Fig. 22 for the simple case

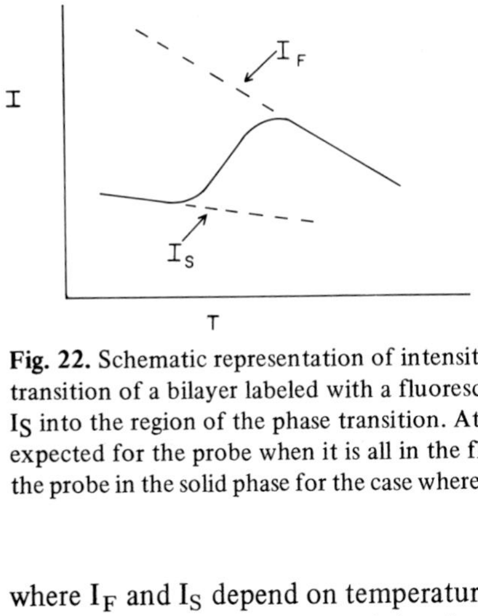

Fig. 22. Schematic representation of intensity (I) versus temperature (T) in the region of the phase transition of a bilayer labeled with a fluorescent probe. *Dashed line:* linear extrapolation of I_F and I_S into the region of the phase transition. At any given temperature I_F is the fluorescence intensity expected for the probe when it is all in the fluid phase, whereas I_S is the intensity expected for all of the probe in the solid phase for the case where ϕ_F and ϕ_S depend on temperature but not composition

where I_F and I_S depend on temperature but not on composition. Linear extrapolation of I_F and I_S into the region of the phase transition allows the left-hand side of Eqs. (137) and (138) to be evaluated. The quantity

$$\overline{I} = \frac{I - I_S}{I_F - I_S} \tag{139}$$

which we shall call a reduced intensity is, according to Eq. (137), essentially a measure of the fraction of bilayer in the fluid phase (in terms of volumes) miltiplied by the partition coefficient. If $K = 1$ and $\beta = 1$, then according to Eq. (138) $\bar{I}$ is a measure of the molar fraction of bilayer in the fluid phase at each temperature T. The temperature T_m at which $\bar{I} = 0.5$ can be considered as the experimental midpoint of the transition. If $K = 1$ and $\beta = 1$, T_m is the temperature at which half the lipid molecules are in the fluid phase and half in the solid phase, i.e., $N_F = N_S$. If K and β are not equal to 1, then at T_m we have $N_S = K\beta N_F$.

The linear extrapolation of I_S and I_F into the region of the phase transition, as shown in Fig. 22, is only approximately correct even in the case where ϕ_S and ϕ_F are independent of composition and P_0 is constant. Thus according to elementary theory the dependence of fluorescence efficiency on temperature in a phase of constant composition is chiefly due to the dependence of the rate of internal conversion k_i on temperature. Assuming that $k_i = k_0 e^{-\epsilon/kt}$, then we can write for the intensity I from a phase with constant composition

$$\frac{1}{I} = \frac{1}{I_0} + b\,e^{-\epsilon/kT} \tag{140}$$

where I_0 is the intensity when the fluorescence efficiency is 1, i.e., when $k_0 e^{-\epsilon/kT} \ll k_e$. This expression indicates that a plot of $\log \frac{1}{I} - \frac{1}{I_0}$ versus 1/T is linear and that such a plot is the proper method for making a linear extrapolation of intensity versus temperature data when composition is constant or fluorescence efficiency is independent of composition and depends only on temperature. Unfortunately I_0 is usually not known, which means that this method of extrapolation is not practical. In view of this limitation, it is common practice to make linear extrapolation of I versus T data using a plot of $\log \frac{1}{T}$ versus 1/T or simply a plot of I versus T, as we have done in Fig. 22.

In the most general case, ϕ_S and ϕ_F are functions of composition as well as temperature. In the absence of any other information concerning ϕ_S and ϕ_F it is convenient to assume that at any temperature T the fluorescence efficiency of the fluid phase ϕ_F is a linear combination of the fluorescence efficiency ϕ_{AF} of the fluorophore in the fluid phase of pure A and the efficiency ϕ_{BF} in the fluid phase of pure B at the temperature T. More specifically, we write

$$I_F = a\,(X_{AF}\phi_{AF} + X_{BF}\phi_{BF})\,P_0 \tag{141}$$

$$= X_{AF}I_{AF} + X_{BF}I_{BF} \tag{142}$$

and similarly for the solid phase we assume

$$I_S = a\,(X_{AS}\phi_{AS} + X_{BS}\phi_{BS})\,P_0 \tag{143}$$

$$= X_{AS}I_{AS} + X_{BS}I_{BS} \tag{144}$$

I_{AF} is the fluorescence intensity at temperature T of the probe in the pure fluid phase A and its value at any temperature can be determined by linear extrapolation in a plot of I versus T for pure A. Similarly for I_{BF}, I_{AS}, and I_{BS}. Experiments in our laboratory indicate that I_F and I_S are practically independent of composition for mixtures of lipids with the same polar heads. For mixtures with different polar heads, e.g., PC and PE, I_S and I_F depend on composition and the more general expression, Eqs. (142) and (143) must be used in the analysis of experimental information.

The profiles of I versus T plots expected for the types of phase diagrams shown in Fig. 20 and specific values of K can be predicted with Eq. (138). As explained above, the values of X_{AF} and X_{AS} for a given value of T and a total mole fraction X_A can be obtained from the fluidus and solidus graphs of the phase diagram. Material balance requires that

$$X_A(N_F + N_s) = X_{AF}N_F + X_{AS}N_S \tag{145}$$

from which we can write

$$\frac{N_F}{N_S} = \frac{X_A - X_{AS}}{X_{AF} - X_A} \tag{146}$$

This equation, together with the phase diagram, a value for β, an assumed dependence of ϕ_F and ϕ_S on temperature and composition and an assumed value for K, allows us to calculate with Eq. (138) values of $\bar{I}$ at different temperatures. Figure 23 shows plots of the reduced intensity $\bar{I}$ versus T calculated for the types of phase diagram of Fig. 20. The important points to be made from theoretical $\bar{I}$ versus T plots are as follows. For lipids that cocrystallize, the breakpoints in the I versus T plots correspond to the onset, T_F, and termination, T_S, temperatures of the phase transition when K = 1. When $K \neq 1$, i.e., when the probe partitions preferentially into the fluid or the solid phase, the width of the transition T_u-T_ℓ in the I versus T plot (where T_u and T_ℓ are the temperatures for the onset and termination of the transition in the I versus T plot as temperature is decreased) is smaller than the true width T_F-T_S and T_u and T_ℓ no longer correspond to T_F and T_S respectively. If the probe partitions preferentially into the fluid phase, T_u moves toward T_S, whereas if it partitions preferentially into the solid phase, T_ℓ moves toward T_F. For the phase diagram for which lipids do not cocrystallize, clear breakpoints may not always occur in the I versus T plots at the onset and at the end of the phase transition, even for the case in which K = 1. In addition, inflections in the I versus T graphs may occur within the range of the transition. This suggests that caution should be exercised in assigning transition temperatures to all temperatures at which there is an inflection in such graphs.

Equation (138) can also be used to obtain values K from I versus T plots for two-component bilayers whose phase diagrams have been independently determined. The procedure involves calculation of $\bar{I}$ versus T plots as described above for different values of K and comparison of these theoretical plots with the experimental ones. The experimental plots of I versus T are converted to plots of $\bar{I}$ versus T by extrapolation of I_S and I_F into the region of the phase transition, assuming a dependence of I_S and I_F on temperature and perhaps also on the composition of the fluid and solid phases as indicated

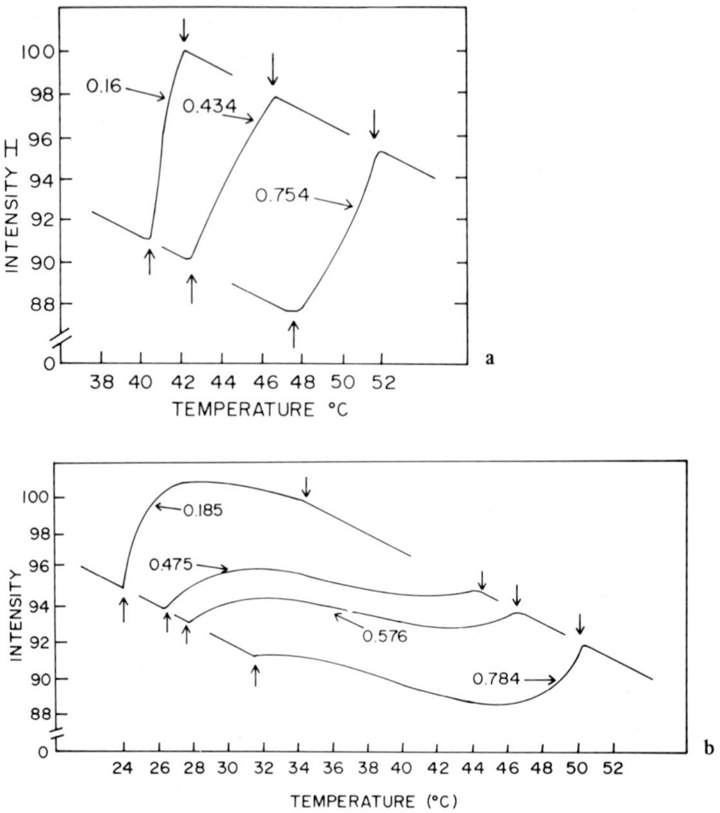

Fig. 23. Examples of theoretical plots of $\bar{I}$ versus T for bilayers with two lipid components for the cases where **(a)** the lipids cocrystallize and **(b)** they do not cocrystallize. Plots in **a** were calculated using the phase diagram for a lipid bilayer composed of dipalmitoyl phosphatidylcholine (DPPC) and distearoyl phosphatidylcholine (DSPC), which cocrystallize. The fractional number next to each plot is the mole fraction of DSPC. *Vertical arrows* show for each mole fraction the temperature at which the transition begins (T_S) and ends (T_L) as determined from the solidus and liquidus graphs in the phase diagram. Plots in **b** were calculated using the phase diagram for a lipid bilayer composed of dimyristoyl phosphatidylcholine (DMPC) and DSPC, which do not cocrystallize. Fractional numbers are for mole fractions of DSPC. For the plots of **a** and **b**, the partition coefficient K was assigned a value of 1 and the slopes of I_S and I_F were set at 0.5% decrease in fluorescence intensity per °C. The magnitude of the transition (change in fluorescence intensity from beginning to end of transition) was selected around 10%. The phase transition temperatures of bilayers of pure DMPC, DPPC, and DSPC are 23°, 41°, and 58°C, respectively [67]

by Eqs. (142) and (143). The value of K is that which gives the best agreement between the theoretical and experimental plots. Figure 24 shows an analysis of experimental data [67].

From analysis of data like that in Fig. 24, it was found that the partition coefficient for perylene in PC and PE bilayers is close to unity. Using a somewhat different method, Thompson et al. [68] found that DPH also partitions between fluid and solid phases with a partition coefficient close to unity. On the other hand, Sklar et al. [69] find that trans-

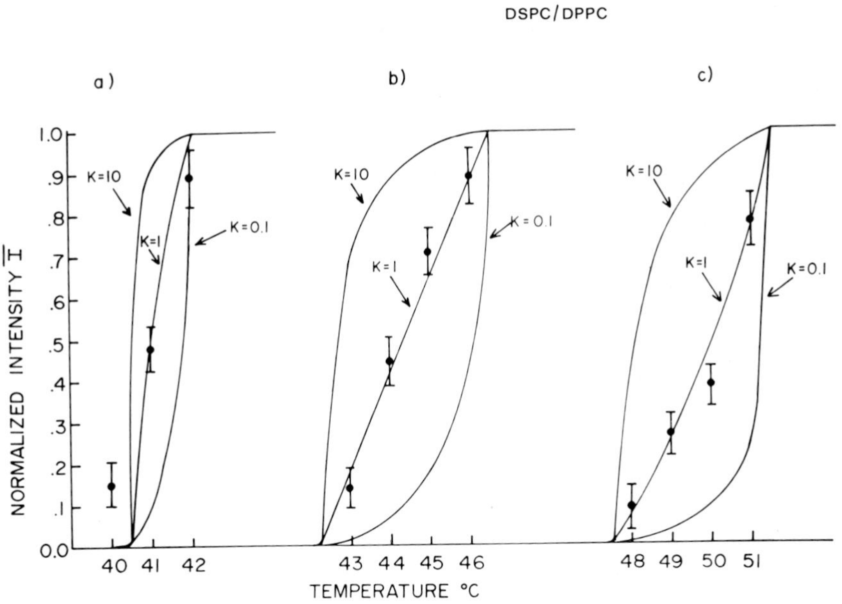

Fig. 24 a-c. Analysis of $\overline{I}$ versus T plots to obtain the partition coefficient K for partitioning of perylene between fluid and solid phases. The experimental points (*black dots* with *error bars*) were taken from I versus T plots for bilayers composed of dipalmitoyl phosphatidylcholine (DPPC) and distearoyl phosphatidylcholine (DSPC). The mole fractions of DSPC are (**a**) 0.16, (**b**) 0.434, and (**c**) 0.754 [67]

parinaric acid preferentially associates with solid phase lipids although cis-parinaric acid partitions with a partition coefficient close to 1. Thus, trans-parinaric acid may be a sensitive probe for detecting solid phases.

Acknowledgment. We would like to express our appreciation to Evangelina Yguerabide for assistance in the preparation of the manuscript. This work was supported by grant numbers USPHY EY 01177-03 (JY) and NSF PCM75-19594 (JY) and USPHY EY-00027-03 (MF).

References

1. Conti, F.: Annu. Rev. Biophys. Bioeng. *4*, 287 (1975).
2. Radda, G.K., Vanderkooi, J.: Biochim. Biophys. Acta *265*, 509 (1972).
3. Weber, G.: Annu. Rev. Biophys. Bioeng. *1*, 553 (1972).
4. Brand, L., Gohlke, J.R.: Annu. Rev. Biochem. *41*, 843 (1972).
5. Azzi, A., in: Methods in Enzymology, Vol. XXXII, part B, p. 234. New York: Academic Press 1974.
6. Becker, R.S.: Theory and Interpretation of Fluorescence and Phosphorescence, New York: Wiley Interscience 1969.
7. Parker, C.A.: Photoluminescence of Solutions, New York: Elsevier 1968.
8. Moore, W.J.: Physical Chemistry, p. 748. New Jersey: Prentice-Hall 1972.

9. Ware, W.R., Lee, S.K., Chow, P.: Chem. Phys. Lett. *2*, 356 (1968).
10. Brand, L., Gohlke, J.R.: J. Biol. Chem. *246*, 2317 (1971).
11. Keuzmann, W.: Quantum Chemistry, Chap. 15. New York: Academic Press 1957.
12. Albrecht, A.C.: J. Mol. Spectroscop. *6*, 84 (1961).
13. McGraw, G.E.: J. Polym. Sci. A-2, *8*, 1323 (1970).
14. Williams, R.J.: J. Chem. Phys. *26*, 1186 (1957).
15. Kauzmann, W.: Quantum Chemistry, Chap. 16. New York: Academic Press 1957.
16. Stryer, L.: Science *162*, 526 (1968).
17. Shinitzky, M., Dianoux, A.A., Gitler, C., Weber, G.: Biochemistry *12*, 521 (1973).
18. Lesslauer, W., Cain, J., Blaise, J.K.: Biochim. Biophys. Acta *241*, 547 (1971).
19. Waggoner, A.S., Stryer, L.: Proc. Natl. Acad. Sci. USA *67*, 579 (1970).
20. Sherwood, D., Yguerabide, J.: unpubl. results.
21. Papahadjopoulos, D., Jacobson, K., Nir, S., Isac, T.: Biochim. Biophys. Acta *311*, 330 (1973).
22. McGrath, A.E., Morgan, C.G., Radda, G.K.: Biochim. Biophys. Acta *426*, 173 (1976).
23. Galla, H., Sackman, E.: Biochim. Biophys. Acta *339*, 103 (1974).
24. Berlin, R.D., Fera, J.P.: Proc. Natl. Acad. Sci. USA *74*, 1072 (1977).
25. Krishnan, K.S., Balaram, P.: FEBS Lett. *60*, 419 (1975).
26. Yguerabide, J., Stryer, L.: Proc. Natl. Acad. Sci. USA *68*, 1217 (1971).
27. Bradley, R.A., Martin, W.G., Schneider, H.: Biochemistry *12*, 268 (1973).
28. Nishijima, Y., Onogi, Y., Asai, T.: J. Polymer. Sci. *15*, Part C, 237 (1966).
29. Desper, C.R., Kimura, I.: J. Appl. Phys. *38*, 4225 (1967).
30. Steinberg, I.Z., in: Biochemical Fluorescence, Chap. 3. New York: Dekker 1975.
31. Yguerabide, J.: unpubl. results.
32. Yguerabide, J.: Methods Enzymol. *26*, 798 (1972).
33. Tao, T.: Biopolymers *8*, 609 (1969).
34. Sherwood, D., Yguerabide, J.: unpubl. results; Sherwood, D.: Ph. D. Thesis, Univ. of California 1974.
35. Kawato, S., Kinosita, K., Jr., Ikegami, A.: Biochemistry *16*, 2319 (1977).
36. Chen, L.A., Dale, R.E., Roth, S., Brand, L.: J. Biol. Chem. *252*, 2163 (1977).
37. Hubbel, W.L., McConnell, H.M.: J. Am. Chem. Soc. *93:2*, 314 (1971).
38. Yguerabide, J.: Spectroscopic and Calorimetric Studies of Viral and Animal Membranes. In: Cell Membranes and Viral Envelopes (eds. Tiffany, J.M., Blough, H.A.). New York: Academic Press 1978.
39. Hare, F., Lussan, C.: Biochim. Biophys. Acta *467*, 262 (1977).
40. Edidin, M.: Annu. Rev. Biophys. Bioeng. *3*, 179 (1974).
41. Singer, S.J.: Annu. Rev. Biochem. *43*, 805 (1974).
42. Yguerabide, J.: Biophys. Soc. Annu. Meet. Abstr. (1971).
43. Freeman, J., Lindenberg, K., Yguerabide, J.: unpubl. results.
44. Yguerabide, J., Walter, R.: to be published.
45. Benson, S.W.: The Foundations of Chemical Kinetics, p. 494. New York, Toronto, London: McGraw-Hill 1960.
46. McGarth, A.E., Morgan, C.G., Radda, G.K.: Biochim. Biophys. Acta *426*, 173 (1976).
47. Rubalcava, B., Munoz, D., Gitler, C.: Biochemistry *8*, 2741 (1969).
48. Fortes, P.A.G., Hoffman, J.F.: J. Membr. Biol. *5*, 154 (1971).
49. Flanagan, M.T., Hesketh, T.R.: Biochim. Biophys. Acta *298*, 535 (1973).
50. Fortes, P.A.G., Yguerabide, J., Hoffman, J.F.: Biophys. Soc. Abstracts, Sixteenth Annual Meeting, 255a. 1972; Fortes, P.A.G.: Ph. D. Thesis, Univ. of Pennsylvania 1972.
51. Yguerabide, J. in: Fluorescence Techniques in Cell Biology (eds. Thaer, A.A., Sernetz, M.), p. 311. New York: Springer 1973.
52. Davies, J.T., Rideal, E.K.: Interfacial Phenomena, p. 75. New York: Academic Press 1963.
53. *Ibid*, p. 183.
54. Haugland, R.P., Yguerabide, J., Stryer, L.: Proc. Natl. Acad. Sci. USA *63*, 23 (1969).
55. Dale, R.E., Eisinger, J. in: Biochemical Fluorescence (eds. Chen, R.F., Edelhoch, H.), Chap. 4. New York: Dekker 1975.

56. For a description of methods for covalent attachment of fluorescent probes to proteins see: Fothergill, J.E., in: Fluorescent Protein Tracing (ed. Nairn, R.C.), 3rd ed., Chap. 2 and 3. Baltimore: Williams and Wilkins 1969.
57. Wu, C.W., Stryer, L.: Proc. Natl. Acad. Sci. USA *69*, 1104 (1972).
58. Wald, G., in: Biochemistry and Physiology of Visual Receptors, p. 1. New York: Springer 1973.
59. Singer, S.J.: Annu. Rev. Biochem. *43*, 805 (1974).
60. Vanderkooi, J.M., Ierokomas, A., Nakamura, H., Martonosi, A.: Biochemistry *16*, 1262 (1977).
61. Veatch, W., Stryer, L.: J. Mol. Biol. *113*, 89 (1977).
62. Bretscher, M.S., Raff, M.C.: Nature *258*, 43 (1975).
63. Shaklai, N., Yguerabide, J., Ranney, H.: Interaction of Hemoglobin with Red Cell Membranes as Shown by a Fluorescent Chromophore. Biochemistry *16*, 5585 (1977).
64. Shaklai, N., Yguerabide, J., Ranney, H.M.: Classification and Localization of Hemoglobin Binding Sites on the Red Blood Cell Membrane. Biochemistry *16*, 5593 (1977).
65. Sheetz, M.P., Singer, S.J.: Proc. Natl. Acad. Sci. USA *71*, 4457 (1974).
66. Chapman, D.: Rev. Biophys. *8*, 2, 185 (1975).
67. Foster, M., Yguerabide, J.: in preparation.
68. Suurkuusk, J., Lentz, B.R., Barenholz, Y., Biltonen, R.L., Thompson, T.E.: Biochemistry *15*, 1393 (1976).
69. Sklar, L.A., Hudson, B.S., Simoni, R.D.: Biochemistry *16*, 819 (1977).
70. Berlman, I.B., in: Handbook of Fluorescence Spectra of Aromatic Molecules, p. 117. New York: Academic Press 1965.
71. Cogan, U., Shinitzky, M., Weber, G., Nishida, T.: Biochemistry *12*, 521 (1973).

Infrared Membrane Spectroscopy

U.P. Fringeli and Hs.H. Günthard

I. Introduction

Application of vibrational spectroscopy to the problem of structure determination of molecules of biological interest goes back to the early uses of raman and infrared spectroscopy in the study of organic molecules. For reviews of earlier work the reader is referred to compilations by Kohlrausch (1943) and by Jones and Sandorfy (1956), whereas more recently a comprehensive discussion has been presented by Bellamy (1975). These compilations accentuate the correlation of vibrational spectra with molecular structure from an essentially empirical point of view and culminate in the establishment of empirical correlation charts. For typical examples the reader is referred to Weast (1974) and Bellamy (1975). There have been many treatments of the theoretical basis of molecular vibrational spectroscopy. Among them the classical work by Herzberg (1945) and by Wilson et al. (1955) should be mentioned. Applications of infrared spectroscopy (IR) to structure problems of biological interest have been summarized by Susi (1969), Fraser and MacRae (1973), and Wallach and Winzler (1974). It was remarked quite eraly that relevant structural information about biological systems often requires study in aqueous solution, which forms the natural environment for most biologically important systems. Besides critical control of experimental conditions and samples the conventional methods of raman spectroscopy may be applied to aqueous solutions in a quite straightforward manner, cf. the contribution by Lord and Mendelson, Chapter 8. The condition of biological environment, i.e., the study in aqueous solutions, by IR spectroscopy is difficult to achieve by conventional absorption technique, since the high absorption coefficient of water in wide regions of the mid and far infrared implies use of thin layers and high concentrations. As a consequence the application of special techniques for measurement of IR spectra of biological material has been a necessity in many cases.

This contribution covers the following topics: (1) specific spectroscopic techniques used in this field, in particular for membrane spectroscopy, (2) discussion of typical results derived from application of IR techniques to model and natural membrane systems and to important constituent molecules of such systems.

Laboratorium für physikalische Chemie, Eidgenössische Technische Hochschule, ETH-Zentrum, 8092 Zürich, Switzerland

II. Experimental Techniques – Spectrophotometric Considerations

Biological applications of IR spectroscopy most often concentrate on absorption measurements of one kind or another, whereas reflection spectroscopy has so far been used only rarely. Absorption measurements are based on determination of transmittance (transmission factor) T or absorbance (absorption factor) A, defined by

$$T = I/I_o, A = 1 - T = (I_o\text{-}I)/I_o \tag{1}$$

where I_o and I denote (IR) radiation intensity incident and transmitted by a layer of sample of thickness ℓ, respectively. For the case of monochromatic radiation of frequency $\tilde{\nu}$, transmittance is related to the absorption coefficient $\alpha(\tilde{\nu})$ and the molar absorption coefficient $\epsilon(\tilde{\nu})$ by the Lambert-Beer Law (c denoting the molarity of the absorbend)

$$T = e^{-\alpha\ell} = e^{-\epsilon c\ell}. \tag{2}$$

Practical measurements often require corrections to obtain true absorption coefficients; for a thorough discussion the reader is referred to Jones and Sandorfy (1956). The nuclear motion of a molecular system consisting of N atoms with fixed orientation in space is in general described by 3N-6 fundamental modes, each of which is approximated by a normal vibration (normal mode). With each normal vibration is associated a normal coordinate, which allows the normal mode to be described as a simple harmonic vibration. Furthermore the interaction of a normal vibration $\tilde{\nu}_s$ with the IR radiation field depends on the quantity $(\partial\vec{\mu}/\partial Q_s)_o$, s = 1, 2, . . . , 3N-6. The vector $(\partial\vec{\mu}/\partial Q_s)$ represents the derivation of the molecular electric dipole moment with respect to the normal coordinate Q_s.

The integrated absorption coefficient of an infrared band

$$A = \int_0^\infty \epsilon(\tilde{\nu})d\tilde{\nu} = -(1/c\ell) \int_0^\infty 1nT(\tilde{\nu})d\tilde{\nu} \tag{3}$$

is related to the "oscillating dipole moment" $(\partial\vec{\mu}/\partial Q)_o$ of the fundamental mode associated with the absorption band by

$$A \propto |\vec{E}_o, (\partial\vec{\mu}/\partial Q)_o|^2 = E_o^2 || (\partial\vec{\mu}/\partial Q)_o|^2 \cos^2(\vec{E}_o, (\partial\vec{\mu}/\partial Q)_o). \tag{4}$$

The vector $\vec{E}_o$ denotes the amplitude of the electric field strength acting on the molecule.[1,2] The last term in Eq. (4) expresses explicitely the dependence of the (integrated) absorption coefficient on the polarization direction of the electric field vector with respect to the oscillating dipole, the latter being conveniently referred to the molecular frame system, i.e., a coordinate system rigidly connected to the nuclear configuration (r_e-structure) of the molecular system. It is customary to relate normal modes (fundamentals) to group vibrations. Group vibrations are normal modes which encompass nu-

1 This expression rests on the assumption that the wavelength of the radiation field is large in comparison with the molecular dimensions.

2 The quantity $|(\partial\vec{\mu}/\partial Q_S)_o|$, denoted here by "oscillating dipole" or "electric dipole moment" derivative with respect to the normal mode s, is often referred to as "transition moment", cf. Zbinden (1964), Susi (1969), and Fraser and MacRae (1973).

clear motions of typical partial nuclear configurations, i.e., atomic groups occurring often in organic molecules like CH_3-, -CH_2-, keto-, carboxylic and amide groups, etc.

Among the 3N-6 normal modes only some may be related to group modes, usually a considerable number are to be regarded as typical for the molecule as an entity. The frequencies of the latter form a unique spectrum characteristic of a molecule and allow simple identification (fingerprint, skeletal modes). They do not in general allow straightforward conclusions regarding molecular structure. In contrast the group modes allow detection of the presence of specific atomic groups. Furthermore the oscillating dipole moment ($\partial\vec{\mu}/\partial Q$) of such modes nearly always bears a fixed relation to the nuclear configuration of the group involved, a property which is often used to derive information about the orientation of specific groups within a complex molecular structure by means of polarization measurements. Information contained in skeletal modes may be obtained to a considerable extent by normal coordinate analysis (NCA). The NCA consists in the calculation of the normal frequencies $\tilde{\nu}_1, \tilde{\nu}_2, \ldots, \tilde{\nu}_{3N-6}$ based on a geometrical molecular model and the harmonic part of the molecular potential function. For molecules of biological interest the NCA may be carried out only numerically by appropriate computer programs, cf. Herzberg (1945); Wilson et al. (1955); Schachtschneider and Snyder (1963); Hunziker (1965); Warschel and Levitt (1975); and Rihak (1979). Both structural information and harmonic force constants are transferred from small molecules with related structural elements. In this contribution NCA results will be referred to in connection with specific membrane constituents. For certain group vibrations the normal mode approximation of isolated molecules has been found inadequate and must be extended in several directions. First, intermolecular interactions often must be taken into account for biological molecules, in particular hydrogen bonding and electric dipole-dipole interactions. In most examples treated so far perturbations of this type may be taken into account within the framework of NCA, if in the latter all noticeable interacting systems are included. Second, Fermi resonance often complicates the NCA concept, a phenomenon related to anharmonicity of the intramolecular potential function, cf. Herzberg (1945). Fermi resonance manifests itself often with typical group vibrations. It may play an important role in the interpretation of both frequency and intensity of vibrational spectra of even simple model systems, cf. Hollenstein and Günthard (1974) and Kühne and Günthard (1976). Unfortunately Fermi resonance in small molecule spectra ($N \leqslant 10$) is still only little studied, in spite of its practical importance. Fermi resonances connected with N-H stretching of the amide group have been discussed by Miyazawa (1960b).

III. Application of IR Spectroscopy to Biological Systems

A. Restrictions

First some facts influencing IR spectroscopy of biological systems should be mentioned:

1. Complexity of Molecules

Since 3N-6 normal vibrations are expected in the general case (C_1-symmetry), a lipid molecule consisting of 130 atoms will result in 384 normal vibrations. Furthermore, this number may be enlarged by overtones, combination tones, and band splitting due to intermolecular interactions. The main problems now arising are: (1) the overlapping of absorption bands, which complicates in many cases the identification of single absorption bands, (2) the complexity of typical molecular motions, which hinders the assignment in terms of group vibrations.

2. Noncrystallinity of Biological Samples

At ambient temperature biological membranes as well as model membranes consisting of natural components are liquid crystalline. This results in a considerable loss of order of the hydrocarbon chains. The influence on the structure of the polar headgroup of phospholipids was found to depend strongly on the nature of this group. The liquid crystalline state may lead to broadening of absorption bands as well as to a certain loss of polarization, which is clearly demonstrated by the spectra of L-α-dipalmitoyl phosphatidylethanolamine (crystalline) and sheep brain cephalin (liquid crystalline) (Fringeli, 1977).

3. Influence of Liquid Water

As has been mentioned above, liquid water has some very strong and broad absorption bands, overlapping a considerable part of the infrared spectral range; see Table 1. The low transmittance of even thin water layers is the principal reason why most infrared investigations of biological materials have been performed with dry samples. However, it

Table 1. Infrared vibrational spectra of H_2O and D_2O (Wallach and Winzler, 1974)

Vibration	H_2O	D_2O
O-X stretching (ν_S)	Very broad band with two main maxima and a shoulder (sh) at 25°C	
	ν = 3920 sh	ν = 2900 sh
	ϵ = 0.83	ϵ = 0.60
	ν = 3490	ν = 2540
	ϵ = 62.7	ϵ = 59.8
	ν = 3280	ν = 2450
	ϵ = 54.5	ϵ = 55.2
Association (ν_A)	ν = 2125	ν = 1555
	ϵ = 3.23	ϵ = 1.74
X-O-X′ bending (ν_2)	ν = 1645	ν = 1215
	ϵ = 20.8	ϵ = 16.1
Libration (ν_L)	Broad band between 300 cm^{-1} and 900 cm^{-1}	
Hindered translation (ν_T)	Prominent shoulder on ν_L band at ~ 190 cm^{-1}; 30°C	

ν, frequency of band maximum in cm^{-1};
ϵ, extinction coefficient x 10^{-3} cm^2 mol^{-1} at absorption maximum

is quite obvious that investigations in aqueous environment are of greater biological relevance, since drastic conformational changes of peptides may occur when the membrane hydration changes (Fringeli and Fringeli, 1979). On the other hand the use of aqueous systems implies measurements with thin-layer cells and high concentrations of solute. Both conditions lead to specific souces of errors. Jones and Sandorfy (1956) have shown that thin-layer measurements may require considerable corrections to yield true absorption coefficients. Furthermore many types of biological molecules tend to association at high concentration.

4. Spectrometer Sensitivity Requirements for Biomembrane Spectroscopy

The following considerations are based on stearic acid mono- and bilayer measurements, cf. Fig. 20. They may be regarded as typical for most lipid mono- and bilayer systems. Since biomembranes are very thin, < 100 Å, the amount of membrane components is so small that the sensitivity requirement for direct absorption measurement is unusually high. From mono- and bilayer spectra of stearic acid (Fig. 20 a-d) one may estimate that the transmission of a monomolecular layer is about 95% for the $\delta(CH_2)$ band, which should be considered as absorption band of medium intensity. This finding is in good agreement with the values derived from Akutsu et al. (1975), where 150 monolayers were used to get sufficient intensity in a transmission experiment. For the sake of comparison it should be mentioned that the transmission accuracy of conventional high-quality spectrometers is about .1%.

B. Application of Special Infrared Techniques

In order to overcome some of the difficulties mentioned in Sect. III.A it is necessary to apply more advanced infrared techniques, which will be depicted below.

1. Attenuated Total Reflection Techniques (ATR)

For a detailed description of this technique the reader is referred to Harrick (1967). Figure 1 shows a schematic comparison of the conventional (a) transmission and the internal reflection (b) technique. The latter is achieved by placing the sample material in close contact with the optically transparent internal reflection element of higher refractive index and working above the critical angle θ_c. From Fresnel's equation it follows that an electromagnetic field exists in the rarer medium beyond the reflecting interface, even under conditions of total reflection. This field exhibits the frequency of the incoming light, but the amplitude falls off exponentially with distance z from the surface, cf. Figs. 1 and 2, i.e.,

$$E = E_o e^{-z/d_p} \quad (5)$$

where d_p denotes the depth of penetration and is given by

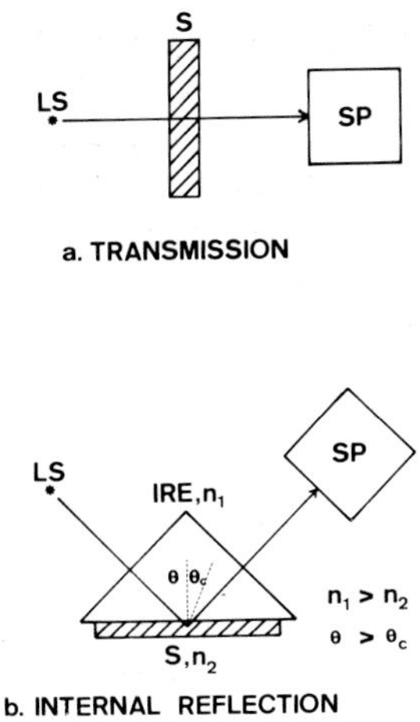

Fig. 1 a and b. Comparison of conventional transmission (**a**) and internal reflection (**b**) technique.
LS, light source; S, sample; SP, spectrometer; IRE, internal reflection element; n_1, refractive index of IRE; n_2, refractive index of S; ϑ, angle of incidence; ϑ_c, critical angle (Harrick, 1967)

$$d_p = \frac{\lambda_1}{2\pi(\sin^2\vartheta - n_{21}^2)^{\frac{1}{2}}} \tag{6}$$

where n_{21} stands for the ratio n_2/n_1 of the refractive indices of the rarer medium (n_2) and of the internal reflection element (n_1). λ_1 denotes the wavelength λ_{vacuum}/n_1 in the latter. According to Eq. (6) the penetration depth amounts to the order of magnitude of the wavelength λ_1 of the infrared radiation. This allows IR absorption measurements of thin layers, provided optical contact between rarer medium (sample) and ATR plate may be established, e.g., by direct deposition of mono- or multilayers. Under this condition

$$\left(\frac{A_{sample}}{A_{H_2O}}\right)_{ATR} \gg \left(\frac{A_{sample}}{A_{H_2O}}\right)_{transmission}.$$

This means that ATR measurements of aqueous systems are less impeded by water absorption than transmission experiments. Recently, Hofer and Fringeli (1979) have shown that in aqueous environment membrane adhesion at a hydrophilic ATR plate (Ge) requires a pretreatment of its surface. Best results have been obtained by coating the surface with a monomolecular layer of propylaminosilane. This layer can also be used for chemical immobilization of enzymes and other membrane constituents. A further advantage of ATR is offered through the possibility of relatively simple multiple-reflection measurement. Depending on incidence angle ϑ, thickness, and length of the ATR element, 30-100 reflections may be achieved by standard-size ATR elements.

Polarized ATR Measurements. In studying oriented layers, ATR technique allows measurements with polarized radiation in a fairly straightforward way. Figure 2 depicts

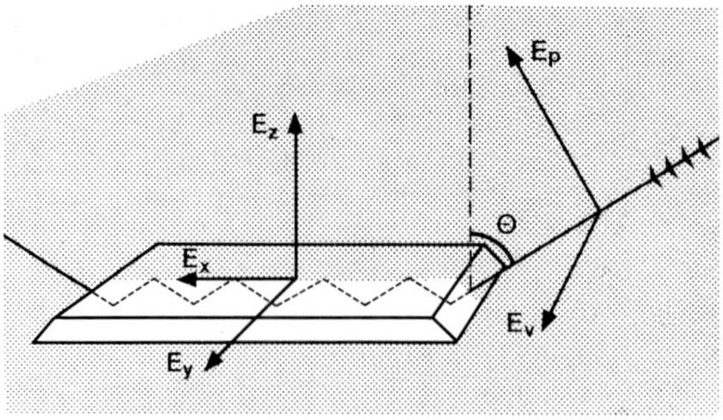

Fig. 2. ATR set-up. ϑ, angle of incidence; E_p, E_v, parallel and perpendicular polarized components of the electric field of incident light; E_x, E_y, E_z, electric field components with respect to the coordinate system corresponding to the internal reflection plate ($E_p \rightarrow E_x, E_z$, $E_v \rightarrow E_y$) (Kopp et al., 1975a)

the experimental set-up schematically. If the incident radiation is ||-(⊥)-polarized, the probe field in the rarer medium is x,z-(y)-polarized. More explicitly the amplitudes of latter fields are approximated in the case of weak absorption for thin rarer-medium layers by: (Harrick, 1967)

$$E_{yo} = E_{\perp} = \frac{2\cos\vartheta}{(1-n_{31}^2)^{\frac{1}{2}}}$$

$$E_{xo} = \frac{2\cos\vartheta\,(\sin^2\vartheta - n_{31}^2)^{\frac{1}{2}}}{(1-n_{31}^2)^{\frac{1}{2}}[(1+n_{31}^2)\sin^2\vartheta - n_{31}^2]^{\frac{1}{2}}} \tag{7}$$

$$E_{zo} = \frac{2\cos\vartheta\, n_{32}^2 \sin\vartheta}{(1-n_{31}^2)^{\frac{1}{2}}[(1+n_{31}^2)\sin^2\vartheta - n_{31}^2]^{\frac{1}{2}}}$$

and for bulk rarer medium

$$E_{yo} = E_{\perp} = \frac{2\cos\vartheta}{(1-n_{21}^2)^{\frac{1}{2}}}$$

$$E_{xo} = \frac{2\cos\vartheta(\sin^2\vartheta - n_{21}^2)^{\frac{1}{2}}}{(1-n_{21}^2)^{\frac{1}{2}}[(1+n_{21}^2)\sin^2\vartheta - n_{21}^2]^{\frac{1}{2}}} \tag{8}$$

$$E_{zo} = \frac{2\sin\vartheta\cos\vartheta}{(1-n_{21}^2)^{\frac{1}{2}}[(1+n_{21}^2)\sin^2\vartheta - n_{21}^2]^{\frac{1}{2}}}\ .$$

n_{ik} denotes the ratio of refractive indices n_i/n_k of medium i and k, where i,k = 1, 2, and 3 stand for ATR plate, thin films, and surrounding medium, respectively. For bulk rarer-medium index 3 must be replaced by 2, thus leading to the expressions (8). The total electric field amplitude for parallel polarization (||,pp) is given by

$$E_{\parallel} = (E_x^2 + E_z^2)^{\frac{1}{2}}\ . \tag{9}$$

Molecular orientation is determined via the measurement of the dichroic ratio R, which is defined as the ratio of the absorption coefficients of ||-(pp)- and ⊥-(vp)-polarized light. Introducing Eq. (4), R is given by

$$R = \frac{A_{\|}}{A_{\perp}} = \frac{(\vec{E}_{\|}, \frac{\partial\vec{\mu}}{\partial Q})^2}{(\vec{E}_{\perp}, \frac{\partial\vec{\mu}}{\partial Q})^2} . \tag{10}$$

For membrane layers the molecular orientation is not distributed statistically. Experience shows that most often one of the molecular axis is symmetrically distributed around one space-fixed direction enclosing an angle γ. Moreover in many cases the oscillating dipole moment deviates by an angle θ from the molecular axis and is cylindrically distributed with regard to the latter. Assuming the z-axis as a space-fixed direction one may derive for the dichroic ratio:

$$R_z^{ATR} = \frac{E_x^2}{E_y^2} + \frac{E_z^2}{E_y^2} \; \frac{2\cos^2\theta + S}{\sin^2\theta + S} . \tag{11}$$

The reader is referred to Zbinden (1964) and Fraser and MacRae (1973) for details of the derivation, which however relates to the case of transmission IR spectroscopy. Extension to ATR is straightforward. It is customary to introduce the order parameter S, defined by

$$S = \frac{F}{N - \frac{3}{2}F} \tag{12}$$

where

$$F = \langle\sin^2\gamma\rangle = \int_0^{\pi/2} f(\gamma)\sin^2\gamma d\gamma, \; N = \langle f(\gamma)\rangle = \int_0^{\pi/2} f(\gamma)d\gamma = 1. \tag{13}$$

$f(\gamma)$ stands for the first density of the distribution of the angle γ. Model distributions $f(\gamma)$ have been discussed by Kratky (1933), Fraser (1958), and Zbinden (1964). The order parameter S becomes zero for perfect ordering and infinity for random distribution. For the sake of comparison it should be mentioned that the parameter S' generally used in EPR and NMR spectroscopy is defined by

$$S' = \frac{3}{2}\int_0^{\pi/2} \cos^2\gamma f(\gamma)d\gamma - \frac{1}{2} . \tag{14}$$

S' becomes unity for perfect ordering and zero for random distribution. However, there is a simple expression relating Eq. (14) to Eq. (12):

$$S' = \frac{1}{1 + \frac{3}{2}S} . \tag{15}$$

For a more general discussion of orientation measurement, the reader is referred to Fringeli (1981b).

2. Modulated Excitation ATR Spectroscopy (ME-ATR)

Application of this special kind of difference spectroscopy leads in many cases to an essential reduction of the diffuculties mentioned in Sects. III.A.1 and 2. The technique of modulation spectroscopy can always be applied provided the system studied allows periodic excitation. It enables selective scanning of those absorption bands of the infrared spectrum resulting from molecules (or parts of them) that are involved in the stimulation process. All absorption bands which are not affected by excitation are suppressed. The dynamics of the periodically excited reaction can be investigated by measuring the phase angle ϕ between stimulated and detected signal. ϕ can easily be calculated from tg $\phi = A_{90}/A_0$, where A_{90} and A_0 denote the absorption coefficients measured in the 90°- and 0°-phase channels, respectively. Modulation spectroscopy has been applied in several fields of physical chemistry by Hexter (1963); Labhart (1964); Johnston et al. (1967); Fringeli (1969); Fringeli and Günthard (1971); and Günthard (1974). A diagram of a ME-ATR spectrometer has been published by Fringeli et al. (1976) in a study of the dynamics of reorientation of liquid crystal molecules under the influence of an external electric field. A detailed kinetic analysis of a photochemical reaction using UV light for excitation was given by Forster et al. (1976). Since the ME spectrum consists exclusively of absorption bands resulting from molecules (or parts of them) which are involved in the excited process, this technique offers a chance for studying complex systems such as biological model membranes or membranes of living cells. The former has been reported by Fringeli and Günthard (1976) and Fringeli and Hofer (1980), the latter by Sherebrin et al. (1972).

C. Summary of Group Vibrations Applicable for Diagnostic Purpose in Membrane Spectroscopy

For a general discussion of group vibrations, the reader is referred to the extensive work by Bellamy (1975) and by Jones and Sandorfy (1956). At this place we list only some selected group vibrations which have turned out to be useful in infrared membrane spectroscopy. Table 2 contains group vibrations of phospholipids, whereas Table 3 deals with group vibrations occurring in polypeptides and proteins.

IV. Review of Results of IR Membrane Spectroscopy

A. Phospholipids

Discussion of spectroscopic results will be presented as follows. The first section will be devoted to various methods for preparation of lipid layers. Stability and structure of such layers largely depend on chemical nature, method of preparation, number of layers,

Table 2. Group vibrations of diagnostic importance in membrane Spectroscopy

Functional group	Description of vibration	Symbol	Wavenumber $[cm^{-1}]$ (intensity)	Estimated direction of dipole moment	Comments
C-CH_3				In the plane	
	Antisym. stretching	ν_{as} (CH_3)	2962 ± 10 (s)	⊥ to the C-C bond	Bellamy (1975)
	Sym. Stretching	ν_s (CH_3)	2872 ± 10 (s)	‖ to the C-C bond	Bellamy (1975)
				In the plane	
	Antisym. bending	δ_{as} (CH_3)	1450 ± 20 (m)	⊥ to the C-C bond	Bellamy (1975)
	Sym. Bending	δ_s (CH_3)	1375 ± 5 (s)	‖ to the C-C bond	Bellamy (1975)
$(-CH_2-)_n$	Antisym. stretching	ν_{as} (CH_2)	2926 ± 10 (s)	⊥ to the bisector of the H-C-H angle	Bellamy (1975)
	Sym. stretching	ν_s (CH_2)	2853 ± 10 (s)	‖ to the bisector of the H-C-H angle	Bellamy (1975)
	CH_2-bending	δ (CH_2)	1465 ± 20 (m)	‖ to the bisector of the H-C-H angle	Bellamy (1975
			⩾ 1470		Hydrocarbon chain in all-trans configuration. Bandsplitting in crystals may occur. Münch et al. (1977)
	CH_2-wagging and CH_2-twisting	γ_w (CH_2) γ_t (CH_2)	1180-1345	‖ to the hydrocarbon chain (all-trans)	Band progression (w) only observed in saturated hydrocarbon chains with all-trans configuration. Number of bands depends on the number of CH_2 groups. Meiklejohn et al. (1957)
	CH_2-rocking	γ_r (CH_2)	720-(~ 1000) (m)	⊥ to the bisector of the H-C-H angle	Band progression is only observed when the hydrocarbon chain assumes all-trans configuration. The 720 cm^{-1} (m) band is always observed, but becomes broader when the chain deviates from all-trans. Bandsplitting in crystals may occur. Münch et al. (1977)

Table 2 cont.

Functional group	Description of vibration	Symbol	Wavenumber $[cm^{-1}]$ (intensity)	Estimated direction of dipole moment	Comments
	C-C-stretching	ν (C-C)	1150 ± 50 1080 ± 50	∥ to the chain direction ⊥ to the chain direction	Only two vibrations are expected to be infrared active in the infinite chain with all-trans conformation. Zbinden (1964)
R(H)C=C(H)R′	C=C-stretching	ν (C=C)	1600-1680 (v)	∥ to the C=C double bond	Bellamy (1975)
	CH antisym. stretching	ν_{as} (CH=CH)	~ 3010 (m)	∥ to the C=C double bond	Bellamy (1975)
$-CH_2-C(=O)-O-R$	CH_2-bending of α-methylene group	δ (α-CH_2)	~ 1420 (w)	∥ to the bisector of the H-C-H angle	Jones (1962)
	C-O single bond stretching	ν (C-O)	1160-1190 (m)	~ ⊥ to the C=O double bond	Band located near 1180 cm^{-1} in the case of a planar C-C(=O)-O-C frame, otherwise shifting to lower frequency occurs. Fringeli et al. (1972)
	C=O double bond stretching	ν (C=O)	1720-1745 (s)	∥ to the C=O double bond	
R-C(=O)-OH	OH-stretching	ν (OH)	3500-3560 (m) 2500-2700 (w)	~ ∥ to the OH-bond	Free carboxylic acid group Hydrogen bonded carboxylic acid group
	C=O double bond stretching	ν (C=O)	1700-1725 (s)	~ ∥ to the C=O double bond	Band near 1700 cm^{-1} in hydrogen-bonded carboxylic acid groups. Bellamy (1975)
	OH-bending	δ (OH)	1210-1320 (s)		Band observed in long-chain fatty acids. OH-bending is involved

R-C(=O)-O$^-$	Antisym. and sym. C=O double stretching	ν_{as} (COO$^-$) ν_s (COO$^-$)	1550-1610 (s) 1300-1420 (m)	~ ⊥ and ~ ∥ to the bisector of the (O=C-O$^-$) angle, respectively	Bellamy (1975)
R-OH	OH-stretching	ν (OH)	3590-3650 (v) 3200-3400 (v.s) 3450-3570 (v)	∥ to the OH-bond	Free hydroxyl group Intermolecular H bonding, broad absorption bands Intramolecular H-bonding. Strong intramolecular H-bonding may shift ν (OH) to considerably lower frequencies. Bellamy (1975)
	OH-bending	δ (OH)	1200-1400 (m)	~ ⊥ to the OH-bond	Strong H-bonding shifts the band to the high wavenumber limit of the range
	CO-stretching	ν (CO)	~ 1050 (m,s) ~ 1100 (m,s) ~ 1150 (m,s)	~ ∥ to the CO-bond	Primary alcohols Secondary alcohols Tertiary alcohols } with δ(OH) partially involved Bellamy (1975)
R-NH$_2$	NH$_2$-stretching	ν_{as} (NH$_2$) and ν_s (NH$_2$)	3000-3500 (m)		Two or more resolved bands in the crystalline state Strong H-bonding results in shifts to lower wave-numbers
	NH$_2$-bending	δ (NH$_2$)	1590-1650 (m)	~ ∥ to the bisector of the NH$_2$-group	
	C-N-stretching	ν (C-N)	1020-1220 (m,w)	~ ∥ to the CN-bond	
R-NH$_3^+$	Antisym. NH$_3^+$-stretching	ν_{as} (NH$_3^+$)	~ 3200 (m)	In the plane ⊥ to the CN-bond	Assignments are mainly based on l-aminohexadecane · HCl. Fringeli (1977)
	Sym. NH$_3^+$-stretching	ν_s (NH$_3^+$)	~ 3020 (m)	∥ to the CN-bond	
	Antisym. NH$_3^+$-bending	δ_{as} (NH$_3^+$)	1570-1620 (w,m)	In the plane ⊥ to the CN-bond	

Table 2 cont.

Functional group	Description of vibration	Symbol	Wavenumber $[cm^{-1}]$ (intensity)	Estimated direction of dipole moment	Comments
	Sym. NH_3^+-bending	$\delta_S (NH_3^+)$	~ 1520 (s)	∥ to the CN-bond	
	C-N-stretching	ν (C-N)	~ 1090 (w)	∥ to the CN-bond	
	Combination of sym. NH_3^+-bending and CCN-bending	$\delta_S (NH_3^+) + \delta$ (CCN)	~ 2050 (w)		This band is typical for the existence of the $-NH_3^+$ group. It is observed in the IR spectra of many aminoacids. Weast (1975)
R(R')NH	NH-stretching	ν (NH)	3300-3500 (m)	∥ to the NH-bond	Strong H-bonding may shift the band to lower wavenumbers. Bellamy (1975)
	NH-bending	δ (NH)	1520-1650 (w)	⊥ to the NH-bond	
	Antisym. (N)-CH_3-bending	$\delta_{as} (CH_3)$	1470-1510 (m)	In the plane ⊥ to the CN-bond	
	Antisym. CNC-stretching	ν_{as} (CNC)	1020-1040 (m)	~ ⊥ to the bisector of the CNC-angle	Dellepiane and Zerbi (1968) Buttler and McKean (1965)
	Sym. CNC-stretching	ν_S (CNC)	920- 990 (m)	~ ∥ to the bisector of the CNC-angle	Fringeli (1977)
R(R')(R'')N	Antisym. (N)-CH_3-bending	$\delta_{as} (CH_3)$	1470-1510 (m)	⊥ or ∥ to the local C_3-axis depending on symmetry type	A_1- and E-type modes for R = R' = R'' = CH_3
	Antisym. CN-stretching	ν_{as} (CN)	940-1040 (m)	~ ⊥ to the local C_3-axis	E-type mode for R = R' = R''
	Sym. CN-stretching	ν_S (CN)	800- 900	∥ to the local C_3-axis	A_1-type mode for R = R' = R''. Each E-type mode splits into two bands, when the C_{3v}-symmetry is lowered

$-N(CH_3)_3^+$	Antisym. CH_3-stretching	ν_{as} (CH_3)	3020-3030 (w,m)		
$(-N(CD_3)_3^+)$	Antisym. CH_3-bending	$\delta_{as}^{A_1}$	1480-1490 (m)	‖ to the local C_3-symmetry axis	A_1-type mode
	Antisym. CH_3-bending	δ_{as}^{E}	1470-1480 (m)	In the plane ⊥ to the local C_3-axis	E-Type mode, Fringeli (1977)
	Sym. CD_3-bending	$\delta_{s}^{A_1}$	965- 980 (w,m)	‖ to the local C_3-axis	A_1-type mode
	Sym. CH_3-bending	$\delta_{s}^{A_1}$	1395-1405 (m)	‖ to the local C_3-axis	A_1-type mode (not observed in lecithin) C_{3v}-sym. of $-N(CH_3)_3^+$ assumed
	Antisym. $N-(CH_3)_3^+$-stretching	δ_{as} $(N-CH_3)$	950- 970 (m)		Two absorption bands
	Sym. $C-N-(CH_3)_3^+$-stretching	ν_s $(N-CH_3)$	920- 930 (w,m) 875- 895 (w,m)	‖ to the local C_3-symmetry axis $(C-N(CH_3)_3^+$-bond)	Trans conformation of O-C-C-N frame Gauche conformation of choline ν(C-N) is in antiphase to the sym. $N-(CH_3)_3^+$-stretching
	Totally sym. C-N-stretching	ν_s (CN)	750- 770 (vw) 695- 715 (vw) 710- 720 (vw) 665- 685 (vw)		Trans conformation d_0-isotope Trans conformation d_9-isotope Gauche conformation d_0-isotope Gauche conformation d_9-isotope Rihak (1979)
R-O-P(=O)(O⁻)-O-R	Antisym. PO_2^--double bond stretching	ν_{as} (PO_2^-)	1220-1260 (s)	⊥ to the bisector of the O . . . P . . . O angle	Shimanouchi et al. (1964)
	Sym. PO_2^--double bond stretching	ν_s (PO_2^-)	1085-1110 (m)	‖ to the bisector of the O . . . P . . . O angle	
	Antisym. PO single bond stretching	ν_{as} $(P(OR)_2)$	815- 825 (m,s)		CO single bond stretching is significantly involved Fringeli (1977)

Table 2 cont.

Functional group	Description of vibration	Symbol	Wavenumber $[cm^{-1}]$ (intensity)	Estimated direction of dipole moment	Comments
	Sym. PO single stretching	ν_S $(P(OR)_2)$	755- 765 (m,s)		
	C-O-stretching	ν (C-O)	1040-1090 (m,s)		PO single bond stretching involved
O II -C-CH-NH- I R	Amide group vibrations				cf. Table 3

Table 3. Amide group vibrations[a] (Fraser and MacRae, 1973)

Designation		Approximate frequency[b] (cm^{-1})	Description
Amide	A	3300	Result of Fermi resonance between the first excited state of the N-H-stretching mode and the second excited state of the amide II vibration[c]
	B	3100	
Amide	I	1650	In-plane modes[d]; the potential energy distributions are such that none is capable of simple description although amide I approximates a C=O-stretching mode
	II	1550	
	III	1300	
	IV	625	
Amide	V	650	Out-of-plane modes[e]; amide V involves an N-H-bending motion, amide VI a C=O-bending motion, and amide VII a torsional motion about the C-N bond
	VI	550	
	VII	200	
		4600	Combination of amide A and amide III[f]
		4860	Combination of amide A and amide II[g]
		4970	Combination of amide A and amide I[f]

[a] Trans configuration
[b] The frequencies are sensitive to environment
[c] Badger and Pullin (1954); Miyazawa (1960a,b); Tsuboi (1964)
[d] Miyazawa et al. (1956, 1958); Miyazawa (1962, 1963, 1967)
[e] Kessler and Sutherland (1953); Miyazawa (1961, 1963, 1967)
[f] Fraser (1956)
[g] Glatt and Ellis (1948)

and further parameters. In a large part of the published work perfect and stable layer structure has apparently been assumed and many studies have been reported without information on preparation and structure. Experience with lipid layer systems has however shown that such information may be relevant. In the remainder of this chapter spectroscopic studies of particular phospholipid layer systems will be discussed. Three different types of ultrastructures of layer systems will be important:

(i) Microcrystalline Ultrastructure (MCU). The layer is composed of submicroscopic domains, each of which is characterized by fixed values of the polar angles γ, φ of molecular orientation with regard to the coordinate system depicted in Fig. 2. The ensemble is described by a fixed value of γ and isotropically distributed angle φ, i.e., isotropic molecular arrangement around the z-axis. Therefore the distribution function becomes $f(\gamma) = \delta(\gamma - \gamma_o)$, where γ_o denotes the angle between the molecular axis and the z-axis. A number of applications to molecular structure determination of biological model membranes are given by Fringeli (1977) and Fringeli and Fringeli (1979).

(ii) Liquid Crystalline Ultrastructure (LCU). The layer is an ensemble of molecules whose long molecular axes feature a narrow distribution with regard to γ and an isotropic distribution with regard to φ. A reasonable distribution function for this case is that used by Kratky (1933) to describe molecular ordering of polymers by stretching. The function may be expressed as below (Zbinden, 1964):

$$f(\gamma) = \frac{v^{3/4} \sin\gamma}{(v^{-3/2}\cos^2\gamma + v^{3/2}\sin^2\gamma)^{3/2}}$$

where v denotes the extension ratio describing the degree of ordering. v = 1 means isotropic distribution, i.e., $f(\gamma) = \sin\gamma$, and $v = \infty$ means perfect ordering.

(iii) Oriented Crystalline Ultrastructure (OCU). The layers are built from an ensemble of molecules with fixed angles γ and φ.

1. Preparation of Oriented Bilayers

The preparation of oriented multilayer assemblies by means of the Langmuir-Blodgett dipping technique has been reviewed, e.g., by Kuhn et al. (1972). Unfortunately this elegant technique may lead to difficulties in many cases where phospholipids should be transferred from the water-air interface to a solid support such as an ATR plate. Using a hydrophilic surface the first layer should attach during withdrawing of the element, the second by redipping, etc. However, in the case of phospholipids the first layer generally detaches by redipping. Nevertheless Procarione and Kauffman (1974) and Akutsu et al. (1975) have reported results obtained from ordered phospholipid multilayers of Langmuir-Blodgett type (y-type layers), cf. Blodgett and Langmuir (1937). On the other hand Levine et al. (1968), Green et al. (1973) and Hitchcock et al. (1975) have built up multilayer systems of phospholipids and polypeptides (Green et al. (1973) by means of Langmuir-Blodgett x-type films, i.e., multimonolayer systems with head-tail-head-tail arrangement of lipid molecules (cf. Blodgett and Langmuir, 1937)). Since X-ray analysis leads to a bilayer structure, recrystallization of the built-up system must necessarily have taken place (cf. Sect. IV.A.10). Furthermore the procedure for monolayer transfer applied by these authors seems to be rather drastic. First, dipping of the solid support through the monomolecular spread film was performed within 1 second and leads unavoidably to local destructions of the spread monolayer. According to the authors this step should produce transfer of one monolayer. Second, withdrawal of the plate took about 10 minutes. No information about monolayer stability at the air-water interface and the transfer ratio was given, i.e., on the ratio of the area of the monolayer on the solid substrate to the area of the corresponding monolayer spread on water under transfer conditions. This ratio should be close to unity for perfect transfer.

In view of the facts mentioned above and of the crystallization phenomena described in Sect. IV.A.10, it must be expected that application of the Langmuir-Blodgett technique has led to microcrystalline systems (MCU) rather than to regular multibilayer assemblies. Another technique has been applied in this laboratory to achieve oriented phospholipid layers. The phospholipid is dissolved ($\sim 10^{-3}$ M) in chloroform. A drop of the solution ($\sim 50\ \mu l$) is put on one side ($\sim 10\ cm^2$) of an ATR reflection plate. Now a small Teflon bar is placed on the plate in such a way that the drop is spread by capillary action between the ATR plate and the bar. Slowly moving the bar along the plate several times until the solvent has evaporated leads to multilayer formation. Homogeneity of the sample can be checked by observing interference colors. Samples spread on a Ge surface, for instance, showed a homogeneous yellow-brownish color over the entire surface, corresponding to a mean thickness of 4-6 bilayers, cf. Fringeli and Günthard (1976) and

Fringeli (1977). However, it should be mentioned that the nature of the solvent from which the membrane is prepared by evaporation may have a significant influence on the molecular structure of the model membrane. This fact must be taken into account, especially in the presence of polypeptides and proteins. Recently Fringeli and Fringeli (1979) have shown that contact with liquid water is a prerequisite for the incorporation of the peptide antibiotic alamethicin into lecithin bilayers.

2. Homologous Series: L-α-Dipalmitoyl Phosphatidylethanolamine to Phosphatidylcholine (L-α-DPPE to L-α-DPPC)

The infrared spectra of these four synthetic phospholipids are shown in Figs. 3-6.

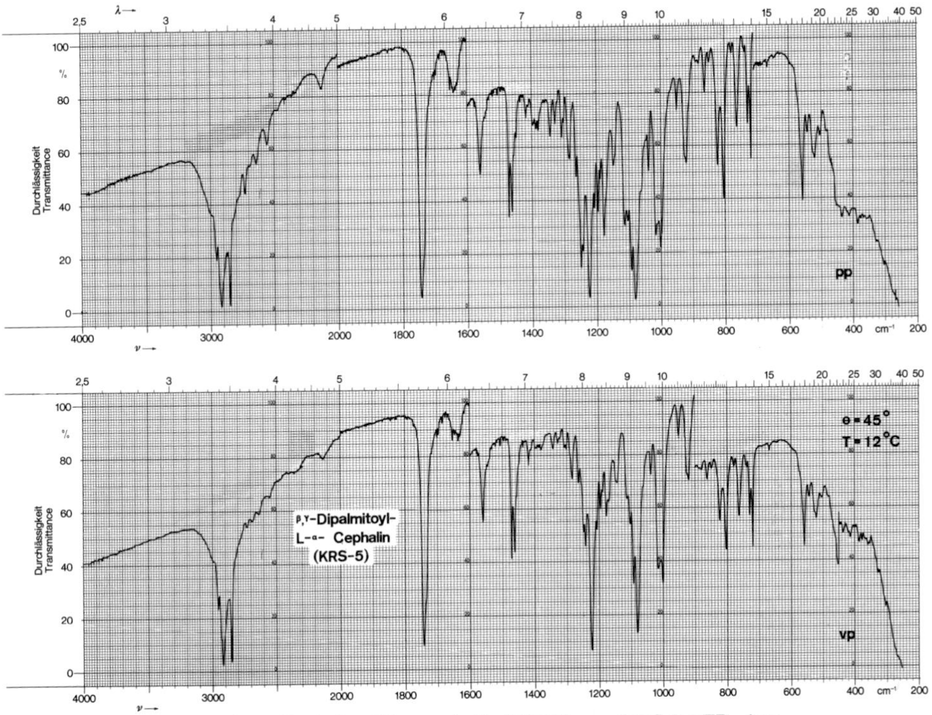

Fig. 3. Oriented layers of β,γ-dipalmitoyl-L-α-cephalin (DPPE) on a KRS-5 ATR plate. pp, parallel polarized; vp, perpendicular polarized; $\vartheta = 45°$; $T = 12°C$; rel. humidity $< 1\%$

a) Crystallinity of Samples

Decreasing crystallinity in the polar part of the molecules with increasing number of N-methyl groups is a typical feature of this homologous series. This behavior is clearly recognized by the varying half-width of typical absorption bands, i.e., crystalline sub-

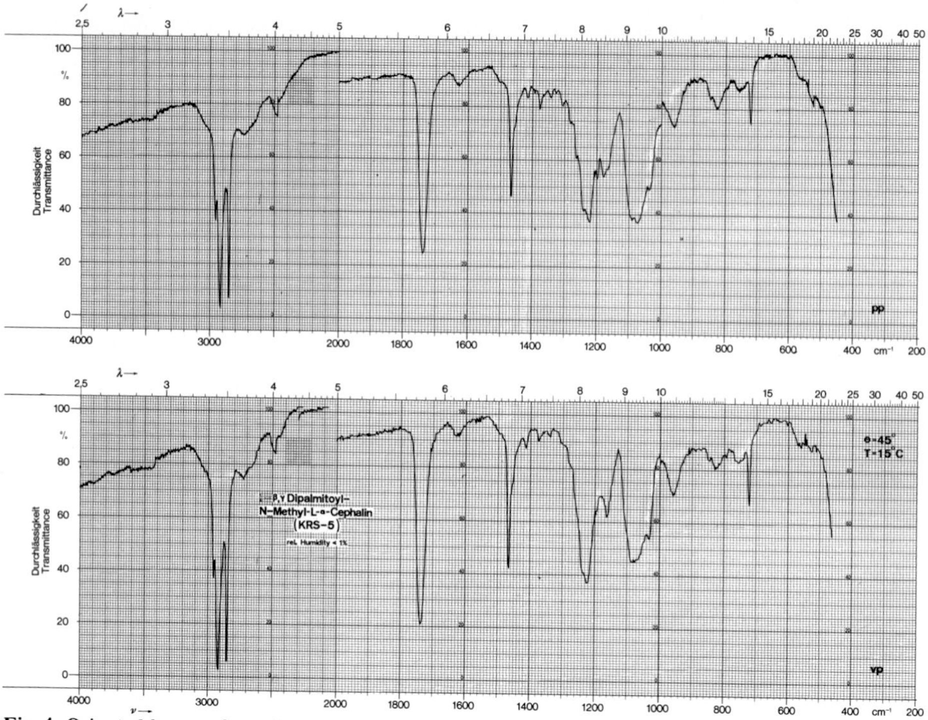

Fig. 4. Oriented layers of β,γ-dipalmitoyl-N-methyl-L-α-cephalin (DPPME) on a KRS-5 ATR plate. pp, parallel polarized; vp, perpendicular polarized; $\vartheta = 45°$; T = 15°C; rel. humidity < 1%

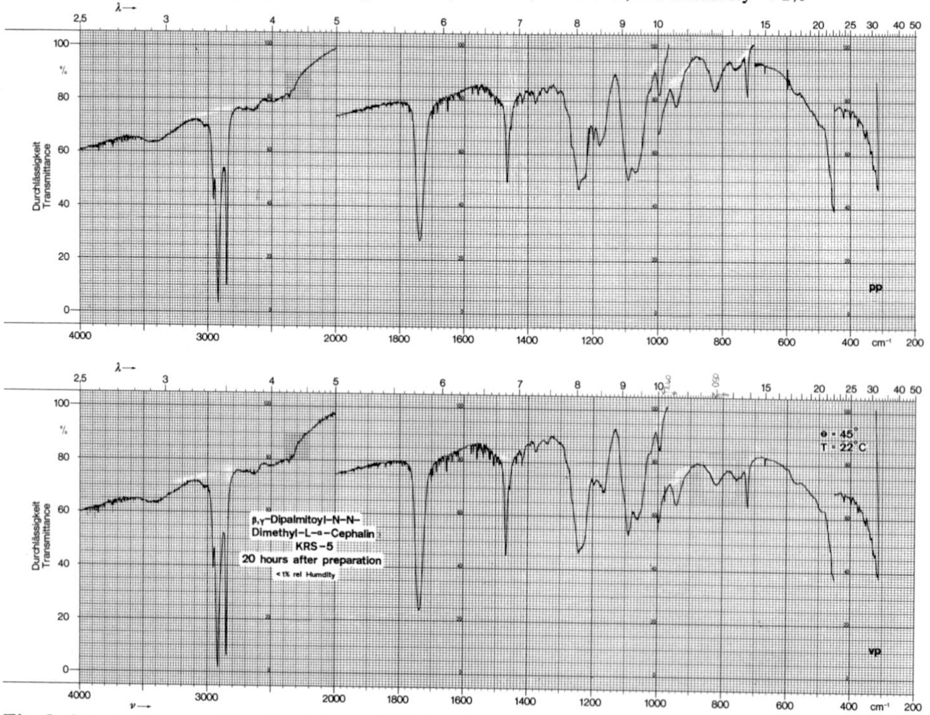

Fig. 5. Oriented layers of β,γ-dipalmitoyl-N-N-dimethyl-L-α-cephalin (DPPDME) on a KRS-5 ATR plate. pp, parallel polarized; vp, perpendicular polarized; $\vartheta = 45°$; T = 22°C; rel. humidity < 1%

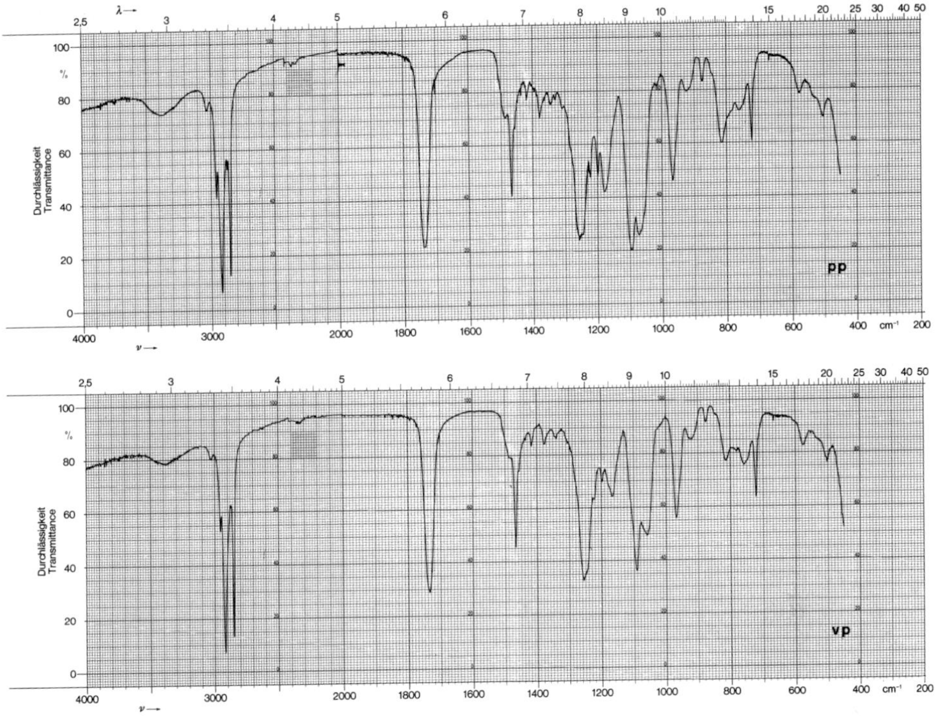

Fig. 6. Oriented layers of β,γ-dipalmitoyl-L-α-lecithin (DPPC) on a CdTe ATR plate. pp, parallel polarized; vp, perpendicular polarized; $\vartheta = 45°$; T = 22°C; rel. humidity < 1%

stance exhibits very sharp absorption bands with strong polarization effects, whereas in the liquid crystalline state, where the molecules may assume different conformations, corresponding absorption bands become considerably broader and show less polarization. All samples of L-α-dipalmitoyl phosphatidylethanolamine (L-α-DPPE) featured microcrystallinity (MCU-type) whereas in both N-methyl (L-α-DPPME) and N-N-dimethyl (L-α-DPPDME) compounds recrystallization from a liquid crystalline to a crystalline state may occur, especially when the multilayers were prepared on a hydrophobic surface such as KRS-5 (a tallium bromide iodide compound) or zinc selenide. This is exemplified by Fig. 7. Similar recrystallization phenomena have been observed by Kopp et al. (1975a,b) with Langmuir-Blodgett-type oriented layers of Ba-stearate and tripalmitin (cf. Sect. IV. A.10).

b) Structure of Hydrocarbon Chains

Saturated normal hydrocarbon chains in all-trans conformation exhibit characteristic sequencies of absorption bands, i.e., the wagging progression, twisting progression, C-C stretching progression, and rocking progression (cf. Table 2). These vibrations have been investigated by many workers, among them Primas and Günthard (1953a,b, 1955, 1956);

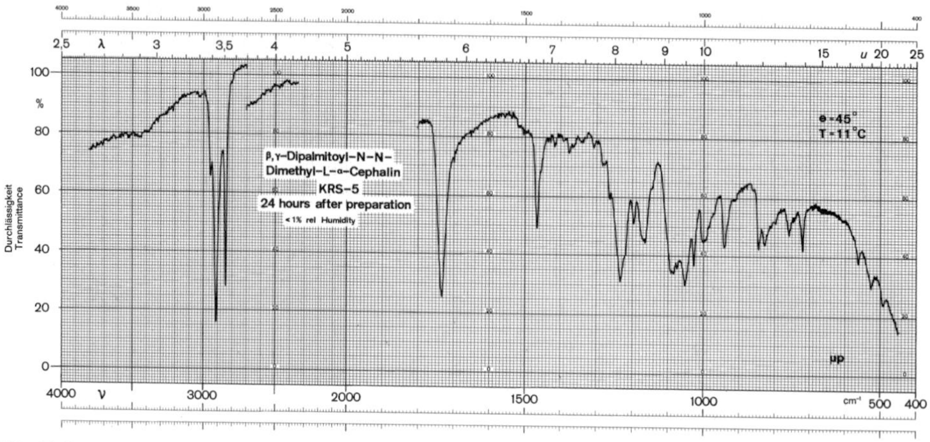

Fig. 7. Oriented layers of β,γ-dipalmitoyl-N-N-dimethyl-L-α-cephalin (DPPDME) on a KRS-5 ATR plate.
Recrystallized after 24 h. up, unpolarized; $\vartheta = 45°$; T = 11°C; rel. humidity $< 1\%$

Sheppard (1959); Snyder (1960, 1967); and Schachtschneider and Snyder (1963). Generally the wagging progression is most prominent. However, its intensity depends strongly on the nature of the endgroups of the corresponding hydrocarbon chain. In n-alkanes γ_w (CH_2) bands have low intensity, whereas polar endgroups such as -COOH or $\text{-}\overset{\text{O}}{\overset{\|}{\text{C}}}\text{-OR}$ increase the intensity considerably. This is due to coupling of γ_w (CH_2) with a vibration of the polar group, which increases the oscillating dipole moment, i.e., the intensity of the absorption band (cf. Sect. II). Deviations of the hydrocarbon chain conformation from all-trans have been found to lead to marked alterations in the wagging progression, such as loss of intensity, band broadening, and finally disappearance of the progression. This behavior is clearly detected with DL-α-dipalmitoyl phosphatidylcholine (DL-α-DPPC) when measured immediately below and above the transition temperature $T_c \approx 41°C$ (cf. Fig. 8a-d). However, little quantitative information is available at present about the behavior of the typical γ_w (CH_2)-, γ_t (CH_2)-, etc., sequences of a given CH_2-chain on the position and number of gauche(g)-type defects. For the case of paraffins, studies of this type have been reported by Snyder (1967). The first gauche defect separates the chain into two parts, each with all-trans conformation, and probably leads to a decoupling of the corresponding γ_w (CH_2) vibrations. Two independent wagging progressions should be expected. The all-trans part adjacent to the polar head should result in the more intensive progression. Aronovic (1957) and Meiklejohn et al. (1957) have studied series of n-aliphatic acids and corresponding salts, respectively. They found that the progressions contain n/2 (n being even) or (n + 1)/2 (n being odd) bands in the frequency range of 1180-1320 cm^{-1}, n denoting the number of CH_2 groups in the chain. One should expect a stochastic appearance of gauche conformations to smear the bands, thus leading to nonresolved infrared spectra. This hypothesis is supported by earlier measurements of the temperature dependence of oriented layers of tripalmitin (Fringeli et al., 1972) as well as by the results from DL-α-DPPC presented in Fig. 8a-d. The intensity behavior of the original γ_w (CH_2) progression enables therefore an estimate of the fraction of satu-

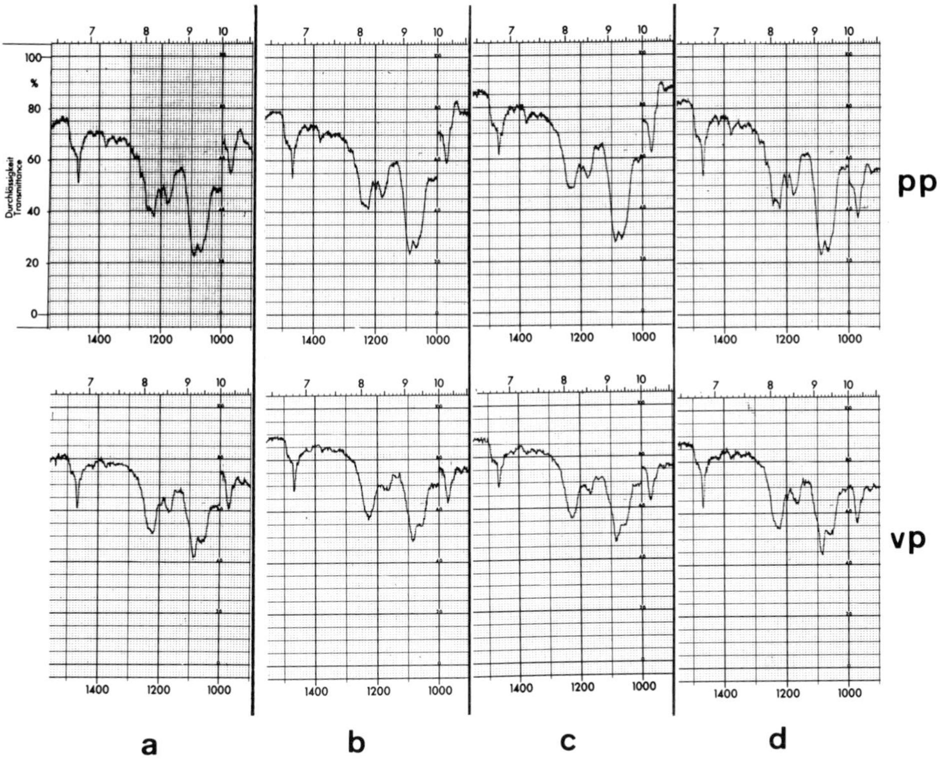

Fig. 8a-d. Homogeneous mixture of 50% β,γ-dipalmitoyl-DL-α-lecithin (DPPC) and 50% water by weight on a germanium ATR plate.
pp, parallel polarized; vp, perpendicular polarized; $\vartheta = 41°$.
a T = 22°C (begin of temperature cycle); **b** T = 41°C; **c** T = 42°C; **d** T = 22°C (end of temperature cycle)

rated hydrocarbon chains in the all-trans conformation. This finding is supported by the fact that both number and position of absorption bands coincide with the corresponding absorption bands in the microcrystal spectrum (Fig. 6). However, knowledge of the molar absorption coefficient $\epsilon_{\gamma_W\,(CH_2)}$ for a methylene group in a saturated hydrocarbon chain assuming all-trans conformation is required in order to determine the amount of all-trans hydrocarbon chains in a given lipid membrane. Because $\epsilon_{\gamma_W\,(CH_2)}$ depends strongly on the nature of the polar endgroup as mentioned above, it seemed reasonable to determine $\epsilon_{\gamma_W\,(CH_2)}$ from crystallized oriented layers of L-α-DPPC. Surprisingly, the amount of all-trans chains was found to be doubled during the transition from the biaxial into the high-crystalline C-phase (Fig. 12) as reported by Fringeli (1981a). A similar effect has been found by Fringeli and Fringeli (1979) upon incorporation of helical alamethicin (linear peptide, 19 amino acids, cf. Sect. IV.C.2). As mentioned below, the fragment $C\text{-}\overset{\overset{O}{\|}}{C}\text{-}O\text{-}C$ of the fatty acid ester group is found to be planar at γ-position and nonplanar at β-position (Hitchcock et al., 1975; Seelig and Seelig, 1975; Fringeli, 1977). Taking this fact into account, one is led to the interpretation that in pure, crystalline

L-α-DPPC only the γ-hydrocarbon chain assumes all-trans conformation. Further support for this opinion is obtained from comparison of γ_w (CH_2) intensities in L-α-DPPC (Fig. 6) and γ-palmitoyl-L-α-phosphatidylcholine (lysolecithin, Fig. 13). Therefore, one has to assume that insertion of helical rods of alamethicin into the membrane as well as the phase-transition into the C-phase lead to a stretching of the β-chain, assuming now all-trans conformation, too. Furthermore polarization measurement with the wagging progression is an efficient tool to determine the mean direction of all-trans hydrocarbon chains, since the oscillating dipole moment of γ_w (CH_2) is directed parallel to the chain (cf. Sect. III.B.1). However, deviation of the direction of the oscillating dipole from the hydrocarbon chain may be induced by certain polar endgroups, and crystal structures. In stearic acid, for instance, Münch et al. (1977) have found the deviation between chain and oscillating dipole moment to be ~30°. There are, however, no indications for a significant deviation in the case of triglycerides (Fringeli et al., 1972) and phospholipids (Fringeli, 1977). All these findings suggest a MCU-type ultrastructure of the homologous series L-α-DPPE to L-α-DPPC. Using Eqs. (11)-(13) with $f(\gamma) = \delta(\gamma\text{-}\gamma_o)$, the mean angle γ_o between hydrocarbon chains and the z-axis can be calculated. Values related to this homologous series are listed in Table 4. As mentioned above, the γ_w (CH_2) progression

Table 4. Direction of hydrocarbon chains in oriented phospholipid multilayers with respect to the plane of the bilayers[a]

Substance[b]	R_z^{ATR} [c]	ϕ_o [d]	Comments	
Tripalmitin	4.0-4.5	71°-73°	Ge-ATR plate,	ϑ = 30°, T = 26°C, < 1% rel. humidity
DPPE	2.4-3.0	63°-67°	Ge-ATR plate,	ϑ = 30°, T = 12°C, < 1% rel. humidity
DPPME	1.9-3.4	59°-68°	KRS-5-ATR plate,	ϑ = 45°, T = 15°C, < 1% rel. humidity
DPPDME	2.3-2.6	62°-65°	KRS-5-ATR plate,	ϑ = 45°, T = 22°C, < 1% rel. humidity
DPPC	2.1-2.9	58°-65°	CdTe-ATR plate,	ϑ = 45°, T = 22°C, < 1% rel. humidity
Lysolecithin	5.7-7.5	74°-77°	Ge-ATR plate,	ϑ = 45°, T = 22°C, ~15% rel. humidity
Lysolecithin-OD	6.9-8.0	76°-78°	Ge-ATR plate,	ϑ = 45°, T = 21°C, < 1% rel. humidity

[a] A microcrystalline ultrastructure (MCU-type) is assumed, cf. Fig. 14 and introduction to Sect. 4.a.
[b] Abbreviations: DPPE, dipalmitoyl phosphatidylethanolamine; DPPME, dipalmitoyl phosphatidylmonomethylethanolamine; DPPDME, dipalmitoyl phosphatidyldimethylethanolamine; DPPC, dipalmitoyl phosphatidylcholine; lysolecithin-OD, O-deuterated lysolecithin.
[c] ATR dichroic ratio of the γ_W (CH_2) band near 1200 cm^{-1}, cf. Sect. 3.b.i.
[d] Angle between hydrocarbon chains and the plane of the bilayers.

vanishes above the transition temperature T_c, i.e., information about chain ordering (conformation and direction) from wagging absorption bands is lost. However, mean order parameter of hydrocarbon chains in the liquid crystalline state can still be determined by means of polarized infrared measurements of other hydrocarbon chain group vibrations, e.g., ν_{as} (CH_2), ν_s (CH_2), ν_{as} (CH=CH), δ (CH_2), δ_s (CH_3), and ν (C-C). Two typical applications should be noted. Firstly, the determination of the order parameters of the rigid part and the flexible hydrocarbon chains of liquid crystal molecules by Fringeli et al. (1976). It was demonstrated that the theoretical calculation of hydrocarbon chains ordering proposed by Marcelja (1974) leads to results which are not consistent with experimental data. Secondly, the mean ordering of the double bonds in egg-lecithin was determined by Fringeli (1977) by means of polarization measurements with antisymmetric CH-stretching of $\overset{H}{R}\!>\!=\!<\!\overset{H}{R'}$. Further use of these data was made by Seelig and Waespe-Šarčevič (1978), since combination of the infrared data with NMR results enabled a more accurate determination of the double bond orientation.

c) Structure of Polar Headgroups

Little information is available concerning group vibrations of the polar headgroup of phospholipids. In order to get more detailed information, two approaches may be applied:

(i) Empirical Method. The infrared spectra of simple model compounds are compared with phospholipid or polypeptide spectra. Such investigations go back to Edsall (1937), who reported raman spectra of $N(CH_3)_4^+$ ions in aqueous solution. A single raman line was observed at 955 cm^{-1}, which was assigned to threefold degenerate (F_2-type) C-N stretching vibrations due to the tetrahedral symmetry (T_d) of the ion in aqueous solution. This band should split into three absorption bands when the symmetry is decreased to C_s or lower.

Three bands are indeed observed in the case of crystalline $N(CH_3)_4Cl$. The space-group is D_{4h}^7 as determined by Wyckoff (1966), thus resulting in a site symmetry of C_2 or C_s, leading to the threefold splitting of the F_2-type vibration of the unperturbed $N(CH_3)_4^+$ tetraeder. The bands are observed at 915 cm^{-1}, 945 cm^{-1}, and 955 cm^{-1}. Corresponding triplets exist in the spectra of CH_3-$(CH_2)_{14}$-$N(CH_3)_3Cl$, choline, acetylcholine, phosphoryl choline, and lecithin, respectively. However, the band situated at ~ 920 cm^{-1} is in some cases shifted to ~ 870 cm^{-1}, depending on the conformation of O-C-C-N in the choline part (Rihak, 1979; cf. NCA below). Assuming local C_s-symmetry in the choline part (trans conformation) the 920 cm^{-1} band may be assigned to a symmetric C-N-stretching mode of the $N(CH_3)_3$ moiety, thus resulting in an oscillating dipole moment along the C_3 symmetry axis of -$N(CH_3)_3^+$, a fact that is clearly demonstrated in the case of oriented layers of hexadecyltrimethylammoniumchloride ($CH_3(CH_2)_{14}N(CH_3)_3Cl$); see Fig. 9. This vibration turned out to be of considerable importance for structural studies of the polar headgroup of lecithin. Furthermore, in view of this approximation the doublet in the 960 cm^{-1} region (Figs. 6, 9) may be assigned to the symmetric and antisymmetric C-N-stretching vibration of the $N(CH_3)_3$ group. The corresponding oscillating dipole moments are expected to be in the plane and perpendicular to the plane of symmetry, respectively. Empirical investigations as discussed above are very useful for finding and assigning group vibrations in complex molecules.

There are of course many other group vibrations found by similar arguments. A number of these are listed in Tables 2 and 3. Furthermore, several are discussed briefly when applied diagnostically in the following sections. A more elaborate but on the other hand more time-consuming technique is normal coordinate analysis of model compounds. Applications of this technique are described below.

(ii) Normal Coordinate Analysis (NCA) (cf. Sect. II). NCA of dimethyl- and diethylphosphate was carried out by Shimanouchi et al. (1964) for three assumed conformations of the C-O-P-O-C frame, the extended all-trans conformation (C_{2v} symmetry) the gauche-gauche conformation (C_2 symmetry), and the unsymmetric trans/gauche confor-

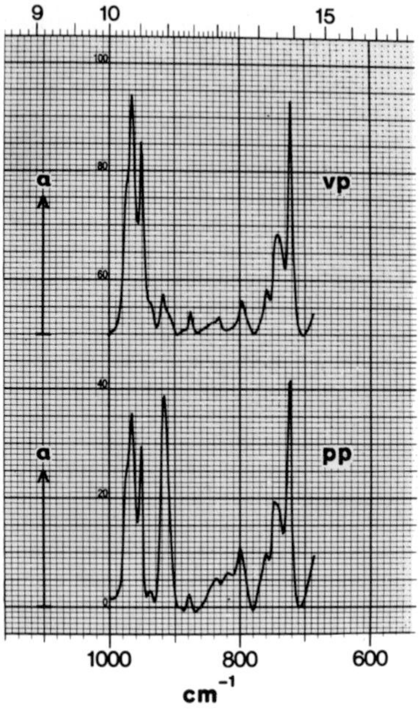

Fig. 9. Oriented layers of hexadecyltrimethylammonium-chloride on a ZnSe ATR plate.
pp, parallel polarized; vp, perpendicular polarized;
$\vartheta = 45°$; T = 25°C

mation. It was found that the $g^{\pm}/g^{\pm}$ conformation results in the best coincidence between calculated and measured vibrational frequencies. Several group vibrations in the 700-1300 cm^{-1} region turned out to be of considerable diagnostic value, especially the four P-O-stretching modes ν_{as} (PO_2^-) (~ 1230 cm^{-1}), ν_s (PO_2^-) (~ 1090 cm^{-1}), ν_{as} ($P(OC)_2$) (~ 820 cm^{-1}), and ν_s ($P(OC)_2$) (~ 760 cm^{-1}) (cf. Table 2). It should be concluded from recent IR polarization measurements with oriented layers of a series of phospholipids that C-O single bond stretching is significantly involved in both $P(OC)_2$ modes (Fringeli, 1977). An NCA was carried out with several isotropic modifications of choline by Rihak et al. (1975) and Rihak (1979). The most important results with respect to the structure of the polar headgroup of DPPC may be summarized as follows:

a. The symmetric N-$(CH_3)_3$ stretching (ν_s (N-$(CH_3)_3$) in choline was found to depend strongly on the torsional angle of the C-C bond. For the trans conformation the absorption band is calculated to be near 920 cm^{-1}, whereas the gauche conformation results in two bands, one near 900 cm^{-1} and the other near 860 cm^{-1}. The calculated potential energy distribution is > 90% for trans (C_s symmetry), however, strong coupling with other vibrations occurs in gauche conformation. The corresponding potential energy distributions are ν_s (N-$(CH_3)_3$) (21%), ν (C-O) (33%), γ_r (α-CH_2) (20%), γ_w (β-CH_2) (14%) for the 895 cm^{-1} band of choline and ν_s (N-$(CH_3)_3$) (67%), ν_s (NC_4) (13%) for the 866 cm^{-1} band. Wychoff (1966) concluded from X-ray data that the conformation is gauche in the crystalline state. Experimentally, as documented by Fig. 10, the calculation of ν_s (N-$(CH_3)_3$) is confirmed. When the sample was prepared on a ZnSe plate by evaporating the solvent (H_2O), the still strongly hydrated choline showed a band at 924 cm^{-1} (Fig. 10), in good agreement with the calculated ν_s (N-$(CH_3)_3$) for the trans conformation. On

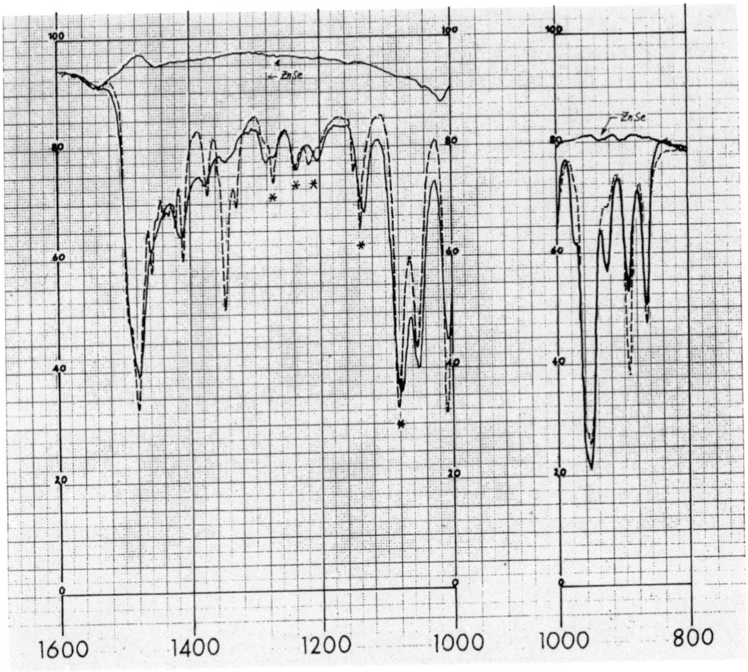

Fig. 10. Dependence of symmetric N-$(CH_3)_3$-stretching (ν_S (N-CH_3)) on O-C-C-N trans-gauche isomerization of choline. Strongly hydrated choline (*solid line*) assumed trans (925 cm^{-1}) and gauche (895 and 866 cm^{-1}) conformation. Loss of hydration has induced a conversion of the choline fraction with trans conformation into gauche conformation (*dashed line*)

drying, this band decreased, while the 895 cm^{-1} and 866 cm^{-1} bands increased synchronously. As mentioned above, the latter are typical for ν_s (N-$(CH_3)_3$) when the O-C-C-N frame assumes gauche conformation. There is no doubt that a direct relation between the three bands exists. Gauche-trans isomerization is the most probable explanation.

b. Another absorption band of diagnostic importance was calculated and observed in the 1020 cm^{-1} region. The oscillating dipole moment is estimated to be parallel to the C-C bond of choline.
c. Concerning the doublet observed at ~ 950 cm^{-1}, NCA confirms the empirical assignment given under (i), especially for C_s symmetry, i.e., trans or cis conformation.
d. The frequency of the totally symmetric N-C-stretching vibration depends strongly on the conformation of the O-C-C-N frame. It is found to be at 720 cm^{-1} for gauche, and near 760 cm^{-1} for trans conformation. However, because the IR intensity of this band is very weak (approximate T_d symmetry), Raman spectroscopy has to be used for conformational analysis via these bands.

(iii) Mole Fractions of Conformers. If reliable assignment of corresponding conformation-dependent vibrations has been achieved via empirical and/or NC analysis, the mole fraction of a conformer can be determined from the integrated intensities of cor-

responding modes in the coexisting conformations. It should be noted, however, that generally corresponding absorption coefficients related to different conformations of a fragment are different, i.e., mole fractions may not be determined by simple comparison of absorption band areas. Corresponding absorption coefficients must be known, which furthermore may depend on the degree of hydration, i.e., on intramolecular interaction. This complicating fact is demonstrated on the one hand by means of Fig. 10, where the area of the 920 cm^{-1} band (solid line, trans conformation of O-C-C-N frame) has to be compared with the differences of corresponding band areas (solid and dashed lines, respectively) of the 895 cm^{-1} and 866 cm^{-1} bands. On the other hand, chlorocholine assumes predominantly trans conformation in the crystal and gauche conformation in aqueous solution. Trans-gauche equilibrium can be controlled in the crystalline state by relative humidity of the gas phase (Rihak, 1979). Chlorocholine turned out to be of particular interest for conformational studies of Cl-C-C-N (corresponding to O-C-C-N in choline or lecithin), because besides vibrations of the quaternary ammonium group (cf. choline above) there is an additional conformation-sensitive vibration, the C-Cl stretching mode (ν(C-Cl)). It is located near 735 cm^{-1} for trans and near 670 cm^{-1} for gauche conformation. Figure 11 a-c demonstrates clearly the appearance of the gauche conformer

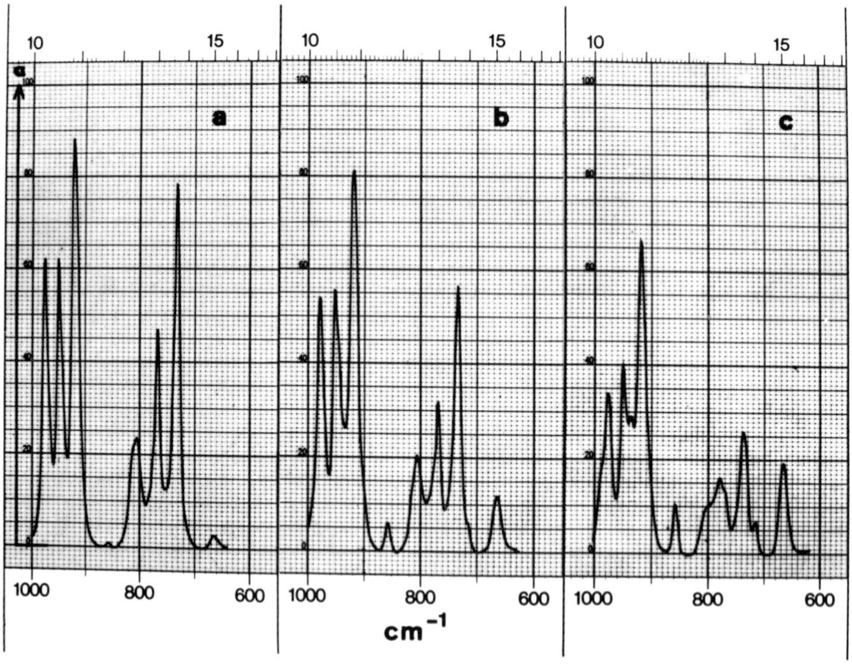

Fig. 11 a-c. Conformation of the Cl-C-C-N fragment of chlorocholine as a function of the relative humidity of the gas phase. **a** 1% rel. humidity; **b** ~ 20% rel. humidity D_2O; **c** 90% rel. humidity D_2O. Typical conformation-sensitive absorption bands are:

ν_S (-N$(CH_3)_3$) : 920 cm^{-1} (trans); ⧺ 910 and 860 cm^{-1} (gauche)
ν (NC_4) : 769 cm^{-1} (trans); 715 cm^{-1} (gauche) (totally symmetric)
ν (C-Cl) : 735 cm^{-1} (trans); 666 cm^{-1} (gauche)
ZnSe ATR plate, ϑ = 45°; parallel polarized incident light; T = 25°C

upon hydration of the crystal (-$N(CH_3)_3^+$: 860 cm^{-1} and ~ 910 cm^{-1}, -C-Cl: 666 cm^{-1}, respectively. The corresponding trans bands are at 920 cm^{-1} (-$N(CH_3)_3$) and 735 cm^{-1} (C-Cl), respectively). Intensities of corresponding modes reflect once more that the absorption coefficients critically depend on the conformation. More explicitly, one may conclude from Figs. 10 and 11 that the absorption coefficient of ν_s ($N(CH_3)_3$) is significantly larger for trans (920 cm^{-1}) than for gauche conformation (860 cm^{-1}). This finding must be considered when gauche/trans mole fractions are estimated by means of the these band areas. The reader is referred to Fringeli (1980a) for a more detailed discussion.

d) Application of Empirical and NCA Assignments from Model Compounds to Phospholipids

(i) Structure of Fatty Acid Ester Groups. The vibrations of the ester groups of fatty acids have been discussed by many authors. For a review the reader is referred to Fringeli (1977). The O-C(=O)-O-C frame exhibits a typical absorption band in the 1160-1180 cm^{-1} region in with C-O single bond stretching involved. It has turned out that in glycerol fatty acid esters the planar arrangement of this fragment leads to an absorption band at 1180 cm^{-1} with predominant polarization along the hydrocarbon chain of the fatty acid. Deviation from planar arrangement results in band shift to ~ 1170 cm^{-1}, band broadening, and loss of the distinct polarization. From this observation one may conclude that for most phospholipids the fatty acid ester group in γ-position assumes planar conformation. The ester group in β-postition, however, is expected to deviate from planarity (Fringeli, 1977). Concerning L-α-DPPE our finding is in good agreement with X-ray results published by Hitchcock et al. (1974, 1975), as well as with NMR data by Seelig and Seelig (1975).

(ii) Structure of Phosphate Group. The polarization of typical phosphate group vibrations, such as ν_{as} (PO_2^-) at ~ 1230 cm^{-1}, ν_s (PO_2^-) at ~ 1090 cm^{-1}, ν_{as} ($P(OR)_2$) at 818 cm^{-1}, and ν_s ($P(OR)_2$) at 760 cm^{-1}, is another common feature of the phospholipids under discussion (cf. Figs. 3-7). Based on NCA results of Shimanouchi et al. (1964) the conformation of C-O-P(O)(O)-O-C is assumed to be g^+/g^+ or g^-/g^-. ν_{as} (PO_2^-) at 1230 cm^{-1} exhibits no distinct polarization, whereas the corresponding symmetric vibration ν_s (PO_2^-) at 1090 cm^{-1} shows typical parallel polarization. This observation leads to the conclusion that the bisector of the O-P-O angle of the $> PO_2^-$ group directs predominantly parallel to the z-axis, i.e., perpendicular to the plane of the bilayers. Furthermore, ν_{as} (PO_2^-) exhibits a typical dependence on both, the degree of crystallinity and the hydration of the phospholipids, see Table 5. Typical polarization behavior is also observed with ν_{as} ($P(OR)_2$) at 818 cm^{-1} and ν_s ($P(OR)_2$ at 760 cm^{-1}; however, less information on the direction of corresponding transition dipole moments is available at present. From comparison of the polarization of corresponding P-O single bond and double bond stretching vibrations one should conclude that C-O single bond stretching is significantly involved in the 760 cm^{-1} and 818 cm^{-1} bands (Fringeli, 1977).

(iii) Structure of Phosphoryl Ethanolamine Group. A detailed discussion of this structure in L-α-DPPE (Fig. 3) and in natural phosphatidylethanolamine from sheep

Table 5. Wavenumbers of the antisymmetric stretching vibration of the $> PO_2^-$ group in some phospholipids

Substance	Wavenumbers [cm^{-1}] 0% Rel. humidity	60% Rel. humidity	90% Rel. humidity	Comments
GPE			1230 (⊥), 1235 (‖)	Two absorption bands with dichroism, hygroscopic
DPPE	1218			Fringeli (1977)
DPPE (recrystallized)	1222		1222	No possibility of hydration from the gas phase, Fig. 3
PE-analog	1228 (‖), 1215 (⊥)			
PE-analog (recryst.)	1218			Fringeli (1977)
Sheep brain cephalin	1230 (‖), 1225 (⊥)		1220 (‖), 1215 (⊥)	Two bands with dichroism, hygroscopic. Fringeli (1977)
DPPME	~ 1230			Fig. 4
PME-analog	1240 (‖), 1230 (⊥)			
DPPDME	1235 ± 5			Fig. 5
DPPDME (recryst.)	~ 1230			Fig. 7
DME-analog	1230 (‖), 1220 (⊥)			Broad absorption bands
DME-analog (recryst.)	1240 (‖), 1230, 1220			Narrow overlapping bands, the 1240 cm^{-1} band is strongly z-polarized
GPC	1240 (‖), 1230 (⊥)	1223 (‖), 1215 (⊥)		
DPPC	1260, 1250, (1225)	1235 ± 5	1235 (‖), 1225 (⊥)	At 0° rel. humidity weak band at 1225 cm^{-1}. Wavenumbers determined at 81°C with molten hydrocarbon chains (0% rel. humidity)
Egg-lecithin	1262 (‖), 1252 (⊥) (1225)	1225	1220	Fig. 17
Lysolecithin	1240-1245	1225		
Lysolecithin-OD	1240-1245	~ 1220		A new band appears at a 1195 cm^{-1}
Lysolecithin-OD (recryst.)	1245 (‖), 1240 (⊥)			Fig. 16
Lysolecithin + $NdCl_3$	~ 1210	~ 1210		Broad band, strong shift due to Nd^{3+}-bonding

Abbreviations: GPE, glycerophosphorylethanolamine; DPPE, dipalmitoyl phosphatidylethanolamine; PE-analog, ketal derivatives of DPPE; DPPME, dipalmitoyl phosphatidylmonomethylethanolamine; DPPDME, dipalmitoyl phosphatidyldimethylethanolamine; GPC, glycerophosphorylcholine; DPPC, dipalmitoyl phosphatidylcholine; lysolecithin-OD, O-deuterated lysolecithin

brain has been given by Fringeli (1977). The main results are summarized as follows:

a. Synthetic L-α-DPPE (Fig. 3) exhibits a high degree of crystallinity, pointing to the existence of preferred conformations in the polar headgroup. In contrast, natural phosphatidylethanolamine exhibits a considerable band broadening. This effect may result from the existence of an equilibrium between different conformers of a given fragment and from altered intermolecular interactions.
b. There is good evidence for a protonated phosphate group (O=P-OH) in L-α-DPPE and for a deprotonated group (O=P-O⁻) in natural phosphatidylethanolamine. The latter is found to be typical for all alkylated ethanolamines.
c. L-α-DPPE cannot be hydrated from the gas phase even at 90% relative humidity (25°C), whereas natural phosphatidylethanolamine and all other alkylated derivatives bind ~ 20% weight of water under the same conditions. This effect gives further evidence for an uncharged polar headgroup in L-α-DPPE and charged polar headgroups in the other phospholipids.
d. The hydrocarbon chains form an angle of 63°-67° with respect to the plane of the bilayer, which is assumed to be parallel to the xy-plane of the ATR plate (Fig. 2).

(iv) N-Methyl and N-N-Dimethyl Phosphatidylethanolamine. A number of special features of the polar headgroup conformation of mono- and dimethyl phosphatidylethanolamine have been discussed by Fringeli (1977).

(v) Special Features of Phosphatidylcholine (Lecithin). Based on NCA data of choline by Rihak et al. (1975), Fringeli (1977) has postulated the choline headgroup of lecithin to assume at least two different conformations with respect to the O-C-C-N frame, i.e., gauche and trans. At the same time Seelig and Gally (1976) have proposed the ethanolamine group of DPPE to undergo fast conformational transitions from gauche$^+$ (g^+) to gauche$^-$ (g^-) and vice versa. In a recent work Seelig (1978) has extended this picture to the phosphocholine group, which is postulated to assume the two enantiomeric conformations g^+ and g^-. These data are in good agreement with X-ray results of glycerphosphorylcholine (Sundaralingam, 1972). If this picture is true then a number of intermediate conformations must be expected. It may be that they are too short-lived to be resolved by NMR spectroscopy. However, since IR spectroscopy has a considerably higher time resolution (~ 10^{-13} s) intermediate conformations must be detected, unless their concentration is too small. IR-ATR spectra reveal that the conformation of the polar headgroup of DPPC is approximately the same in the microcrystalline state (biaxial phase) as in the liquid crystalline state (L_α phase). Because it is unlikely that the whole phosphocholine group would be able to undergo all-g^+ to all-g^- transitions in the crystalline state, one should moreover assume that certain intermediate conformations could be frozen, i.e., would represent "stable" states of the polar headgroup. Based on the general remarks on the O-C-C-N structure at the beginning of this chapter, one may estimate from the IR-ATR spectrum of microcrystalline L-α-DPPC (Fig. 6) that approximately 10% of the choline headgroup of DPPC assumes trans conformation. Our postulation of a nonuniform conformation of the polar head of L-α-DPPC in all phases described so far (Powers and Pershan, 1977) obtains further support by the discovery of a new highly-ordered phase of L-α-DPPC monohydrate (Fringeli, 1981a; Fig. 12). Obviously there must be considerable differences between the molecular structures of

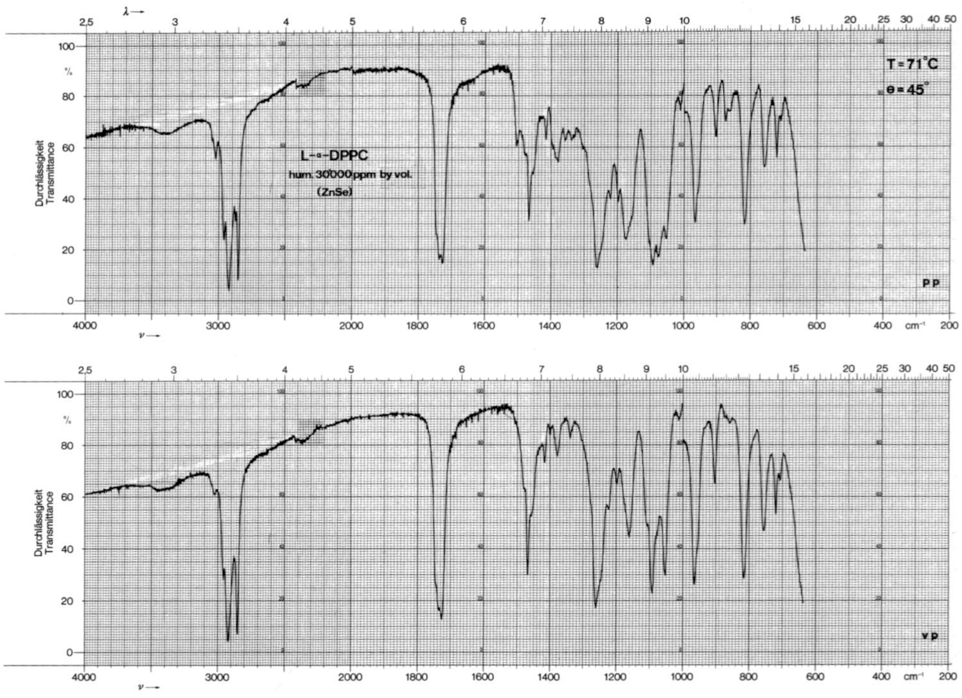

Fig. 12. Oriented layers of β,γ-dipalmitoyl-L-α-lecithin (DPPC) monohydrate recrystallized at 75°C (phase C) on a ZnSe ATR plate. ‖, parallel polarized; ⊥, perpendicular polarized; $\vartheta = 45°$

L-α-DPPC in the new phase and in the biaxial or liquid crystalline phases, as easily concluded from a comparison of Fig. 12 with Figs. 6 and 8d, respectively. The main features of the new phase, we call it C, may be summarized as follows:

a) The phase transition biaxial → C of the L-α-DPPC monohydrate occurs at ~ 71°C
b) Phase C is also stable in dry atmosphere at temperatures below 71°C; however, sudden transition to the biaxial phase occurs in humid atmosphere
c) The conformation of the O-C-C-N frame of choline seems to be uniformly gauche, as concluded from ν_S $(N(CH_3)_3)$ (gauche) at 905 cm^{-1} and 875 cm^{-1} (cf. *Structure of Phosphate Group* above). These two bands in the spectrum of L-α-DPPC have been shown to result from the choline headgroup by substituting the N-methyl groups by ^{13}C-labeled methyl groups (Fringeli, 1981a)
d) The totally symmetric NC_4-stretching mode is clearly resolved at 710 cm^{-1}, indicating gauche conformation of O-C-C-N, too
e) The orientation of the $C\text{-}N(CH_3)_3$ bond (axis of local C_3 symmetry) is perpendicular to the bilayer plane, as concluded from the distinct polarizations of ν_S $(N(CH_3)_3)$ (gauche) at 875 cm^{-1} and of δ_{as} $(N)\text{-}CH_3)$ (A-type) at 1507 cm^{-1} (symmetric combination of antisymmetric N-methyl CH_3-bending modes, cf. Fringeli, 1977). In the biaxial phase, however, the corresponding orientation of the $C\text{-}N(CH_3)_3$ bond was reported to form an angle of 52°-54° with respect to the bilayer plane (Fringeli, 1977)
f) The C-O-P(O)(O)-O-C conformation is concluded from the narrow and symmetric shape of the antisymmetric and symmetric POC-stretching modes (820 cm^{-1} and 758 cm^{-1}, respectively) to be uniform. According to Shimanouchi et al. (1964) and Forrest and Lord (1977) gauche-gauche conformation seems to be most probable from the point of view of vibrational spectroscopy Furthermore comparing the POC-stretching bands of L-α-DPPC in the biaxial phase (Fig. 6) with those of the C phase (Fig. 12) one has to conclude that the former does not exhibit a uniform conformation of C-O-P(O)(O)-O-C
g) L-α-DPPC exists in the C phase most probably as monohydrate, as concluded from the OH-stretching band of bound water at 3400 cm^{-1}

(For a discussion of the main features of L-α-DPPC in the biaxial phase, the reader is referred to Fringeli, 1977).

(vi) Special Features of Lysolecithin. In comparison with L-α-DPPC (Fig. 6) for γ-palmitoyl-β-lyso-α-lecithin (Fig. 13) no significant differences with respect to the orientation of the $> PO_2^-$ group as well as to the existence of gauche and trans isomers of the O-C-C-N moiety were found. However significant deviation of the "C_3"-axis of $\text{-}N(CH_3)_3^+$ and of the C-C bond of choline from the xy-plane is found, the latter probably resulting from a rotation of the polar headgroup around the adjacent P-O bond. The O-H bond of the β-hydroxyl group is approximately parallel to the plane of the bilayers. Strong hydrogen bonding of this group was found to be intramolecular rather than intermolecular. H bond formation to one O-atom of the $> PO_2^-$ group, which seems most reasonable, is assumed. Thus, based on polarization measurements of ν (OH) and δ (OH) the fully extended conformation of the whole molecule may be excluded. Rotation around the $C_\beta\text{-}C_\gamma$ bond of glycerol is proposed in order to orient the O-H bond parallel to the plane of the bilayers.

Direction of C_3-Symmetry Axis of $\text{-}N(CH_3)_3$.

a. ν_s $(N\text{-}CH_3)_3$: 930 cm^{-1} (trans), 905 cm^{-1} and 875 cm^{-1} (gauche).

From inspection of Fig. 13 it follows that both absorption bands under discussion show more z-polarization than the corresponding bands in DPPC. The dichroic ratios are found to be R^{ATR}_{trans} (930 cm^{-1}) = 1.40 ± 0.05 (Ge) and R^{ATR}_{gauche} (875

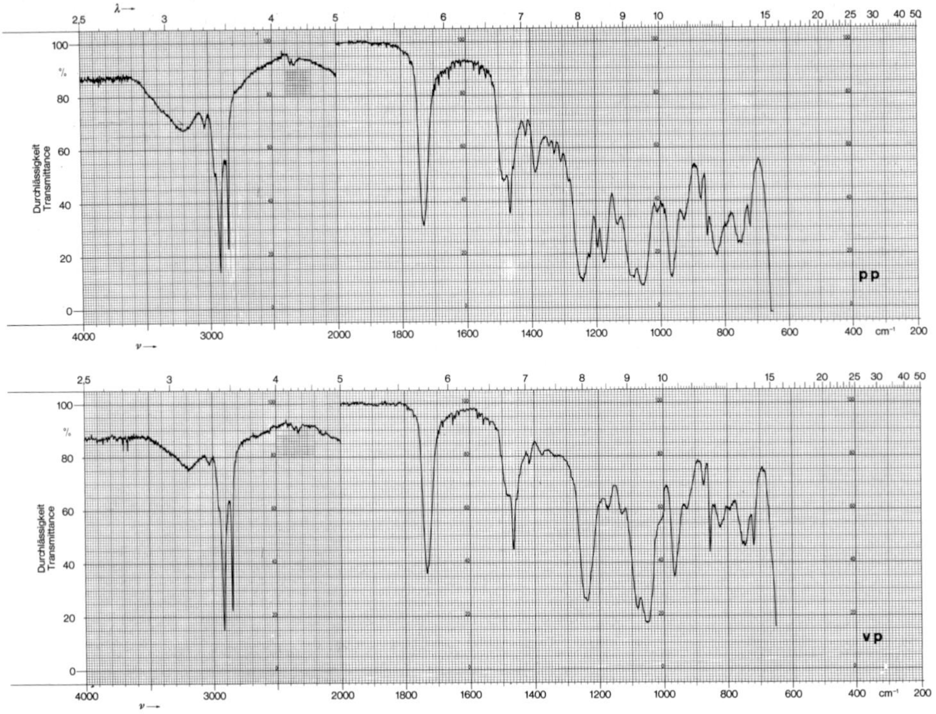

Fig. 13. Oriented layers of L-γ-palmitoyl-lyso-α-lecithin (lysolecithin) on a germanium ATR plate. pp, parallel polarized; vp, perpendicular polarized; $\vartheta = 45°$; T = 22°C; rel. humidity ~ 15%

cm^{-1}) = 1.25 ± 0.03 (Ge), indicating a significant conformational change with respect to L-α-DPPC (biaxial phase).

b. $\delta_{as}^{A_1}$ *(N-CH_3):* 1488 cm^{-1}.
The finding of (a) is confirmed by the polarization of the 1490 cm^{-1} band, which indicates the mean direction of the C_3-axis of -$N(CH_3)_3$ (cf. Fringeli, 1977). This direction is estimated to devitate more than 45° from the xy-plane.

Direction of C-C Bond in the Choline Part. In analogy to the bands just mentioned, the oscillating dipole moment of this vibration directs also toward the z-axis. The corresponding dichroic ratio was determined to R^{ATR} (1014 cm^{-1}) = 2.30 ± 0.50 (Ge). The change in polarization of this band and of the ν_s (N-CH_3) absorption band with respect to DPPC could result from a rotation of the choline moiety around adjacent C-O and P-O bonds. This conjecture is supported by the pair of absorption bands assigned to ν (P-O), 823 cm^{-1} (asym.) and 750 cm^{-1} (sym.). The relative polarization of the pair changes significantly in going from DPPC to lysolecithin.

β-OH-Bending Vibration: δ(OH): 1380 cm^{-1}. This vibration results in a medium absorption band at 1380 cm^{-1}. The characteristic feature of this band is its predominant z-polarization, indicating an oscillating dipole moment directed practically parallel to the z-axis. Using a simple oscillating dipole model on postulating its direction to be perpendicular to the OH bond direction, one is led to locate the latter in the xy-plane.

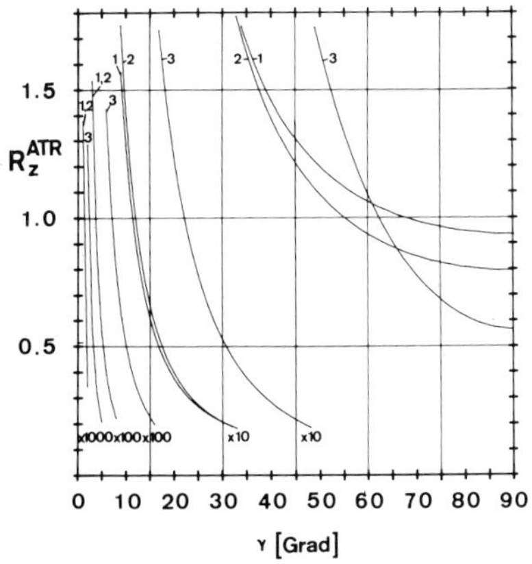

Fig. 14. Relationship between ATR-dichroic ratio R_Z^{ATR} and angle γ between the molecular axis and the z-axis for oriented layers with microcrystalline ultrastructure (MCU-type, cf. introduction to Sect. IV.a).
Calculation is based on Eqs. (7) and (11)-(13) with $f(\gamma) = \delta(\gamma-\gamma_O)$. The angle of incidence was 45°. 1, $n_1 = 4.0$ (germanium), $n_2 = 1.55$ (lipid), $n_3 = 1.0$ (air); 2, $n_1 = 2.4$ (zinc selenide)[a], $n_2 = 1.55$ (lipid), $n_3 = 1.0$ (air); 3, $n_1 = 2.4$ (zinc selenide)[a], $n_2 = 1.55$ (lipid), $n_3 = 1.33$ (water)

[a] Or KRS-5, CdTe

β-OH-Stretching Vibration: ν (OH), 3200 cm⁻¹. The conclusion drawn from δ (OH) is supported by the observed polarization of ν (OH). The corresponding dichroic ratio was determined from dehydrated oriented layers to be R^{ATR} (3200 cm^{-1}) = 0.90 ± 0.10, thus confirming the paralleleity of O-H bond and xy-plane (KRS-5, $\vartheta = 45°$).

Hydrogen Bonding in Alcohols. Hydrogen bonding may affect strongly frequency, half-width, and intensity of both OH-stretching (ν(OH)) and O-H-bending (δ(OH)) modes. Band broadening and shifting are special features of intermolecular H bonds, whereas intramolecular H bonds result in only slight increase of the band width, combined with smaller frequency shifts (Pimentel and McClellan, 1960). Typical behavior of the OH-stretching motion in a system with intermolecular H bonding is shown in Fig. 15. On the other hand the frequency of the δ (OH) mode is shifted to higher values. Stuart and Sutherland (1956) have studied many aliphatic alcohols in dilute solution. They have proposed δ (OH) to occur in the range 1200-1330 cm^{-1}. The pure alcohol on the other hand exhibits a broad band, which has two maxima near 1410 and 1330 cm^{-1}. In the light of these observations the relatively narrow ν (OH) and δ (OH) bands in oriented layers of lysolecithin point rather to strong intramolecular hydrogen bonding than to intermolecular bonding. Hydrogen bonding to one oxygen atom of the PO_2^- groups is proposed. In this case molecular model considerations show the OH bond to be parallel to the hydrocarbon chain for the fully extended lysolecithin molecule. This conformation however must be excluded from polarization measurements of ν (OH) and δ (OH) (see above). Rotation around the C_β-C_γ bond of glycerol is required in order to orient the OH bond parallel to the xy-plane, i.e., the plane of the bilayers.

Finally, it should be mentioned that model considerations also show other types of hydrogen bonds to be feasible. A molecular arrangement involving intermolecular H bonding has been proposed by Hauser (1976).

O-Deuteration. The protons of the β-OH group and of adsorbed water can be exchanged by hydration of lysolecithin multilayers with D_2O from the gas phase. Thereby the typical shifts of ν (OH) from the 3300 cm^{-1} to the 2500 cm^{-1} region are observed (Pimentel and McClellan, 1960). The δ (OH) absorption band at 1390 cm^{-1} vanishes, however the δ (OD) band could not be assigned because of overlapping with other bands.

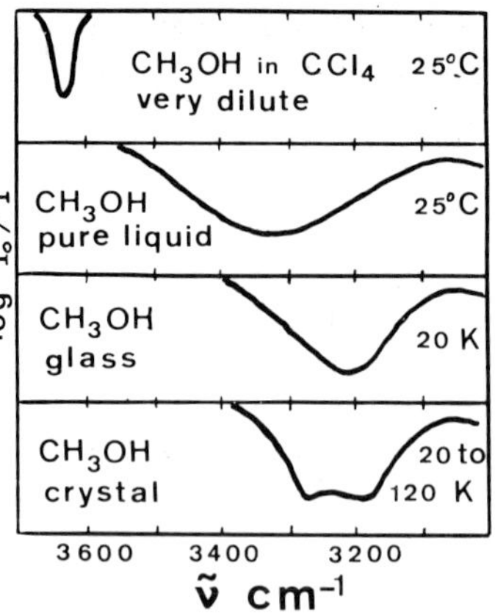

Fig. 15. Infrared spectra of OH-stretching vibration of methanol in the condensed phase (Pimentel and McClellan, 1960)

Crystallization of Oriented Layers after Deuteration. Exposure of oriented lysolecithin layers to an atmosphere with 60% relative humidity of D_2O at 11°C generally results in only alteration of ν (OH) and δ (OH) absorption bands (cf. above). However it may happen that drastic changes of most absorption bands occur due to recrystallization of the samples (Fig. 16). It was reported earlier by Gallagher (1959) that H→D isotopic

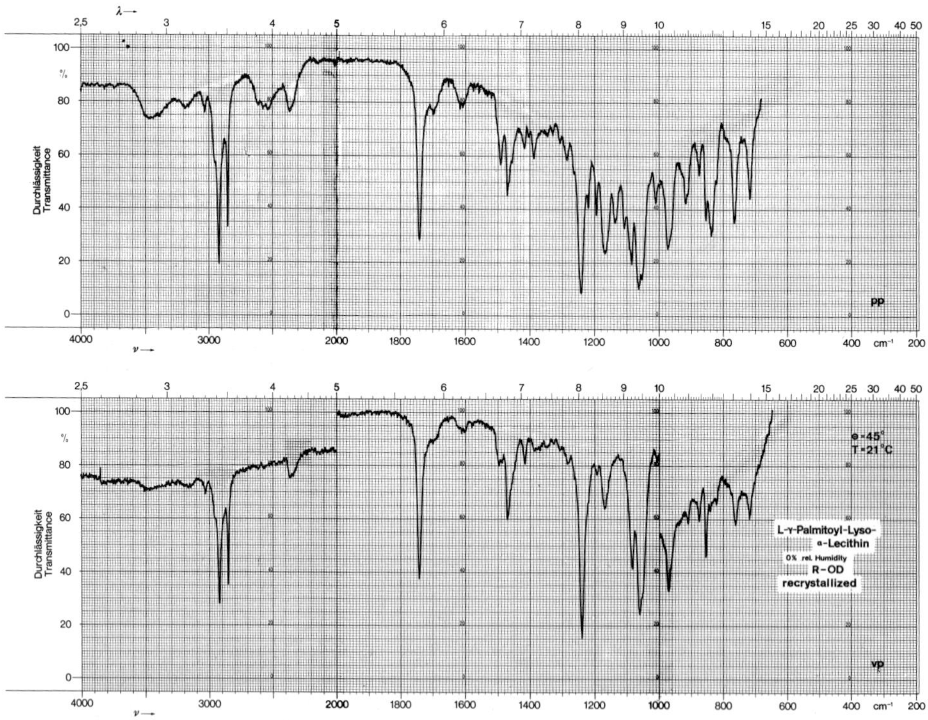

Fig. 16. Oriented layers of L-γ-palmitoyl-lyso-α-lecithin, recrystallized after O-deuteration (lysolecithin-OD, R-OD) on a germanium ATR plate.
pp, parallel polarized; vp, perpendicular polarized; $\vartheta = 45°$; T = 21°C; rel. humidity < 1%

exchange may lead to alteration of interatomic distances in crystals. The recrystallization must therefore be understood as an isotope effect. Indeed reexposure to H_2O vapor reconverts the spectra of Fig. 16 into the initial form, Fig. 13.

3. Influence of Chemical Modifications

a) Unsaturated Hydrocarbon Chains

Since most naturally occurring lipids contain unsaturated hydrocarbon chains, the question arises how double bonds influence the structure of hydrocarbon chains as well as the structure of the polar headgroup.

a. IR absorption bands from double bonds. For a comprehensive summary of double bond group vibrations, the reader is referred to Bellamy (1975). The C=C-stretching vibration is expected to absorb in the 1620-1680 cm^{-1} range with variable intensity. L-α-l-palmitoyl-2-palmitoleoyl-lecithin shows a weak absorption band near 1660 cm^{-1}. However, it may occur that natural phospholipids exhibit an intense absorption band near 1610 cm^{-1} as observed, e.g., in the case of egg-phosphatidylcholine, Fig. 17. It has turned out that this band only appears upon slight oxidation of the double bonds in unsaturated hydrocarbon chains. A reasonable explanation of this effect is the intermediate formation of an enolized β-diketone during the oxidation process.

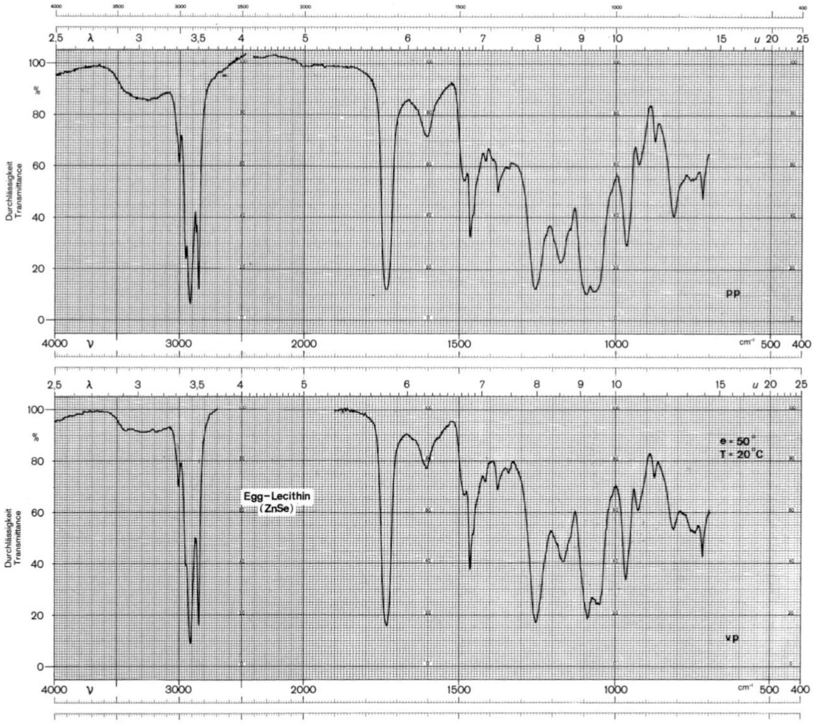

Fig. 17. Oriented layers of egg-lecithin on a ZnSe ATR plate. pp, parallel polarized; vp, perpendicular polarized; ϑ = 50°; T = 20°C

This group assumes an ionic resonance structure, which probably is the reason for the tremendously high absorption coefficient (100 or even more times the intensity of a usual ketone C = O band) commonly observed (Rasmussen et al., 1949). Therefore the 1610 cm^{-1} band may be used as very sensitive probe for the detection of slightly oxidized unsaturated hydrocarbon chains.

b. Influence of C=C double bonds on the structure of hydrocarbon chains. The decrease in hydrocarbon chain ordering by introducing double bonds is drastically demonstrated by the loss of the CH_2-wagging sequence, i.e., the loss of the all-trans conformation of the hydrocarbon chains (cf. Sect. IV.A.2.b and Figs. 6, 12, and 17).

c. Influence of unsaturated hydrocarbon chains on the polar headgroup. In the case of phosphatidylcholine (Figs. 6, 17) no significant influence could be detected. Phosphatidylethanolamine however showed drastic alterations depending on whether the hydrocarbon chains are saturated or not (Fringeli, 1977). It should be mentioned, however, that small amounts of impurities could also influence the structure of the polar part.

b) Isotopic Substitution

Isotopic modifications of a molecule generally exhibit the same molecular geometry and potential. The vibrational frequencies, however, may be altered due to change of the atomic masses. Under certain conditions conformational changes may occur, especially due to H↔D exchanges in crystals (cf. Sect. IV.A.2.d, *Special Features of Lysolecithin*, and Figs. 13, 16). In most cases, however, isotopic modifications facilitate the assignment of vibrational modes and are furthermore a prerequisite for reliable NCA calculations (Rihak, 1979).

c) Replacement of Both Fatty Acid Ester Groups by a Ketal Group

The main feature of this substitution is the loss of the wagging sequence γ_w (CH_2), although the hydrocarbon chains are expected to assume all-trans conformation already a few C-atoms after branching at the ketal group. A reasonable explanation is the lacking polar headgroup. As a consequence there is no enhancement of the oscillating dipoles of γ_w (CH_2) by inductive effects resulting in low intensity of the γ_w (CH_2) progression. No definite answer concerning the influence of the ketal group on the structure of the polar headgroup can be given at present, since overlapping with ketal group vibrations and crystallization phenomena complicate the interpretation.

d) Lysolecithin

Significant alterations of the structure of the polar headgroup are induced by hydrolization of the fatty acid ester in β-position. This subject is discussed in Sect. IV.A.2.d part (vi).

4. Influence of Temperature on Lipid Multilayers

a) Temperature-Induced Conformational Changes in Tripalmitin Oriented Multilayers

The structure of tripalmitin multilayers prepared by the Langmuir-Blodgett technique (Blodgett and Langmuir, 1937) has been investigated between room temperature and

62°C by means of IR-ATR spectroscopy (Fringeli et al., 1972). It was found that the bilayer structure exhibits crystalline features with pseudo-hexagonal structure. The methylene groups of the hydrocarbon chains are in the all-trans conformation. Increasing temperature produces at first a continuous conformational change in the glycerol part of the molecule. This melting process starts about 10°C below the melting of hydrocarbon chains. The latter is found to result in sudden disordering of the chains within 1°C. Later infrared and electron microscopic studies by Kopp et al. (1975a, 1975b) revealed that the results by Fringeli et al. (1972) have to be related to tripalmitin microcrystals and not to the initial Langmuir-Blodgett layers. The latter have turned out to be unstable (cf. Sect. IV.A.10.a).

b) DL-α-Dipalmitoyl Phosphatidylcholine

The sample consisted of 50% DL-α-DPPC and 50% H_2O by weight. Spectra were measured at 22°C and near the transition temperature of 41°C (Fig. 8 a-d). It was found that most hydrocarbon chains remain in the all-trans conformation until the transition temperature is reached. Then a sudden transition from all-trans to disorder is observed, indicated by the vanishing γ_w (CH_2) band progression (cf. Sect. IV.A.2.b). This finding is in analogy to tripalmitin oriented layers, however, no drastic conformational change in the polar part of lecithin could be detected.

5. Influence of Hydration on the Structure of Lecithin Multilayers

The main receptors for water molecules in both lecithin and lysolecithin are found to be the phosphate group, the carbonyl group and the choline group. Weak influence of hydration on the structure of hydrocarbon chains has been observed.

At ambient temperature solid oriented layers of lecithin bind water molecules with a wide range of association energies. Based on experimental evidence, loosely bound H_2O molecules associate in the solid with lecithin with reaction half-times in the order of seconds. This relatively fast kinetics allows use of conventional phase-sensitive detectors to rectify the periodic infrared signals resulting from a periodic variation of water vapor concentration in the surrounding atmosphere. This technique has been used recently to study the hydration sites of lecithin (Fringeli and Günthard, 1976). This modulation technique results in more detailed information of kinetics and molecular aspects than conventional spectroscopy (cf. Sect. III.B.2).

a) Hydration Modulation Spectroscopy of L-α-Dipalmitoyl Phosphatidylcholine

Oriented layers of DPPC, prepared on an ATR reflection plate, have been exposed to a mean relative humidity of 60%, 81%, and 85% at 30°C. The modulation amplitude was estimated to be smaller than ± 2% relative humidity at a frequency of 0.5 Hz. Typical modulation spectra are shown in Fig. 18 a-c. It should be noted that the reference phase for the zero degree channel ($\phi = 0°$) of the phase-sensitive detector was set in such a

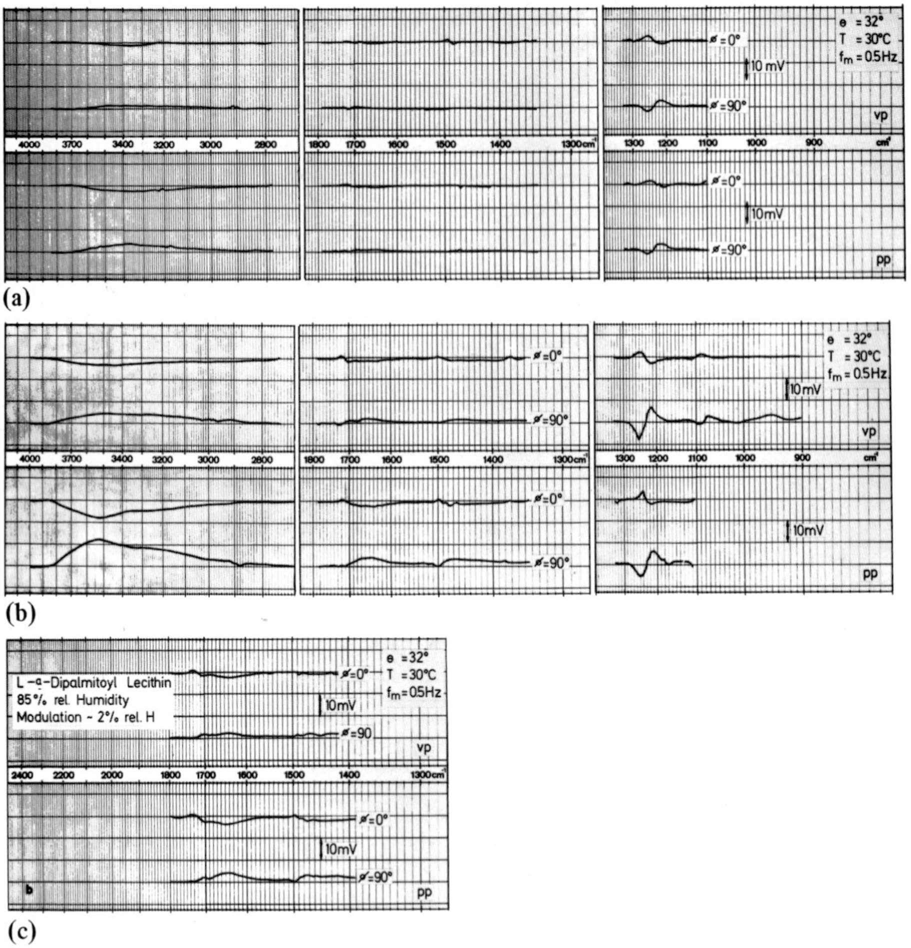

Fig. 18 a-c. Hydration-modulated excitation ATR infrared spectra of L-α-dipalmitoyl lecithin (DPPC). Modulation amplitude = ~ 2% rel. humidity; modulation frequency = 0.5 Hz; Φ, phase angle between stimulation and modulated infrared signal: Φ = 0°, Φ = 90°; sensitivity = 10 mV/div; pp, parallel polarized; vp, perpendicular polarized; ϑ = 32°; T = 30°C; germanium ATR plate. Mean relative humidities: 60% (**a**), 81% (**b**), 85% (c)

way that absorption bands appearing under the influence of increased humidity (in phase), e.g., bound H_2O absorbing in the 3400 and 1650 cm^{-1} region, direct downward, whereas absorption bands decreasing under the influence of increasing humidity (180° phase), e.g., a phosphate group vibration which is shifted to lower wavenumbers under the influence of hydration, direct upward. Overlapping bands with slightly different absorption frequencies and 0° and 180° phase may then give rise to dispersion-like band shapes ("biphasic" shape). The principle results of this study may be summarized as follows:

a. *Hydration of the phosphate group.* The phosphate group turned out to be the main receptor of water molecules. This is demonstrated by the biphasic modulation bands near 1220 and 1090 cm^{-1}.

b. *Hydration of the carbonyl group.* This remains unaltered when the relative humidity is modulated at 60%, since no ν (C=O) modulation band is observed in Fig. 18a. At 81% mean relative humidity however C=O group hydration is indicated by a weak biphasic band in the 1730 cm^{-1} region (Fig. 18b). The hydration effect is further enhanced when modulation occurs at a mean value of 85% relative humidity (Fig. 18c). This finding however does not imply that the carbonyl group is completely dehydrated below 80% relative humidity, since the binding kinetics of the water molecules bound already below 80% relative humidity could be so slow that no significant change is achieved within the modulation period of 2 seconds. From stationary IR measurements it is known that the lecithin sample has to stay several hours at elevated temperature under dry nitrogen flow for complete dehydration.

c. *Hydration of fatty acid ester groups.* The shoulder at 1180 cm^{-1} observed in MEIR spectra measured at relative humidities above 80% indicates that this group vibration is affected synchronously with the carbonyl-stretching vibrations ν (C=O).

d. *Hydration of the choline group.* There is no doubt that the choline group is also affected by hydration, since weak modulation bands are observed in spectral regions where typical choline absorption bands are expected. These regions are 1490, 1050, and 960 cm^{-1}. No definite explanation is available of whether choline modulation bands result from direct hydration or from hydration-induced conformational changes.

e. *Influence on the structure of hydrocarbon chains.* The broad band in the 1220 cm^{-1} region is weakly structured by MEIR bands of the wagging progression. There are several explanations for this observation:
i) Weak modulation of the wagging absorption coefficients by periodic hydration of the ester group, i.e., modulation of the vibrational coupling between ester group and CH_2-wagging motion. It was mentioned earlier (Sect. IV.A.2.b) that wagging modes obtain almost all intensity by coupling with a polar headgroup. If this explanation is true, no conformational changes in the hydrocarbon chains would be expected.
ii) Spatial reorientation of the hydrocarbon chains under the influence of hydration, without changing the initial chain conformation.
iii) Hydration-induced disturbance of all-trans structure near the ester group.
There is some evidence that (ii) or (iii) is more probable than (i), because the amplitude ratios $A_{90°} / A_{0°}$ (ν_{as} (PO_2^-)) and $A_{90°} / A_{0°}$ (ν (C=O)) are different from $A_{90°} / A_{0°}$ (γ_W (CH_2)), i.e., different phase responses to the periodic stimulation exhibit different kinetics at the processes. If (i) were true, no phase difference should occur.
Thus a weak hydration-induced effect on the hydrocarbon chains may be assumed, e.g., due to a biaxial → $L_{\beta'}$ phase transition (Powers and Pershan, 1977).

b) Egg-Lecithin

Analogous experiments to those in Sect. IV.A.5.a have been carried out by Fringeli and Günthard (1976) with oriented multilayers of egg-lecithin. The mean relative humidity was 75%. Apart from the lacking γ_W (CH_2)-MEIR bands, which are mostly absent in natural lipids at ambient temperature (cf. IV.A.3.a (b)), no principal differences concerning the hydration sites are detected with respect to DPPC. However, it should be mentioned that the ν (C=O) modulation band in the 1730 cm^{-1} region has significantly more intensity than the corresponding band in DPPC. This fact could result from a slightly different structure of the polar headgroup of synthetic and natural phosphatidyl choline, allowing water molecules to penetrate more easily to the carbonyl groups in the latter case. Corresponding information was obtained from stationary spectra of synthetic dipalmitoyl phosphatidylethanolamine (Fig. 3) and sheep brain cephalin. DPPE did not bind significant amounts of water in atmospheres of 90% relative humidity, whereas the natural cephalin could be hydrated easily.

c) Lysolecithin

Lysolecithin turned out to be significantly more hygroscopic than DPPC or egg-lecithin. Already at 50% mean relative humidity quite intense MEIR absorption bands are obtained (Fig. 19b). The intensity of the MEIR bands associated with δ (OH) (1390 cm^{-1}) and ν (OH) (3200-3300 cm^{-1}) is very weak. This points to strong intramolecular hydrogen bonding. D_2O modulation experiments support this finding since H→D exchange is found to be too slow to result in a measurable MEIR signal at a modulation frequency of 0.5 Hz. Furthermore there is good evidence for a disturbance of hydrocarbon chains induced by hydration of the polar part of the molecules since distinct MEIR bands of ν (CH) are observed at 2855 cm^{-1} and 2925 cm^{-1}. On the other hand the δ_{as} (CH_3), δ_{as} (N-CH_3), and δ_s (N-CH_3) region shows an unusually broad and unstructured absorption. Both features may be related to overlapping spectra resulting from different conformations of the choline headgroup. No clear information via γ_w (CH_2) is available since the progression is overlapped by the prominent ν_{as} (PO_2^-) MEIR band.

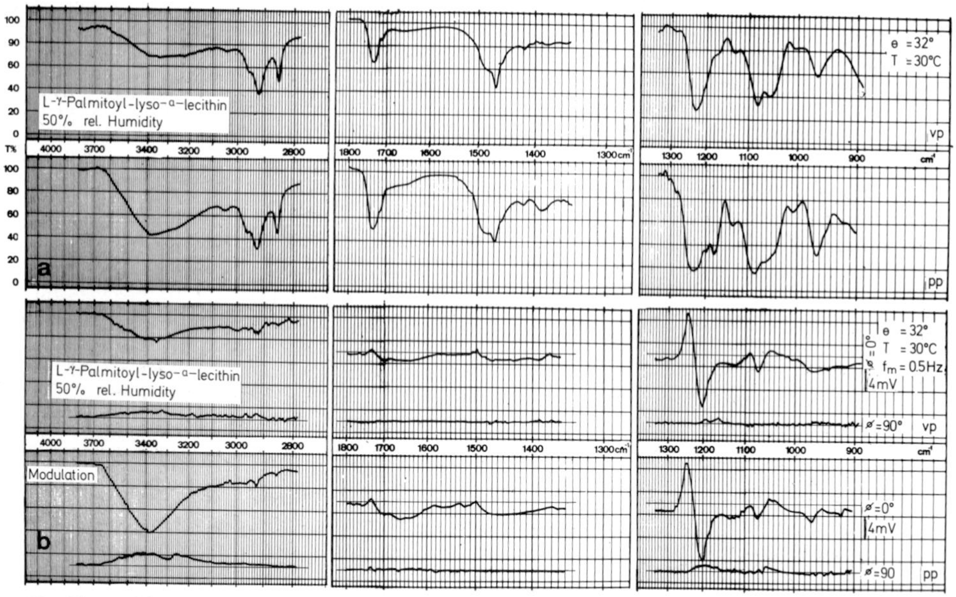

Fig. 19 a and b. Oriented layers of L-γ-palmitoyl-lyso-α-lecithin (lysolecithin) on a germanium ATR plate.
pp, parallel polarized; vp, perpendicular polarized; ϑ = 32°; T = 30°C; mean relative humidity 50%. **a** Conventional ATR infrared spectra; **b** hydration-modulated excitation ATR infrared spectra. Modulation amplitude = ~ 2% rel. humidity; modulation frequency = 0.5 Hz; phase angle between stimulation and modulated infrared signal: Φ = 0°, Φ = 90°; sensitivtiy = 4mV/div

6. Bilayer Spectra of Stearic Acid

The spectra presented in Fig. 20 a-c result from one double layer with head-head arrangement. A stearic acid monolayer was spread on aqueous subphase of pH 3.0-3.5. The pres-

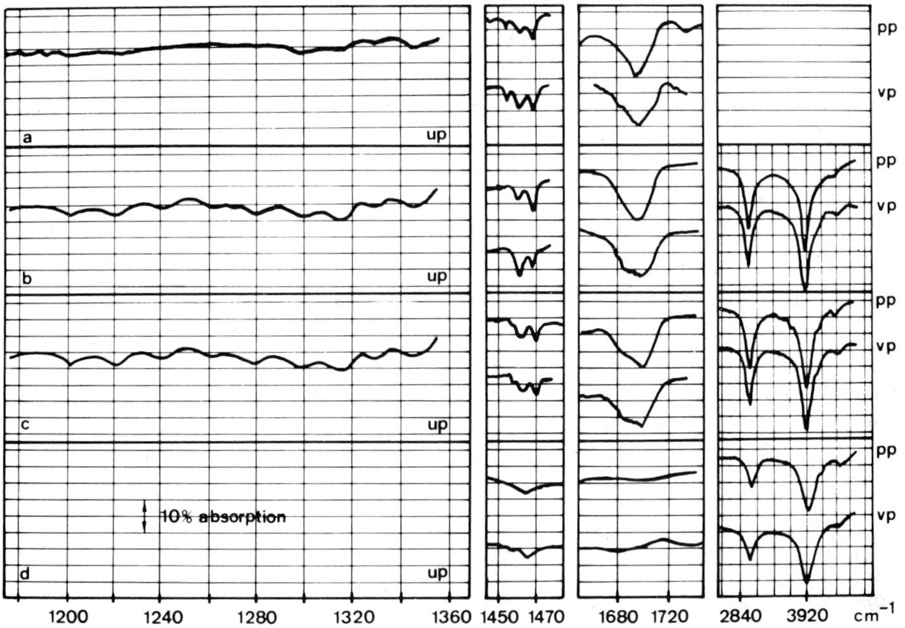

Fig. 20 a-d. Stearic acid bilayer (Langmuir-Blodgett-type, head-head arrangement) (**a-c**), and stearic acid monolayer (**d**) on a germanium ATR plate.
pp, parallel polarized; vp, perpendicular polarized; up, unpolarized; $\vartheta \cong 25°$; T = 30°C.
a Double layer transferred at 20 dynes/cm; **b** double layer transferred at 30 dynes/cm; **c** rectystallized double layer. Final state of (**a**) and (**b**) reached after ~ 24 h; **d** first monolayer with carboxylic acid group attached to the germanium plate

sure-area diagram exhibits a discontinuity at a film pressure of ≈ 25 dynes/cm (Gaines, 1966). Three monolayers were transferred to a germanium ATR plate by means of the technique of Blodgett and Langmuir (1937). One sample was prepared at 20 dynes/cm, i.e., below the discontinuity (Fig. 20a), the other at 30 dynes/cm, above it (Fig. 20b). The reference plate was coated with one monolayer in order to compensate the first monolayer on the sample plate. A marked dependence of the bilayer structure on film pressure is observed, namely:

1. *ν (C=O).* No polarization is observed in the low pressure (LP) film (Fig. 20a), however strong y-polarization is exhibited by the high pressure (HP) film (Fig. 20b). It should be noted that y-polarization of this extent requires anisotropic molecular arrangement in the xy-plane, which is achieved by withdrawing and dipping the ATR plate along the x-direction during the film preparation (cf. Fig. 2).
2. *δ (CH_2).* The LP film exhibits three unpolarized bands in the 1456-1470 cm^{-1} region, whereas in the HP film two polarized CH_2-bending absorption bands appear. The same polarization was observed in stearic acid single-crystal ATR spectra by Münch et al. (1977).
3. *γ_w (CH_2).* No wagging progression could be observed in the LP film, but weak wagging bands appear in the HP film. Thus one may conclude that the hydrocarbon

chains of the spread monolayer exhibit all-trans conformation, when the film pressure is greater than 25 dynes/cm.

4. *Instability of Langmuir-Blodgett-layers.* Twenty-four hours after preparation of both LP and HP films, both spectra have become equal (Fig. 20c). This phenomenon most probably results from recrystallization of the Langmuir-Blodgett-type layers (cf. Sect. 4.a.x).
   Figure 20d shows the ATR spectrum of the first stearic acid monolayer attached to the germanium plate. It should be mentioned that the δ (CH_2) band at $\approx$ 1465 cm^{-1} shows no splitting as observed in the bilayer spectra Fig. 20 a-c. Furthermore no ν (C=O) band could be detected at $\approx$ 1700 cm^{-1}, probably due to the interaction of the carboxylic acid group with the hydrophilic germanium surface.

7. Infrared Spectra of Natural Lipids

a) Sheep Brain Cephalin

The features of cephalin from sheep brain with respect to DPPE have been discussed in section IV.A.2.d part (ii).

b) Egg-Lecithin

For comparative discussion of the structures of egg-lecithin and DPPC, the reader is referred to sections IV.A.2.b and IV.A.3.a.

c) Mixture of Natural Lipids

The lipids from the electric organ of *Torpedo marmorata* have been extracted by Hopff (1975). The mixture consisted of 3.5% sphingomyeline, 26% lecithin, 13% phosphatidyllinositol, 22% phosphatidylethanolamine and phosphatidylserine, 3% polyglycerol, 16% cholesteryl triglycerol, and 16.5% of unidentified components. The spectrum of the dry lipid extract (Fig. 21a) shows the typical features of natural lipids at ambient temperature: broad, weakly structured absorption bands and low polarization, which means low degree of molecular ordering. Weakly polarized absorption bands are observed near 830 cm^{-1} (probably ν_{as} (P-O)), 1090 cm^{-1} (ν_s (P=O)), 1120 cm^{-1} (probably ν (C-C)). This last band could be used to determine the mean direction of hydrocarbon chains. Furthermore distinct polarization is shown by the asymmetric CH-stretching vibrations of the double bonds at 3010 cm^{-1}. The dichroic ratio is R^{ATR} (ν_a (CH=CH)) = 1.70 ± 0.10, i.e., the z-axis is the predominant mean direction of the double bonds. The ν (C=C) absorption band at 1650 cm^{-1} exhibits a dichroic ratio of R^{ATR} (ν (C=C)) = 1.40 ± 0.15, which is in agreement with the behavior of ν_{as} (CH=CH). One should expect that ν_a (CH=CH) is a better normal coordinate than ν (C=C), which probably is coupled with adjacent ν (C-C) modes. The 3010 cm^{-1} band should therefore be used to determine the mean direction of C=C double bonds. Furthermore the ν (C=C) band is ob-

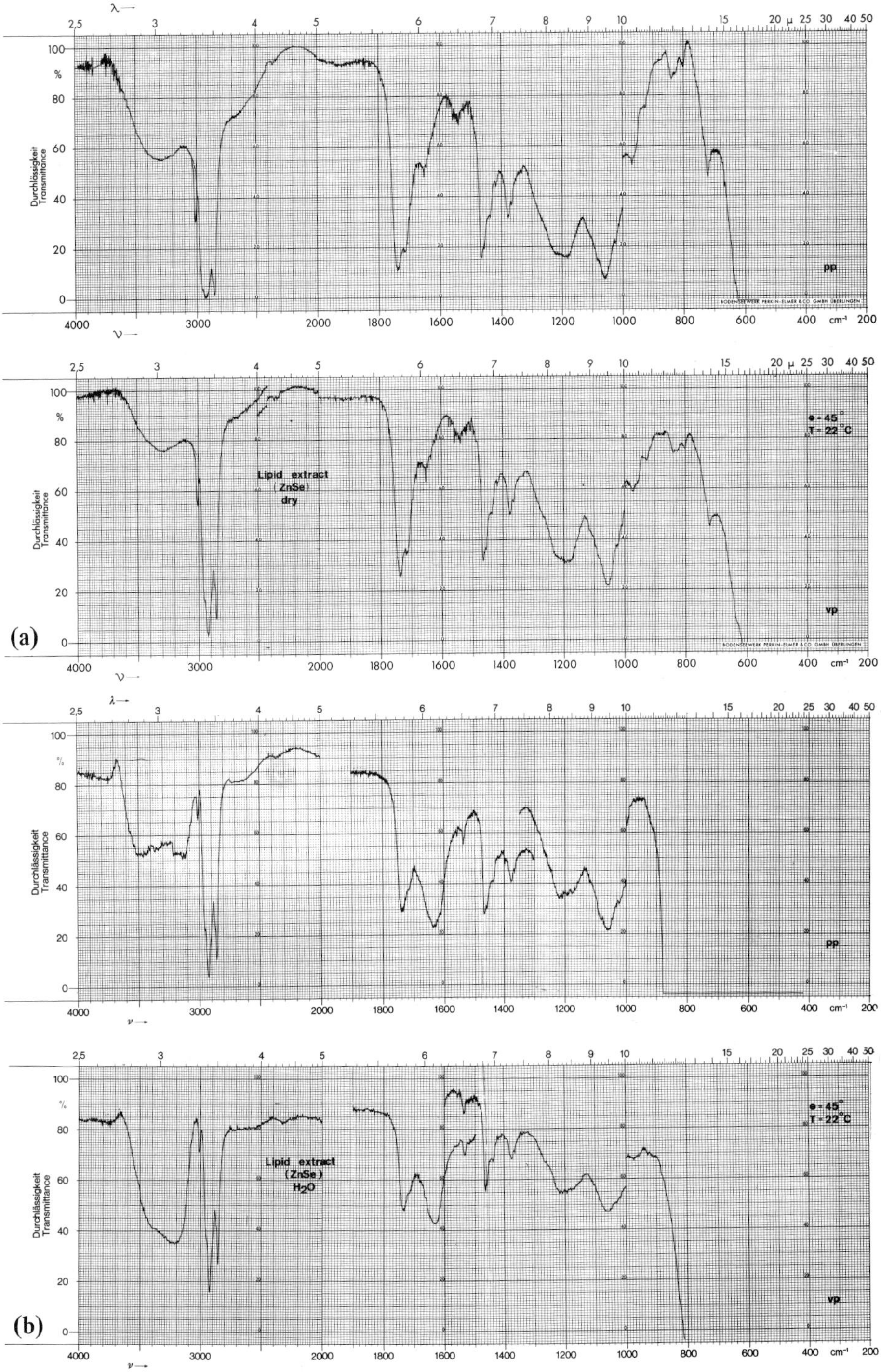

Fig. 21 a and b. Oriented layers of the lipid extract from the electric organ of *Torpedo marmorata* on a zinc selenide ATR plate.
pp, parallel polarized; vp, perpendicular polarized; $\vartheta = 45°$; T = 22°C.
a Layers exposed to dry nitrogen gas; **b** layers exposed to liquid water. The broad absorption bands in the 3300 cm^{-1} and 1640 cm^{-1} region result from incomplete compensation of liquid water absorption

scured by δ (H_2O) as soon as the sample is hydrated from the gas phase or is in contact with liquid water. The latter is the case in Fig. 21b. The principal amount of liquid water absorption has been compensated in the reference beam. A complete compensation of H_2O absorption bands is, however, not possible, since water adsorbed to the lipid layers exhibits a slightly different absorption spectrum than water in the bulk phase (Zundel, 1969).

8. Interaction of Lecithin with Neodymium Ions

Lanthanide ions are often used as shift reagents in NMR spectroscopy (Hauser, 1976; Hauser et al., 1976). Since these ions bind to the phosphate group the principal question arises whether induced conformational changes occur or not, i.e., whether the structure of the polar headgroup derived from such NMR experiments corresponds to the natural structure or to a structure which is altered due to lanthanide ion complexation.

IR-ATR measurements with dry oriented layers of lysolecithin show drastic spectral changes of typical phosphate group, hydroxyl group, and hydrocarbon chain absorption bands. Hydration in an atmosphere of 60% relative humidity restores the original conformation to a small extent. La (III) cations are obviously chelated by lysolecithin with the $> PO_2^-$ and -OH groups acting as ligands. No infrared data obtained under the same experimental conditions of Hauser (1976) and Hauser et al. (1976) are available at this time. However, in view of the results discussed below one should be aware that structural changes induced by La (III) cation binding could occur even under the conditions used by these authors. Finally it is surprising that the precision claimed by these authors did not provide information about the existence of choline-part trans and gauche conformers. Such conformers were detected in bilayers and should exist also apparently in vesicles.

Lysolecithin-Nd^{3+} – Interaction. Lysolecithin was dissolved in water (5% weight) containing $NdCl_3$ at a molar ratios of 10:1, 2:1, and 1:1 (lipid: Nd^{3+}). Oriented layers were obtained by evaporation of the solvent. IR-ATR spectra were recorded at 25°C and relative humidities of 15% and 60%, respectively, Fig. 22.

The structure of undisturbed layers of lysolecithin (Fig. 13) was discussed earlier (cf. Sect. IV.A.2.a to IV.A.2.d). Nd^{3+}-doped layers at 15% relative humidity (Fig. 22a) show alterations in the phosphate group. ν_{as} (PO_2^-) is shifted from ≈ 1240 to ≈ 1210 cm^{-1} and ν_s (PO_2^-) from ≈ 1090 to ≈ 1110 cm^{-1}. The most striking observation, however, is the loss of the wagging progression and the strong decrease of the OH-bending absorption band at 1390 cm^{-1}. A reasonable explanation is complexation of the Nd^{3+} ion by both phosphate group and hydroxyl group. Evidently this complexation requires a conformational change of the glycerol part which also affects the all-trans conformation of the γ-hydrocarbon chain, thus erasing the γ_w (CH_2) progression. On the other hand no significant conformational change of the choline part could be detected. Increasing hydration of the sample partially restores the initial lysolecithin conformation in the phosphate, glycerol, and hydrocarbon chain region (Fig. 22e,f). This fact may result from a partial substitution of the β-OH group as ligand of Nd^{3+} by H_2O.

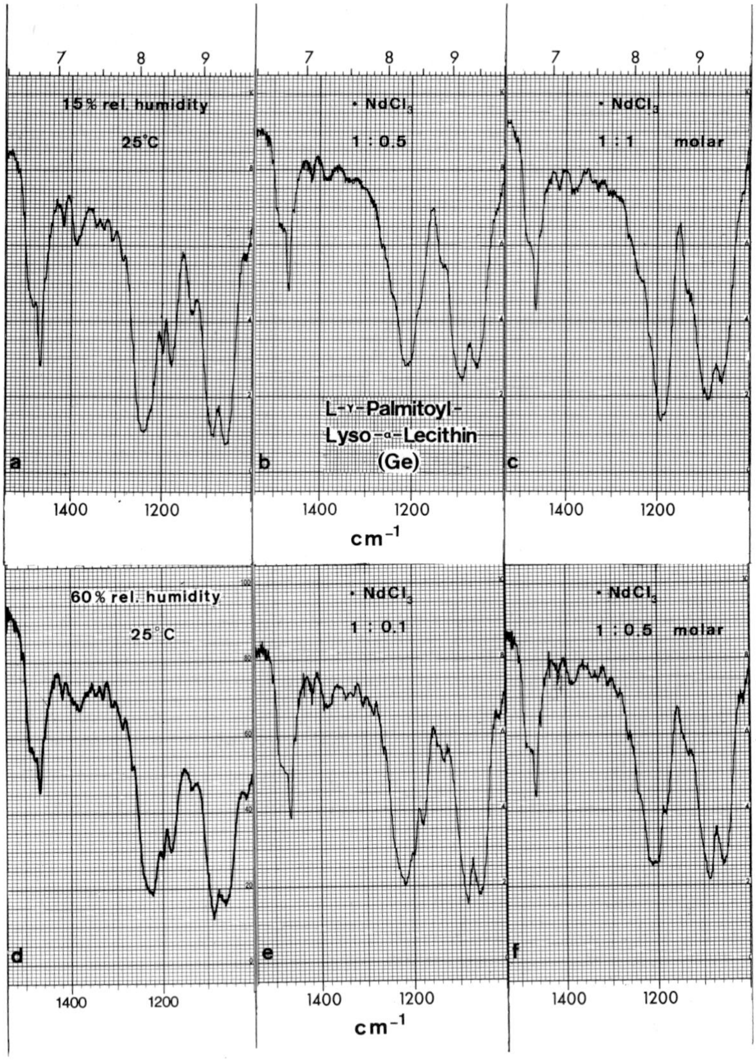

Fig. 22 a-f. Oriented layers of L-γ-palmitoyl-lyso-α-lecithin interacting with Nd^{3+}-ions in dry (b,c) and humid (e,f) atmosphere. Germanium ATR plate. Parallel polarized indicdent light, $\vartheta = 45°C$, $T = 22°C$. Relative humidity ~ 15%: **a** pure lysolecithin; **b** lysolecithin and $NdCl_3$, molar ratio 1:0.5; **c** lysolecithin and $NdCl_3$, molar ratio 1:1. Relative humidity ~ 60%: **d** pure lysolecithin; **e** lysolecithin and $NdCl_3$, molar ratio 1:0.1; **f** lysolecithin and $NdCl_3$, molar ratio 1:0.5

9. Conformational Changes in Phospholipids under the Influence of an Electric Field

Little work has been done in this field so far. The principal reason could be insufficient spectrometer sensitivity, since the signals to be expected are of slight intensity, when compared with the undisturbed absorption bands. May et al. (1970) have studied the influence of electric fields on brain cephalin and lecithin films. The applied field strengths were varied in the range of 10^4-10^6 V/cm. Resulting spectral changes, although very weak signals, were interpreted as conformational changes ocurring first in the phosphate group and at higher field strength in the ammonium group region. It should be noted that no spectral changes could be observed in the case of synthetic cephalin and lecithin, both containing saturated hydrocarbon chains.

Application of ATR modulation technique with polarized IR light should enhance the sensitivity considerably.

10. Rearrangement Phenomena of Lipid Multilayer Assemblies

a) Langmuir-Blodgett Layers of Ba-Stearate and Tripalmitin

The stability of mono- and multilayers of barium stearate, cadmium arachidate, and tripalmitin has been studied by means of electron microscopy and ATR spectroscopy by Kopp et al. (1975a,b). The layers were found to rearrange spontaneously from regular films into ultrastructures with the shape of irregular, flat islands of varying thickness. The kinetics of the phase transformation of the first layer depends on the substrate, that of successive layers appears to depend on the total number of layers, the temperature, the surrounding medium, and small amounts of impurities.

In the case of tripalmitin, infrared spectra revealed that Langmuir-Blodgett layers transferred at 15 dynes/cm are in a liquid crystalline state exhibiting disordered hydrocarbon chains and furthermore some flexibility in the glycerol part. The typical features of liquid crystalline layers are the lacking wagging progression γ_w (CH_2), a broad methylene bending band δ (CH_2) at 1469 cm^{-1}, (1473 cm^{-1}), a broad unstructured carbonyl stretching band ν (C=O) at 1735 cm^{-1}, (1736, 1728 cm^{-1}), as well as a typical shift and broadening of the ν (C-O) ester group vibration at 1169 cm^{-1} (1182 cm^{-1}). The positions of the corresponding bands in the microcrystalline sample are given in brackets. Recently the kinetics of the rearrangement of 4 tripalmitin layers has been studied based on line shape analysis of the δ (CH_2) and ν (C-O) bands (Kopp et al., 1975b).

In a recent X-ray study, on the other hand, Lesslauer (1974) has reported stable Ba-stearate multilayers. His findings are supported by optical investigations of Kuhn (1972).

In the view of the IR spectroscopic observations made with tripalmitin layers (see above) as well as with staric acid Langmuir-Blodgett layers (cf. Sect. IV.A.6), one may conclude that this type of layer exhibits an intrinsic instability, which may be more or less evident, depending on the conditions of preparation and storage as well as on the nature of lipids.

b) Rearrangement of Phospholipid Multilayers

Many synthetic phospholipid multilayers prepared by evaporation of the solvent (cf. Sect. IV.A.1) show recrystallization phenomena similar to tripalmitin (cf. IV.A.10.a). Although DPPC turned out to exist in stable multilayers, crystallization was observed in the case of DPPE, DPPME, DPPDME, and some phospholipid analogs such as ketal compounds as well as 1-oleoyl-2-n-hexadecyl-2-deoxy-lecithin. It was found that the crystallization tendency depends strongly on the nature of the solid support (KRS-5, ZnSe, Ge) in analogy to electronmicroscopic findings by Kopp et al. (1975a).

11. Radiation Damage in Lipid Multilayers Produced in High-Resolution Electron Microscopy

Progress in molecular electron microscopy is limited by radiation damage. Molecular degradation of organic materials in the electron microscope has been investigated by numerous authors, among them the works of Stenn and Bahr (1970) and of Isaacson (1975) should be mentioned. Loss of mass, loss of specific elements, loss of crystallinity, changes in the optical and in the energy loss spectra, and changes in the electron microscopic image are methods often applied in studying radiation damage. ATR-IR spectroscopy has turned out to be an efficient tool to study radiation damage, because very thin layers (less than 100 Å) can be investigated. This has the advantage of similarity of the test specimen to the real electronmicroscopic preparations, especially membrane-like layer systems. Preliminary experiments with Langmuir-Blodgett-type tripalmitin layers have been reported by Hahn et al. (1974). More recently Baumeister et al. (1976a) have used the same tripalmitin model membranes, which were irradiated with 100-keV electrons in a microscope. The intensity decay of typical group vibrations such as ν_{as} (CH_3), ν_s (CH_2), δ (α-CH_2), δ (CH_2), δ_s (CH_3), γ_w (CH_2), ν (C-O), and ν (C-C) has been determined quantitatively. The C-C backbone showed a significant latency effect up to doses of 0.6 $e^-/Å^2$ and was completely disordered at 3 $e^-/Å^2$, corresponding to about three inelastic processes per molecule. The decay of the C-C frame is initiated by conformational changes in the hydrocarbon chains monitored through γ_w (CH_2) and δ (CH_2) and by H-abstraction monitored through ν_s (CH_2). It has been shown recently that the randomization of the secondary structure of β-lactoglobulin (Baumeister et al., 1976b) and of Langmuir-Blodgett-type polypeptide monolayers (Baumeister et al., 1976c) occurs also in the dose range between 0.5 and 2.0 $e^-/Å^2$.

Finally it should be mentioned that the required dose for electron micrography with molecular resolution must be orders of magnitude higher than the doses applied in these experiments.

B. Infrared Spectroscopy of Polypeptides and Proteins

For a comprehensive discussion of conformational analysis of polypeptides and proteins the reader is referred to the reviews of Susi (1969) and Fraser and MacRae (1973). Typical amide group vibrations were found to be of diagnostic interest. The results of extensive investigations in this field are summarized in Table 3. If polarized light is applied, the analysis requires knowledge of the direction of the oscillating dipole moment. Since the amide group is assumed to have C_s symmetry (trans conformation) the transition moments are expected to be oriented parallel or perpendicular to the plane of symmetry. More detailed information on the transition moment direction of in-plane modes is available from Fraser and MacRae (1973).

1. Effect of Coupling of Amide Groups

In proteins and polypeptides the vibration in different amide groups are coupled through the main chain and through the formation of hydrogen bonds. Both the frequencies of the normal modes and the direction of the transition moments are affected by coupling. Since different types and degrees of chain ordering generally occur in polypeptides and proteins, spectral analysis may become very complicated.

Therefore most investigations have concentrated on spectral analysis of the regular portions of the main chain, assuming that regular and irregular portions may be treated separately. Vibrational analysis of the regular conformation is generally approached through analysis of the modes of an isolated chain of infinite length. The procedure has been described by Liang et al. (1956), Liang and Krimm (1956), Krimm (1960), and Elliott (1969). Miyazawa (1960c) has shown that the vibrational modes of the amide groups in an infinite chain may be approximated by

$$\nu(\delta) = \nu_o + \sum_j D_j \cos(j\delta) \tag{16}$$

where ν_o is the corresponding frequency of the isolated amide group, and D_j is a coupling constant between the reference amide group and the jth group along the chain. δ denotes the phase difference between the corresponding atomic motions of successive groups.

Equation (16) considers only intrachain interactions, however the expression may be completed for intra- and interchain interactions (Miyazawa, 1960c), which is of importance in the case of parallel or antiparallel β-structures, where intermolecular hydrogen bonding plays an important role (Pauling et al., 1951; Pauling and Corey, 1951). The amide group frequency is then given by

$$\nu(\delta,\delta') = \nu_o + \sum_j D_j \cos(j\delta) + \sum_j D'_j \cos(j\delta') \tag{17}$$

where D'_j and δ' are interchain coupling constant and interchain phase angle of successive groups, respectively.

α-Helix. Equation (16) may be applied in the case of helical conformation, since intramolecular coupling is predominant. Higgs (1953) has derived the IR selection rules for an isolated infinite α-helical chain. He has found that there are only two infrared active vibrations, corresponding to the phase angles $\delta = O$ and $\delta = \pm\chi$, respectively. χ denotes the angle between two successive groups relative to the helix axis, which was found to be $\chi = \frac{2\pi}{3.6}$ (Pauling et al., 1951). The transition moment associated with $\delta = O$ is parallel to the helix axis and the transition moments for the modes with $\delta = \pm\chi$ are mutually perpendicular and normal to the helix axis. Since hydrogen bonds usually are formed between every third group, one may expect that apart from the nearest-neighbor coupling constand D_1 also D_3 must be taken into account, however Krimm (1962) reported that D_2 is also significant.

Thus two frequencies should be observed:

$$\begin{aligned} \nu(o) &= \nu_o + D_1 + D_2 + D_3 \ (\|) \\ \nu(\chi) &= \nu_o + D_1 \cos\chi + D_2 \cos(2\chi) + D_3 \cos(3\chi) \ (\perp) \end{aligned} \tag{18}$$

With the latter being twofold degenerate.

Recently Nevskaya and Chirgadze (1976) have reported an analysis of amide I and II vibrations, based on perturbation theory in a dipole-dipole approximation. The authors have calculated amide I/II vibrations of helices of finite length. The results may be of importance for structural analysis of small peptides ($n \leqslant 18$) by means of IR spectroscopy.

Parallel-Chain β Conformation. The unit cell contains two succeeding groups of the same chains which can move either in phase ($\delta = 0$) or out of phase ($\delta = \pi$). Coupling to adjacent groups in a neighboring chain occurs through intermolecular H bonds. Only the intermolecular in-phase vibrations ($\delta' = 0$) lead to infrared active modes (Miyazawa, 1960c). Thus each amide band should split into a doublet, exhibiting parallel (||) and perpendicular (⊥) polarization, respectively. The frequencies are expected at

$$\begin{aligned} \nu(0,0) &= \nu_o + D_1 + D_1' \ (\|) \\ \nu(\pi,0) &= \nu_o - D_1 + D_1' \ (\perp). \end{aligned} \tag{19}$$

Antiparallel-Chain β Conformation. This arrangement involves segments of two chains with 4 peptide groups per unit cell. Four vibrations are expected, one of them infrared inactive (Miyazawa, 1960c), namely:

$$\begin{aligned} \nu(0,\pi) &= \nu_o + D_1 - D_1' \ (\|) \\ \nu(\pi,0) &= \nu_o - D_1 + D_1' \ (\perp, \text{in plane}) \\ \nu(\pi,\pi) &= \nu_o - D_1 - D_1' \ (\perp, \text{out of plane}) \\ \nu(0,0) &= \nu_o + D_1 + D_1' \ (\text{inactive}). \end{aligned} \tag{20}$$

The amide I vibrations of both β conformations have been studied recently by Chirgadze and Nevskaya (1976a,b). Results for finite polypeptide chain length have been derived.

2. Analysis of Amide Bands in Polypeptides and Proteins

Numerical values of amide frequencies are summarized in review articles by Susi (1969), Fraser and MacRae (1973), and by Wallach and Winzler (1974). Only two typical examples will be mentioned here:

(i) pH-Induced Conformational Change of Poly-L-Lysine. Poly-L-Lysine assumes random coil conformation in aqueous solution of pH 9. Increase of pH to 11.9 induces a time-dependent conformational change to antiparallel β-structure where the α-helical form appears as transient (Susi et al., 1967). The corresponding time-dependent spectra measured in D_2O solution are shown in Fig. 23. This example demonstrates that under favorable conditions a clear distinction between β-structure and α-helix or random coil is possible. A distinction between corresponding amide I bands of α-helix and random coil, respectively, becomes difficult. In the case of oriented fibrous proteins and polypeptides however, a distinction appears possible by means of the application of both polarized light and band shape analysis. A typical example is given below.

(ii) Separation of the Components of the Amide I Band of Bombyx Mori Fibroin. This investigation by Suzuki (1967) yielded a nice confirmation of theoretical consider-

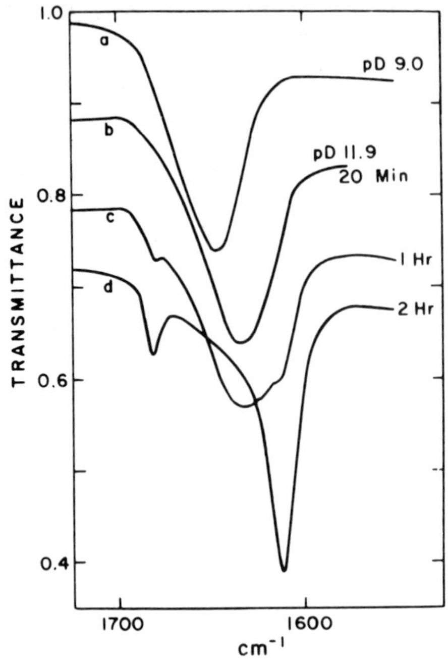

Fig. 23 a-d. Rearrangement of poly-L-lysine in D_2O solution as reflected in the infrared transmission spectrum. Path length, 0.1 mm; concentration, 0.4 wt.%; pure D_2O as reference; consecutive spectra displaced by 0.1 scale units (upper curve is on scale). Curves: a, random conformation, pD 9.0; b, α-helix, pD 11.9, 20 min; c, transition between α-helix and antiparallel-chain β-conformation, pD 11.9, 1 h; d, antiparallel-chain β-conformation, pD 11.9, 2 h (Susi et al., 1967)

ations reported in Sect. IV.B.1. The amide I bands shown in Fig. 24 are consistent with a fibrous protein consisting of parts exhibiting β-structure and noncrystalline regions, i.e., flexible regions and/or regions where corresponding chain segments assume different conformations. The results of this study are collected in Table 6. Obviously the application of polarized infrared light and band shape analysis turned out to be an efficient tool for secondary structure determination in oriented proteins. Among many applications the works of Loeb and Baier (1968), Loeb (1969) and Malcolm (1970) should be mentioned. In these studies synthetic polypeptides or proteins which had been oriented by spreading on a water surface followed by compression of the film were investigated. Transfer of the oriented layers was performed by the technique of Blodgett and Langmuir (1937).

Further extensive use of line shape analysis has been made by Rüegg et al. (1975) in the infrared study of the amide I, II, and III band regions of ribonuclease, lysozyme, chymotrypsin, myoglobin, alpha-casein, alpha-lactoalbumin, beta-lactoglobulin A, and beta-lactoglobulin. The fraction of β-structure determined in these proteins was found to be in good agreement with X-ray data.

3. Enzyme-Substrate Interaction

Very little is kown concerning the nature of enzyme-substrate complexes not only from infrared spectroscopy but also from all the other physical techniques applicable to such problems.

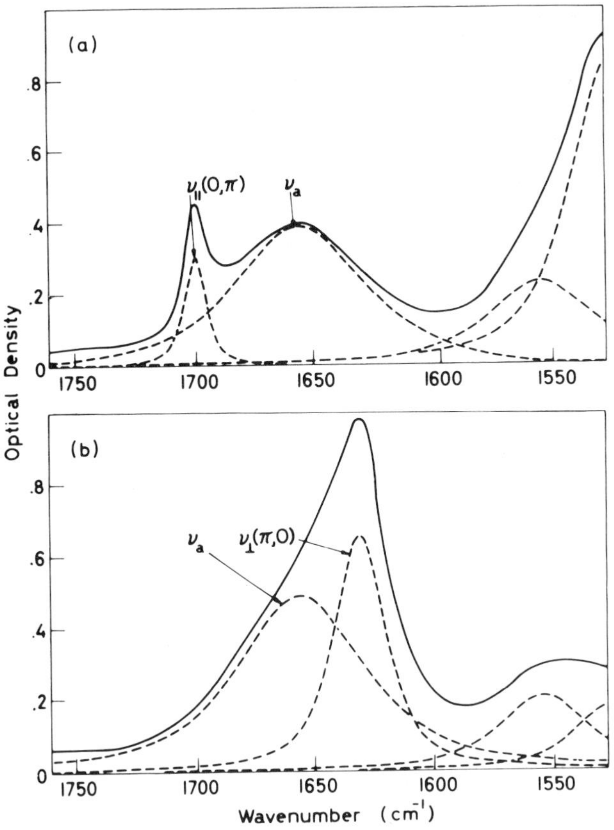

Fig. 24 a and b. Separation of the components of the amide I band in the infrared transmission spectrum of *Bombyx mori* silk.
a Spectrum obtained with the electric vector vibrating parallel to the fiber axis; **b** spectrum obtained with the electric vector vibrating perpendicular to the fiber axis (Suzuki, 1967)

Table 6. Band parameters of amide I components in the infrared spectrum of *Bombyx mori* fibroin (Suzuki, 1967)

Component[a]	Frequency of maximum (cm^{-1})	Half bandwidth (cm^{-1})	Peak absorbance Parallel	Peak absorbance Perpendicular
$\nu_{\perp}(\pi,0)$	1630	25.9	0.000	0.655
$\nu_{\parallel}(0,\pi)$	1699	11.3	0.300	0.005
ν_r	1656	66.5	0.380	0.490

[a] $\nu_{\perp}(\pi,0)$ and $\nu_{\parallel}(0,\pi)$ are vibrational modes of the antiparallel-chain pleated sheet (cf. Eq. [20]), ν_r is associated with noncrystalline material.

One may expect that application of chemical modulation technique in IR spectroscopy should result in details about the nature of the enzyme-substrate complex. Such experiments may be carried out analogous to those performed for hydration site determination of oriented lipid layers (cf. Sect. IV.A.5).

Recently, Fringeli and Hofer (1980) have got some evidence for electrostatic control of the activity of membrane bound acetylcholinesterase.

C. Lipid-Protein Interaction

Few reliable infrared studies have been reported in this field. However, ATR infrared technique may result in promising information of lipid-protein interaction on a molecular level. Some examples will be depicted below.

1. Interaction of Hemoglobin with Arachidic Acid and Methyl Arachidate

Fromherz et al. (1972) reported on arachidic acid and methyl arachidate spread on a water surface. Hemoglobin was then adsorbed from the aqueous subphase, followed by a transfer of four double layers to a germanium ATR plate by means of the Langmuir-Blodgett technique (Sect. IV.A.1). The principal features of this study may be summarized as follows:

1. The arachidic acid-hemoglobin system exhibits predominantly hydrophilic or electrostatic interaction between the carboxyl group and hydrophilic sites of the hemoglobin. No significant alteration was observed in the amide I region, leading to the assumption that the predominant α-helical conformation of native hemoglobin remains unaltered. However, chain restructuring to random coil cannot be excluded.
   The nature of protein-lipid interaction was mainly concluded from the appearance of a band at ~ 1540 cm^{-1} leading to a skewing of the amide II band. The new 1540 cm^{-1} band was assigned to ν (C=O) of $-COO^-$. Since layers were prepared at pH ~ 4, the formation of the ionized carboxyl group could only result from a proton transfer to the protein, thus indicating ionic protein-lipid interaction. However, it should be mentioned that the same skewing of the amide II band of hemoglobin is also observed in the absence of arachidic acid (Fringeli, unpublished work). Furthermore the lack of the δ (CH_2) band-splitting after protein binding indicates an induced alteration of the arachidic acid crystal structure. Whether this effect results from hydrophilic and/or hydrophobic protein-lipid interaction could be concluded from the behavior of the CH_2-wagging progression γ_w (CH_2), but unfortunately the spectra presented by Fromherz et al. (1972) did not cover this interesting range (1200-1300 cm^{-1}).
2. The methyl arachidate-hemoglobin system exhibits predominantly hydrophobic interaction, since the carbonyl stretching mode ν (C=O) at 1735 cm^{-1} of the fatty acid ester remains unaffected upon protein binding. Furthermore prominent changes in the amide I region are observed. A possible explanation of the appearance of the band at 1620 cm^{-1} is restructuring of a considerable number of chain fragments from the α-helical and/or random coil into the β-structure.

2. Interaction of Alamethicin with L-α-Dipalmitoyl Phosphatidylcholine

Alamethicin is a peptide antibiotic isolated from the fungus *Trichoderma viride*. Under proper conditions it can produce action potentials similar to those of nerve axons (Müller and Rudin, 1968). After these early investigations there was considerable confusion in the literature with respect to the primary structure of the interaction with lipid bilayers, and the mechanism of pore formation; e.g., alamethicin was believed to be a ring peptide with 18 amino acids until Martin and Williams (1976) showed that the molecule is linear and consists of 19 amino acids. The following primary structure of the alamethicin R_F30 (ALA) component is taken from Gisin et al. (1977):

1 5 10
Ac-Aibu-Pro-Aibu-Ala-Aibu-Ala-Gln-Aibu-Val-Aibu-Gly-Leu-

15 19
Aibu-Pro-Val-Aibu-Aibu-Glu-Gln-Phol

where Aibu is α-aminoisobutyric acid and Phol is phenylalaninol. A distinct hydrophobicity of the peptide is expected from the amino acid composition. Concerning the molecular mechanism of pore formation, most models presented so far are based on the assumption that alamethicin is adsorbed at the membrane/water interface. Insertion of alamethicin into the membrane is reported to occur under the influence of an electric field (Boheim and Kolb, 1978). This molecular mechanism was derived predominantly from kinetic data of membrane conductance measurements, i.e., from information on a macroscopic level. Alternatively, Fringeli and Fringeli (1979) have shown recently by means of IR-ATR spectroscopy that in the absence of an electric field alamethicin is already located predominantly in the hydrophobic phase of the bilayer membrane and not adsorbed at its interface as commonly assumed. Therefore field-induced pore formation cannot be initiated by insertion of peptide molecules from the hydrophilic surface into the interior of the membrane, but rather by opening of preformed alamethicin aggregates in the membrane. Some interesting features of the L-α-DPPC/alamethicin membrane should be noted. Firstly, as documented by Fig. 25d and g, alamethicin is incor-

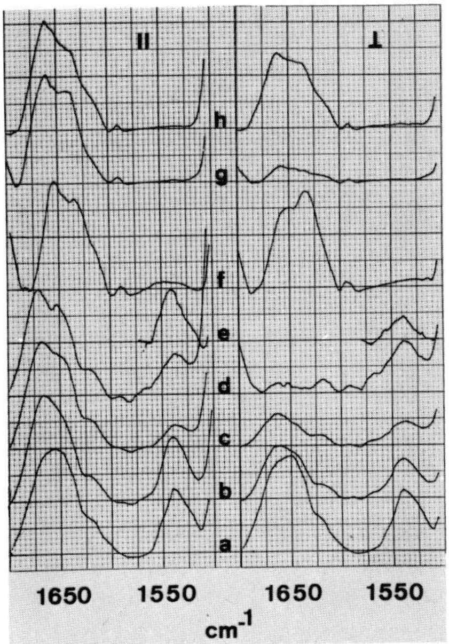

Fig. 25 a-f. Structural changes of alamethicin in DPPC multilayer membranes. Changes depending upon sample preparation and environment are reflected by amide I and amide II IR-ATR absorption bands (absorbance plot, $\alpha = -\ln T$, in which T is transmittance). ∥ and ⊥, parallel and perpendicular polarized incident light. Samples: a, sum spectrum of pure DPPC and alamethicin, dry N_2, 25°C; b, model membrane prepared from $CHCl_3$ solution (10 mM DPPC, 0.2 mM alamethicin), dry N_2, 25°C; c, sample after 30 min contact with liquid H_2O, dry N_2, 50°C after adequate drying; d, model membrane prepared as for (b), 77.5 h in contact with liquid H_2O, dry N_2, 50°C after adequate drying, molar ratio 80:1; e, sample (d) in liquid H_2O before drying, 25°C, molar ratio 80:1; f, sample (d) after addition of liquid ^2H_2O containing 90 nM alamethicin (equilibrium concentration), 25°C, molar ratio 80:1; g, sample (f) after adequate drying, kept in N_2 (gas) with 30,000 ppm ^2H_2O (~ 90% relative humidity), 25°C, molar ratio 80:1; h, sample (f) after sudden drying with dry N_2 (gas), 50°C, molar ratio 80:1

porated into the membrane, assuming helical conformation. The helix axis is oriented parallel to the hydrocarbon chains of DPPC, resulting in distinct z-polarization of amide I and xy-polarization of amide II. Helical peptide conformation, however, is only typical for dry and hydrated membranes. Addition of liquid H_2O (Fig. 25e) or liquid D_2O (Fig. 25f) leads to helix unfolding and association of extended alamethicin molecules to dimers and multimers. These aggregates are considered preformed pores. No definite answer can be given at present as to, how pore opening is achieved by the influence of a properly applied electric field. Fringeli and Fringeli (1979) have proposed a field-induced conformational change of the peptide, however this mechanism has to be verified, e.g., by electric-field-modulated IR-ATR spectroscopy. Secondly, it should be mentioned that alamethicin incorporation into the membrane could not be achieved from chloroform solution, contact with liquid water turned out to be a prerequisite for peptide incorporation, cf. Fig. 25 a-e. Furthermore Fringeli (1980) has shown that the experimental conditions used in IR-ATR studies of alamethicin interaction with dipalmitoyl phosphatidylcholine bilayers are similar to those used in black film experiments. For that purpose the equilibrium distribution of alamethicin between the bilayer membrane and the surrounding aqueous phase was determined. It turned out that alamethicin concentration in the aqueous phase typical for IR-ATR experiments is between 25 nM and 200 nM. On the other hand this concentration range is also typical for black film experiments with alamethicin and a number of its analogs (Boheim et al., 1978). Furthermore the corresponding molar ratios lipid:peptide in the membrane were found to be 600:1 and 50:1 respectively. Figure 26 shows the results of a typical equilibration experiment. A

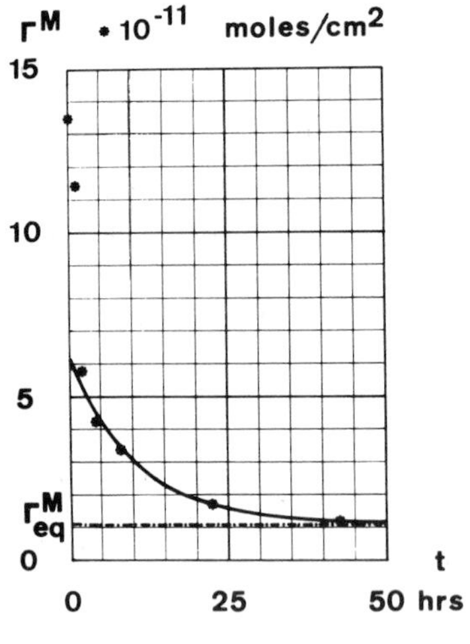

Fig. 26. Typical plot of alamethicin surface concentration Γ^M against the time of liquid water exposure at 25°C. The fast decrease during the first two hours results from alamethicin F50 (Gln^{18}) exhibiting a diffusion rate about 10 times higher than that of alamethicin F30 (Glu^{18}). The solid line is obtained by a least squares fit with respect to a single exponential decay. It levels off at the equilibrium surface concentration Γ^M_{eq}. Mean thickness of the membrane: 5.8 bilayers

model membrane with a mean thickness of 5.8 DPPC bilayers contained $13.5 \cdot 10^{-11}$ moles/cm^2 alamethicin (i.e., $2.3 \cdot 10^{-11}$ moles/(cm^2 · bilayer)). Now a given volume of pure H_2O was pumped in a closed cycle through the ATR cell, resulting in dissolution of alamethicin from the membrane until the equilibrium concentration in the aqueous phase was reached. This concentration can be calculated from the difference of initial and equilibrium surface concentration, the total membrane area, and the volume of circulating water. Furthermore the time course of the equilibration process can be used to determine the transfer time constant τ_0 for alamethicin diffusion across a DPPC bilayer membrane. Diffusion coefficients in the interior of the membrane, D_M, and across the membrane/water interface, D_I, have been estimated. A number of typical results are presented in Table 7. The application of IR-ATR spectroscopy for the determination of distribution coefficients of membrane constituents between the membrane and the aqueous surrounding, as well as the determination of trans-membrane diffusion coefficients, demonstrates once more the versatility of ATR techniques.

3. Interaction of Acetylcholinesterase with the Lipids Extracted from the Electric Organ of *Torpedo marmorata*

This experiment was performed in aqueous environment at pH 7. The same lipid samples as in Sect. IV.A.7.c were used. However, the surrounding liquid water was replaced by a dilute aqueous solution of acetylcholinesterase. Protein penetration into the lipid membrane could be monitored through ATR spectra. The penetration process lasted several hours until equilibrium was reached. Protein binding was indicated by the appearance of amide I and amide II bands, the former, however, being strongly overlapped by the H_2O bending absorption at $\sim 1640 \text{ cm}^{-1}$. A biologically interesting feature of this system is that bound ACHE keeps its enzymatic activity for several days at room temperature. This fact should enable the performance of substrate-enzyme (acetylcholine (ACH)-ACHE) interaction studies by means of chemical modulation technique, cf. Sect. IV.B.3.

D. Infrared Spectroscopy of Living Nerve Cells

IR measurements with living cells are confronted with considerable experimental difficulties: first because biomembranes consist of a large variety of substances which form only a minor amount of the whole cell; and second because the experiments have to be performed under biological conditions, i.e., in aqueous anvironment. Nevertheless Sherebrin et al. (1972) and Sherebrin and MacClement (1973) have reported a series of experiments performed on nerves of frog, lobster, and garfish. A simple type of modulation technique was applied in combination with transmission infrared spectroscopy in order to obtain difference spectra of the nerve between resting and activated state. Although digital signal processing was used to enhance sensitivity the measured signals must be considered as being at the limit of significance, since the difference in transmission was in most cases smaller than 0.05%.

Acknowlegments. The authors wish to thank Mrs. M. Fringeli and Miss. S. Wiedemann for technical assistance. Stimulating discussion with Dr. P. Rihak from the Laboratory for Physical Chemistry and Dr. F. Kopp from the Institute of Cell Biology are acknowledged. The following chemicals were kindly provided: lecithin-d9 by Prof. J. Seelig and egg phosphatidylcholine by Dr. D. Walz from the Biocenter, Basel; lipid extract and acetylcholineesterase from *Torpedo marmorata* by PD Dr. W. Hopff from the Institute of Pharmacology, University of Zurich; ketal phospholipid analogous by Dr. H. Eibl from the Max Planck Institute for Biophysical Chemistry, Göttingen. ^{13}C-labeled dipalmitoyl phosphatidylcholine was obtained from R. Berchtold, Biochemical Laboratory, Bern.

Table 7. ALA distribution and diffusion in DPPC bilayers: aqueous environment, 25°C[a]

Substance	Γ_{eq}^{o} [b] (moles/cm²)	K[c]	τ[d] (hours)	$\bar{n}$[e] (bilayers)	$\tau_o = \tau/\bar{n}$[f] (hours)	D_M/D_I[g] (cm)	D_M[h] (cm²/s)	D_I[i] (cm/s)	τ_M[j] (hours)
ALA F30	$8.4 \cdot 10^{-12}$	$4.17 \cdot 10^{-6}$	48.8	15.65	3.12	$\geqslant 2.09 \cdot 10^{-12}$	$\geqslant 3.0 \cdot 10^{-17}$	$\leqslant 1.4 \cdot 10^{-5}$	$\leqslant 1.2$
ALA F30	$7.7 \cdot 10^{-12}$	$3.95 \cdot 10^{-6}$	90.9	28.63	3.18	$\geqslant 1.98 \cdot 10^{-12}$	$\geqslant 3.0 \cdot 10^{-17}$	$\leqslant 1.5 \cdot 10^{-5}$	$\leqslant 1.2$
ALA F30	$1.1 \cdot 10^{-12}$	$9.76 \cdot 10^{-6}$	8.8	3.38	2.60	$\geqslant 4.88 \cdot 10^{-12}$	$\geqslant 3.6 \cdot 10^{-17}$	$\leqslant 7.4 \cdot 10^{-6}$	$\leqslant 0.95$
ALA F30	$7.6 \cdot 10^{-13}$	$1.03 \cdot 10^{-5}$	12.9	5.76	2.24	$\geqslant 5.15 \cdot 10^{-12}$	$\geqslant 4.2 \cdot 10^{-17}$	$\leqslant 8.2 \cdot 10^{-6}$	$\leqslant 0.83$
ALA F50	$1.7 \cdot 10^{-12}$	$\sim 10^{-5}$	1.67	6.83	0.245	$\geqslant 5 \cdot 10^{-12}$	$\geqslant 3.8 \cdot 10^{-16}$	$\leqslant 7.6 \cdot 10^{-5}$	$\leqslant 0.09$ (5.5 min)

[a] The statistical error of all data is < 10%. However, there could be a shift up to 20% resulting from systematic errors of amide I absorption coefficient and of the electric field components E_{ox}^2, E_{oy}^2, E_{oz}^2 [cf. Eq. (7)].

[b] ALA surface concentration per monolayer.

[c] Distribution coefficient, $K = c_{eq}$ (water) / c_{eq} (membrane).

[d] Time constant for ALA diffusion across $\bar{n}$ DPPC bilayers.

[e] Mean number of DPPC bilayers.

[f] Time constant for ALA diffusion across one bilayer. ALA diffusion along the water layer between successive DPPC bilayers can be neglected, i.e., $\tau_o = \tau/\bar{n}$.

[g] $D_M/D_I \geqslant K \cdot \ell_o$, ℓ_o = thickness of a DPPC bilayer.

[h] Diffusion coefficient of ALA across the hydrophobic region of the bilayer.

[i] Film diffusion coefficient of ALA at the membrane water interface (ALA incorporation) related to D_M^{min}.

[j] Mean transfer time for ALA diffusion across the hydrocarbon chain region of the bilayer membrane, according to the Einstein-Smoluchowsky equation: $\tau_M = \ell_o^2/2\,D_M$.

Furthermore we express our gratitude to Mr. O. Diener, Mr. E. Pleisch, Mr. R. Gunzinger, Mr. M. Andrist, and Mr. W. Jäggi for skilful help with mechanical, optical, and electronic problems related to ATR spectroscopic techniques used in this work.

This work was supported by the Swiss National Foundation (Project No. 3.521-0.75 and 3.192.0.77) and by the Fritz Hoffmann-La Roche Foundation (Project No. 127).

References

Akutsu, H., Kyogoku, Y., Nakahara, H., Fukuda, K.: Conformational analysis of phospahtidylethanolamine in multilayers by infrared dichroism. Chem. Phys. Lipids *15*, 222-242 (1975).

Aronovic, S.M.: Ph. D. Thesis, University of Wisconsin (1957).

Badger, R.M., Pullin, A.D.E.: The infrared spectrum and structure of collagen. J. Chem. Phys. *22*, 1142 (1954).

Baumeister, W., Fringeli, U.P., Hahn, M., Kopp, F., Seredynski, J.: Radiation damage in tripalmitin layers studied by means of infrared spectroscopy and electron microscopy. Biophys. J. *16*, 791-810 (1976a).

Baumeister, W., Fringeli, U.P., Hahn, M., Seredynski, J.: Radiation damage of protein in the solid state: changes of β-lactoglobulin secondary structure. Biochim. Biophys. Acta *453*, 289-292 (1976b).

Baumeister, W., Hahn, M., Fringeli, U.P.: Electron-beam induced conformational changes in polypeptide layers: an infrared study. Z. Naturfosch. *31c*, 746-747 (1976c).

Bellamy, L.J.: The Infrared Spectra of Complex Molecules. London: Methuen 1975.

Blodgett, K.B., Langmuir, I.: Built-up films of barium stearate and their optical properties. Phys. Rev. *51*, 964-982 (1937).

Boheim, G., Kolb, H.-A.: Analysis of the multi-pore system of alamethicin in a lipid membrane. I. Voltage-jump current relaxation measurements. J. Membr. Biol. *38*, 99-150 (1978).

Boheim, G., Irmscher, G., Jung, G.: Trichotoxin A-40, a new membrane-exciting peptide. Part B: Voltage dependent pore formation in bilayer lipid membranes and comparison with other alamethicin analogues. Biochim. Biophys. Acta *507*, 485-506 (1978).

Buttler, M.J., McKean, D.C.: Infrared spectra and vibrational assignments in dimethylamines. Spectrochim. Acta *21*, 465-483 (1965).

Chirgadze, Yu.N., Nevskaya, N.A.: Infrared spectra and resonance interaction of amide-I vibration of the antiparallel-chain pleated sheet. Biopolymers *15*, 607-625 (1976a).

Chirgadze, Yu.N., Nevskaya, N.A.: Infrared spectra and resonance interaction of amide-I vibration of the parallel-chain pleated sheet. Biopolymers *15*, 627-636 (1976b).

Dellepiane, G., Zerbi, G.: Normal coordinate calculations as a tool for vibrational assignments. I. Fundamental vibrations of simple aliphatic amines, J. Chem. Phys. *48*, 3573-3583 (1968).

Edsall, J.T.: Raman spectra of amino acids and related substances. III. Ionization and methylation of the amino group. J. Chem. Phys. *5*, 225-237 (1937).

Elliott, A.: Infrared Spectra and Structure of Organic Long-Chain Polymers. London: Arnold 1969.

Forrest, G., Lord, R.G.: Laser raman spectroscopy of biomolecules X. J. Raman Spectrosc. *6*, 32-37 (1977).

Forster, M., Fringeli, U.P., Günthard, Hs.H.: Kinetics of photochemical systems with modulated optical excitation. Helv. Chim. Acta *56*, 389-407 (1973).

Forster, M., Loth, K., Andrist, M., Fringeli, U.P., Günthard, Hs.H.: Kinetic study of the photooxidation of pyrocatechol by modulated electronic excitation IR and ESR spectroscopy (MEIR and ESR). Chem. Phys. *17*, 59-80 (1976).

Fraser, R.D.B.: Interpretation of infrared dichroism in fibrous proteins the 2μ region. J. Chem. Phys. *24*, 89-95 (1956).

Fraser, R.D.B.: Interpretation of infrared dichroism in axially oriented polymers. J. Chem. Phys. *28*, 1113-1115 (1958).

Fraser, R.D.B., MacRae, T.P.: Conformations in Fibrous Proteins and Related Synthetic Polypeptides. Chap. 5. New York: Academic Press 1973.

Fringeli, U.P.: Bau eines Infrarot-Modulations-Spektrometers zur Messung der Absorption und der Lebenszeit elektronisch angeregter Zustände. Thesis Nr. 4366. ETH Zürich: 1969.

Fringeli, U.P.: The structure of lipids and proteins studied by attenuated total reflection (ATR) infrared spectroscopy. II. Oriented layers of a homologous series: Phosphatidylethanolamine to phosphatidylcholine. Z. Naturforsch. *32c*, 20-45 (1977).

Fringeli, U.P.: A new crystalline phase of L-α-Dipalmitoylphosphatidylcholine monohydrate. Biophys. J., (in press) (1981a).

Fringeli, U.P.: Quantitative analysis of polarized infrared ATR spectra. J. Chem. Phys. (1981b).

Fringeli, U.P.: Distribution and Diffusion of alamethicin in a lecithin/water model membrane system. J. Membr. Biol. *54*, 203-212 (1980).

Fringeli, U.P., Fringeli, M.: Pore formation in lipid membranes by alamethicin. Proc. Natl. Acad. Sci. USA *76*, 3852-3856 (1979).

Fringeli, U.P., Günthard, Hs.H.: Modulated excitation infrared spectrophotometer. Appl. Opt. *10*, 819-824 (1971).

Fringeli, U.P., Günthard, Hs.H.: Hydration sites of egg lecithin determined by means of modulated excitation infrared spectroscopy. Biochim. Biophys. Acta *450*, 101-106 (1976).

Fringeli, U.P., Müldner, H.G., Günthard, Hs.H., Gasche, W., Leuzinger, W.: The structure of lipids and proteins studied by Attenuated Total Reflection (ATR) infrared spectroscopy. I. Oriented Layers of Tripalmitin. Z. Naturforsch. *27b*, 780-796 (1972).

Fringeli, U.P., Schadt, M., Rihak, P., Günthard, Hs.H.: Hydrocarbon chain ordering in liquid crystals investigated by means of infrared attenuated total reflection (IR-ATR) spectroscopy. Z. Naturforsch. *31a*, 1098-1107 (1976).

Fringeli, U.P., Hofer, P.: Electric-field induced effects in acetylcholinesterase. Neurochemistry International *2*, 185-192 (1980).

Fromherz, P., Peters, J., Müldner, H.G., Otting, W.: An infrared spectroscopic study on the lipid-protein interaction in an artificial lamellar system. Biochim. Biophys. Acta *274*, 644-648 (1972).

Gaines, G.L., Jr.: Insoluble Monolayers at Liquid-Gas Interfaces, p. 165. New York: Interscience 1966.

Gallagher, K.H.: The isotope effect in relation to bond length in hydrogen bonds in crystals. In: Hydrogen Bonding (ed. Hadzi, D.), pp. 45-54. New York: Pergamon Press 1959.

Gisin, B.F., Kobayashi, S., Hall, J.E.: Synthesis of a 19-residue peptide with alamethicin-like activity. Proc. Natl. Acad. Sci. USA *74*, 115-119 (1977).

Glatt, L., Ellis, J.W.: Near infrared absorption of nylon films: dichroism and rupture of N-H...O bonds on melting. J. Chem. Phys. *16*, 551 (1948).

Green, J.P., Phillips, M.C., Shipley, G.G.: Structural investigations of lipid, polypeptide and protein multilayers. Biochim. Biophys. Acta *330*, 243-253 (1973).

Günthard, Hs.H.: Modulations spectroscopy. Ber. Bunsenges. Phys. Chem. *78*, 1110-1115 (1974).

Hahn, M., Fringeli, U.P., Baumeister, W.: Radiation damage studies of membrane model systems by means of ATR-IR spectroscopy. Proc. 8th Int. Congr. Electron Microsc. Canberra. Australian Academy of Science, Vol. II, 672-673, 1974.

Harrick, N.J.: Internal Reflection Spectroscopy. New York: Wiley 1967.

Hauser, H.: The conformation of the polar group of lecithin and lysolecithin. J. Colloid Interface Sci. *55*, 85-93 (1976).

Hauser, H., Phillips, M.C., Levine, B.A., Williams, R.J.P.: Conformation of the lecithin polar group in charged vesicles. Nature *261*, 390-394 (1976).

Herzberg, G.: Infrared and Raman Spectra of Polyatomic Molecules. New York: Van Nostrand 1945.

Hexter, R.M.: Excitation-modulation spectroscopy: a technique for obtaining vibrational spectra of excited electronic states. J. Opt. Soc. Am. *53*, 703-709 (1963).

Higgs, P.W.: The vibration spectra of helical molecules: infrared and raman selection rules, intensities and approximate frequencies. Proc. Soc. London Ser. A *220*, 472-485 (1953).

Hitchcock, P.B., Mason, R., Thomas, K.M., Shipley, G.G.: Structural chemistry of 1,2 dilauroyl-

DL-Phosphatidylethanolamine: molecular conformation and intermolecular packing of phospholipids. Proc. Natl. Acad. Sci. USA *71*, 3036-3040 (1974).

Hitchcock, P.B., Mason, R., Shipley, G.G.: Phospholipid arrangements in multilayers and artificial membranes: quantitative analysis of the x-ray diffraction data from a multilayer of 1,2-dimyristoyl-DL-phosphatidylethanolamine. J. Mol. Biol. *94*, 297-299 (1975).

Hofer, P., Fringeli, U.P.: Structural investigation of biological material in aqueous environment by means of infrared ATR spectroscopy. Biophys. Struct. Mech. *6*, 67-80 (1979).

Hollenstein, H., Günthard, Hs.H.: Gas and matrix infrared spectra of n-methyl-ethylidenimine and valence force field. Chem. Phys. *4*, 368-389 (1974).

Hopff, H.: Isolierung von Acetylcholinesterase und Charakterisierung der katalytisch aktiven Stelle. Habilitationsschrift, p. 18. Universität: Zürich 1975.

Hunziker, H.: On the numerical solution of the normal coordinate problem. J. Mol. Spectrosc. *17*, 131-135 (1965).

Isaacson, M.S.: Specimen damage in the electron microscope. In: Principles and Techniques of Electron Microscopy, Vol. VI (ed. Hayat, M.A.), New York: Van Nostrand 1975.

IUPAC Manual of Symbols and Terminology for Physicochemical Quantities, p. 16. London: Butterworth 1971.

Johnston, H.S., Graw, G.E., Paukert, T.T., Richards, L.W., Bogaerde, T., van den: Molecular modulation spectrometry, 1. New method for observing infrared spectra of free radicals. Proc. Natl. Acad. Sci. USA *57*, 1146-1153 (1967).

Jones, R.N.: The infrared absorption spectra of deuterated esters. III. Methyl laurate. Can. J. Chem. *40*, 301-320 (1962).

Jones, R.N., Sandorfy, C.: The application of infrared and raman spectroscopy to the elucidation of molecular structure. In: Technique of Organic Chemistry, Vol. IX (ed. Weissberger, A.), pp. 247-580. New York: Interscience 1956.

Kessler, H.K., Sutherland, G.B.B.M.: The out-of-plane deformation frequency of the NH group in the peptide link. J. Chem. Phys. *21*, 570-571 (1953).

Kohlrausch, K.W.F.: Ramanspektren, Hand- und Jahrbuch der Chemischen Physik, Vol. IX. Leipzig: Becker und Erler 1943.

Kopp, F., Fringeli, U.P., Mühlethaler, K., Günthard, Hs.H.: Instability of Langmuir-Blodgett layers of barium stearate, cadmium arachidate and tripalmitin, studied by means of electron microscopy and infrared spectroscopy. Biophys. Struct. Mech. *1*, 75-96 (1975a).

Kopp, F., Fringeli, U.P., Mühlethaler, K., Günthard, Hs.H.: Spontaneous rearrangement in Langmuir-Blodgett layers of tripalmitin studied by means of ATR infrared spectroscopy and electron microscopy. Z. Naturforsch. *30c*, 711-717 (1975b).

Kratky, O.: Zum Deformationsmechanismus der Faserstoffe, I. Kolloid Z. *64*, 213-222 (1933).

Krimm, S.: Infrared spectra of high polymers. Adv. Polym. Sci. *2*, 51-172 (1960).

Krimm, S.: Infrared spectra and chain conformation of proteins. J. Mol. Biol. *4*, 528-540 (1962).

Kühne, H., Günthard, Hs.H.: Spectroscopic study of the ozoneethylene reaction, matrix IR spectra of 3 isotopic ethylene ozonides. J. Phys. Chem. *80*, 1238-1247 (1976).

Kuhn, H.: Electron tunneling effects in monolayer assemblies. Chem. Phys. Lipids *8*, 401-404 (1972).

Kuhn, H., Möbius, D., Bücher, H.: Spectroscopy of monolayer assemblies. In: Techniques of Chemistry (eds. Weinberger, A., Rossiter, B.), pp. 577-702, Vol. I, Part IIIb, Chap. VII. New York: Wiley 1972.

Labhart, H.: Eine experimentelle Methode zur Ermittlung der Singulett-Triplett-Konversionswahrscheinlichkeit und der Triplett-Spektren von gelösten organischen Molekeln. Helv. Chim. Acta *47*, 2279-2288 (1964).

Lesslauer, W.: X-ray diffraction from fatty-acid multilayers. Angular width of reflexions from systems with few unit cells. Acta Cryst. *B30*, 1932-1937 (1974).

Levine, Y.K., Bailey, A.I., Wilkins, M.H.F.: Multilayers of phospholipid bimolecular leaflets. Nature *220*, 577-578 (1968).

Liang, C.Y., Krimm, S.: Infrared spectra of high polymers. III. Polytetrafluoroethylene and polychlorotrifluoroethylene. J. Chem. Phys. *25*, 563-571 (1956).

Liang, C.Y., Krimm, S., Sutherland, G.B.B.M.: Infrared spectra of high polymers. I. Experimental methods and general theory. J. Chem. Phys. *25*, 543-548 (1956).

Loeb, G.I.: Infrared spectra of protein monolayers: paramyosin and β-lactoglobulin. J. Colloid Interface Sci. *31*, 572-574 (1969).

Loeb, G.I., Baier, R.E.: Spectroscopic analysis of polypeptide conformation in polymethyl glutamate monolayers. J. Colloid Interface Sci. *27*, 38-45 (1968).

Lord, R.C., Mendelsohn, R.: Raman spectroscopy of membrane constituents and related molecules. In: Molecular Biology, Biochemistry and Biophysics, Vol. XXXI: Membrane Spectroscopy (ed. Grell, E.), Berlin-Heidelberg-New York: Springer 1980.

Malcolm, B.R.: Surface chemistry of poly (β-benzyl L-aspartate). Biopolymers *9*, 911-922 (1970).

Marčelja, S.: Chain ordering in liquid crystals. II. Structure of bilayer membranes. Biochim. Biophys. Acta *367*, 165-176 (1974).

Martin, D.R., Williams, R.J.P.: Chemical nature and sequence of alamethicin. Biochem. J. *153*, 181-190 (1976).

May, L., Kamble, A.B., Acosta, I.P.: The effect of electric fields on brain cephalin and lecithin films. J. Membr. Biol. *2*, 192-200 (1970).

Meiklejohn, R.A., Meyer, R.J., Aronovic, S.M., Schuette, H.A., Meloch, V.W.: Characterization of long-chain fatty acids by infrared spectroscopy. Anal. Chem. *29*, 329-334 (1957).

Miyazawa, T.: Normal vibrations of monosubstituted amides in the cis configuration and infrared spectra of diketopiperazine. J. Mol. Spectrosc. *4*, 155-167 (1960a).

Miyazawa, T.: The characteristic band of secondary amides at 3100 cm^{-1}. J. Mol. Spectrosc. *4*, 168-172 (1960b).

Miyazawa, T.: Perturbation treatment of the characteristic vibrations of polypeptide chains in various configurations. J. Chem. Phys. *32*, 1647-1652 (1960c).

Miyazawa, T.: Internal rotation and low frequency spectra of esters, monosubstituted amides and polyglycine. Bull. Chem. Soc. Jp. *34*, 691-696 (1961).

Miyazawa, T., in: Polyamino Acids, Polypetides and Proteins (ed. Stahmann, M.A.), p. 201. Madison, Wisconsin: University of Wisconsin Press 1962.

Miyazawa, T.: Infrared studies of the conformations of polypeptides and proteins. In: Aspects of Protein Structure (ed. Ramachandran, G.N.), pp. 257-264. New York: Academic Press 1963.

Miyazawa, T.: Infrared spectra and helical conformation. In: Poly-α-Amino Acids (ed. Fasman, G.D.), pp. 69-103. New York: Dekker 1967.

Miyazawa, T., Shimanouchi, T., Mizushima, S.: Characteristic infrared bands of monosubstituted amides. J. Chem. Phys. *24*, 408-418 (1956).

Miyazawa, T., Shimanouchi, T., Mizushima, S.: Normal vibrations of N-methylacetamide. J. Chem. Phys. *29*, 611-616 (1958).

Müller, P., Rudin, D.O.: Action potentials induced in biomolecular lipid membranes. Nature *217*, 713-719 (1968).

Münch, W., Fringeli, U.P., Günthard, Hs.H.: Polarized IR-ATR-single crystal spectra of stearic acid d_0 and d_{35}. Spectrochim. Acta *33A*, 95-109 (1977).

Nevskaya, N.A., Chirgadze, Yu.N.: Infrared spectra and resonance interactions of amide I and II vibrations of α-helix. Biopolymers *15*, 637-648 (1976).

Pauling, L., Corey, R.B.: The pleated sheet, a new layer configuration of polypeptide chains. Proc. Natl. Acad. Sci. USA *37*, 251-256 (1951).

Pauling, L., Corey, R.B., Branson, H.R.: The structure of proteins: two hydrogen-bonded helical configurations of the polypeptide chain. Proc. Natl. Acad. Sci. USA *37*, 205-211 (1951).

Pimentel, G.C., McClellan, A.L.: The Hydrogen Bond. San Francisco: Freeman 1960.

Powers, L., Pershan, P.S., Monodomain samples of dipalmitoyl phosphatidylcholine with varying concentrations of water and other ingredients. Biophys. J. *20*, 137-152 (1977).

Primas, H., Günthard, Hs.H.: Die Infrarotspektren von Kettenmolekeln der Formel $R'CO(CH_2CH_2)_n COR''$. 1. Rocking und Twisting-Grundtöne. Helv. Chim. Acta *36*, 1659-1670 (1953a).

Primas, H., Günthard, Hs.H.: Die Infrarotspektren von Kettenmolekeln der Formel $R'CO(CH_2CH_2)_n COR''$. 2. Die Normalschwingungen des Symetrietypus B_u. Helv. Chim. Acta *36*, 1791-1803 (1953b).

Primas, H., Günthard, Hs.H.: Theorie der Intensitäten der Schwingungsspektren von Kettenmolekeln. 1. Allgemeine Theorie der Berechnung von Intensitäten der Infrarotspektren von großen Molekeln. Helv. Chim. Acta *38*, 1254-1262 (1955).

Primas, H., Günthard, Hs.H.: Theorie der Intensitäten der Schwingungsspektren von Kettenmolekeln. 2. Zur Berechnung der Intensitäten der Infrarotspektren von freien Kettenmolekeln der Symmetrie C_{2h}. Helv. Chim. Acta *39*, 1182-1192 (1956).

Procarione, W.L., Kauffman, J.W.: The electrical properties of phospholipid bilayer Langmuir films. Chem. Phys. Lipids *12*, 251-260 (1974).

Rasmussen, R.S., Tunnicliff, D.D., Brattain, R.R.: Infrared and ultraviolet spectroscopic studies of ketones. J. Am. Chem. Soc. *71*, 1068-1072 (1949).

Rihak, P.: Raman and Infrarot-Spektroskopie von Cholin und Lezithin. ETH-Thesis No. 6393, Zürich 1979.

Rihak, P., Fringeli, U.P., Günthard, Hs.H.: The structure of the polar headgroup of lecithin studied by means of infrared ATR-spectroscopy. V^{th} Biophys. Congr. Copenhagen (1975).

Rüegg, M., Metzger, V., Susi, H.: Computer analysis of characteristic infrared bands of globular proteins. Biopolymers *14*, 1465-1471 (1975).

Schachtschneider, J.H., Snyder, R.G.: Vibrational analysis of the n-paraffins-II. Normal co-ordinate calculations. Spectrochim. Acta *19*, 117-168 (1963).

Seelig, A., Seelig, J.: Bilayers of dipalmitoyl-3-sn-phosphatidylcholine conformational differences between the fatty acyl chains. Biochim. Biophys. Acta *406*, 1-5 (1975).

Seelig, J.: Phosphorus-31 nuclear magnetic resonance and the head group structure of phospholipids in membranes. Biochim. Biophys. Acta *505*, 105-141 (1978).

Seelig, J., Gally, H.-U.: Investigation of phosphatidylethanolamine bilayers by deuterium and phosphorus-31 nuclear magnetic resonance. Biochemistry *15*, 5199-5204 (1976).

Seelig, J., Waespe-Šarčević, N.: Molecular order in cis and trans unsaturated phospholipid bilayers. Biochemistry *17*, 3310-3315 (1978).

Sheppard, N.: Rotational isomerism about C-C bonds in saturated molecules as studied by vibrational spectroscopy. Adv. Spectrosc. *1*, 288-353 (1959).

Sherebrin, M.H., MacClement, B.A.E.: Modification of an infrared spectrophotometer to study nerve excitation. IEEE Transactions *BM 20*, 63-65 (1973).

Sherebrin, M.H., MacClement, B.A.E., Franko, A.J.: Electric-field-induced shift in the infrared spectrum of conducting nerve axons. Biophys. J. *12*, 977-989 (1972).

Shimanouchi, T., Tsuboi, M., Kyogoku, Y.: Infrared spectra of nucleic acids and related compounds. In: Advances in Chemical Physics, Vol. VII, pp. 435-496. New York: Interscience 1964.

Snyder, R.G.: Vibrational spectra of crystalline n-paraffins. Part. I. Methylene rocking and wagging modes. J. Mol. Spectrosc. *4*, 411-434 (1960).

Snyder, R.G.: Vibrational study of the chain conformation of the liquid n-paraffins and molten polyethylene. J. Chem. Phys. *47*, 1316-1350 (1967).

Stenn, K., Bahr, G.F.: Specimen damage caused by the beam of the transmission electron microscope, a correlative reconsideration. J. Ultrastruct. Res. *31*, 526-550 (1970).

Stuart, A.V., Sutherland, G.B.B.M.: Effect of hydrogen bonding on the deformation frequencies of the hydroxyl group in alcohols. J. Chem. Phys. *24*, 559-570 (1956).

Sundaralingam, M.: Molecular structures and conformation of the phospholipids and sphingomyelins. Ann. N.Y. Acad. Sci. *195*, 324-355 (1972).

Susi, H.: Infrared spectra of biological macromolecules and related systems. In: Structure and Stability of Biological Macromolecules (eds. Timasheff, S.N., Fasman, G.D.), pp. 575-663. New York: Dekker 1969.

Susi, H., Timasheff, S.N., Stevens, L.: Infrared spectra and protein conformations in aqueous solutions. J. Biol. Chem. *242*, 5460-5466 (1967).

Suzuki, E.: A quantitative study of the amide vibrations in the infrared spectrum of silk fibroin. Spectrochim. Acta *23A*, 2303-2308 (1967).

Tsuboi, M.: Some problems in the infrared spectra of polypeptides and polynucleotides. Biopolymers Symp. *1*, 527-547 (1964).

Wallach, D.F.H., Winzler, R.J.: Evolving Strategies and Tactics in Membrane Research. Berlin-Heidelberg-New York: Springer 1974.

Warschel, A., Levitt, H.: A general program for consistent force field evaluation of equilibrium geometry and vibrational frequencies of ground and excited states of conjugated molecules and of ground state of an arbitrary molecule. Quantum Chemistry Program Exchange. Program No. 247. Bloomington/Indiana: Indiana University, Chemistry Department (1975).

Weast, R.C.: Handbook of Physics and Chemistry, 55th edn., pp. 224. Cleveland/Ohio: CRC Press 1974-75.

Wilson, E.B., Decius, J.C., Cross, P.C.: Molecular Vibrations. New York: McGraw-Hill 1955.

Wyckoff, R.W.G.: Crystal Structures. 2nd. edn., Vol. V, pp. 254-256. New York: Interscience 1966.

Zbinden, R.: Infrared Spectroscopy of High Polymers. New York: Academic Press 1964.

Zundel, G.: Hydration and Intermolecular Interaction. New York: Academic Press 1969.

Chemical Relaxation Spectrometry

H. Ruf and E. Grell

I. Introduction

This article essentially deals with the application of chemical relaxation spectrometry to the investigation of membrane processes and includes a section related to some applications of dielectric relaxation.

Kinetic studies are carried out for the detailed investigation of chemical and biochemical reactions as a function of time. Note that the stationary static behaviour of a reversible equilibrium represents the overall sum of the component reaction steps. Thus, kinetic studies provide information concerning the reaction mechanism or reaction path involved and permit the detection and characterization of reaction intermediates and allow the determination of the kinetic parameters (rate constants). Nevertheless, it should be noted that kinetic studies can never verify a reaction mechanism. They can only provide supporting evidence for or against a particular model. Therefore, the general strategy for the evaluation of reaction mechanisms is always based on the exclusion principle.

The classic experimental techniques require that the reactants are mixed and the concentration changes due to the reaction are then detected. Formerly, the time range accessible to the experimental observation was limited by a mixing time of about 1 s. In order to reduce the mixing time, stopped-flow methods have been developed that enable the determination of half-lives of chemical species down to about 1 ms (Gibson and Milnes, 1964). A further reduction of the dead time has been achieved by introducing a continuous flow method coupled with integrating observation along the axis of the flow tube (Gerischer and Heim, 1965). However, this has the serious disadvantage that very large reactant volumes are required and this has considerably reduced the numer of possible applications. For these reasons, flow methods generally can not be employed for the investigation of very rapid reaction steps such as diffusion-controlled processes (second order reactions) or interconversion steps (first order reactions).

For the study of very rapid reactions in solution which are characterized by a reversible equilibrium system, chemical relaxation methods have been developed in which resolution times of about 10^{-9} s or better have been achieved (Eigen et al., 1953; Eigen, 1954; Eigen and De Maeyer, 1963). These methods, therefore, allow the evaluation of reaction mechanisms in terms of individual elementary steps.

As far as the scope of this article is concerned, individual binding to the membrane and rearrangement and transport steps in membranes are considered in terms of the rele-

---

Max-Planck-Institute for Biophysics, Heinr. Hoffmannstr. 7, 6000 Frankfurt 71, FRG

vant elementary processes. This is schematically illustrated in Fig. 1, where a carrier molecule C is bound to the surface region of the membrane and subsequently moves into the nonpolar region. The latter elementary process can be considered as a transport step.

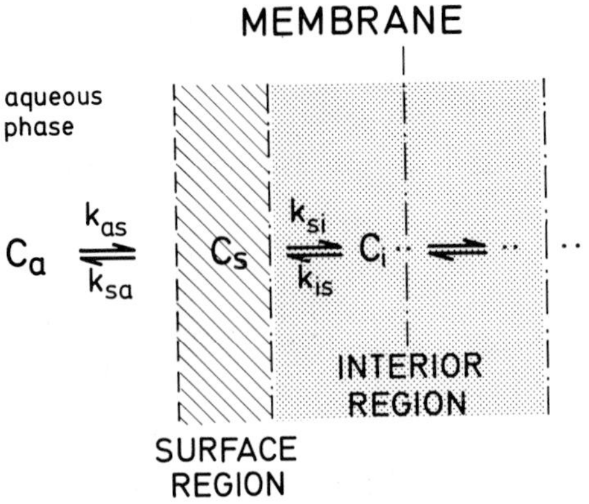

Fig. 1. Schematic illustration of the elementary steps involved in the transport of a carrier molecule C across a membrane: Binding from the aqueous phase (C_a) to the polar surface region (C_s) and movement into the nonpolar interior region

The general principle of the chemical relaxation methods relies on a shift of the chemical equilibrium due to a controlled perturbation of one of the intensive properties such as temperature T (temperature jump method), pressure P (pressure jump method) and electric field E (field jump method). As a consequence of such a perturbation experiment the concentrations of the chemical species involved vary as a function of time. After the variation of an intensive property, the readjustment of the chemical system to the new equilibrium condition is observed experimentally, employing suitable detection systems. It is evident that only those readjustment or relaxation processes can be resolved that are slower than the time required to vary the intensive property. The duration of typical stepwise perturbation ranges from ns to μs and depends on the particular method employed.

In the case of small perturbations the concentration shift of a species, A, involved in a one-step reaction, can be expressed by the following linearized differential equation:

$$\delta \dot{c}_A = -\frac{1}{\tau}[c_A(t) - \bar{c}_A] = -\frac{1}{\tau}\delta c_A \tag{1}$$

where $c_A(t)$ denotes the time-dependent concentration, $\bar{c}_A$ the equilibrium concentration of the final state after the perturbation and $\delta c_A = c_A(t) - \bar{c}_A$ the time-dependent concentration shift. τ is the relaxation time of the particular one-step equilibration process and is related to its kinetic parameters.

The stepwise shift of the intensive property leads to a response which is characterized by an exponential time dependence of the concentrations as shown in Fig. 2. The solution of Eq. (1) is

$$\delta c_A = \delta_o c_A \, e^{-t/\tau} \tag{2}$$

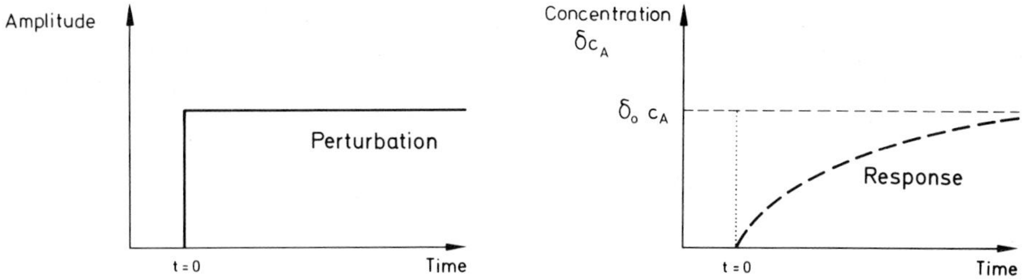

Fig. 2. Relaxation process: Stepwise perturbation and exponential response of the chemical equilibrium system

where $\delta_0 c_A$ represents the maximum concentration shift due to the readjustment of the equilibrium to the new conditions.

For multi-step reactions, coupled linear differential equations have to be considered. In this case, a spectrum of relaxation times is obtained. The corresponding relaxation times depend on the equilibrium concentrations of all species and the rate constants of all steps involved. For details of the general treatment as well as of relaxation spectra in terms of mean relaxation times, special texts may be consulted (Castellan, 1963; Eigen and De Maeyer, 1963, 1973; Schwarz, 1968; Czerlinski, 1969).

In addition to the nonperiodic, transient perturbations discussed above, mention should also be made of the stationary methods. A periodic change of an intensive property, e.g., due to a sinusoidal forcing function, leads also to a periodic shift of the concentrations of the species involved. In addition to the frequency dependence there is a phase relation between the periodic perturbation and the response of the chemical system. The relation for the frequency-dependent concentration shift of species A in a one-step process is:

$$\delta_\omega c_A = \frac{\delta_0 c_A}{\sqrt{1 + \omega^2 \tau^2}} \, e^{-i\varphi} \tag{3}$$

The square root term represents the amplitude decrease of the periodic process as a function of the angular frequency, ω. φ is the phase angle (φ = arc tg($\omega\tau$)). The term $\delta_0 c_A$ is the maximum concentration shift (i.e., $\omega \to 0$). Ultrasonic and dielectric absorption and dispersion methods are typical examples of stationary techniques.

The parameters of a relaxation experiment include one or more relaxation times and the corresponding shift amplitudes which evidently depend on the magnitude of the variation of the intensive property. From the number of relaxation times of an equilibrium system the minimum number of individual steps can be deduced. The values of the relaxation times enable an evaluation of the kinetic parameters of the particular model under consideration. The relaxation amplitudes allow the determination of thermodynamic parameters, e.g., of individual elementary processes in the case of multi-step equilibrium systems. The following general correlation holds between the concentration shift of species A and the amplitude of the intensive paramter z, where ΔZ is the perturbation amplitude for transient or periodic forcing functions:

$$\delta_o c_A = \left(\frac{\delta c_A}{\partial \ln K}\right)_z \left(\frac{\partial \ln K}{\partial z}\right) \delta z = \Gamma \frac{\Delta Z}{RT} \delta z \tag{4}$$

K is the stability constant of the chemical equilibrium, $\Gamma = (\partial c_A/\partial \ln K)_z$ is the amplitude factor and $\Delta Z = RT\partial \ln K/\partial z$ defines the conjugated extensive reaction property corresponding to z. R is the gas constant. According to van't Hoff's equations, the following correlations hold:

$$\Delta Z = \frac{\Delta H}{T} \quad \text{for } z{=}T \text{ at constant pressure} \tag{5}$$
(ΔH: reaction enthalpy)

$$\Delta Z = -\Delta V \quad \text{for } z{=}P \text{ at constant temperature} \tag{6}$$
(ΔV: reaction volume)

$$\Delta Z = \Delta M \quad \text{for } z{=}E \text{ at constant temperature and pressure} \tag{7}$$
(ΔM: electrical moment)

Equations (5) and (6) are related to temperature and pressure jump experiments respectively. Equation (7) holds for field jump experiments and for dielectric measurements at high electric fields.

For adiabatic pressure disturbances which occur at constant entropy (e.g., ultrasonic measurements), Eq. (6) is modified to

$$\Delta Z = -\Delta V + \frac{\alpha_p}{c_p \rho} \Delta H \quad \text{for } z{=}P \text{ at constant entropy} \tag{8}$$

where α_p is the thermal expansion coefficient, ρ the density and c_p the specific heat.

According to the equations presented above it is evident that relaxation experiments on a given system can only be carried out if at least one of the thermodynamic parameters (ΔH, ΔV or ΔM) is not negligible.

To experimentally monitor the readjustment of the reversible equilibrium system to the new external conditions, fast and sensitive detection systems are required. Measurements of absorbance in the UV and visible range, and of fluorescence, may be used to follow a specific reaction, as described in the following section.

Although chemical relaxation methods are considered to represent the major technique for the investigation of fast biochemical reactions (cf. Pecht and Rigler, 1977), some other techniques (cf. Hiromi, 1979) should also be considered. Measurements of fluorescence life times (cf. Chap. 5 of this volume) or of brillouin scattering of light by thermal phonons, pulse radiolysis, and flash photolysis belong to the relatively fast kinetic techniques. On the other hand, polarographic and magnetic resonance studies generally allow kinetic studies only down to the time range of about 1 ms (cf. Hammes, 1973a).

The recent introduction of pulsed flow techniques (Owens and Margerum, 1980; Owens et al., 1980) wich allow the measurements of minimal half-life times of species down to about 50 to 100 μs is expected to stimulate considerably the investigation of fast reaction in solution.

II. Experimental Methods

A brief summary will be given of the relaxation techniques which have so far been successfully applied to the investigation of membrane processes. The experimental techniques used in the study of homogeneous (or quasi-homogeneous) solutions differ from those used in the study of heterogeneous systems (such as the black lipid bilayer membrane separating two aqueous phases according to Mueller and Rudin, 1967). In the former optical detection systems are generally used whereas the latter is mainly based on current detection.

A. Techniques for the Investigation of Single Lipid Bilayer Membranes

Temperature Jump Relaxation. Knoll and Stark (1977) were the first to report temperature jumps of 0.1° to 0.5°C induced by the absorption of an intensive light flash acting as the energy source. In this method the flash is focussed on the membrane employing a hollow mirror. In order to increase light absorption, commercial ink is added to both aqueous compartments. The response is followed by current detection with light-shielded Ag-AgCl electrodes. The time resolution is about 100 μs. It is limited by the HF-noise in the detection system accompanying the discharge and by the charging time of the circuit. An improved method (Brock et al., 1979) is based on the fast heating by a Nd-glass-laser beam (wavelength 1.06 μm) with a pulse energy of 17 Joule and a pulse duration of 500 μs, enabling temperature jumps of 0.4° to 0.8°C.

Electric Field Jump Relaxation. This was the first relaxation method applied to the investigation of a heterogeneous membrane system (Ketterer et al., 1971; Stark et al., 1971).

The experimental arrangement consists of a generator and an electronic switch producing a voltage pulse of 10 to 40 mV of variable duration which is applied to the black lipid bilayer membrane via electrodes in both compartments. The rise time of the electronic switch is about 1 μs. The current is measured with Ag-AgCl electrodes with large surface areas. The time resolution of this method is about 1 μs, being limited by the spike resulting from the charging of the membrane capacitance. The time-dependent response of the current is measured as the voltage drop across an external resistor as indicated on an oscilloscope. A schematic illustration of this method is shown in Fig. 3. This technique can easily be applied to repetitive measurements.

Field jump relaxation experiments can be performed either by applying a stepwise voltage for a given time period or by switching off an applied voltage. Only in the latter case is relaxation towards true equilibrium conditions observed. In order to reduce the charging period of the bilayer membrane to less than 1 μs, independently of the magnitude of the series resistances in the compartments, a special voltage clamp method has been developed (Sandblom et al., 1975).

Charge-Pulse Technique. Although it can be considered to be related only to the relaxation methods above, mention should be made of the charge-pulse technique (Feldberg and Kissel, 1975; Benz and Läuger, 1976) which is characterized by a high time resolution. By this technique the black lipid bilayer membrane is charged up to a voltage of 10 mV through a pair of platinized platinum electrodes, within 20 to 50 ns, employing

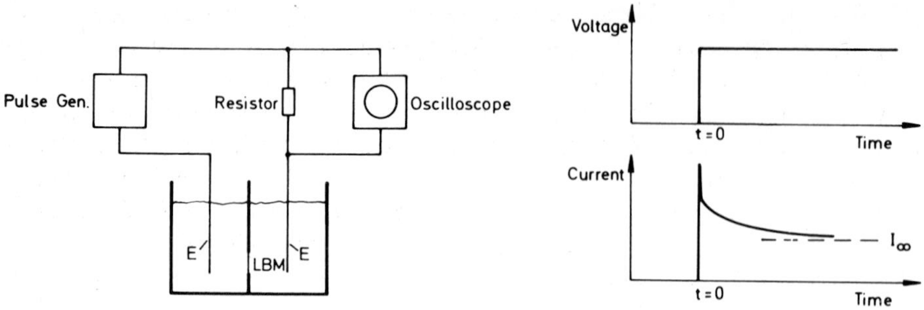

Fig. 3. Schematic illustration of the electric field jump technique with current detection (*E* electrodes) for measurements with lipid bilayer membranes (*LBM*); the lipid membrane separates the two aqueous compartments. (From Grell et al., 1974)

a fast FET-switch of very high impedance. The charge-pulse technique requires the measurement of the time course of the voltage across the membrane. The time resolution of this method is about 400 ns. In order to increase the sensitivity pulse sequences are applied.

B. Techniques for the Investigation of Membranes in Homogeneous or Quasi-Homogeneous Solutions

In order to apply techniques that have been developed for kinetic studies in homogeneous solutions to the investigation of membrane processes, it is required to use nonheterogeneous membrane systems. Of course, membrane components such as nonclosed membrane fragments, micelles, and dissolved membrane proteins would fulfill the necessary requirements. However, investigation of such components would not provide direct information concerning the elementary steps and mechanisms of action of transport processes. The desired information may be usefully obtained from studies of small closed vesicles either naturally occurring or prepared from lipids or from biological membranes. Such vesicles constitute a pseudo-homogeneous solution which may be studied with classical relaxation methods and in addition exhibit transport properties. Three types of simple phospholipid vesicles (cf. Fig. 4) are available for such studies. These are, namely, monolayer prevesicles where an inner aqueous compartment is separated from the outer organic solvent phase (Träuble and Grell, 1971), monolayer vesicles where the single lipid layer separates the inner organic phase from the outer aqueous medium (Sackmann and Träuble, 1972; Ruf, 1974) and finally bilayer vesicles which may be prepared according to the methods of Huang (1969), Seufert (1970) and Brunner et al. (1976). Apart from their remarkable stability these vesicular dispersions exhibit a total surface area which is considerably larger than that of planar bilayer membranes. This property of the vesicular system facilitates the application of such optical detection methods as are commonly used in chemical relaxation methods. In the following account, some aspects of the various techniques will be mentioned briefly. Details can be obtained from several extensive review articles (cf. Eigen and De Maeyer, 1963; Czerlinski, 1966; Hammes, 1973a).

Temperature Jump Relaxation Spectrometry. The temperature jump method employing spectrophotometric detection with fast photomultipliers was developed by Eigen and De Maeyer (1963) and is the most widely used transient chemical relaxation technique. The rapid temperature rise of up to 5°C occurs within microseconds and is performed by Joule heating of a sample containing an inert electrolyte. The fast heating is achieved by discharging a HV-capacitance C_{HV} (e.g., 5 x 10^{-8} F) through the conducting solution of the sample of resistance R_{cell} (typical values are 20 to 200 Ω). The capacitor can be loaded with a voltage, U_{HV}, which may be as high as 40 kV. The discharge is initiated by firing the spark gap (cf. Fig. 5). The time constant, τ_{heat}, characterizing the heating process is given by

$$\tau_{heat} = \frac{1}{2} R_{cell} C_{HV} \tag{9}$$

The magnitude of the temperature jump in the sample, δT, can be estimated according to Eq. (10),

$$\delta T = g_{cell}^{-1} \frac{C_{HV} U_{HV}^2}{2\rho c_p} \tag{10}$$

where g_{cell} is a cell constant, e.g., the effective volume fraction that has been heated up. The cells contain platinum or gold electrodes (cf. Fig. 5). Their geometry enables uniform heating. The new equilibrium condition, due to the temperature rise, can be maintained for 0.2 to 2 s depending on the size and geometry of the cell, as well as on the efficiency of the thermostatic control of the cell.

The change of concentration of the equilibrium components subsequent to the temperature jump is measured spectrophotometrically employing absorption or fluorescence detectors as indicated in Fig. 5 (Eigen, De Maeyer, 1963; Rabl, 1973; Rigler and Ehrenberg, 1973; Hammes, 1973b; Rigler et al., 1974; Jovin, 1975). The light beam from the lamp passes through the monochromator, the cell, and several optical components, and finally reaches the sample photomultiplier. In order to compensate for light intensity fluctuations of the lamp, dual beam arrangements, comprising a beam splitter and a reference multiplier have been incorporated. The balanced differential signal is amplified, filtered and displayed on an oscilloscope. In order to improve the signal-to-noise ratio special signal filter systems have been introduced.

Only those relaxation processes can be resolved which are slower than the heating process. The time range generally available is in the region of 2 x 10^{-6} to 1 s. The sensitivity of the absorption detection system extends to about 10^{-4} extinction units. The volume of the cells made out of Plexiglas, Dynal, or Teflon varies from 0.3 to 1 ml. In order to avoid cavitation effects the samples have to be degassed prior to the experiments.

In addition to the classic instrumentation discussed above, nanosecond temperature jump techniques for small volumes of about 0.1 ml have been developed. A sample, containing high electrolyte concentrations, is also resistively heated (up to about 10°C) by the discharge of a coaxial cable (Hoffman, 1971). A different approach is to use intensive pulses of an iodine laser beam (wavelength 1.315 μm) to obtain jumps of several degrees (Holzwarth et al., 1977). This technique does not require high ionic strength.

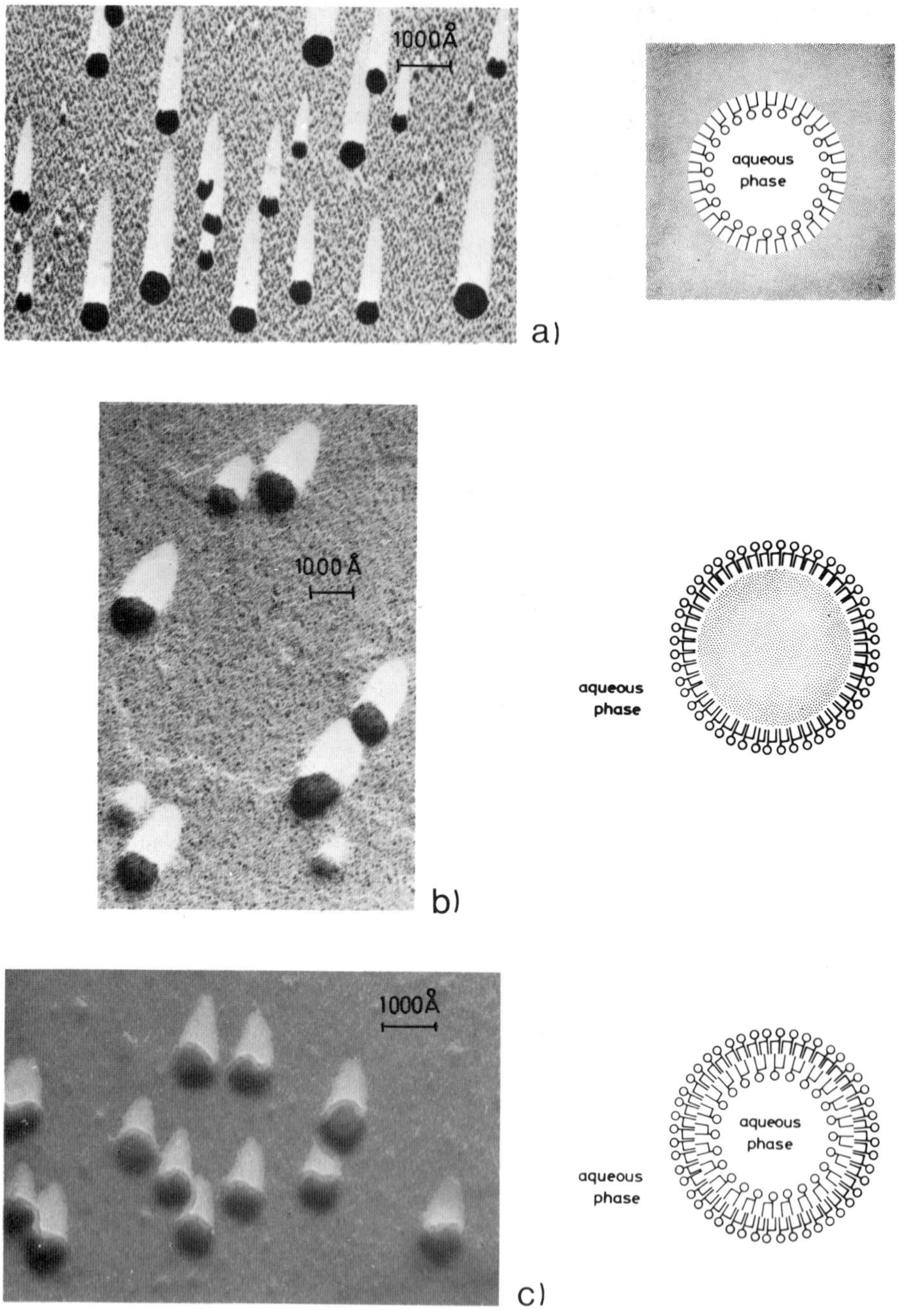

Fig. 4 a-c. Electron micrographs (shadow casting with Ge or Pt/Pd) and schematic illustration of egg lecithin vesicles. **a** Monolayer prevesicles in an organic phase (di-n-butylether); prevesicles contain 1 M CsCl. (From Träuble and Grell, 1971). **b** Monolayer vesicles in aqueous solution; the phospholipid monolayer separates the inner organic phase (chlorobenzene). (From Ruf, 1974). **c** Bilayer vesicles. (From Grell et al., 1975)

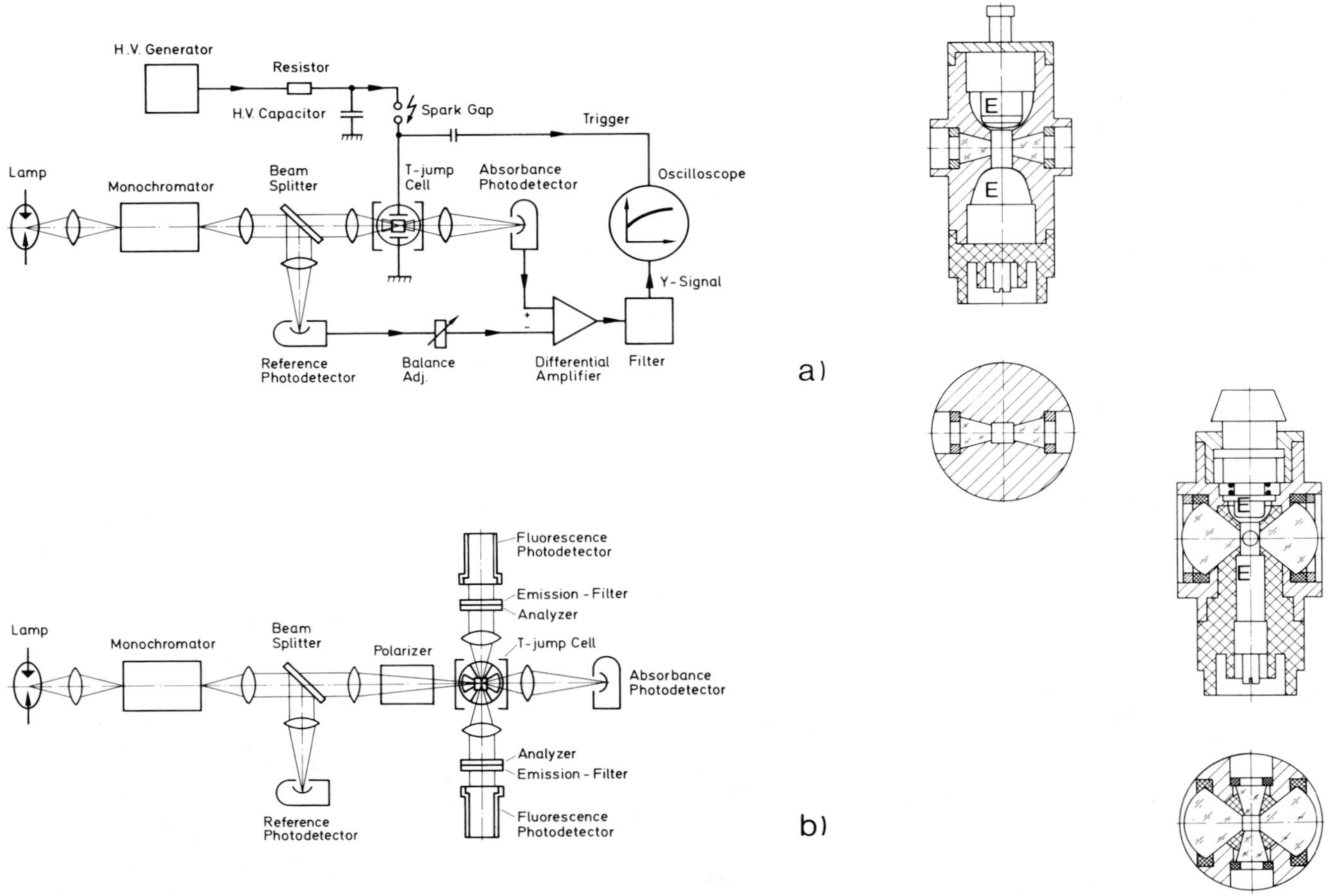

Fig. 5 a and b. Schematic arrangement of temperature jump relaxation spectrometers and sample cells (*E* electrode) for absorption (**a**) and fluorescence (**b**) detection as developed at the Max-Planck-Institute for Biophysical Chemistry, Göttingen

Pressure Jump Method. This method is based on a perturbation of the chemical equilibrium by a pressure change. Since chemical equilibrium systems are generally not very sensitive to pressure changes, either very high pressures or repetitive techniques have to be applied. The present systems use bursting membranes (Knoche, 1973), shock-wave techniques or compression of solutions by voltage-dependent extension of piezoelectric crystal stacks (Clegg et al., 1975). Considering typical thermodynamic parameters for an equilibrium system (e.g., $\Delta V = 10$ ml/mol, $\Delta H = 10$ kcal/mol) at about 25°C, a temperature change of 1°C is equivalent to a pressure change of about 140 atm with respect to the change in equilibrium constant. Relaxation times in the range from 30 μs to 20 s can be measured with these techniques, employing conductometric or optical detection systems.

Electric Field Jump Method. The perturbation of chemical equilibrium systems by an electric field can be interpreted on the basis of the dissociation field effect and the dipole field effect [cf. Eq. (7)]. The dissociation field effect describes, for example, the decrease of the stability constant of a weak electrolyte in the presence of an electric field. For high fields the shift of the equilibrium constant depends linearly on the field strength. In the case of the dipole field effect, however, the shift of equilibrium constant depends on the square of the field strength. The former mechanism is predominantly responsible for the equilibrium shift in aqueous solution induced by field jump experiments.

For the investigation of very fast reactions in the time range of about 5×10^{-8} to 10^{-5} s the discharge of a coaxial cable is used for the generation of high voltage pulses. Concentration changes are also measured spectrophotometrically and the cells resemble those used for the temperature jump studies. This method allows the application of pulses of up to 100 kV/cm. In contrast to the temperature jump experiments, the conductivity of the sample must be low to avoid superimposed temperature effects. Details are described elsewhere (Eigen and De Maeyer, 1963; Ilgenfritz, 1971; De Maeyer and Persoons, 1973; Grünhagen, 1973, 1974).

Ultrasonic Absorption Methods. Kinetic information can be obtained from ultrasonic velocity dispersion and from absorption measurements. The latter technqiue offers fewer experimental difficulties. As typical stationary techniques, both methods operate at fixed frequencies and allow the use of a narrow bandwidth even at the highest frequencies.

The sound wave produces periodic fluctuations of temperature and pressure, thus perturbing the chemical equilibrium system [cf. Eq. (8)]. If the angular frequency ω is considerably lower than the reciprocal relaxation time, τ^{-1}, the response of the equilibrium, in terms of concentration changes, is in phase with the periodic forcing function and reaches its full amplitude. In case of angular frequencies which are very high compared with τ^{-1}, the phase delay of the response of the chemical reaction is 90° and only a diminished amplitude can be expected. Optimal amplitude conditions are reached if the value of the frequency is identical with that of τ^{-1} for the chemical process, where the phase delay is 45°C. The quantitative description for a one-step process is given by

$$\alpha\lambda = H_o \frac{\omega\tau}{1 + (\omega\tau)^2} + L\omega \qquad (11)$$

$\alpha\lambda$ is the absorption per wavelength which is proportional to the absorption per cycle. L represents the contribution of effects which cannot be attributed to the chemical relaxation process. A plot of $\alpha\lambda$ versus logf ($\omega = 2\pi f$) for a single equilibrium is shown in Fig. 6. The determination of the frequency exhibiting the maximum amplitude, f_{max}, in the

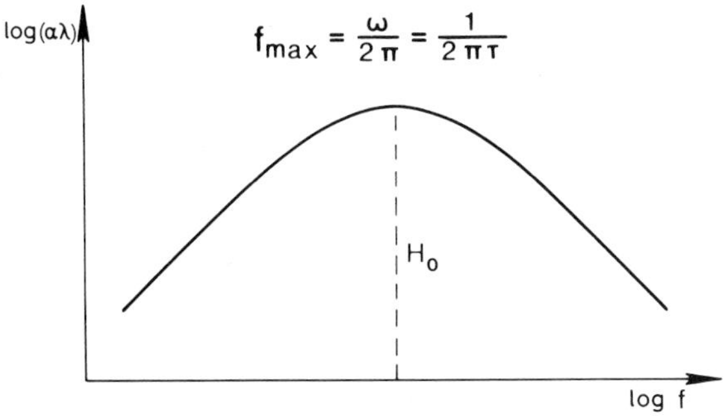

Fig. 6. Ultrasonic absorption: Double logarithmic plot of absorption per wavelength ($\alpha\lambda$) versus frequency (f) for a single relaxation process

plot of Fig. 6 (Debye curve) allows the calculation of the corresponding relaxation time. The maximum absorption amplitude (cf. Fig. 6) is given by

$$H_o = \frac{\pi\Gamma}{\kappa_s RT} (\Delta V - \frac{\alpha_p}{\rho c_p} \Delta H)^2 \tag{12}$$

where κ_s denotes the adiabatic compressibility. If, for example, the reaction enthalpy of an equilibrium system or step is known, the reaction volume can be calculated from the maximum ultrasonic absorption amplitude according to Eq. (12).

For the investigation of membranes and membrane components, two different absorption methods have been used, namely an acoustic resonator technique in the frequency range 0.2 to 30 MHz and a pulse technique between 15 and 150 MHz (Eggers, 1967/68; Eggers and Funck, 1973). In the case of the resonator method, the half power bandwidths, Δf, of the standing sound waves are determined for the sample and the corresponding solvent reference system separately. The absorption per wavelength due to the chemical system is calculated according to

$$\alpha\lambda = \pi \frac{\Delta f_{sample} - \Delta f_{solvent}}{f} \tag{13}$$

The ultrasonic wave acts as a disturbance as well as a sensor of the chemical equilibrium system. Therefore, no optical detection systems are required for the measurement of the response. No conductive solutions are required, thus the ultrasonic absorption methods are suitable for the investigation of fast relaxation processes in nonpolar solvents. Unfortunately, these methods are comparatively insensitive, and rather high concentrations of solutes (10^{-3}-10^{-1} M) are required. The typical time range for the determination of relaxation times in sample volumes of 1 to 3 ml varies from about 5 ns to 5 μs. Details of ultrasonic methods are discussed in several articles (Eigen and De Maeyer, 1963; Stuehr, 1973).

Dielectric Methods. If the electric moment of an equilibrium system is finite, the concentrations can be perturbed by the application of an electric field [cf. Eq. (7)]. Static electric field strengths as high as 100 kV/cm are required. Methods for the investigation of relaxation processes of dipolar equilibria have been developed and applied by Bergmann et al. (1963), Bergmann (1963), De Maeyer et al. (1968), De Maeyer and Persoons (1973).

Dielectric relaxation of phospholipid membranes in the absence of a static electric field has been studied by various methods in which the complex permittivity is determined as a function of the frequency in the range of 50 MHz to 60 GHz (cf. Kaatze et al., 1975; Kaatze et al., 1979b).

III. Theoretical Aspects

This section deals with the evaluation of kinetic and thermodynamic parameters of the experimentally observed relaxation times and amplitudes. The relaxation time, τ, is a measure of the time required to reach equilibrium after a perturbation due to a change of an intensive quantity, and is defined in Eq. (1). The amplitude factor, Γ, has been introduced as a measure of the chemical turnover (Eigen and De Maeyer, 1963). Generally speaking, the relaxation time provides information concerning rate and equilibrium constants and the relaxation amplitude characterizes also equilibrium constants and thermodynamic parameters such as the reaction enthalpy. The investigation of relaxation times and amplitudes as a function of temperature allows a detailed determination of kinetic and thermodynamic activation parameters of single elementary steps.

A. General Considerations

One-Step Equilibria. If the equilibrium of a simple, one-step binding reaction, at temperature T, is perturbed by a sudden rise in temperature, δT, the forward and backward rate constants, k_{12} and k_{21}, are defined for the temperature $T + \delta T$, as indicated below

$$A + B \underset{k_{21}}{\overset{k_{12}}{\rightleftharpoons}} AB \tag{14}$$

The equilibrium constant (neglecting activity coefficients) at $T + \delta T$ based on the equilibrium concentration, $\bar{c}_i$, of species i is given by

$$K = \frac{k_{12}}{k_{21}} = \frac{\bar{c}_{AB}}{\bar{c}_A \ \bar{c}_B} \tag{15}$$

The following differential equation holds

$$-\dot{c}_A = k_{12} c_A c_B - k_{21} c_{AB} \tag{16}$$

where c_i represents the time-dependent concentration of species i. These concentrations can be expressed in terms of equilibrium concentrations, $\overline{c}_i$, and changes in concentration, δc_i, induced by the temperature jump and due to the mass conservation $\delta c_A = \delta c_B = -\delta c_{AB} = \delta c_i$

$$c_A = \overline{c}_A + \delta c_A = \overline{c}_A + \delta c_i \tag{17}$$

$$c_B = \overline{c}_B + \delta c_B = \overline{c}_B + \delta c_i$$

$$c_{AB} = \overline{c}_{AB} - \delta c_{AB} = \overline{c}_{AB} - \delta c_i$$

The consideration of $\dot{\overline{c}}_i = 0$ leads to

$$\dot{c}_A = \frac{d(\overline{c}_A + \delta c_i)}{dt} = \dot{\overline{c}}_A + \delta \dot{c}_i = \delta \dot{c}_i \tag{18}$$

Substituting Eq. (17) into (16) and taking the equilibrium relation, Eq. (15), into account yields

$$\begin{aligned} \delta \dot{c}_i &= -k_{12}(\overline{c}_A + \delta c_i)(\overline{c}_B + \delta c_i) + k_{21}(\overline{c}_{AB} - \delta c_i) = \\ &= -[k_{12}(\overline{c}_A + \overline{c}_B) + k_{21}]\, \delta c_i - k_{12}(\delta c_i)^2 \end{aligned} \tag{19}$$

Under the conditions of small perturbations ($\delta c_i \ll \overline{c}_i$) the above differential equation can be linearized [$(\delta c_i)^2 = 0$] without significant error. According to the definition of the relaxation time [Eq. (1)] it follows

$$\delta \dot{c}_i = -[k_{12}(\overline{c}_A + \overline{c}_B) + k_{21}]\, \delta c_i = -\tau^{-1}\, \delta c_i \tag{20}$$

The expression for the reciprocal relaxation time of the one-step binding equilibrium is

$$\tau^{-1} = k_{12}(\overline{c}_A + \overline{c}_B) + k_{21} \tag{21}$$

Equation (21) predicts a linear dependence of the reciprocal relaxation time on the sum of the equilibrium concentrations. The rate constants can be calculated from the slope and intercept. The evaluation of the relaxation time according to Eq. (21) assumes that the corresponding equilibrium constant is known. If this is not the case, τ^{-1} can be expressed in terms of total concentrations, c_i°, allowing the determination of k_{12} and K according to

$$\tau^{-1} = k_{12}\sqrt{(c_A^{\circ} + c_B^{\circ} + K^{-1})^2 - 4c_A^{\circ}c_B^{\circ}} \tag{22}$$

The amplitude factor Γ for the above equilibrium [cf. Eq. (14)] is

$$\Gamma = \left(\frac{1}{\overline{c}_A} + \frac{1}{\overline{c}_B} + \frac{1}{\overline{c}_{AB}}\right)^{-1} \tag{23}$$

or expressed in total concentration according to Winkler-Oswatitsch and Eigen (1979).

$$\Gamma = \frac{K^{-1}}{2} \left[\frac{1}{1-4c_A^\circ c_B^\circ/(c_A^\circ + c_B^\circ + K^{-1})^2} - 1 \right] \tag{24}$$

The equilibrium constant, K, for the one-step binding equilibrium [Eq. (14)] can be determined from relaxation titration experiments according to Eqs. (22) and (24) (cf. Winkler, 1969; Winkler-Oswatitsch and Eigen, 1979).

Typical plots of $k\tau$ versus the total concentration of B, c_B° ($c_B^\circ = \bar{c}_B + \bar{c}_{AB}$), and of the amplitude factor versus c_B° are shown in Fig. 7 a and b, respectively, for a hypothet-

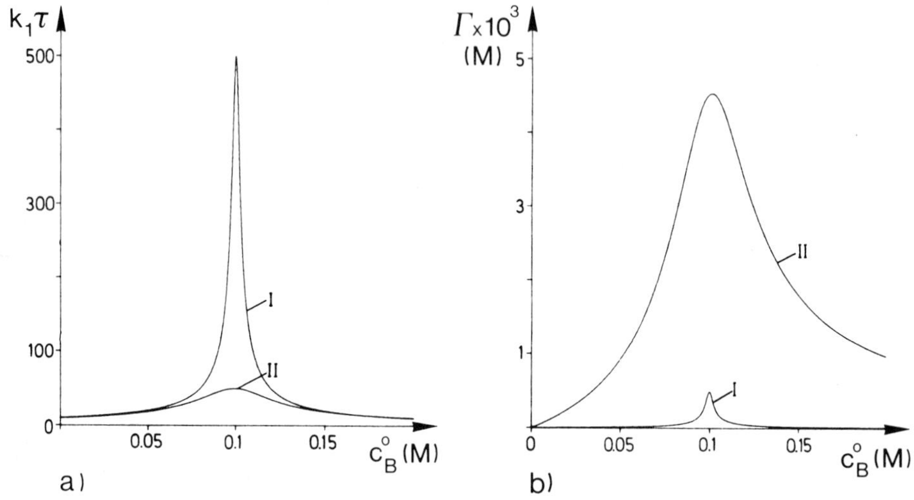

Fig. 7 a and b. Titration experiments employing chemical relaxation methods: Evaluation of relaxation time (**a**) and amplitude (**b**) (Parameters c_A° = 0.1 M; K = $10^5 M^{-1}$ (*I*), K = $10^3 M^{-1}$ (*II*). (From Strehlow and Knoche, 1977)

ical example under conditions where the total concentration of A, c_A°, is kept constant. Clear maxima are observed. The maximum relaxation amplitude is obtained if all three equilibrium concentrations are equal to the value of the reciprocal equilibrium constant ($\bar{c}_A = \bar{c}_B = \bar{c}_{AB} = K^{-1}$). The relaxation amplitude are valuable tools for the determination of equilibrium constants (cf. Winkler-Oswatitsch and Eigen, 1979).

The amplitude factors represent the ratio of the experimental relaxation amplitude to the total signal. The following relationship between the change, δI, of an optical absorption signal (light intensity I) and the concentrations of species i holds for small concentration changes:

$$\delta I = -2.303\, I\, \delta E; \qquad \delta E = \sum_i \epsilon_i\, \delta c_i\, l \tag{25}$$

where E is the extinction, ϵ_i is the decadic extinction coefficient and l is the light path. A similar expression has been derived for the fluorescence signal (Rigler et al., 1974).

Applying van't Hoff's relation [Eq. (4)] to the binding equilibrium [Eq. (14)] and by making the assumption that only species A and AB are absorbing, gives

$$\delta E = \Gamma \, \Delta\epsilon \, l \, (\Delta H \frac{\delta T}{RT^2}) \tag{26}$$

where $\Delta\epsilon = \epsilon_{AB} - \epsilon_A$ and δE is linearly dependent on Γ. According to Eq. (26), a plot of δE versus Γ allows the determination of the reaction enthalpy, ΔH, from the corresponding slope. In addition, a plot of the relaxation amplitude as a function of the relaxation time has been introduced by Jovin (1975) for the determination of ΔH.

Winkler-Oswatitsch and Eigen (1979) have suggested an elegant procedure by which the kinetic and thermodynamic parameters of an equilibrium such as in Eq. (14), may be experimentally determined. It is based on the expression

$$\tau = \frac{1}{k_{12}(c_A^o + c_B^o + K^{-1})} + \frac{2}{k_{12}(c_A^o + c_B^o + K^{-1})\, K^{-1}\, \Delta Z} (\Gamma \Delta Z) \tag{27}$$

where ΔZ is the extensive property [cf. Eqs. (5-7)].

In contrast to the above consideration of the simplest possible binding equilibrium, the expression for the simplest one-step rearrangement process

$$A \underset{k_{21}}{\overset{k_{12}}{\rightleftharpoons}} B \tag{28}$$

is

$$\tau^{-1} = k_{12} + k_{21} \tag{29}$$

From Eq. (29) it is concluded that the reciprocal relaxation time of a single first-order reaction step is concentration-independent. The expression for the amplitude is given by

$$\Gamma = (\frac{1}{\bar{c}_A} + \frac{1}{\bar{c}_B})^{-1} \tag{30}$$

Two-Step Equilibrium. The simplest two-step binding equilibrium is indicated by the following scheme:

$$\underset{}{A + B} \underset{\substack{k_{21}\\ \text{(I)}}}{\overset{k_{12}}{\rightleftharpoons}} AB \underset{\substack{k_{32}\\ \text{(II)}}}{\overset{k_{23}}{\rightleftharpoons}} C \tag{31}$$

Since two equilibrium steps are involved, two relaxation times (τ_1 and τ_2) should be expected. If both steps (I and II) are characterized by similar reaction rates, τ_1 and τ_2 are solutions of a quadratic equation. Plots of $(\tau_1^{-1} \cdot \tau_2^{-1})$ and of $(\tau_1^{-1} + \tau_2^{-1})$ versus concentration terms are required for the evaluation of all rate constants. This is rather elaborate.

In most of the cases, experimental conditions can be found, where the rate of equilibration of step (I) is much faster than step (II). The expressions for the relaxation times are then:

$$\tau_1^{-1} = k_{12}(\bar{c}_A + \bar{c}_B) + k_{21} \tag{32}$$

and

$$\tau_2^{-1} = \frac{k_{12}(\bar{c}_A + \bar{c}_B) \cdot k_{23}}{k_{12}(\bar{c}_A + \bar{c}_B) + k_{21}} + k_{32} \tag{33}$$

The determination of the parameters k_{12} and k_{21} is identical to that for the simple equilibrium [Eq. (14)]. The reciprocal of the second relaxation time is also plotted versus $(\bar{c}_A + \bar{c}_B)$. The corresponding plot shows the more complex concentration dependence and exhibits saturation behaviour for large values of $(\bar{c}_A + \bar{c}_B)$. If k_{12} and k_{21} or alternatively k_{12}/k_{21} is known, the kinetic parameters k_{23} and k_{32} can be determined. From Eq. (33) it follows that τ_2 depends not only on the rate constants of step (II) but also on those of the faster step (I).

If step (I) in the following two-step rearrangement system

$$\underset{(I)}{A \underset{k_{21}}{\overset{k_{12}}{\rightleftharpoons}} B} \underset{(II)}{\underset{k_{32}}{\overset{k_{23}}{\rightleftharpoons}} C} \tag{34}$$

is assumed to be fast compared to step (II), the following expressions for the reciprocal relaxation times hold:

$$\tau_1^{-1} = k_{12} + k_{21} \qquad \tau_2^{-1} = \frac{k_{23}}{1 + k_{21}/k_{12}} + k_{32} \tag{35}$$

For details concerning the determination of kinetic and thermodynamic parameters from the relaxation times and amplitudes of two-step and multi-step reactions, special texts may be consulted (Eigen and De Maeyer, 1963, 1973; Castellan, 1963; Hammes and Schimmel, 1966; Winkler, 1969; Schimmel, 1971; Thusius, 1972; Thusius et al., 1973; Jovin, 1975; Ilgenfritz, 1977; Winkler-Oswatitsch and Eigen, 1979).

In the following section special consideration will be given to the analysis of the relaxation times of membrane reactions.

B. Special Considerations of Membrane-Related Processes: Interaction of Small, Neutral Molecules with Spherical Membrane Vesicles

Dispersions of vesicles can be considered as quasi homogeneous solutions of membrane particles, and the binding of small molecules to these vesicles can be treated by analogy with chemical reactions.

The membrane of such a vesicle forms a spherical shell, enclosing a small volume of the aqueous bulk phase. Molecules adsorbed by the membrane are assumed to be bound either at the surfaces or within the nonpolar interior of the membrane. In the case of lipid bilayer vesicles, the surface part corresponds to the region of the polar head groups, while the nonpolar interior is formed by the hydrocarbon chains of the lipid molecules. Schematically, the vesicular membrane can be subdivided into three shells: A thick shell representing the nonpolar part, which is enclosed by two thin shells representing the surface parts (Fig. 8).

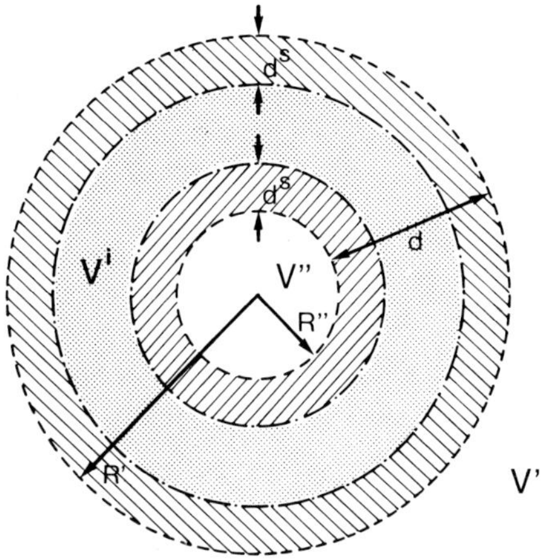

Fig. 8. Schematic representation of a vesicle. R′ and R″ are the radii of outer and inner membrane surface, respectively. V′ and V″ are the volumes of outer and inner aqueous bulk phase, and V^1 is the volume of the nonpolar part of the membrane. d^S ($d^S \ll R', R''$) is the thickness of the surface shells, and d is the total thickness of the membrane

Consequently we have to consider two different processes: The exchange of molecules between the membrane surfaces and their adjacent aqueous bulk phases, and the exchange between the nonpolar interior and the surfaces.

Adsorption to a surface from either side may be described by a two-step mechanism

$$A + S \underset{k_2}{\overset{k_1}{\rightleftharpoons}} \overline{AS} \underset{k_4}{\overset{k_3}{\rightleftharpoons}} AS$$

A particle A in the bulk phase approaches the surface S due to its diffusional motion and forms an encounter complex $\overline{AS}$, which then transforms into the adsorbed state by entering the surface region. Desorption of a particle is the reverse process. k_1, k_2, k_3, and k_4 are the rate constants. The reciprocal value of the rate of the overall reaction

$$A + S \underset{k_2^*}{\overset{k_1^*}{\rightleftharpoons}} AS$$

is obtained from the sum of the reciprocal velocities of the individual steps. If one step is much faster than the other, the overall rate is determined by the slower one. When the second step is fast compared to the first, the rate of adsorption is diffusion-controlled ($k_1^* = k_1$). The rate constant of desorption ($k_2^* = k_2 K^S$) includes the equilibrium constant $K^S = k_4/k_3$ of the rapidly equilibrating process $\overline{AS} \rightleftharpoons AS$. If the transition $\overline{AS} \rightleftharpoons AS$ is rate limiting, analogous expressions for the overall rate constants are obtained ($k_1^* = k_3K^\circ$, $k_2^* = k_4$, $K^\circ = k_1/k_2$).

Here, the special case is treated in which the adsorption of small particles from the aqueous bulk phase to the membrane is diffusion-controlled, whereas the rate of adsorption from the non-polar interior is determined by an energy-barrier located in the interface. It is assumed that the particle exchange between the membrane surface and the aqueous phases proceeds much faster than between the surface and the nonpolar interior. This implies that after a rapid disturbance of the equilibrium, practically no particle transfer takes place across the membrane while the adsorption-desorption processes in the two aqueous subsystems equilibrate, or that the latter are uncoupled during that period. To describe the relaxational behaviour of our system, it is necessary to solve a system of three coupled linear differential equations. The latter are obtained by linearizing the rate equations near the equilibrium. However, according to the above assumptions, the problem need not be solved in its general form (see Eigen and De Maeyer, 1963). The fast equilibration of the diffusion-controlled processes will not be affected by the much slower exchange of molecules between the surface of the membrane and its nonpolar interior. Thus, the rate equations for the diffusion-controlled reactions can be solved separately without considering the latter process. In order to solve the rate equation for the slow exchange processes between the surface and the nonpolar interior, on the other hand, the diffusion-controlled reactions have to be taken into account with respect to their equilibrium constants.

The basic method of this treatment of the diffusion-controlled adsorption from the aqueous bulk phases is the same as in von Smoluchowski's (1916), and many subsequent works (Debye, 1942; Umberger and La Mer, 1945; Eigen, 1954; Weller, 1957), i.e., the diffusion flux of particles into and out of a sphere is calculated. Two spheres are considered, which correspond to the inner and outer surface of the membrane.

The rate of adsorption from the nonpolar interior, on the other hand, is determined by the probability of an encounter complex crossing the energy barrier in the interface. Thus, it is given by the number of encounter complexes times this probability (k_3). The concentration of the encounter complexes is nearly in equilibrium with the concentration of the particles in the bulk. This implies that the number or encounter complexes is simply proportional to the bulk concentration and to the area of the surface in question. It should be mentioned that the concentration in the interior bulk phase, as used in this treatment, is only a mean concentration. The actual distribution of particles within the nonpolar interior needs not be specified in this approach, since all processes regulating this distribution are assumed to be fast compared to the absorption step.

Concerning the state of the particles adsorbed in the surface parts of the membrane, the two limiting cases will be considered, in which the mutually non-interacting molecules may diffuse freely within the surface area, like molecules in a two-dimensional gas, or in which the molecules are bound to fixed binding sites. Which one of the two cases, termed nonlocalized and localized binding (see Aveyard and Haydon, 1973) applies to a given system depends on the lateral mobility of an adsorbed molecule in relation to its life-time in this particular state.

1. Nonlocalized Binding

a) Diffusion-Controlled Adsorption of Neutral Molecules from the Aqueous Phases

Due to the independent diffusion of each molecule, the exchange between the membrane surface and the aqueous bulk phase may be divided into an adsorption and a desorption process. Diffusion-controlled adsorption implies that every encounter between a molecule and a vesicle results in binding. A vesicle, therefore, can be considered to be a sink for molecules diffusing within the aqueous bulk phase. Conversely, with respect to the desorption from the surface into the bulk phase, the vesicle represents a source of molecules. In order to calculate the fluxes it is necessary to solve one diffusion problem, in which the vesicle is a sink for molecules, and one, in which it is a source.

The diffusion equation, which has to be solved for both aqueous subsystems takes, for this particular problem, the form

$$\frac{\partial (r\, n)}{\partial t} = D\, \frac{\partial^2 (r\, n)}{\partial r^2} \tag{36}$$

indicating that the particle concentration n depends only on the distance r from the centre of the vesicle and the time t. Since the relatively large vesicle is practically immobile, D is, to a good approximation, the diffusion coefficient of the small particle. The solution of Eq. (36) is given by

$$n(r,t) = C + \frac{B}{r} + \frac{2A}{r\sqrt{\pi}} \left(\int_0^w e^{-w^2}\, dw \right) \tag{37}$$

where $w = (r\text{-}R) / (2\sqrt{Dt})$. A, B and C are integration constants, which are determined by the appropriate starting and boundary conditions.

In order to calculate the influx from the bulk phase the concentration of the free particles is assumed to be zero at the boundary of the sphere and equal to the mean bulk concentration at large distances from the boundary. Formally this reads for the outer bulk phase

$$\begin{aligned} &1.\ n = n_o' \quad \text{for } t = 0 \text{ and } r > R' \\ &2.\ n = 0 \quad \text{for } r = R' \text{ and } t > 0 \end{aligned} \tag{38}$$

and for the inner aqueous bulk phase

$$\begin{aligned} &1.\ n = n_o'' \quad \text{for } t = 0 \text{ and } r < R'' \\ &2.\ n = 0 \quad \text{for } r = R'' \text{ and } t > 0 \end{aligned} \tag{39}$$

where n_o' and n_o'' denote the corresponding mean bulk concentrations.

In order to calculate the efflux the concentration of bound particles is presumed to remain constant at the boundary and is considered to be zero at large distances from the surface (Eigen, 1954). Thus, for desorption into the outer aqueous bulk phase

$$\begin{aligned} &1.\ n = 0 \quad \text{for } t = 0 \text{ and } r > R' \\ &2.\ n = n'_R \quad \text{for } r = R' \text{ and } t > 0 \end{aligned} \tag{40}$$

and correspondingly into the inner aqueous bulk phase

$$\begin{aligned} &1.\ n = 0 \quad \text{for } t = 0 \text{ and } r < R'' \\ &2.\ n = n''_R \quad \text{for } r = R'' \text{ and } t > 0 \end{aligned} \tag{41}$$

n'_R and n''_R are the concentrations of the encounter complex at the corresponding surface, which are related to the numbers of the surface-bound molecules $N^{s\prime}$ and $N^{s\prime\prime}$ by

$$n'_R = K^s N^{s\prime} / (4\pi R'^2 d^s) \tag{42}$$

$$n''_R = K^s N^{s\prime\prime} / (4\pi R''^2 d^s) \tag{43}$$

$K^s N^{s\prime}$ and $K^s N^{s\prime\prime}$ are the statistical probabilities of finding the particles, which are absorbed in the surface shells of thickness d^s, at the boundaries towards the aqueous bulk phases (the occurrence of the equilibrium constant K^s is a direct consequence of the assumption that the actual binding step is fast compared to the diffusional process).

In the general case the number of particles adsorbed per second at a sphere of radius R may be considered as the number of binding reactions per second a vesicle performs with the particles. This may be expressed by

$$\dot{N}_{ad} = 4\pi R^2 D \left(\frac{\partial n}{\partial r}\right)_{r=R} = k^{as} \frac{1000\, n_o}{L} \tag{44}$$

k^{as} is the molar rate constant for adsorption, L is Avogadro's number, and n represents the mean aqueous bulk concentration. Correspondingly, the rate of desorption from the surface into the aqueous bulk phase can be considered as the rate of dissociation of vesicle-particle complexes, which is given by

$$N_{de} = 4\pi R^2 D \left(\frac{\partial n}{\partial r}\right)_{r=R} = k^{sa} N^s \tag{45}$$

where k^{sa} is the rate constant for the desorption process. For our two subsystems we have to introduce into Eqs. (44) and (45) the corresponding primed or double-primed notation, in which the subphases are specified.

The calculations are then straightforward. Introducing the above starting and boundary conditions into Eq. (37), and using Eqs. (43-45) we obtain

$$k^{as\prime} = \frac{4\pi D L R'}{1000} \left(1 + \frac{R'}{\sqrt{\pi D t}}\right) \tag{46}$$

$$k^{sa\prime} = \frac{K^s D}{d^s R'} \left(1 + \frac{R'}{\sqrt{\pi D t}}\right) \tag{47}$$

$$k^{as''} = \frac{4\pi DLR''}{1000} (1 + \frac{R''}{\sqrt{\pi Dt}}) \tag{48}$$

$$k^{sa''} = \frac{K^s D}{d^s R''} (1 + \frac{R''}{\sqrt{\pi Dt}}) \tag{49}$$

In addition the four kinetic parameters shown above include a time-dependent term which becomes negligible after a time $t \gg R/D$. The processes then become stationary and these parameters become real constants. This is the approximation used in the treatments of diffusion-controlled chemical reactions. Neglecting the time-dependent term in Eqs. (48) and (49) gives a relationship analogous to that derived by Woolley and Diebler (1979), who showed that the binding between a small neutral dye molecule and a spherical vesicle is diffusion-controlled. However, as will be shown below, this time-dependent term generally cannot be omitted, especially for the equilibration of the processes in subsystem (″).

A vesicular dispersion contains a number of vesicles m. To simplify matters, it is assumed that these are all of the same size. The total volume V of such a dispersion then consists of V' and $m(V'' + V^i)$. The rate equations for the processes in the two subsystems (′) and (″) are obtained as usual taking into account all reactants and products, respectively.

There is a marked difference between the two subsystems. A particle in the outer aqueous bulk phase may react with each vesicle of the dispersion, whereas a particle in the volume V'' may be adsorbed only at the inner surface of the vesicle in which it is confined. Thus in the first case the probability of an encounter is proportional to the vesicle concentration, while in the second it is independent of the number of vesicles present in a dispersion. In the latter case this is completely determined by the inner volume and the diffusion coefficient.

The concentrations in a vesicular dispersion, as measured by kinetic methods, normally refer to the total volume. Therefore, the actual concentrations in the two subsystems have to be transformed. This is done by calculating the number of molecules and the adsorption and desorption rates in the corresponding subsystem and dividing them by the total volume and Avogardo's number. The index v indicates that a concentration refers to the total volume. The rate equations for the two subsystems then become

$$\dot{c}_v^{s'} = k^{as'} \frac{V}{V'} c_v' c_{VES} - k^{sa'} c_v^{s'} \tag{50}$$

and

$$\dot{c}_v^{s''} = k^{*as''} c_v'' - k^{sa''} c_v^{s''} \tag{51}$$

where c_v^s, c_v and c_{VES} denote the concentrations of surface-bound particles, free particles and vesicles, respectively.

The quantity $k^{*as''}$ includes the volume of the aqueous bulk phase $V'' = 4\pi R''^3/3$ enclosed by a vesicle. This is given by

$$k^{*as''} = \frac{3D}{R''^2} (1 + \frac{R''}{\sqrt{\pi Dt}}) \tag{52}$$

For small deviations from the equilibrium the rate equations may be linearized. Introducing the differences of the actual concentration from the equilibrium concentration $\delta c_V = c_V - \bar{c}_V$ and $\delta c_V^s = c_V^s - \bar{c}_V^s$, and taking into account that in each subsystem the total particle number is constant, or

$$\delta c_V^s = -\delta c_V \tag{53}$$

for both subsystems (′) and (″), we obtain the differential equation

$$\delta \dot{c}_V^{s'} = -\frac{1}{\tau'}\left(1 + \frac{R'}{\sqrt{\pi D t}}\right)\delta c_V^{s'} \tag{54}$$

and

$$\delta \dot{c}_V^{s''} = -\frac{1}{\tau''}\left(1 + \frac{R''}{\sqrt{\pi D t}}\right)\delta c_V^{s''} \tag{55}$$

where the proportionality factors are given by

$$\frac{1}{\tau'} = \frac{4\pi D L R'}{1000}\,\frac{V}{V'}\,c_{VES} + \frac{K^s D}{d^s R'} \tag{56}$$

and

$$\frac{1}{\tau''} = \frac{D}{R''}\left(\frac{3}{R''} + \frac{K^s}{d^s}\right) \tag{57}$$

τ' and τ'' are the relaxation times. The above equations indicate that only τ' is dependend on the vesicle concentration.

If the diameter of vesicles in a uniform population is known, the vesicle concentration can be determined via the lipid concentration according to the equation

$$c_{VES} = c_{LIP}^{o}\,\frac{a_L}{4\pi(R'^2 + R''^2)} \tag{58}$$

where c_{LIP}^{o} is the lipid concentration and a_L the area a lipid molecule covers in the bilayer membrane. For this the reciprocal relaxation time is given by

$$\frac{1}{\tau'} = \frac{D L\, a_L}{1000}\,\frac{R'}{(R'^2 + R''^2)}\,\frac{V}{V'}\,c_{LIP}^{o} + \frac{K^s D}{d^s R'} \tag{59}$$

If typical values are assigned to these parameters ($D = 10^{-5}$ cm^2 s^{-1}, $K^s = 10^{-4}$, d^s = 8 Å, R' = 250 Å, R'' = 200 Å, a_L = 60 Å, $c_{LIP}^{o} = 10^{-3}$M and $V/V' \sim 1$) estimates for the values of the relaxation times of $\tau' = 7.3\ 10^{-5}$ s and $\tau'' = 1.3\ 10^{-7}$ s can be obtained. After those times the values of the time-dependent term in the corresponding equation [Eqs. (54) and (55)] are about 0.05 and 1, respectively. Therefore, this term can be neglected only in Eq. (54).

Integration of the differential equations yields

$$\delta c_V^{s'} = (\delta c_V^{s'})_o\, e^{-\frac{1}{\tau'}\left(t + 2R'\sqrt{\frac{t}{\pi D}}\right)} \tag{60}$$

and

$$\delta c_V^{s''} = (\delta c_V^{s''})_o \, e^{-\frac{1}{\tau''}\left(t + 2R''\sqrt{\frac{t}{\pi D}}\right)} \tag{61}$$

where $(\delta c^{s'})_o$ and $(\delta c^{s''})_o$ are the maximum deviations from the corresponding intermediate equilibrium states to which the two fast processes adjust prior to relaxation of the slow process. As mentioned above, the two aqueous subsystems are uncoupled during the initial time period, since there is practically no particle transfer across the membrane. Therefore, there is a conservation of mass for each subsystem [Eq. (53)]. The concentrations in this intermediate equilibrium situation are determined by this conservation of mass and, of course, by the intensive properties of the system (temperature, pressure, etc.). However, on a larger time scale, while the processes in the nonpolar subsystem equilibrate, this separate conservation of mass is no longer valid. Consequently, the concentrations in the aqueous subsystems vary with time. The corresponding fast equilibria, however, are always adjusted. This is what is termed "pre-equilibrated".

Equations (60) and (61) indicate that equilibration of the diffusion-controlled processes occurs according to a complicated time function. Only if the time-dependent term is neglected can the relaxation process be described by a single exponential. It would then follow from Eq. (54)

$$\delta c_V^{s'} = (\delta c_V^{s'})_o \, e^{-\frac{t}{\tau'}} \tag{62}$$

For illustration, the difference between exact and approximate solution are shown in Fig. 9 (D, R′ etc. as above).

The equilibration in both systems is actually faster than expected from a treatment based on stationary rates. This is a consequence of the nonstationary part of the adsorption or desorption rate, which at the beginning contributes to a greater or lesser extent to the relaxation towards equilibrium. The difference is not severe with respect to the binding from the outer aqueous bulk phase under the conditions outlined above. For large vesicles, high vesicle concentrations or weakly bound molecules, however, the approximate solution [Eq. (62)] should be used with caution.

The above estimation indicates that the process in subsystem (″) characterized by the relaxation time τ'' is too fast to be resolved employing normal temperature jump methods. If the vesicle are very much larger it would be possible to measure it by these techniques. It is important to note here that under appropriate conditions the processes in the two subsystems equilibrate on two different time scales. Therefore, they can be analysed separately. It should be mentioned that the assumptions made on the coupling of the subsystems may be violated for very large vesicles. For this case a system of coupled differential equations has to be solved.

b) Non-Diffusion-Controlled Adsorption from the Nonpolar Interior

The rate of adsorption from the nonpolar interior of a vesicle to the two surfaces $AS' \rightleftharpoons AI \rightleftharpoons AS''$ is assumed to be controlled by an energy barrier at the corresponding interface. The number of particles adsorbed per second is then proportional to the num-

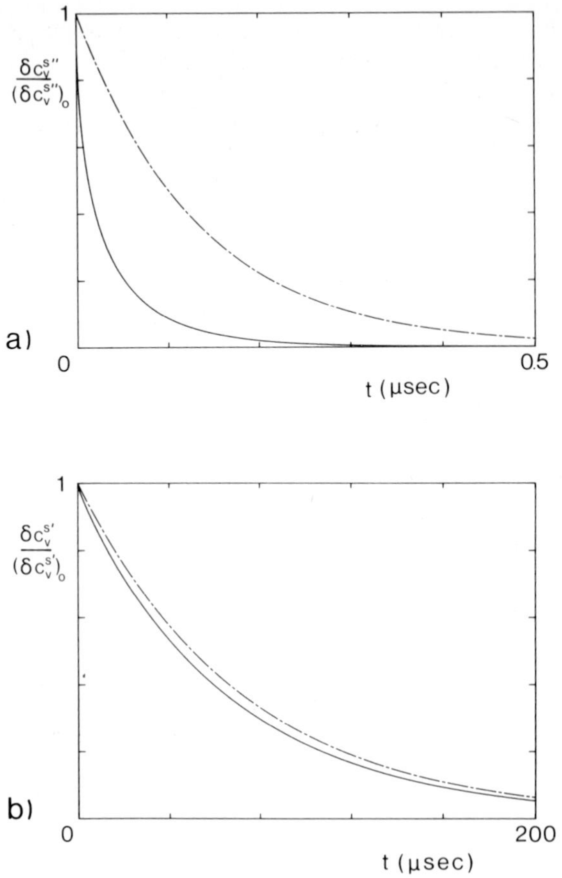

Fig. 9 a and b. Diffusion-controlled adsorption of neutral molecules to the inner (**a**) and outer (**b**) surface region of a vesicle: Theoretical relaxation curves for the equilibration in the two aqueous subsystems (') and (''). The exact solutions are represented by *solid lines* and the approximate solutions, where the time-dependent term in Eqs. (54) and (55) has been neglected, are represented by *dashed lines*

ber of encounter complexes, which in turn is proportional to the concentration in the bulk phase and to the area of the surface. The desorption rate, on the other hand, is simply proportional to the number of particles adsorbed at the surfaces. Thus the rate constants can be written:

$$k^{is} = k^{is'} + k^{is''} = \frac{4\pi L(R'^2 + R''^2)}{1000} k_3^i \tag{63}$$

and

$$k^{si} = k^{si'} + k^{si''} = k_4^i \tag{64}$$

k_3^i and k_4^i are the rate constants for the crossing of the energy barrier. k_3^i here has the dimension cm s^{-1} since it includes the thickness of the thin layer of nonpolar phase in which the particles form an encounter complex.

In a similar way the rate equation is obtained in terms of the overall concentrations measured by appropriate means:

$$\dot{c}_v^i = - k_v^{is} c_v^i + k^{si} c_v^s \tag{65}$$

where c_v^s is the sum of the concentrations of the subsystems ($c_v^s = c_v^{s'} + c_v^{s''}$) and c_v^i is the concentration of particles being in the nonpolar bulk phase. The rate constant

$$k_v^{is} = 3 \frac{(R'^2 + R''^2)}{(R'^3 - R''^3)} k_3^i \tag{66}$$

includes the volume of the nonpolar part of a vesicle.

The adsorption rate from the nonpolar interior is proportional to the area of the suface and so, at equilibrium, the mean numbers of particles bound at the two surfaces of a vesicle are in the same ratio as the squares of the corresponding radii, or $N^{s'} / N^{s''} = R'^2 / R''^2$. At equilibrium, the number of particles bound to a surface is also proportional to the surface. It follows, therefore, that at equilibrium the concentrations in the aqueous bulk phases inside and outside a vesicle are the same ($c' = c''$).

At equilibrium, therefore, we have to consider only three different particle species: particles in the aqueous bulk phase, particles in the nonpolar bulk phase and surface-bound particles. These are related by

$$\overline{c}_v^s = K_v^{is} \overline{c}_v^i \tag{67}$$

and

$$\overline{c}_v^s = K_v^{as} \overline{c}_v c_{VES} \tag{68}$$

where $\overline{c}_v$ is the sum of the equilibrium concentrations of the two subsystems ($\overline{c}_v = \overline{c}_v' + \overline{c}_v''$).

The equilibrium constants are given by

$$K_v^{is} = 3 \frac{(R'^2 + R''^2)}{(R'^3 - R''^3)} \frac{k_3^i}{k_4^i} \tag{69}$$

and

$$K_v^{as} = \frac{4\pi d^s L (R'^2 + R''^2)}{1000 K^s} \frac{V}{(V' + mV'')} \tag{70}$$

respectively.

To calculate the relaxation behaviour of the particle exchange between the nonpolar interior and the surface of the membrane the difference of the actual concentration from its equilibrium concentration is introduced as a new variable and the rate equation [Eq. (65)] is linearized for near-equilibrium conditions. Here, in addition, the two pre-equilibrium exchange processes between the surfaces and their adjacent aqueous bulk phases have to be taken into account. The relation $N^{s\prime} / N^{s\prime\prime} = R'^2 / R''^2$, also applies approximately to the intermediate equilibrium state (in our numerical example for concentrations $c \lesssim 10^{-3}$ M). Then the two mass action equations describing the pre-equilibria may be combined as in Eq. (68). The time-dependent concentration differences then shift according to

$$\delta c_v^s = K_v^{as} \, c_{VES} \, \delta c_v \tag{71}$$

According to the mass conservation law the concentration differences are related by

$$\delta c_v^s + \delta c_v^i + \delta c_v = 0 \tag{72}$$

Combining these relations with the linearized rate equation yields a linear differential equation

$$\delta \dot{c}_v^i = -\frac{1}{\tau^i} \, \delta c_v^i \tag{73}$$

of which the solution is a single exponential function, namely

$$\delta c_v^i = (\delta c_v^i)_o \, e^{-\frac{t}{\tau^i}} \tag{74}$$

where $(\delta c_v^i)_o$ is the maximum deviation from equilibrium. The reciprocal relaxation time is given by

$$\frac{1}{\tau^i} = k_v^{is} + k^{si} \frac{K_v^{as} \, c_{VES}}{1 + K_v^{as} \, c_{VES}} \tag{75}$$

This equation indicates a nonlinear dependence of the relaxation time upon the vesicle concentration. For high vesicle concentration ($K_v^{as} \, c_{VES} \gg 1$) the relaxarion time reaches a constant value.

2. Localized Binding

Localized binding refers to substrate molecules that are bound at the membrane to fixed binding sites. Such a binding site might be a membrane protein, for example. As mentioned earlier the applicability of this model will depend on the life-time of the complex relative to the lateral diffusion of the binding site in the membrane.

In this treatment it is assumed that a binding site exhibits 1:1 stoichiometry and that the binding sites are distributed homogeneously over the membrane surface. The surfaces of a vesicle, therefore, consist of numbers of binding sites, $N_o' = 4\pi R'^2/a_o$ and

$N_o'' = 4\pi R''^2/a_o$, where a_o is the area covered by a binding site. $N^{s'}$ and $N^{s''}$ denote the number of occupied binding sites. A substrate molecule striking an occupied binding site cannot be bound. Therefore, unlike the model given in the previous section, the probability of an impinging molecule being bound is less than one. θ represents the surface coverage, that is, the ratio of occupied sites to the total number of binding sites at a surface. The probability of a molecule from the aqueous bulk phase impinging on a free site is given by $(1\text{-}\theta')$ or by $(1\text{-}\theta'')$, respectively. The adsorption rates are obtained by multiplying the rate parameters [Eqs. (46, 48 and 63)] by the corresponding probabilities.

In the previous section concerning nonlocalized binding, the area covered by an adsorbed molecule has been disregarded. The relationships for nonlocalized binding apply only to relatively small surface coverages. For higher surface coverages, however, it would be necessary to introduce a corresponding expression for the probability of striking an unoccupied area at the surface.

A consequence of taking into account the area covered by an adsorbed molecule, or the area of a binding site, is that the maximum number of molecules which may be bound at the membrane is finite. Therefore, an additional equation for the conservation of mass is necessary, which accounts for the sum of free and occupied binding sites.

The derivation of the rate equations for the localized binding model is completely analogous to that given above. It provides very similar results and the same relaxation spectrum is obtained.

The reciprocal relaxation time for the adsorption from the outer aqueous bulk phase is given as

$$\frac{1}{\tau'} = \frac{k^{as'}}{N_o'} \frac{V}{V'} (\overline{c}_B' + \overline{c}_V') + k^{sa'} \tag{76}$$

where $\overline{c}_B'$ is the concentration of free binding sites (referred to the total volume). The relaxation time exhibits a dependence on both the vesicle and the particle concentration. Correspondingly, the expression for the relaxation time for the adsorption from the inner aqueous bulk phase reads

$$\frac{1}{\tau''} = \frac{k^{as''}}{N_o'' \, c_{VES}} (\overline{c}_B'' + \overline{c}_V'') + k^{sa''} \tag{77}$$

As in the nonlocalized model, the relaxation time is independent of the vesicle concentration, since both $\overline{c}_B''$ and $\overline{c}_V''$ are proportional to this concentration, but here the relaxation time exhibits a dependence on the particle concentration.

The expression for the relaxation time has been derived for the same conditions as above, which means that the surface coverage is assumed to be nearly the same at both surfaces when the system is in the intermediate equilibrium state. Thus

$$\frac{1}{\tau^i} = \frac{k_\theta^{is}}{c_{VES}} \left(\overline{c}_B + \overline{c}_V^i \frac{K_\theta^{as}\,\overline{c}_B}{1 + K_\theta^{as}(\overline{c}_B + \overline{c}_V)}\right) + k^{si} \frac{K_\theta^{as}\,\overline{c}_B}{1 + K_\theta^{as}(\overline{c}_B + \overline{c}_V)} \tag{78}$$

where $\overline{c}_B$ and $\overline{c}_V$ are the sums of the corresponding concentrations in the subsystems. The rate constant k_θ^{is} and the equilibrium constant K_θ^{as}, including the number of binding sites of a vesicle, are given by

$$k_\theta^{is} = \frac{3\,a_o}{4\pi\,(R'^3 - R''^3)}\;k_3^i \tag{79}$$

and

$$K_\theta^{as} = \frac{d^s L\,a_o}{1000\,K^s}\;\frac{V}{(V' + mV'')} \tag{80}$$

In the limiting case, where the concentration of unoccupied binding sites is large, compared to the concentrantions of free and bound molecules in the corresponding subsystem, the above expressions for the relaxation times become simplified to those of the nonlocalized binding model. This can easily be seen by introducing $\overline{c}'_B \simeq N'_o\,c_{VES}$, $\overline{c}''_B \simeq N''_o\,c_{VES}$ and $\overline{c}_B = \overline{c}'_B + \overline{c}''_B$ into Eqs. (76), (77), and (78), and by neglecting the corresponding concentrations of unbound molecules.

3. Discussion of the Models

The above two models treat two limiting cases in which the surface-bound substrate molecules have a high or a low lateral mobility. Associated with this is a difference in the probabilities of an impinging substrate molecule being adsorbed, which becomes appreciable for higher surface coverages. Correspondingly, differences arise in the expressions relating the relaxation times to the concentrations of vesicles and substrate molecules. In other respects the expressions are similar.

The relaxation time, τ'', in the nonlocalized model is independent of the vesicle concentration and of the concentration of substrate molecules. In the localized model τ'' is also independent of the vesicle concentration but not of the concentration of the substrate molecules. The reciprocal relaxation time, τ'^{-1}, is linearly dependent on the vesicle concentration in the nonlocalized case, and on the sum of the concentrations of vacant binding sites and unbound molecules in the localized case. In both models, the relaxation time, τ^i, reaches a constant value for high vesicle concentrations. This can be illustrated by the following. At high vesicle concentrations practically all substrate molecules are bound to the vesicles. Adding more vesicles will no longer affect the ratio of surface-bound molecules and molecules in the nonpolar interior and therefore has no effect on the relaxation time. Formally, the slow exchange of molecules between the surface and the nonpolar interior of the membrane can be treated in the same way as a mono-molecular reaction under these conditions.

The main difference is exhibited in the different relationships between the relaxation time and the concentration of the substrate molecules, and this may serve as a tool to discriminate between the two models. However, it should be noted that in the case where the surface coverages are small, the expressions for the two models become identical. In order to choose between one or other model, additional methods have to be used.

Lipid molecules arranged in a bilayer have a rather high lateral mobility, as determined by fluorescence correlation spectroscopy (Fahey et al., 1977). The constants for the binding of small organic molecules to lipid vesicles, which are in the range $10\text{-}10^5\ M^{-1}$ (litres per mol lipid) (Haynes, 1973; Ruf, 1974, in preparation; Woolley, 1979), indicate that the binding would be determined predominantly by hydrophobic interactions. This suggests that the surface-bound molecules also have a high mobility. The nonlocalized model, therefore, seems to be more appropriate for these systems. If membrane proteins or lipid molecules associated with proteins form the binding sites, the lateral mobility of the bound molecule may be reduced considerably, such that the model of localized binding applies.

The models presented here involve a four-step mechanism for the transport of a molecule across the membrane, in which the rate-limiting step is the transfer from the membrane surface into the nonpolar interior. The membrane permeability P is then given by the product of the partition coefficient β (which relates the concentration of surface-bound molecules to the concentration of substrate molecules in the aqueous bulk phase) and the rate constant of the rate-limiting step:

$$P = \beta\, k^{si} \tag{81}$$

The partition coefficient is given according to the relation $\beta = 1000\, k^{as'} / (4\pi R'^2 L k^{sa'})$, the ratio $V/(V' + mV'')$ has been set to 1, which is valid for most experimental conditions.

IV. Applications to Membrane Studies

A. Studies of Membrane-Active Compounds in Homogeneous Solution

1. Peptides

Chemical relaxation methods have been extensively applied to the characterization of macrocyclic compounds (molecular weight range about 600-1200), including peptides, depsipeptides and related compounds (lactones and ethers), which act as carriers for alkali ions in black lipid bilayer membranes. Prior to studies of the compounds in the membrane-bound state, investigations were carried out in homogeneous solution (Winkler, 1972; Grell et al., 1975; Grell and Oberbäumer, 1977; Chock et al., 1977; Burgermeister and Winkler-Oswatitsch, 1977). The following experimental approaches have been employed.

Ultrasonic adsorption studies (cf. Sect. II.B.) of fast conformational rearrangements of the uncomplexed macrocyclic ligands in the frequency range 0.2-150 MHz have been performed in nonpolar solvents. The kinetics of complex formation between these ligands and alkali ions, usually in methanol as the solvent, have been studied employing ultrasonic absorption techniques and temperature jump relaxation spectrometry (cf. Sect. II.B.).

For the neutral and flexible ligands (e.g., n-actins, valinomycin and crown-ethers) the highest formation rate constants found are around $5 \times 10^8\ M^{-1}\ s^{-1}$, close to those of diffusion-controlled reactions. It has been concluded that complex formation of alkali ions with these macrocyclic ligands must involve a subsequent stepwise ligand exchange process. According to this interpretation all solvate molecules of the inner coordination shell are substituted in a stepwise manner, by the coordinating groups of the ligand. Since the rate constant for a diffusion-controlled encounter between a solvated alkali ion and such a ligand would be around $3 \times 10^9\ M^{-1}\ s^{-1}$ (Grell et al., 1975), it is possible to estimate that about 10^{-9} s are required for a single substitution process, provided that the required conformational rearrangements of the ligand during the coordination process are sufficiently fast. This suggests that solvate substitution is the rate-limiting step of complex formation.

In the case of valinomycin intermediate state of complex formation have been experimentally observed. The rate-limiting step of cation coordination is a conformational change of the ligand which occurs during the stepwise substitution process (Grell et al., 1975). It has been further shown that the overall rate constant of complex formation between valinomycin and alkali ions is very sensitive to the polarity of the medium and can vary by four orders of magnitude between methanol and water (Grell and Oberbäumer, 1977). This is due to the solvent dependence of a pre-equilibrium of the uncomplexed valinomycin. In the case of these flexible ligands, the alkali ion specificity is mainly determined by the overall dissociation rate constants.

A different mechanism for complex formation between alkali ions and sterically nonflexible ligands exhibiting a narrow cavity has been suggested, in which the rate-limiting step is attributed to the dissociation of more than one solvate molecule from the inner coordination sphere of the cation in a single reaction step (Grell and Oberbäumer, 1977). In this case not only the overall dissociation rate constants but also the overall formation rate constants depend on the solvation energy and thus on the size of the cation.

Similar results have been obtained from extensive studies on synthetic macrocyclic ligands which do not, however, exhibit the same degree of membrane activity as the compounds mentioned above (Tümmler et al., 1977; Tümmler et al., 1979).

Temperature jump studies on virginiamycin S, which acts as a proton carrier in black lipid bilayer membranes, have provided some information on the elementary steps of proton transfer (Grell and Oberbäumer, 1977; Ruf et al., 1978).

The first relaxation studies of a membrane-bound model carrier were directed toward a dynamic characterization of the conformational equilibria of valinomycin in the uncomplexed state, bound to a phospholipid membrane (Grell et al., 1975). The unfavorable optical properties of the membrane-active ligands in which amide, ester and ether groups are the main chromophors, and the low stabilities of the corresponding alkali ion complexes in aqueous medium, have not favored a successful application of the temperature jump technique to the study of the kinetics of complex formation with the membrane-bound carrier molecules.

2. Proteins

Only a small number of contributions related to the investigation of membrane or membrane-related proteins in homogeneous or quasi-homogeneous solution are available.

The kinetics of complex formation between specific and fluorescent ligands (N-methylacridinium, 1-methyl-7-hydroxyquinolinium) and the 11S acetylcholinesterase isolated from the electric organs of the eel *Electrophoros electricus* have been studied by temperature jump relaxation spectrometry (Rosenberry and Neumann, 1977). Unexpectedly high rates of complex formation of about 2 x 10^9 M^{-1} s^{-1} were found, closely approaching the estimated value for diffusion-controlled reactions. However, no evidence was found to suggest that the high rate constants were associated with a fast initial binding of the ligand to an unspecific binding site, with subsequent surface diffusion to the specific site of the enzyme. The effective charge on the active site region of acetylcholinesterase has been deduced from the ionic strength dependence of the formation rate constant with N-methylacridinium (Nolte et al., 1980). The relevance of these findings with respect to the physiological action of acetylcholine has been discussed.

Tsong and Yang (1978) have investigated the dynamic properties of the benzphetamine complex of microsomal cytochrome P-450 from rat liver by temperature jump spectrometry. The experimentally observed relaxation times in the 50 ms and 300 ms time region were attributed to conformational rearrangements of the protein after substrate binding. Addition of dimyristoyl-L-α-lecithin liposomes and vesicles affected both relaxation times. It has been concluded that the physical state of the lipid envelope around the enzyme regulates the rate of the substrate-induced conformational changes.

The dynamic propertieş of the isolated acetylcholine receptor from *Torpedo california* has been investigated by the temperature jump technique, employing murexide as a spectroscopic Ca^{2+} indicator to monitor concentration changes in the receptor-bound Ca^{2+} (Neumann and Chang, 1976). Since acetylcholine and Ca^{2+} compete for the same binding site, this method permits an indirect measurement of the acetylcholine binding. The acetylcholine binding process is characterized by at least a two-step reaction. The formation rate constant is 2.4 x 10^7 M^{-1} s^{-1}, which is considerably smaller than that expected for the binding of acetylcholine to the acetylcholinesterase. The biological relevance of these results has been discussed on the basis of a theoretical model (Neumann, 1979).

B. Model Membrane Systems

1. Micelles

The kinetics of micelle formation (e.g., sodium dodecyl sulphate) have been investigated by ultrasonic absorption and pressure jump (cf. Sect. II.B.) studies (Herrmann and Kahlweit, 1973; Strehlow and Knoche, 1977). Ultrasonic absorption studies demonstrate fast single steps (in the time range of diffusion-controlled reactions) corresponding to the collection of monomers into small aggregates. Pressure jump studies (cf. Sect. II.B.) reveal a much slower process in the millisecond region, which is attributed to the formation of micelles from monomers and oligomers. The results have been interpreted on the basis of a detailed theoretical model.

Electrical field jump experiments (cf. Sect. II.B.) have been carried out on cetylpyridinium iodide micelles at just above the critical micelle concentration (Grünhagen, 1975a,b). An increase of the electric field induces chemical relaxation processes including a fast dissociation in the 50 ns range, an association in the 1 μs range and a slower dissociation in the 50 μs range. The fast dissociation is attributed to diffusion-controlled counter ion binding with a formation rate constant of 1.5×10^{10} M^{-1} s^{-1}. The association process, which is very sensitive to the electrical field, is interpreted as being a special structural field effect and the slowest process reflects the dynamics of the equilibration between monomers and micelles. Although the electric field jump technique may be considered to represent the most suitable one for the investigation of field-dependent elementary steps relevant to membrane processes, no further applications exist in homogeneous or quasi-homogeneous solution.

2. Mobility of Membrane Constituents

The mobility of membrane constituents, such as lipid molecules, or certain parts of them, have been characterized employing magnetic resonance techniques (cf. Chaps. 1 and 2). The mobility can be determined from dielectric relaxation studies (Kaatze et al., 1975; Pottel et al., 1978; Kaatze et al., 1979a; Kaatze et al., 1979b). Although dielectric relaxation in the absence of an electric field (cf. Sect. II.B.) does not strictly belong to the chemical relaxation methods, this technique should be mentioned since some relevant applications come under the scope of this article.

The complex permittivity in the frequency range of 10^5 to 6×10^{10} Hz has been measured with dispersed phospholipids at several temperatures and with micelles of lysolecithin or simple surfactants. These studies have enabled a direct determination of the mobility of the dipolar zwitterionic head groups of the lipid molecules. The reorientational motion of the head groups of dispersed synthetic lecithins is characterized by a time constant of around 30 ns. From the studies at various temperatures in the range of the gel-liquid crystalline transition of the phospholipids a high degree of cooperativity is found for the diffusive motions of the phosphorylcholine groups at the bilayer interface. In addition, it has been clearly demonstrated that the mobility of the trimethylammonium headgroup is considerably smaller than in the case of the micellar arrangement of lipid or detergent molecules. Despite the interesting possibilities of this method for the investigation of protein-lipid interactions, no such applications are yet available.

3. Lipid Dispersions and Phase Transitions

Relaxation studies have been performed on phospholipid dispersions and on suspensions of vesicles, prepared from natural lipid mixtures. The complex nature of the processes has been revealed by temperature jump relaxation spectrometry, employing indirect detection using a pH indicator (Hammes and Tallmann, 1970), or light scattering (Owen et al., 1970), and also by ultrasonic attenuation measurements (Hammes and Roberts, 1970).

Temperature jump studies employing Joule heating have already indicated a disadvantage of this technique, namely the possibility of partially disrupting the bilayer membrane as a consequence of the electrical discharge under conditions where large temperature changes occur. In addition, ion permeations and fusion processes may be indiced by large temperature jumps. If elaborate control experiments are to be avoided, it is recommended that temperature jumps do not exceed 1°C.

Since 1971, much effort has been devoted to the study of the kinetics of the gel-liquid crystalline phase transition of synthetic phospholipids containing one particular fatty acid. Employing the temperature jump technique with an indirect detection system (1-anilino-8-naphthalenesulphonate, bromothymol blue), Träuble (1971) noted that there is a single relaxation process with a maximum relaxation time of about 1200 ms, for a degree of transition around 0.8, for dipalmitoyl-L-α-lecithin dispersions. Similar kinetic studies of Fischer (1973) on dimyristoyl-L-α-lecithin dispersions employing direct observation of the intensity of the scattered light indicated that the response could not be described by a single exponential. However, the dependence of the main time constant on temperature was similar to the above and exhibited a clear maximum at the transition temperature. In addition, Fischer (1973) was able to demonstrate that the change of the scattering intensity and also the change in the refractive index occurring during the phase transition, are quantitatively related to the corresponding change of the specific volume of the lipid. A very profound and thorough study of the scattering properties of lipid dispersions was later published by Chong and Colbow (1976). The 10 to 1000 ms time range for phospholipid phase transitions was confirmed by Strehlow (1978) employing stopped flow measurements and by Clegg et al. (1975) employing pressure jump measurements with direct turbidity detection.

Tsong (1974) performed temperature jump experiments on sonicated and unsonicated dispersions of dimyristoyl-L-α-lecithin and was able to resolve at least two relaxation times in the 1 to 100 ms time range for the phase transition. The results have been interpreted in terms of nucleation and propagation processes. An extensive temperature jump study of Tsong and Kanehisa (1977) revealed two concentration-independent relaxation times, one in the 10 ms and the other in the 1000 ms time range for multilayered dimyristoyl- and dipalmitoyl-L-α-lecithin dispersions. Both relaxation times clearly exhibited maxima near the transition temperature. Qualitatively similar results were obtained for bilayer vesicles, although the corresponding relaxation times were one order of magnitude faster. An interpretation based on a two-dimensional Ising lattice has been suggested.

Besides the transitions with relaxation times in the 1 to 1000 ms range, considerably faster nanosecond processes have been found by various groups employing ultrasonic studies between 0.5 and 300 MHz (Eggers and Funck, 1976; Gamble and Schimmel, 1978; Mitaku et al., 1978; Harkness and White, 1979). These fast processes have also been related to the lipid phase transition. The amplitudes in the lower frequency range exhibited the expected temperature dependence and showed a maximum near the transition temperature. Sonicated dispersions exhibit a lower transition temperature and a broader transition range than unsonicated ones. There is good agreement of experimental observations on sonicated dispersions between Eggers and Funck (1976) and Harkness and White (1979) who demonstrated that the volume change, for the phase transition, calculated from the ultrasonic absorption amplitude is considerably larger than that obtained with static methods. Eggers and Funck (1967) suggested an interesting molecular interpretation of the ultrasonic absorption data. The observed 100 ns relaxation are attributed to an increased rotational mobility, near the transition temperature, of the polar headgroups, which under these conditions are assumed to be freely

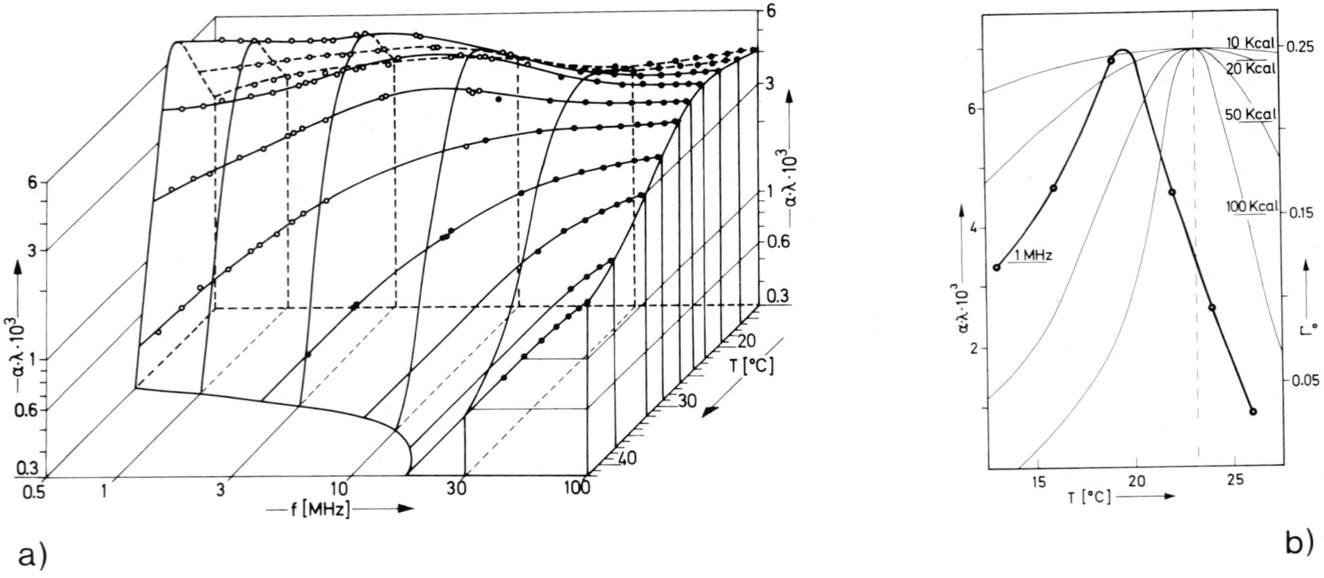

Fig. 10 a and b. Ultrasonic spectra of lecithin vesicle suspensions: **a** Ultrasonic excess absorption ($\alpha\lambda$) at different temperatures of a dispersion of dimyristoyl-L-α-lecithin (0.036 M) in aqueous tetramethylammonium chloride (0.02 M). **b** Plot of temperature dependence of $\alpha\lambda$ at 1 MHz together with calculated temperature profiles for different ΔH values. (From Eggers and Funck, 1976)

rotating and more strongly hydrated in the form of cooperative units (of 10 to 20 head-groups). The report of Gamble and Schimmel (1978) is not consistent with this interpretation and disregards the results of Eggers and Funck (1976).

The kinetic investigations on sonicated and unsonicated dispersions of dimyristoyl- and dipalmitoyl-L-α-lecithin in both principal time ranges have been reexamined extensively by Gruenewald et al. (1980) employing the pressure jump method with optical detection in the millisecond time range and by Gruenewald et al. (1980) employing a laser temperature jump method in the nanosecond time range. Both methods were based on turbidity detection.

The observed nanosecond relaxation time showed no significant temperature dependence in the range of the transition temperature. This indicated that the fast nanosecond process does not directly represent the phase transition process. Two distinct concentration-independent relaxation times were found by the pressure jump technique; both time constants exhibit a pronounced maximum at the transition midpoint. For both lecithins the relaxation times were observed in the 1 and 10 ms time range for unsonicated dispersions and in the 100 μs and 1 ms range for sonicated ones. These results are similar to those reported by Tsong and Kanehisa (1977), who employed temperature jump studies, although their time constants were about two orders of magnitude smaller. The study of Gruenewald et al. (1980) suggests conclusively that the millisecond processes represent the main change of the degree of transition with respect to the gel-liquid crystalline phase transition. Although this work has considerably helped to clarify the general situation, a consistent model, encompassing all reported effects, is not yet available.

The inherent experimental difficulties in the preparation of stable sonicated vesicular dispersions which exhibit suitable properties for optical detection methods may be responsible for some of the inconsistencies of the results obtained in different laboratories. Further experimental studies on very well defined systems, and the development of suitable models to explain them, are desirable.

4. Binding of Small Molecules to Membranes

The initial temperature studies related to the binding of small molecules to vesicular lipid systems were evaluated on the basis of suitable models. Thus, the numerical values of the rate constants of the earlier reports cannot directly be compared to those of Ruf (1974; in prep.) and Woolley and Diebler (1979).

Haynes (1973) published the first results on the binding of 1-anilino-8-naphthalene-sulphonate (ANS) to a monolayer vesicle system. A formation rate constant of 2.1 x 10^8 M^{-1} s^{-1} and a dissociation rate constant of 4.5 x 10^3 s^{-1}, at 31°C, was reported. The dynamic properties of the membrane-bound ANS have been further characterized by Haynes and Staerk (1974). The binding of Ca^{2+} to negatively charged phosphatidic acid (Haynes, 1977) and of the model carrier X537A (Haynes et al., 1980) to dimyristoyl-L-α-lecithin in vesicular model systems using the temperature jump technique have been investigated. Formation rate constants of the order of magnitude of 10^7 M^{-1} s^{-1} have been reported.

A single relaxation process, detected by temperature jump relaxation spectrometry, has also been found for the binding of the neutral N-phenyl naphthylamine (NPN) to bilayer vesicles composed of the negatively charged C_{12}-methyl phosphatidic acid (Woolley and Diebler, 1979). The membrane-bound NPN is assumed to be preferentially located between the hydrocarbon chains of the inner region of the bilayer membrane (Woolley, 1978). The formation rate constant is 5×10^6 M^{-1} s^{-1} expressed in terms of monomeric lipid, or 10^{11} M^{-1} s^{-1} expressed in terms of vesicles. It has been shown that the binding is due to a diffusion-controlled process, which is consistent with the corresponding model (cf. Woolley and Diebler, 1978).

A very detailed analysis of the kinetics of the binding of a small molecule to sonicated phospholipid dispersions, characterized by a two-step binding process, has been reported by Ruf (1974; in prep.) and Ruf et al. (1978). The corresponding theoretical aspects have been given in Section III.B. of this article.

The azo-dye o-methyl red (o-dimethylaminophenylazo benzoic acid), a well-known pH indicator (pK_{II} = 4.9), binds to lipid bilayer vesicles (egg lecithin) both in its anionic and in its neutral (zwitterionic) form. The apparent binding constants, which have been determined also by spectrophotometric titrations (Ruf, in prep.), indicate that the affinity for neutral forms to the membrane is nearly two orders of magnitude larger than that for the anion. The values for these binding constants compare fairly well with those calculated from the rate constants. Binding to the membrane is accompanied by remarkable changes in the corresponding absorption spectra. In order to investigate the kinetics of binding, a temperature jump method with absorbance detection has been employed.

The temperature jump experiments reveal a relaxation spectrum consisting of at least three relaxation times. The fastest processes could not be resolved in time by this method. These processes are observed additively, as a substantial shift of the relaxation amplitude in the initial phase. Part of this shift results from the very fast change of the protolytic equilibrium of o-methyl red, in the aqueous bulk phase, due to the new conditions. This has been measured by temperature jump experiments in the absence of vesicles. Another part, however, arises from the rapid equilibration of the binding from the aqueous bulk phase enclosed by the vesicles, as has been discussed in the theoretical Section III.B. Therefore, the actual number of relaxation times is more than three.

The two processes resolved by the method are associated with the binding of o-methyl red to the vesicles (Fig. 11). The processes proceed on rather different time scales and can each be described, with sufficient accuracy, by a single exponential. This

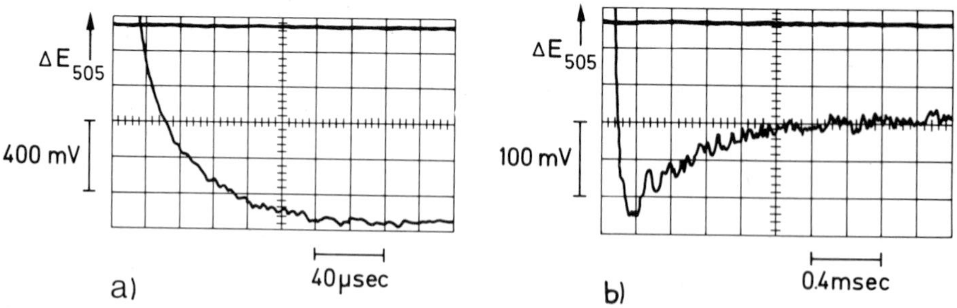

Fig. 11 a and b. Oscillograms of temperature jump relaxation experiments of 5×10^{-5} M o-methyl red in the presence of egg lecithin bilayer vesicles (5×10^{-5} M lecithin) and citrate buffer pH 5.08 containing 0.06 M KCl. The initial, fast relaxation process (**a**) is followed by a slower process (**b**), which is indicated on an expanded scale. (From Ruf, in preparation)

means that the kinetics of the system can be treated as in the theoretical part, i.e., the processes faster than the process of interest can be trated as pre-equilibrated. The relaxation times of the two processes exhibit, within the experimentally accessible concentration range, no dependence on the o-methyl red concentration. This is in agreement with the idea of nonlocalized binding for these systems.

The faster of the two relaxation processes (Fig. 11a) is associated with the binding of o-methyl red from the aqueous bulk phase to the membrane surface. Consistent with the theoretical expectations for this "bimolecular reaction", the relaxation time, τ_1, is linearly dependent on the lipid concentration as shown in Fig. 12a for various pH values. However, the relaxation time, τ_1, represents here the combined binding of the two

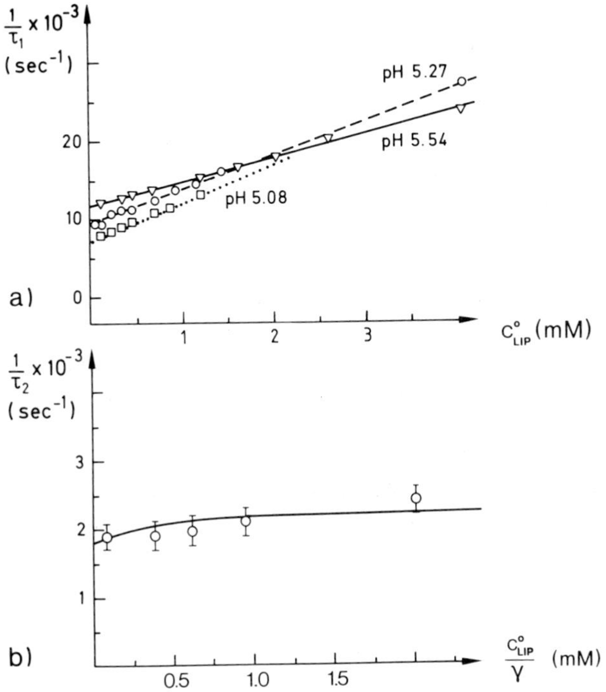

Fig. 12 a and b. Evaluation of relaxation times: Plot of the reciprocal relaxation time, $1/\tau_1$, versus the total lipid concentration, $\overset{\circ}{c}_{LIP}$, at pH 5.08, 5.27 and 5.54 (**a**) and plot of $1/\tau_2$ versus $\overset{\circ}{c}_{LIP}/\gamma$ at pH 5.08 (**b**). The *solid line* in (**b**) indicates the theoretical curve. The factor γ is given by $\gamma = 1 + 1/(K_I\bar{c}_H)$, where K_I denotes the association constant of the protolytic equilibrium of o-methyl red in the aqueous phase and $\bar{c}_H$ is the proton activity. Other experimental conditions are as given in the legend of Fig. 11. (From Ruf, in preparation)

o-methyl red forms, which interchange by protolytic reactions both in the aqueous bulk phase and at the membrane surface. The protolytic reactions are fast compared to the binding processes. Such a coupled reaction system explains the findings that the relaxation time, τ_1, for small lipid concentrations decreases with increasing pH value (i.e., the relaxation process becomes faster the fewer neutral o-methyl red molecules are present in the solution) and that for high lipid concentrations the reverse is valid (Fig. 12a). The characteristics of the pH dependence of τ_1 provide direct evidence for the existence of the protolytic reaction of the surface-bound species, which is often termed a heterogeneous reaction.

The slower relaxation process (Fig. 11b) represents the equilibration of the processes by which neutral o-methyl red molecules are exchanged between the surface and the nonpolar interior. The dependence of the relaxation time, τ_2, on the lipid concentration is shown in Fig. 12b. The theoretical curve, which has been calculated on the basis of the model in the theoretical section, fits fairly well to the experimental data.

The rate constants, evaluated from the experimental data are summarized in Table 1, where AH stands for the neutral form and A for the anionic form of o-methyl red.

Table 1. Binding of o-methyl red to egg lecithin vesicles (20°C): Kinetic parameters

$k_{AH}^{as'} = 8.9 \times 10^6\ M^{-1}\ s^{-1}$	$k_{AH}^{si} = 5 \times 10^2\ s^{-1}$
$k_{AH}^{sa'} = 4.9 \times 10^3\ s^{-1}$	$k_{AH}^{is} = 2 \times 10^3\ s^{-1}$
$k_{A}^{as'} = 1.3 \times 10^6\ M^{-1}\ s^{-1}$	
$k_{A}^{sa'} = 1.0 \times 10^5\ s^{-1}$	

The theoretical value for the adsorption rate constant, $k^{as'}$, according to Eq. (59) ($R' = 250$ Å, $R'' = 200$ Å, $a = 60$ Å^2, $D = 10^{-5}\ cm^2\ s^{-1}$) is given by $k^{as'} = 8.78 \times 10^6\ M^{-1}\ s^{-1}$. Comparison with the experimentally determined value for the neutral o-methyl red species shows a surprisingly good agreement. This is rather coincidental, since the vesicles in a dispersion are not of uniform size, and the values of the parameters for the theoretical value are based on estimates. However, this agreement indicates that the adsorption to the membrane of neutral o-methyl red molecules proceeds in a diffusion-controlled manner. The smaller value of the corresponding rate constant of the anionic form could arise from electrostatic interactions, since the vesicles cannot be considered as being completely uncharged.

The rate constants for the exchange between the membrane surface and the nonpolar interior the existence of an energy barrier, which is rate-limiting for the overall transport of the neutral species across the membrane. The permeability, calculated according to Eq. (81), becomes $P = 0.3\ cm\ s^{-1}$ which compares well with values determined for other small neutral molecules.

These results clearly demonstrate the powerful advantages of the application of chemical relaxation methods such as the temperature jump technique to the investigation of single elementary steps involved in membrane processes. It is emphasized that even permeabilities of individual species can be estimated from kinetic parameters under certain conditions.

5. Transport

Since the present article deals mainly with the application of chemical relaxation methods to homogeneous or quasi-homogeneous solutions, employing optical detection systems, the results of transport studies, involving mainly single planar membranes, will be presented only briefly here.

The initial kinetic study of transport processes by the field jump technique on black lipid membranes (cf. Sect. II.A.), employing current detection, has been devoted to the permeation of hydrophobic ions such as dipicrylamine and tetraphenyl borate (Ketterer et al., 1971). This technique has also been successfully applied to the kinetic investigation of valinomycin in phosphatidylinositol bilayer membranes (Stark et al., 1971). The overall rate constant for complex formation between valinomycin and K^+ at the membrane interface was found to be 5×10^4 M^{-1} s^{-1}. This value is consistent with that expected for complex formation in homogeneous aqueous medium (Grell and Oberbäumer, 1977). The rate constants for the diffusion step of the valinomycin-K^+ complex, and of the back diffusion of the unloaded carrier are both 2×10^4 s^{-1}.

The development of the charge pulse technique (cf. Sect. II.A.) with high time resolution (Benz and Läuger, 1976), and the temperature jump method for the investigation of single bilayer membranes (Knoll and Stark, 1977), has enabled a reinvestigation of the carrier-mediated alkali ion transport by valinomycin. In the case of the charge pulse experiments three relaxation processes related to the valinomycin-induced cation transport have been resolved, thus allowing the determination of all rate constants for complex formation and transport of different alkali ions for different lipid compositions of the membrane (Benz et al., 1977). In addition, the charge pulse method has also been successfully applied to the elucidation of the mechanism of action of alamethicin in black lipid membranes (Boheim and Benz, 1978). Alamethicin forms voltage-dependent ion conductance channels in lipid membranes.

The comparatively slow transport of ANS and X537A across vesicular phospholipid bilayer membranes could also be detected in quasi-homogeneous solution employing stopped-flow techniques (Haynes and Simkowitz, 1977; Haynes et al., 1980). Molecular models have been described in these reports to explain the results.

C. Biological Systems

Despite the large number of possible applications of interest, only a few applications of chemical relaxation methods to the field of natural membrane systems are available.

The binding of ANS to membrane fragments, isolated from the innervated, excitable face of electroplax cells, has been investigated by temperature jump spectrometry (Jovin, 1975). The observed relaxation phenomenon is attributed to an isomerization reaction between two forms of the bound aromatic molecule.

A detailed study of the kinetics of permeability changes of the soluble contents in chromaffin granules induced by electric impulses of 20 to 50 kV/cm and a duration of 10 to 40 μs has been carried out (Neumann and Rosenheck, 1972; Lindner et al., 1977). The transient permeability change, as well as the resealing of the perturbed membrane, was monitored directly.

Relaxation phenomena investigated by the temperature jump technique employing Joule heating and turbidity detection have also been observed in erythrocyte suspensions (Tsong et al., 1976). Two fast processes have tentatively been assigned to water relaxations in the membrane structure, whereas two slower processes are attributed to membrane ruptures and to hemolysis caused by the rapid dielectric perturbation of the erythrocyte membrane.

Moore et al. (1972) and Moore (1975) have introduced the temperature jump method to the investigation of relaxation processes associated with changes in the membrane conductance of myelinated nerves. The local temperature jumps are performed by heating with pulses from a laser beam.

The small number of such investigations clearly indicates that numerous fundamental studies especially in the field of biological membranes remain for future investigations. Some of the required theoretical aspects have been presented in this article.

Acknowledgements. We wish to thank Dr. S. Nandi for invaluable assistance in the preparation of this manuscript and Dr. F. Sauer for very stimulating and helpful discussions.

References

Benz, R., Läuger, P.: Kinetic analysis of carrier-mediated ion transport by the charge-pulse technique. J. Membr. Biol. *27*, 171-191 (1976)

Benz, R., Fröhlich, O., Läuger, P.: Influence of membrane structure on the kinetics of carrier-mediated ion transport through lipid bilayers. Biochim. Biophys. Acta *464*, 465-481 (1977)

Bergmann, K.: Dielektrische Absorption als Folge chemischer Relaxation. II. Ausführung der Experimente. Ber. Bunsenges. Phys. Chem. *67*, 826-832 (1963)

Bergmann, K., Eigen, M., De Maeyer, L.: Dielektrische Absorption als Folge chemischer Relaxation. Ber. Bunsenges. Phys. Chem. *67*, 819-826 (1963)

Boheim, G., Benz, R.: Charge-pulse relaxation studies with lipid bilayer membranes modified by alamethicin. Biochim. Biophys. Acta *507*, 262-270 (1978)

Brock, W., Jordan, P.C., Stark, G.: Laser-temperature-jump method for the determination of the rate of desorption of hydrophobic ions in lipid membranes. In: Annual Meeting of the Deutsche Gesellschaft für Biophysik. Abstracts of Poster Presentations (eds. Adam, G., Stark, G.), p. 46. Heidelberg: Springer 1979

Brunner, J., Skrabal, P., Hauser, H.: Single bilayer vesicles prepared without sonication. Physicochemical properties. Biochim. Biophys. Acta *455*, 322-331 (1976)

Burgermeister, W., Winkler-Oswatitsch, R.: Complex formation of monovalent cations with biofunctional ligands. In: Topics in Current Chemistry (ed. Boschke, F.L.), Vol. 69, pp. 91-196. Berlin, Heidelberg, New York: Springer 1977

Castellan, G.W.: Calculation of the spectrum of chemical relaxation times for a general reaction mechanism. Ber. Bunsenges. Phys. Chem. *67*, 898-908 (1963)

Chock, P.B., Eggers, F., Eigen, M., Winkler, R.: Relaxation studies on complex formation of macrocyclic and open chain antibiotics with monovalent cations. Biophys. Chem. *6*, 239-251 (1977)

Chong, C.S., Colbow, K.: Light scattering and turbidity measurements on lipid vesicles. Biochim. Biophys. Acta *436*, 260-282 (1976).

Clegg, R.M., Elson, E.L., Maxfield, B.W.: A new technique for optical observation of the kinetics of chemical reactions perturbed by small pressure changes. Biopolymers *14*, 883-887 (1975)

Czerlinski, G.H.: Chemical Relaxation. New York: Dekker 1966

Debye, P.: Reaction rates in ionic solutions. Trans. Electrochem. Soc. *82*, 265-272 (1942)

De Maeyer, L., Persoons, A.: Electric field methods. In: Techniques of Chemistry (ed. Weissberger, A.), Vol. VI, Part 2 (ed. Hammes, G.G.), pp. 211-236. New York: Wiley 1973

De Maeyer, L., Eigen, M., Suarez, J.: Dielectric dispersion and chemical relaxation. J. Am. Chem. Soc. *90*, 3157-3161 (1968)

Eggers, F.: Eine Resonatormethode zur Bestimmung von Schall-Geschwindigkeit und Dämpfung an geringen Flüssigkeitsmengen. Acustica *19*, 323-329 (1967/68)

Eggers, F., Funck, Th.: Ultrasonic measurements with milliliter liquid samples in the 0.5-100 MHz range. Rev. Sci. Instrum. *44*, 969-977 (1973)

Eggers, F., Funck, Th.: Ultrasonic relaxation spectroscopy in liquids. Naturwissenschaften *13*, 280-285 (1976)

Eigen, M.: Über die Kinetik sehr schnell verlaufender Ionenreaktionen in wässriger Lösung. Z. Phys. Chem. Frankfurt am Main *1*, 176-200 (1954)

Eigen, M., De Maeyer, L.: Relaxation methods. In: Technique of Organic Chemistry (ed. Weissberger, A.), Vol. VIII, Part II, pp. 895-1054. New York: Interscience 1963

Eigen, M., De Maeyer, L.: Theoretical basis of relaxation spectrometry. In: Techniques of Chemistry (ed. Weissberger, A.), Vol. VI, Part II (ed. Hammes, G.G.), pp. 63-146. New York: Wiley 1973

Eigen, M., Kurtze, G., Tamm, K.: Reaktionsmechanismus der Ultraschallabsorption in wäßrigen Elektrolytlösungen. Z. Elektrochem. *57*, 103-118 (1953)

Fahey, P.F., Koppel, D.E., Barak, L.S., Wolf, D.E., Elson, E.L., Webb, W.W.: Lateral diffusion in planar lipid bilayers. Science *195*, 305-306 (1977)

Feldberg, S.W., Kissel, G.: Charge pulse studies of transport phenomena in bilayer membranes. J. Membr. Biol. *20*, 269-300 (1975)

Fischer, U.: Die kristallin-flüssig-kristalline Umwandlung von Dimyristoyllecithin in wässriger Dispersion. Optischer Nachweis und Kinetik. Diplomarbeit Techn. Univ. München 1973

Gamble, R.C., Schimmel, P.R.: Nanosecond relaxation processes of phospholipid bilayers in the transition zone. Proc. Natl. Acad. Sci. USA *75*, 3011-3014 (1978)

Gerischer, H., Heim, W.: Eine Strömungsmethode mit integrierender Anzeige zur Untersuchung schneller Reaktionen in Lösung. Z. Phys. Chem. Frankfurt am Main *46*, 345-352 (1965)

Gibson, Q.H., Milnes, L.: Apparatus for rapid and sensitive spectrophotometry. Biochem. J. *91*, 161-171 (1964)

Grell, E., Oberbäumer, I.: Dynamic aspects of carrier-mediated cation transport through membranes. In: Molecular Biology, Biochemistry and Biophysics, Vol. XXIV: Chemical Relaxation in Molecular Biology (eds. Pecht, I., Rigler, R.), pp. 371-413. Berlin, Heidelberg, New York: Springer 1977

Grell, E., Funck, Th., Eggers, F.: Structure and dynamic properties of ion-specific antibiotics. In: Membranes (ed. Eisenman, G.), Vol. III, pp. 1-126. New York: Dekker 1975

Gruenewald, B., Blume, A., Watanabe, F.: Kinetic investigations on the phase transition of phospholipid bilayers. Biochim. Biophys. Acta *597*, 41-52 (1980)

Gruenewald, B., Frisch, W., Holzwarth, J.F.: Nanosecond relaxation in lipid bilayers. Experienta *36*, 722 (1980)

Grünhagen, H.H.: Fast spectrophotometric detection system for coupled physical and chemical field effects in solution. Biophysik *10*, 347-354 (1973)

Grünhagen, H.H.: A high power square wave pulse generator for the investigation of fast electric field effects in solution. Meßtechnik 19-23 (1974)

Grünhagen, H.H.: Electric field effects in systems of associating amphiphilic electrolytes. Chem. Phys. Lipids *14*, 201-210 (1975a)

Grünhagen, H.H.: Chemical relaxation of cetylpyridinium iodide micelles in high electric fields. J. Colloid Interface Sci. *53*, 282-295 (1975b)

Hammes, G.G. (ed.): Investigation of elementary reaction steps in solution and very fast reactions. In: Techniques of Chemistry (ed. Weissberger, A.), Vol. VI, Part 2. New York: Wiley 1973a

Hammes, G.G.: Temperature-jump methods. In: Techniques of Chemistry (ed. Weissberger, A.), Vol. VI, Part 2 (ed. Hammes, G.G.); pp. 147-186. New York: Wiley 1973b

Hammes, G.G., Roberts, P.B.: Ultrasonic attenuation measurements in phospholipid dispersions. Biochim. Biophys. Acta *203*, 220-227 (1970)

Hammes, G.G., Schimmel, P.R.: Chemical relaxation spectra: Calculation of relaxation times for complex mechanisms. J. Chem. Phys. *70*, 2319-2324 (1966)

Hammes, G.G., Tallmann, D.E.: Application of the temperature-jump technique to the study of phospholipid dispersions. J. Am. Chem. Soc. *92*, 6042-6046 (1970)

Harkness, J.E., White, R.D.: An ultrasonic study of the thermotropic transition of dipalmitoyl phosphatidylcholine. Biochim. Biophys. Acta *552*, 450-456 (1979)

Haynes, D.H.: Studien der Bindung und des Transportes von Ionen und Molekülen an Phospholipid-Membranen. In: Technische Biochemie; DECHEMA Monographien, Bd. 71 (ed. Rehm, H.J.), pp. 119-133. Weinheim: Verlag Chemie 1973

Haynes, D.H.: Divalent cation-ligand interactions of phospholipid membranes. Equilibria and kinetics. In: Metal-Ligand Interactions in Organic Chemistry and Biochemistry, Part 2 (eds. Pullmann, B., Goldblum, N.), pp. 189-212. Dodrecht: Reidel 1977

Haynes, D.H., Simkowitz, P.: 1-Anilino-8-naphthalenesulfonate: A fluorescent probe of ion and ionophore transport kinetics and trans-membrane asymmetry. J. Membr. Biol. *33*, 63-108 (1977)

Haynes, D.H., Staerk, H.: 1-Anilino-8-naphthalenesulfonate: A fluorescent probe of membrane surface, structure, composition and mobility. J. Membr. Biol. *17*, 313-340 (1974)

Haynes, D.H., Chiu, V.C.K., Watson, B.: Study of the Ca^{2+} transport mechanism of X537A in phospholipid membranes using fluorescence and rapid kinetic techniques. Arch. Biochem. Biophys. *203*, 73-89 (1980)

Herrmann, U., Kahlweit, M.: On the kinetics of the formation of NaDS micelles. Ber. Bunsenges. Phys. Chem. *77*, 1119-1121 (1973)

Hiromi, K.: Kinetics of Fast Enzyme Reactions. New York: Halsted 1979.

Hoffmann, G.W.: A nanosecond temperature-jump apparatus. Rev. Sci. Instrum. *42*, 1643-1647 (1971)

Holzwarth, J.F., Schmidt, A., Wolff, H., Volk, R.: Nanosecond temperature-jump technique with an iodine laser. J. Phys. Chem. *81*, 2300-2301 (1977)

Huang, C.: Studies on phosphatidylcholine vesicles. Formation and physical characteristics. Biochemistry *8*, 344-352 (1969)

Ilgenfritz, G.: The electric field jump relaxation method. In: Probes of Structure and Function of Macromolecules and Membranes. Vol. 1 (eds. Chance, B., Yonetani, R.), pp. 505-510. New York: Academic Press 1971

Ilgenfritz, G.: Theory and simulation of chemical relaxation spectra. In: Molecular Biology, Biochemistry and Biophysics, Vol. XXIV: Chemical Relaxation in Molecular Biology (eds. Pecht, I., Rigler, R.), pp. 1-42. Berlin, Heidelberg, New York: Springer 1977

Jovin, T.M.: Fluorimetric kinetic techniques; chemical relaxation and stopped-flow. In: Biochemical Fluorescence. Concepts (eds. Chen, R.F., Edelhoch, H.), pp. 305-374. New York: Dekker 1975

Kaatze, U., Henze, R., Seegers, A., Pottel, R.: Dielectric relaxation in colloidal phospholipid aqueous solutions. Ber. Bunsenges. Phys. Chem. *79*, 42-53 (1975)

Kaatze, U., Henze, R., Eibl, H.: Motion of the lengthened zwitterionic head groups of C_{16}-lecithin analogues in aqueous solutions as studied by dielectric relaxation measurements. Biophys. Chem. *10*, 351-362 (1979a)

Kaatze, U., Henze, R., Pottel, R.: Dielectric relaxation and molecular motions in C_{14}-lecithin-water systems. Chem. Phys. Lipids *25*, 149-177 (1979b)

Ketterer, B., Neumcke, B., Läuger, P.: Transport mechanism of hydrophobic ions through lipid bilayer membranes. J. Membr. Biol. *5*, 225-245 (1971)

Knoche, W.: Pressure-jump methods. In: Techniques of Chemistry (ed. Weissberger, A.), Vol. VI, Part 2 (ed. Hammes, G.G.), pp. 187-210. New York: Wiley 1973

Knoll, W., Stark, G.: Temperature-jump experiments on thin lipid membranes in the presence of valinomycin. J. Membr. Biol. *37*, 13-28 (1977)

Lindner, P., Neumann, E., Rosenheck, K.: Kinetics of permeability changes induced by electric impulses in chromaffin granules. J. Membr. Biol. *32*, 231-254 (1977)

Mitaku, S., Ikegami, A., Sakanishi, A.: Ultrasonic studies of lipid bilayer. Phase transition in synthetic phosphatidyl-choline liposomes. Biophys. Chem. *8*, 295-304 (1978)

Moore, L.E., Holt, J.P., Jr., Lindley, B.D.: Laser temperature jump technique for relaxation studies of the ionic conductances in myelinated nerve fibres. Biophys. J. *12*, 157-174 (1972)

Moore, L.E.: Mambrane conductance changes in single nodes of Ranvier, measured by laser-induced temperature-jump experiments. Biochim. Biophys. Acta *375*, 115-123 (1975)

Mueller, P., Rudin, D.O.: Developement of K^+-Na^+ discrimination in experimental bimolecular lipid membranes by macrocyclic antibiotics. Biochem. Biophys. Res. Commun. *26*, 398-404 (1967)

Neumann, E.: Membrane permeability control by the acetylcholine gating system. In: Molecular Mechanism of Biological Recognition (ed. Balaban, M.), pp. 449-463. Amsterdam: Elsevier 1979

Neumann, E., Chang, H.W.: Dynamic properties of isolated acetylcholine receptor protein: Kinetics of the binding of acetylcholine and Ca ions. Proc. Natl. Acad. Sci. USA *73*, 3994-3998 (1976)

Neumann, E., Rosenheck, K.: Permeability changes induced by electric impulses in vesicular membranes. J. Membr. Biol. *10*, 279-290 (1972)

Nolte, H.-J., Rosenberry, T.L., Neumann, E.: Effective charge on acetylcholinesterase active sites determined from the ionic strength dependence of association rate constants with cationic ligands. Biochemistry *19*, 3705-3711 (1980)

Owen, J.D., Hemmes, P., Eyring, E.M.: Light scattering temperature jump relaxation in mixed solvent suspensions of phosphatidylcholine vesicles. Biochim. Biophys. Acta *219*, 276-282 (1970)

Owens, G.D., Margerum, D.W.: Pulsed-flow spectroscopy. Analyt. Chem. *52*, 91A-106A (1980)

Owens, G.D., Taylor, R.W., Ridley, T.Y., Margerum, D.W.: Pulsed-flow instrument for measurement of fast reactions in solution. Anal. Chem. *52*, 130-138 (1980)

Pecht, I., Rigler, R. (eds): Chemical Relaxation in Molecular Biology; Molecular Biology, Biochemistry and Biophysics, Vol. XXIV. Berlin, Heidelberg, New York: Springer 1977

Pottel, R., Kaatze, U., Müller, St.: Dielectric relaxation in aqueous solutions of zwitterionic surfactant micelles. Ber. Bunsenges. Phys. Chem. *82*, 1086-1093 (1978)

Rabl, C.R.: Relaxationsmeßtechnik. In: Technische Biochemie; DECHEMA Monographien, Bd. 71 (ed. Rehm, H.J.), pp. 187-205. Weinheim: Verlag Chemie 1973

Rigler, R., Ehrenberg, M.: Molecular interactions and structure as analyzed by fluorescence relaxation spectroscopy. Q. Rev. Biophys. *6*, 139-199 (1973)

Rigler, R., Rabl, C.R., Jovin, T.M.: Temperature-jump apparatus for fluorescence measurements. Rev. Sci. Instrum. *45*, 580-588 (1974)

Rosenberry, T.L., Neumann, E.: Interaction of ligands with acetylcholinesterase. Use of temperature-jump relaxation kinetics in the binding of specific fluorescent ligands. Biochemistry *16*, 3870-3878 (1977)

Ruf, H.: Mechanismus des Carrier-induzierten Protonentransports durch Lipid-Membranen. Diss. Techn. Univ. Braunschweig 1974

Ruf, H.: The mechanism of carrier-mediated proton transport across lipid bilayer membranes. I. The thermodynamics and kinetics of the binding of the model compound o-methyl red to liposomes. In prep.

Ruf, H., Oberbäumer, I., Grell, E.: Chemical relaxation spectrometry: Investigation of mechanisms involved in membrane processes. In: Transport by Proteins (eds. Blauer, G., Sund, H.), pp. 27-44. Berlin: De Gruyter 1978

Sackmann, E., Träuble, H.: Studies of the crystalline-liquid crystalline phase transition of lipid model membranes. I. Use of spin labels and optical probes as indicators of the phase transition. J. Am. Chem. Soc. *94*, 4482-4491 (1972)

Sandblom, J., Hägglund, J., Erikson, N.E.: Electrical relaxation processes in black lipid membranes in the presence of a cation-selective ionophore. J. Membr. Biol. *23*, 1-19 (1975)

Seufert, W.D.: Model membranes: Spherical shells bounded by one bimolecular layer of phospholipids. Biophysik 7, 60-73 (1970)

Smoluchowski, M.v.: Drei Vorträge über Diffusion, Brownsche Molekularbewegung und Koagulation von Kolloidteilchen. Physik. Z. *17*, 557-599 (1916)

Schimmel, P.R.: On the calculation of chemical relaxation amplitudes. J. Chem. Phys. *54*, 4136-4137 (1971)

Schwarz, G.: Kinetic analysis by chemical relaxation methods. Rev. Mod. Phys. *40*, 206-218 (1968)

Stark, G., Ketterer, B., Benz, R., Läuger, P.: The rate constants of valinomycin-mediated ion transport through thin lipid membranes. Biophys. J. *11*, 981-994 (1971)

Strehlow, H., Knoche, W.: Fundamentals of chemical relaxation. In: Monographs in Modern Chemistry (ed. Ebel, H.F.), Vol. X. Weinheim: Verlag Chemie 1977

Strehlow, U.: Untersuchungen zur Kinetik des elektrostatisch ausgelösten Phasenüberganges in Modellmembranen. Diss. Univ. Göttingen 1978

Stuehr, J.: Ultrasonic methods. In: Techniques of Chemistry (ed. Weissberger, A.), Vol. VI, Part 2 (ed. Hammes, G.G.), pp. 237-284. New York: Wiley 1973

Thusius, D.: Relaxation amplitudes for systems of two coupled equilibria. J. Am. Chem. Soc. *94*, 356-363 (1972)

Thusius, D., Foucault, G., Guillain, F.: The analysis of chemical relaxation amplitudes and some applications to reactions involving macromolecules. In: Dynamic Aspects of Conformation Changes in Biological Macromolecules (ed. Sadron, C.), pp. 271-284. Doodrecht: Reidel 1973

Träuble, H.: Phasenumwandlungen in Lipiden. Mögliche Schaltprozesse in biologischen Membranen. Naturwissenschaften *58*, 277-284 (1971)

Träuble, H., Grell, E.: The formation of asymmetrical spherical lecithin vesicles. Neurosci. Res. Program Bull. *9* (No. 3), 373-380 (1971)

Tsong, T.Y.: Kinetics of the crystalline-liquid crystalline phase transition of dimyristoyl-L-α-lecithin bilayers. Proc. Natl. Acad. Sci. USA *71*, 2684-2688 (1974)

Tsong, T.Y., Kanehisa, M.I.: Relaxation phenomena in aqueous dispersions of synthetic lecithins. Biochemistry *16*, 2674-2680 (1977)

Tsong, T.Y., Yang, C.S.: Rapid conformational changes of cytochrome P-450: Effect of dimyristoyl-lecithin. Proc. Natl. Acad. Sci. USA *75*, 5955-5959 (1978)

Tsong, T.Y., Tsong, T.-T., Kingsley, E., Siliciano, R.: Relaxation phenomena in human erythrocyte suspensions. Biophys. J. *16*, 1091-1104 (1976)

Tümmler, B., Maass, G., Weber, E., Wehner, W., Vögtle, F.: Noncyclic crown-type polyethers, pyridinophane cryptands, and their alkali metal ion complexes: Synthesis, complex stability, and kinetics. J. Am. Chem. Soc. *99*, 4683-4690 (1977)

Tümmler, B., Maass, G., Vögtle, F., Sieger, H., Heimann, U., Weber, E.: Open-chain polyethers. Influence of aromatic donor end groups on thermodynamics and kinetics of alkali metal ion complex formation. J. Am. Chem. Soc. *101*, 2588-2598 (1979)

Umberger, J.K., La Mer, V.K.: The kinetics of diffusion controlled molecular and ionic reactions in solution as determined by measurements of the quenching of fluorescence. J. Am. Chem. Soc. *67*, 1099-1109 (1945)

Weller, A.: Eine verallgemeinerte Theorie diffussionsbestimmter Reaktionen und ihre Anwendung auf die Fluoreszenzlöschung. Z. Phys. Chem. Frankfurt am Main *13*, 335-352 (1957)

Winkler, R.: Kinetik und Mechanismus der Alkali- und Erdalkalimetallkomplexbildung in Methanol. Diss. Techn. Univ. Wien 1969

Winkler, R.: Kinetics and mechanism of alkali ion complex formation in solution. In: Structure and Binding, Vol. X, pp. 1-24. Berlin, Heidelberg, New York: Springer 1972

Winkler-Oswatitsch, R., Eigen, M.: Die Kunst zu Titrieren. Angew. Chemie *91*, 20-51 (1979)

Woolley, P.: The binding of a neutral aromatic molecule to a negatively-charged lipid membrane. I. Thermodynamics and mode of binding. Biophys. Chem. *10*, 289-303 (1979)

Woolley, P., Diebler, H.: The binding of a neutral aromatic molecule to a negatively-charged lipid membrane. II. Kinetics and mechanism. Biophys. Chem. *10*, 305-318 (1979)

Raman Spectroscopy of Membrane Constituents and Related Molecules

R.C. Lord[1] and R. Mendelsohn[2]

I. Introduction

The Raman effect was discovered a half century ago, but only in the last ten years has it been applied effectively to biological problems. The development of the continuous-wave laser has been responsible for most of these applications. For a time after this development it was fashionable to denote Raman spectra produced by laser excitation as "laser Raman spectra". Because it seems unlikely that Raman spectra will be excited by any other than laser sources in the future, in the present chapter the terms "Raman effect", "Raman spectra", and the like will not be modified by the prefix "laser", and it will be understood that lasers are used as the source of excitation. The rapid growth of biological Raman spectroscopy that followed the development of the laser has resulted in a bibliography with few references prior to 1967.

When light is propagated through a uniform material, part of the light is absorbed and part transmitted by the material. In addition a part, generally a small part, is scattered by means of various scattering processes. This chapter is concerned only with the very weak scattering process called Raman scattering, and, specifically, only with Raman scattering due to molecular vibrations.

Raman scattering is best studied when produced by a narrow intense beam of monochromatic radiation and observed at right angles to the beam, so that the beam's intensity does not overwhelm the much less intense scattered radiation. When the light scattered by a simple pure liquid such as water or benzene is analyzed by a spectrometer, the spectrum obtained consists of a relatively strong line – the Rayleigh line – at the frequency of the incident monochromatic radiation, together with much weaker companion lines on either side of it. The pattern of these companion lines, called Raman lines or Raman bands, is a property of the scattering material.

Conventionally a Raman line is characterized by its frequency, measured as wavenumbers in cm^{-1}, expressed in terms of the difference in wavenumber $\Delta\sigma$ between that of the incident beam (the so-called exciting line) and that of the Raman line.[3] Raman

---

[1] Spectroscopy Laboratory and Department of Chemistry, Massachusetts Institute of Technology Cambridge, Massachusetts 02139, USA

[2] Department of Chemistry, Rutgers University, Newark, New Jersey 07102, USA

[3] For electromagnetic radiation travelling in free space, the velocity of propagation c in cm/s is equal to the product of the frequency ν in Hz (one Hz is one cycle per s) and wavelength λ in cm: $c = \nu\lambda$. The wavenumber σ in cm^{-1} is $\sigma = \nu/c = 1/\lambda$. It is standard practice to convert wavelengths measured in air to their slightly larger values in free space before calculating $\Delta\sigma$.

lines on the low wavenumber side of the exciting line are called Stokes lines and those on the high wavenumber side anti-Stokes lines. The wavenumber pattern of the Stokes and anti-Stokes lines is symmetrical about the exciting line but the intensity pattern is not, the Stokes lines being more intense (see Sect. III). Thus for vibrational Raman spectroscopy the Stokes lines are universally used.

The differences in wavenumber $\Delta\sigma$ between the Stokes lines and the exciting line are sometimes referred to as "Raman shifts", but more often simply as "Raman frequencies". Since the wavenumber scale is the same as the frequency scale except for the factor $1/c$, and since the range of $\Delta\sigma$ for molecular vibrations is approximately 10-4000 cm^{-1} it is considerably more convenient to use the wavenumber scale in cm^{-1} than the frequency scale in Hz. Thus Raman spectroscopists often speak in loose fashion of "Raman frequency in wavenumbers". There is as yet no official term for the unit cm^{-1}, though the name Kayser has been proposed.

The numerical values of the molecular Raman shifts correspond to the actual values of molecular vibrational frequencies in cm^{-1}. These latter also correspond to the characteristic frequencies of the mid and far infrared regions of the spectrum. Thus information about molecular vibrational frequencies that is furnished by infrared absorption spectroscopy is of the same kind as that provided by the $\Delta\sigma$s of the Raman lines. This point is illustrated by Figure 1, which shows the Raman scattering spectrum of liquid N-methylacetamide plotted on the same wavenumber scale as that of the infrared transmission spectrum. It can be seen that the frequencies of the minima in infrared transmission (i.e., the infrared absorption maxima) are in many cases the same as those of maximum intensity of Raman scattering. However, there are often examples of infrared minima for which there is no Raman maximum (for instance, the band marked "amide II" in Fig. 1), and vice versa. More importantly, the strongest lines in the Raman spectrum do not unusually correspond to the strongest infrared bands. Hence the two kinds of spectra are complementary.

It is clear from Figure 1 that the intensities of Raman lines and infrared bands are a significant property of the two kinds of spectra. However, the relationship of intensity to the characteristics of the sample being studied is quite different for the two spectra. Under the usual operating conditions of an infrared spectrometer, transmission of a collimated monochromatic beam of infrared radiation of wavenumber σ depends exponentially on the sample's thickness b and its absorptivity a_σ:

$$\% T_\sigma = 100(I/I_o)_\sigma = 100 \exp(-a_\sigma b) \tag{1}$$

where I_o and I are the intensities of the beam before and after transmission through the sample. No such simple relationship exists for Raman intensities but it can be said that the intensity of Raman scattering is proportional to the intensity of the exciting radiation and also to the number of molecules irradiated. The proportionality constant for any given line in the Raman spectrum depends on variables that have to do with the experimental arrangement, which affects all lines equally, and those that concern the individual line only. The former include the volume of the irradiated sample, the solid angle over which the scattered radiation is collected, and attenuation of scattering by reflection losses, sample inhomogeneity and the like; the latter consist of the frequency of the molecular vibration producing the line and the first derivative

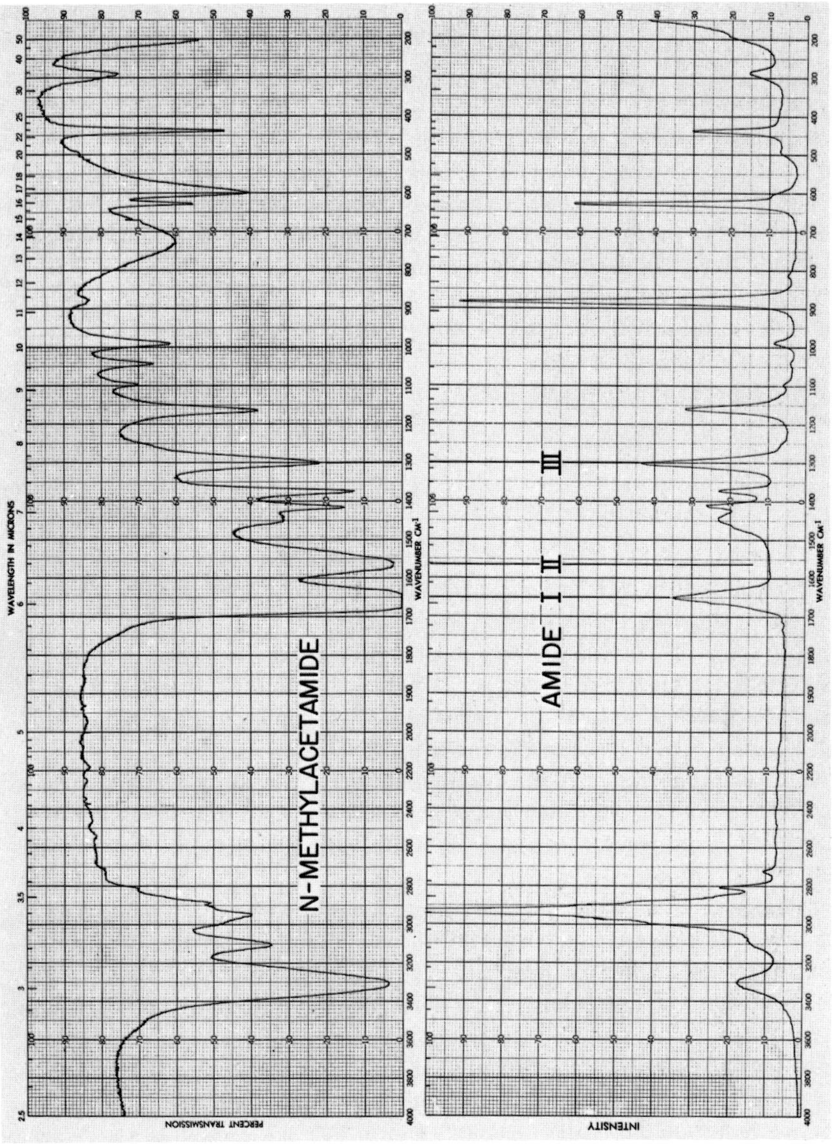

Fig. 1. Infrared transmission spectrum (*upper*) and Raman spectrum (*lower*) of liquid N-methylacetamide. The two spectra were recorded by commercial instruments to the same scale and have been directly reproduced from the original records. Zero wavenumber on the Raman scale corresponds to the wavenumber of the laser radiation used for excitation (Rayleigh line)

of the molecular polarizability with respect to the vibrational normal coordinate. The latter quantity is the analog of the first derivative of the molecular dipole moment with respect to the normal coordinate, whose square determines the infrared absorptivity a_σ.

The accurate measurement of the intensity of scattered radiation is not an easy matter. There are many instrumental problems in addition to complications inherent in the spectra. Instrumental factors can often be cancelled out by measuring intensities with respect to an "internal standard" whose scattering properties are known. For example, carbon tetrachloride can be added to organic liquids and sulfate ion to aqueous solutions to a known volume concentration.

Intensity measurement is of special importance in the determination of the state of polarization of the Raman scattering. This state is defined by a depolarization factor ρ, which for scattering at an angle of 90° to the beam of exciting radiation is given by:

$$\rho = I_{\perp}/I_{\parallel} \tag{2}$$

Here $I_{\perp}$ and $I_{\parallel}$ are the intensities of Raman radiation of a given frequency that is polarized respectively perpendicular and parallel to the plane normal to the incident beam. It can be shown (see, for example, the monographs of Placzek (1934); Herzberg (1945); Konigstein (1972), that for randomly oriented molecules such as those in a liquid or solution, ρ must have a value between 0 and 6/7 for unpolarized exciting radiation, or between 0 and 3/4 when the incident radiation is polarized, as is usual for laser radiation. The components $I_{\perp}$ and $I_{\parallel}$ for the Raman spectrum of aqueous lysozyme are shown in Figure 2. The direction of the electric vector of the laser beam is arranged to be normal to the direction in which the scattered radiation is observed. In special cases involving resonant scattering, ρ may have values larger than 6/7 (Placzek, 1934; Spiro and Strekas, 1972b).

The numerical value of the depolarization factor ρ of a given Raman line is important because it is related to the nature of the vibration that produces the line. Vibrations that preserve molecular symmetry have ρ less than 3/4 for polarized incident radiation (usually substantially less), while vibrations that are antisymmetrical in any way have ρ = 3/4.

When the molecules of a sample have a regular orientation, as in a single crystal, the depolarization factor can be replaced by other ratios of a more diverse nature. These ratios arise from the variety of ways in which the electric vector of the incident light and the direction of observation of the scattered radiation can be oriented with respect to the crystal axes. By suitable permutation of these variables the individual components of the scattering polarizability tensor discussed in Section IIIA may be evaluated. However, single crystals of biological macromolecules are not often available for such studies, and this kind of study will not be discussed further. An example is the investigation of oriented polyalanine by Fanconi and co-workers (1969).

Because the kind of information provided by Raman spectra is so closely similar to that of infrared spectra, it is useful to compare the two techniques for problems related to membrane structure. The advantages of Raman spectroscopy are:

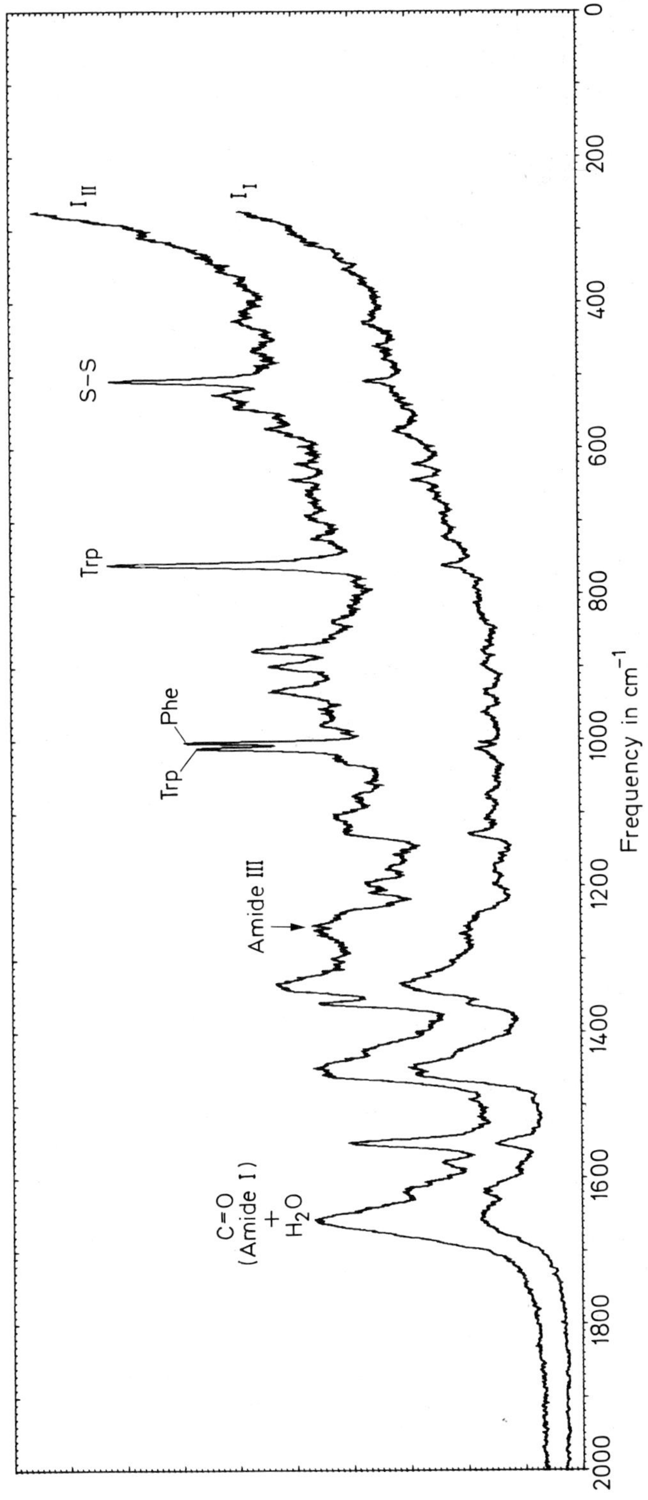

Fig. 2. Polarized components of the Raman spectrum of 30% aqueous lysozyme (courtesy of Dr. J.R. Scherer). $I_I = I_\perp$; $I_{II} = I_{\parallel}$

1. The relatively weak Raman scattering of liquid water, which enables the investigation of aqueous solution of moderate dilution (~1%) by nonresonant Raman spectroscopy, and of solutions of considerable dilution (0.001%) by resonant Raman methods. Liquid water absorbs infrared radiation strongly throughout the range of interest (10-4000 cm^{-1}) and the study of even concentrated aqueous solutions is troublesome except in certain narrow wavenumber regions (see chapter 6 of Hs.H. Günthard and U.P. Fringeli).
2. The high intensities of Raman lines due to vibrations that stretch homopolar chemical bonds, for example, S-S, C-C, C=C, and C-S bonds (see Fig. 2). In the infrared, bands due to these vibrations may be weak or missing altogether.
3. The wide wavenumber range covered in the usual span of commercial Raman spectrometers. Modern infrared instruments may cover a large part of this range, but usually two separate instruments are needed to measure infrared spectra over the entire range 10-4000 cm^{-1}.

The disadvantages of the Raman technique are:
1. The importance of the optical uniformity of the sample. When the sample is optically inhomogeneous, the intensity of excitation by the laser beam within the sample may be reduced and much of the scattered Raman radiation may be deflected from the collecting system. In the infrared, sample inhomogeneity is less important and techniques for minimizing it are well established.
2. Impairment of the quality of the spectrum by fluorescent impurities. The intensity of fluorescence (scattered light emitted after absorption of the laser beam) from a molecule is usually several orders of magnitude higher than that of Raman scattering. Thus a trace of fluorescent impurity (<0.01%) may produce a characteristically broad spectrum which obscures the Raman effect. The only satisfactory cure for this difficulty is purification of the sample. Infrared radiation does not produce fluorescence and small amounts of impurities do not affect the infrared spectrum adversely.
3. Vulnerability of the sample to the power and photochemical activity of the laser beam. Laser radiation has high intensity and actinic power, so that alteration of the sample is a matter of concern. There are techniques for avoiding such effects (see Sect. IIB) but they require larger amounts of sample. Infrared radiation, on the other hand, has little photochemical effect and normally does not damage the sample either by this means or by simple heating.
4. Insensitivity to dilute solutions. Except for the resonant Raman effect (discussed in detail later), it is not usually possible to obtain satisfactory Raman spectra of solutions whose concentrations are much less than 1% by weight. In the infrared, absorption spectra of dilute solutions can be obtained, when the solvent is sufficiently transparent, by increasing the length of the absorbing path to make up for low concentration. However, the infrared opacity of liquid water precludes such a procedure for aqueous solutions.

Except for the foregoing major differences, infrared and Raman spectroscopy are comparable in sample size, availability and cost of instruments of high performance, convenience, speed, precision of spectral measurement, and similar practical factors. It is frequently desirable to use both techniques for a given problem when the considerations outlined above do not dictate one or the other.

It should be mentioned at the outset that relatively little work has been done to date in membrane research by either Raman or infrared spectroscopy. Nonetheless the promise shown by the two techniques in the papers discussed in this chapter and Chapter 6 of Günthard and Fringeli is high, and it is certain that much use will be made of these new approaches to the investigation of membrane structure and function.

II. Experimental Methods

A. Instrumentation

The design of efficient instrumentation for Raman spectroscopy is governed by the fact that the effect is normally extremely weak. The intensity of Raman scattering is $10^{-2}-10^{-5}$ times weaker than Rayleigh scattering for liquids or solutions, which in turn is many orders of magnitude weaker than the incident power level. Consequently, intense sources are required to excite the Raman spectrum. In addition, double or triple monochromators are necessary for analysis of the spectrum in order to minimize stray light (light of frequency different from that which the monochromator is set to pass but which nevertheless appears at the exit slit). Removal of stray light due to Rayleigh scattering or reflection of the laser line by the sample is especially important. Finally, an efficient detection system is required.

In this section, we outline briefly the nature of the instrumentation used in modern Raman spectroscopic research. A photograph and optical diagram of a spectrometer are given in Figure 3. For further details the reader is referred to several monographs in the field (Gilson and Hendra, 1970; Loader, 1970; Szymanski, 1970; Tobin, 1971; Durig and Harris, 1972).

1. Sources

Laser sources are the most suitable and are nearly universally used for exciting Raman spectra. They offer an intense, monochromatic, well-collimated beam of light which can easily be focused to a small diameter. The most widely used are of the continuous-wave (cw) rare gas variety that emit in the visible region of the spectrum. The type, principle emission wavelengths, and typical power of several commonly used lasers are given in Table 1.

Historically, the He-Ne laser (632.8 nm) was the first to be widely employed for Raman work, although it suffers from the two disadvantages that its power output is relatively low and photomultiplier tubes required for detection in the red spectral region have a much reduced sensitivity. In spite of this, He-Ne excitation has been used in the past to obtain Raman spectra of a large variety of molecules. In particular, substances that absorb in the 400-600 nm range and may be decomposed by such radiation can be studied with 632.8 nm excitation.

Argon ion lasers (488 nm and 514.5 nm) offer several advantages for Raman work. In addition to the high power available (Table 1), the higher frequency gives

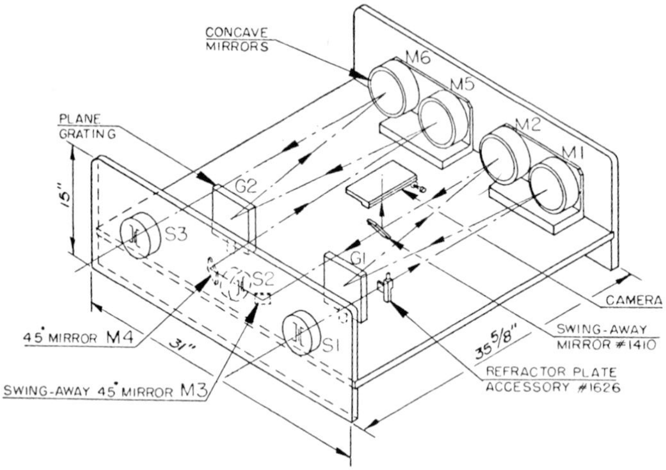

Fig. 3 a,b. (a) Photograph of Spex Raman spectrometer. (b) Optical diagram of the Spex 1401 double monochromator. Scattered radiation enters through S1 entrance slit, is reflected to grating G1 via mirror M1, is reflected by mirror M2 through intermediate slit S2 into a similar monochromator. The radiation then passes through the exit slit S3 and into the detection system. The gratings are connected by a bar (Spex Industries, Metuchen, New Jersey, USA)

a higher intensity of scattering in proportion to the fourth power of the frequency. For example, a factor of 2.8 is gained by utilizing 488 nm as opposed to 632.8 nm radiation. These gains can be used to improve the signal-to-noise ratio or to obtain spectra at lower concentrations or amounts of sample.

For resonance Raman studies (Sect. IIA5 and IV.C), sources are required that lie within molecular absorption bands, and it is essential to have a wide variety of exciting wavelengths. The development of continuously tunable dye lasers will aid in this area. In such devices the fluorescent emission of a dye solution is excited with a "pump" laser. The resulting broad-band output can be tuned within a cavity resonator to produce lasing action at any wavelength within the fluorescence band. For example, the dye Rhodamine 6G in methanol-water solution absorbs in the 480-530 nm region. When excited by Ar^+ radiation at 514.5 nm, it provides an output tunable from 550 to 620 nm. The power levels available depend on the power of the pump laser but 10%-20% conversion should be achieved. Various dyes with differing emission characteristics can produce lasing action from 530 nm to 700 nm (Snavely, 1973; Hall, 1978). To produce radiation in the range 400-600 nm nitrogen lasers, which emit at 337 nm, can be coupled with appropriate dyes.

2. Spectrometers

As Raman-scattered radiation is very weak, the monochromator used for analysis must discriminate strongly against Rayleigh scattering and reflected laser radiation from the

Table 1. Commonly used laser wavelengths

Type	Wavelength in air (Å)	Typical power (mw)
Ar^+	4579	100
	4658	75
	4727	75
	4765	500
	4880	1500
	4965	400
	5017	400
	5145	1500
	5287[a]	200
Ar^{++}	3511[b]	variable
	3638[b]	variable
Kr^+	5208	150
	5308	150
	5682	150
	6471	500
	6764	50
He-Ne	6328	100
He-Cd	4416	75
N_2	3371	variable

[a] Requires special optics.

[b] Requires quartz optics on an Ar^+ laser.

sample. For a single monochromator, the discrimination ratio I_{Raman}/I_{σ_0}, where I_{σ_0} is the intensity of light at the incident frequency σ_0, and I_{Raman} is the intensity of the Raman radiation, is often quoted for a frequency 50 cm^{-1} away from the exciting line. A typical value of this ratio for a good monochromator is of the order 10^{-5}. Since the ratio of Raman to Rayleigh intensity approaches this value, it is necessary to employ two monochromators in tandem in a spectrometer to reject the stray light effectively. A rejection of 10^{-10} or better may be achieved thereby. Most commercial instruments incorporate such a design. In fact, a number of commercial Raman instruments have triple monochromators which enable the observation of vibrations at small frequency shifts (<10 cm^{-1}) from the exciting line, even for highly inhomogeneous samples.

Many monochromators employ the Czerny-Turner design, which uses a plane grating and identical spherical mirrors for collimation and focusing. The optical diagram of the Spex "Ramalog" instrument, for example, in which the Czerny-Turner design is used, is shown in Figure 3b.

3. Detection Systems

The radiation appearing at the exit slit of the monochromator is detected photoelectrically in all commercial Raman instruments. Three types of detection system are commonly used in Raman spectroscopy (Bulkin, 1969; Tobin, 1971; Durig and Harris, 1972):

1. *Direct Current (DC) Amplification.* The photocurrent developed in the photomultiplier produces a voltage drop across a load resistor which is measured with a sensitive voltmeter. DC detection suffers from zero drift and is generally poor at low signal levels.
2. *Photon Counting.* Each pulse that arrives at the photomultiplier anode is amplified, passed through a discriminator and used to trigger a standard pulse. The great advantage of this technique is that the output pulses are amenable to direct computer processing. It suffers at high signal levels because of the finite time resolution of the process (i.e., if two photoelectrons appear at the anode within a time shorter than the system's resolving time, they will be observed as a single pulse). This disadvantage is minimal at low signal levels and photon counting is the most commonly used detection system for biological samples.
3. *Synchronous Detection with a Lock-in Amplifier.* The laser beam is chopped periodically and part of it is made to produce a reference signal, which is compared with the photomultiplier output at the chopping frequency by means of a lock-in amplifier. With this technique, it is possible to compensate easily for variation of laser power during an experiment. The main drawback with synchronous detection is that one-half of the laser power is lost in the chopping process, and the resultant signal-to-noise ratio is reduced by a factor of $1/\sqrt{2}$.

Several types of phototube may be employed successfully to detect the Raman radiation. Two requirements which must be met are those of high quantum efficiency and low dark current. Phototubes which possess an extended S-20 response curve are eminently suitable for Raman spectroscopy because of their enhanced sensitivity in the red region of the spectrum. More recently, photomultipliers with photocathodes made of semiconductors such as gallium arsenide have been developed which are sensitive up to 850-900 nm.

4. Measurement of the Depolarization Ratio

There are several experimental methods for evaluating the depolarization ratio ρ defined earlier (Eq. (2) above; Gilson and Hendra, 1970; Tobin, 1971; Koningstein, 1972), but only the one commonly used will be described here. The scattered radiation is collected in a direction (taken as the y-axis) perpendicular to that of the incident laser beam (x-axis). The laser beam is linearly polarized in the z-direction. A suitable polarization analyzer is placed between the sample and the entrance slit of the spectrometer. The analyzer is mounted on a rotating mechanism so that it can be set to pass light that is polarized in either the x- or the z-direction. The intensitiy of the former, as measured at a particular wavenumber by the detection system, is $I_\perp$; the latter gives $I_\parallel$ at the same wavelength.

The spectrometer itself is selective in the efficiency with which it passes polarized light, and the selectivity unfortunately varies with wavenumber. Hence it is essential to neutralize the effect of the spectrometer by inserting a "scrambler" between the analyzer and the spectrometer. The scrambler convertes the parallel and perpendicularly polarized beams to nonpolarized form without altering the relative values of $I_\perp$ and $I_\parallel$, so that the spectrometer affects both beams equally. A wedge of calcite or crystal quartz placed in front of the slit at an appropriate angle is normally used as a scrambler.

Several techniques have been described for single-scan automatic recording of the depolarization ratio throughout the entire range of the Raman spectrum, which avoids the necessity of recording the two components $I_\perp$ and $I_\parallel$ separately (Proffitt and Porto, 1973; Kiefer and Topp, 1974).

B. Sampling Procedures

Raman spectroscopy possesses the substantial advantage that samples may be examined in any physical state. Spectra can be obtained from pure liquids, solutions, crystals, polycrystalline powders, fibers, and surface films.

Samples in the liquid state or in solution may be studied in ordinary melting-point capillaries containing 2-20 μl. Temperature in such systems can be controlled by a stream of nitrogen flowing over the cell (Miller and Harney, 1970) or by utilizing thermostatted sample holders (e.g., Thomas and Barylski, 1970).

Capillary cells are most conveniently illuminated with transverse excitation. The laser beam impinges on the curved surface of the cell and is focused inside. A fluorescent background sometimes arises from the glass itself, but this can usually be minimized by careful focusing of the laser beam or by using more expensive quartz capillaries. Alternatively, the capillary tube can be illuminated coaxially, for which an internal diameter large enough to accommodate the laser beam and a flattened end to the capillary (to serve as a window) are needed.

The concentration of material required to obtain Raman spectra in solution is naturally quite sample-dependent. Spectra have been reported for proteins and nucleic acids at concentrations of 20 mg/ml (Small and Peticolas, 1971; Bellocq et al., 1972). Thus 0.2 mg of scattering substance is needed for a 10-μl sample. For solutes of lower molecular weight, concentrations of the order 0.1 M are probably sufficient, although this figure is extremely variable. Aromatic and conjugated olefinic groups scatter more strongly than mono-olefins and aliphatic groups. Saturated polar compounds with only first-row atoms, e.g., carbohydrates, scatter rather poorly, while compounds containing atoms of high atomic number give rise to strong scattering.

Polycrystalline materials may be examined in capillaries or as pressed pellets. A milligram or two of material suffices for the spectrum. Solid substances must be pure and free from surface contaminants in order to yield satisfactory Raman spectra; a trace of impurity which absorbs the laser light almost always leads to intense local heating and sample destruction. Fibers may be stretched and examined as oriented specimens, while single crystals are usually mounted for study on a goniometer of the type commercially available for X-ray crystallography.

The recent emergence of resonance Raman scattering (Sect. IV.C) has created new possibilities for the study of colored molecules, but also has required modified sample-handling techniques. When molecules are irradiated with light that is strongly absorbed, sample destruction due to local heating or photolysis is quite possible. Several methods have been developed to circumvent this, all of which involve relative motion of the sample and the laser beam. A simple and extremely useful approach involves rotation of the sample in the laser beam, which avoids local heating. Rotating holders have been described for solutions (Kiefer and Bernstein, 1971a) and for powders (Kiefer and Bernstein, 1971b) with sample requirements of 2 cc for the former and 10 mg for the latter. Another arrangement (Koningstein and Gachter, 1973) keeps the sample fixed and "wobbles" the point of illumination by a slightly eccentric motion of a plane mirror in the path of the laser beam. A semimicro arrangement for powders that employs 1 mg of material has also been described (Long et al., 1974; Carey, 1978).

In order to reduce the volume of solution required for resonance Raman work, Spiro and co-workers have utilized a capillary combined with a peristaltic pump and closed loop to Tygon tubing to circulate the sample through the cell (Woodruff and Spiro, 1974). In this fashion, the volume required was reduced to 0.8 ml and temperature control was simply maintained. The current methods of resonance Raman spectroscopy have been summarized by Kiefer (1974).

Sample concentrations for resonance Raman spectroscopy are greatly reduced from those for nonresonant scattering. Spectra have been reported for hemoglobin and cytochrome c derivatives at concentrations of 10^{-4} M (Spiro and Strekas, 1972a,b; Strekas and Spiro, 1972) and β-carotene in a 10^{-6} M solution (Gill et al., 1970). The intensity of resonance Raman scattering is not necessarily a monotonically increasing function of sample concentration, as is ordinary Raman intensity, but instead passes through a maximum value determined by sampling geometry. The optimum concentrations for various standard arrangements have been calculated in terms of the molar extinction coefficients of the colored material at the exciting frequencies utilized and at the resultant frequencies of Raman scattering (Lippincott et al., 1959; Shriver and Dunn, 1974; Strekas et al., 1974).

III. The Theoretical Basis of Raman Spectroscopy

A. Theory of Raman Scattering

The Raman effect is a special kind of light scattering and its theoretical basis is a branch of the general theory of scattering phenomena. There are two main parts to the general theory. One concerns the production of secondary radiation by the action of the incident light on an isolated particle such as a molecule; the other deals with the consequences of the phase relationships among the waves of secondary radiation procuded by the totality of molecules in the scattering system. For the Raman effect our concern is chiefly with the first of these and we give here a qualitative summary of this part of the theory. For more detailed accounts the reader is referred to the treatises listed in the bibliography, Part A, particularly those of Placzek (1934), the founder of the theory, and Herzberg (1945).

1. Molecular Polarizability and Light Scattering

A light wave is a travelling wave of electric and magnetic fields, of which only the electric component gives rise to Raman scattering. When a light wave passes over a molecule composed of electrons and nuclei, the electric field of the wave at any given instant will be the same throughout the molecule, because the wavelength of the light (~500 nm) is so much longer than the size of the molecule (~1 nm). Hence the field will exert the same force on all the electrons in the molecule and will act to displace them from their average positions around the positively charged atomic nuclei. The displacements will produce a dipole moment **M** in the molecule that is to a good approximation proportional to the electric field strength **E**:

$$\mathbf{M} = \boldsymbol{\alpha}\,\mathbf{E}. \tag{3}$$

The proportionally factor $\boldsymbol{\alpha}$, called the polarizability, relating the two vector quantities **M** and **E**, is a 3 x 3 tensor. For example, M_x, the x-component of **M**, is given bv

$$M_x = \alpha_{xx}E_x + \alpha_{xy}E_y + \alpha_{xz}E_z \tag{4}$$

where E_x, E_y, E_z are the components of **E**. Analogous equations hold for M_y and M_z. Thus two subscripts are needed for each element of $\boldsymbol{\alpha}$, the first showing the component of **M** and the second the component of **E** related by that element.

The electric field of the light wave is of course alternating rapidly with time. If a fixed molecule is irradiated with the sharply monochromatic, plane-polarized radiation of wavenumber σ in cm^{-1} from a laser, the time variation $E_z(t)$ is described by

$$E_z(t) = E_{max} \cos 2\pi c\sigma t \tag{5}$$

where E_{max} is the value of E_z at its maximum, t is the time in seconds measured from a suitable starting time, c is the velocity of light in free space, and the direction of polarization is taken as the z-direction. A consequence of this is that **M** also varies with time. If we assume a time-independent $\boldsymbol{\alpha}$, we have for the z-component of **M**

$$M_z(t) = \alpha_{zz} E_{max} \cos 2\pi c\sigma t. \tag{6}$$

That is, the induced dipole moment oscillates sinusoidally with the same frequency $c\sigma$ as the incoming light. The oscillating dipole moment in turn emits radiation at the same frequency with a radiant power proportional to $\overline{M_z^2}$, the time average of $M_z(t)^2$.

Because M_z depends on α_{zz} as well as E_z, the properties of the molecule enter the scattering theory through α. For example, if $\boldsymbol{\alpha}$ depends on the time, this will affect the time dependence of the induced moment **M**. $\boldsymbol{\alpha}$ may vary with time, for example, by virtue of the vibrations of the molecule, because the ease with which electrons may be displaced by an external field depends on how tightly they are held to the nuclei, which in turn depends on the internuclear separations r_{ij}.

In a diatomic molecule AB, which has one internuclear separation r_{AB}, the molecular vibration consists of a sinusoidal variation of r_{AB} with time:

$$\Delta r(t) = r_{AB}(t) - r^{\circ}_{AB} = \Delta r_{max} \cos 2\pi c\sigma_{vib} t \tag{7}$$

Here $\Delta r(t)$ is the displacement of the atoms at time t from their equilibrium separation r°_{AB}, Δr_{max} is the maximum displacement during the vibration, and σ_{vib} is the vibrational frequency of the molecule in cm^{-1}. The time t will be taken to have the same starting point as the time scale for the light wave in Eq. (2). This ignores the fact that the vibrations of different molecules in an assembly have different phases, but simplifies the discussion for present purposes.

We now suppose that the polarizability of the diatomic molecule depends linearly on the displacement Δr. For the component α_{zz} of $\boldsymbol{\alpha}$, for example, we postulate that

$$\alpha_{zz}(t) = \alpha^{\circ}_{zz} + \alpha'_{zz}\Delta r(t) = \alpha^{\circ}_{zz} + \alpha'_{zz}\Delta r_{max} \cos 2\pi c\sigma_{vib} t. \tag{8}$$

Here α°_{zz} and α'_{zz} are constants, the former being a polarizability element of the nonvibrating molecule and the latter the rate at which this element changes with r_{AB}, i.e., $d\alpha_{zz}/dr_{AB}$.

When this expression for α_{zz} is substituted in Eq. (6) we find

$$M_z(t) = \alpha^{\circ}_{zz} E_{max} \cos 2\pi c\sigma t + \alpha'_{zz}\Delta r_{max} E_{max} \cos 2\pi c\sigma t \cos 2\pi c\sigma_{vib} t. \tag{9}$$

The product of cosines in Eq. (9) gives the result

$$M_z(t) = \alpha^{\circ}_{zz} E_{max} \cos 2\pi c\sigma t + \tfrac{1}{2}\alpha'_{zz}\Delta r_{max} E_{max} \cos 2\pi c(\sigma+\sigma_{vib})t \\ + \tfrac{1}{2}\alpha'_{zz}\Delta r_{max} E_{max} \cos 2\pi c(\sigma-\sigma_{vib})t. \tag{10}$$

This basic equation says that when light falls on a vibrating diatomic molecule, the induced dipole moment (as exemplified by M_Z) has a complicated time dependence consisting of three components. The first term on the right is a component vibrating with the frequency σ of the incoming light and with a magnitude determined by the polarizability α°_{ZZ} and the field strength of the light. The second term is a component vibrating at a frequency which is the sum of the frequencies of the light and the molecular vibration. The third term is a component vibrating at a frequency given by that of the light wave minus that of the molecule. Both these components have magnitudes dependent on α' (i.e., the polarizability derivative $d\alpha_{ZZ}/dr_{AB}$), the amplitude of the vibration, and the field strength of the light.

The different components of M_Z in Eq. (10), having different time dependences, give rise to scattering at different frequencies. The first term on the right side produces scattered radiation with the frequency of the incoming light, which is called Rayleigh scattering. The second term gives rise to scattering at a higher frequency, $\sigma + \sigma_{vib}$, and the third to scattering at a lower one, σ-σ_{vib}. These latter two, the Raman scattering, appear as a pair of lines in the spectrum of the scattered light at frequencies that are spaced symmetrically above and below σ by the amount σ_{vib}. It is the ability to measure this molecular property σ_{vib} that gives the Raman effect its importance.

2. Quantized Molecular Energies and the Raman Effect

The preceding discussion has been based on classical electromagnetic theory, which provides a convenient and intuitive approach to Raman scattering. The quantum theoretical view of the scattering process as quite different: it takes into account the quantization of molecular levels and provides a formal procedure (not to be discussed here) for the calculation of $\boldsymbol{\alpha}$ in terms of the electronic properties of molecules.

To a good approximation, for a gaseous diatomic molecule, the molecular energy E_{mol} is

$$E_{mol} = E_{elec} + E_{vib} + E_{rot} + E_{trans} \tag{11}$$

where the respective subscripts refer to the electronic, vibrational, rotational and translational degrees of freedom. In condensed phases the latter two are turned into quasi-harmonic oscillations, and for present purposes only the electronic and vibrational motions need be considered. The energies of the former, E_{elec}, are quantized at higher values (10,000-100,000 cm^{-1}) than the former (100-4000 cm^{-1})[4]. No closed algebraic expression can be written for E_{elec} but for E_{vib} in a diatomic molecule one has, to a good approximation:

$$E_{vib} \text{ in } cm^{-1} = (v + \tfrac{1}{2})\sigma_{vib} \qquad v = 0, 1, 2, \ldots \tag{12}$$

[4] Frequency in cm^{-1} is convertible to energy in ergs/molecule by the factor hc, h being Planck's constant

The vibrational quantum number v has only integral values, so that the energy levels are equally spaced by the amount σ_{vib}.

The quantum mechanical description pictures light scattering as a two-photon process. In the first step of the process a photon of the incoming light combines with the scattering molecule to raise it to a higher energy state of short lifetime. This state, which may or may not correspond to one of the quantized levels E_{mol}, is reached by the upward arrows of Figure 4. The analogous process in the classical description is the induction of the alternating dipole moment **M** by the field **E** of the light wave. The second step is the release of a photon after a very short period of time ($<10^{-11}$ s), which is indicated by the downward arrows of Figure 4. The release of the photon has as its classical analog the emission of radiation by the oscillating induced dipole moment.

The energy of the released photon depends on the energy difference between the upper level reached by the molecule (the horizontal dotted lines of Fig. 4) and the final level reached on the downward transition. If the downward arrow and the upward arrow have the same length (that is, the energy differences, apart from sign, are the same), the incoming and emitted photons will have the same wavelength. This is the process for Rayleigh scattering (left-hand side of Fig. 4), in which the scattered light suffers no change in wavelength.

If the downward transition terminates on a vibrational energy level that is higher than the starting level (Stokes transition in Fig. 4), the molecule gains energy in the

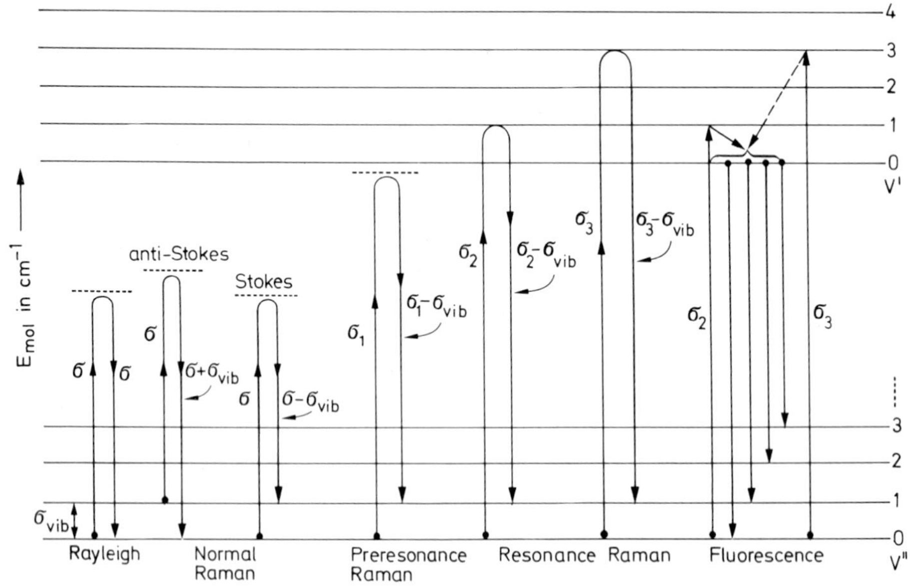

Fig. 4. Energy and frequency relationships in molecular light-scattering processes. The lengths of the upward arrows are proportional to the frequencies of the incoming light, lengths of the downward arrows to the frequencies of scattered radiation. Only Stokes transitions are shown except for the one labeled "anti-Stokes". v', v" are the respective values of the vibrational quantum number (Eq. (12)) in the upper and lower electronic states of the molecule

process while the emitted photon has less energy than the incoming one (shorter downward arrow). Conversely, if the downward transition terminates on a vibrational level that is lower in energy than the starting level (anti-Stokes transition), the molecule loses energy in the process and the emitted photon has more energy than the incoming one (longer downward arrow). These two processes are represented respectively by the last and next to last terms in Eq. (10) and correspond to Raman lines at frequencies lower by σ_{vib} and higher by σ_{vib} than the exciting frequency σ.

So far as Eq. (10) is concerned, the Stokes and anti-Stokes transitions are expected to be of equal intensity, since the coefficients of the two terms in the equation are the same. However, the anti-Stokes transitions can only take place when there are molecules that have energy corresponding to the first level ($v''=1$) above the lowest level $v''=0$ in Figure 4. The number N_1 of such molecules is

$$N_1 = N_o \exp(-hc\sigma_{vib}/kT) \tag{13}$$

when N_o is the number in the lowest level, T is the absolute temperature and k is the Boltzmann constant (gas constant per molecule, R/N). When T = 300K and σ_{vib} = 480 cm^{-1}, N_1 is 0.1 N_o. Thus the anti-Stokes line for a frequency of 480 cm^{-1} will be only one-tenth as intense as the Stokes line. Because of the exponential nature of Eq. (13), an anti-Stokes line at 2x480, or 960 cm^{-1}, would be 0.01 times as strong as the corresponding Stokes line, and so on. For this reason it is customary to record only the Stokes lines in vibrational Raman spectroscopy (compare Figs. 1 and 2).

From Figure 4 it might be concluded that other downward transitions in the Stokes lines are possible, terminating on levels v''= 2, 3, 4, etc. However, a selection rule can be derived from Eq. (10) which says that v can only change by one unit up (Stokes) or down (anti-Stokes) from its original value. Thus the classical Eq. (10) correctly shows that light scattered by the Raman effect contains frequency differences from the incoming light that correspond directly to the molecular vibrational frequencies.

Figure 4 also shows several other kinds of Raman scattering. The transitions labeled "Preresonance Raman" are those excited by a laser frequency that has almost enough energy to produce direct electronic absorption by the molecule. In such a case the selection rules are the same as for the normal Raman scattering but the intensity of scattering increases more than expected by the nonresonant scattering law. This law says that the intensity of scattering is proportional to the fourth power of the frequency of the scattered light. However, as the frequency of the exciting radiation approaches the energy of the upper electronic energy level expressed in frequency units, the intensity of scattering goes up much more rapidly than σ^4.

If the exciting frequency actually equals or exceeds somewhat the energy of the upper electronic level, the radiation can be absorbed and three processes can ensue. For our purposes the most significant is the prompt reemission (in 10^{-11} s or less) of the absorbed radiation to give rise to the resonance Raman effect. However, if the radiation is not reemitted for a longer time (10^{-8} s or more), two other processes may occur. the molecule may go to the lower electronic state by nonradiative paths that convert the absorbed energy to heat, or the molecule may emit the absorbed radiation as fluorescence. In the latter case the time delay of 10^{-8} seconds or more is enough for the excited molecule, which has acquired vibrational as well as electronic energy from the incoming radiation (σ_2 and σ_3 giving v′=1 and v′=3 respectively in Fig 4), to lose that vibrational energy (dashed lines in Fig. 4). Hence the fluorescent scattering transitions start at the same upper level (v'=0), irrespective of the frequency of the exciting radiation σ_2, σ_3, etc., thereby giving scattered radiation at fixed frequencies. This distinguishes resonance Raman from fluorescent scattering, because the former occurs not at a fixed place in the spectrum but at a place dependent on the exciting frequency, i.e., at σ_2-σ_{vib}, σ_3-σ_{vib}, etc.

Which of the three processes is most likely depends on the molecule, but even if resonance Raman effect is the least likely, it is still many times more probable than the normal Raman effect, so that the intensity of resonance Raman scattering is several orders of magnitude greater than the normal effect. This has the result mentioned earlier that concentrations 10^{-3} times smaller may give satisfactory spectra. However, the vibrations that appear in the resonance Raman effect are only those in which the atoms of the absorbing chromophore participate strongly. For example, in the resonance Raman effect of hemoglobin (Spiro and Strekas, 1972a, b) or rhodopsin (Callender et al., 1976), the spectrum contains only frequencies of the heme or retinal moiety, respectively. At the concentration levels used for these materials, the contribution of the globin or opsin part of the molecule is much too feeble to be seen.

3. Polarization Characteristics of Raman Scattering.

The polarizability tensor $\boldsymbol{\alpha}$ (defined in Eq. (4) above) has nine elements, but for any given molecule there can be found a system of internal coordinate axes such that all elements are zero except α_{xx}, α_{yy}, and α_{zz}. When the molecule has axes and planes of symmetry, the coordinate system is easily found because the symmetry determines it. This system is of course fixed to the molecule. In fluids or random solid materials the molecules and their polarizability tensors are randomly oriented with respect to the directions of the incoming exciting radiation and the outgoing scattered light. However, there are certain properties of the tensor that permit the scattering by randomly oriented molecules to be computed. These are the isotropic property A, defined as

$$A = (\alpha_{xx} + \alpha_{yy} + \alpha_{zz})/3 \tag{14}$$

and the anisotropic property γ defined as

$$\gamma^2 = \tfrac{1}{2} [\alpha_{xx}-\alpha_{yy})^2 + (\alpha_{xx}-\alpha_{zz})^2 + (\alpha_{yy}-\alpha_{zz})^2 . \tag{15}$$

Clearly if all the tensor elements are equal, γ is zero.

To calculate scattering intensities from a sample of randomly oriented molecules, we illuminate the sample with linearly polarized radiation. Let the direction of incoming radiation be the x-direction, the direction of polarization (i.e., the direction of the electric vector **E** of the incoming light wave) be the z-direction, and the direction in which the scattering is measured the y-direction. If the intensity $I_\perp$ of scattered light polarized in the x-direction is computed, it is found to be

$$I_{\perp} = \text{const.}\ 3\gamma^2 \overline{E_Z{}^2} \tag{16}$$

while that polarized in the z-direction, $I_{\parallel}$, is

$$I_{\parallel} = \text{const.}\ (45A^2 + 4\gamma^2)\overline{E_Z{}^2}. \tag{17}$$

The constant in the two equations is the same and depends on the number of scattering molecules per unit volume and other numerical quantities that need not concern us here. $\overline{E_Z{}^2}$ is the average value of the square of the electric vector, to which the intensity of the incoming radiation is proportional.

The depolarization factor ρ defined earlier (Eq. (2)) as $I_{\perp}/I_{\parallel}$ is thus given by

$$\rho = 3\gamma^2/(45A^2 + 4\gamma^2). \tag{18}$$

Equation (18) applies either to Rayleigh or Raman scattering, the former coming from the first term on the right of Eq. (8) and the latter from the second term. There is an important difference between the quantities A and γ for the two kinds of scattering. For Rayleigh scattering, the tensor elements are all positive. For Raman scattering, however, as Eq. (8) shows, the elements α' represent changes $d\alpha/dr$ in the polarizability due to molecular vibration, and they can be either positive or negative. Thus the range in values of the ratio γ^2/A^2 is much greater in principle for Raman than for Rayleigh scattering. As a result, ρ for Rayleigh scattering of polarized light is necessarily small ($<1/3$) whereas for Raman scattering ρ can have any value from 0 to 3/4.

Detailed study of the effect of molecular symmetry on molecular vibrations shows that unless a molecule preserves all its symmetry during a given vibration, the isotropic property A of the polarizability changes α' is zero. Thus from Eq. (18) ρ for nontotally symmetrical vibrations is 3/4. For totally symmetrical vibrations both A and γ are nonzero, with ρ falling between 0 and 3/4, except for molecules of cubic and icosahedral symmetry. In these relatively unusual molecules, the totally symmetrical vibrations have $\alpha'_{xx} = \alpha'_{yy} = \alpha'_{zz}$, so that $\gamma = 0$ and $\rho = 0$. Thus a measurement of ρ enables one to decide whether the vibration that produces a Raman line is totally symmetrical or not: if $\rho = 3/4$, it is not; for $\rho < 3/4$, it is. For intermediate values of ρ there may be an empirical usefulness in knowing the values of ρ for various Raman lines of model molecules. Such information is an addition to the evidence relating the presence of a given Raman line to the functional group to which it belongs.

The foregoing discussion applies to nonresonance Raman scattering of radiation by randomly oriented molecules. When the molecules are not randomly oriented, as in single crystals, it becomes possible to measure tensor elements individually by appropriate arrangement of the direction of the incoming electric vector and the direction fo viewing the scattered light. Since single-crystal studies will rarely be useful in the Raman spectroscopy of membrane materials we do not pursue the topic further (see Fanconi et al., 1969, for details).

Finally a feature of resonance scattering should be mentioned. The isotropic and anisotropic scattering properties A and γ give the intensities shown in Eqs. (16) and (17), provided the scattering tensor is symmetric, that is, $\alpha'_{xy} = \alpha'_{yx}$, $\alpha'_{xz} = \alpha'_{zx}$ and

$\alpha'_{yz} = \alpha'_{zy}$. This is true for nonresonance scattering, but for resonance scattering there can be certain vibrational modes for which the nondiagonal tensor elements may have the relation $\alpha'_{xy} = -\alpha'_{yx}$, etc. When this is the case, the expression for ρ alters. Its range is no longer $0 \leq \rho \leq 3/4$, but may extend above unity, in fact very much above unity. Raman lines in the resonance spectrum of hemoglobin have been found with ρ much larger than 3/4 (Spiro and Strekas, 1972b), thereby verifying after forty years the prediction of Placzek (1934) that such inverse polarization should exist in resonance scattering.

B. Vibrational Frequencies and Molecular Conformations

The evaluation of molecular structure from the data of vibrational spectroscopy, that is, from the frequencies, intensities and polarization states observed in the Raman effect and infrared absorption, is based on two different but related approaches. One is the relatively rigorous analysis of the spectroscopic data for small symmetrical molecules and the other is a more general, "vigorous" analysis, by methods of analogy. The latter must be used with molecules too large and unsymmetrical to be treated rigorously.

The spectra of small molecules can be interpreted by means of selection rules derived from molecular symmetry (Herzberg, 1945), with the help of theoretical treatment of the vibrational motions (so-called "normal-coordinate calculations"). For molecules that can be studied in the vapor state, it is further possible to make accurate spectroscopic measurements of the interatomic distances. With large molecules in which n, the number of atoms, is more than, for example 20, approximations are necessary. The number of atomic coordinates is 3n, and of internal vibrations 3 n-6, so that a formidable number of coordinates and frequencies are involved. Moreover, most biological molecules have no symmetry, and methods based on symmetry arguments cannot be applied.

The steps in the interpretation of the spectra of small molecules are:

1. Determination of the infrared and Raman spectra of the molecule. This includes measurement of frequencies, intensities and states of polarization. Additional information may be obtained from the spectra of isotopic modifications, in which spectral changes are produced by mass effects only, and from the spectra observed in different states of aggregation.
2. Interpretation of the spectra on the basis of an assumed geometric model of the molecule. If the structure is known from X-ray or other studies, that is taken as the model. The spectra are then interpreted as completely as possible with the help of selection rules and other theoretical arguments. If the model is clearly inconsistent with the totality of the data, it is rejected and a more suitable structure is assumed for the interpretation.
3. Calculation of the normal modes of vibration on the basis of the observed frequencies assigned to them in step (2). Usually the amount of data is insufficient for a rigorous calculation and additional assumptions must be made for this purpose. When the calculation is completed, the nature of each molecular vibration – that is, the amplitudes and phases of the displacements of every atom on the molecule during the vibration – is known.

By carrying out this program for a large number of representative small molecules, a fund of information useful in understanding the spectra of large molecules has been obtained. For example, it has been found that atoms tend to move along chemical bonds (in stretching vibrations) or perpendicular to them (bending vibrations). Because bonds are usually easier to bend than to stretch, frequencies of stretching vibrations are two to three times higher than those of bending vibrations involving the same atoms. Some vibrations are found to involve only a few, closely connected atoms. In this case the group of atoms has a characteristic vibrational frequency known as a group frequency.

Group frequencies in the infrared spectrum have been known empirically for almost a century, and many such frequencies were catalogued for the Raman effect in the decade after its discovery (1928-1938). The most useful group frequencies are those that are intense enough to be easily observed, occur in a spectral range that is reasonably free from other intense frequencies, and show small variations (~1%-10%) that can be correlated with geometrical and other parameters. In favorable cases the polarization characteristics of the infrared absorption (see Chapter 6) or Raman scattering due to a group vibration may give direct information about the orientation of the group in an ordered sample such as a crystal or a fiber.

For biological systems such as membranes, group frequencies can be used for the following purposes: (1) identification and quantitative measurement of the groups present; (2) determination of the environment of the groups, that is, such factors as the nature of adjacent chemical groups, the orientation of the group, the geometry of group's environment, etc. The usefulness of this kind of information depends on the group frequencies present and on the extent to which analogous simpler systems have been characterized. If model systems have been thoroughly studied, for example, by correlating the values of group frequencies with structural parameters of interest from X-ray diffraction or other sources, it is often possible to draw specific structural conclusions from the spectroscopic data for biopolymers (Lord, 1977). Examples of this procedure are given in the following section.

IV. Applications to Membrane Studies

Although Raman spectroscopy has been applied extensively to problems in molecular biology (see review articles by Lord, 1971, 1977; Thomas, 1971; Koenig, 1972; Peticolas, 1972b, 1975; Fawcett and Long, 1973; Lewis and Spoonhower, 1974; Spiro, 1974a, b; Frushour and Koenig, 1975; Spiro and Stein, 1977; Yu, 1977), it is only beginning to be used in the study of membranes. However, the results achieved so far show that Raman spectra can complement and extend the information obtainable by other techniques, and much activity in this field is to be expected in the near future. In this section we discuss the Raman spectra of the classes of biomolecules that are important in the structure and function of membranes. No discussion of nucleic acids is given because of their lesser importance in this connection. The infrared and Raman spectra of nucleic acids and polynucleotides have been extensively reviewed by Hartmann et al. (1973) and Tsuboi et al. (1973).

A. Raman Spectra of Polypeptides and Proteins

Information about the configurations of proteins can be obtained from their Raman spectra with the help of the spectra of amino acids and polypeptides of known structure (Lord, 1971, 1977; Koenig, 1972; Frushour and Koenig, 1975). As discussed above, this procedure works through relationships, quantitative where possible, between the geometry of these model molecules and spectroscopic data. Such relationships are almost always limited to small groups of atoms, and frequencies directly characteristic of larger structural units (tertiary and quaternary structure) have not yet been identified in infrared and Raman spectra. Frequencies characteristic of helical structure in a polypeptide, for example, are essentially those of the peptide group of atoms in the special conformation it assumes in the helical chain.

1. Amino Acids

The vibrations of amino acids, $H_3\overset{+}{N}$-CHR-CO_2^-, divide into those of the side groups R and those of the $H_3\overset{+}{N}$-CH-CO_2^- moiety. When the amino acid is polymerized, the latter are superseded by the peptide vibrations as the degree of polymerization (DP) increases. The side-chain frequencies remain, though they may alter if polymerization alters the conformation of the side chain. It is thus important to establish, by examining the spectra of oligopeptides, which frequencies are characteristic of the side chains in their various conformations and which arise from the polypeptide chain.

Reviews of the spectra of amino acids have been given (Lord and Yu, 1970a; Koenig, 1972; Simons et al., 1972), but only a few relationships between conformations of the side chains and their characteristic frequencies have been worked out. Some of these frequencies can be illustrated with the help of Figure 5, which shows the Raman spectrum of aqueous ribonuclease A compared with that obtained by adding up with proper weighting the spectra of the amino acids in the enzyme (Lord and Yu, 1970b). The two spectra have numerous points of agreement; these are the prominent characteristic frequencies of the side chains. When the agreement is quantitative, it implies either that the side-chain conformations are the same in the amino acid and the protein or that the characteristic frequency and intensity of the side-chain contributions are independent of its conformation. The latter appears to be the case, for example, with the phenylalanine peak at 1004 cm^{-1} in Figure 5. A case of disagreement because of conformational differences is visible at 647 and 659 cm^{-1}, where the intensity ratio of the line at the former position (due to tyrosine) to that of the latter (C-S vibrations in cystine and methionine) is different in the two spectra. The effect of conformation in sulfides and disulfides on Raman intensities in this frequency range has been discussed by Sugeta et al. (1973) and Van Wart et al. (1976).

On the other hand, there are striking differences in the two spectra at those frequencies characteristic respectively of $H_3\overset{+}{N}$-CH-COOH (for example, 825 and 1720 cm^{-1}) and -HN-CH-CO- (for example, the amide I region, 1650-1680 cm^{-1}, and the amide III region, 1230-1300 cm^{-1}), as is to be expected.

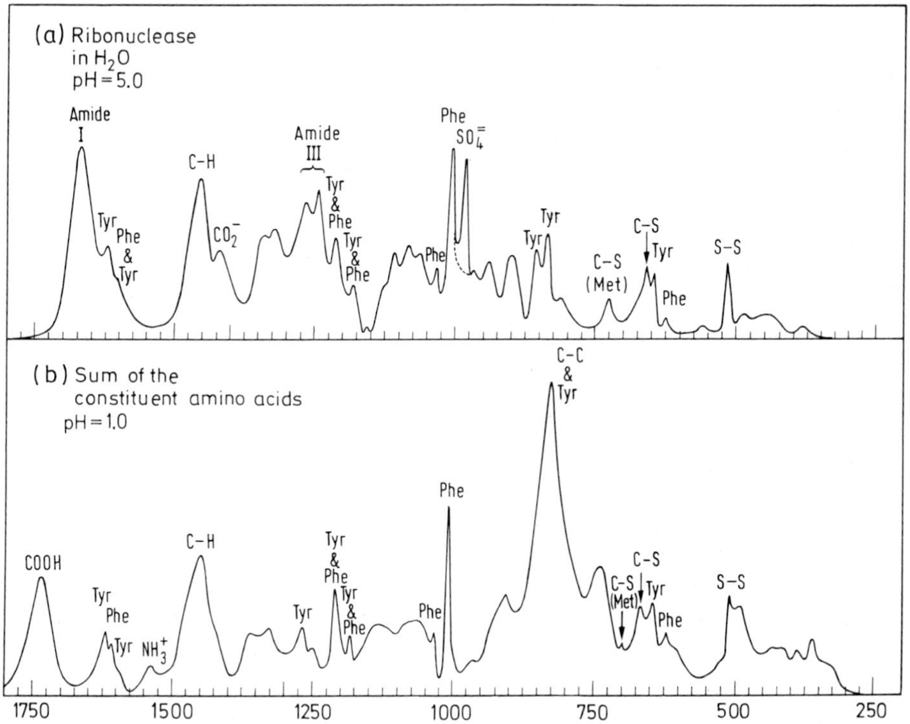

Fig. 5 a, b. Comparison of the Raman spectrum of native ribonuclease A with the sum of the spectra of the constituent amino acids. (a) Spectrum of ribonuclease A (20% in water at pH 5) redrawn to eliminate water background. (b) Sum of the aqueous spectra of the amino acids at pH 1.0 (concentrations adjusted to relative abundance in the protein) (Lord and Yu, 1970b)

2. Polypeptides

Polypeptides form relatively simple and useful models of proteins for Raman study. The peptide group being the same in both, the difference lies in the fact that there is only one kind of side chain in a homopolypeptide, whereas the side chain varies from one residue to the next in a protein. By synthesis and appropriate processing of polypeptides, model materials for different secondary structures (α-helical, antiparallel β-pleated sheet, etc.) may be obtained. In this way relationships have been established between frequencies in the amide I and III regions on the one hand and the geometry of the secondary structure on the other (Table 2). An extensive review of these relationships has been given by Frushour and Koenig (1975).

The conformation of the side chain may also vary in oligomers of different DPs. It is therefore of considerable importance to explore this variation by examination of the spectra of oligomers of different secondary structures. The effect of the secondary structure on side-chain conformations and relations between different side-chain geometries and their Raman frequencies can thus be established. These relations are of obvious importance in the interpretation of the spectra of proteins.

Table 2. Amide I and III frequencies observed in proteins and polypeptides

	Amide I (cm^{-1})		Amide III (cm^{-1})	
Structure	Raman	IR	Raman	IR
α-Helix	1650-55 S	1650-1655 VS 1685 W	1260-1290 M	1260-1290 M
β-Sheet (antiparallel)	1665-1670 S	1630 S	1229-1240 M	1235 M
		1685 W	1290 W	
Intermediate	1665-1670 M	1665 M	1240-1255 M	1250 M

As an example, Figure 6 shows the Raman spectra of solid L-valine, some oligomers of L-valine, and poly-L-valine (Fasman et al., 1978a). The lines in the monomer due to the terminal NH_3^+ and CO_2^- groups, such as the symmetric stretching vibration of the CO_2^- at about 1420 cm^{-1}, disappear as the DP increases. Conversely, lines associated with the peptide chain, such as the amide I near 1660 cm^{-1}, grow more intense. However, some side-chain frequencies, for example, that at 846 cm^{-1} in the monomer, show relatively little change in intensity with DP. The spectrum of the hexapeptide is much simpler than that of the monomer, dimer, and trimer, and quantitatively almost identical with those of poly-L-valine with DPs ranging from 26 to 500. The uniformity and simplicity of the spectra of polymers from DP 6 to DP 500 indicates a constancy of side-chain as well as backbone geometry in these materials.

While Raman spectra of homopolypeptides as a function of DP have not yet been published for all the amino acids, it is expected that such data will soon be forthcoming, and that the interpretation of the spectra of proteins will be made more reliable thereby.

3. Proteins

Just as with polypeptides, the Raman frequencies in protein spectra are characteristic of either the backbone or the side chains. The most important backbone frequencies are those mentioned above for the polypeptides, namely the amide I and III frequencies (Table 2). The other frequencies of the amide group are less important in the Raman effect, amide II at 1550 cm^{-1} because it is usually very weak or missing (though strong in the infrared spectrum) and amide IV and lower frequencies (<800 cm^{-1}) because they are not very intense and are often overlapped by stronger side-chain lines. The amide I vibration, mainly due to the carbonyl stretching motion, occurs quite reliably in the 1650-1680 cm^{-1} range and is usually free from interference by other lines in proteins. However, it may be overlapped by the broad water band at ~1640 cm^{-1} in aqueous solutions, in which case the replacement of H_2O by ^2H_2O (whose corresponding band occurs at ~1210 cm^{-1}) will uncover the amide I′ frequency (amide I′ refers to the carbonyl vibration in the group $\text{-C}^{\alpha}\text{HR-(CO)-N}^2\text{H-C}^{\alpha'}\text{HR}'$, whose frequency is shifted to lower values, 1610-1650 cm^{-1}, by ^{2}H-substituion).

The amide III frequency is at least twice as sensitive as the amide I in the Raman effect to conformational changes, since it is found to range from about 1230 cm^{-1} in pure β-pleated-sheet structures to as high as 1300 cm^{-1} or thereabouts in α-helical

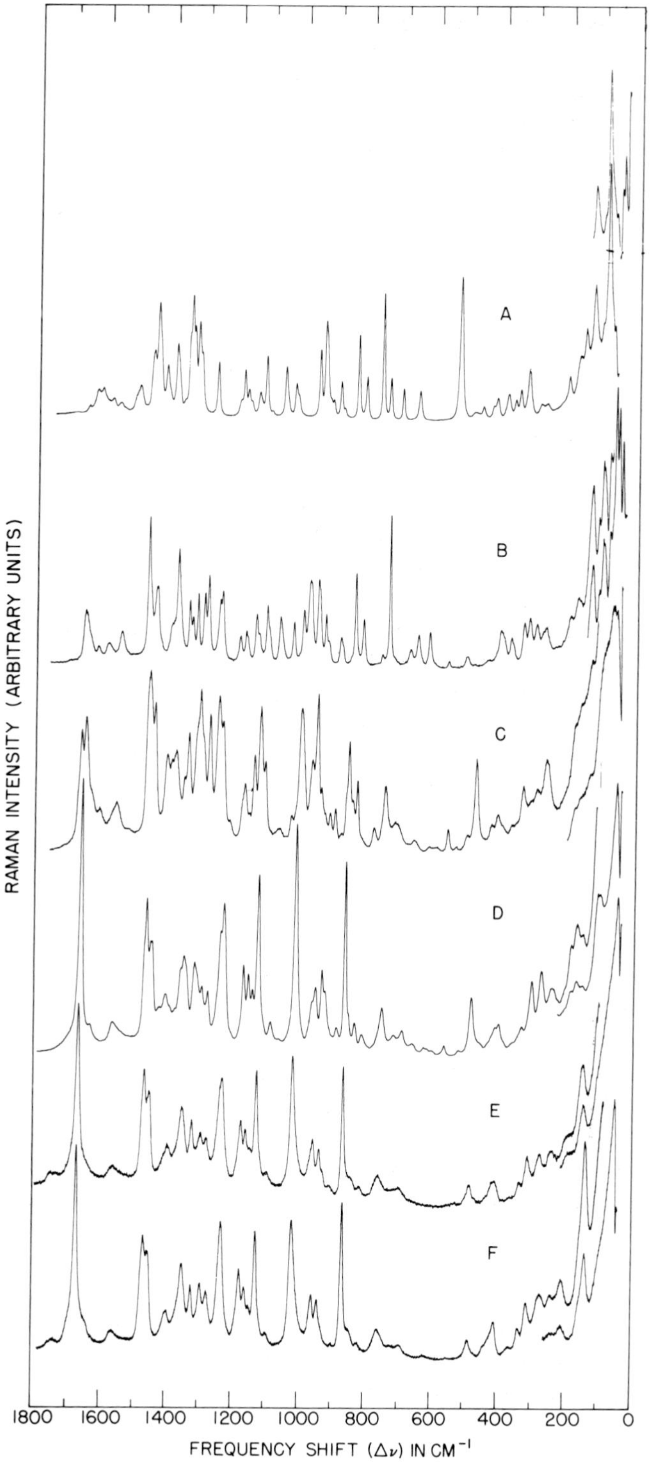

Fig. 6. Raman spectra of: (A) L-valine; (B) di-L-valine; (C) trimer; (D) tetramer; (E) hexamer; (F) polymer (DP=26), all in solid state (Fasman et al., 1978a)

conformations. A number of authors (Lord and Yu, 1970a; Koenig, 1972; Chen and Lord, 1974; Yu, 1972) have suggested a correlation between the amide III frequency and the backbone conformation in proteins. The most detailed of these, which has yet to be tested extensively, is that of Lord (1977), who proposed a curvilinear relation between the amide III frequency and the torsional angle, ψ, which measures the rotation of the backbone about the C^{α}HR-CO bond of a given peptide group. The other torsional angle, ϕ, about the NH-$C^{\alpha'}$HR′ bond appears to have less effect on the amide III frequency. Since both angles ψ and ϕ are well defined in many homopolypeptide structures, it is possible to establish points on the curve from such models. In proteins, however, the group R and R′ vary from one peptide link to the next and the actual ϕ, ψ angles depart substantially from their ideal values in the homopolymers. Thus a continuous range of these is possible in proteins, which leads to a continuous range of amide III frequencies in their Raman spectra.

If the contour of the amide III region can be made to yield a conformational distribution in terms of the angle ψ, it will be useful to those concerned with the structures of oligopeptides and proteins incorporated in membranes. In the first place, as is now well recognized, one may be able to say qualitatively, or even semiquantitatively, whether such a protein exists mainly in the helical conformation, the pleated-sheet conformation, some intermediate conformation, or some kind of mixture of these. Secondly, it may be possible to follow conformational changes in such proteins as a function of environmental factors at work upon the membrane: pH, concentrations of various ions, reagents such as detergents, and temperature. The extent to which this kind of information may be obtained in quantitative form from the amide III alone or in combination with the amide I and possibly other characteristic peptide frequencies remains to be seen. Enough is now known, however, to enable substantial qualitative conformational information to be obtained from the Raman spectra of proteins.

As an example of the use of the amide III intensity in the Raman spectrum to follow conformational change in a protein, we cite the variation with temperature of intensity at 1250 cm^{-1} in the spectrum of ribonuclease A (Chen and Lord, 1976). This frequency is intermediate between those of the α-helix and β-pleated-sheet conformations, and its intensity increase with temperature measures changes in peptide geometry from those conformations to others of intermediate values of the ϕ, ψ angles (Fig. 7). The "melting temperature," T_m, taken as the point half-way between the plateaus of the native and denatured forms, is found at 60°C and is in accord with the values obtained by other methods.

Some of the frequencies characteristic of the side groups R in proteins are also useful in revealing the conformations of these groups. As mentioned earlier, however, not all of these frequencies are sensitive to the molecular environment. Among those that do show dependence are the S-S vibrations at 510 cm^{-1} (see Van Wart and Scheraga, 1976) and the tyrosine doublet at 830-850 cm^{-1} (Siamwiza et al. 1975). In the thermal denaturation of ribonuclease A both the frequency and the half-width of the S-S line yield a value of T_m in agreement with that obtained from Figure 7, as does the intensity ratio of the tyrosine doublet (Chen and Lord, 1976). The latter shows in addition that the strongly hydrogen-bonded tyrosine residues, presumed to be numbers 25, 92, and 97, remain strongly bonded until a minimum temperature of about 48°C is reached, at which they begin to detach from their acceptors.

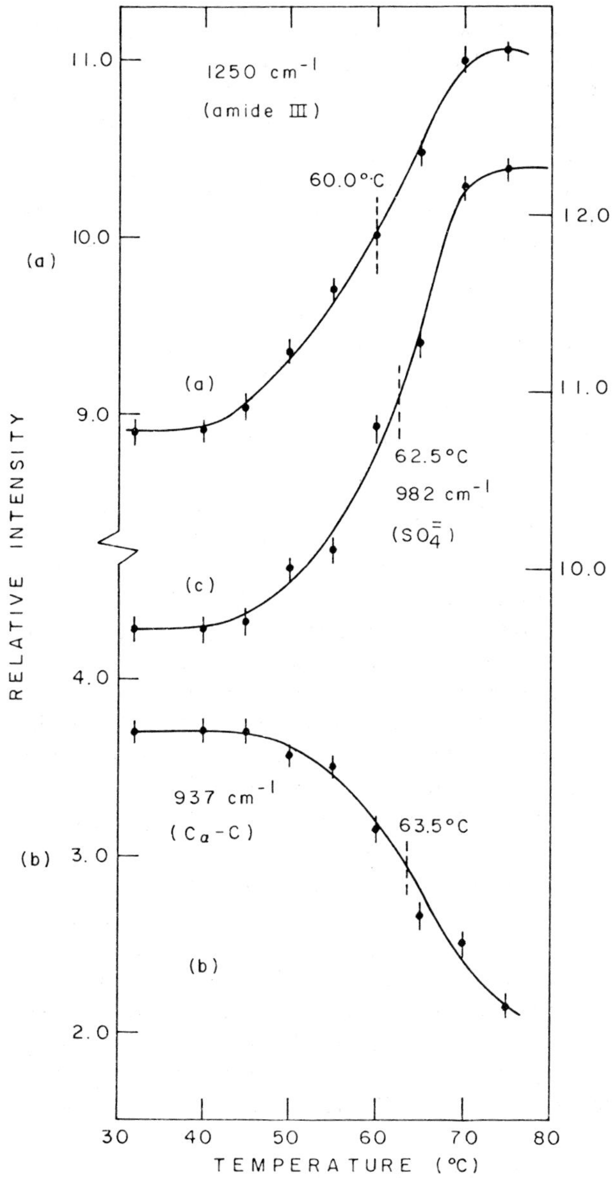

Fig. 7. Temperature dependence of the amide III intensity at 1250 cm^{-1} in 7% aqueous ribonuclease A (*top curve,* a) (Chen and Lord, 1976)

To date little work of the above sort has been done on the conformations of proteins incorporated in membranes. The structure of the protein component of erythrocyte ghosts has been investigated by Bulkin (1972) and later by Lippert et al. (1975), who obtained excellent spectra of those membranes in aqueous suspension (Fig. 8). They used the amide I, amide I ′, and amide III lines to assess the configuration, which was determined to be 40%-55% α-helix and little β-pleated-sheet, the remainder having intermediate structure. The strong lines at 1527 and 1160 cm^{-1} in Figure 8 have now been identified as resonance-enhanced lines due to a small amount of β-carotene present in the sample (see Sect. IV. B. 5).

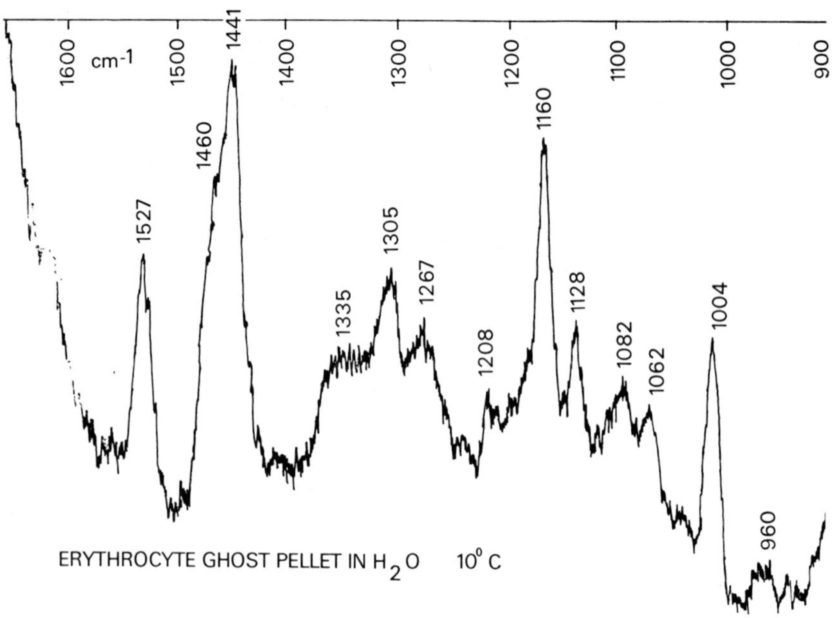

Fig. 8. Raman spectrum of erythrocyte ghosts suspended in water at 10°C (Lippert et al., 1975)

Another study of the conformation of a particular membrane protein is that of Rothschild et al. (1976) on the visual protein opsin. Calf opsin membranes were isolated and their normal (that is, nonresonance) Raman spectra obtained with 514.5 nm excitation after bleaching of the retinal to eliminate the resonance Raman effect. The protein spectrum was overlaid by that of the phospholipid in the membrane, but this did not prevent the measurement of enough of the protein spectrum for the authors to conclude that opsin has mainly an α-helical structure with little if any β-pleated-sheet. A substantial fraction of the tyrosines in the opsin appears to be strongly hydrogen-bonded from the 850:830 intensity ratio of less than unity (this ratio is 1.5 or more when tyrosine is weakly H-bonded; Siamwiza et al., 1975). The phospholipid hydrocarbon chains were said to be in a fluid state at room temperature from the higher intensity of the lipid line at 1082 compared to that at 1068 cm^{-1} (see next section).

Although only a few studies such as those just described have been made thus far, it is clear that both resonance and nonresonance Raman spectroscopy will be valuable in the study of membrane-associated proteins.

B. Raman Spectra of Lipids and Model Membrane Systems

1. Introduction

The most prominent features in the Raman spectra of model membranes arise from vibrations of the lipid hydrocarbon chains. Analysis of the spectroscopic results is simplified by the extensive work available on the vibrational spectra of paraffins as model compounds (Schachtschneider and Snyder, 1963; Snyder and Schachtschneider, 1963; Schaufele and Shimanouchi, 1967; Snyder, 1967; Schaufele, 1968). The sensitivity of Raman spectroscopy to changes of phase has been shown by Schaufele (1968), who obtained spectra for a number of straight chain hydrocarbons in both solid and liquid states. The molecules investigated were of structure $CH_3(CH_2)_{n-2}CH_3$ with n = 13-36. Furthermore, several detailed analyses of the spectrum of polyethylene have been performed (cf. Tasumi et al., 1962; Schaufele and Shimanouchi, 1967), so that the origin of the vibrational frequencies of the all-trans chain conformation present in that system is well understood. In addition, as discussed below, dispersion curves of frequency versus phase angle have been calculated for the polyethylene normal modes (Lippert and Peticolas, 1972, based on Snyder, 1967).

For a polymer with n atoms per translational repeat unit and M repeat units, there are 3nM vibrations which occur in 3n frequency branches. Within each branch, the frequency of each of the M possible modes depends on the relative phase of the nuclear displacements in the other translational unit cells, each mode being characterized by a phase angle ϕ_k:

$$\phi_k = k\pi/M \text{ with } k = 0, 1, 2, \ldots M\text{-}1 \qquad (19)$$

The dispersion curve of a polyethylene zigzag chain is shown in Figure 9.

For an infinite polymer (e.g., polyethylene), only those vibrations with k = 0, that is, those with identical nuclear displacements in each unit cell, are allowed in the Raman spectrum. For segments of chain that are not infinite in length, the selection rule k = 0 is not rigorous and vibrations at other k values – usually k = 1 or 2 – may also be observed. The frequency of any vibration with k = 0 will be a sensitive function of the all-trans segment length present in the CH_2 oligomer. Hence, those vibrations for which there is sufficient variation of frequency with phase angle will be useful as structural probes in membrane systems, in which the hydrocarbon chains undergo gel-liquid crystal phase transitions, and the all-trans segment length alters.

2. The Spectral Regions Below 600 cm^{-1} and 1000-1150 cm^{-1}

Two regions of the spectrum have proved useful for monitoring changes in the effective all-trans segment length in model membrane systems, that below 600 cm^{-1} and that between 1000 and 1150 cm^{-1}. The region below 600 cm^{-1} shows a series of bands (fundamentals and odd-numbered overtones) for short-chain crystalline hydrocarbons whose frequency varies inversely with chain length. The vibration involved is a C-C chain longitudinal acoustical mode (LAM) which has been termed the accordion mode by Schaufele and Shimanouchi (1967) because of its characteristic motion. These authors have fitted the measured Raman frequencies to a power series expansion in terms of the number of CH_2 units in the hydrocarbon chain. These frequencies have been used for quantitative analysis of mixtures of fatty acids in the C_{12}-C_{24} range (Warren and Hooper, 1973). In addition, the LAMs for a variety of sodium alkyl sulfates and potassium aliphatic carboxylates in the solid state and aqueous solution have been observed and analyzed in order to compare the conformation of the hydrocarbon

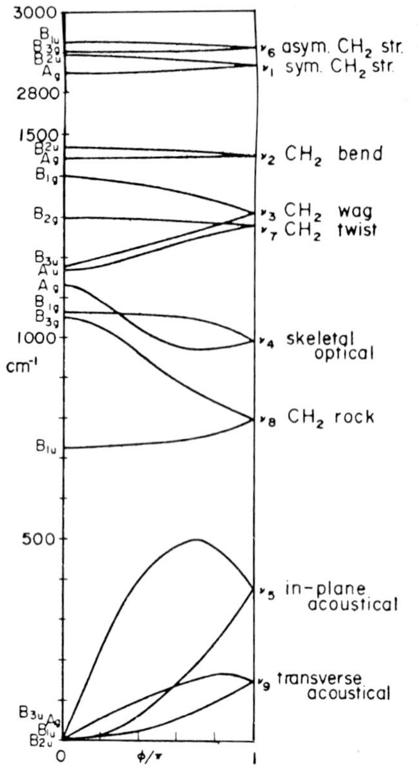

Fig. 9. Phonon dispersion curve of a polyethylene zigzag chain. The Brillouin zone is that of the translational group $-CH_2-CH_2-$. The modes ν_4 and ν_5 are most important for following conformational changes in model membrane systems (Lippert and Peticolas, 1972)

chains in the two states. For chain lengths greater than eight carbon atoms, the presence of gauche conformations in solution is indicated (Okabayashi et al., 1974).

Brown et al. (1973) have reported a Raman spectroscopic study of a sonicated dispersion of dipalmitoylphosphatidylethanolamine. At temperatures below the melting temperature, T_m, a sharp band is observed at 161 cm^{-1}, exactly the same frequency as in palmitic acid. Above T_m the mode broadens and shifts to higher frequencies (Fig. 10), as expected for the shorter effective chain lengths of the many possible gauche structures that can be formed in the liquid-crystal phase. In general there is some difficulty in observing this region of the spectrum, since scattering of the exciting line by turbid suspensions is quite intense and the Raman spectrum at small frequency shifts is often obscured. However, the sensitivity of the spectrum to conformational change (vide infra) and to alterations in the morphology of the crystalline forms of hydrocarbons (cf. Khoury et al., 1973; Olf et al., 1974) indicates that it will be useful in future studies.

The region 1000-1150 cm^{-1} contains vibrations in which alternate carbon atoms move in opposite directions along the chain axis. The first observation of the sensitivity of this region to conformational change in model membrane systems was made by Lippert and Peticolas (1971). They monitored spectral changes that occur in the Raman spectrum during the gel-liquid crystal phase transition of a sonicated suspension of dipalmitoyl lecithin. Below T_m, two strong lines were observed at 1128 cm^{-1} and 1064 cm^{-1} and a weaker one at 1100 cm^{-1} (Fig. 11). Initially, all three vibrations

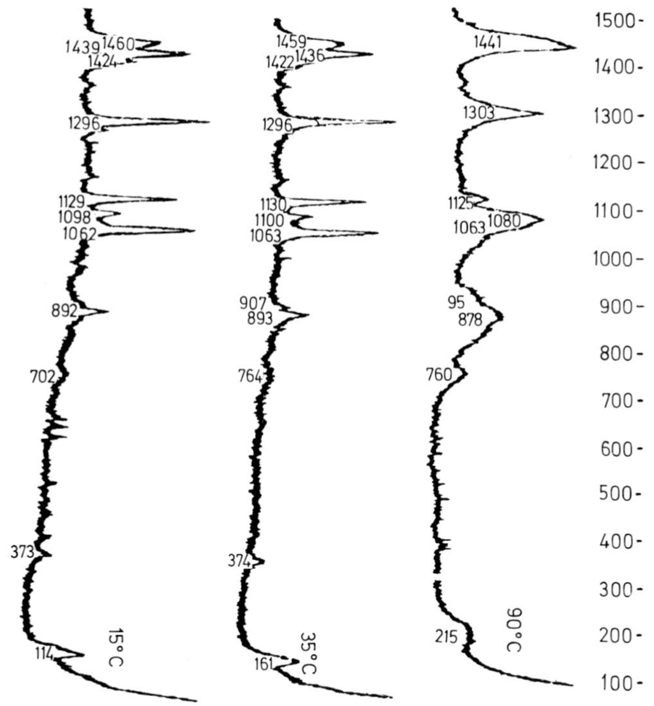

Fig. 10. Raman spectra of dipalmitoylphosphatidylethanolamine above and below T_m (Brown et al., 1973)

were assigned to the aforementioned skeletal optical modes of the hydrocarbon chain. The assignment of the 1100 cm^{-1} band was later changed to a stretching vibration of the PO_2^- group by Lippert and Peticolas (1972), a change confirmed by Mendelsohn et al. (1975a) and Spiker and Levin (1975). No alterations in the overall conclusions were required thereby.

The spectrum of dipalmitoyl lecithin in this region is quite similar to that of polycrystalline n-hexadecane, and appears characteristic of a highly ordered linear hydrocarbon structure. The intensities of the two strong lines were found to decrease abruptly as T_m (41°C) was approached. Above T_m, a line appeared at 1089 cm^{-1} whose intensity increased with temperature. This feature was assigned to C-C vibrations of random liquid-like configurations. The band merged with the PO_2^- symmetric stretching vibration, which shifted downward in frequency from 1100 cm^{-1}. The transition was found to be highly cooperative. Similar results were noted for dipalmitoyl lecithin monohydrate (T_m = 69°C). The addition of cholesterol in a 1:1 mole ratio causes the phase transition to become noncooperative, as does dehydration of the lecithin (Fig. 12). The results indicate that cholesterol rigidifies the lipid above T_m, but fluidizes the close-packed structure below T_m. The role of cholsterol in regulating the fluidity in membrane systems can therefore be directly demonstrated from the Raman spectra (Faiman et al., 1976).

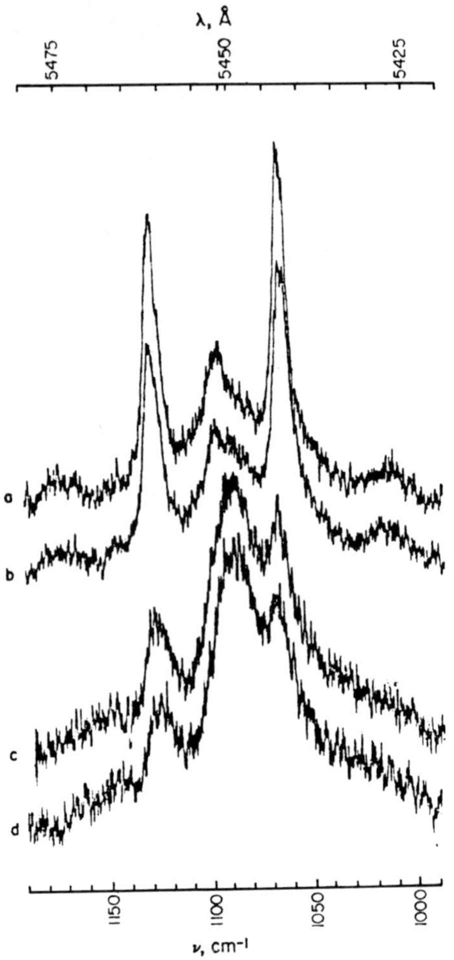

Fig. 11. Raman spectra in the region of 1100 cm^{-1} for sonicates of 20% D, L-dipalmitoyl lecithin in water. (a) At 20°C; (b) 30°C; (c) 40°C; (d) 50°C (Lippert and Peticolas, 1971)

Similar experiments have been performed for egg lecithin and egg-lecithin-cholesterol mixtures above T_m (Mendelsohn, 1972). The Raman spectra showed the hydrocarbon chains to be in random configuration and the cholesterol to cause a marked increase in chain rigidity. Spectra were obtained both from sonicated and unsonicated suspensions and from films cast on an aluminum surface. The temperature dependence of the 1000-1150 cm^{-1} region has been studied in several other phospholipid-water mixtures (Bulkin and Krishnamachari, 1971).

Lippert and Peticolas (1972) have reported Raman spectra of the 1000-1150 cm^{-1} region for a variety of saturated and unsaturated fatty acids. Regular changes in the frequencies of the saturated chains as a function of chain length were noted. For unsaturated molecules containing one double bond it was possible to distinguish vibrations arising from the acid terminal and methyl terminal segments. This observation led to a detailed analysis of the phase transition of sonicated L-α dioleoyl lecithin vesicles. The results showed that below T_m, the interior of the bilayer is more fluid than

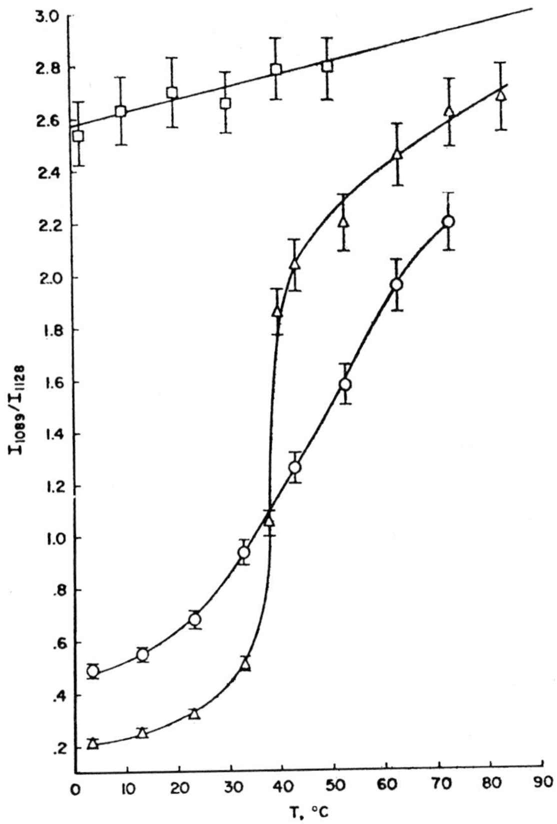

Fig. 12. Temperature dependence of the ratio I_{1089}/I_{1129} for D, L-dipalmitoyl lecithin (DPL). (Δ) Sonicate of 20% DPL in water; (○) sonicate of 20% 1:1 cholesterol: DPL in water; (□) 10% w/w solution of DPL in chloroform (Lippert and Peticolas, 1971)

the region closer to the polar surface. Above T_m, both regions are found to be in a liquid-like conformation.

Gaber and Peticolas (1977) have proposed a quantitative correlation of this spectral region with the phospholipid hydrocarbon-chain conformation. It is assumed that the skeletal optical mode near 1130 cm^{-1} is the sum of intensities from individual all-trans chain segments. In this linearizing assumption the Raman intensity, I_R, is given in terms of the probability of an all-trans n-mer, P(n), by

$$I_R = I_o \sum_{n=8}^{N} nP(n) \quad (20)$$

where Σ P(n) = 1 and n is the number of carbon atoms in the all-trans sequence within a molecule of N carbon atoms. I_o is the intensity per trans unit. The minimal segment for which this description is assumed to apply is n $\geq$ 8. In practice, this limitation must be overlooked as it is not possible to deduce the number of carbon atoms in a given all-trans sequence.

To correlate data from different phospholipid molecules, a longitudinal order parameter, S_{trans}, which is simply a relative measure of the all-trans chain probability (and cannot be related to order parameters derived from ESR or ^{2}H NMR measurements) is defined as:

$$S_{trans} = \frac{(I_{1133}/I_{ref})\text{ observed}}{(I_{1133}/I_{ref})\text{ solid DPPC}} \qquad (21)$$

I_{ref} is usually the band intensity at 1090 cm^{-1}; however for precise work on phosphatidylcholines, Gaber and Peticoles (1977) recommend the use of a head group vibration at 722 cm^{-1}. For purposes of the normalization procedure, it is assumed that all the CH_2 units in solid DPPC are in the trans form. The use of this analysis has been called into question recently (Karvaly and Loshchilova, 1977); however its semiquantitative utility has been shown by comparisons of Raman and ^{2}H NMR results (Gaber and Peticolas, 1977).

3. The C-H Stretching Region

The above spectral regions yield structural information concerned primarily with the conformations of the hydrocarbon chains themselves. On the other hand Larsson (1973) has demonstrated that the region 2750-3200 cm^{-1}, which contains the C-H stretching vibrations of the methyl and methylene groups, is sensitive to intermolecular interactions between the chains. The relative intensities of Raman frequencies in this range vary for different phases of lipid-water systems. In addition, different close packing arrangements for the hydrocarbon chains are distinguishable. The utility of the C-H stretching frequencies for monitoring of chain melting in phosphatidylcholine and phosphatidylethanolamine vesicles has been demonstrated (Brown et al., 1973; Akutsu and Kyogoku, 1975).

Detailed vibrational assignments and the origins of the spectral changes observed on melting have been discussed by several groups of workers. Sunder et al. (1976) have compared the spectra of stearic acid and its 18-d_3 derivative in the solid state and have identified the terminal methyl group C-H stretching vibrations as weak features at 2967, 2954, and 2937 cm^{-1}. Their work confirmed earlier studies of Spiker and Levin (1975) on the phospholipids. Bunow and Levin (1977b) investigated the broad, structure-sensitive feature which appears at 2930 cm^{-1} in phospholipid liquid-crystal phases, and showed that it originates from antisymmetric C-H stretching vibrations of chain methylene groups that are inactive in the Raman spectrum of the all-trans chain conformation. The mode becomes increasingly intense due to loss of symmetry as chain disorder increases. Several other studies of the vibrational assignments in the 2800-3200 cm^{-1} region are more or less consistent with the aforementioned (Bulkin and Krishnan, 1971; Bulkin, 1976; Faiman et al., 1976; Verma and Wallach, 1977).

Three recent studies have attempted to determine quantitatively the origin of the sensitivity of this spectral region to conformational change (Mendelsohn et al., 1975a; Gaber and Peticolas, 1977; Snyder et al., 1978). Mendelsohn et al. found that changes in the C-H stretching region, occurring 100°C below T_m for sphingomyelin, could be identified with thermal motions of the hydrogen atoms and concomitant disorder in their (geometric) positions. Gaber and Peticolas (1977) realized that the spectral parameter used to characterize the C-H stretching region (I_{2885}/I_{2850}) responded both to changes in the number of gauche rotamers in the chains and to alteration in interchain vibrational coupling. On the basis of isotopic dilution experiments with a mixture of hexadecane and perdeuterated hexadecane, they showed that half of the observed variation in I_{2885}/I_{2850} is derived from loss of vibrational coupling. The spectroscopic origin of the phenomenon of vibrational coupling and positional disorder has been elegantly clarified by Snyder et al. (1978). Broad features underlying the antisymmetric C-H stretching mode at 2885 cm^{-1} are responsible for the sensitivity of its intensity to chain packing. The broad band arises from Fermi resonance between the symmetric methylene C-H stretching mode and appropriate binary combinations involving methylene bending modes near 1450 cm^{-1}. Due to the behavior of the hydrocarbon chain as a line group, more than one binary combination is possible and broad spectral features result. For a detailed discussion, the reader is referred to the original study.

The effect of sonication on the hydrocarbon chain conformation in model membrane systems has recently been probed in three independent studies by Raman spectroscopy. The technique of sonication is widely used to transform the large coarse aggregates which form initially on phospholipid suspension in water into smaller (and often single-walled) vesicles convenient for the study of membrane transport phenomena. Initial Raman studies of egg-lecithin and dipalmitoyl lecithin multilayers subject to a gentle bath type of sonication (Mendelsohn et al., 1976a; Spiker and Levin, 1976b) revealed changes in the C-H stretching regions of the molecule due to disruptions of lateral interactions in the hydrocarbon chains. This is a result of the small radius of curvature and imperfect chain packing in vesicle systems. No changes were observed in the 1100 cm^{-1} region of egg lecithin, which suggests that for a system already in the liquid-crystal state, sonication does not alter the trans/gauche population ratio.

The effect of sonication on the trans/gauche population ratio in phospholipid gel phases depends significantly on the type of vesicle preparation used and the particle size of the resultant suspension. Mendelsohn et al. (1976a) found minimal changes for DPPC at 23°C subjected to bath sonication, while Spiker and Levin (1976a) observed a slight broadening in the melting profiles for vesicles compared with multilayers.

Using probe sonication, Gaber and Peticolas (1977) prepared single-walled vesicles of uniform size according to the method of Huang (1969). In addition to the loss of interchain order observed by the other workers, they noted a significant decrease in the number of trans units in the phospholipid chains below T_m. In sonicated vesicles the main DPPC transition was shifted to a value of 37°C from 41.5°C in multilamellar systems. In summary, the Raman studies indicate that interchain interactions are strongly perturbed by sonication procedures of even the gentlest type, while significant changes in trans/gauche ratios may be induced in gel phase hydrocarbons by the production of single-walled bilayer species.

The sensitivity of the Raman technique to slight structural changes in the phospholipid gel or solid state was further demonstrated by Mendelsohn et al. (1975a). Relatively high resolution spectra of sphingomyelin show striking changes in the 1100 cm^{-1} region which are consistent with the formation of gauche rotamers in the solid state well before the onset of melting. A wax-like structure was postulated. In an important contribution, Yellin and Levin (1977a) examined multilayers of dimyristoyl, dipalmitoyl, and distearoyl lecithin below T_m. Again, the formation of gauche rotamers well below the gel-liquid crystal phase transition temperature was noted. It was found possible to estimate enthalpy differences between chains in the all-trans chain conformation and chains containing gauche rotamers. It was further possible (Yellin and Levin, 1977b) to show that the number of CH_2 groups in the gauche conformation per lipid molecule was two for the dipalmitoyl and dimyristoyl lecithin gels and six for the distearoyl derivative. It is evident from the above studies that a wealth of detail is available from Raman spectra of simple phospholipid systems.

The most detailed study of phospholipid phase transitions in one-component lipid systems has been reported by Gaber et al. (1978a, b, c) for dipalmitoyl lecithin using Raman difference spectroscopy. Detailed conformational information about the molecule was obtained for each of three well-defined temperature ranges. Below the pretransition temperature, the occurrence of certain spectral features in the 2860 cm^{-1} range led to the suggestion of the triclinic crystal structure. Between the pretransition and the main chain melt, approximately 1-2 gauche rotamers per chain were formed and the lateral interactions were disrupted between chains. It was further suggested that the gauche rotamers are highly restricted in this temperature domain, and are found only at the ends of long, all-trans segments. Above T_m, the number of gauche rotamers is drastically increased.

Lis et al. (1975) have examined structural changes induced in lipid bilayers by various ions. From the intensity patterns in the 1100 cm^{-1} region it was inferred that dipositive ions decrease the proportions of gauche rotamers in lecithin-water systems, the effect being in the order $Ba^{2+} < Mg^{2+} < Cd^{2+}$. Unipositive cations as well as anions exhibit little effect. In a further study of conformational changes induced by external agents, Szalontai (1976) has shown that benzene is able to promote a gel-liquid transition in dipalmitoyl lecithin multilayers.

4. Lipid-Protein Interactions

Lipid-protein interactions in model systems have been studied by various workers. Chen et al. (1974) have obtained the Raman spectrum of a coprecipitate of sodium dodecyl sulphate (SDS) and lysozyme. The spectrum of pure solid SDS is that of a rigid all-trans structure of the molecule. Upon coprecipitation with the enzyme, spectral changes were observed that indicate the formation of a number of bends in the hydrocarbon chain (see Fig. 13). It appears that protein-SDS interaction requires an alteration in SDS geometry and that the backbone conformation of the protein is also substantially altered by SDS coprecipitation. The helical regions of the lysozyme are disrupted and random-coil regions appear to predominate in the coprecipitate.

Larsson and Rand (1973) have reported spectra of an insulin-sodium hendecylphosphate coprecipitate. Changes in the 1000-1150 cm^{-1} region in the spectrum of the lipid were observed that are consistent with the formation of gauche structures in the hydrocarbon chains upon coprecipitation. A tentative model was suggested for the interaction in which chain-protein contacts predominate. Comparison of the spectra of sodium hendecylphosphate-insulin and sodium hexadecylphosphate-insulin showed them to be similar in the C-C stretching region but significantly different in the C-H stretching regions. It was postulated that the sodium hexadecylphosphate chains are surrounded by other chains and that they crystallize when the lipid-protein complex is dried, in contrast to the sodium hendecylphosphate-insulin interaction.

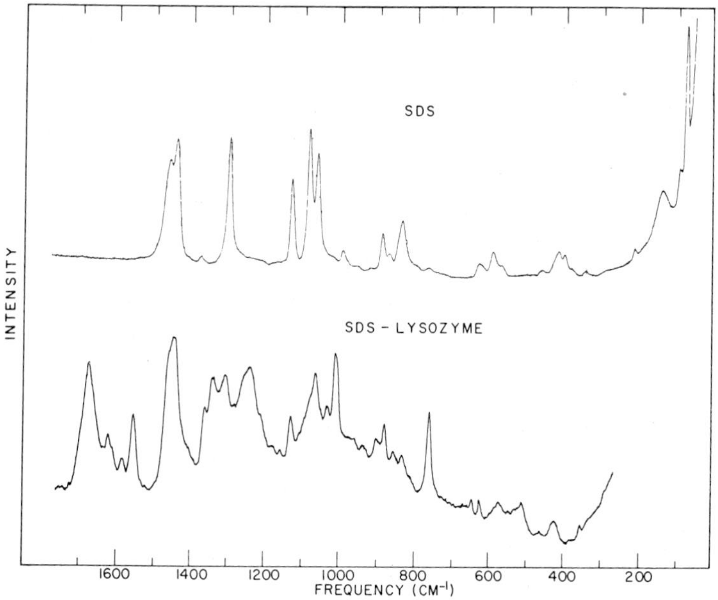

Fig. 13. Raman spectra of solid sodium dodecyl sulfate (SDS) and a precipitated SDS-lysozyme complex (Chen et al., 1974)

Raman studies have been reported for several other systems that serve as prototypes of lipid-protein interaction. Lis et al. (1976a) have examined the effect of the ionophores alamethicin and valinomycin on membrane organization. Both peptides tend to fluidize dimyristoyl lecithin but produce little alteration in chain conformation of dipalmitoyl lecithin. A detailed temperature profile would serve as a useful adjunct to this work. On the other hand Verma and Wallach (1976a) suggest that the peptide mellitin tends to rigidify dipalmitoyl lecithin multilayers.

Lis et al. (1975) have examined the effect on lipid structure of a variety of intrinsic and extrinsic membrane proteins. Human fibrinogen and bovine serum albumin induce intensity changes in dimyristoyl lecithin, a fact which suggests structural modification in the lipids. Additional alterations were noted for cytochrome c and cytochrome oxidase but no detailed analysis was undertaken.

Several further studies have been reported with reconstitututed lipid-protein systems. Curatolo et al. (1978) obtained spectra of myelin apoprotein proteolipid reconstituted with dimyristoyl lecithin and egg lecithin. Spectra in the C-H stretching region showed that the DMPC hydrocarbon chains possessed some residual "solid" character at temperatures as high as 18°C above T_m for the lipid. When the apoprotein was combined with egg lecithin, the main chain melting was still observed below 0°C. However, a new structural transition characteristic of the lipid-protein complex was observed at 12°C. The Raman and supporting calorimetric results indicated that some of the phospholipid was sequestered by the protein into regions that can undergo a cooperative thermal transition higher than the normal and presumably consists of saturated chains.

Most recently, Mendelsohn and Taraschi (1979) have examined complexes of the polypeptide hormone glucagon with DMPC. This lipid-protein complex is unusual in that it forms a soluble species only at temperatures where the phospholipid is in its gel state and precipitates out of solution above the phospholipid T_m. Spectra for the soluble complex showed that, at 7.5°C, lateral interactions between phospholipid molecules were completely disrupted and the Raman spectrum in the C-H stretching region resembled that of sonicated DMPC vesicles, even though sonication is not used in the preparation of the complex. Spectra in the C-C stretching region indicated that at 7.5°C, an additional 2 gauche rotamers per phospholipid hydrocarbon chain occurred in the complex, as compared with DMPC in either multilamellar or small unilamellar vesicle form. However, the chains appeared to be immobilized in the complex, as the noncooperative formation of gauche rotamers normally noted below T_m (e.g., Yellin and Levin, 1977a) for the pure phospholipids is not observed in the complex.

In a related study, Chapman et al. (1977) have investigated the interactions of gramicidin A with DMPC and DPPC. Raman spectroscopic studies showed that above the transition temperature for the lipid, the gramicidin A causes a marked decrease in the number of gauche rotamers of lipid hydrocarbon chains. Below the lipid T_m the data suggest a slight fluidization of the chains. Similar conclusions were reached by Weidekamm et al. (1977), who showed in addition that a mixture of lecithin and cholesterol is not affected by the peptide.

A serious impediment to using the C-C and C-H stretching vibrations as probes of lipid conformation in model systems arises from the scatttering by nonlipid components in these spectral regions. An accurate determination of the lipid component is thus difficult. Mendelsohn et al. (1976b) indicated a way to avoid this difficulty by inserting a completely deuterated fatty acid into a model membrane system and following the C-D stretching vibrations during the membrane gel-liquid crystal transition. The linewidth of the C-D stretching vibrations of the bound fatty acid was shown to be a sensitive probe of membrane hydrocarbon chain order (Fig. 14). Since the C-D stretching vibrations occur in a spectral region free of interference from nonlipid components, lipid conformation can be unambiguously monitored in multicomponent systems.

The availability of deuterated phospholipids has led to a variety of interesting extensions of the above approach. Gaber et al. (1978b, c) and Bunow and Levin (1977c) assigned spectral features of 1,2-dipalmitoyl-d_{62} phosphatidylcholine, and two groups (Gaber et al., 1978b,c; Sunder et al., 1978) obtained melting curves for several features in the Raman spectrum of the molecule. The results indicate that deuterated phospholipids provide spectral probes in the form of isolated conformation-sensitive vibrations, especially in the C-D stretching region. It is therefore feasible to monitor the molecular order and hydrocarbon chain conformation of each component in mixed phospholipid systems where one component is deuterated.

The first application along these lines was reported by Mendelsohn and Maisano (1978), in studies of DMPC (and its -d_{54} derivative) with DSPC. Two distinct melting regions were observed for the 1:1 mole ratio mixture of DMPC/DSPC. Use of DMPC-d_{54} as one component of the binary mixture permitted identification of the lower transition (22°C) with the melting primarily of the shorter chain component. An interesting observation in the work was the response of the C-H stretching vibrations

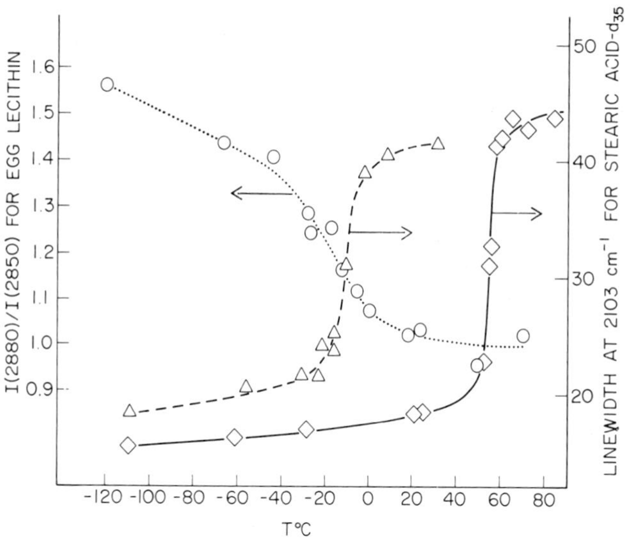

Fig. 14. Temperature dependence of linewidth (full width at half maximum intensity) of the line at 2103 cm^{-1} in the Raman spectrum of stearic acid-d_{35}, pure (◇) and in lecithin multilayers (Δ). Temperature dependence of intensity ratio of lines at 2880 and 2850 cm^{-1} in the spectrum of egg lecithin multilayers (○) (Mendelsohn et al., 1976b)

of the distearoyl component to the melting of the dimyristoyl chains, a consequence of restoration of lateral interactions in the former upon the change of phase of most of the latter component. The general conclusion from the work is that lateral phase separation processes can be directly monitored in a complex system. The overall results were in good agreement with calorimetric studies. Further experiments along these lines have been reported for the DPPC-d_{62}/DPPE system (Mendelsohn and Taraschi, 1978). The conformation of each phospholipid was monitored in a binary mixture and in a ternary mixture containing cholesterol as the third component. The addition of cholesterol produced additional phases in the system. Recent studies (R. Mendelsohn and C. Koch, unpublished work) have led to the production of phase diagrams for binary lipid mixtures as well as determination of the conformational variation of each component throughout the phase separation region.

A novel approach to the use of deuterated phospholipids has been taken by Gaber et al. (1978c). The compounds 1-palmitoyl, 2-palmitoyl-d_{31}-3-sn-phosphatidylcholine, and 1-palmitoyl-d_{31}, 2 palmitoyl-3-sn-phosphatidylcholine were synthesized. The conformations of chains 1 and 2 of each molecule were simultaneously and independently monitored by Raman spectroscopy. Differences in the spectral characteristics of the two compounds at a given temperature were attributed to nonequivalent conformations of the fatty acyl chains at positions 1 and 2. It was found that below the pretransition, the conformation of chain 2 departs slightly more from the all-trans structure than chain 1. In general, the use of deuterated phospholipids promises to be of considerable importance in studies of lipid-protein interaction, where the conformation of lipid can be determined free from interference of overlapped protein molecules.

5. Naturally Occurring Membranes

The acquisition of Raman data from naturally occurring membranes presents some technical difficulties which have yet to be completely solved. Typical of the problems is the occurrence of trace amounts of fluorescent components in most biomembranes. Another problem not widely appreciated has been described recently by Forrest (1976, 1978). In a detailed study of fat-globule membranes isolated from whole milk, he demonstrated that the Raman data are strongly dependent on the thermal history of the sample. The rates of cooling or heating a membrane sample were found to have a remarkable influence on the ordering of the acyl chains. It is clear from this work that the effect of temperature on the Raman spectrum of a natural membrane must be investigated before reliable conclusions can be drawn from any spectrum measured at a single temperature. Despite these difficulties several naturally occurring membranes have been examined by Raman scattering. It seems fair to note that the signal-to-noise ratios obtained thus far in all but a few cases are such that the conclusions drawn must be considered tentative.

The most thoroughly studied system for which Raman spectra have been reported to date is that of hemoglobin-free erythrocyte ghosts, as discussed above in Section IV. A. 3. In addition to their study of the protein component of this system, Lippert et al. (1975) showed from an analysis of the 1000-1150 cm^{-1} region that the phospholipid component has about 60% of its hydrocarbon chains in the all-trans conformation. Interestingly, their spectra contained intense resonance-enhanced bands due to β-carotene (Fig. 8) at 1160 and 1527 cm^{-1}. The presence of this molecule was confirmed subsequently by Verma and Wallach (1975), who also attempted to use the intense C-C and C=C stretching bands of the carotenoid as a structural probe. They further studied the temperature dependence of the C-H stretching region and were able to observe the main gel-liquid crystal transition of the lipids at about $-8°C$ (Verma and Wallach, 1976b, c). They also reported an irreversible conformational change at 40°C, which was attributed to alterations at sites of lipid-protein interaction.

Several other membrane systems have been examined by Raman spectroscopy. Milanovich et al. (1976) have investigated sarcoplasmic reticulum membranes and concluded that the lipids are in a more fluid state than those in erythrocyte ghosts. However, no study of the temperature dependence of the Raman spectrum was made, so that detailed structural information is not available. Finally, Schmidt-Ullrich et al. (1975, 1976) have obtained data on plasma membranes from resting rabbit thymocytes and from cells mitogenically stimulated with concanavalin A. Several differences were seen in the two data sets, apparently corresponding to changes in lipid structure.

C. Chromophoric Systems Studied by Resonance Raman Spectroscopy

As discussed above in Section III, resonance enhancement of the intensity of Raman lines in a chromophoric system may occur when the Raman spectrum is excited by radiation whose wavelength lies within a molecular absorption band. Since the intensities of Raman lines due to vibrations coupled to the electronic transition of the chromophore may be several orders of magnitude greater than those of nonresonant lines, the effect provides a structural probe at biologically relevant concentrations (10^{-3}-10^{-7} M). The resonance Raman technique is clearly useful for a wide variety of biological applications, since the structure of a particular region of a biomolecule (the immediate vicinity of the chromophore) may be selectively probed. Lines that are not enhanced by resonance get lost in the background noise. The theory of resonance Raman scattering has been considered by several authors (Koningstein, 1972; Behringer, 1974), and applications to biological systems have been reviewed (Lewis and Spoonhower, 1974, Spiro, 1974a, b, 1975a, b; Carey, 1978).

The potential of resonance Raman scattering for the study of biological systems was first demonstrated by Rimai and co-workers (Gill et al., 1970) in their investigations of molecules related to β-carotene. They observed resonance-enhanced vibrations arising primarily from C=C and C-C stretching modes of the polyene chains in live tissues of plants containing these chains (Fig. 15). The spectra of the chromophores in the live tissue differed from those in solution, showing that the technique is sensitive to changes in the environment of the chromophore.

Detailed studies have been carried out on a series of molecules related to vitamin A aldehyde (retinal). The most intense vibrations of the spectrum are observed near 1550 and 1150 cm^{-1} and, as in the carotenoids, arise from C=C and C-C stretching respectively. The various isomers of retinal (9-, 11-, and 13-cis and all-trans) are easily distinguished through their resonance Raman spectra (Gill et al., 1971; Heyde at al., 1971; Rimai et al., 1971, Rimai et al., 1973; Cookingham et al., 1976). The biological importance of retinal derives in part from the fact that its 11-cis isomer complexes with a protein, opsin, to form the basic photoreceptor pigment in mammalian systems, rhodopsin. In rhodopsin, the retinal is covalently bound to the protein as a Schiff's base. An additional noncovalent interaction occurs between the retinal and opsin which perturbs the absorption spectrum of the polyene from 380 nm in retinal to 500 nm in bovine rhodopsin.

Several resonance Raman studies have been reported for rhodopsin and its bacterial counterpart, bacteriorhodopsin (Lewis et al., 1973; Mendelsohn, 1973, 1976; Lewis et al., 1974; Oseroff and Callender, 1974; Callender et al., 1976). The noncovalent interaction between retinal and opsin is shown in the spectra by a lowering of the C=C frequency in the polyene, in direct proportion to the magnitude of the shift to lower frequency in the visible absorption spectra of the pigments. The Raman data also demonstrate that retinal and opsin are joined in situ through a protonated Schiff's base. It has further been possible to observe Raman spectra arising from particular intermediates in the rhodopsin bleaching cycle (Lewis, 1976; Mathies et al., 1976, Mendelsohn, 1976). This approach may permit evaluation of any slight conformational changes that occur in these chromophoric systems during the photocycle.

Recent technical developments have led to the observation of time-resolved resonance Raman spectra. These have been of use in the study of intermediates in the bacteriorhodopsin system on the millisecond (Marcus and Lewis, 1977; Terner et al., 1977) and even nanosecond scales (Campion et al., 1977). It was possible to show that deprotonation of the Schiff's base in bR 570 occurs before the formation of the bM 412 intermediate.

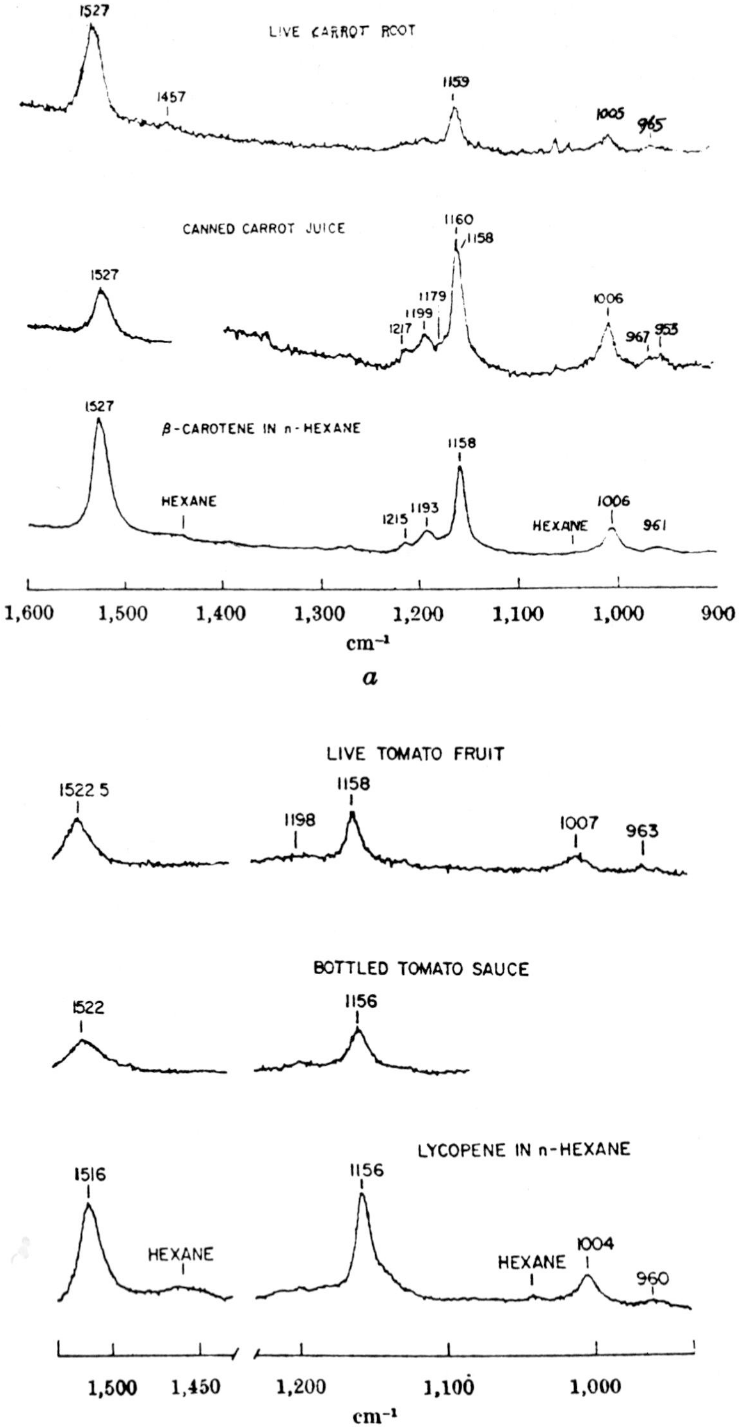

Fig. 15. Resonance Raman spectra of β-carotene and lycopene from different sources (Gill et al., 1970)

The resonance Raman effect has been extensively used to probe the structure of the chromophore in heme proteins (Brunner et al., 1972; Spiro and Strekas, 1972a, b, 1974; Strekas and Spiro, 1972, 1973, 1974; Brunner, 1973; Brunner and Sussner, 1973; Collins et al., 1973; Nestor and Spiro, 1973; Pezolet et al., 1973; Strekas et al., 1973; Yamamoto et al., 1973; Loehr et al., 1974; Sussner et al., 1974; Adar, 1975; Kitagawa et al., 1975a; Felton et al., 1976; for reviews see Spiro, 1975a, b; Spiro and Stein, 1977). Spectra have been reported for a variety of derivatives of hemoglobin and cytochrome c, in which most of the observed vibrations arise from the iron complex of protoporphyrin IX (heme).

The electronic spectra of most heme proteins and metalloporphyrins from which the resonant Raman intensity is derived display a similar pattern (Gouterman, 1959). Very intense absorption (the Soret band) appears at about 400 nm, while a pair of weaker bands, termed α and β in order of increasing frequency, are observed between 500 and 600 nm. Different sets of vibrational modes of the heme moiety are brought into resonance, depending upon whether laser excitation is in the Soret or α-β region. If D_{4h} (square planar) point group symmetry is assumed for the heme chromophore, a detailed analysis (Spiro and Strekas, 1972b; Pezolet et al., 1973) shows that totally symmetric A_{1g} modes appear through excitation in the Soret region, while nontotally symmetric B_{1g}, B_{2g}, and A_{2g} vibrations appear through excitation in the α-β region. The A_{2g} vibrations, which are inactive in the normal Raman effect, can be identified from measurements of the depolarization ratio ρ. The A_{2g} vibrations have values for ρ of infinity, whereas in the normal Raman effect, values greater than 3/4 are not allowed (Sect. III. A. 3). The A_{1g} modes can be distinguished from the B_{1g} or B_{2g}, as ρ for A_{1g} is expected at ~1/8 and for the latter pair ρ = 3/4.

The complexity of the resonance Raman spectra of various hemoglobin derivatives is illustrated in Figure 16, where the spectra of oxy-, carboxy-, deoxy-, azido-, and aquomethemoglobin, excited at 568.2 nm, are shown. An analysis of the spectral features has been reported by Spiro and Strekas (1974), and correlations of resonance Raman bands with heme structure have been made for the set of derivatives studied. Several vibrations of the porphyrin skeleton in the range 1000-1650 cm^{-1} are sensitive to changes in the spin and oxidation state of the central iron atom. The spectra of molecules with known spin and oxidation states have been used to identify unambiguously the state of oxyhemoglobin as low spin Fe^{3+}-O_2^- (Yamamoto et al., 1973; Spiro and Strekas, 1974). In addition, the structure of the heme in cytochrome c′ from *Rhodopseudomonas palustris* has been shown to be planar with the spin state of the central iron = 3/2 (Strekas and Spiro, 1974). Raman data have also been obtained for hemoglobin derivatives in which the central iron atom has been replaced by cobalt (Woodruff et al., 1974). An upper limit for the displacement of the cobalt atom out of the ring plane was given as 0.15 A.

Preliminary investigations of the effect of protein on heme structure have been reported by Salmeen et al. (1973) and Rimai et al. (1975). Spectral differences were observed between the carbon monoxide complexes of myoglobin and hemoglobin which were attributed to this effect. Finally, various hydroperoxidases were examined by Felton et al. (1976). Evidence was presented in support of the hypothesis that the iron atom is coplanar with the porphyrin ring in these molecules.

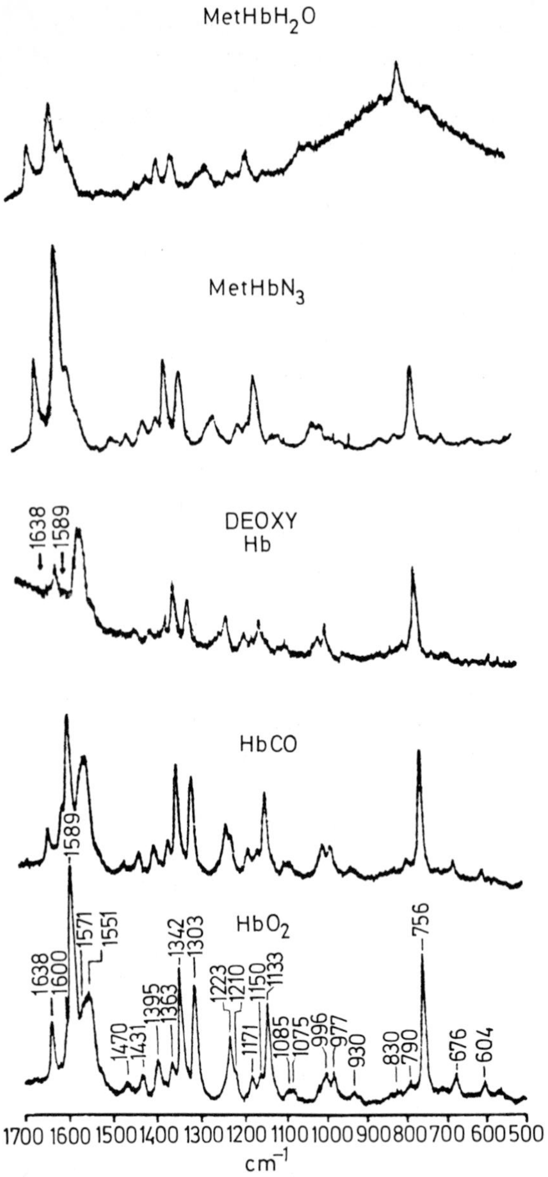

Fig. 16. Resonance Raman spectra of various derivatives of human hemoglobin. Excitation was by 568.2 nm at aqueous concentrations of about 10^{-4} M Fe (Strekas and Spiro, 1972)

To obtain a detailed understanding of the heme protein resonance Raman spectra, several studies of simpler protein-free, metallo and nonmetallo porphyrins have been undertaken (Burger et al., 1970). The effect of asymmetric peripheral substituents in reducing the symmetry of the chromophore from its idealised D_{4h} has been studied for the Ni, Co, and Cu derivatives of mesoporphyrin IX dimethyl ester (Verma et al., 1974). In addition, the effect of changing the peripheral substituent pattern has been probed in a study of metal-free porphin and a variety of derivatives with substituted side chains (Plus and Lutz, 1974; Verma and Bernstein, 1974a, b, c; Kitagawa et al., 1975b). The utility of resonance Raman spectroscopy in qualitative analysis has been demonstrated by studies of the closely related position isomers Cu-etioporphyrin I, II III, and IV, which were easily distinguished by means of their spectra (Mendelsohn et al., 1975b; Sunder et al., 1975). Finally, the spectra of hemin and its imidazole complex have been studied as a model for the hemoglobin chromophore itself (Verma and Bernstein, 1974b).

The structural origin of the frequency shifts in those bands which serve as spin and/or oxidation state markers has been variously interpreted as arising from porphyrin doming (Stein et al., 1975) or enlargement of the central porphyrin core (Yu, 1977). Recent chemical applications of resonance Raman studies in this area include determination of pH-induced structural alterations in *Rhodospirullum rubrum* (Kitagawa et al., 1977) and observation of two distinct species of ferric myoglobin at alkaline pH (Ozaki et al., 1976).

Several systems with chromophores related to the porphyrin structure have been investigated by resonance Raman spectroscopy. Spectra of spinach chloroplasts have been described for which variation of the excitation wavelength leads to selective enhancement of chlorophyll a, chlorophyll b, or the carotenoid constituents (Lutz and Breton, 1973). Selective enhancement of the Raman spectra of bacteriochlorophyll a and bacteriopheophytin a was observed in reaction centers of *Rhodopseudomonas spheroides* (Lutz and Kleo, 1976). The interactions of the various components in the reaction centers could be monitored, although no detailed interpretations of the complex spectra were undertaken. Spectra have also been reported for vitamin B_{12} and several derivatives in which the observed vibrations arise from the corrin ring system (George and Mendelsohn, 1973; Mayer et al., 1973; Wozniak and Spiro, 1973).

The resonance Raman technique is potentially useful as a structural probe of the metal binding site in nonheme metallo proteins. The mode of binding of molecular oxygen to two molecules, hemocyanin with a Cu(II) site, and hemerythrin with a central iron atom, has been investigated (Dunn et al., 1973, 1975; Freedman et al., 1975). In each study the O_2 stretching vibration was observed and isotopic shifts from $^{16}O_2$ to $^{18}O_2$ in the metal-oxygen stretching vibrations of the complexes were used to confirm the structure as an O_2^- peroxide type of linkage to the central metal. The oxygen binding sites have been further probed by use of ^{16}O-^{18}O binding. Spectra for the mixed isotope bound to hemerythrin show that the two oxygen atoms are in different environments (Kurtz et al., 1976).

Siiman et al. (1974, 1976) and Miskowski et al. (1975) have obtained resonance Raman spectra of several copper-containing proteins (laccase, stellacyanin and ceruloplasmin) which possess intense visible absorption bands ("blue" copper proteins). These data were interpreted by comparison with the Raman spectra of copper ovotransferrin and of amino acids and some of their copper complexes. Low-lying vibrations in the 250-540 cm^{-1} region were identified as arising from stretching modes of Cu-O or Cu-N bonds. Studies with model compounds suggested that amide nitrogen and carbonyl oxygen of the protein backbone are coordinated to the central copper site.

Amino acid vibrations of residues complexed to the central iron atom have been observed for Fe(III) ovotransferrin and Fe(III) human serum transferrin. Resonance-enhanced modes arising from the p-hydroxy phenyl ring vibrations of tyrosine or the imidazole ring of histidine attached to the protein chain were noted (Tomimatsu et al., 1973; Carey and Young, 1974; Gaber et al., 1974). Finally resonance Raman spectra have been obtained for the protein ferredoxin, which contains an iron-sulphur cluster (Tang et al., 1975). Evidence was found for distortion of the Fe_4S_4 structure from the tetrahedral geometry observed in simpler model systems.

An inherent difficulty with the resonance Raman technique is the requirement that suitable chromophoric residues be present in the system of interest. While new laser sources are steadily extending the spectral region available for investigation (Sect. II.A.1), a different approach to the problem has been developed, especially by Carey and co-workers. This approach is based on changes observed in the resonance Raman spectra of small colored molecules when they bind to proteins. Such reporter molecules are selected to yield structural information about the nature of the macromolecular binding site. The technique appears to have general utility for providing an appropriate chromophore to a macromolecular system lacking one.

The sensitivity of the technique was demonstrated in studies on the binding of methyl orange to bovine serum albumin, where the dye molecule was shown to be bound in a hydrophobic site on the protein (Carey et al., 1972). Further experiments have been reported on the nature of the interaction between 2, 4-dinitrophenyl haptens and rabbit antibodies. Binding by the antibody of the azo haptens causes many changes in the hapten Raman spectra. The C-N bonds of the azo group appear to twist upon binding, and the double bond character of the N-O bonds increases (Carey et al., 1973).

The detailed nature of the information available is illustrated in a study of the binding of aromatic sulfonamides to the metallo-protein zinc carbonic anhydrase (Kumar et al., 1974). Changes in the structure of the sulfonamide upon binding could be explained by comparison of the spectra of the bound molecules with the free sulfonamide and the free ionized form of the sulfonamide (Fig. 17). The close resemblance of the spectra in the region 1350-1450 cm^{-1} between the bound and ionized forms suggests that the bound sulfonamide group is present in the ionized state. Carey and co-workers (Carey and Schneider, 1974, 1976; Carey et al., 1976) have also introduced resonance Raman labels at the active site of the enzymes chymotrypsin and papain. Spectra of the active acyl enzyme complex of papain differ markedly from those of the free substrate or the product of enzymatic hydrolysis (Fig. 18). The spectral data imply a structural change in the acyl enzyme which precedes the rate-determining step in the reaction. The band at 1570 cm^{-1} appearing in the enzyme-substrate complex has been attributed to a mode arising from a conjugated double bond system (Carey et al., 1978). A series of model compounds has been synthesized which mimic the spectral properties of the enzyme-substrate complex. It is proposed that a highly polarized electron system is set up at the active site of the enzyme. The entire area of resonance Raman labels has been reviewed recently by Carey (1978), where further details of this extremely useful application of the Raman effect are to be found.

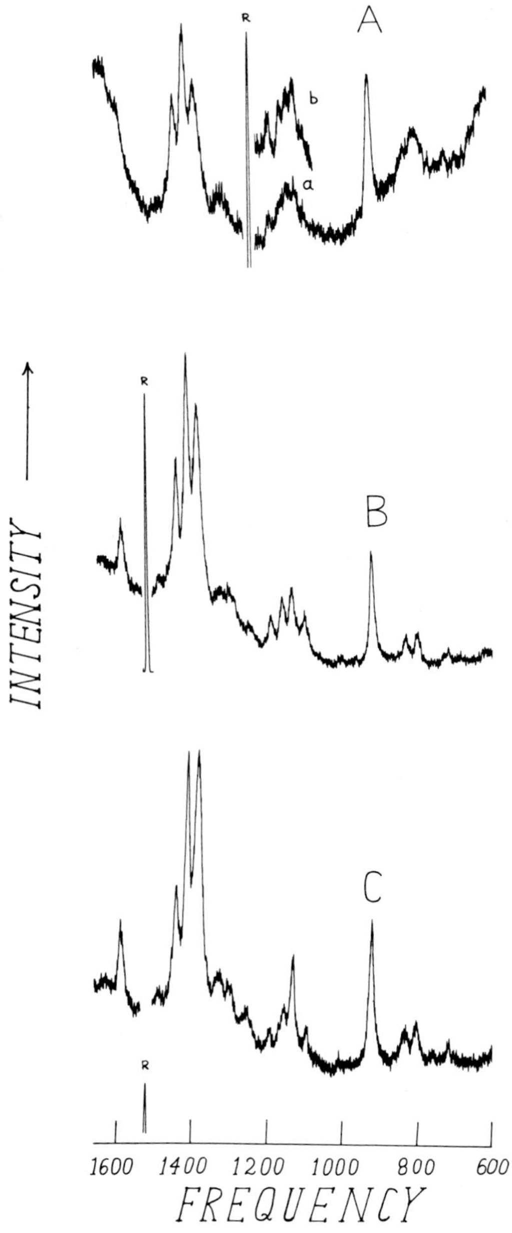

Fig. 17 a-c. Resonance Raman spectra of 4-sulfonamido-4′-hydroxyazobenzene (441.6 nm excitation). (A) Bound to Zn carbonic anhydrase at pH 8 at two different sulfonamide concentrations: (a) 1.6×10^{-5} and (b) 5×10^{-5} M, with respective enzyme concentrations 3.5 and $6.6 \times 10.^{-5}$ M. (B) Anionic form at 3.9×10^{-5} M, pH 13. Free sulfonamide at 11×10^{-5} M, pH 8.2 (Kumar et al., 1974)

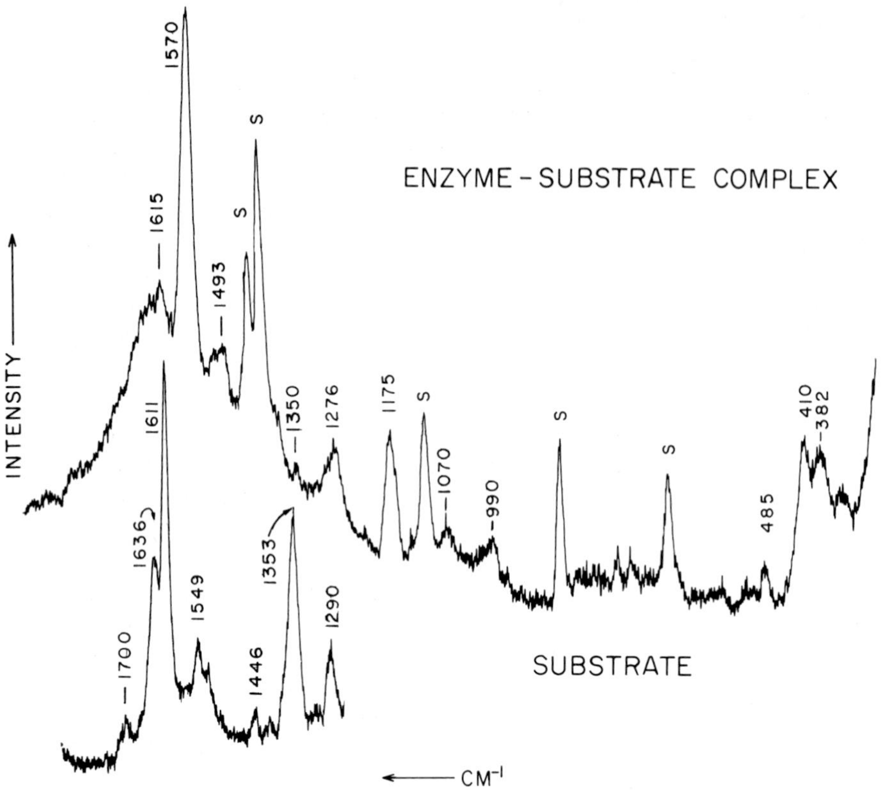

Fig. 18. Resonance Raman spectra of 4-dimethylamino-3-nitro (α-benzamido)cinnamoyl-papain complex (*top*) and acyl substrate alone (*bottom*). S: solvent bands (due to residual dimethylformamide). Note absence of the strong substrate bands at 1631, 1611, and 1353 cm^{-1} from the top spectrum and the new band at 1570 cm^{-1} (Carey et al., 1976)

V. Future Trends

As the utilization of Raman spectroscopic techniques in membrane research goes through its present rapid expansion, two kinds of trends are evident. The first consists of advances in Raman technology that will permit the investigator to obtain suitable spectra under even more demanding sample requirements. The second is based on developments in membrane research that will provide improved isolation, purification and characterization of the important structural components of biological membranes. The most successful applications of Raman scattering to problems of membrane structure and function should result from the interaction of these two trends, that is, from collaboration between the membrane biologist and the spectroscopist. The former will provide guidance as to the urgent problems as well as sample material that will yield meaningful spectra, while the latter's contribution will be the technical expertise and understanding of molecular spectra essential for the acquisition and interpretation of the spectra.

A. Advances in Raman Technology

Several groups, particularly Delhaye and co-workers (Bridoux et al., 1974; Delhaye, 1976) have demonstrated the feasibility of obtaining Raman spectra in times of a millisecond or less. The techniques, which use image intensifiers and television camera tubes, were reviewed earlier by Gilson and Hendra (1970). They should prove useful in studies of the structure of short-lived species such as enzymatic reaction intermediates, and of the dynamics of conformational changes that occur, for example, in membrane components during transport processes.

Two recent developments will permit the observation of Raman spectra in the presence of intense fluorescent backgrounds. One approach takes advantage of the differing lifetimes of Raman scattering ($<10^{-11}$ s) and fluorescent emission (10^{-8} s) by utilizing picosecond laser pulses synchronized with rapid detection. Several preliminary ventures into this field have been described (Van Duyne et al., 1974; Wilbrandt et al., 1975; Woodruff and Atkinson, 1976). The second approach depends on the phenomenon of coherent anti-Stokes Raman scattering (CARS) (Maker and Terhune, 1965; Begley et al., 1974). The phenomenon is observed when two nearly collinear laser beams of frequencies σ_1 and $\sigma_2 (\sigma_1 > \sigma_2)$ are directed on to a sample that has a Raman-active vibration at a frequency $\sigma_1 - \sigma_2$. Intense coherent scattering is observed at $2\sigma_1 - \sigma_2$. Since the scattered light is on the anti-Stokes side of the exciting lines and is coherent, possible interference from fluorescence, which is incoherent and on the Stokes side, is minimized. Although intense laser light is needed to observe the phenomenon, and although the possibility of sample deterioration in the laser beams is significant, recent studies have demonstrated the feasibility of the approach for biological materials (Nestor et al., 1976).

The utilization of currently available laser and electronics technology in biological Raman spectroscopy should greatly increase its applications. For example, ultraviolet laser excitation will allow extension of the resonance Raman effect into that spectral region (Pezolet et al., 1975). Interfacing of computers and Raman spectrometers makes automatic the collection, storage, and reduction of the spectroscopic data, and facilitates substraction of a solvent background and improvement of the signal-to-noise ratio by repetitive scanning and smoothing. A variety of applications, such as following by Raman spectroscopy the concentration and conformational changes in temperature-jump kinetics (Peticolas, 1976) or studying refrigerated microsamples (Delhaye and Dhamelincourt, 1975), require computer control. Recent developments in the determination of Raman spectra of microscopic samples are described by Dhamelincourt (1978) and Blaha et al. (1978).

B. Advances in Membrane Research

Several recent approaches to the understanding of membrane structure could profitably be augmented by Raman spectroscopic studies. For example, the reconstitution of specific membrane functions by insertion of membrane proteins into artificial lipid systems is under intensive current investigation (see, for example, Estrada-O and Gitler, 1974). Raman spectrocopic experiments could monitor conformational changes that occur during this process. Spectra of active and inactive membrane proteins could be combined with spectra of the lipid bilayers before and after insertion of the membrane protein. In this fashion, the conformation of both membrane constituents could be followed.

Resonance Raman scattering has already been profitably used to study structural alterations at particular regions of large macromolecular systems, as discussed in Section III.C. This technique can clearly be extended to biological membranes. In fact, if the measurements can be made on a rapid timescale, it will become feasible to follow structural changes that occur during processes such as active transport and catalysis. In this case, the Raman technique would provide a novel approach by which membrane function may be probed.

The promise of Raman spectroscopy as a tool in the study of biological membrane structure clearly is the result of the versatility of the method. The components of a complex system may be separated and their conformations individually investigated in a variety of physical states. The structural changes that occur upon the mutual interaction of the components may then be deduced and the effects of external influences easily evaluated. In this fashion, some general features of membrane structure may be elucidated. Finally, the resonance Raman technique may be used to probe specific regions of a complex system, and thus yield insight into the mechanism of membrane function.

Acknowledgements. We wish to thank the various authors and publishers for permission to use figures from published work, as indicated in each case. We also thank Dr. Ira W. Levin for preprints of papers in press. R.C.L. wishes to acknowledge financial support from the National Science Foundation for the work from his laboratory.

References

A. Monographs and review articles

Carey, P.R.: Resonance Raman spectroscopy in biochemistry and biology. Q. Rev. Biophys. *11*, 309-370 (1978).

Durig, J.R., Harris, W.C.: Raman spectroscopy. In: Physical Methods of Chemistry, Part IIIB, Optical, Spectroscopic and Radioactivity Methods (eds. Weissberger, A., Rossiter, B.W.,) Chap. 2. New York: Wiley-Interscience 1972.

Fawcett, V., Long, D.A.: Vibrational Spectroscopy of Macro-molecules, Vol. I, Chap. 7 of Molecular Spectroscopy. London: Specialist Per. Rep. Chem. Soc. 1973.

Frushour, B.G., Koenig, J.L.: Raman spectroscopy of proteins. In: Advances in Infrared and Raman Spectroscopy (eds. Clark, R.J.H., Hester, R.E.), Vol. I, Chap. 2, pp. 35-97. London: Heyden 1975.

Gilson, T.R., Hendra, P.J.: Laser Raman Spectroscopy. New York: Wiley-Interscience 1970.

Hall, J.L.: Stabilized lasers and precision measurements. Science *202*, 147-156 (1978).

Hartmann, K.A., Lord, R.C., Thomas, G.J.Jr.: Structural studies on nucleic acids and polynucleotides by infrared and Raman spectroscopy. In: Physical Chemistry of Nucleic Acids (ed. Duchesne, J.) Vol. II, Chap. 10, pp. 1-89 New York: Academic Press 1973.

Herzberg, G.: Infrared and Raman Spectra. New York: Van Nostrand 1945.

Koenig, J.L.: Raman Spectroscopy of biological molecules: A review. J. Polymer Sci. Part D *6*, 59-177 (1972).

Koningstein, J.A.: Introduction to the Theory of the Raman Effect. Dordrecht. Reidel 1972.

Lewis, A., Spoonhower, J.: Tunable laser resonance Raman spectroscopy in biology. In: Neutron, X-Ray Laser Spectroscopy in Biology and Chemistry (eds. Yip, S., Chen, S.) Chap. 2. New York: Academic Press 1974.

Loader, J.: Basic Laser Raman Spectroscopy. London: Heyden 1970.

Lord, R.C.: Laser Raman spectroscopy of biological macromolecules. Pure Appl. Chem. Suppl. *7*, 179-190 (1971).

Lord, R.C.: Strategy and tactics in the Raman spectroscopy of biomolecules. Appl. Spectrosc. *31*, 187-194 (1977).

Peticolas, W.L.: Inelastic light scattering and the Raman effect. Annu. Rev. Phys. Chem. *23*, 93-116 (1972a).

Peticolas, W.L.: Inelastic laser light scattering from biological and synthetic polymers. Adv. Polymer Sci. *9*, 285-333 (1972b).

Peticolas, W.L.: Applications of Raman spectroscopy to biological macromolecules. Biochimie *57*, 417-428 (1975).

Placzek, G.: Die Rayleigh- und Ramanstreuung. In: Handbúch der Radiologie (ed. Marx, E.), Vol. VI, 2, pp. 209-374. Leipzig: Akademische Verlagsges. 1934. English translation (UCRL-Trans-526(L) available from Clearing House for Federal Scientific and Technical Information, Springfield, Virginia 22151.)

Snavely, B.B.: Tunable Dye Lasers. In: Organic Molecular Photophysics, Vol. I (ed. Birks, J.B.) Chap. 5 New York: Wiley 1973.

Spiro, T.G.: Raman spectra of biological materials. In: Chemical and Biochemical Applications of Lasers (ed. Moore, C.B.) Chap. 2. New York: Academic Press 1974a.

Spiro, T.G.: Biological applications of resonance Raman spectroscopy: Haem proteins. Proc. R. Soc. London Ser. *A345*, 89-105 (1975a).

Spiro, T.G., Stein, P.: Resonance effects in vibrational scattering from complex molecules. Annu. Rev. Phys. Chem. *28*, 501-521 (1977).

Szymanski, H.A. (ed.): Raman Spectroscopy. New York: Plenum Press, Vol. 1, 1967; Vol. II, 1970.

Thomas, G.J., Jr.. Infrared and Raman spectroscopy. In: Physical Techniques in Biological Research 2nd edn. (ed. Weissberger, A.) Vol. 1A, Chap. 4. New York: Academic Press 1971.

Tobin, M.C.. Laser Raman Spectroscopy. New York: Wiley-Interscience 1971.

Tsuboi, M., Takahashi, S., Harada, I.: Infrared and Raman spectra of nucleic acids – Vibrations in the base residues. In: Physical Chemistry of Nucleic Acids (ed. Duchesne, J.) Vol. II, Chap. 11, pp. 91-145. New York: Academic Press 1973.

Yu, N.-T.: Raman spectroscopy: A conformational probe in biochemistry. Crit. Rev. Biochem. *4*, 229-280 (1977).

B. References to the original literature

Adar, F.: Resonance Raman spectra of cytochrome b_5 and its mesoheme and deuteroheme modifications. Arch. Biochem. Biophys. *170*, 644-650 (1975).

Akutsu, H., Kyogoku, Y: Infrared and Raman spectra of phosphatidyl ethanolamine and related compounds. Chem. Phys. Lipids *14*, 113-122 (1975).

Begley, R.F., Harvey, A.B., Byer, R.L.: Coherent anti-stokes Raman spectroscopy. Appl. Phys. Lett. *25*, 387-390 (1974).

Behringer, J.: The relation of resonance Raman scattering to resonance fluorescence. J. Raman Spectrosc. *2*, 275-300 (1974).

Bellocq, A.M., Lord, R.C., Mendelsohn, R.: Laser-excited Raman spectroscopy of biomolecules. III. Native bovine serum albumin and beta-lactoglobulin. Biochim. Biophys. Acta *257*, 280-287 (1972)

Blaha, J.J., Rosasco, G.J., Etz, E.S.: Raman microprobe characterization of residual carbonaceous material associated with urban airborne particulates. Appl. Spectrosc. *32*, 292-297 (1978).

Bridoux, M., Chapput, A., Delhaye, M., Tourbez, H., Wallart, F.: Rapid and ultrarapid Raman spectroscopy. In: Laser Raman Gas Diagnostics (eds. Lapp, M., Penney, C.M.) pp. 249-258. New York, London: Plenum Press 1974.

Brown, K.G., Peticolas, W.L., Brown, E.: Raman studies of conformational changes in model membrane systems. Biochem. Biophys. Res. Commun. *54*, 358-364 (1973).

Brunner, H.: Resonance Raman scattering on the haem group of cytochrome C. Biochem. Biophys. Res. Commun. *51*, 888-893 (1973).

Brunner, H., Sussner, H.: Resonance Raman scattering on haemoglobin. Biochim. Biophys. Acta *310*, 20-31 (1973).

Brunner, H., Mayer, A., Sussner, H.: Resonance Raman scattering on hemoglobin and deoxy-hemoglobin. J. Mol. Biol. *70*, 153-156 (1972).

Bulkin, B.J.: Raman spectroscopy. J. Chem. Educ. *46*, A781-A800; A859-868 (1969).

Bulkin, B.J.: Raman spectroscopic study of human erythrocyte membranes. Biochim. Biophys. Acta *274*, 649-651 (1972).

Bulkin, B.J.: Infrared and Raman spectroscopy of liquid crystals. Appl. Spectrosc. *30*, 261-268 (1976).

Bulkin, B.J., Krishnamachari, N.: Vibrational spectra of liquid crystals. IV. Infrared and Raman spectra of phospholipid-water mixtures. J. Am. Chem. Soc. *94*, 1109-1112 (1971).

Bulkin, B.J., Krishnan, K.: Vibrational spectra of liquid crystals. III. Raman spectra of crystal, cholesteric and isotropic cholesterol esters, 2800-3100-cm^{-1} region. J. Am. Chem. Soc. *93*, 5998-6004 (1971).

Bunow, M.R., Levin, I.W.: Vibrational Raman spectra of lipid systems containing amphotericin B. Biochim. Biophys. Acta *464*, 202-216 (1977a).

Bunow, M.R., Levin, I.W.: Comment on the carbon-hydrogen stretching region of vibrational Raman spectra of phospholipids. Biochim. Biophys. Acta *487*, 388-394 (1977b).

Bunow, M.R., Levin, I.W.: Raman spectra and vibrational assignments for deuterated membrane lipids: 1,2-dipalmitoyl phosphatidylcholine-d_9 and -d_{62}. Biochim. Biophys. Acta *489*, 191-206 (1977c).

Burger, H., Burczyk, K., Buchler, J.W., Fuhrhop, J.H., Hofler, F., Schrader, B.: Das Raman-Spektrum eines Metalloporphyrins. Inorg. Nucl. Chem. Lett. *6*, 171-176 (1970).

Callender, R.H., Doukas, A., Crouch, R, Nakanishi, K.: Molecular flow resonance Raman effect from retinal and rhodopsin. Biochemistry *15*, 1621-1629 (1976).

Campion, A., Terner, J., El-Sayed, M.A.: Time-resolved resonance Raman spectroscopy of bacteriorhodopsin. Nature *265*, 659-661 (1977).

Carey, P.R., Schneider, H.: Resonance Raman spectra of chymotrypsin acyl enzymes. Biochem. Biophys. Res. Commun. *57*, 831-837 (1974).

Carey, P.R., Schneider, H.: Evidence for a structural change in the substrate preceding hydrolysis of a chymotrypsin acyl enzyme: Application of the resonance Raman labelling technique to a dynamic biochemical system. J. Mol. Biol. *102*, 679-693 (1976).

Carey, P.R., Young, N.M.: The resonance Raman spectrum of the metalloprotein ovotransferrin. Can. J. Biochem. *52*, 273-280 (1974).

Carey, P.R., Schneider, H., Bernstein, H.J.: Raman spectroscopic studies of ligand-protein interactions: The binding of methyl orange by bovine serum albumin. Biochem. Biophys. Res. Commun. *47*, 588-595 (1972).

Carey, P.R., Froese, A., Schneider, H.: Resonance Raman spectroscopic study of 2,4-dinitrophenyl hapten antibody interactions. Biochemistry *12*, 2198-2208 (1973).

Carey, P.R., Carriere, R.G., Lynn, K., Schneider, H.: Resonance Raman evidence for substrate reorganization in the active site of papain, Biochemistry *15*, 2387-2393 (1976).

Carey, P.R., Carriere, R.G., Phelps, D.J., Schneider, H.: Charge effects in the active site of papain: Resonance Raman and absorption evidence for electron polarisation occurring in the acyl group of some acyl papains. Biochemistry *17*, 1081-1087 (1978).

Chapman, D., Cornell, B.A., Eliasz, A.W., Perry, A.: Interactions of helical polypeptide segments which span the hydrocarbon region of lipid bilayers. Studies of the gramicidin A lipid-water system. J. Mol. Biol. *113*, 517-538 (1977).

Chen, M.C., Lord, R.C.: Laser-excited Raman spectroscopy of biomolecules. VI. Some polypeptides as configurational models. J. Am. Chem. Soc. *96*, 4750-4752 (1974).

Chen, M.C., Lord, R.C.: Laser-excited Raman spectroscopy of biomolecules. IX. Study of thermal unfolding of ribonuclease A. Biochemistry *15*, 1889-1897 (1976).

Chen, M.C., Lord, R.C., Mendelsohn, R.: Laser-excited Raman spectroscopy of biomolecules. V. Conformational changes associated with the chemical denaturation of lysozyme. J. Am. Chem. Soc. *96*, 3038-3042 (1974).

Chen, M.C., Giegé, R., Lord, R.C., Rich, A.: Laser-excited Raman spectroscopy of biomolecules. VII. Yeast phenylalanine transfer RNA in the crystalline state and in solution. Biochemistry *14*, 4385-4391 (1975).

Chen, M.C., Giegé, R., Lord, R.C., Rich, A.: Laser-excited Raman spectroscopy of biomolecules. XIV. Spectra of ten native tRNAs and 5S RNA from *E. Coli*. Biochemistry *17*, 3134-3138 (1978).

Collins, D.W., Fitchen, D.B., Lewis, A.: Resonant Raman scattering from cytochrome C: Frequency dependence of the depolarization ratio. J. Chem. Phys. *59*, 5714-5719 (1973).

Cookingham, R.E., Lewis, A., Collins, D.W., Marcus, M.A.: Preresonance Raman spectra of crystals of retinal isomers. J. Am. Chem. Soc. *98*, 2759-2963 (1976).

Curatolo, W., Verma, S.P., Sakura, J.D., Small, D.M., Shipley, G.G., Wallach, D.F.H.: Structural effects of myelin proteolipid apoprotein on phospholipids: A Raman spectroscopic study. Biochemistry *17*, 1802-1807 (1978).

Delhaye, M.: Pulsed lasers and imaging techniques in Raman spectroscopy. In: Proc. 5th. Int. Conf. Raman Spectrosc. (ed. Schmid, E.D.) Freiburg i. Br.: Schulz 1976.

Delhaye, M., Dhamelincourt, P.: Raman microprobe and microscope with laser excitation. J. Raman Spectrosc. *3*, 33-43 (1975).

Dhamelincourt, P.: Raman microprobe techniques – some analytical applications. In: Proc. 6th Int. Conf. Raman Spectrosc., Vol. I, pp. 399-408. London: Heyden (1978).

Dunn, J.B.R., Shriver, D.F., Klotz, I.M.: Resonance Raman studies on the electronic state of oxygen in hemerythrin. Proc. Natl. Acad. Sci. USA *70*, 2582-2584 (1973).

Dunn, J.B.R., Shriver, D.F., Klotz, I.M.: Resonance Raman studies of hemerythrin-ligand complexes. Biochemistry *14*, 2689-2695 (1975).

Duyne, R.P. Van, Jeanmaire, D.L., Shriver, D.F.: Mode-locked laser Raman spectroscopy-a new technique for the rejection of interfering background luminescence signals. Anal. Chem. *46*, 213-222 (1974).

Estrada-O, S., Gitler, C. (eds.): Perspectives in Membrane Biology. New York: Academic Press 1974.

Faiman, R., Larsson, K.: Assignment of the C-H stretching vibrational frequencies in the Raman spectra of lipids. J. Raman Spectrosc. *4*, 387-394 (1976).

Faiman, R., Long, D.A.: Raman spectroscopic studies of the effect of some of the gramicidin antibiotics on the phospholipid bilayer. J. Raman Spectrosc. *5*, 87-92 (1976).

Faiman, R., Larsson, K., Long, D.A.: A Raman spectroscopic study of the effect of hydrocarbon chain length and chain unsaturation on lecithin-cholesterol interaction. J. Raman Spectrosc. *5*, 3-7 (1976).

Fanconi, B., Tomlinson, B., Nafie, L.A., Small, W., Peticolas, W.L.: Polarized laser Raman studies of biological polymers. J. Chem. Phys. *51*, 3993-4008 (1969).

Fasman, G.D., Itoh, K., Liu, C.S., Lord, R.C.: Laser-excited Raman spectroscopy of biomolecules. XI. Conformational study of poly(L-valine) and copolymers of L-valine and L-alanine. Biopolymers *17*, 125-143 (1978a).

Fasman, G.D., Itoh, K., Liu, C.S., Lord, R.C.: Laser-excited Raman spectroscopy of biomolecules. XII. Thermally induced conformational changes in poly(L-glutamic acid). Biopolymers *17*, 1729-1746 (1978b).

Felton, R.H., Romans, A.Y., Yu, N.T., Schonbaum, G.R.: Laser Raman spectra of oxidized hydroperoxidases. Biochim. Biophys. Acta *434*, 82-89 (1976).

Forrest, G.: Cyclic Phosphates, Dinucleotide Coenzymes and Lipids. Ph. D. Thesis, Massachusetts Institute of Technology 1976.

Forrest, G.: Raman spectroscopy of the milk globule membrane and triglycerides. Chem. Phys. Lipids *21*, 237-252 (1978).

Forrest, G., Lord, R.C.: Laser-excited Raman spectroscopy of biomolecules. X. Frequency and intensity of the phosphodiester stretching vibrations of cyclic nucleotides. J. Raman Spectrosc. *6*, 32-37 (1977).

Freedman, T.B., Loehr, J.S., Loehr, T.M.: A resonance Raman study of the copper protein hemocyanin. New evidence of the oxygen-binding site. J. Am. Chem. Soc. *98*, 2804-2815 (1975).

Gaber, B.P., Peticolas, W.L.: On the quantitative interpretation of biomembrane structure by Raman spectroscopy. Biochim. Biophys. Acta *465*, 260-274 (1977).

Gaber, B.P., Miskowski, V., Spiro, T.G.: Resonance Raman scattering from iron III- and copper II transferrin and an iron III model compound. A spectroscopic interpretation of the transferrin binding site. J. Am. Chem. Soc. *96*, 6868-6873 (1974).

Gaber, B.P., Yager, P., Peticolas, W.L.: Interpretation of biomembrane structure by Raman difference spectroscopy: Nature of the endothermic transitions in phosphatidylcholines. Biophys. J. *21*, 161-176 (1978a).

Gaber, B.P., Yager, P., Peticolas, W.L.: Deuterated phospholipids as nonperturbing components for Raman studies of biomembranes. Biophys. J. *22*, 191-207 (1978b).

Gaber, B.P., Yager, P., Peticolas, W.L.: Conformational non-equivalence of chains 1 and 2 of dipalmitoyl phosphatidycholine as observed by Raman spectroscopy. Biophys. J. *24*, 677-689 (1978c).

George, W.O., Mendelsohn, R.: Resonance Raman spectrum of cyanocobalamin. Appl. Spectrosc. *27*, 390-391 (1973).

Gill, D., Kilponen, R., Rimai, L.: Resonance Raman scattering of laser radiation by vibrational modes of carotenoid pigment molecules in intact plant tissues. Nature *227*, 743-744 (1970).

Gill, D., Heyde, M.E., Rimai, L.: Raman spectrum of the 11-cis isomer of retinaldehyde. J. Am. Chem. Soc. *93*, 6288-6289 (1971).

Gouterman, M.: Study of the effects of substitution on the absorption spectra of porphin. J. Chem. Phys. *30*, 1139-1161 (1959).

Heyde, M.R., Gill, D., Kilponen, R.G., Rimai, L.: Raman spectra of Schiff bases of retinal (models of visual photoreceptors). J. Am. Chem. Soc. *93*, 6776-6780 (1971).

Huang, C.: Studies on phosphatidylcholine vesicles. Formation and physical characteristics. Biochemistry *8*, 344-352 (1969).

Karvaly, B., Loshchilova, E.: Comments on the quantitative interpretation of biomembrane structure by Raman spectroscopy. Biochim. Biophys. Acta *470*, 492-496 (1977).

Khoury, F., Fanconi, B., Barnes, J.O., Bolz, L.H.: Effects of polymorphism on the Raman-active longitudinal acoustical mode frequencies of n-paraffins. J. Chem. Phys. *59*, 5849-5857 (1973).

Kiefer, W.: Laser-excited resonance Raman spectra of small molecules and ions. Appl. Spectrosc. *28*, 115-135 (1974).

Kiefer, W., Bernstein, H.J.: A cell for resonance Raman excitation with lasers in liquids. Appl. Spectrosc. *25*, 500-501 (1971a).

Kiefer, W., Bernstein, H.J.: Rotating Raman sample technique for coloured crystalline powders. Resonance Raman effect in solid $KMnO_4$, Appl. Spectrosc. *25*, 609-613 (1971b).

Kiefer, W., Topp, J.A.: A method for the automatic scanning of the depolarization ratio in Raman spectroscopy. Appl. Spectrosc. *28*, 26-34 (1974).

Kitagawa, T., Kyogoku, Y., Iisuka, T., Ikeda-Saito, M., Yamanaka, T.: Resonance Raman scattering from hemoproteins. Effects of ligands upon the Raman spectra of various C-type cytochromes. J. Biochem. (Japan) *78*, 719-728 (1975).

Kitagawa, T., Ogoshi, H., Watanabe, E., Yoshida, Z.: Resonance Raman scattering from metalloporphyrins: Metal and ligand dependence of the vibrational frequencies of octaethylporphyrins. J. Phys. Chem. *79*, 2629-2635 (1975).

Kitagawa, T., Ozaki, Y., Teraoka, J., Kyogoku, Y., Yamanaka, T.: The pH dependence of the resonance Raman spectra and structural alterations at heme moieties of various c-type cytochromes. Biochim. Biophys. Acta *494*, 100-114 (1977).

Koningstein, J.A., Gachter, B.F.: Surface-scanning technique for Raman spectroscopy of highly coloured crystals. J. Opt. Soc. Am. *63*, 892-893 (1973).

Koyama, Y., Toda, S., Kyogoku, Y.: Raman spectra and conformation of the glycerophosphorylcholine headgroup. Chem. Phys. Lipids *19*, 74-92 (1977).

Kumar, K., King, R.W., Carey, P.R.: Carbonic anhydrase-aromatic sulfonamide complexes, A resonance Raman study. FEBS Lett. *48*, 283-287 (1974).

Kurtz, D.M., Shriver, D.F., Klotz, I.M.: Resonance Raman spectroscopy with unsymmetrically isotopic ligands. Differentiation of possible structures of hemerythrin complexes. J. Am. Chem. Soc. *98*, 5033-5055 (1976).

Larsson, K.: Conformation-dependent features in the Raman spectra of simple lipids. Chem. Phys. Lipids *10*, 165-176 (1973).

Larsson, K., Rand, R.P.: Detection of changes in the environment of hydrocarbon chains by Raman spectroscopy and its application to lipid-protein systems. Biochim. Biophys. Acta *326*, 245-255 (1973).

Lewis, A.: Tunable-laser resonance Raman spectroscopic investigations of the transduction process in vertebrate rod cells. Fed. Proc. *35*, 51-53 (1976).

Lewis, A., Fager, R.S., Abrahamson, E.W.: Tunable laser resonance Raman spectroscopy of the visual process. I: The spectrum of rhodopsin. J. Raman Spectrosc. *1*, 465 (1973).

Lewis, A., Spoonhower, J., Bogomolni, R.A., Lozier, R.H., Stoeckenius, W.: Tunable laser resonance Raman spectroscopy of bacteriorhodopsin. Proc. Natl. Acad. Sci. USA *71*, 4462-4466 (1974).

Lippert, J.L., Peticolas, W.L.: Laser Raman investigation of the effect of cholesterol on conformational changes in dipalmitoyl lecithin multilayers. Proc. Natl. Acad. Sci. USA *68*, 1572-1576 (1971).

Lippert, J.L., Peticolas, W.L.: Raman-active vibrations in long chain fatty acids and phospholipid sonicates. Biochim. Biophys. Acta *282*, 8-17 (1972).

Lippert, J.L., Gorczyca, L.E., Meiklejohn, G.: A laser-Raman-investigation of phospholipid and protein configurations in hemoglobin free erythrocyte ghosts. Biochim. Biophys. Acta *382*, 51-57 (1975).

Lippincott, E.R., Sibilia, J.P., Fisher, R.D.: Optimum concentration effect in Raman spectroscopy. J. Opt. Soc. Am. *49*, 83-89 (1959).

Lis, L.J., Kaufmann, J.W., Shriver, D.F.: Effect of ions on phospholipid layer structures as indicated by Raman spectroscopy. Biochim. Biophys. Acta *406*, 453-464 (1975).

Lis, L.J., Kauffman, J.W., Shriver, D.F.: Raman spectroscopic detection and examination of the interaction of amino acids, polypeptides and proteins with the phosphatidylcholine lamellar structure. Biochim. Biophys. Acta *436*, 513-522 (1976a)

Lis, L.J., Goheen, S.C., Kauffman, J.W., Shriver, D.F.: Laser-Raman spectroscopy of lipid-protein systems. Differences in the effect of intrinsic and extrinsic proteins on the phosphatidylcholine Raman spectrum. Biochim. Biophys. Acta *443*, 331-338 (1976b).

Lis, L.J., Goheen, S.C., Kauffman, J.W.: Raman spectroscopy of fatty acid Blodgett-Langmuir multilayer assemblies. Biochem. Biophys. Res. Commun. *78*, 494-497 (1977).

Loehr, J.S., Freedman, T.B., Loehr, T.M.: Oxygen binding to hemocyanin: A resonance Raman spectroscopic study. Biochem. Biophys. Res. Commun. *56*, 510-515 (1974).

Long, G.J., Basile, L.J., Ferraro, J.R.: A semimicro sampling technique for resonance Raman spectroscopy. App. Spectrosc. *28*, 73-74 (1974).

Lord, R.C., Yu, N.T.: Laser-Raman spectroscopy of biomolecules I. Native lysozyme and its constituent amino acids. J. Mol. Biol. *50*, 509-524 (1970a).

Lord, R.C., Yu, N.T.: Laser-Raman spectroscopy of biomolecules. II. Native ribonuclease and alpha-chymotrypsin. J. Mol. Biol. *51*, 203-213 (1970b).

Loshchilova, E., Karvaly, B.: Laser Raman studies of molecular interactions with phosphatidylcholine multilayers. II. Effect of mono and divalent ions on bilayer structure. Biochim. Biophys. Acta *514*, 274-285 (1978).

Lutz, M., Breton, J.: Chlorophyll association in the chloroplast: Resonance Raman spectroscopy. Biochem. Biophys. Res. Commun. *53*, 413-418 (1973).

Lutz, M., Kleo, J.: Resonance Raman scattering of bacteriochlorophyll, bacteriopheophytin and spheroidene in reaction centers of *Rhodopseudomonas spheroides.* Biochem. Biophys. Res. Commun. *69*, 711-717 (1976).

Maker, P.D., Terhune, R.W.: Study of optical effects due to an induced polarization of third order in the electric field strength. Phys. Rev. *137*, A801-818 (1965).

Marcus, M.A., Lewis, A.: Kinetic resonance Raman spectroscopy. Dynamics of deprotonation of the Schiff base of bacteriorhodopsin. Science *195*, 1328-1330 (1977).

Mathies, R., Oseroff, A.R., Stryer, L.: Rapid-flow resonance Raman spectroscopy of photolabile molecules: Rhodopsin and isorhodopsin. Proc. Natl. Acad. Sci. USA *73*, 1 (1976).

Mayer, E., Gardiner, D.J., Hester, R.E.: Resonance Raman spectra of vitamin B_{12} and dicyanocobalamin. Biochim. Biophys. Acta *297*, 568-570 (1973).

Mendelsohn, R.: Laser-Raman spectroscopic study of egg lectihin and egg lecithin-cholesterol mixtures. Biochim. Biophys. Acta *290*, 15-21 (1972).

Mendelsohn, R.: Resonance Raman spectroscopy of the photoreceptor-like pigment of *Halobacterium halobium.* Nature *243*, 22-24 (1973).

Mendelsohn, R.: Thermal denaturation and photochemistry of bacteriorhodopsin from *Halobacterium cutirubrum* as monitored by resonance Raman spectroscopy. Biochim. Biophys. Acta *427*, 295-301 (1976).

Mendelsohn, R., Maisano, J.: Use of deuterated phospholipids in Raman spectroscopic studies of membrane structure. I. Multilayers of dimyristoyl phosphytidylcholine (and its -d_{54} derivative) with distearoyl phosphatidylcholine. Biochim. Biophys. Acta *506*, 192-201 (1978).

Mendelsohn, R., Taraschi, T.: Deuterated phospholipids as Raman spectroscopic probes of membrane structure: Dipalmitoylphosphatidylcholine-dipalmitoylphosphatidylethanolamine multilayers. Biochemistry *17*, 3944-3949 (1978).

Mendelsohn R., Taraschi, T.: Raman scattering from glucagon-dimyristoyl lecithin complexes: A model system for serum lipoproteins. J. Am. Chem. Soc. *101*, 1050-1052 (1979).

Mendelsohn, R., Sunder, S., Bernstein, H.J.: Structural studies of biological membranes and related model systems by Raman spectroscopy: Sphingomyelin and 1, 2-dilauroyl phosphatidylethanolamine. Biochim. Biophys. Acta *413*, 329-340 (1975a)

Mendelsohn, R., Sunder, S., Verma, A.L., Bernstein, H.J.: Resonance Raman spectra of Cu-etioporphyrins I, I-meso-d_4 and IV. J. Chem. Phys. *62*, 37-44 (1975b)

Mendelsohn, R., Sunder, S., Bernstein, H.J.: The effect of sonication on the hydrocarbon chain conformation in model membrane systems: A Raman spectroscopic study. Biochim. Biophys. Acta *419*, 563-569 (1976a).

Mendelsohn, R., Sunder, S., Bernstein, H.J.: Deuterated fatty acids as Raman spectroscopic probes of membrane structure. Biochim. Biophys. Acta *443*, 613-617 (1976b).

Milanovich, F.P., Yeh, Y., Baskin, R.J., Harney, R.C.: Raman spectroscopic investigations of sarcoplasmic reticulum membranes. Biochim. Biophys. Acta *419*, 243-250 (1976).

Miller, F.A., Harney, B.M.: Variable temperature sample holder for Raman spectroscopy. Appl. Spectrosc. *24*, 291-292 (1970).

Miskowski, V., Wang, S.-P. W., Spiro, T.G., Shapiro, E., Moss, T.H.: The copper coordination group in "blue" copper proteins: Evidence from resonance Raman spectra. Biochemistry *14*, 1244-1250 (1975).

Nestor, J., Spiro, T.G.: Circularly polarized Raman spectroscopy: Direct determination of antisymmetric scattering in the resonance Raman spectrum of ferrocytochrome c. J. Raman Spectrosc. *1*, 539-550 (1973).

Nestor, J., Spiro, T.G,., Klauminzer, G.: Coherent anti-stokes Raman scattering (CARS) spectra, with resonance enhancement of cytochrome c and vitamin B_{12} in dilute aqueous solution. Proc. Natl. Acad. Sci. USA *73*, 3329-3332 (1976).

Okabayashi, H., Okuyama, M., Kitagawa, T., Miyazawa, T.: The Raman spectra and molecular conformations of surfactants in aqueous solution and crystalline states. Bull. Chem. Soc. Jpn. *47*, 1075-1077 (1974).

Olf, H.G., Peterlin, A., Peticolas, W.L.: Laser-Raman study of the longitudinal acoustic mode in polyethylene. J. Polymer Sci. (Polymer Phys.) *12*, 359-384 (1974).

Oseroff, A.R., Callender, R.H.: Resonance Raman spectroscopy of rhodopsin in retinal disk membranes. Biochemistry *13*, 4243-4248 (1974).

Ozaki, Y., Kitagawa, T., Kyogoku, Y.: Raman study of the acid-base transition of ferric myoglobin: Direct evidence for the existence of two molecular species at alkaline pH. FEBS Lett. *62*, 369-372 (1976).

Peticolas, W.L.: New approaches to Raman spectroscopy of biomolecules. In: Proc. 5th Int. Conf. Raman Spectrosc. (ed. Schmid, E.D.) Freiburg i. Br.: Schultz, 1976.

Pezolet, M., Nafie, L.A., Peticolas, W.L.: Complete polarization measurements for non-symmetric Raman tensors. J. Raman Spectrosc. *1*, 455-464 (1973).

Pezolet, M., Yu, T., Peticolas, W.L.: Resonance and preresonance Raman spectra of nucleotides using ultraviolet lasers. J. Raman Spectrosc. *3*, 55-61 (1975).

Plus, R., Lutz, M.: Resonance Raman spectroscopy of porphin. Spectrosc. Lett. *7*, 133-145 (1974).

Proffitt, W., Porto, S.P.S.: Depolarization ratio in Raman spectroscopy as a function of frequency. J. Opt. Soc. Am. *63*, 77-80 (1973).

Rimai, L., Gill, D., Parsons, J.L.: Raman spectra of dilute solutions of some stereoisomers of vitamin A type molecules. J. Am. Chem. Soc. *93*, 1353-3158 (1971).

Rimai, L., Heyde, M.E., Gill, D.: Vibrational spectra of some carotenoids and related linear polyenes. A Raman spectroscopic study. J. Am. Chem. Soc. *95*, 4493-4501 (1973).

Rimai, L., Salmeen, I., Petering, D.H.: Comparison of the resonance Raman spectra of carbonmonoxy and oxyhemoglobin and myoglobin. Similarities and differences in heme electron distribution. Biochemistry *14*, 378-382 (1975).

Rothschild, K.H., Andrew, J.R., DeGrip, W.J., Stanley, H.E.: Opsin structure probed by Raman spectroscopy of photoreceptor membranes. Science *191*, 1176-1178 (1976).

Salmeen, I., Rimai, L., Gill, D., Yamamoto, T., Palmer, G., Hartzell, C.R., Beinert, H.: Resonance Raman spectroscopy of cytochrome c oxidase and electron transport particles with excitation near the Soret band. Biochem. Biophys. Res. Commun. *52*, 1100-1107 (1973).

Schachtschneider, J.H., Snyder, R.G.: Vibrational analysis of the n-paraffins II. Normal coordinate calculations. Spectrochim. Acta *19*, 117-168 (1963).

Schaufele, R.F.: Chain shortening in polymethylene liquids. J. Chem. Phys. *49*, 4168-4175 (1968).

Schaufele, R.F., Shimanouchi, T.: Longitudinal acoustical vibrations of finite polymethylene chains. J. Chem. Phys. *47*, 3605-3610 (1967).

Schmidt-Ullrich, R., Verma, S.P., Wallach, D.F.H.: Anomalous side chain amidation in plasma membrane proteins of simian virus 40 transformed by lymphocytes indicated by isoelectric focusing and laser Raman spectroscopy. Biochem. Biophys. Res. Commun. *67*, 1061-1069 (1975).

Schmidt-Ullrich, R., Verma, S.P., Wallach, D.H.F.: Concanavalin A stimulation modifies the lipid-protein structure of rabbit thymocyte plasma membranes: A laser Raman study. Biochim. Biophys. Acta *426*, 477-488 (1976).

Shriver, D.F., Dunn, J.B.R.: The backscattering geometry for Raman spectroscopy. Appl. Spectrosc. *28*, 319-323 (1974).

Siamwiza, M.N., Lord, R.C., Chen, M.C., Takamatsu, T., Harada, I., Matsuura, H., Shimanouchi, T.: Interpretation of the doublet at 850 and 830 cm^{-1} in the Raman spectra of tyrosyl residues in proteins and certain model compounds. Biochemistry *14*, 4870-4876 (1975).

Siiman, O., Young, N.M., Carey, P.R.: Resonance Raman studies of "blue" copper proteins. J. Am. Chem. Soc. *96*, 5583-5585 (1974).

Siiman, O., Young, N.M., Carey, P.R.: Resonance Raman spectra of "blue" copper proteins and the nature of their copper sites. J. Am. Chem. Soc. *98*, 744-748 (1976).

Simons, L., Bergström, G., Blomfeldt, G., Forss, S., Stenbäck, H., Wansen, G.: Laser Raman spectroscopy of amino acids, oligopeptides and enzymes. Comment. Phys.-Math. *42*, 125-207 (1972).

Small, E.W., Peticolas, W.L.: Conformational dependance of the Raman scattering intensities from polynucleotides. Biopolymers *10*, 69-88 (1971).

Snyder, R.G.: Vibrational study of the chain conformation of the liquid n-paraffins and molten polyethylene. J. Chem. Phys. *47*, 1316-1360 (1967).

Snyder, R.G., Schachtschneider, J.H.: Vibrational analysis of the n-paraffins I. Assignments of infrared bands in the spectra of C_3H_8. Spectrochim. Acta *19*, 85-116 (1963).
Snyder, R.G., Hsu, S.L., Krimm, S.: Vibrational spectra in the C-H stretching region and the structure of the polymethylene chain. Spectrochim. Acta *34*A, 395-406 (1978).
Spiker, R.C., Levin, I.W.: Raman spectra and vibrational assignments for dipalmitoyl phosphatidylcholine and structurally related molecules. Biochim. Biophys. Acta *388*, 361-373 (1975).
Spiker, R.C., Levin, I.W.: Phase transitions of phospholipid singlewall vesicles and multilayers. Measurement by vibrational Raman spectroscopic frequency differences. Biochim. Biophys. Acta *433*, 457-468 (1976a).
Spiker, R.C., Levin, I.W.: Effect of bilayer curvature on vibrational Raman spectroscopic behavior of phospholipid-water assemblies. Biochim. Biophys. Acta *455*, 560-575 (1976b).
Spiro, T.G.: Resonance Raman spectroscopy: A new structure probe for biological chromophores. Acc. Chem. Res. *7*, 339-344 (1974b).
Spiro, T.G.: Resonance Raman spectroscopic studies of heme proteins. Biochim. Biophys. Acta *416*, 169-189 (1975b).
Spiro, T.G., Strekas, T.C.: Hemoglobin: Resonance Raman spectra. Biochim. Biophys. Acta *263*, 830-833 (1972a).
Spiro, T.G., Strekas, T.C.: Resonance Raman spectra of hemoglobin and cytochrome c: Inverse polarization and vibronic scattering. Proc. Natl. Acad. Sci. USA *69*, 2622-2626 (1972b).
Spiro, T.G., Strekas, T.C.: Resonance Raman spectra of heme proteins. Effects of oxidation and spin state. J. Am. Chem. Soc. *96*, 338-345 (1974).
Stein, P., Burke, J.M., Spiro, T.G.: Structural interpretation of heme protein resonance Raman frequencies. Preliminary normal coordinate analysis results. J. Am. Chem. Soc. *97*, 2304-2305 (1975).
Strekas, T.C., Spiro, T.G.: Cytochrome c: Resonance Raman spectra. Biochim. Biophys. Acta *278*, 188-192 (1972).
Strekas, T.C., Spiro, T.G.: Hemoglobin resonance Raman excitation profiles with a tunable dye laser. J. Raman Spectrosc. *1*, 387-392 (1973).
Strekas, T.C., Spiro, T.G.: Resonance Raman evidence for anomalous heme structure in cytochrome c′ from *Rhodopseudomonas palustris.* Biochim. Biophys. Acta *351*, 237-245 (1974).
Strekas, T.C., Packer, A.J., Spiro, T.G.: Resonance Raman spectra of ferri-hemoglobin fluoride: Three scattering regimes. J. Raman Spectrosc. *1*, 197-206 (1973).
Strekas, T.C., Adams, D.H., Packer, A., Spiro, T.G.: Absorption corrections and concentration optimization for absorbing samples in resonance Raman spectroscopy. Appl. Spectrosc. *28*, 324-327 (1974).
Sugeta, H., Go, A., Miyazawa, T.: Vibrational spectra and molecular conformations of dialkyl disulfides. Bull. Chem. Soc. Jpn. *46*, 3407-3411 (1973).
Sunder, S., Mendelsohn, R., Bernsein, H.J.: Resonance Raman spectroscopy as an analytical probe for biological chromophores: Spectra of four Cu-etioporphyrins. Biochem. Biophys. Res. Commun. *62*, 12-16 (1975).
Sunder, S., Mendelsohn, R., Bernstein, H.J.: Raman studies of the C-H and C-D stretching region in stearic acid and some specifically deuterated derivatives. Chem. Phys. Lipids *17*, 456-465 (1976).
Sunder, S., Cameron, D., Mantsch, H.H., Bernstein, H.J.: Infrared and laser Raman studies of deuterated model membranes: Phase transition in 1,2-perdeuterodipalmitoyl-sn-glycero-3-phosphocholine. Can. J. Chem. *56*, 2121-2126 (1978).
Sussner, H., Mayer, A., Brunner, H., Fasold, H.: Raman study on the two quaternary states of unligated hemoglobin. Eur. J. Biochem. *41*, 465-469 (1974).
Szalontai, B.: Phase transitions in lipid multibilayers induced by benzene: A Raman spectroscopic study. Biochem. Biophys. Res. Commun. *70*, 947-950 (1976).
Tang, S.-P.W., Spiro, T.G., Antanaitis, C., Moss, T.H., Holm, R.H., Herskovitz, T., Mortensen, L.E.: Resonance Raman spectroscopic evidence for structural variation among bacterial ferredoxin, HiPIP and $Fe_4S_4(SCH_2Ph_4)_4{}^{2-}$. Biochem. Biophys. Res. Commun. *62*, 1-6 (1975).

Tasumi, M., Shimanouchi, T., Miyazawa, T.: Normal vibrations and force constants of polymethylene chain. J. Mol. Spectrosc. *9*, 261-287 (1962).

Terner, J., Campion, A., El-Sayed, M.A.: Time-resolved resonance Raman spectroscopy of bacteriorhodopsin on the millisecond time scale. Proc. Natl. Acad. Sci. USA *74*, 5212-5216 (1977).

Thomas, G.J. Jr., Barylski, J.R.: Thermostatting capillary cells for a laser-Raman spectrophotometer. Appl. Spectrosc. *24*, 463-464 (1970).

Tomimatsu, Y., Kint, S., Scherer, J.R.: Resonance Raman spectroscopy of iron (III)-ovotransferrin and iron (III)-human serum transferrin. Biochem. Biophys. Res. Commun. *54*, 1067-1074 (1973).

Verma, A.L., Bernstein, H.J.: Resonance Raman spectra of protohemin and protohemin-imidazole complex. J. Raman spectrosc. *2*, 163-173 (1974a).

Verma, A.L., Bernstein, H.J.: Resonance Raman spectra of metal-free porphin and some porphyrins. Biochem. Biophys. Res. Commun. *57*, 255-262 (1974b).

Verma, A.L., Bernstein, H.J.: Resonance Raman spectra of Cu-porphin. J. Chem. Phys. *61*, 2560-2565 (1974c).

Verma, A.L., Mendelsohn, R., Bernstein, H.J.: Resonance Raman spectra of the nickel, cobalt and copper chelates of mesoporphyrin IX dimethyl ester. J. Chem. Phys. *61*, 383-390 (1974).

Verma, S.P., Wallach, D.F.H.: Carotenoids as Raman active probes of erythrocyte membrane structure. Biochim. Biophys. Acta *401*, 168-176 (1975).

Verma, S.P., Wallach, D.H.F.: Effect of melittin on thermotropic lipid state transitions in phosphatidylcholine liposomes. Biochim. Biophys. Acta *426*, 616-623 (1976b).

Verma, S.P., Wallach, D.F.H.: Multiple thermotropic state transitions in erythrocyte membranes: A laser Raman study of the C-H stretching and acoustical regions. Biochim. Biophys. Acta *436*, 307-318 (1976b).

Verma, S.P., Wallach, D.F.H.: Erythrocyte membranes undergo cooperative, pH-sensitive state transitions in the physiological temperature range: Evidence from Raman spectroscopy. Proc. Natl. Acad. Sci. USA *73*, 3558-3561 (1976c).

Verma, S.P., Wallach, D.F.H.: Raman spectra of some saturated, unsaturated and deuterated C_{18} fatty acids in the HCH-deformation and C-H stretching regions. Biochim. Biophys. Acta *486*, 217-227 (1977).

Verma, S.P., Wallach, D.F.H., Schmidt-Ullrich, R.: The structure and thermotropism of thymocyte plasma membranes as revealed by laser Raman spectroscopy. Biochim. Biophys. Acta *394*, 633-645 (1975).

Wallach, D.H:F., Verma, S.P.: Raman and resonance Raman scattering by erythrocyte ghosts. Biochim. Biophys. Acta *382*, 542-551 (1975).

Warren, C.H., Hooper, D.L.: Chain length determination of fatty acids by Raman spectroscopy. Can. J. Chem. *51*, 3901-3904 (1973).

Wart, H.E. Van, Scheraga, H.A.: Raman spectra of strained disulfides: Effect of rotation about S-S bonds on S-S stretching frequencies. J. Phys. Chem. *80*, 1823-1832 (1976).

Wart, H.E. Van, Cardinaux, F., Scheraga, H.A.: Low frequency Raman spectra of dimethyl, methyl ethyl and diethyl disulfides and rotation about their carbon-sulfur bonds. J. Phys. Chem. *80*, 625-630 (1976).

Weidekamm, E., Bamberg, E., Brdiczka, D., Wildermuth, G., Macco, F., Lehmann, W., Weber, R.: Raman spectroscopic investigation of the interaction of gramicidin A with dipalmitoyl phosphatidylcholine liposomes. Biochim. Biophys. Acta *464*, 442-447 (1977).

Wilbrandt, R., Pagsberg, P., Hansen, K.B., Weisberg, C.V.: Fast resonance Raman spectroscopy of a free radical. Chem. Phys. Lett. *36*, 76-81 (1975).

Woodruff, W.H., Atkinson, G.H.: Vidicon detection of resonance Raman spectra: Cytochrome c. Anal. Chem. *48*, 186-189 (1976).

Woodruff, W.H., Spiro, T.G.: A circulating sample cell for temperature control in resonance Raman spectroscopy. Appl. Spectrosc. *28*, 73-74 (1974).

Woodruff, W.H., Spiro, T.G., Yonetani, T.: Resonance Raman spectra of cobalt-substituted hemoglobin: Cooperativity and displacement of the cobalt atom upon oxygenation. Proc. Natl. Acad. Sci. USA *71*, 1065-1069 (1974).

Wozniak, W.T., Spiro, T.G.: Resonance Raman spectra of vitamin B_{12} derivatives. J. Am. Chem. Soc. *95*, 3402-3404 (1973).
Yamamoto, T., Palmer, G., Gill, D., Salmeen, I.T., Rimai, L.: The valence and spin state of iron in oxyhemoglobin as inferred from resonance Raman spectroscopy. J. Biol. Chem. *248*, 5211-5213 (1973).
Yellin, N., Levin, I.W.: Cooperative unit size in the gel-liquid crystalline phase transition of dipalmitoyl phosphatidylcholine-water multilayers: An estimate from Raman spectroscopy. Biochim. Biophys. Acta *468*, 490-494 (1977a).
Yellin, N., Levin, I.W.: Hydrocarbon chain disorder in lipid bilayers. Temperature dependent Raman spectra of 1,2-diacyl phosphatidylcholine-water gels. Biochim. Biophys. Acta *489*, 177-190 (1977b).
Yellin, N., Levin, I.W.: Hydrocarbon chain trans-gauche isomerization in phospholipid bilayer gel assemblies. Biochemistry *16*, 642-647 (1977c).
Yu, N.T., Liu, C.S., O'Shea, D.C.: Laser Raman spectroscopy and the conformation of insulin and proinsulin, J. Mol. Biol. *70*, 117-132 (1972).

Low-Angle X-Ray Diffraction

N.P. Franks[1] and Y.K. Levine[2]

I. Introduction

The first X-ray diffraction patterns that were recorded from biological membranes were published at about the same time (the early 1930s) as the first patterns from protein crystals. Growth in the two fields, however, has been very different. Protein crystallography has been an area of intensive research and represents one of the great successes of molecular biology. Membrane diffraction attracted relatively few workers until the late 1960s, when research in this area expanded rapidly. In 1973 the field was reviewed separately by Levine and Shipley and Worthington and in this article we will concentrate mainly, although not exclusively, on work since this time, with emphasis on those areas in which there has been either particular activity or success.

X-rays are scattered from electrons, and a successful analysis of the diffracted intensity will give the distribution of electron density through the structure. The amount of information that can be obtained in a diffraction experiment is limited by the degree of disorder in the structure averaged over the time that it takes to record the diffraction pattern. Consequently, since most biological membranes have no regular structure over their surfaces, and since the time-scale for molecular motion within the membrane plane is very short compared with the time needed to record the diffracted intensity, these membranes can only be characterized by electron density profiles along the one dimension perpendicular to the membrane plane. Useful information can sometimes be obtained by simply measuring the positions of the intensity maxima in a diffraction pattern. For example, if the specimen is chemically well-defined, the partial thicknesses of the various components (protein, lipid and water) can be calculated without a complete analysis of the diffracted intensity.

If a complete analysis of the intensity distribution is to be attempted, then there are essentially two main problems. The first is that in order to reconstruct the electron density profile, both the amplitudes of the scattered waves and their relative phases are required, yet only intensities can be observed. This means that the phase information is lost. The second problem is that, even if the correct phases can be deduced and an electron density profile derived, there is no simple way of assigning the various levels of electron density to the chemical constituents; hydrated protein and lipid headgroups, for example, have about the same electron density.

1 Biophysics Section, Department of Physics, Imperial College of Science and Technology, Prince Consort Road, London SW7 2AZ, Great Britain

2 Rijksuniversiteit Utrecht, Fysisch Laboratorium, Princetonplein 5, Postbus 80.000, 3508 TA Utrecht, The Netherlands

One other difficulty, peculiar to membranes, is that the diffraction pattern is dominated by the structure of the lipid molecules, while the interesting functions of biological membranes reside mainly with the proteins. Even for membranes which consist of only about 25% by weight lipids, it is the bilayer structure formed by these lipids which largely determines the distribution of scattered intensity. This is because the variations in electron density over the lipid molecules (between about 0.45 electrons/Å^3 for the headgroups and 0.17 electrons/Å^3 for the terminal methyl groups) are considerably greater than the variations within protein molecules at the resolution of most membrane profiles. Consequently it has often been difficult to derive information on the locations of the protein molecules.

Despite these problems, the technique of X-ray diffraction can provide valuable information about the structure of both biological membranes and model systems which cannot be obtained using other methods.

II. Experimental Methods

The X-ray diffraction experiment consists of placing a membrane specimen in a collimated beam of X-rays (which are more or less monochromatic) and measuring the intensity of the scattered radiation as a function of angle. Laboratory sources produce X-rays by bombarding a metal target (almost always copper) with electrons which have been accelerated through a voltage of about 30-40 kV. The targets are water-cooled and usually made to rotate so as to improve the efficiency of heat dissipation. The X-rays that are produced consist of a few sharp spectral lines characteristic of the anode material (for copper about 85% of this energy is in the Kα line at about 1.54 Å and most of the rest in the Kβ line at about 1.39 Å) and a broad background whose wavelength distribution is determined mainly by the accelerating voltage. Much more intense sources are now available from syncrotrons (Rosenbaum et al., 1971) and, in the future, perhaps also from laser-produced plasmas (Frankel and Forsyth, 1979), although radiation damage may prove to be a serious problem.

The X-rays are collimated into either a fine line (about 100 microns x a few millimeters) or a point (about 100 microns in diameter) using various methods ranging from simple pin-holes to systems which use a combination of reflecting mirrors and crystal monochromators (see Witz, 1969, for a review). The degree of monochromatization that is achieved depends largely on the type of collimating system although thin metal foils, which absorb some wavelengths far more than others, are also used. With copper radiation, for example, a nickel foil 0.018 mm thick will practically eliminate the Kβ line with a tolerable reduction in the intensity of the Kα radiation.

The membrane specimens can either be sealed in thin-walled (about 10 microns) glass or quartz capillary tubes or mounted in small chambers with thin plastic or beryllium windows. Beryllium is to be preferred since it does not produce any coherent scatter at low angles. The specimen chambers can usually be temperature-controlled. For all membrane specimens, the optimum thickness for Cu Kα radiation is about 1 mm. The optimum occurs for a thickness roughly equal to the reciprocal of the linear absorption coefficient; too thin a specimen does not scatter strongly enough,while too thick a specimen absorbs too much. To prevent intense air scatter, the path between the specimen and the film or counter is either evacuated or filled with helium. Specimen-to-film distances vary but are of the order of 10 cm. This distance has to be measured accurately so that distances on the film can be converted into scattering angles. A small metal beam-stop prevents the film being fogged by the rescattering of the main beam, whose intensity is many orders of magnitude greater than the intensity of the X-rays scattered by the specimen. The diffraction is recorded by either a stack of films or a position-sensitive detector (Gabriel and Dupont, 1972). The detector is considerably faster than film because it has essentially no background; however, because only one-dimensional counters are commonly available at present, both methods are used. Exposure times vary enormously, depending upon the type of specimen and the diffraction equipment, and can range from days to seconds or less. There has even been a recent report of a diffraction pattern from myelin being recorded in about one nanosecond using a laser-produced plasma source (Frankel and Forsyth, 1979). Typical exposure times in most experiments, however, are a few hours with film or minutes with a counter. When film is used, the intensity of the scattered X-rays is determined by measuring the optical density of the film. Scanning densitometers which can be programed to integrate over defined areas of the diffraction pattern are now widely used.

The preparation of membrane specimens depends, of course, on the type of membrane being studied. The simplest form of specimen consists of a dispersion of membrane sheets (e.g., the purple membrane) or vesicles (e.g., erythrocyte ghosts) in buffer. For technical reasons, however (see Sec. III), there are important advantages in using membrane stacks. Certain biological membranes occur naturally in stacks (e.g. myelin and the disk membranes in the rod outer segment) and can be X-rayed intact and even sometimes in vivo (Webb, 1972). Attempts have been made to stack membrane vesicles artificially by centrifugation and/or controlled dehydration. These attempts have had mixed success since the structural integrity of membranes requires a high water-activity and in some cases even moderate drying can lead to the separation of membrane components. In particular, pure lipid phases readily separate out if the relative humidity is lowered too much. Another technique which has been used to produce membrane stacks consists of repeatedly dipping a substrate through a monolayer of purified membrane components spread at an air-water interface (Levine and Wilkins, 1971; McIntosh et al. 1976, 1977). Oriented multilayers of pure lipids can also be formed simply by allowing a solution of the lipids in organic solvents to evaporate slowly under a gentle stream of humid nitrogen.

III. Theory

A. Origin of the Diffraction Pattern

X-rays interact with electrons in a structure and these electrons act as secondary X-ray sources. The amplitude of the scattered radiation from each region of the structure is directly proportional to the local electron density, and the interference between the waves scattered by different regions gives rise to the diffraction pattern. The interference arises because of the phase differences between the scattered waves and thus reflects the geometric arrangement in space of the scattering centers. The diffraction pattern therefore reflects the time-averaged distribution of electron density throughout the structure.

Consider the interaction between a plane wave of X-radiation and two scattering centers, one of which is at an arbitrary origin in space while the other is at a point P, some distance from it, whose position is described by the vector r (Fig. 1a). We can characterize the incident X-radiation in terms of a wave vector k_i which lies in the direction of propagation and has modulus $|k_i| = 1/\lambda$, where λ is the wavelength of the radiation. It can be seen from the figure that, in the direction of the scattered wave (represented by the vector k_s), the path difference between a ray which has passed through the origin and one which has passed through the point at r is $\lambda r \cdot (k_s - k_i)$. The phase difference in radians is obtained by multiplying by $2\pi/\lambda$ which gives

$$\text{phase difference} = 2\pi r \cdot (k_s - k_i) = 2\pi\,(r \cdot R) \tag{1}$$

where we have defined a new vector $R = (k_s - k_i)$. This vector has a simple geometric interpretation since it is perpendicular to imaginary planes which contain the scattering centers and from which the X-rays can be considered to have been 'reflected'.

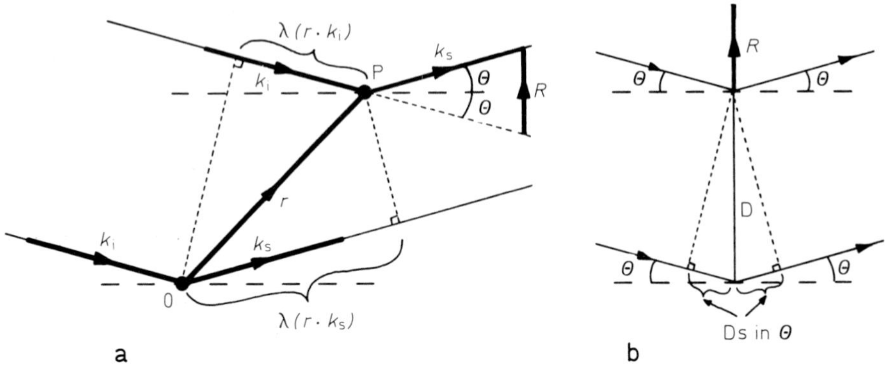

Fig. 1a and b. A plane wave of X-radiation incident upon (**a**) two scattering centers (at O and P) (**b**) two planes of constant electron density (*horizontal dashed lines*)

W.L. Bragg was the first to point out that X-ray scattering from crystals could be thought of in terms of 'reflections' from planes of constant electron density. The construction he used is shown in Fig. 1b. For planes a distance D apart, total constructive interference occurs only when the angle of incidence is such that the path difference between the two rays is an integral number of wavelengths, i.e.,

$$2\,D \sin\theta = h\lambda \tag{2}$$

which is the well-known Bragg's law. The integer h is called the order of the 'reflection'. The analogy with normal reflection is not complete since X-ray 'reflections' only occur at specific angles for a given D.

Experimentally one observes the angular deviation of the scattered X-rays from the main beam direction and this scattering angle (2θ) can be related to distances between planes of constant electron density in the structure using Bragg's law. It can be seen from Eq. (2) that there is a reciprocal relationship between the Bragg angle θ and distances in the structure. For example, when X-rays are diffracted from a stack of membranes, the long-range structural periodicity gives rise to diffraction at low angles, whereas the finer details of the membrane structure itself will give rise to diffraction at higher angles. Thus X-ray diffraction patterns are recorded out to as high an angle as possible so that these fine details may be resolved. The Bragg spacing D corresponding to the highest angle at which diffraction is observed is usually said to be the *resolution* of the pattern. A few workers, presumably with Rayleigh's criterion in mind, divide this value by two, which makes their data sound twice as good!

The reciprocal relationship between distances in the diffraction pattern and distances in the structure can be seen more clearly by relating Bragg's law to the vector R. The modulus of R can be seen from Fig. 1a to be

$$|R| = 2 \sin \theta/\lambda \tag{3}$$

(The amplitude of the wave vector does not change during the diffraction process.) Combining Eqs. (2) and (3) gives

$$|R| = \mathrm{h}/\mathrm{D}. \tag{4}$$

The vector R is called the *reciprocal space* vector and is the coordinate used to describe the positions of reflections in the diffraction pattern; distances in the structure, on the other hand, are said to be in *real space.*

We will now consider the diffraction from a number of membranes stacked regularly, one on top of the other, separated by fluid layers of constant width. Although there are a few notable exceptions, most membranes have no well-ordered repeating structures over their surfaces. We can therefore consider our membrane stack to have a characteristic time-averaged electron density distribution only in the one dimension perpendicular to the membrane plane. This one-dimensional projection can be thought of as a continuous distribution of electron density which repeats periodically. We define the electron density $\rho(\mathrm{x})$, at a distance x from some arbitrary origin, relative to the electron density of the fluid. The amplitude of the scattered wave from this point will be proportional to $\rho(\mathrm{x})$ and its phase relative to the origin will be $2\pi\mathrm{Rx}$ where R is the component of R along the perpendicular (see Eq. [1]). To obtain the total resultant amplitude scattered from the whole membrane array we integrate over the thickness of the array, thus

$$\mathrm{F(R)} = \int_{\mathrm{x}} \rho(\mathrm{x}) \exp (2\pi \mathrm{iRx})\mathrm{dx} \quad . \tag{5}$$

F(R) is called the *structure factor* or *structure amplitude* and defines both the amplitude and phase of the scattered radiation (i is the square root of minus one).

The integral of Eq. (5) has the form of a Fourier integral and therefore defines a unique and reciprocal relationship between the electron density distribution $\rho(\mathrm{x})$ and the structure factor F(R). That is to say that F(R) is the *Fourier transform* of $\rho(\mathrm{x})$ so that if F(R) is known at every value of R then the electron density distribution can be obtained by using the inverse relationship

$$\rho(x) = \int_R F(R) \exp(-2\pi iRx)\, dR. \tag{6}$$

(Although it is an arbitrary distinction, most workers in fact refer to $\rho(x)$ as the Fourier transform of F(R) and call F(R) the inverse Fourier transform of $\rho(x)$. However, it is common to talk of either as simply the Fourier transform of the other.)

F(R) in Eq. (5) relates to the entire stack of membranes; we still have to extract the structure factor of a single membrane from it. Suppose our stack consists of N membranes each of thickness D_m separated by fluid layers each of thickness D_f. The electron density distribution will then be periodic in x, repeating itself every distance $D = D_m + D_f$, i.e.,

$$\rho(x + nD) = \rho(x);\ 0 \leqslant x \leqslant D \quad n = 1,2 \ldots N-1 \ . \tag{7}$$

We can therefore rewrite Eq. (5) as a sum of integrals which can be then factorized to give

$$\begin{aligned} F(R) &= \left\{ \sum_{n=0}^{N-1} \exp(2\pi iRnD) \right\} \int_0^D \rho(x) \exp(2\pi iRx)dx \\ &= \left\{ \sum_{n=0}^{N-1} \exp(2\pi iRnD) \right\} F_u(R) \end{aligned} \tag{8}$$

where

$$F_u(R) = \int_0^D \rho(x) \exp(2\pi iRx)dx \tag{9}$$

is the structure factor of an isolated repeat unit of the array. It is thus apparent that the structure factor of the array can be factorized into a product of the structure factor of a single repeat unit and a sum of terms which represent the phase differences introduced into the scattered waves by the spatial separation of the repeat units. This sum is a geometric series and is called the interference function G(R):

$$\begin{aligned} G(R) &= \sum_{n=0}^{N-1} \exp(2\pi iRnD) \\ &= \exp[\pi iR(N-1)D] \frac{\sin(N\pi RD)}{\sin(\pi RD)} \end{aligned} \tag{10}$$

The interference function has maxima whenever all the terms in the series have the same value of +1, and this happens only when

$$RnD = m\ ;\ m = 0, \pm 1, \pm 2 \ldots \quad . \tag{11}$$

Since both n and m are integers, then Eq. (11) will be satisfied for all n if R times D is also an integer, i.e.,

$$RD = h\ ;\ h = 0, \pm 1, \pm 2 \ldots \tag{12}$$

which is a restatement of Bragg's law. In other words, the interference function has peaks in reciprocal space at positions where reflections are predicted to occur by Bragg's law.

We will now consider the diffraction from two types of membrane specimen, both of which are commonly encountered in practice. In the first case we consider a specimen which consists of a relatively large number of membranes stacked regularly, and in the second case specimens which are dispersions of membrane sheets or large vesicles where there is no fixed phase relationship (i.e. geometry) between the individual membranes.

The interference function for the first type of specimen will consist of a series of sharp peaks at points $R = \pm h/D$ in reciprocal space. Since the amplitude of the interference function is

$$|G(R)| = \frac{\sin(N\pi RD)}{\sin(\pi RD)} \tag{13}$$

(the exponential factor in Eq. [10] giving the phase) then the maxima in the interference function will have height $|G(h/D)|=N$. The half-widths will be of order $1/(ND)$ so that the widths of the Bragg reflections will be inversely proportional to the number of coherently diffracting units in the array. Thus for large N the interference function exists only in the neighborhood of the points $R = \pm h/D$ and effectively samples the function $F_u(R)$, the Fourier transform of the single repeat unit, at equal intervals of reciprocal space. Because diffracted X-rays are concentrated into small regions of reciprocal space (see Fig. 4), diffraction patterns from membrane stacks can be recorded out to quite high angles and can therefore give relatively high resolution information about membrane structure.

We can consider the second type of specimen also to contain the same number of membranes N, but since these membranes are dispersed at random in a fluid, they will have no fixed phase relationship to each other and they can be considered to scatter X-rays independently. We can therefore work out the diffraction pattern from a single membrane (i.e., $N = 1$) and add up N such identical patterns. We can see from Eq. (10) that the interference function G(R) for $N = 1$ is simply unity over all of reciprocal space. The total structure factor F(R) thus simply reduces to the structure factor of the single repeat unit and the diffracted amplitude is the continuous function $F_u(R)$. Although the total intensity scattered from N membranes dispersed at random is identical to that scattered from the same N membranes placed in a regular stack, the fact that the diffracted X-rays are now spread out over reciprocal space (see Fig. 2), rather than being concentrated in narrow regions, means that the patterns from membrane dispersions are usually observed only to moderate or low resolution. As we shall see, however, the continuous nature of the diffraction sometimes simplifies the analysis.

So far we have distinguished those specimens in which the membranes are stacked regularly from those in which there is no fixed spatial relationship between the membranes (i.e., on the basis of the degree of *order*). There is another useful distinction that can be made on the basis of the degree of *orientation* of the membrane planes about an axis parallel to the main beam direction. If the membranes can be oriented, then the scattering from the membrane plane can be separated from the scattering due to the variation in electron density in the direction perpendicular to the membrane plane (see Fig. 4). The degree of order in a specimen need not be related to the degree of disorientation. Regular stacks of membranes can be disoriented while membrane dispersions can, sometimes, be oriented.

As discussed above, there is a unique relationship between the structure factor F(R) and the electron density distribution $\rho(x)$ through the Fourier transformation, so that if one is known then the other can be calculated. Unfortunately, in a diffraction experiment we do not observe the function F(R) but rather the diffracted intensity I(R). The observed intensity distribution is related to the diffracted amplitude by

$$I(R) = F^*(R) \cdot F(R) = |F(R)|^2 \tag{14}$$

where $F^*(R)$ is the complex conjugate of F(R). Thus from the observed intensity we can only directly calculate the modulus of the structure factor while we lose the phase information. This is the well-known phase problem and constitutes the principal difficulty in the complete analysis of X-ray diffraction patterns.

It should be mentioned here that the actual diffracted intensity which is measured, either with a counter or by densitometering a film, has to be corrected by a number of geometrical factors. The most important of these is a factor which takes into account the disorientation of the membranes with respect to the X-ray beam (often called the Lorentz factor by analogy with a correction made in crystallography). For a point-focus beam the diffracted intensity will be spread out in reciprocal space at constant R so that the measured intensities will be reduced by a factor proportional to R^2. If the intensities on a film are integrated around an arc or a circle then they need only be multiplied by R rather than R^2. A minor correction is needed because oscillating electrons, which act as the secondary X-ray sources, do not radiate energy isotropically and the measured intensities have to be divided by $[1 + \cos^2 2\theta]/2$ to take this effect into account, although it is only a small correction at low angles. Another correction due to preferential absorption by the specimen at different angles is also sometimes made. Throughout this discussion, when we refer to the intensity distribution, we mean one for which these geometrical corrections have already been made.

One other important point that should be appreciated is that, almost invariably, only relative intensities are recorded due to the technical difficulties in measuring absolute intensities (Luzzati, 1960; Kratky, 1967). Thus the electron density profiles can only be directly calculated on a relative scale, which has sometimes greatly hindered their interpretation. Absolute electron density scales can in some cases be indirectly inferred (see later).

The phase problem is enormously simplified if the electron density profile of the structure is, or at least is assumed to be, centrosymmetric, i.e., $\rho(x) = \rho(-x)$. We can rewrite Eq. (9) as

$$\begin{aligned} F_u(R) &= \int_{-D/2}^{D/2} \rho(x) \exp(2\pi i R x) dx \\ &= \int_{-D/2}^{D/2} \rho(x) \cos(2\pi R x) dx + i \int_{-D/2}^{D/2} \rho(x) \sin(2\pi R x) dx \ . \end{aligned} \tag{15}$$

The sine transform, the second term in Eq. (15), is identically zero since the sine function is antisymmetric ($\sin(x) = -\sin(-x)$). In general, we can also write the structure factor $F_u(R)$ in terms of its amplitude $|F_u(R)|$ and phase $\alpha(R)$ as

$$F_u(R) = |F_u(R)|\exp(i\alpha) = |F_u(R)|\cos\alpha + i|F_u(R)|\sin\alpha \qquad (16)$$

and, since Eq. (15) shows that $F_u(R)$ is purely real for a centrosymmetric structure, we see that for this case

$$\sin\alpha = 0 \text{ so that } \alpha = n\pi;\ n = 0, \pm 1, \pm 2\ldots \quad . \qquad (17)$$

i.e., the phase angles of the structure factor $F_u(R)$ can only be integral multiples of π. These phase angles correspond to the structure factor having either positive or negative values and thus the phase problem reduces to simply determining the signs associated with the observed structure factor amplitude $|F_u(R)|$. Furthermore the signs can only change when the amplitude (and therefore the observed intensity) goes to zero.

We will now discuss in detail the analysis of membrane diffraction patterns and review the various methods that have been used to solve the phase problem. Once again we will consider our two types of membrane specimen; a random dispersion of membrane sheets or large vesicles in a fluid, and a regular multimembrane stack.

B. Analysis of Diffraction Patterns from Membrane Dispersions

A typical diffraction pattern from a membrane dispersion is shown in Fig. 2. The pattern is circularly symmetric and shows three broad bands. Wilkins et al. (1971) identified these bands with the squared Fourier transform of the individual membranes and interpreted the patterns from a number of biological membranes in terms of a basic lipid bilayer structure. They showed that the bands had peaks at reciprocal space values which were approximately multiples of $1/D_{hg}$, where D_{hg} is the distance between the electron-dense headgroups (see Fig. 3). A rough thickness for the membrane could thus be obtained directly from the diffraction pattern. Blaurock (1973a) has since calculated how the positions and shapes of the bilayer bands would be affected due to the presence of surface layers of protein.

If the membrane is assumed to be centrosymmetric, then, in principle, all that has to be done in order to calculate the electron density distribution $\rho(x)$ is to identify the zeros in intensity where the signs change, use these signs together with the observed $|F_u(R)|$, and apply the Fourier cosine transform. One problem that immediately arises, however, is that, since there is a practical limit to how close to the main beam direction scattered X-rays can be observed, then if the continuous curve is to be transformed, the region close to and including the origin of reciprocal space has to be extrapolated from the observed data. The value of the transform at the origin has a simple physical interpretation. From Eq. (9) we can see that

$$F_u(O) = \int_0^D \rho(x)dx \quad . \qquad (18)$$

Thus, remembering that $\rho(x)$ is a contrast function defined relative to the electron density of the fluid, we see that $F_u(O)$ is simply the total number of electrons in the structure in excess of the number in an equivalent volume of the fluid. In most mem-

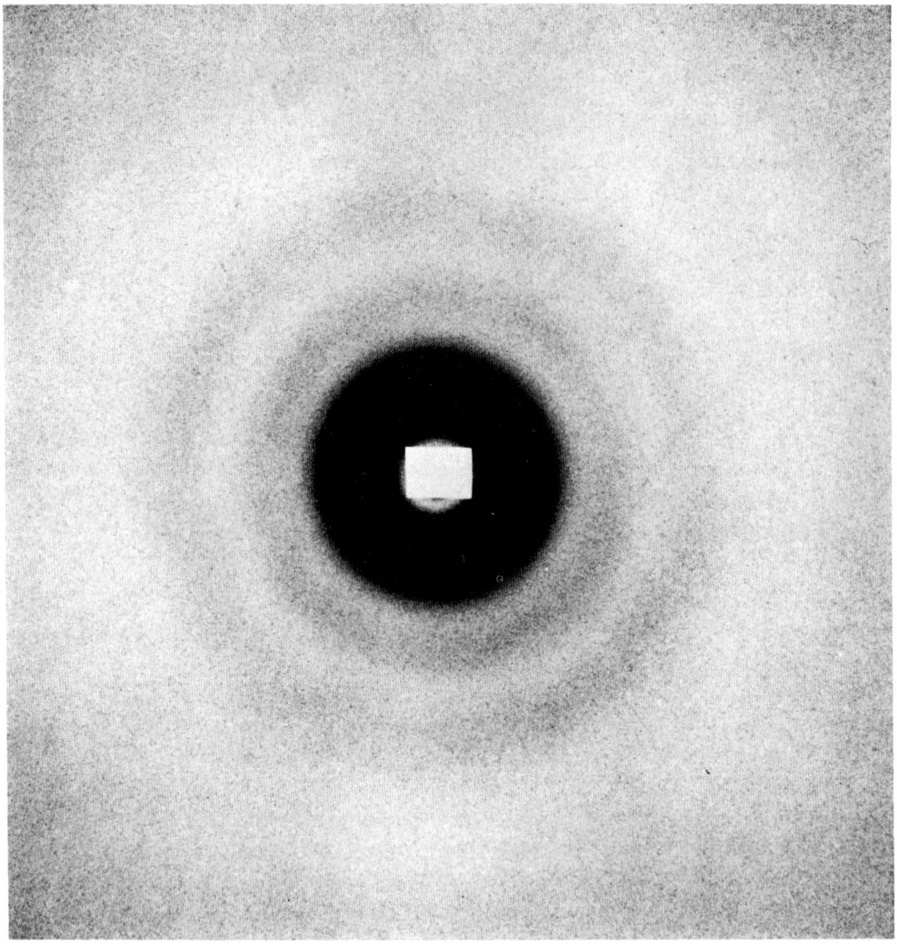

Fig. 2. The low-angle pattern from a dispersion of membrane vesicles

brane experiments $F_u(O)$ will be positive. If the sign is not known then, strictly speaking, even if the relative signs of the bands could be deduced, this would still leave an ambiguity between the electron density distribution and its negative. In practice, however, this choice is not difficult if a lipid bilayer-type profile can be assumed. A more serious difficulty is that the observation of zero intensity is only a necessary but not a sufficient condition for the sign to change, and rather accurate intensity data are required to determine whether or not a sign change in fact occurs. If zeros are not observed between the bands then this may mean that the membrane electron density profile is asymmetric, although there are a number of other factors which can result in diffraction being observed where there should be zeros. For example, if the incident beam is not sufficiently monochromatic, if the membrane thickness is not uniform or if there is scattering from structures within the plane of the membrane or indeed from nonmembraneous structures in the sample, then clear zeros will not be observed. Also, if the membranes are not large vesicles or sheets but have significant curvature, then transform zeros will not be expected either (Moody, 1975).

One alternative to the approach of simply assigning phases to the diffraction bands by inspection is to apply the Fourier transform to the intensity distribution itself. The resulting function has a direct interpretation in terms of the electron density distribution $\rho(x)$. We can see this by writing the intensity as

$$I(R) = F^*(R) \cdot F(R)$$
$$= \int_{-\infty}^{\infty} \int_{-\infty}^{\infty} \rho(u)\rho(x) \exp [2\pi i R(u-x)]\, du dx \tag{19}$$

where we have used the dummy variable u so we could write the double integral. If we now make the substitution $x' = u - x$ then we have

$$I(R) = \int_{-\infty}^{\infty} \left\{ \int_{-\infty}^{\infty} \rho(x)\rho(x + x')\, dx \right\} \exp (2\pi i R x')dx' \tag{20}$$

(since $du = dx'$ in the u integral). We see that the intensity distribution I(R) is the inverse Fourier transform of the function in the brackets. This function, obtained by applying the Fourier transform to I(R), is called the *autocorrelation* or *self-correlation* function of $\rho(x)$ and is usually represented by

$$Q(x') = \rho(x) \otimes \rho(-x) = \int_{-\infty}^{\infty} \rho(x)\rho(x + x')dx \tag{21}$$

where the symbol $\otimes$ means the function $\rho(x)$ is *convoluted* with $\rho(-x)$. In general, the autocorrelation function $Q(x')$ will be symmetric about its maximum value, which will occur at the origin and will only exist between $\pm D_m$, where D_m is the thickness of the membrane (see Fig. 3b,d). $Q(x')$ will also show peaks at positions corresponding to the separation of particularly electron-dense or electron-deficient regions in the structure. Some information can therefore be derived by the interpretation of the Fourier transform of the intensity function I(R). For example Blaurock and Wilkins (1969) identified a peak in the transform of the intensity scattered from intact retina as arising from correlations between the two membranes of the disk which formed the basic repeat unit (see also Fig. 17). Because of the relatively low resolution of membrane diffraction patterns, the information that can be obtained by a direct interpretation of the autocorrelation function is very limited.

It is possible however in principle, and sometimes in practice, actually to deconvolute the autocorrelation function to obtain the electron density profile. It was first shown by Hosemann and Bagchi (1962) that a unique solution could be obtained if the electron density function was centrosymmetric, although the same ambiguity discussed above between $\rho(x)$ and its negative remained. Such a deconvolution has subsequently been attempted by a number of workers (e.g., Lesslauer et al., 1971). The method that was first used for the deconvolution is based on a recursion formula. In this method $Q(x')$ is divided into a number of small intervals and the electron density values at an equivalent number of points are successively determined by solving a set of equations. The accurate determination of each electron density value, however, requires the determination of the previous value, so that errors in the observed intensity distribution (and therefore in $Q(x')$) will quickly propagate and lead to an unreliable electron density profile.

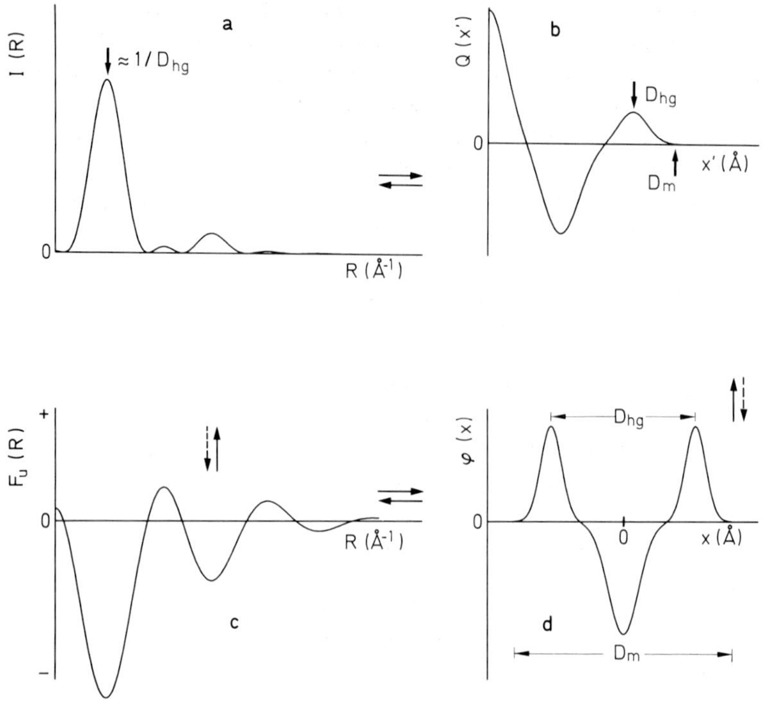

Fig. 3a–d. Diffraction from a random dispersion of centrosymmetric membranes **a** intensity distribution; **b** autocorrelation function; **c** structure factor; **d** electron density distribution. *Solid arrows*: steps that can always be made; *dashed arrows*: steps that are required in the structure analysis, and which cannot always be made

Since this method of deconvolution and the approach of inferring the sign changes by the behavior of the intensity distribution near zeros both require accurate intensity data, the involved deconvolution of the autocorrelation function has little to offer over the simple method of assigning phases to the diffraction bands by inspection and applying the Fourier transform.

Various alternatives to the recursion method for the deconvolution of $Q(x')$ have since been proposed. These methods include an iterative model-building approach (the relaxation method of Worthington et al., 1973), a Fourier series technique (Pape, 1974) and a procedure which relates the moments of $\rho(x)$ and $Q(x')$ (Moody, 1974).

So far we have only discussed methods for solving the phase problem when the membranes in the dispersion are assumed to be centrosymmetric. Recently, an interesting new method has been used to analyze the diffraction from dispersions of the asymmetric membranes of the acetylcholine receptor (Ross et al., 1977; Stroud and Agard, 1979). An essentially identical procedure has also been developed independently by Makowski (personal communication) for analyzing diffraction patterns from the filamentous bacteriophage Pf1. Central to the method is the obviously reasonable assumption that the membrane has a finite thickness. This assumption effectively limits the number of possible phase functions $\alpha(R)$ which are consistent with the observed intensity distribution. This can be seen to be so since, as mentioned above, the autocorrelation function, which is directly related to the observed intensity distribution, will only exist between $\pm D_m$, where D_m is the membrane thickness. Thus only phase functions that result in electron density profiles of thickness D_m will be acceptable solutions. The procedure of Stroud and Agard consists of guessing a solution for $\rho(x)$, Fourier-transforming (using Eq. [9]), combining the calculated phase function with the observed $|F_u(R)|$, and then back-transforming. This cycle is repeated until convergence is reached. All the solutions that result, and in general there will be more than one, will be electron density profiles of the same thickness, the constraint effectively imposed by using a transform (Eq. [9]) with a finite limit; convergence occurs provided the chosen limit is greater than the actual membrane thickness. Stroud and Agard argue, on the basis of model calculations, that there will usually be only a small number of solutions and sometimes only one. The procedure is formally the same as the refinement of atomic models in crystallography, where usually a model that is quite close to reality has to be used if the refinement is to converge. Here on the other hand, where continuous diffraction data are used, the final solutions are apparently independent of the starting model. The constraint of a finite membrane thickness is equivalent to the refinement of a flat solvent region, a technique which has been used to improve phases in crystallography.

This new method is demonstrated by Stroud and Agard using the diffraction data collected by Blaurock from a dispersion of the purple membranes of *Halobacterium halobium.* From these data two asymmetric electron density profiles were derived which were shown to account for the observed intensity distribution. An identical pair of profiles had previously been published by Blaurock and King (1977) who had also come to the conclusion that these were the only two electron density profiles that were consistent with the observed intensity data (see Fig. 23).

Another method which has been used to analyze diffraction from dispersions, similarly applicable to asymmetric structures, has been proposed independently by both King (1975) and Mitsui (1978). Their method involves the analytical extension of the intensity distribution into the complex plane. They show that if the positions of all of the zeros in the complex plane are known, then a unique phase solution is possible. In practice, of course, only the zeros along the real axis can be observed so that a number of solutions exist. Both workers show, however, that there are a relatively small number of solutions possible and Mitsui (1978), for example, arrives at essentially the same two profiles for the purple membrane that Blaurock and King (1977) and Stroud and Agard (1979) determined.

C. Analysis of Diffraction Patterns from Membrane Stacks

The methods for analyzing the patterns from membrane stacks are usually quite different to those used to analyze the patterns from membrane dispersions. This is because, as outlined above, the patterns themselves show very different features. Instead of a few broad bands which are observed with membrane dispersions (see Fig. 2), a number of sharp Bragg reflections, sometimes a dozen or more, can be recorded from membrane stacks (see Fig. 4). Although the problem is essentially the same, that of assigning phase angles to each of the reflections (instead of to each of the bands), the task is clearly very much greater due to the large number of reflections. What is worse is that, since the reflections are sharp, a single series of lamellar reflections gives relatively little information about the shape of the Fourier transform which they are sampling. So far it has only been possible to analyze the diffraction from stacks of centrosymmetric structures, although it should be stressed that this does not exclude asymmetric membranes since often pairs of such membranes occur naturally, or can be stacked, back-to-back, to form centrosymmetric units. While it is certainly true that the analysis of patterns from membrane stacks presents greater problems than the analysis of those from dispersions, the rewards are greater too because of the higher resolution that can usually be obtained.

Because of the repeating nature of the membrane stack, it is convenient to represent the relationship between the electron density distribution $\rho(x)$ and the structure factors $F_u(h/D)$ in terms of a Fourier series rather than a Fourier transform. The Fourier series for a centrosymmetric electron density function is

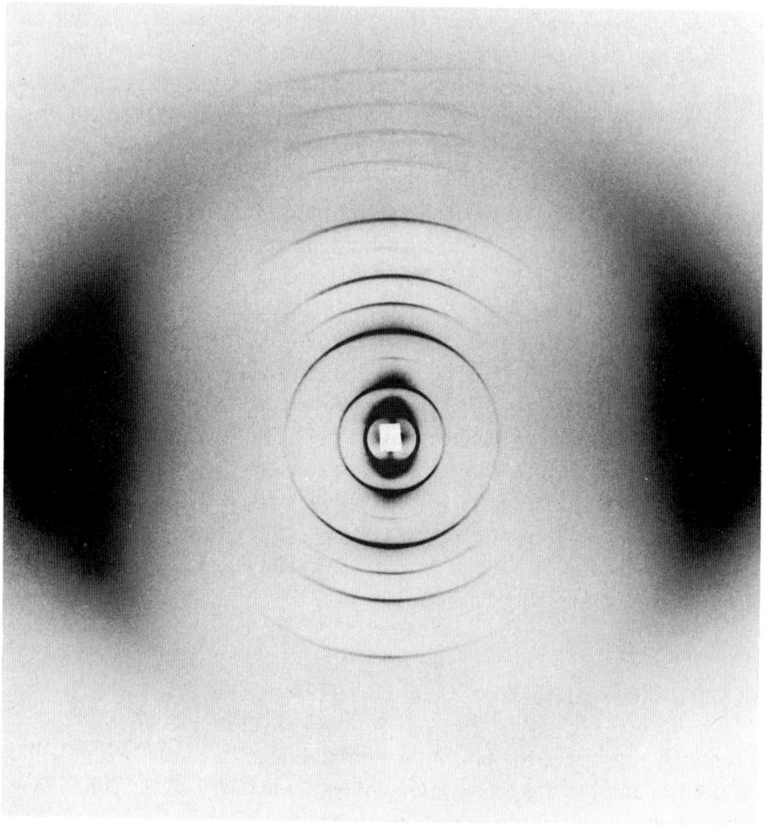

Fig. 4. Diffraction pattern from a stack of lipid bilayers. The pattern consists of a series of lamellar reflections along the meridian together with a broad, equatorially oriented band which shows that the hydrocarbon region of the bilayer is in a fluid state (Franks and Lieb, 1979)

$$\rho(x) = \frac{F_u(O)}{D} + \frac{2}{D} \sum_{h=1}^{h_{max}} F_u(\frac{h}{D}) \cos(\frac{2\pi hx}{D}) \qquad (22)$$

where h_{max} is the order of the highest observed reflection. $\rho(x)$ therefore consists of a sum of harmonics whose amplitudes are the structure factor amplitudes $|F_u(h/D)|$ and whose frequency increases with increasing Bragg order. This is consistent with what has been said earlier in that as we add higher orders to the Fourier synthesis finer detail can be resolved in the structure. The first term in Eq. (22) is a constant and defines the mean value of $\rho(x)$; note that if the electron density is defined relative to the fluid level, it does not represent the mean absolute electron density. In general it is more difficult to assign phases to the weaker reflections but, happily, these terms have the least effect in the Fourier series. The origin of the unit cell can be shifted from O to D/2 simply by changing the signs of the odd orders, leaving the other signs unchanged.

A word of warning here. It is not permissable to calculate a Fourier synthesis using only the low orders if higher terms of significant amplitude exist, and consider this to be an accurate representation of the structure at lower resolution. The exclusion of the higher terms introduces the truncation artifact of a ripple in the synthesis at about the frequency of the highest order included. A procedure that can be used is to apply a smoothing function to the observed data. For example, the observed amplitudes can be multiplied by a Gaussian which will reduce the higher orders relative to the lower orders. The characteristic width of the Gaussian can be chosen so that only the structure factors which are to be included in the synthesis have significant amplitudes.

One approach to the phase problem, championed particularly by Worthington and his co-workers (e.g., Worthington, 1969, 1973a,b) is the use of electron density models. A model consisting of a number of strips, each with constant electron density, is proposed, and the intensities predicted from this model compared to the observed intensities. The model parameters (the widths and heights of the strips) are then adjusted until the observed and predicted intensities agree. The number of independent variables can be altered by changing the number of strips in the model. This procedure is open to some objections. Step-function models will inevitably predict high-angle diffraction beyond the limits of the observed data, and since these terms are ignored when model and observed intensities are compared, then this can affect the optimization of the model parameters. A more serious problem is that even if a model can be derived, there is no way of knowing whether it is either correct or unique and the initial starting model may well influence the final result. The problem of uniqueness, of course, can never be solved since it is obviously not possible to try all conceivable models. One advantage of the model-building approach is that the model can provide a useful and simple physical interpretation of the structure in terms of chemically distinct regions provided, of course, that the model is correct.

In the previous section we discussed methods of analysis which involved the Fourier transform of the continuous intensity distribution. The equivalent function when discrete reflections are observed is a Fourier series with the intensities instead of the structure factors used as the coefficients. This is the Patterson function used extensively in crystallography:

$$P(x) = \frac{I(O)}{D} + \frac{2}{D} \sum_{h=1}^{h_{max}} I(\frac{h}{D}) \cos (\frac{2\pi hx}{D}) \quad . \tag{23}$$

Although, like the autocorrelation function, the Patterson function P(x) can be interpreted in terms of correlations between electron-dense and electron-deficient regions in the structure, the interpretation is more difficult since P(x) is in fact Q(x′) convoluted with the autocorrelation of the lattice function. (The lattice function is the transform of the interference function and is therefore a series of sharp peaks a distance D apart, where D is the unit cell size.) Thus the Patterson function repeats every distance D, while Q(x′) is twice the width of the structure, so there will usually be some overlap of neighboring autocorrelation functions (compare Figs. 3b and 5b). It was pointed out by Worthington et al. (1973), however, that if the repeat units could be moved far enough apart by swelling, then the autocorrelation function could be separated from the Patterson function. It is also possible to extract the autocorrelation function from the observed intensities if the number of repeat units in the stack is sufficiently small. This was first suggested by Hosemann and Bagchi (1962) and performed by Lesslauer and Blasie (1972) for a specimen which consisted of a few bilayers of barium stearate. Both of these methods for extracting the autocorrelation function suffer from the same drawback, that is, they are only applicable to systems where the maximum resolution tends to be somewhat limited (either because stacking disorder occurs at high swelling states or because the Bragg peaks broaden when N is small). Alternatively, Moody (1974) has shown that Q(x′) can be calculated even if the repeat units are not swollen far apart, but again only at lower resolution unless a number of swelling states can be measured. This method though has not yet been applied.

If the repeat units can be swollen apart, provided the structure (or strictly its electron density profile) does not change, then this offers a powerful and widely used method of obtaining the signs of the structure factors. Since the values of $F_u(h/D)$ represent sampled points on the continuous Fourier transform (see Fig. 5c) then, if enough swelling states can be measured, this transform can be mapped out and the zeros where sign changes occur identified. Before the data can be plotted, the intensities from different swelling states have to be scaled together. Blaurock (1967) showed that the intensities in any one set at a spacing D should be scaled so that

$$\sum_{h=-\infty}^{\infty} I(h/D) = D/D_{min} \tag{24}$$

where D_{min} is the minimum spacing in the swelling series. He pointed out that since in practice I(O) is not observed and that only a finite number (h_{max}) of reflections are recorded, only an inexact version of Eq. (24) can in fact be used. These points were subsequently stressed by Stamatoff and Krimm (1976).

The principle of minimum wavelength (Bragg and Perutz, 1952; Perutz, 1954) can be used as a guide as to where sign changes might occur in the transform. This principle states that, for a structure of maximum width W, $F_u(R)$ can only have components with a minimum wavelength of 2/W, so that two adjacent peaks in the intensity

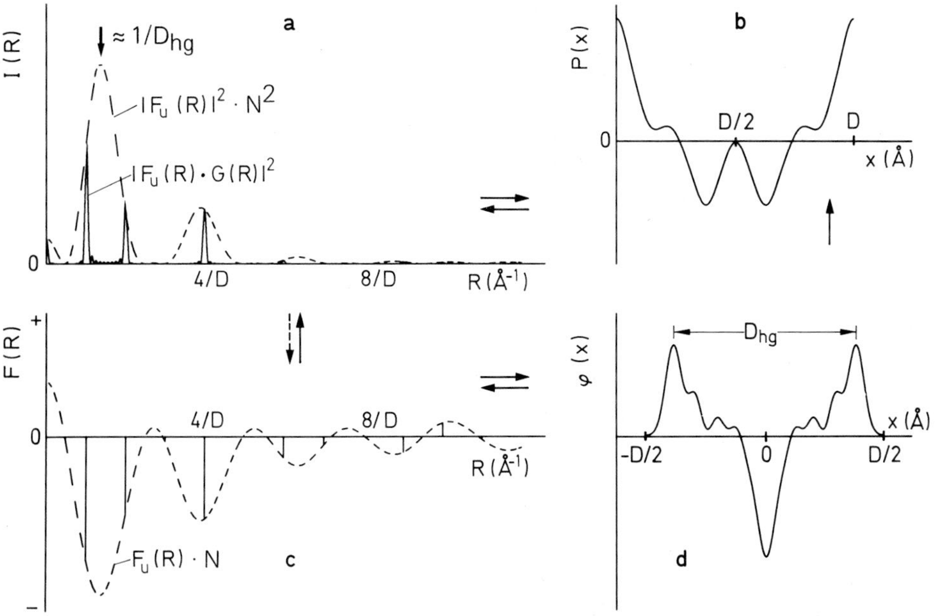

Fig. 5a-d. Diffraction from a regular stack of centrosymmetric membranes **a** intensity distribution. *Dashed line* proportional to intensity with no stacking, and this curve is sampled by the square of the interference function. For clarity, the interference function was calculated for a small number (10) of membranes; **b** Patterson function; **c** structure factor; **d** electron density distribution. D is the unit cell dimension

distribution with a separation less than this value are likely to have opposite sign. The rule is only approximate, since peaks in the transform can be closer together than the minimum wavelength due to the superposition of low and high frequency components in the transform. Nonetheless, the minimum wavelength principle is an important guide if used with care.

More rigorously, Shannon's sampling theorem from communication theory can be used. The application to crystallography was suggested by Sayre (1952) and used in membrane diffraction for the first time by Moody (1963). The theory states that the complete continuous Fourier transform $F_u(R)$ can be reconstructed using just the set of observed $|F_u(h/D)|$ together with their chosen signs, and a value of $F_u(O)$. Thus

$$F_u(R) = \sum_{h=-h_{max}}^{h_{max}} F_u\left(\frac{h}{D}\right) \frac{\sin[\pi(RD-h)]}{\pi(RD-h)} \quad . \tag{25}$$

This means that, in principle, only two sets of intensities are needed since, for a constant structure, only the correct combination of signs for the two sets of amplitudes will result in the same continuous Fourier transform calculated using Shannon's theorem (Worthington et al., 1973). The use of just two data sets, however, relies heavily on both the accuracy of the experimental measurements and the assumption that the structure does not change during swelling. It is always more prudent to measure as many sets of data as possible so as to reduce and quantify experimental random errors.

It is commonly stated that if the observed $|F_u(h/D)|$ fall on a smooth curve during swelling, then this means that the membrane structure does not change, so that these values must be sampling the modulus of the continuous Fourier transform. This is not so. The observation that the measured structure factors fall on a smooth curve means only that there has been no discontinuous change in structure. If a continuous change in the structure does occur during swelling then the data sample a set of Fourier transforms and it may not be evident from these data alone that there has been any change in structure (see Fig. 6). If, on the other hand, the data points for each order vary smoothly but these individual curves do not make up a smooth function, then this is clear evidence of a structural change although once again a continuous one. This can be very striking (see Fig. 6 of Franks and Lieb, 1979) but does not mean, we believe, that the swelling technique cannot be used. What is important is that usually the structural changes that occur are fairly small and continuous in nature, increasing disorder being the most common, so that the phases can still be assigned on the assumption that the reconstructed Fourier transforms should change in a reasonable and consistent way with swelling.

Assigning phases on the assumption that the minimum change in structure corresponds to the correct solution (Moody, 1963), can be misleading. For example, Torbet and Wilkins (1976) showed that for a multilayer system where the bilayer thinned during swelling (dipalmitoyl lecithin), the correct set of phases corresponded

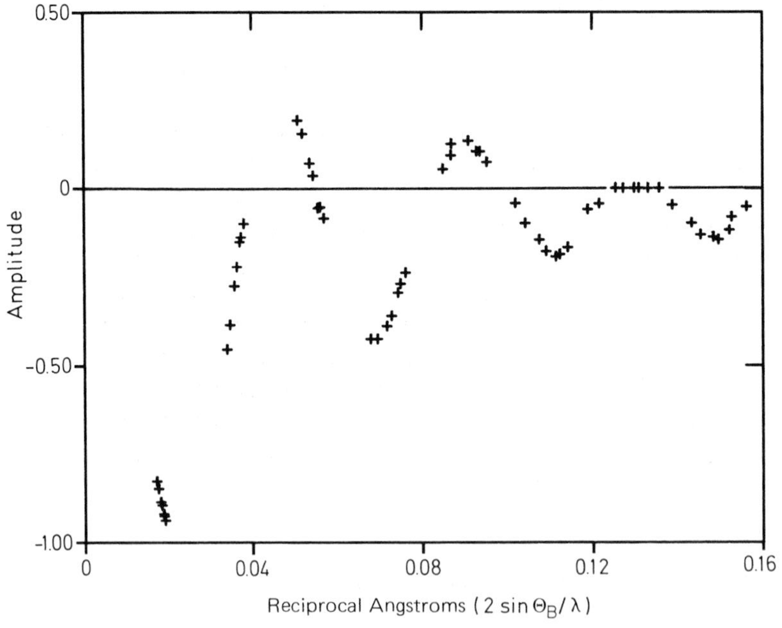

Fig. 6. An example of a swelling series. The data points lie on a smooth curve which is similar to, but not exactly, the Fourier transform of the membrane electron density profile. The structure changes slightly during swelling, so the data points sample a set of transforms (Franks, 1976)

to the set of transforms that behaved reasonably (i.e., the peaks in the transforms moved progressively to higher angles) rather than the set that were most similar to one another.

There are some other methods, apart from model-building, which attempt the determination of the phases using only a single set of data. The most thorough of these is the pattern recognition approach of Luzzati et al. (1972). In this procedure, all possible phase combinations are considered and the correct solution chosen on the basis of known or postulated properties of the electron density profile, such as levels of electron density and partial specific thicknesses of particular components. Another method which has been used (Worthington and Khare, 1978) for lipid multilayers that do not swell, assumes the bilayer profile has a total headgroup region less than or equal to half the bilayer thickness and a relatively flat hydrocarbon region. Such methods are only as good as the assumptions on which they are based and it is important to appreciate that the validity of the assumptions cannot be proven a posteriori; only self-consistency can be hoped for.

The use of heavy atom methods has had limited success with biological membranes (see Akers and Parsons, 1970; Harker, 1972; Blaurock, 1973b). However, phases have been determined for bilayers of fatty acids associated with a series of alkaline earth cations (McIntosh et al., 1976), using a method similar to that of Hargreaves (1946). More recently, Franks et al. (1978) showed that cholesterol analogs, halogenated at carbon 26, could be isomorphously exchanged with normal cholesterol in a bilayer and that this exchange could be used to determine the signs of the lamellar reflections (see Fig. 7). The method was tested on bilayers composed of a mixture of lecithin and cholesterol and the phases of the first seven orders (equivalent to a resolution of about 7 Å) were determined. In principle, this method can be used with protein-containing model systems and, with some modifications, biological membranes.

One important advantage that heavy atom methods have is that, if the number of heavy atoms present in the membrane is known, then an absolute scale of electron density can be derived (e.g., McIntosh et al., 1976, 1977; Franks et al., 1978). Absolute scales can also be derived if the electron density of the fluid can be changed. In some cases, where large changes can be made without changing the membrane structure (e.g., Blaurock 1972) the absolute scale can be reliable. In many other cases, however, only small changes in electron density can be made compared to the range of electron density in the structure. Moreover, because of termination ripples, the levels of fluid density in the profiles can be difficult to define. Consequently, small errors which are made in the assignment of the fluid electron density to a particular point in the profile can propagate seriously and lead to quite large errors in the densities calculated for the rest of the membrane profile.

D. The Effects of Thermal Motion and Stacking Disorder

Thermal motion is, of course, always present and its effect on the diffraction pattern is to reduce the intensities of the Bragg orders, particularly at higher angles. Debye (1914) showed that the intensities of the reflections were reduced by a factor

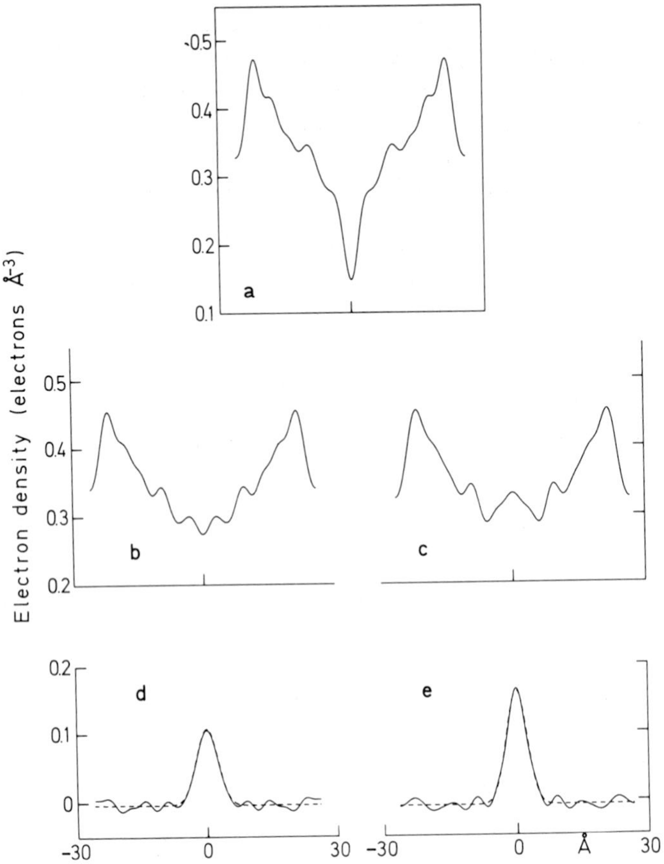

Fig. 7a-e. Electron density profiles of bilayers of lecithin and 40% (molar) of (**a**) normal cholesterol, (**b**) 26-bromocholesterol and (**c**) 26-iodocholesterol. The difference electron density profiles are shown in (**d**) and (**e**). The presence of halogen atoms in the membrane can be used to determine the signs of the lamellar reflections (because the increase in electron density is confined to a narrow region near the unit cell origin) and to derive absolute electron density scales (Franks et al., 1978)

$$\exp(-2M) \tag{26}$$

where

$$M = 8\pi^2\overline{u_0^2}\,\sin^2\theta/\lambda^2$$

and $\overline{u_0^2}$ is the mean square displacement of the repeating units in a lattice about their mean positions, in the direction of R. The multiplication of the intensity distribution by this factor is equivalent to convoluting the electron density function by a Gaussian. In other words, the effect of thermal motion is to blur out detail in the structure. The widths of the Bragg reflections are not affected by thermal motion.

Since the total intensity scattered from the specimen is independent of the location of the atoms, the intensity that is lost from the sharp Bragg reflections is transferred to the background. If the vibrations of neighboring units in the array are independent, then this diffuse scatter is a monotonically increasing function, rising from zero intensity at the reciprocal space origin. If, however, the motion of one unit is coupled to its neighbors, then the diffuse scatter is not a simple smooth curve. If thermal motion is considered in terms of a set of plane waves moving through the lattice causing coupled displacement of the lattice points, then the resulting diffuse scatter occurs as broad bands centered at the Bragg reflections (James, 1962); this is commonly observed in diffraction patterns from crystals. Despite the more complicated distribution of the thermal diffuse scatter, the intensities of the Bragg reflections are still reduced by the same factor (Eq. [26]).

Although stacking disorder can take many forms, a broad distinction between two classes is usually made. Stacking disorder of the first kind is said to occur if, despite the short-range disorder, there is sufficient long-range order for a regular lattice to be laid down over large distances. If this long-range order does not exist, the lattice is said to display disorder of the second kind. For both types of disorder, the intensities of the Bragg reflections are reduced with increasing angle. In addition, a diffuse background is predicted which rises from zero intensity at the reciprocal space origin. For disorder of the first kind the widths of the reflections do not change. With disorder of the second kind, however, the widths of the Bragg peaks progressively increase with increasing scattering angle.

Recently, several workers have tried to take account of the stacking disorder in membrane preparations using simple models for the disorder (Schwartz et al., 1975; Blaurock and Nelander, 1976; Pape et al., 1977; Nelander and Blaurock, 1978). It is clear from this work that a large amount of the diffuse scatter which is observed in some membrane diffraction patterns can be accounted for in terms of stacking disorder and it seems likely that the disorder is characteristic of fresh material (Blaurock, 1978). It is sometimes difficult, however, to quantify the contribution of oriented, nonmembrane structures to the diffraction. Another problem that arises, as pointed out by Melchior et al. (1979), is that if the electron density profile is 'sharpened' by deconvoluting the disorder from the observed diffraction pattern, then this can introduce truncation ripples in the profile. This can occur because the deconvolution tends to increase the intensities of the highest orders while higher, unobserved reflections remain at zero intensity, thus causing an abrupt change in intensity at the limit of the observed diffraction.

IV. Applications to Membrane Systems

A. Lipids

It has long been appreciated that lipids play a central role in membrane structure. Over the last ten years or so, in particular, this has stimulated a vast literature on the structure of lipid/water phases. The variety of structures that have been observed is remarkable, and Luzzati and his colleagues have repeatedly stressed the extraordinary polymorphism that lipids display over a relatively narrow range of conditions. This work has been reviewed several times and we refer interested readers to these reviews (Luzzati, 1968; Levine, 1973; Shipley, 1973; Luzzati and Tardieu, 1974). Of the large number of long-range organizations that occur, the one-dimensional lamellar phase is clearly the most relevant to the structure of biological membranes. In this phase the lipids form a bimolecular layer and thus a hydrocarbon core is sandwiched between two planes of hydrated polar headgroups. All of the biological membranes studied so far using X-ray diffraction have been shown to include a basic lipid bilayer structure. While not forgetting that other lipid structures exist, we will restrict our discussion to the properties of the lamellar phase.

A further classification (Luzzati, 1968) distinguishes different lamellar systems on the grounds of the short-range order of the hydrocarbon chains. If the chains are fluid and give a broad diffraction band centered at about 4.6 Å, similar to diffraction from liquid paraffins, then the phase is called Lα (see Fig. 4). If the chains are stiff and parallel then they pack hexagonally (with a lattice constant of about 4.85 Å) and give a sharp reflection at about 4.2 Å; when the chains are perpendicular to the bilayer plane the phase is called Lβ. Another phase has now been recognized to consist of bilayers which contain regions of fluid and crystalline chains (Ranck et al., 1974). This structure has been called, not surprisingly, L$\alpha\beta$.

Under normal physiological conditions the diffraction from biological membranes shows that the chains are in a fluid state, although diffraction from crystalline chains can sometimes be observed in membranes which have been cooled or in the membranes of certain bacteria which have been supplemented by various fatty acids (Esfahani et al., 1971). The X-ray diffraction technique is sensitive to quite small amounts of crystalline chains, a few percent being fairly easily identifiable (e.g., Davis et al., 1976; Ranck et al., 1974). While it remains quite possible that small clusters of more ordered lipids might occur in biological membranes, and these clusters may be involved in structural transitions, it seems unlikely that extensive regions of membrane lipids will exist in the crystalline state. We will therefore only consider structural studies of lamellar systems which contain lipid bilayers whose hydrocarbon regions are in a fluid state.

Levine and Wilkins (1971) studied the structures of oriented multilayers of egg lecithin and mixtures of egg lecithin and cholesterol. They observed that upon increasing hydration, the thickness of the pure lecithin bilayer was reduced with an accompanying increase in the cross-sectional area occupied by the lipid molecules. When cholesterol was present in the bilayer, the thickness remained roughly constant. Electron density profiles were derived, and the effects of adding cholesterol were clearly seen. In particular, the addition of cholesterol caused a localization of the terminal methyl groups in a region around the center of the bilayer. Some localization of the methyl groups was also evident from the profiles with egg lecithin alone. The interaction of cholesterol with lipids has been reviewed by Phillips (1972).

Torbet and Wilkins (1976) extended these earlier studies on egg lecithin by recording and analyzing higher resolution diffraction patterns. They studied the effects of hydration upon the bilayer electron density profiles in more detail and also observed a shoulder on the inner side of the headgroup peak. By a comparison with much higher resolution profiles, obtained when the hydrocarbon chains were in a crystalline state, and using model-building calculations, Torbet and Wilkins concluded that this shoulder corresponded to the fatty acid ester groups of the lecithin molecule.

Using a combination of X-ray and neutron diffraction, Franks (1976) and Worcester and Franks (1976) analyzed the structure of egg lecithin/cholesterol bilayers more fully. From these studies the positions of the principal molecular groups in the bilayer were determined. An extensive X-ray study on the effects of cholesterol on lecithin bilayers has been performed by McIntosh (1978). He found that cholesterol tended to stabilize the structure of the bilayer over a range of different temperatures and hydrations. This was also discussed by Franks and Lieb (1979), who investigated the structure of bilayers of dimyristoyl lecithin and cholesterol. They pointed out that membranes which shared common transmembrane proteins (for example the sodium pump) might be expected to have the same hydrocarbon thickness, and suggested that cholesterol may act as a 'thickness buffer'. Franks and Lieb derived electron density profiles at 4 Å resolution, which is remarkably high for a bilayer with fluid chains. This high resolution revealed features in the bilayer profiles which could be directly attributed to variations in the electron density of the cholesterol molecule itself. This is illustrated in Fig. 8, where the bilayer profile is compared to the electron density distribution along a cholesterol molecule, calculated using the single crystal data of Craven (1976). The two peaks in the cholesterol profile match peaks in the bilayer profile when the cholesterol hydroxyl group is close to the position of the fatty acid ester groups of the lecithin molecule. By a comparison with results obtained using neutron diffraction, Franks and Lieb showed that the cholesterol molecule was immersed in the hydrocarbon region, with its hydroxyl group at the water interface, confirming earlier results (Franks, 1976; Stockton and Smith, 1976; Worcester and Franks, 1976).

The interactions between cholesterol and both sphingomyelin and lecithin have recently been studied by Calhoun and Shipley (1979), and diffraction studies on oriented multilayers of sphingomyelin/cholesterol have also been reported (Khare and Worthington, 1977). Khare and Worthington concluded that part of the hydrophobic cholesterol nucleus was associated with the polar region of the sphingomyelin molecule. The same conclusion had been drawn previously by Rand and Luzzati (1968)

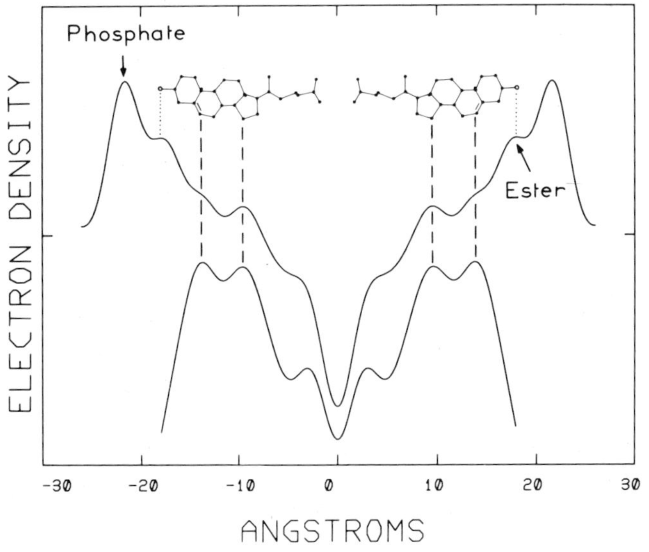

Fig. 8. The position of cholesterol in the bilayer. *Top*: cholesterol atomic coordinates; *middle*: lecithin/cholesterol bilayer electron density profile; *bottom*: cholesterol profile calculated using the atomic coordinates supplied by Dr. B.M. Craven (Franks and Lieb, 1979)

from studies of extracted red blood cell lipids which, interestingly, also contain some sphingomyelin. Both of these studies were at relatively low resolution, however, and it remains to be seen whether or not the apparent difference in the position of cholesterol in bilayers of lecithin and sphingomyelin will be confirmed by higher resolution data.

B. Protein/Lipid Model Systems

The obvious advantage in studying protein/lipid systems composed of purified membrane components, rather than the intact membrane itself, is that the interpretation of the diffraction results is enormously simplified and quite firm conclusions can often be drawn about the structure of the protein/lipid phase. An equally obvious disadvantage is the lack of certainty that the configuration and disposition of the protein molecules are the same in the model system as they are in vivo. This is particularly so for protein molecules which have a clear functional asymmetry, such as transport systems, since in general it will be difficult to reconstitute a model system and ensure that all the protein molecules are facing the same way across a lipid bilayer. In some cases the retention of enzyme activity or transmembrane pumping can be used as an assay in the reconstituted system. Several workers have shown the importance of correlating X-ray diffraction results with electron microscopy whenever possible (see, for example, Ranck et al., 1974; Gulik-Krzywicki, 1975).

Because of the relative chemical homogeneity of model systems compared to intact membranes, where dozens of different protein and lipid species may be present, useful information can be obtained simply by measuring the lamellar repeat distance, D. If the relative weight concentrations and mass densities of the constituents are known, then their effective partial thicknesses can be calculated (see Gulik-Krzywicki et al., 1969). The partial thickness of a given component is the length that it would occupy in the one-dimensional unit cell if it were compressed into a single slab which excluded the other components. Obviously this distance need not (and indeed usually will not) correspond to any real width in the structure, nonetheless a comparison of the relative partial thicknesses of protein, lipid and water for different systems can give useful information about the interactions between the components. For example, the partial thickness of lipid, d_ℓ, is given by

$$d_\ell = \phi_\ell D \tag{27}$$

where ϕ_ℓ is the volume fraction of the lipid and is given by

$$\phi_\ell = \frac{C_\ell \bar{V}_\ell}{C_\ell \bar{V}_\ell + C_p \bar{V}_p + C_w \bar{V}_w} \tag{28}$$

where C is the weight concentration, $\bar{V}$ is the partial specific volume, and the subscripts p,ℓ and w refer to protein, lipid, and water. (In pure lipid systems it is more convenient (Levine, 1973) to use the partial thickness of hydrocarbon since this will equal the actual hydrocarbon thickness as it is naturally a water-excluding region.)

The effect that the presence of protein has on the value of the lipid partial thickness can be interpreted in terms of two classes of protein/lipid interactions. This was shown by Gulik-Krzywicki et al. (1969), who correlated the results from X-ray diffraction and other spectroscopic measurements (see also Gulik-Krzywicki et al., 1972). If the partial thickness of the lipid does not change when protein is added, then this implies that the interaction is rather superficial; for example, an extrinsic membrane protein associating with the lipid headgroups through mainly electrostatic interactions. If, on the other hand, the addition of the protein results in a substantial reduction in the lipid partial thickness, then this indicates that the hydrophobic regions of the protein are interacting with the lipid hydrocarbon chains; for example, an intrinsic membrane protein penetrating into the bilayer. Both types of interactions have been observed with purified protein/lipid systems.

Gulik-Krzywicki et al. (1969) studied mixtures of both lysozyme and cytochrome c with a number of different lipids under a variety of conditions. Many different phases were observed with lysozyme. Amongst them were two lamellar phases which illustrated the distinction outlined above between electrostatic and hydrophobic interactions. At higher lipid concentrations, d_ℓ remained the same as it was in the absence of protein, whereas at higher protein concentrations, d_ℓ was reduced when protein was added. With cytochrome c, two lamellar phases were described in detail, both of which involved purely electrostatic protein/lipid interactions. One phase had a repeat period of 78.5 Å with a protein partial thickness of 23 Å, while the other had a repeat period of 112 Å and d_p was 42.9 Å. In both cases the lipid partial thickness was the same, and equal to that in a pure lipid system. It was proposed that the structures consisted of layers of cytochrome c, one or two molecules thick, sandwiched between the lipid bilayers (see Fig. 9). Essentially identical structures were also proposed by Shipley et al. (1969) on the basis of their X-ray studies on cytochrome c/phospholipid mixtures.

The structure proposed for the lamellar phase with the smaller repeat period was subsequently confirmed by Blaurock (1973c), who derived one-dimensional electron density profiles in the presence and absence of cytochrome c (see also Gulik-Krzywicki, 1975). Blaurock found that the addition of the protein to small single-walled lipid ves-

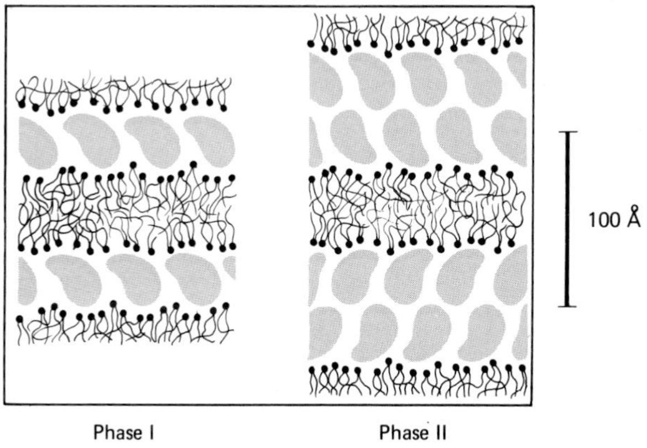

Fig. 9. Two phases found with mixtures of cytochrome c/phosphatidyl inositol/water. The bilayer thickness is the same in both structures (Gulik-Krzywicki et al., 1969)

icles caused the formation of some multishelled structures, resulting in the superposition of sharp reflections on top of a broad diffraction band. He analyzed his results in the light of a theory which had been developed in the preceding paper (Blaurock, 1973a), which predicted the changes that would occur in an X-ray diffraction pattern when protein is added to the surface of a lipid bilayer. Two electron density profiles were derived, one using the intensities of the sharp reflections alone and the other including the broad nonmultilayer diffraction, and these were compared to the profile for the lipids alone. As can be seen from Fig. 10, the two profiles can be interpreted as a similar structure but with differing amounts of protein bound to the bilayer surface. It is clear that the addition of cytochrome c has not affected the dimensions of the lipid bilayer itself.

Although the interaction between cytochrome c and lipids is certainly of interest, since it is known to be associated with the inner mitochondrial membrane, it seems likely that it interacts specifically with a number of other membrane proteins in vivo, so that it may be difficult to relate the results discussed above to the structure of the intact membrane. An extrinsic membrane protein that may interact mainly with lipids is the basic protein in myelin. Mixtures of this basic protein, extracted from either the central nervous system (one protein termed A1) or the peripheral nervous system (two proteins, P1 and P2), with various fractions of myelin lipids were studied by Mateu et al. (1973). Of the three lipid fractions used, acidic, amphoteric, and neutral, only the acidic fraction gave well-organized phases when basic proteins were added. One particularly interesting phase, which is described in some detail, had an unusually large repeat period and gave a sharp reflection at 4.2 Å as well as a broad band at about 4.6 Å, indicating two different populations of hydrocarbon chains. The structural parameters

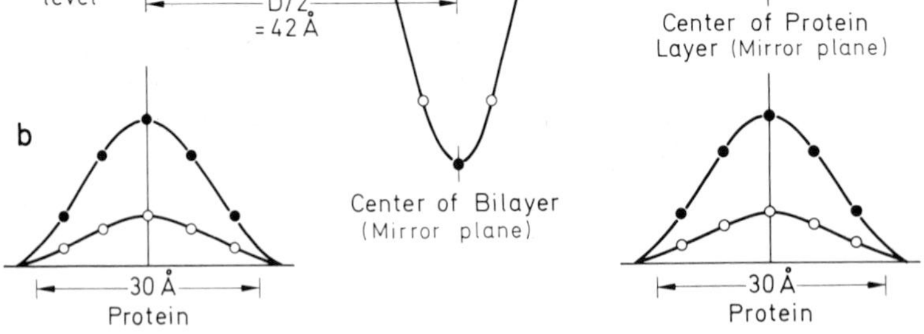

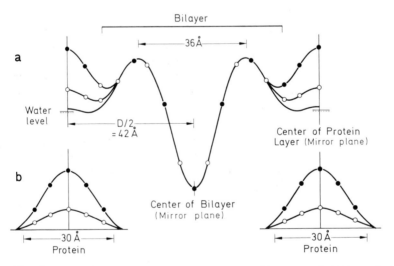

Fig. 10 a and b. Centrosymmetric electron density profiles across a lipid bilayer in the absence (*lower curve*) and in the presence of different amounts (○ and •) of cytochrome c. The differences between the upper two profiles and the lower one in (**a**) are shown below in (**b**). Compare phase I of Fig. 9 (Blaurock, 1973c)

Table 1. Structural parameters of two phases of acidic lipids and myelin basic proteins: A1 from CNS and Pl from PNS. D is the unit cell dimension, c_p, d_ℓ and c_w are the weight concentrations and d_p, d_ℓ and d_w the partial thicknesses of protein, lipid, and water. d_{hc} is the partial thickness of hydrocarbon. The temperature was 20°C (Mateu et al., 1973)

	D (Å)	c_p/c_ℓ	c_w	d_p (Å)	d_ℓ (Å)	d_w (Å)	d_{hc} (Å)
CNS	154	0.32	0.10	25.0	113.0	16.0	81.0
PNS	175	0.42	0.15	33.8	113.0	28.2	81.0

for the phases observed with two different basic proteins, one from the peripheral nervous system (PNS) and the other from the central nervous system (CNS), are given in Table 1. What is striking is the large value of d_ℓ, which suggests that the phase contains two lipid bilayers in the repeat period. Using the constraints imposed by the calculated partial thicknesses, and the pattern recognition procedure of Luzzati et al. (1972), a small number of possible electron density profiles were derived. The final choice between these profiles was made by comparing the hydrocarbon regions of the PNS and CNS structures and picking the sign combination that gave the most similarity in this region. The resulting electron density profile and the molecular interpretation is shown in Fig. 11. It was proposed that the lipids segregated into two bilayers, one of mainly sulphatides with predominantly frozen chains, and the other of mainly phospholipids with fluid chains. The asymmetry in the headgroup peak widths indicates a preferential affinity of the protein for the sulphatides.

The authors point out the amusing correlation between the repeat periods which they observe (listed in Table 1) and the repeat periods of intact PNS and CNS myelin (see Fig. 13). They also point out, however, the important differences between their system and the intact membranes. Namely, myelin contains one type of asymmetric membrane whereas their phase contains two types of symmetric membrane. Also their system does not contain cholesterol (myelin lipids have about 40% molar cholesterol), and if cholesterol is added then the repeat period halves.

Rhodopsin can be extracted from retinal rod outer segments using various detergents, and these protein/detergent complexes have been studied by low-angle X-ray diffraction (Sardet et al., 1976) and used to produce a rhodopsin/lipid lamellar phase by gradually removing the detergent molecules in the presence of lipid (Chabre et al., 1972). In the low-angle study, absolute intensities were recorded and a number of parameters which characterized the rhodopsin/detergent complex were derived. It was concluded that the hydrated rhodopsin molecule must be extremely elongated (perhaps as long as 95 Å) and span a rather flat detergent micelle whose thickness is approximately that of a lipid bilayer. The lamellar rhodopsin/lipid phase produced by Chabre et al. (1972) had a repeat period which varied between about 110 Å and 70 Å depending on the water content (70% - 35% by weight). The electron density profile at about 30 Å resolution closely resembled a pure lipid bilayer with peaks at about ± 20 Å from the bilayer center. The partial thickness for the lipids, however, was calculated to be only about 18 Å, compared to about 45 Å for pure lipids, indicating there

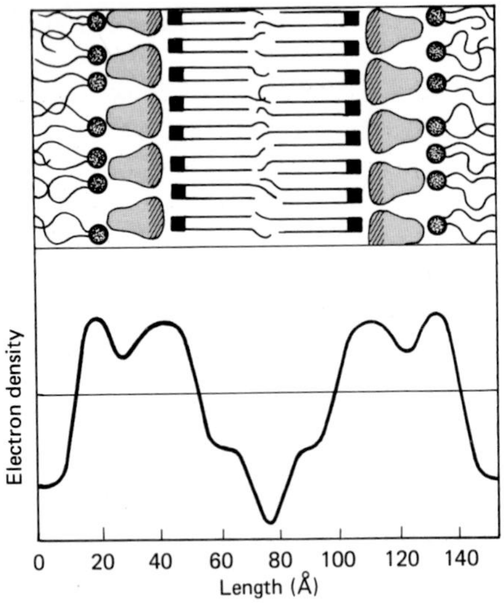

Fig. 11. Proposed structure and electron density profile for a phase composed of an acidic fraction of myelin lipids and a basic protein (A1) from central nervous system myelin (Mateu et al., 1973)

must be substantial penetration of the rhodopsin into the hydrocarbon region of the lipid bilayer. The observation that the electron density profiles with rhodopsin present so closely resemble that of a pure lipid bilayer is, of course, consistent with the protein contributing a fairly uniform level of electron density across the unit cell.

Interestingly, Chabre et al. (1972) reported that they observed additional reflections indicating lipid segregation, and therefore presumably disruption of the protein/lipid phase, at a repeat period close to 75 Å. If the disruption of the rhodopsin/lipid phase occurred due to the interaction between rhodopsin molecules in neighboring bilayers, then this implies a minimum length for the molecule of about 75 Å. Alternatively, if the lipid bilayer is about 55 Å in extent, then a rhodopsin molecule 95 Å long would begin to penetrate neighboring bilayers at a repeat period about 75 Å, which could also explain the observed disruption.

Brady and his co-workers (Brady et al., 1979a, b; Brady and Fein, 1979) have recently applied the methods used to analyze the diffraction from liquids to data obtained from mixtures of egg lecithin and an intrinsic myelin protein, N2. The method relies on the recognition and isolation of the various contributions to the scattering curve. If the various transforms can be separated, then they can be deconvoluted to give electron density profiles. One clear result from this work was the Guinier plots of the low-angle intensity data whose slopes corresponded to a radius of gyration of 41.5 Å. The shape of the scattering curve was found to be consistent with the protein being in the form of a sphere with a diameter of 108 Å. Such a molecule would be sufficiently long not only to span the lipid bilayer in intact myelin but also to interact with protein molecules in neighboring bilayers across the extracellular and cytoplasmic spaces (see Fig. 13). This sort of intermembrane interaction has been observed in myelin by Pinto da Silva and Miller (1975) using freeze-fracture electron microscopy.

C. Myelin

There have been more X-ray diffraction studies on the myelin sheath which surrounds nerve axons than on any other natural membrane. Since the pioneering work of Schmitt et al. (1935, 1941) this has been an area of intensive research which has been reviewed a number of times (e.g., Levine, 1973; Shipley, 1973; Worthington, 1973a; Kirschner and Caspar, 1977). We will therefore only briefly outline the results obtained prior to 1973 and then go on to discuss more recent work.

Up until 1973 there had been many attempts to determine the appropriate set of signs for the first few orders of myelin lamellar diffraction, and several different solutions had been proposed (for details see the reviews cited above). A rigorous proof that the correct choice had been made was not obtained until Blaurock's analysis (1971) of the intensity changes that occurred as a result of swelling at both the extracellular and cytoplasmic boundaries. The phase combination was subsequently confirmed by McIntosh and Worthington (1974) using methods described by Worthington et al. (1973). Low-resolution (about 30 Å) electron density profiles for swollen and compacted myelin are shown in Fig. 12. Although Stamatoff and Krimm (1976) have since proposed that the signs of certain weak orders in patterns from swollen myelin should be different, Blaurock (1979) has convincingly reaffirmed the original set of signs. Moreover, Blaurock suggested that Stamatoff and Krimm had been misled because at the lowest states of swelling there may be some interdigitation of material from neighboring membranes (see Fig. 12) so that their assumption of a constant membrane profile was not valid, even though the membrane structure itself may remain unaltered.

While the signs for the first six orders of nerve myelin are now generally agreed, the correct set of signs for the higher orders is still a matter for debate. There are essen-

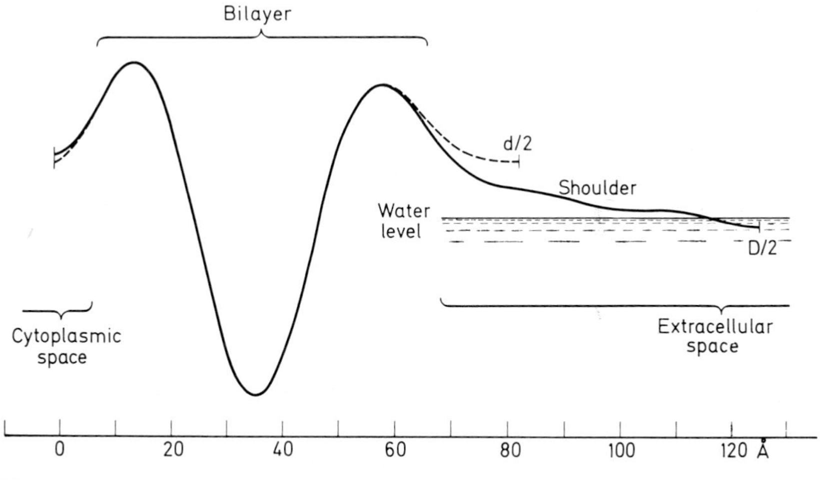

Fig. 12. Low resolution (30 Å) profiles for frog sciatic-nerve myelin. The solid curve is for myelin swollen in water (D = 252 Å). The *shoulder* is interpreted as material extending from the membrane surface; *dashed curve* is for myelin first swollen, and then compacted by 1 mM $CaCl_2$ (d = 166 Å). It was suggested that the elevated density at the extracellular boundary resulted from interdigitation of material from the neighboring membrane (Blaurock, 1979)

tially two different solutions which result in quite different electron density profiles and therefore different molecular interpretations. Caspar and Kirschner (1971) determined the signs of the higher orders by comparing the diffraction from optic and sciatic nerve myelin. They then assumed that each of the membrane units was approximately centrosymmetric and considered the correct combination of signs was that which gave the greatest similarity between the optic and sciatic myelin profiles. The two profiles are shown in Fig. 13. The profiles were interpreted as showing an asymmetric distribution of cholesterol between the two halves of the bilayer, there being a greater concentration on the extracellular side. The protein was said to be associated with the lipid headgroups and distributed in a weak solution in the fluid spaces. It was thought probable that there would be some protein extending across the hydrocarbon layer, but an upper limit of about 15% was put on the space that it could occupy in this region.

An alternative set of signs was proposed by Worthington and King (1971) on the basis of model-building and virtually the same set (the weak seventh order had a different sign) was found by Worthington and McIntosh (1973, 1974) using both sampling methods and deconvolution of the autocorrelation function. The electron density profile preferred by Worthington and McIntosh (1974) is shown in Fig. 14. It is clearly quite different to the Caspar and Kirschner profiles (Fig. 13) and was interpreted as showing a symmetric lipid bilayer (with a symmetric cholesterol distribution) with substantial interdigitation of the hydrocarbon chains, and layers of protein on the surfaces, there being more on the cytoplasmic side.

Recent results now allow a definitive choice to be made between the two proposed solutions. In 1976 Blaurock noted the similarity between the high-angle continuous diffraction from swollen myelin and a calculated single-membrane transform. This point was stressed further by Nelander and Blaurock (1978), who showed that if there was substantial variability in the width of the cytoplasmic space, then the higher-angle diffraction data would be dominated by the Fourier transform of the single-membrane unit and not that of the membrane pair. They observed some variability in the cytoplasmic space in normal myelin and argued that in swollen myelin this variability would be greater. Since the methods used by Worthington and McIntosh (1974) assume that the higher-angle data correspond to diffraction from the membrane pair, the results of Nelander and Blaurock cast serious doubt on the validity of their phase assignment.

Other evidence, which comes from recent studies with extracted myelin lipids, also strongly supports the phase solution of Caspar and Kirschner. In their method of determining the phases they made the a priori assumption that the structures of optic and sciatic nerve myelin were very similar. The similarity between the derived electron density profiles of the intact membranes (Fig. 13) is necessarily consistent with, but does not prove, the validity of the original assumption. Independent information has been obtained by studying the structures of bilayers of extracted lipids. Because of the much smaller repeat period observed with the extracted lipids, electron density profiles can be derived at a higher resolution than the intact myelin profiles using only a few reflections whose phases can be determined unambiguously. Moreover, since the dry weight of myelin consists of about 75% lipid, it is reasonable to expect that the

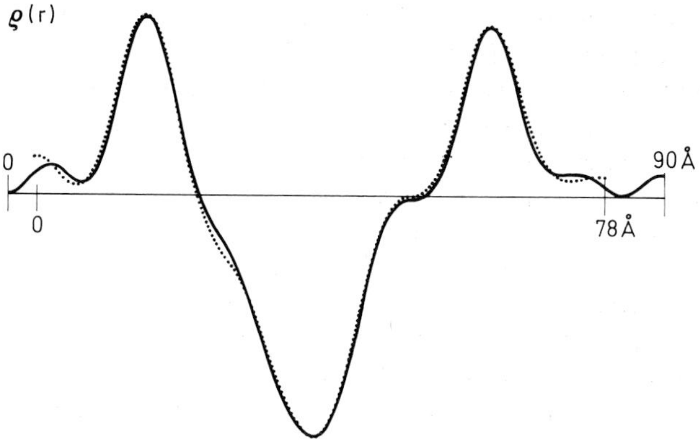

Fig. 13. High resolution (10 Å) profiles for rabbit sciatic-nerve (*solid curve*) and rabbit optic-nerve (*dotted curve*) myelin according to Caspar and Kirschner (1971). The extracellular space is on the right (Kirschner and Caspar 1972)

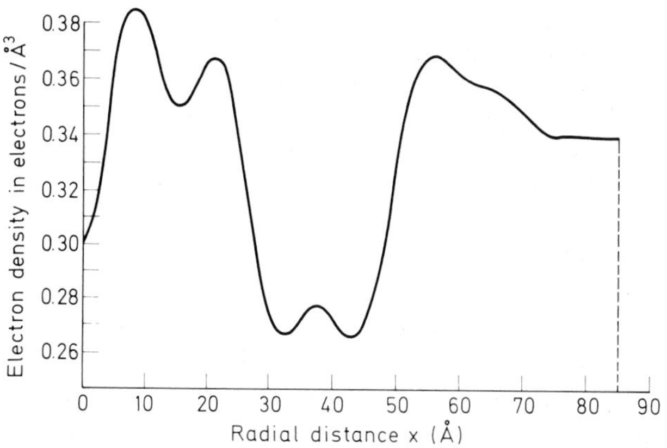

Fig. 14. Absolute electron density profile for frog sciatic-nerve myelin (D = 171 Å) at 14 Å resolution, according to Worthington and McIntosh (1974). The extracellular space is on the right (Worthington and McIntosh, 1974)

shape of the electron density profile of the intact membrane will largely reflect the structure of the bilayer that the extracted lipids form, although, of course, the lipid profiles will be centrosymmetric.

Electron density profiles for lipids extracted from peripheral (intradural root) and central (brain stem) nervous system myelin are shown in Fig. 15. These profiles are part of an extensive study on the structure of oriented multilayers of myelin lipids (Franks, Melchior, Kirschner and Caspar, in preparation). The first point to be made is the close similarity between the profiles for the lipids from PNS and CNS myelin, which strengthens the assumption made by Caspar and Kirschner in their analysis of the intact membranes. Secondly, there is a strong resemblance between the lipid profiles of Fig. 15 and the profiles of Fig. 13 and it is quite easy to account for the profile of the intact membrane in terms of the bilayer that the extracted lipids form and the addition of small amounts of protein. On the other hand, it would be much more difficult to account for the shape of the profile in Fig. 14 in terms of the bilayer of extracted lipids. (We note that the myelin used by Worthington and McIntosh was from frog, while the profiles in Fig. 13 are for rabbit, however, when their phase assignment is applied to the rabbit data, a very similar profile to Fig. 14 results.) Invoking Occam's razor therefore would strongly favor the phase assignment of Caspar and Kirschner for the higher orders of nerve myelin.

Nelander and Blaurock (1978) have reexamined possible phase assignments for the higher orders in the light of their analysis of stacking disorder in myelin (Blaurock and Nelander, 1976). They had previously shown that a large part of the oriented diffuse scatter observed in myelin diffraction patterns could be accounted for by the lack of long-range order in the myelin stack. They derived parameters which characterized

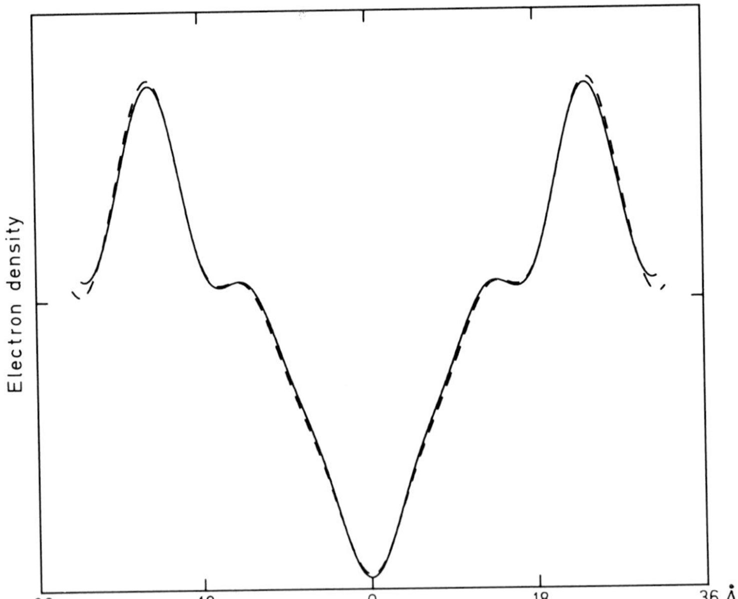

Fig. 15. Centrosymmetric profiles for extracted myelin lipids. *Solid curve*: lipids from central nervous system myelin (D = 61.17 Å); *dashed curve*: lipids from peripheral nervous system myelin (D = 62.21 Å). For each profile, 7 orders were used in the Fourier synthesis. The data are taken from Franks, Melchior, Kirschner and Caspar (in preparation). See also Melchior et al. (1979)

this disorder and found, surprisingly, that the cytoplasmic space was more variable than the larger extracellular space. Nelander and Blaurock (1978) then compared the observed diffuse scatter with that predicted, using each of the possible phase combinations for the higher orders. They concluded that, in addition to the Caspar and Kirschner set of phases, there were three other sets (the signs of either or both of the 8th and 15th orders were reversed) which also accounted for the diffuse scatter, so they could not distinguish between these phase solutions. Since the intensities of the 8th and 15th orders are relatively weak, however, all four electron density profiles are broadly similar. The calculated profiles were compared (Blaurock and Nelander, 1979) to the electron density distribution across a bilayer of egg lecithin and 40% molar cholesterol. This was a sensible comparison to make since myelin lipids also have about this proportion of cholesterol. It was concluded that the extracellular side of the intact myelin membrane could not be accounted for by the lipids alone and that there must be some penetration of protein into this half of the hydrocarbon region. This conclusion was based partly on the observation that the height of the shoulder in the low-resolution lecithin/cholesterol profile was about midway between the peak and trough densities, while in the intact membrane it was significantly closer to the peak density. The profiles of Fig. 15, however, show that for myelin lipids the cholesterol shoulder is indeed that much closer to the headgroup density and thus the extracellular half of the intact myelin bilayer may, in fact, be accounted for in terms of the lipid alone without the need for protein penetration. This does not exclude the possibility that a uniform layer of protein extends across the hydrocarbon region.

In an attempt to characterize the forces which stabilize the myelin sheath, Kirschner and Caspar (1975) studied the reversible structural transformation that occurs when myelin is exposed to dimethlysulfoxide (DMSO). They observed the formation of a compacted phase (with a repeat period of about two-thirds that of the native period) which coexisted with the native myelin, but whose amount varied with DMSO concentration. Above about 40% DMSO only the compacted phase remained. The authors concluded that the myelin sheath was stabilized by both long-range forces between the membranes and short-range interactions with the interstitial protein. A similar compaction has been reported by Melchior et al. (1979) for either central or peripheral nerve myelin when the nerve was exposed to at least 10 mM Ca^{2+}. An electron density profile for the compacted myelin was compared with the profile for Ca^{2+}-flocculated myelin lipids and found to correspond closely over the hydrocarbon region. Large differences were apparent in the headgroup region where protein had been compacted. A series of striking electron micrographs showed that the compacted phase corresponded to membranes which excluded the particles observed by freeze-fracture. It was suggested that this lateral segregation of the particles is driven by the attraction of the lipid bilayers.

In an interesting and promising approach to correlating chemical composition and structure, Kirschner and Sidman (1976) have compared the myelin from different neurological mutants and from animals of different ages. They observed significant differences in the diffraction patterns. For example, they found that the membranes of immature myelin were about 2 Å thinner than the mature membranes, possibly reflecting the lower protein content. In one mutant, the quaking mouse, the electron density level of the extracellular fluid was significantly higher than that of the normal membrane.

D. Disk Membranes of the Rod Outer Segment

Retinal rod outer segment is another system, like myelin, where membranes occur naturally in relatively well-ordered stacks. This has been exploited by several groups who have recorded the lamellar diffraction from intact retina and analyzed the data to give one-dimensional electron density profiles across the membrane pair (see, for example Blaurock and Wilkins, 1969; Corless, 1972; Worthington, 1973b; Chabre and Cavaggioni, 1973; see also Shipley, 1973, for other references). By 1973 there was broad agreement about the appropriate set of signs for the first eleven reflections, although the phases for a couple of the weak orders were ambiguous. Consequently, the electron density profiles all looked rather similar and it was generally accepted that the membrane contained a basic lipid bilayer structure. The profiles of Worthington (1973b), however, were rather more asymmetric with a higher peak on the side of the membrane facing the inside of the disk. This was largely due to the fact that no explicit Lorentz correction was applied to the intensities. Worthington interpreted this asymmetry as showing that the rhodoposin molecules were located on the inside of the disk membrane.

Both Corless (1972) and Chabre and Cavaggioni (1973) observed small, but significant, changes in the relative intensities of the lamellar reflections after total bleaching of the photopigment. Both groups found that these changes in intensity corresponded to an increase in electron density on the cytoplasmic side of the membrane (i.e., the outside of the disk). In addition, exploiting the greater time resolution available with a position-sensitive detector, Chabre and Cavaggioni found a sudden decrease in repeat period (of about 1/2%) and an increase in long-range disorder upon bleaching. After some time (depending upon the temperature), the reflections sharpened again and the repeat period slowly increased. In the absence of Ca^{2+} ions, the sudden shrinkage was enhanced, and it was suggested that the reduction in repeat period was a result of switching off the sodium ion current which flows into the outer segment in the dark. Blasie (1972a, b) also investigated changes in the diffraction pattern during bleaching and proposed a model in which the rhodopsin molecules moved into the membrane when the photopigment was bleached.

The effects of illumination upon the membranes were reinvestigated by Chabre (1975) and Chabre and Cavaggioni (1975). They used both intact retina and isolated rod outer segments which had been detached from the retina. The isolated outer segments were oriented by a magnetic field and allowed to sediment under gravity. Chabre confirmed the previous conclusion that, upon bleaching, there was an immediate increase in electron density on the cytoplasmic side of the membrane. The electron density profiles before and after bleaching are shown in Fig. 16. The difference between the two profiles (the dotted curve) has been magnified five-fold. The increase in the long-range disorder and subsequent slow swelling that was observed with intact retina (Chabre and Cavaggioni, 1973) did not occur with the isolated rod preparation. Chabre questioned the conclusions of Blasie (1972a, b) about the movement of the rhodopsin molecule into the membrane upon bleaching. It was shown that the specimen geometry used by Blasie (isolated disks were stacked by centrifugation and the X-ray beam was oriented parallel to the sedimentation axis) could result in contamination of the true equatorial scattering by strong lamellar diffraction because of imperfect orientation of the membranes. Chabre found no measurable change in the true equatorial diffraction (at about 55 Å) as a result of bleaching.

The similarity of the electron density profile across the rod outer segment membrane to that of a pure lipid bilayer was stressed by Chabre (1975), although he suggested that the whole profile was elevated with respect to the water level due to the presence of protein. It is worth noting that the comparison was made with a bilayer of synthetic lipids whose chains were in a frozen state (Luzzati et al., 1972) and that the obviously more appropriate comparison with bilayers of lipids extracted from the rod outer segment has not yet been made.

A small transient decrease in the repeat period was observed upon bleaching (Chabre and Cavaggioni, 1975) with the isolated rods. The explanation of the similar, although larger, shrinkage observed with the intact retina was that the sodium dark current was switched off. This could not account for the observation in the isolated rod preparation since very few inner segments or cell bodies would be present. The authors therefore suggested that an active ion pump might exist in the outer segment membrane and that an active efflux of ions was causing the observed shrinkage with the isolated rods.

Using the data of Chabre and Cavaggioni, Schwartz et al. (1975) reanalyzed the lamellar diffraction and attempted to account for the stacking disorder in the specimen by assuming a Gaussian distribution of distances between the membranes across both the interdisk and intradisk spaces. As a result of the analysis, the characteristic widths of these distributions were found to be 19 Å and 8 Å respectively, for a repeat period of 295 Å and a mean distance between the membranes across the intradisk space of 88 Å. The combination of signs proposed by Schwartz et al. for the lamellar diffraction was the same as Chabre's except for a different sign for the very weak fifth order. The resulting electron density profile was very similar to those published previously.

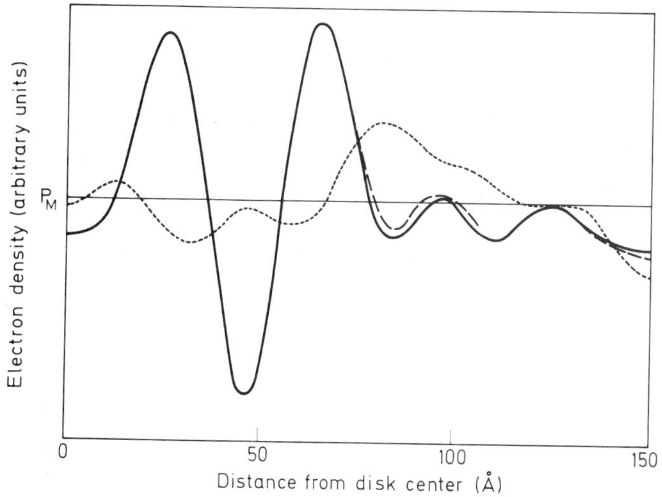

Fig. 16. Electron density profile across one membrane in the rod outer segment disk. (——), dark adapted; (- - - -), bleached; (· · · ·), difference profile enlarged fivefold (Chabre, 1975)

E. Sarcoplasmic Reticulum

The membranes of the sarcoplasmic reticulum regulate the contractile cycle of muscle by the passive release and active transport of calcium ions. The calcium ATPase polypeptide has a molecular weight of approximately 100,000 and constitutes the majority of the protein in the membrane. In purified fractions, the Ca^{2+}ATPase can make up as much as 90% of the total protein, which makes this membrane system a particularly attractive choice for low-angle X-ray diffraction studies.

Using a purified membrane fraction, Dupont et al. (1973) prepared oriented stacks of membranes by centrifuging a dispersion of vesicles and allowing the resultant pellet to air-dry in the cold for varying times. This technique of centrifugation and controlled drying had previously been used by Coleman et al. (1969) for a variety of membranes. The diffraction patterns showed up to 13 orders of a lamellar repeat which varied between 170 Å and 270 Å (corresponding to about 25% - 50% by weight water) although a 60 Å reflection, characteristic of a separate lipid phase, appeared if the specimens were allowed to dry too much. Since the specimen consisted of a stack of vesicles, the repeat unit was a pair of membranes placed back-to-back forming a centrosymmetric pair.

The Patterson function was calculated for each swelling state and showed a constant peak at about 37 Å, which was interpreted as the distance between the electron-dense lipid headgroups across the bilayer, and a peak which varied between about 60 Å and 90 Å during swelling, which was interpreted as the distance between the two membranes across the inside of the vesicle (see Fig. 17). The difference between this value and the total repeat period was then taken to be the separation of the membranes between neighboring vesicles. It was found that there were two distinct regions of swelling; at lower water contents, the space within each vesicle remained constant so that during swelling the collapsed vesicles simply moved apart. At higher water contents (at lamellar spacings beyond about 220 Å) swelling occurred both inside and between the vesicles with the space inside the vesicles changing more rapidly (see Fig. 17b).

The data were analyzed by plotting two distinct swelling series, the first (low water contents) assuming a constant structure with respect to an origin at the center of a vesicle, and the second (at higher water contents) assuming a constant structure with respect to an origin at a point between the vesicles. While the first assumption is sound, since the intravesicle space remained constant with swelling for the 'dry' transform, the second is less secure, since some swelling was observed between the vesicles in the higher water content range. Nonetheless, both sets of data showed a number of well-defined loops and nodes so that the relative signs of the bands could be fairly easily determined on the basis of the minimum wavelength principle. The data beyond about 0.035 $Å^{-1}$ could not be given signs relative to the bands at lower angles, however, due to the presence of an extended zero in the data near this region.

The electron density profiles that were derived (at a resolution of about 30 Å) are shown in Fig. 18 for three different states of swelling. The changes seen in the profiles are consistent with a rather constant membrane structure and variable fluid spaces and thus, to a large extent, justify the choice of phases. Each membrane shows a characteristic bilayer profile, with electron-dense peaks at about ± 20 Å from the bilayer

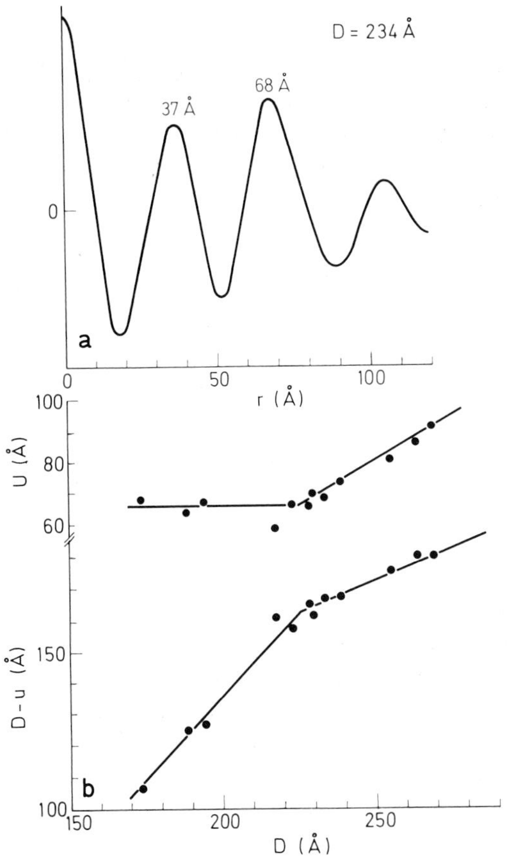

Fig. 17 a and b. One-dimensional Patterson function for a pair of sarcoplasmic reticulum membranes. The peak at 37 Å corresponds to the distance between the bilayer headgroups and the peak at 68 Å is the distance between the membrane centers across the inside of the vesicle. **b** Intravesicle (u) and extravesicle (D - u) distances plotted versus the repeat period (D) (Dupont et al., 1973)

center. In addition, there is a significant peak of density about 54 Å from the center of the bilayer which is identified with the projections seen at the surfaces of sarcoplasmic reticulum vesicles using the electron microscope. The partial thickness of the lipids was calculated to be about 26 Å which indicates, as discussed in Sec. IV.B, that a substantial portion of the protein molecule penetrates the lipid bilayer.

In a later paper, Dupont and Hasselbach (1973) showed that as accessible SH groups were blocked (there are about 10-12 groups accessible per 10^5 daltons of protein) a linear decrease in Ca^{2+} ATPase activity was accompanied by a progressive decrease in the feature at about 54 Å from bilayer center seen in the electron density profiles. When four or more groups were blocked there was no measurable Ca^{2+}-activated ATPase activity and the electron density profiles had become symmetric, with a halving of the fundamental repeat unit.

At about the same time, Worthington and Liu (1973) published a rather different electron density profile for the sarcoplasmic reticulum membrane. Oriented stacks of membranes were also prepared by centrifugation and lamellar spacings between about 230 Å and 270 Å were recorded. In these experiments it was found that the space inside the vesicles remained relatively unchanged, so that the origin was chosen to be inside the vesicle. The electron density profile that was calculated differs from

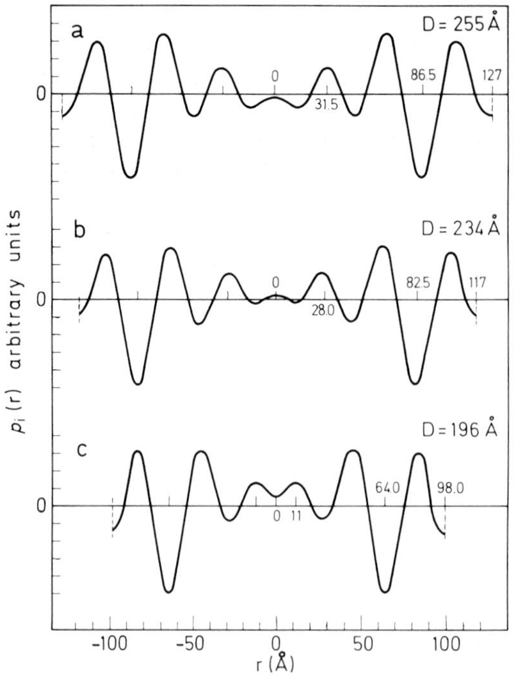

Fig. 18 a-c. Sarcoplasmic reticulum profiles for three different swelling states. Note the electron dense peak at about 54 Å from the bilayer center. The origin is in the center of the extra-vesicle space (Dupont et al., 1973)

those of Dupont et al. for two reasons: firstly, a different sign is chosen for the first band, and secondly, it is argued that a Lorentz factor is unnecessary. The resulting profile is very asymmetric, with the highest peak on the inside of the vesicle which is interpreted as being the site for the majority of the membrane protein, the opposite conclusion to that drawn by Dupont et al. (1973). In a subsequent paper, Liu and Worthington (1974) argue in favor of their choice of sign for the first band on the basis of experiments using various concentrations of glycerol in the suspending medium. It may not be valid, however, to compare the signs for the two sets of data directly, since the sign of the first band depends upon the sign of F(O), which can be either positive or negative depending on whether the mean electron density of the membrane is greater than, or less than, that of the suspending medium. In the specimens of Dupont et al., the electron density of the fluid between membranes is not known, since the membranes were centrifuged in buffer and the resulting pellets partially dehydrated.

More recent work (Herbette et al., 1977) has arrived at the same choice of signs and has supported the general conclusions of Dupont et al. (1973). Using a similar membrane fraction and much the same technique for preparing specimens, Herbette et al. recorded diffraction patterns which showed only three sharp reflections (as opposed to the 13 or so which Dupont et al. observed), with the higher angle diffraction consisting of broad bands. The data were analyzed using the method of Schwartz et al. (1975) which attempts to account for the stacking disorder in the specimen, although it was assumed in this case that there was no significant variation in the intravesicle space. The data were phased unambiguously out to about 20 Å resolution and the resulting membrane profile looks rather like those of Dupont et al., though with the detail somewhat smoothed out. It is possible that this smoothing is a consequence of there being some variabilty in the intravesicle space, although Herbette et al. argue that this disorder is small. As with the profiles of Dupont et al., there is clear evidence of significant electron density extending beyond the lipid bilayer and this density is interpreted as protein. The data observed at higher angles (out to about 11 Å) could not be phased unambiguously and two alternative profiles were presented. Neither of these profiles would weaken the conclusion that there is a substantial amount of protein on the external surface of the vesicle.

Oriented bands of diffraction were also observed at about 10 Å on the equator and about 5.2 Å on the meridian, which were interpreted as arising from α-helices oriented perpendicular to the membrane plane (see also Sec. IV.H and Fig. 22).

Although it is true that the earlier workers (i.e., Worthington and Liu [1973], and Dupont et al. [1973]) did not take into account the stacking disorder in their specimens, we feel that this omission is not as serious as it may seem, since it is clear from the published patterns that the disorder present in the specimens of Herbette et al. is very much greater than that observed by either of the other two groups.

Both Dupont et al. and Herbette et al. performed the important measurement of the enzymatic activity of the membranes after specimen preparation and X-ray exposure. While the former group measured ATPase activity and Ca^{2+} uptake after resuspending the pellets, Herbette et al. employed the added refinement of monitoring the Ca^{2+} uptake spectrophotometrically, using arsenazo III as a calcium indicator. Both groups found that a substantial fraction of the normal enzyme activity and Ca^{2+} pumping remained after centrifugation, partial drying, and X-ray exposure.

F. Red Blood Cell Membranes

In a series of papers, Finean and his co-workers studied a number of different erythrocyte membrane preparations under a variety of conditions, using both electron microscopy and X-ray diffraction (e.g., see Knutton et al., 1970, and Shipley, 1973, for other references). Their principal conclusion was that the membrane was an asymmetric structure, about 100 Å thick. They also stressed the lability of some preparations and showed that even moderate dehydration of hemoglobin-free membranes resulted in the separation of the membrane components into different phases.

Wilkins et al. (1971) recorded patterns from dispersions of hemoglobin-free erythrocyte membranes and interpreted the observed bands as showing that the membrane must contain a basic lipid bilayer structure whose thickness is about 45 Å.

More recently, Stamatoff et. al. (1975) attempted to stack hemoglobin-free erythrocytes by centrifugation and controlled drying. The patterns that they recorded showed a surprisingly small periodicity (between about 55 Å and 70 Å) and the resulting electron density profile showed a single symmetric membrane with electron-dense peaks at about ± 23 Å from the bilayer center. The striking resemblance of this profile to those obtained with pure lipids (e.g., lecithin/cholesterol, Levine and Wilkins, 1971) was noted by Blaurock and Lieb (1975), who suggested that the process of centrifugation and partial drying had resulted in the separation of a pure lipid phase. This conclusion was reinforced by Finean et al. (1975), who once again emphasized the ease with which phase separation can occur upon dehydration.

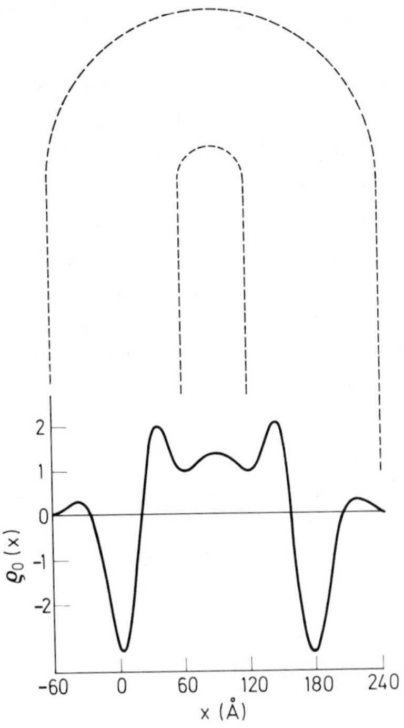

Fig. 19. Electron density distribution across a flattened red blood cell ghost. The repeat period was 600 Å and only the profile across the membrane themselves is shown (Pape et al., 1977)

Using a new technique, Lesslauer (1976) prepared oriented stacks of erythrocyte membranes by agglutinating the cells with phytohemagglutinin M. A series of sharp lamellar peaks was observed with spacings ranging from about 160 Å - 186 Å, and this period was shown by electron microscopy to contain two membranes. On the basis of a qualitative comparison between the intensities of the lamellar orders and the diffraction observed by Wilkins et al. (1971) from dispersions, it was argued that the membranes must be asymmetric. In a subsequent paper, Lesslauer (1978) used the oriented stacks of membranes to record a meridionally oriented band at 1.5 Å and an equatorially oriented band at about 10.5 Å, which were taken as evidence for transmembrane α-helices, packed side-by-side.

Stacks of red blood cell membranes have also been obtained by Pape et. al. (1977) by centrifugation in the presence of sorbitol, which was found to improve the order in the stacking. The specimens were, nonetheless, very disordered and only three discrete reflections of an unusually large repeat period (600 Å) were observed, with the higher-angle reflections merging into a broad band. Because of the large spacing, the autocorrelation function could be directly extracted from the intensity data (see Sec. IIIc). An attempt was made to account for the disorder and a 'sharpened' autocorrelation function, predicted in the absence of intravesicle disorder, was then deconvoluted to give the electron density distribution.

The derived profile is shown in Fig. 19. It shows a pair of asymmetric membranes placed back-to-back. What is most striking is the large asymmetry in the bilayer, there being considerably more density on the cytoplasmic side, and an unusually large distance of about 75 Å between the electron-dense peaks on each side of the low-density trough. The density between the membranes, in the compacted cytoplasmic space, is interpreted as the proteins spectrin and actin which together constitute about 35% of the total membrane protein.

At first sight there appears to be a discrepancy between these results and the data of Wilkins et al. (1971), which were interpreted as showing a bilayer thickness of about 45 Å, and the results of Stamatoff et al. (1975), which also indicate the red blood cell lipids form a bilayer with about this thickness (see also Rand and Luzzati, 1968). If spectrin was present in the specimens of Wilkins et al., and there is no reason to suppose that it was not, then it must have formed a rather diffuse and extensive layer so as not to modulate the intensity distribution over the first bilayer band (Blaurock, 1973a). The main effect of such a protein distribution would be to increase the low-angle scatter, consistent with the relatively high intensity at the first minimum observed by Wilkins et al. The electron micrographs published by Tilney and Detmers (1975), which show an amorphous material (probably mainly spectrin) extending between 100 Å and 200 Å from the cytoplasmic surface of the membrane, would support the above interpretation of the X-ray data. When the membranes are stacked, then the cytoplasmic surfaces come together (in Fig. 19 the inner peaks are about 100 Å apart) and the amorphous material within the cytoplasm will build up to give a significant peak of electron density which would tend to exaggerate the asymmetry and may distort the positions of the bilayer peaks.

G. Gap Junctions

The gap junction is a specialized region where the plasma membranes of two cells are fused together. Preliminary X-ray diffraction patterns (Goodenough and Stoeckenius, 1972) confirmed earlier electron microscopic evidence (Robertson, 1963) that this region contained a regular hexagonal structure in the planes of the plasma membranes.

Using a combination of electron microscopy, X-ray diffraction, and chemical analysis, the three-dimensional structure of the gap junction has been investigated further (Caspar et al., 1977; Makowski et al., 1977). Both the electron microscopy and the X-ray diffraction patterns showed that the hexagonal array in the membrane plane had a lattice constant which varied between about 80 Å and 90 Å and was remarkable in that it exhibited a high degree of long-range order coupled with substantial short-range disorder. Meridional diffraction from oriented gap junctions consisted of a series of broad bands separated by fairly well-defined minima. Assuming the pair of membranes in the gap junction contained a center of symmetry, the observed diffraction bands were given signs on the basis of the minimum wavelength principle. Once again the familiar problem of an extended region of low intensity in the data (see also Dupont et al., 1973; Franks, 1976; and Fig. 6) resulted in an ambiguity in the phasing of the higher bands. This was neatly resolved by comparing the predicted diffraction in this region for the two alternative phase choices with the low intensity observed experimentally. The analysis was slightly complicated by the finding (also observed by electron microscopy) that the gaps had a tendency to stack, so that the transform of a single gap was modulated by an interference function. This interference function was determined, and the electron density profile refined, by imposing a real space constraint based on the maximum thickness of the gap junction which could be defined from

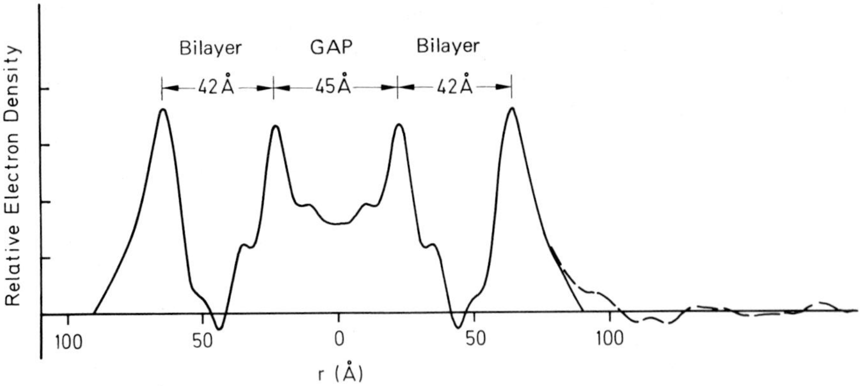

Fig. 20. Gap junction profile. *Solid curve:* profile calculated after correcting for the interference effects due to the partial stacking of the gap junctions; *dashed curve*: ripple that occurs before correction. Note that the electron density at the bilayer center is close to that of water, indicating substantial protein penetration (Makowski et al., 1977)

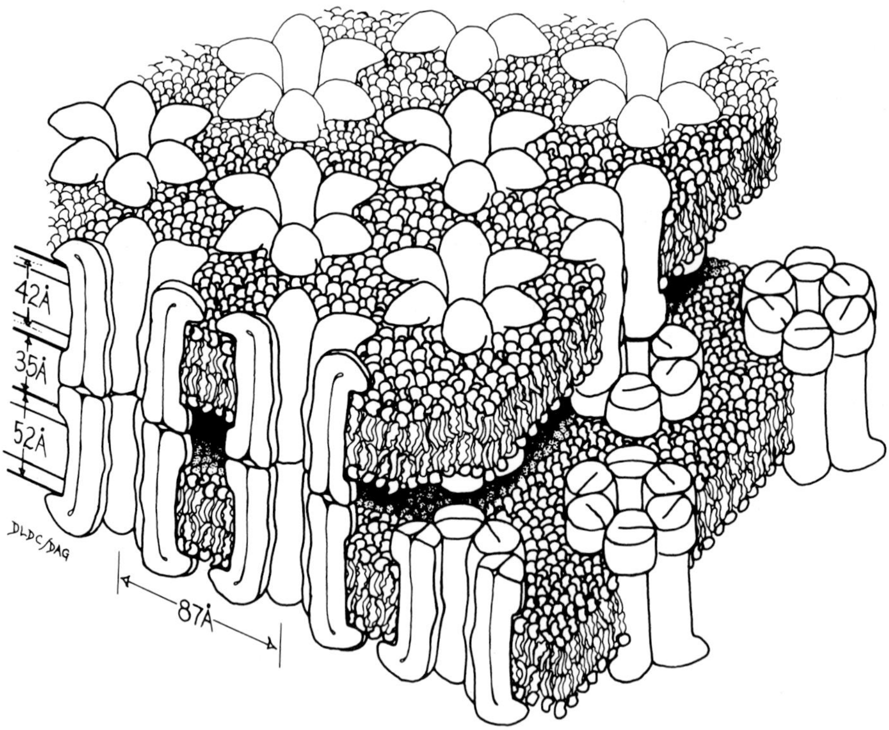

Fig. 21. A three-dimensional representation of a gap-junction structure based on X-ray diffraction, electron microscopy, and chemical analysis. Hexamers of protein molecules form aqueous channels across a pair of lipid bilayers and these hexamers are themselves centered on a hexagonal grid, with a lattice constant of 87 Å. Compare the electron density profile of Fig. 20 (Makowski et al., 1977)

the electron microscopy. Although this refinement, in fact, had little effect on the profile, the pair correlation function corresponding to the interference function was derived and showed that the gaps had a strong tendency to pair, with a nearest-neighbor distance of about 160 Å.

The electron profile across the gap junction is shown in Fig. 20. The two bilayer profiles are greatly elevated with respect to the solvent density, indicating a substantial amount of protein crosses the gap. Also the asymmetric step heights in the hydrocarbon region are reminiscent of the myelin profiles (see Fig. 13). In both cases the higher step is on the extracellular side of the membrane.

The hexagonal reflections on the equator of the X-ray pattern were analyzed to give a two-dimensional projection of the junction. Since, in projection, the junction has P6m symmetry, the structure factors are real so that only signs had to be chosen for the reflections. These signs were chosen so that the electron density map was consistent with the filtered images obtained from the electron microscope using negative stain. It was found that only one sign combination was consistent. The images seen by the two techniques provided complementary information yet also showed significant differences. Both images were interpreted as showing a hexagonal array of protein cylinders with a central channel perhaps extending all of the way across the gap. The diameter of the cylinder in the electron density map, however, was significantly less than the stain-excluding region within the gap which is emphasized in the image from the electron microscope. This was interpreted as showing that protein occupies a larger fraction of the surface area within the gap than it does within the bilayers. A three-dimensional representation of the gap-junction structure corresponding to the electron density profile in Fig. 20 is shown in Fig. 21.

The variations in lattice constant of the hexagonal array, which were observed between specimens, were found to be correlated with the width of the gap; as the gap narrowed (with reducing lattice constant), the protein within the gap spread out to occupy a greater area while apparently maintaining a constant volume. The authors speculated that the structural variations they observed for different specimens may be reflecting conformational changes that occur in vivo during the regulation of cell-cell communication.

H. The Purple Membrane of *Halobacterium Halobium*

The purple membrane of the bacterium *Halobacterium halobium* is now probably the best-known example of a membrane which contains a crystalline arrangement of protein molecules within the plane. These molecules of bacteriorhodopsin are now known to act as light-driven proton pumps (Oesterhelt and Stoeckenius, 1973).

X-ray reflections from an hexagonal array in the membrane plane were first recorded by Blaurock and Stoeckenius (1971), who measured a lattice constant of 63 Å and indexed the reflections out to about 7 Å resolution (see Fig. 22). They also derived an electron density profile from the meridional diffraction (assuming a center of symmetry) which showed a bilayer-type profile with a mean density considerably greater than the solvent density, indicating a substantial amount of protein in the bilayer.

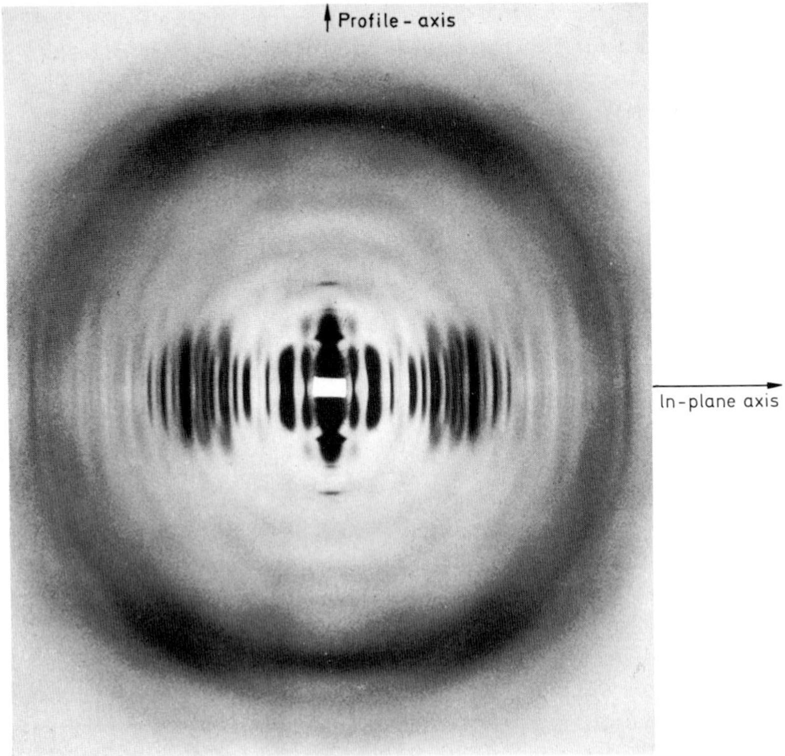

Fig. 22. Low-angle X-ray diffraction pattern of dried purple membranes. The hexagonal reflections along the equator arise from the crystalline array of bacterio-rhodopsin molecules in the membrane plane. Reflections along the meridian profile axis show a membrane stacking distance of about 49 Å (Blaurock, 1975)

Diffraction experiments were continued independently by both Blaurock (1975) and Henderson (1975). Henderson concentrated on the equatorial diffraction and showed that the reflections could be indexed out to at least 4 Å resolution. He concluded that the lattice had P3 symmetry (only one unit cell thick) and that each unit cell contained a cluster of three molecules, i.e., one per asymmetric unit. Diffraction recorded on the meridian at about 5.1 Å and 1.5 Å was interpreted as arising from a coiled-coil arrangement of α-helices, oriented perpendicular to the membrane plane. On the basis of the lengths of the equatorial reflections in the meridional direction, it was shown that these α-helices must extend across most of the membrane. It was found that the relative intensities of the equatorial reflections only changed slightly when the electron density of the suspending medium was increased, implying there is little protrusion of the protein molecules from the surfaces, which must therefore be relatively flat. Also, when the membranes were exposed to a few mM uranyl acetate, only the low orders of the hexagonal reflections were affected. Henderson interpreted this observation as meaning that the lipids were in a disordered state, since it was most likely that the heavy atoms were binding on to the lipid phosphate groups.

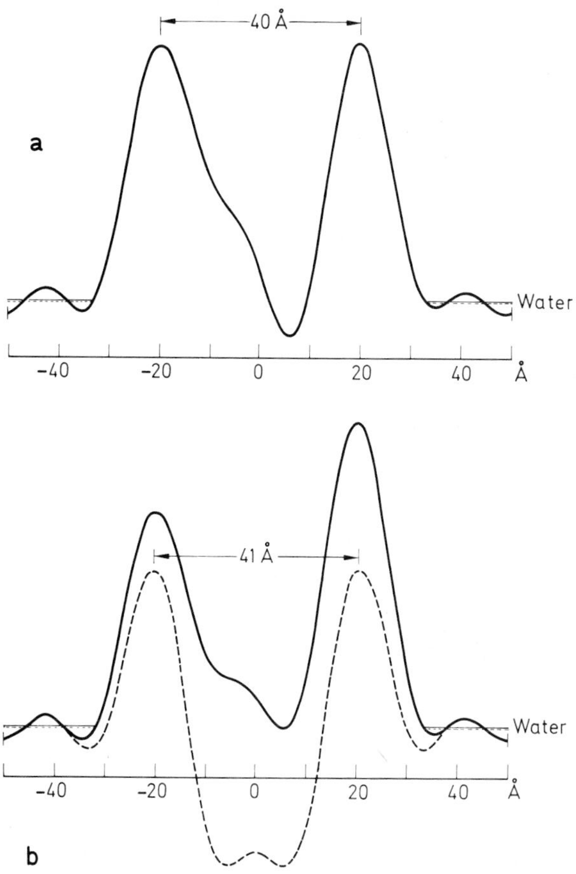

Fig. 23 a and b. Two possible profiles for the intact purple membrane. *Dashed curve*: a symmetric profile for the extracted lipids (Blaurock and King, 1977)

Blaurock (1975) came to essentially the same conclusions about the presence of α-helical segments placed perpendicular to the membrane plane and the symmetry relating the bacteriorhodopsin molecules. In addition, he recorded diffraction patterns from dispersions of lipids of both the purple membrane and red membrane fractions of the bacterium. The bilayer profiles for the two lipid fractions were found to be very similar, and this profile was used in model calculations to try and define a distribution of protein and lipid across the membrane which would account for the observed meridional diffraction from the intact purple membrane. Blaurock concluded that there must be at least some lipid present on both sides of the membrane, in a bilayer-like arrangement, and that a continuous monolayer on one side, with only protein on the other surface, was inconsistent with the observed meridional intensity.

The meridional diffraction was analyzed still further by Blaurock and King (1977). They showed that only two electron density profiles were consistent with the observed intensity distribution. Both profiles are asymmetric, although one more so than the other (Fig. 23). Below the two profiles in the figure is the electron density distribution for the extracted lipids. It is striking that, despite the unusually high protein content (about 75% of the dry weight), the thickness of the intact membrane is very close to that of the extracted lipids. These results establish the presence of a bilayer of lipids beyond doubt.

The three-dimensional structure of the purple membrane has been determined at about 7 Å resolution by low-dose electron microscopy (Unwin and Henderson, 1975; Henderson and Unwin, 1975) and shows clearly that the bacteriorhodopsin molecule is composed largely of α-helical segments, oriented more or less perpendicular to the membrane plane. It should be noted, however, as stressed by Blaurock and King (1977), that the X-ray diffraction results provide additional information about the purple membrane structure that is not redundant with the electron microscope images, particularly information about the distribution of the lipid molecules.

V. Conclusions

Over at last ten years or so, there has been considerable debate and controversy in the field of membrane diffraction. For almost every biological membrane that has been studied alternative structural solutions have been proposed. For nondiffractionists the debates about the correct methods of structural analysis must be bewildering, and it would not be surprising if doubts had been raised about what contributions X-ray diffraction can make to the study of biological membranes. Part of the difficulty, perhaps, has been the tendency to ascribe the same degree of certainty to a membrane electron density profile as it is usually reasonable to expect in models resulting from X-ray crystallography. It is now clear that there is no single, correct way to analyze low-angle membrane diffraction patterns, and that self-consistency rather than rigorous proof should be aimed for. Most importantly, it is crucial not to rely solely on the diffraction technique, and other information, particularly from electron microscopy and chemical analysis, should be used whenever possible. On the other hand, it should be said that the value of diffraction techniques, which can often provide direct and reliable information about membrane structure, has tended to be underestimated by other spectroscopists.

We feel that a good measure of concensus is now being reached about the best ways of analyzing the diffraction patterns and, perhaps most of all, about what membrane profiles should look like. Indeed, the unbiased observer cannot fail to have been struck by the rather monotonous appearance of the membrane electron density profiles which we have illustrated. While this measure of agreement is certainly encouraging, it also highlights the major difficulty with the technique. It is clearly the bilayer of lipids which dominates the diffraction pattern, so that detailed information on the distribution of the protein molecules has been hard to come by. A few specialized membranes which contain high concentrations of a single protein species (e.g., the purple membrane and gap junctions) are notable exceptions, and it is the considerable success that has been achieved with these chemically relatively homogeneous membranes that, in our opinion, points the way ahead. Since there are only a few examples of such specialized membranes which occur naturally, we feel that the study of well-defined model systems composed of purified membrane components is likely, in the long term, to provide the most definitive information about the distribution of protein molecules in membranes. The principal problem with such an approach is likely to be the preparation of suitable specimens, rather than difficulties in analyzing the X-ray diffraction patterns. The relative ease with which the conditions in model systems can be changed (e.g., varying protein/lipid ratios) greatly facilitates both the determination of the phase angles and the interpretation of the resulting electron density profiles in terms of the chemical composition.

Acknowledgments. We are grateful to Dr. W.R. Lieb for his careful reading, and helpful criticism of the manuscript. One of us (NPF) thanks the British Oxygen Corporation for support.

References

Akers, C.K., Parsons, D.F.: Diffraction of Myelin Membrane II, Determination of the Phase Angles of the Frog Sciatic Nerve by Heavy Atom Labeling and Calculation of the Electron Density Distribution of the Membrane. Biophys. J. **10**, 116–136 (1970).

Blasie, J.K.: Location of Photopigment Molecules in the Cross-section of Frog Retinal Receptor Disk Membranes. Biophys. J. **12**, 191–204 (1972a).

Blasie, J.K.: Net Electric Charge on Photopigment Molecules and Frog Retinal Receptor Disk Membrane Structure. Biophys. J. **12**, 205–213 (1972b).

Blaurock, A.E.: Ph. D. Thesis, University of Michigan (1967).

Blaurock, A.E.: Structure of the Myelin Membrane: Proof of the Low Resolution Profile. J. Mol. Biol. **56**, 35–52 (1971).

Blaurock, A.E.: Locating Proteins in Membranes. Chem. Phys. Lipids **8**, 285–291 (1972).

Blaurock, A.E.: X-ray diffraction pattern from a bilayer with protein outside. Biophys. J. **13**, 281–289 (1973a).

Blaurock, A.E.: Some Comments on 'Myelin Structure as Revealed by X-ray Diffraction' by David Harker. Biophys. J. **13**, 1261–1262 (1973b).

Blaurock, A.E.: The structure of a lipid-cytochrome c membrane. Biophys. J. **13**, 290-297 (1973c).

Blaurock, A.E.: Bacteriorhodopsin: a transmembrane pump containing α-helix. J. Mol. Biol. **93**, 139-158 (1975).

Blaurock, A.E.: Myelin patterns reconciled. Biophys. J. **16**, 491–501 (1976).

Blaurock, A.E.: Disorder is characteristic of nerve myelin. Biochim. Biophys. Acta **510**, 11–17 (1978).

Blaurock, A.E.: On phasing the small-angle X-ray diffraction pattern from nerve myelin. Biophys. J. **26**, 147–155 (1979).

Blaurock, A.E., King, G.I.: Asymmetric structure of purple membrane. Science **196**, 1101–1104 (1977).

Blaurock, A.E., Lieb, W.R.: X-ray diffraction studies of biomembranes. Nature **255**, 370–371 (1975).

Blaurock, A.E., Nelander, J.C.: Disorder in nerve myelin: analysis of the diffuse X-ray scattering. J. Mol. Biol. **103**, 421–431 (1976).

Blaurock, A.E., Nelander, J.C.: Locating the P_0 Protein in the X-ray profile of frog sciatic-nerve myelin. J. Neurochem. **32**, 1753–1760 (1979).

Blaurock, A.E., Stoeckenius, W.: Structure of the purple membrane. Nature New Biol. **233**, 149–154 (1971)

Blaurock, A.E., Wilkins, M.H.F.: Structure of frog photoreceptor membranes. Nature **223**, 906–909 (1969).

Brady, G.W., Birnbaum, P.S., Moscarello, M.A.: The model membrane system: egg lecithin + myelin protein (N2). Effect of solvent density on the X-ray scattering. Biophys. J. **26**, 49–60 (1979a).

Brady, G.W., Birnbaum, P.S., Moscarello, M.A., Papahadjopoulos, D.: Liquid diffraction analysis of the model membrane system: egg lecithin + myelin protein (N2). Biophys. J. **26**, 23–42 (1979b).

Brady, G.W., Fein, D.B.: The effect of added protein on the interchain X-ray peak profile in egg lecithin. Biophys. J. **26**, 43–48 (1979).

Bragg, Sir Lawrence, Perutz, M.F.: The structure of haemoglobin. Proc. R. Soc. London Ser. A. **213**, 425–435 (1952).

Calhoun, W.I., Shipley, G.G.: Sphingomyelin–Lecithin bilayers and their interaction with cholesterol. Biochemistry **18**, 1717–1722 (1979).

Caspar, D.L.D., Kirschner, D.A.: Myelin membrane structure at 10 Å resolution. Nature New Biol. **231**, 46–52 (1971)

Caspar, D.L.D., Goodenough, D.A., Makowski, L., Phillips, W.C.: Gap junction structures I. Correlated electron microscopy and X-ray diffraction. J. Cell Biol. **74**, 605–628 (1977).

Chabre, M.: X-ray diffraction studies of retinal rods I. Structure of the disc membrane: effect of illumination. Biochim Biophys. Acta **382**, 322–335 (1975).

Chabre, M., Cavaggioni, A.: Light-induced changes of ionic flux in retinal rod. Nature New Biol. **244**, 118–120 (1973).

Chabre, M., Cavaggioni, A.: X-ray diffraction studies of retinal rods II. Light effects on osmotic properties. Biochim. Biophys. Acta **382**, 336–343 (1975).

Chabre, M., Cavaggioni, A., Osborne, H.B., Gulik-Krzywicki, T.; A rhodopsin-lipid-water lamellar system: Its characterization by X-ray diffraction and electron microscopy. FEBS Lett. **26**, 197–202 (1972).

Coleman, R., Finean, J.B., Thompson, J.E.: Structural and functional modifications induced in muscle microsomes by trypsin. Biochim. Biophys. Acta **173**, 51–61 (1969).

Corless, J.M.: Lamellar structure of bleached and unbleached rod receptor membrane. Nature **237**, 229–231 (1972).

Craven, B.M.: Crystal structure of cholesterol monohydrate. Nature **260**, 727–729 (1976).

Davis D.G., Inesi, G., Gulik-Krzywicki, T.: Lipid molecular motion and enzyme activity in sarcoplasmic reticulum membrane. Biochemistry. **15**, 1271–1276 (1976).

Debye, P.: Interferenz von Röntgenstrahlen und Wärmebewegung. Ann. Phys. **43**, 49–95 (1914).

Dupont, Y., Hasselbach, W.: Structural changes in sarcoplasmic reticulum membranes induced by SH reagents. Nature **246**, 41–44 (1973).

Dupont, Y., Harrison, S.C., Hasselbach, W.: Molecular organization in the sarcoplasmic reticulum membrane studied by X-ray diffraction. Nature **244**, 555–558 (1973).

Esfahani, M., Limbrick, A.R., Knotton, S., Oka, T., Wakil, S.J.: The molecular organization of lipids in the membrane of Escherichia coli: Phase transitions. Proc. Natl. Acad. Sci. USA **68**, 3180–3184 (1971).

Finean, J.B., Freeman, R., Coleman, R.: X-ray diffraction patterns from haemoglobin-free erythrocyte membranes. Nature **257**, 718–719 (1975).

Frankel, R.D., Forsyth, J.M.: Nanosecond X-ray Diffraction from Biological Samples with a Laser-Produced Plasma Source. Science **204**, 622–624 (1979).

Franks, N.P.: Structural analysis of hydrated egg lecithin and cholesterol bilayer I. X-ray diffraction. J. Mol. Biol. **100,** 345–358 (1976).

Franks, N.P., Arunachalam, T., Caspi, E.: A direct method for determination of membrane electron density profiles on an absolute scale. Nature **276**, 530–532 (1978).

Franks, N.P., Lieb, W.R.: The structure of lipid bilayers and the effects of general anaesthetics: an X-ray and neutron diffraction study. J. Mol. Biol. **133**, 469–500 (1979).

Gabriel, A., Dupont, Y.: A position-sensitive proportional detector for X-ray crystallography. Rev. Sci. Instrum. **43**, 1600–1602 (1972).

Goodenough, D.A., Stoeckenius, W.: Isolation of mouse hepatocyte gap junctions. J. Cell Biol. **54**, 646–656 (1972).

Gulik-Krzywicki, T.: Structural studies of the associations between biological membrane components. Biochim. Biophys. Acta **415,** 1–28 (1975).

Gulik-Krzywicki, T., Shechter, E., Luzzati, V., Faure, M.: Interactions of proteins and lipids: structure and polymorphism of protein-lipid-water phases. Nature **223**, 1116–1121 (1969).

Gulik-Krzywicki, T., Shechter, E., Luzzati, V., Faure, M.: In: Biochemistry and Biophysics of Mitochondrial Membranes (eds. Azzone, G.F., Carafoli, E., Lehninger, A.L., Quagliariello, E., Siliprandi, N.) New York: Academic Press 1972.

Hargreaves, A.: Crystal structure of zinc P-toluenesulphonate. Nature **158**, 620–621 (1946).

Harker, D.: Myelin membrane structure as revealed by X-ray diffraction. Biophys. J. **12**, 1285–1295 (1972).

Henderson, R.: The structure of the purple membrane from Halobacterium Halobium: analysis of the X-ray diffraction pattern. J. Mol. Biol. **93**, 123–138 (1975).

Henderson, R., Unwin, P.N.T.: Three-dimensional model of purple membrane obtained by electron microscopy. Nature **257**, 28–32 (1975).

Herbette, L., Marquardt, J., Scarpa, A., Blasie, J.K.: A direct analysis of lamellar X-ray diffraction from hydrated oriented multilayers of fully functional sarcoplasmic reticulum. Biophys. J. **20**, 245–272 (1977).

Hosemann, R., Bagchi, S.N.: Direct Analysis of Diffraction by Matter. Amsterdam: North-Holland 1962.

James, R.W.: The Optical Principles of the Diffraction of X-Rays. London: Bell 1962.

Khare, R.S., Worthington, C.R.: An X-ray diffraction study of sphingomyelin-cholesterol interaction in oriented bilayers. Mol. Cryst. Liq. Cryst. **38**, 195–206 (1977).

King, G.I.: Direct structure determination of asymmetric membrane systems from X-ray diffraction. Acta Cryst. **A31**, 130–135 (1975).

Kirschner, D.A., Caspar, D.L.D.: Comparative diffraction studies on myelin membranes. Ann. N.Y. Acad. Sci. **195**, 309–320 (1972).

Kirschner, D.A., Caspar, D.L.D.: Myelin structure transformed by DMSO. Proc. Natl. Acad. Sci. USA **72**, 3513–3517 (1975).

Kirschner, D.A., Caspar, D.L.D.: Diffraction studies of molecular organization. In: Myelin (ed. Morell, P.). New York: Plenum Press 1977.

Kirschner, D.A., Sidman, R.L.: X-ray diffraction study of myelin structure in immature and mutant mice. Biochim. Biophys. Acta **448**, 73–87 (1976).

Knutton, S., Finean, J.B., Coleman, R., Limbrick, A.R.: Low-angle X-ray diffraction and electron microscopy studies of isolated erythrocyte membranes. J. Cell Sci. **7**, 357–371 (1970).

Kratky, O.: Adaptation of the technique of diffuse small-angle X-ray scattering to extreme demands. In: Small-angle scattering (ed. Brumberger,H.). pp. 63–120. New York: Gordon and Breach 1967.

Lesslauer, W.: On the structure of agglutinated sheep red blood cell membranes. Biochim. Biophys. Acta **436**, 25–37 (1976).

Lesslauer, W.: Phytohemagglutinin and transmembrane proteins in agglutinated sheep erythrocyte ghost membranes. Biochim. Biophys. Acta **510**, 264–269 (1978).

Lesslauer, W., Blasie, J.K.: Direct determination of the structure of barium stearate multilayers by X-ray diffraction. Biophys. J. **12**, 175–190 (1972).

Lesslauer, W., Cain, J., Blasie, J.K.: On the location of 1-Anilino-8-Naphtalene Sulfonate in lipid model systems. Biochim Biophys. Acta **241**, 547–566 (1971)

Levine, Y.K.: X-ray diffraction studies of membranes. Prog. Surf. Sci. **3** (4) 279–352 (1973).

Levine, Y.K., Wilkins, M.H.F.: Structure of oriented lipid bilayers. Nature New Biol. **230**, 69–72 (1971).

Liu, S.C., Worthington, C.R.: Electron density levels of sarcoplasmic reticulum membranes. Arch. Biochem. Biophys. **163**, 332–342 (1974).

Luzzati, V.: Interpretation des Mesures absolues de diffusion centrale des rayons X en collimation ponctuelle ou lineaire: solutions de particules globulaires et de bâtonnets. Acta Cryst. **13**, 939–945 (1960).

Luzzati, V.: X-ray diffraction studies of lipid-water systems. In: Biological Membranes (ed. Chapman, D.), Chapt. 3. New York–London: Academic Press 1968.

Luzzati, V., Tardieu, A.: Lipid phases: structure and structural transitions. Ann. Rev. Phys. Chem. 79–94 (1974).

Luzzati, V., Tardieu, A., Taupin, D.: A pattern recognition approach to the phase problem: application to the X-ray diffraction study of biological membranes and model systems. J. Mol. Biol. **64**, 269–286 (1972).

Makowski, L., Caspar, D.L.D., Phillips, W.C., Goodenough, D.A.: Gap junction structures II. Analysis of the X-ray diffraction data. J. Cell. Biol. **74**, 629–645 (1977).

Mateu, L., Luzzati, V., London, Y., Gould, R.M., Vosseberg, F.G.A., Olive, J.: X-ray diffraction and electron microscope study of the interaction of myelin components. The structure of a lamellar phase with a 150 to 180 Å repeat distance containing basic proteins and acidic lipids. J. Mol. Biol. **75**, 697–709 (1973).
McIntosh, T.J.: The effect of cholesterol on the structure of phosphatidylcholine bilayers. Biochim. Biophys. Acta **513**, 43–58 (1978).
McIntosh, T.J., Worthington, C.R.: Direct determination of the lamellar structure of peripheral nerve myelin at low resolution. (17 Å). Biophys. J. **14**, 363–385 (1974).
McIntosh, T.J., Waldbillig, R.C., Robertson, J.D.: Lipid bilayer ultrastructure. Electron density profiles and chain tilt angles determined by X-ray diffraction. Biochim. Biophys. Acta **448**, 15–33 (1976).
McIntosh, T.J., Waldbillig, R.C., Robertson, J.D.: The molecular organization of asymmetric lipid bilayers and lipid-peptide complexes. Biochim. Biophys. Acta **466**, 209–230 (1977).
Melchior, V., Hollingshead, C.J., Caspar, D.L.D.: Divalent cations cooperatively stabilize close membrane contacts in myelin. Biochim. Biophys. Acta **554**, 204–226 (1979).
Mitsui, T.: X-ray diffraction studies of membranes. Adv. Biophys. **10**, 97–135 (1978).
Moody, M.F.: X-ray diffraction pattern of nerve myelin: a method for determining the phases. Science **142**, 1173–1174 (1963).
Moody, M.F.: Structure determination of membranes in swollen lamellar systems. Biophys. J. **14**, 697–702 (1974).
Moody, M.F.: Diffraction by dispersions of spherical membrane vesicles. I. The basic equations. Acta. Cryst. **A31**, 8–15 (1975).
Nelander, J.C., Blaurock, A.E.: Disorder in nerve myelin: phasing the higher order reflections by means of the diffuse scatter. J. Mol. Biol. **118**, 497–532 (1978).
Oesterhelt, D., Stoeckenius, W.: Functions of a new photoreceptor membrane. Proc. Natl. Acad. Sci. USA **70**, 2853–2857 (1973).
Pape, E.H.: X-ray small-angle scattering: a new deconvolution method for evaluating electron density distributions from small-angle scattering diagrams. Biophys. J. **14**, 284–294 (1974).
Pape, E.H., Klott, K., Kreutz, W.: The determination of the electron density profile of the human erythrocyte ghost membrane by small-angle X-ray diffraction. Biophys. J. **19**, 141–161 (1977).
Perutz, M.F.: The structure of haemoglobin, III. Direct determination of the molecular transform. Proc. R. Soc. London Ser. A **225**, 264–286 (1954).
Phillips, M.C.: The physical state of phospholipids and cholesterol in monolayers, bilayers, and membranes. Prog. Surf. Sci. **5**, 139–221 (1972).
Pinto da Silva, P., Miller, R.G.: Membrane particles on fracture faces of frozen myelin. Proc. Natl. Acad. Sci. USA **72**, 4046–4050 (1975).
Ranck, J.L., Mateu, L., Sadler, D.M., Tardieu, A., Gulik-Krzywicki, T., Luzzati, V.: Order–disorder conformational transitions of the hydrocarbon chains of lipids. J. Mol. Biol. **85**, 249–277 (1974).
Rand, R.P., Luzzati, V.: X-ray diffraction study in water of lipids extracted from human erythrocytes. Biophys. J. **8**, 125–137 (1968).
Robertson, J.D.: The occurence of a sub-unit pattern in the unit membrane of club ending in Mauthner cell synapses in goldfish brains. J. Cell Biol. **19**, 201–221 (1963).
Rosenbaum, G., Holmes, K.C., Witz, J.: Synchroton radiation as a source for X-ray diffraction. Nature **230**, 434–437 (1971).
Ross, M.J., Klymkowsky, M.W., Agard, D.A., Stroud, R.M.: Structural studies of a membrane-bound acetylcholine receptor from Torpedo californica. J. Mol. Biol. **116**, 635–659 (1977).
Sadet, C., Tardieu, A., Luzzati, V.: Shape and size of bovine rhodopsin: a small-angle scattering study of a rhodopsin–detergent complex. J. Mol. Biol. **105**, 383–407 (1976).
Sayre, D.: Some implications of a theory due to Shannon. Acta Cryst. **5**, 843 (1952).
Schmitt, F.O., Bear, R.S., Clark, G.L.: X-ray diffraction studies on nerve. Radiology **25**, 131–151 (1935).

Schmitt, F.O., Bear, R.S., Palmer, K.J.: X-ray diffraction studies on the structure of the nerve myelin sheath. J. Cell Comp. Physiol. **18**, 31–42 (1941).

Schwartz, S., Cain, J.E., Dratz, E.A., Blasie, J.K.: An analysis of lamellar X-ray diffraction from disordered membrane multilayers with application to data from retinal rod outer segments. Biophys. J. **15**, 1201–1233 (1975).

Shipley, G.G.: Recent X-ray diffraction studies of biological membranes and membrane components. In: Biological Membranes, Vol. II (eds. Chapman, D., Wallach, D.F.H.), pp. 1–89. New York: Academic Press 1973.

Shipley, G.G., Leslie, R.B., Chapman, D.: X-ray diffraction study of the interaction of phospholipids with cytochrome c in the aqueous phase. Nature **222**, 561–562 (1969).

Stamatoff, J.B., Krimm, S.: Phase determination of X-ray reflections for membrane type systems with constant fluid density. Biophys. J. **16**, 503–516 (1976).

Stamatoff, J.B., Krimm, S., Harvie, N.R.: X-ray diffraction studies of human erythrocyte membrane structure. Proc. Natl. Acad. Sci. USA **72**, 531–534 (1975).

Stockton, G.W., Smith, I.C.P.: A deuterium nuclear magnetic resonance study of the condensing effect of cholesterol on egg phosphatidylcholine bilayer membranes. I. Perdeuterated fatty acid probes. Chem. Phys. Lipids **17**, 251–263 (1976).

Stroud, R.M., Agard, D.A.: Structure determination of asymmetric membrane profiles using an iterative Fourier method. Biophys. J. **25**, 495–512 (1979).

Tilney, L.G., Detmers, P.: Actin in erythrocyte ghosts and its association with spectrin. Evidence for a nonfilamentous form of these two molecules in situ. J. Cell Biol. **66**, 508–520 (1975).

Torbet, J., Wilkins, M.H.F.: X-ray diffraction studies of lecithin bilayers. J. Theoret. Biol. **62**, 447–458 (1976).

Unwin, P.N.T., Henderson, R.: Molecular structure determination by electron microscopy of unstained crystalline specimens. J. Mol. Biol. **94**, 425–440 (1975).

Webb, N.G.: X-ray diffraction from outer segments of visual cells in intact eyes of the frog. Nature **235**, 44–46 (1972).

Wilkins, M.H.F., Blaurock, A.E., Engelman, D.M.: Bilayer structure in membranes. Nature New Biol. **230**, 72–76 (1971).

Witz, J.: Focusing monochromators. Acta Cryst. **A25**, 30–42 (1969).

Worcester, D.L., Franks, N.P.: Structural analysis of hydrated egg lecithin and cholesterol bilayers II. Neutron diffraction. J. Mol. Biol. **100**, 359–378 (1976).

Worthington, C.R.: The interpretation of low-angle X-ray data from planar and concentric multilayered structures. The use of one-dimensional electron density strip models. Biophys. J. **9**, 222–234 (1969).

Worthington, C.R.: X-ray diffraction studies on biological membranes. In: Current Topics in Bioenergetics, Vol. V (eds. Sanadi, D.R., Packer, L.), pp. 1-39. New York: Academic Press 1973a.

Worthington, C.R.: X-ray analysis of retinal photoreceptor structure. Exp. Eye Res. **17**, 487–501 (1973b).

Worthington, C.R., Khare, R.S.: Structure determination of lipid bilayers. Biophys. J. **23**, 407–425 (1978).

Worthington, C.R., King, G.I.: Electron density profiles of nerve myelin. Nature **234**, 143–145 (1971).

Worthington, C.R., Liu, S.C.: Structure of sarcoplasmic reticulum membranes at low resolution (17 Å). Arch. Biochem. Biophys. **157**, 573–579 (1973).

Worthington, C.R., McIntosh, T.J.: Direct determination of the electron density profile of nerve myelin. Nature New Biol. **245**, 97–99 (1973).

Worthington, C.R., McIntosh, T.J.: Direct determination of the lamellar structure of peripheral nerve myelin at moderate resolution (7 Å). Biophys. J. **14**, 703–729 (1974).

Worthington, C.R., King, G.I., McIntosh, T.J.: Direct structure determination of multilayered membrane-type systems which contain fluid layers. Biophys. J. **13**, 480–494 (1973).

Subject Index

Molecular Biology Biochemistry and Biophysics

Editors: A. Kleinzeller, G. F. Springer, H. G. Wittmann

Volume 22: H. J. Fromm

Initial Rate Enzyme Kinetics

1975. 88 figures, 19 tables. X, 321 pages
ISBN 3-540-07375-2

This book fills a gap in the literature on enzyme kinetics. It outlines how one may reach definitive conclusions regarding enzyme and substrate interaction, the functional groups of enzymes involved in catalysis, mapping of the active site, intermediates involved in enzyme catalysis, allosteric mechanisms and evaluation of kinetic parameters that are essential for an understanding of enzyme regulation. The text covers such topics as kinetic nomenclature, systematic procedures for the derivation of rate equations, reversible enzyme inhibition, isotope exchange, temperature and pH effects on catalysis, and allostery. The experimental protocol involved in many kinetic procedures is discussed in detail. Specific examples from the literature are used to illustrate each topic.

Volume 23: M. Luckner, L. Nover, H. Böhm

Secondary Metabolism and Cell Differentiation

1977. 52 figures, 7 tables. VI, 130 pages
ISBN 3-540-08081-3

The volume is based on the concept that the biosynthesis of secondary natural substances is a widely found characteristic of cell specialization in almost all organisms. The central theme together with chapters on coordinated enzyme synthesis and possible effectors of gene expression in this field of metabolism, is the analysis of differentiating programs, including the formation of enzymes of secondary metabolism. Particular aspects of the formation of secondary natural substances in plant cell culture are treated in a separate chapter.

Springer-Verlag
Berlin
Heidelberg
New York

Volume 24

Chemical Relaxation in Molecular Biology

Editors: I. Pecht, R. Rigler

1977. 141 figures, 50 tables. XVI, 418 pages
ISBN 3-540-08173-9

The purpose of this monograph is to give a representative cross section of the current research activities dedicated to the analysis of essential steps in biological reactions. This covers the range of the following topics: hydrogen-bond formation, nucleotide base pairing, protein folding, isomerisation of protein and nucleic acid conformations, interactions between protein and proteins, nucleic acid and proteins, enzymes and substrates, antibody and haptens or ionic transport through membranes. A common denominator in these studies is the search for an understanding of the laws that govern the dynamic behaviour of living systems.

Volume 25

Advanced Methods in Protein Sequence Determination

Editor: S. B. Needleman

1977. 97 figures, 25 tables. XII, 189 pages
ISBN 3-540-08368-5

The determination of protein sequences has become so commonplace that, as more laboratories have entered this area of study, the sophistication of the technology has, in fact, progressed to make use of physical properties not previously utilized for this purpose. Earlier manual techniques have become automated; current instrumentation operates at higher parameters and with greater precision than before. Thus the present volume supplements the earlier one in presenting details of the more advanced technologies (optical, high pressure, X-ray, immunology etc.) being used in sequence determination today.

Volume 26: A. S. Brill

Transition Metals in Biochemistry

1977. 49 figures, 18 tables. VIII, 186 pages
ISBN 3-540-08291-3

This monograph is a concise review of the current state and developments in the field, with emphasis upon the application of physical methods to the investigation of metal coordination. Molecular functions of proteins containing transition metal ion prosthetic groups are summarized. Where established by X-ray diffraction, the three-dimentional structures of relevant metal binding sites in proteins are described. Light absorption and electron paramagnetic resonance are treated in depth. Those aspects of the theory are presented which can be directly employed in the quantification and interpretation of experimental data. The monopraph provides a basis for closer communication between scientists of different backgrounds with a common interest in the biochemistry of transition metal ions.

Volume 27

Effects of Ionizing Radiation on DNA

Physical, Chemical und Biological Aspects

Editors: A. J. Bertinchamps (Coordinating Editor), J. Hüttermann, W. Köhnlein, R. Téoule
1978. 74 figures, 48 tables. XXII, 383 pages
ISBN 3-540-08542-4

For the first time, the three essential approaches to research on the effects of ionizing radiation on DNA and its constituents have been described together in one book, providing an overall view of the fundamental problems involved. A result of the European study group on "Primary Effects of Radiation on Nucleic Acids", this book contains the current state of knowledge in this field, and has been written in close collaboration by 27 authors.

Volume 28: A. Levitzki

Quantitative Aspects of Allosteric Mechanisms

1978. 13 figures, 2 tables. VIII, 106 pages
ISBN 3-540-08696-X

This book provides a concise but comprehensive treatment of the basic regulatory phenomena of allostery and cooperativity. It critically evaluates the differences between the allosteric models and their applicability to real situations. For the first time the full analysis of the different allosteric models is given, and compared with the pure thermodynamic approach. The treatment of the subject of allostery in this book is of great value to enzymologists, receptorologists, pharmacologists and endocrinologists, as it provides the basic rules for the study of ligand-protein and ligand-receptor interactions.

Volume 29: E. Heinz

Mechanics and Energetics of Biological Transport

1978. 35 figures, 3 tables. XV, 159 pages
ISBN 3-540-08905-5

This book presents the interrellations of mechanistic models on the one hand and the kinetic and energetic behavior of transport and permeatin processes on the other, using the principles of irreversible thermodynamics. The advantages of each method are compared. The special aim is to show how to appropriate formulas can be transformed into each other, in order to recognize in what way the kinetic parameters correspond to those of irreversible thermodynamics.

Volume 30: D. Vázquez

Inhibitors of Protein Biosynthesis

1979. 61 figures, 13 tables. X, 312 pages
ISBN 3-540-09188-2

This is the first comprehensive treatment of how antibiotics and other compounds inhibit protein biosynthesis. Various antibiotics and compounds are analyzed to illustrate the mode of action and selectivity of a number of drugs used medically as antibacterial or antitumor agents. Antibiotics are also studied to shed light on ribosomal structure and the process of translation.
This research offers valuable information for general microbiologists, pharmacologists, biochemists, molecular biologists and all specialists working on the problems of protein biosynthesis and ribosomal structure.

Volume 32

Chemical Recognition in Biology

Editors: F. Chapeville, A.-L. Haenni
1980. 190 figures, 39 tables. Approx. 450 pages
ISBN 3-540-10205-1

This volume is a collection of papers presented at the Symposium on Chemical Recognition in Biology organized in Grignon (France) in July 1979 on the occasion of the 80th anniversary of Fritz Lipmann, one of the most outstanding figures of modern biochemistry.
The topics covered in this book extend from precise enzymatic systems to highly complex cellular organisms. Special emphasis is laid on recognition of ligands, enzymic catalysis, enzyme regulation, nucleic acid-protein interactions, mutagenesis, and protein biosynthesis. The volume concludes with philosophical reflections on molecular biology, culture and society, and personal recollections of Fritz Lipmann.

Springer-Verlag
Berlin
Heidelberg
New York